For Frank
with best
regards
Dick

LEGAL STRATEGIES FOR INDUSTRIAL INNOVATION

Richard A. Givens, Editor
Botein, Hays, Sklar & Herzberg
New York, New York

Edward S. Binkowski
Bruce M. Kramer
Frank J. Mauro
Lloyd McAulay
Kenneth Ross
Richard H. Rowe
Frank F. Skillern
Robert S. Taft

Shepard's/McGraw-Hill
P.O. Box 1235, Colorado Springs, Colorado 80901

McGraw-Hill Book Company

New York • St. Louis • San Francisco • Colorado Springs
Auckland • Bogota • Hamburg • Johannesburg • London
Madrid • Mexico • Montreal • New Delhi • Panama • Paris
Sao Paulo • Singapore • Sydney • Tokyo • Toronto

12345678910 SHHA 891098765432

Library of Congress Cataloging in Publication Data
Main entry under title:

Legal strategies for industrial innovation.

Includes index.
Contents: Legal issues affecting innovation / Richard A. Givens — Innovation and labor relations / Richard A. Givens — Antitrust and innovation / Richard A. Givens — [etc.]
1. Technological innovations—Law and legislation—United States. 2. Industrial laws and legislation—United States. I. Givens, Richard A.
KF4270.L43 1982 344.73'095 82-10247

ISBN 0-07-023331-4 347.30495

Preface

This book seeks to highlight strategic options available to innovators in dealing with the legal system as both support system and as obstacle, in the light of the increasing importance of innovation and its wise management for business survival and also for the survival of our open society.

I am grateful to my wife, Janet Eaton Givens, and my daughters, Susan Ruth and Jane Lucile, for endless patience and encouragement and for a generalist, nonlegal perspective on the problems touched upon here. I am grateful to my parents, Meredith B. Givens and Ruth Wheelock Ayres Givens, both economists, for sensitizing me to the importance of many of the issues raised here.

Richard A. Givens
April, 1982

The Authors

Richard A. Givens, Editor and Contributor

A.B., Columbia College; M.S., Econ., University of Wisconsin; LL.B., Columbia Law School; Private law practice, 1959-1961; Assistant U.S. Attorney, Criminal Division, Southern District of New York, 1961-1971; Regional Director, Federal Trade Commission, 1971-1977; Private law practice, since 1977; Member of the New York Bar; Author of professional book and journal articles.

Contributors

Lloyd McAulay

A.B., College of the City of New York; LL.B., Yale Law School; Electrical engineer, U.S. Navy Department, 1950-1956; Private law practice, since 1960; Member of the New York Bar; Admitted to practice before the United States Patent Office; Author of professional publications.

Robert S. Taft

B.A., Dartmouth College; LL.B., Columbia University; LL.M., New York Law School; Professor of Law, New York Law School, since 1973; Private law practice, since 1960; Member of the New York Bar; Author of numerous professional publications and journal articles.

Richard H. Rowe

A.B., Bates College; J.D., Harvard Law School; Trial Attorney, Branch of Administrative Proceedings and Investigations, Division of Corporation Finance, SEC, 1964-1967; Legal Assistant to an SEC Commissioner, 1967-1969; Staff Member, Special Disclosure Policy Study, SEC, 1968-1969; Special Counsel, Office of the Director, Division of Corporation Finance, SEC, 1969; Assistant Vice President, Vice President, Shareholders Management Company, 1969-1970; Special Counsel, Office of the Chief Counsel, Division of Corporation Finance, SEC, 1970; Trial Attorney and Assistant Director, Division of

Corporation Finance, SEC, in charge of Office of Administrative Proceedings and Investigations, 1970-1972; SEC staff member, Industrial Issuers Advisory Committee, 1972; Associate Director/Disclosure Policy, Division of Corporation Finance, SEC, 1972-1976; Director, Division of Corporation Finance, SEC, 1976-1979; Adjunct Professor of Law, Georgetown University Law Center, Fall, 1977; Private law practice, since 1979; Member of New York and Washington, D.C. Bars; American Bar Association; Federal Bar Association; Author of numerous professional publications and journal articles.

Frank F. Skillern

A.B., University of Chicago; J.D., University of Denver; LL.M., University of Michigan; Private law practice, 1967; General Attorney, Maritime Administration, 1967-1968; Assistant Professor of Law, Ohio Northern University, 1969-1971; Assistant, Associate, and Professor of Law, School of Law, Texas Tech University, since 1971; Visiting Professor, University of Texas Law School, Summer, 1978, University of Arkansas Law School, 1979-1980; University of Tulsa College of Law, 1981-1982; Editor, *Natural Resources Lawyer;* Member of Colorado and Texas Bars; American Bar Association; Author of numerous professional books and journal articles.

Bruce M. Kramer

A.B., J.D., University of California at Los Angeles; LL.M., University of Illinois, College of Law; Private law practice, 1972-1973; Assistant, Associate, and Professor of Law, School of Law, Texas Tech University, since 1974; Visiting Professor, School of Law, University of Indiana, Fall 1979; Lewis & Clark Law School, Summer 1980; University of Florida, Fall 1982-Spring 1983; Member of California Bar; Author of numerous professional publications.

Frank J. Mauro

A.B., Union College, M.P.A., Syracuse University; Assistant Director, Federal-State Liaison Office, New York State Senate, 1969-1971; Assistant and Acting Director, Municipal Assistance and Policy Development Division, New York State Commission on Cable Television, 1973-1975; Director, Office of Research and Analysis, New York State Assembly, 1975-1976; Deputy Director, Program and Committee Staff, New York State Assembly, 1976-1978; Director, Program Development Group, New York State Assembly, since January 1979.

Kenneth Ross

A.B., University of California at Riverside; J.D., University of California at Los Angeles; Private law practice, 1973-1976; Counsel for Westinghouse Electric Corporation, 1976-1982; Product Liability Counsel for Emerson Electric Company, since 1982; Editor-in-chief, Journal of Products Liability; Board of Editors, Product Liability Newsletter; Member of California Bar; Author of numerous professional journal articles and publications.

Edward S. Binkowski

A.B., Ph.D., Princeton University; J.D., Fordham University Law School; Member of Research Staff, Applied Mathematics Group, Western Electric Engineering Research Center, 1970-1974; Instructor, Biology Department of Cornell University Graduate School of Medical Sciences, and Research Associate, Sloan-Kettering Institute, Biostatistics Laboratory, 1974-1975; Supervisor of Corporate Planning, AT&T, Corporate Planning Division, since 1975; member of New York Bar; Author of professional journal articles, in several fields.

Contents

Summary

Detailed

Legal Issues Affecting Innovation

Richard A. Givens

1

§1.01 Introduction

Attention to innovation can permit an organization to seize the initiative instead of merely reacting to changed circumstances[1] in ways that many times

---

[1] Common events that compel changes include:

Introduction of new products, services or techniques by competitors. For illustrative controversies, *see, e.g.,* Berkey Photo, Inc v Eastman Kodak Co, 603 F2d 263 (2d Cir 1979), *cert denied,* 444 US 1093 (1980); SCM v Xerox Corp, 645 F2d 1195 (2d Cir 1981), *cert denied,* — US — (1982), *affg* 463 F Supp 983 (D Conn 1978). *See also* Riegel Textile Corp v Celanese Corp, 649 F2d 894 (2d Cir 1981).

Shortages or increases in cost of goods or services needed by the organization, such as labor or raw materials. *E.g.,* for those caused by raw material cartels, *see* IAM v OPEC, 649 F2d 1354 (9th Cir 1981).

Changes in transportation, communication or industry structure which introduce new competitor. See, for historical comparison, Commons, *American Shoemakers, 1648-1895,* in Labor and Administration 219 (Commons ed 1913). For juridical consequences, *see, e.g.,* United States v Women's Sportswear Mfrs Assn, 336 US 460, 464 (1949); Hunt v Washington Apple Advertising Commn, 432 US 333 (1977).

Deterioration in infrastructures, such as transportation facilities or training in needed skills by the general educational system. *E.g.,* loss of local feeder air and rail transportation to regional centers or in problem-solving ability of graduates of the educational system. *Compare,* as to the latter, Wertheimer, Productive Thinking (1959); Fantini & Weinstein, Toward a Contact Curriculum (B'nai

leave the organization at a disadvantage. The legal system, as part of the environment within which any organization must function, furnishes both opportunities and obstacles as to innovation. Awareness of the opportunities is important. Dealing with the obstacles with adroitness can itself yield a competitive advantage.[2] These legal challenges can be confronted more effectively with advance attention before a crisis stage is reached, and with flexibility once an issue becomes concrete.

§1.02 Legal Promotion of Innovation

Promotion of innovation by the legal system occurs because innovation is the major source of increased productivity, and productivity furnishes the economic pie which delimits individual and societal attainment.[3]

Among the critical consequences of a lag in the rate of technological innovation are:

1. Lack of substitutes for scarce nonrenewable resources[4]
2. Decreased productivity resulting in an inability to pay for the cost of meeting additional requirements imposed by changing circumstances. These may include:
    a. international commitments[5]
    b. costs of dealing with the problems caused by development of substitutes for techniques causing, for example, adverse effects as to the environment, health, safety and resource consumption[6]
    c. reduction of taxes to encourage incentives for later productivity gains from enhanced effort[7]
    d. increases in compensation to offset inflation, provide incentives or keep pace with assumed or expected productivity gains[8]
    e. expansion of social services by the society as a whole

---

B'rith 1969); Hechinger, *Psychologist Sees a Key to Learning in Managing "Unsolvable" Problems,* NY Times, Aug 18, 1981, at C5; Fiske, *Reading Analysis Is Called Lacking,* NY Times, Dec 14, 1980, at F1, Browne, *Is Problem Solving America's Lost Art?,* NY Times, July 29, 1980, at C1; Baker, *Terminal Education,* NY Times Magazine, Nov 9, 1980, at 29.

[2] *See generally* Givens, Advocacy (Shepard's/McGraw-Hill 1980).

[3] *See generally* Worsinger, *New Technologies and Antitrust,* 47 NY St BJ 651 (1975), *reprinted in* 81 Case & Com 33 (1976).

[4] On the social importance of substitutes, compare the consequences involved in IAM v OPEC, 649 F2d 1354 (9th Cir 1981). On the economic effects, *see* United States v EI duPont de Nemours & Co (Cellophane), 351 US 377 (1956).

[5] *See* as to the full range involved, Gardner, Sterling-Dollar Diplomacy (1956).

[6] See generally **ch 6.**

[7] See generally **ch 5;** Mansfield, *Tax Policy and Innovation,* 215 Sci 1365 (Mar 12, 1982).

[8] See **ch 12.**

   f. care for increased numbers of retired citizens or workers
3. Erosion of the industrial and technological basis for national security
4. Erosion of international competitive position
5. Cost-push inflation leading to erosion of value of the currency
6. Increased unemployment as a direct result of higher costs. These make it more difficult to maintain or increase sales of goods or services
7. Higher costs also tend to require restriction of demand as the only means to limit inflationary pressure, thus curtailing economic activity

Policies designed to encourage innovation are often subject to great expansion once individuals become aware of the available opportunities.[9] Such policies are reflected in laws relating to patents,[10] taxation[11] and funding sources for high-risk, socially vital forms of innovation.[12] Policies in support of innovation also are reflected in laws which have overlapping objectives, such as those concerned with antitrust,[13] labor,[14] products liability[15] and government procurement.[16]

§1.03 —Individual Enhancement of Pro-Innovation Policies

Pro-innovation policies can be enhanced by individual efforts to trigger the availability of laws which promote, aid or even finance innovation. Such activities are likely to be most successful when: (1) the innovation sought is of vital social importance; for example, because it will deal with actual or impending scarcities of critical resources, needs for a strategic industrial base relevant to national survival or serious environmental, safety or other social problems; and (2) private investment encouraged by then-existing market incentives is insufficient because of a combination of one or more of the following factors:

a. There is a high risk involved, which carries a negative investment premium for the very reason individuals and businesses buy insurance, and which can be offset by spreading the risk

---

[9] E.g., Apollo Project to place astronauts on the moon, Manhattan Project. See Sullivan, Build Brother Build (1969), on economic aspects of the Apollo effort.

[10] See **ch 4.** *See also* Practising Law Institute, Domestic and International Licensing of Technology (1980).

[11] See **ch 5.**

[12] See **chs 6** and **12.**

[13] See **ch 3.**

[14] See **ch 2.**

[15] See **ch 10;** Sherman, Products Liability for the General Practitioner (Shepard's/McGraw-Hill 1981).

[16] Not covered here in detail, but see generally options discussed in **ch 12.**

b. The profitability of the effort is likely to be postponed, also a negative factor in obtaining private capital
c. The requirements are lumpy, i.e., the project cannot be assured to go forward at all unless a large amount of capital is amassed, making all initial investments even riskier and indeed contingent on more being obtained
d. Legal barriers pose serious roadblocks
e. Much of the benefit from the innovation will escape the innovator and accrue to the public at large[17]

§1.04 Legal Obstacles to Innovation

The legal system impedes innovation in numerous ways. Planning to deal with barriers is thus at least as critical as planning to take advantage of opportunities offered by the legal system. The obstacles are essentially of four major types.

The first type is those legal policies designed to promote other objectives, such as the rights of employees, eradication of invidious discrimination, protection of safety[18] and supervision of practice in regulated fields.[19]

The second type is those policies designed to slow down or stop types of innovation regarded as actually or potentially harmful. These policies are reflected in numerous laws such as the National Environmental Policy Act, Consumer Product Safety Act, food and drug laws, and so forth.[20] They may be broadly categorized as outgrowths of an accelerating technology assessment[21] movement. Ultimately, however, this movement can promote desirable technologies as well as impede those considered harmful by decision makers.[22]

The third obstacle is the inertia of the legal system itself. This is due to multiple layers of approvals and reviews for the same proposed course of action, both within agencies[23] and involving multiple agencies, multiple levels of government and different branches of the same or different levels, and the resulting susceptibility of our procedural structure to frequent reversals of

---

[17] See Terleckyj, Effects of R&D on the Productivity Growth of Industries (National Planning Assn 1974).

[18] See **ch 2.**

[19] See **chs 7-11;** *see generally* Schwartz & Schwartz, Engineering Evidence (Shepard's/McGraw-Hill 1981).

[20] See **ch 7.**

[21] *See* Technology Assessment Act, 2 USC §471 *et seq; see also* Long, Appropriate Technology and Social Values (1978), *reviewed,* 212 Sci 38 (Apr 3, 1981); Moesteler, *Innovation and Evaluation,* 211 Sci 881 (Feb 27, 1981).

[22] See **ch 12.**

[23] Compare references cited in notes to Givens, Advocacy ch 14 (Shepard's/McGraw-Hill 1980).

decisions once seemingly made one way or the other.[24]

A final obstacle is legal barriers deliberately created at the behest of actual or potential competitors.[25]

§1.05 —Strategies for Dealing with Obstacles

Strategies for dealing with legal barriers to innovation may, depending on the circumstances, include:

1. A *fulfillment* strategy of seeking to meet not only the letter but also the underlying objectives of a legal requirement so as to reduce the likelihood of founded or unfounded objections to a planned innovation[26]
2. A *consultation* strategy involving the earliest possible dialogue with agencies and interest groups potentially involved[27]
3. A *confrontation* strategy contemplating a challenge if necessary to the legal or constitutional basis of anticipated barriers should accommodation fail[28]
4. A *detour* strategy involving either alternate technologies or alternate legal categories which may supersede, or be able to be amended to supersede, the original legal barriers[29]

§1.06 The Increasing Relevance of Innovation

Issues concerning innovation are likely to be of enhanced importance in the future for several reasons. First, physical resources necessary to sustain our industrialized society are proving to be limited and thus increasingly expensive and subject to interruptions of supply.[30] Innovation can help solve problems in this area.

---

[24] For one example of resulting delay, *see* Federal Maritime Commn v Aktiebolagiet Svenska Amerika Linien, 390 US 238 (1968) (involving only one level of government yet taking nearly a decade).

[25] *See, e.g.*, Eastern RR Presidents Conference v Noerr Motor Freight Inc, 365 US 127 (1961); California Motor Transp, Inc v Trucking Unlimited, 404 US 508 (1972); LG Balfour Co v FTC, 442 F2d 1 (7th Cir 1971); *see also* Sullivan, Antitrust §238 (1977); Givens, *Litigation as a Means of Economic Competition,* 27 Rec AB City NY 572 (1972).

[26] On purpose interpretations, *see, e.g.*, Broadcast Music, Inc v CBS, 441 US 1 (1979); Silver v New York Stock Exch, 373 US 341 (1963); references cited in Givens, Advocacy ch 7 (Shepard's/McGraw-Hill 1980).

[27] See references cited in Givens, *supra* note 26, ch 16 at 392-93; and see **ch 7-8.**

[28] For a successful challenge, *see, e.g.*, Raymond Motor Transp, Inc v Rice, 434 US 429 (1978).

[29] *See generally* Schelling, The Strategy of Conflict (1960).

[30] For an extreme view, see Meadows, Meadows, Randers & Behrens, The Limits to Growth (1972); Taylor, Rethink: A Paraprimitive Solution (1973); *but see* O'Neill, The High Frontier (1977); Clarke, The Promise of Space (1968); Berry, The Next Ten Thousand Years (1974); Feinberg, The Prometheus Project (1969).

In addition, the significant growth of scientific and technological knowledge is in itself inherently exponential: the development of each new area results in unanswered questions which in turn, afford new opportunities for innovation.[31] If some enterprises or societies fail to grasp these opportunities, others will.[32] This, then, may create crises for those who failed to act in the first instance.[33]

Further, as long as contending superpowers exist, their need for protection against each other will escalate. By increasing productivity, innovation can strengthen the economy. The stronger the civilian economy, the stronger the strategic position of the society,[34] and also the smaller the proportion of its resources which must be devoted to direct defense. Ideally, scientific and technological innovation on an unprecedented scale could generate sufficient power for a coalition of open societies to cause totalitarian adversaries to prefer to dismantle the balance of terror in order to win admission into a new consortium.[35]

Even that desirable outcome, however, would not relax the urgency of still further innovation. Indeed, entire industries are devoted to the carrying on of innovation at many levels of implementation, so that a slackening of the rate of innovation would itself have far-reaching negative consequences.[36]

§1.07 Approach of This Volume

The purpose of this volume is to call attention to legal opportunities and obstacles for innovators. The focus is upon alternate strategies which may be available to deal with challenges confronting the innovator in each area, not on providing encyclopedic coverage of the numerous areas of law involved. The references in each chapter are designed to provide a starting point for evaluating the strategic options discussed; limited introductory material is included early in each chapter for orientation of those not already familiar with the areas involved.

Since every area of law both promotes and retards innovation, only those believed to be most commonly pivotal are discussed here. While the primary

---

[31] See, e.g., Mendelssohn, The Search for Absolute Zero (1966); *compare* Starr & Rudman, *Parameters of Technological Growth,* 182 Sci 358 (Oct 26, 1973).

[32] *Compare* Speer, Inside the Third Reich (1970).

[33] *Compare, e.g.,* Shirer, The Collapse of the Third Republic (1969) *with* Thomson, Vote of Censure (1968); Churchill, Memoirs of the Second World War (1978).

[34] *Compare* Sakharov, Progress, Coexistence and Intellectual Freedom (1968); Heisenberg, Physics and Beyond (1971); Orwell, 1984 (1949); J.Q. Adams, 1 State of the Union Messages of the Presidents 246 (Chelsea House 1966); Thomson, Vote of Censure (1968).

[35] See **ch 12.**

[36] See Slichter, *The Industry of Discovery,* 128 Sci 1610 (Dec 22, 1968); Schumpeter, Capitalism, Socialism and Democracy (3d ed 1950).

emphasis is on dealing with the legal structure as it is already developing, some major policy options are discussed within each chapter and also at the conclusion of this book.[37]

[37] Ch 12.

Innovation and Labor Relations

2

Richard A. Givens

§2.01 Impact of Innovation on Labor Relations

Innovation affects labor by opening up new jobs,[1] by creating demands for new skills and indeed entire new industries and by displacing formerly needed employees and, frequently, their skills as well. These multifaceted potential gains and losses furnish fertile ground for legal and industrial strategy, conflict, compromise, confrontation and synthesis of interests.[2]

One of the first imperatives for an innovator is to analyze the interests likely to be affected by options under consideration. The potential outcome can then be considered, and alternative approaches assessed.

The earlier these approaches are considered, the wider the possible options, including modification of technologies to avoid dealing with some groups or to reduce conflict with others, as well as consideration of various types of benefits and the estimation of probable results of possible conflicts.

Failure to anticipate labor conflict can often block innovations or lead to

[1] *See, e.g.*, Sullivan, Build Brother Build (1969); Adams, *The Automobile—A Luxury Becomes A Necessity*, in Hamilton, Price and Price Policies 27-82 (1938).

[2] *See generally* on such "non-zero sum games," Schelling, The Strategy of Conflict (1960); Axelrod, *The Evolution of Cooperation*, 211 Sci 1309 (Mar 27, 1981).

years of expensive and uncertain judicial and industrial relations battles. Accommodations that mortgage the future of an entire industry by including permanent requirements that generate unnecessary work can also result.[3] On the other hand, creative solutions which protect existing employees in ways that permit innovation to go forward, often expanding an industry so as to create a large number of new jobs, can be successfully developed—sometimes in the heat of conflict[4] but also, much more successfully, through advance planning.

A key element in the latter context is consultation fairly far in advance with interest groups which must inevitably be dealt with,[5] and placement of the largest possible number of options on the table,[6] so that discussions do not become deadlocked unnecessarily by brinkmanship[7] over specific demands of either party even though alternative accommodations may be possible.[8]

---

[3] *See, e.g.,* American Newspaper Publishers Assn v NLRB, 345 US 100 (1953); NLRB v Gamble Enters, 345 US 117 (1953). *Compare* Order of Ry Telegraphers v Chicago & NW Ry, 362 US 330 (1960). On the strong motives underlying this, *see, e.g.,* Perlman, A Theory of the Labor Movement (1928); Zausner, Unvarnished (1941); Porter, Job Property Rights (1954); NY Times, Dec 19, 1959, at 50, col 1; Dulles, Labor in America (1949); Rottenberg, *Property in Work,* 15 Indus & Lab Rel Rev 402 (1962); *see also* Jou-Jou Designs, Inc v ILGWU, 643 F2d 905 (2d Cir 1981); Danielson v Joint Board, 494 F2d 1230 (2d Cir 1974); Note, 73 Yale LJ 812 (1964).

[4] *See, e.g.,* Brotherhood of Local Firemen v Chicago, B & QRR, 225 F Supp 11 (DDC), *affd,* 331 F2d 1020 (DC Cir), *cert denied,* 377 US 918 (1964) (*upholding* 77 Stat 132 (1963); see S Rep No 459, 88th Cong, 1st Sess (1963)), discussed in Shils, *Industrial Unrest in the Nation's Rail Industry,* 15 Labor LJ 81, 99-104 (1964). *See also* Adamson Eight-Hour Day Act, 39 Stat 721 (1916) (codified at 45 USC §65 *et seq*), upheld in Wilson v New, 243 US 332 (1917); 44 Stat 577 (1926), *as amended,* (codified at 45 USC §§151-188), discussed in Murphy, *Agreement on the Railroads: The Joint Railway Conference of 1926,* 11 Labor LJ 823 (1960); 92 Cong Rec 5754-62, 5780-81, 5843-58, 5968-75, 5988 (1946) (discussion of railway strike of 1946); Givens, *National Emergency Labor Disputes,* 34 Temple LQ 17 (1960); Marshall, *New Perspectives on National Emergency Disputes,* 18 Labor LJ 451 (1967).

[5] Ewing, The Human Side of Planning (1969); see Drucker, Concept of the Corporation (1953). For interesting examples in the 1930s and 1940s, see Perkins, The Roosevelt I Knew (1946).

[6] See Givens, Advocacy ch 11 (Shepards/McGraw-Hill 1980). On problems with one available alternative, see Committee on Corporation Law, *Special Interest Directors,* 35 Rec AB City NY 26 (1980).

[7] Forkosch, *"Take It or Leave It" as a Bargaining Technique,* 18 Labor LJ 676 (1967); *see* South Shore Hosp v NLRB, 630 F2d 40, 45 (1st Cir 1980); Note, 76 Harv L Rev 807 (1963). On dealing with such tactics, see Bramson, Coping With Difficult People (1981). On the legal issues involved, *see e.g.,* Pease Co v NLRB, 666 F2d 1044 (6th Cir 1981) and cases cited therein; NLRB v Herman Sausage Co, 275 F2d 229, 231 (5th Cir 1960); NLRB v United Clay Mines, 219 F2d 120 (6th Cir 1955); NLRB v Insurance Agents, 361 US 477 (1960); American Ship Bldg v NLRB, 380 US 300 (1965).

[8] For a gold mine of folklore on successful negotiation strategy, see Eisenberg, *Private Ordering Through Negotiation,* 89 Harv L Rev 673 (1976); *see generally* Nierenberg, The Art of Negotiating (1968); Nirenberg, Getting Through to People (1963). Special problems, of course, arise in public-sector or subsidized fields where bargaining involves outside funding sources or potential sources. *See* Division 1235 v Metropolitan Transit

§2.02 Options Concerning Employees Threatened with Displacement

Numerous alternatives exist for treatment of employees who may be displaced by technological innovation, [9] including the following among others. First, nothing at all may be done. This may not produce conflict if there are ample alternatives open to the employees affected, or if the enterprise is sufficiently small.

Second, work rules may be negotiated requiring additional employees beyond those believed needed by management, at times partially for safety or similar reasons, but primarily to protect jobs.[10] This commonly adopted approach results in locking additional costs into the industry or firm structure on a long-term basis. This often guarantees future conflict when these costs become difficult to support, and in some cases can contribute to the demise of large enterprises and even major segments of entire industries.

Third, retraining or seniority transferable to new jobs may be offered, but this is rarely a complete solution since not all employees may be able to take advantage of the jobs offered, or the jobs may be few in number.[11]

Other alternatives include provision of financial payments such as severance pay or supplemental unemployment benefits (SUB).[12] Further, agreements can be made that any reductions of the total work force or in a specific category

---

Auth, 477 F Supp 1027 (MD Tenn 1979); Comment, 78 Colum L Rev 1332 (1978). As to bankrupt entities, see Note, 81 Colum L 391 (1981).

[9] *See generally,* note 4 supra; Barnett, Machinery and Labor (1926); Bachman, *Cushioning the Impact of Technological Change,* 13 Labor LJ 731 (1962); Gomberg, *The Work Rules and Work Practices Problem,* 12 Labor LJ 643 (1961); Cox, *Use and Abuses of Union Power,* 35 Notre Dame Law 624 (1960); Fanning, *The Challenge of Automation in the Light of the Natural Law,* 11 Labor LJ 875 (1960); NY Times, Apr 11, 1960, at 18, col 1; Hommers, *Protection of Employees Affected by Railroad Consolidations,* 15 Labor LJ 736 (1964). *Compare* Pittsburgh Plate Glass Co, 33 Lab Arb (BNA) 614 (1960); Dayton, *Work Rules in Industry,* 12 Labor LJ 380 (1961); Note, 32 Mercer L Rev 833 (1981).

[10] See note 3 supra; *e.g.,* NLRB v ILA, 447 US 490 (1980); National Wookwork Mfrs Assn v NLRB, 386 US 612 (1967); United States v Petrillo, 75 F Supp 176 (ND Ill 1948); Westinghouse Broadcasting Co v Dukakis, 412 F Supp 580 (D Mass 1976) (decided under 60 Stat 89 (1946) (formerly codified at 47 USC §506), *repealed by* 94 Stat 2747 (1980)); National Assn of Window Glass Mfrs v United States, 263 US 403 (1923).

[11] See Aaron, *Reflections on the Legal Nature and Enforceability of Seniority Rights,* 75 Harv L Rev 1532 (1962); Blumrosen, *Seniority Rights and Industrial Change,* 47 Minn L Rev 505 (1963); Givens, *The Search for an Alternative to Protection,* 30 Fordham L Rev 17 (1961).

[12] For expansion of this approach, see the United Automobile Workers' Ford and General Motors agreements of 1982, described in UAW-Ford Report (spec ed Feb 1982); Holusha, *Ford Workers Accept New Contract By Wide Margin,* NY Times, Mar 1, 1982, at 1, 15; see Raskin, *The Cooperative Economy,* NY Times, Feb 14, 1982, §3, at 1.

For an early background, see Chernick & Naeb, *Legal and Political Aspects of Integration of Unemployment Insurance and SUB Plans,* 12 Indus & Lab Rel Rev 20 (1958); Note, 1962 Duke LJ 605. *see also* Carmichael v Southern Coal & Coke Co, 301 US 495 (1937).

For contract issues, *see* Heheman v EW Scripps Co, 661 F2d 1115 (6th Cir 1981).

will be by attrition only.[13] This option, although attractive to management, perhaps combined with severance pay or SUB payments, may not meet the institutional interests of trade unions which would lose membership as the number of employees in an enterprise or category dwindled should attrition occur.[14] Similarly, it cannot replace work rules to the extent actually based on safety or related concerns.[15]

An employer may want to consider yet another option, that of making contributions to industry funds, as in the musical recording industry, designed to create new jobs. A large multi-employer group is often necessary to make this effective. In addition, annual or other long-term employment may be guaranteed.[16] Such an approach is the easiest to implement, but also the least necessary, if employment in the industry is expanding.[17]

Finally, it may be agreed that some job-threatening practices, such as subcontracting, will not be utilized in specified fields.[18]

Changes in the workplace due to technological innovation may raise collective bargaining issues. Bargaining may not be legally required over whether to close down an operation entirely, as distinct from how to handle the dislocation.[19] If the employees are represented by a union, however, bargaining over the options to be considered is necessary under labor law and is also normally highly desirable as a practical measure to minimize conflict entirely apart from

---

[13] See note 4 *supra.*

[14] See Perlman, A Theory of the Labor Movement (1928); Zausner, Unvarnished (1941).

[15] This factor is undoubtedly of increased importance due to enhanced awareness of job risks. Occupational Safety and Health Act, 84 Stat 1590 (1970), *as amended* (codified at 29 USC §651 *et seq*); *e.g.*, Gomes et al, At Work in Copper (INFORM, Inc 1979); *See also* Note, 49 Fordham L Rev 432 (1980).

[16] See Raskin, *Pay by the Hour? The Week? The Year? For Life?*, NY Times Magazine, Sept 4, 1966, at 6; Chernick, Economic Effects of Steady Employment and Earnings (1942); Chernick & Hellickson, Guaranteed Annual Wages (1945); Roosevelt, 3 State of the Union Messages of the Presidents 2840-41 (Chelsea House 1966); Jones, *Guaranteed Pay Proposed by United States*, NY Times July 27, 1966, at 35, col 1.

[17] *See* NY Times, Feb 4, 1966, at 14, col 1; Federal Trade Commission, New York Regional Office, Report on Hearings on Obstacles to Expansion of the Building and Construction Industry (1973); Editorial, *New Deal for Construction*, 119 America 93 (Aug 17, 1968); HR 1445, 96th Cong, 1st Sess (1979).

[18] *E.g.*, note 12 supra; *see generally* on validity, when enforced by pressure against third parties, Local 189, Amalgamated Meat Cutters v Jewel Tea Co, 381 US 676 (1965); Handler & Zifchak, *Collective Bargaining and the Antitrust Laws*, 81 Colum L Rev 495 (1981). In the garment industry involving a special exemption, *see* Jou-Jou Designs, Inc, v ILGWU, 643 F2d 905 (2d Cir 1981); Danielson v Joint Board, 494 F2d 1230 (2d Cir 1974). *See generally* Berman Enters Inc v Local 333, Intl Longshoremen's Assn, 644 F2d 930 (2d Cir 1981); Equitable Gas Co v NLRB, 637 F2d 980 (3d Cir 1981); Naylor, *Subcontracting, Penalty Clauses and Plant Removals*, 30 Drake L Rev 203 (1980).

[19] First Natl Maintenance Corp v NLRB, — US —, 101 S Ct 2573, 69 L Ed 2d 318 (1981); NLRB v Robin Am Corp, 667 F2d 1170 (5th Cir 1981); Heinz, *The Partial Closing Conundrum*, 1981 Duke LJ 71. See Note, *Labor Law Problems in Plant Relocation*, 77 Harv L Rev 1100 (1964).

its legal necessity. Unions and employees may themselves suggest options to deal with plant closing or other displacements, ranging from contract modifications in return for greater job security[20] to purchase of the plant.[21] For these reasons, as well as to minimize inevitable conflict, consultations with employees even in the absence of union representation may be important.

Conflicts also often arise between employee groups as to who will handle new work, and between trade unions as to who will represent employees doing it.[22] Those previously handling related work may have a greater stake and, hence, a greater likelihood of success in such cases, but experience indicates that the decision of management as to who should do the work will be of great importance in determining the ultimate result.[23]

§2.03 Conflicts over Job Security Arrangements

Direct combat over job security between employers and employees is common. In recent decades, these direct labor-management disputes have rarely resulted in successful legal intervention unless a national emergency dispute was created.[24] Efforts to ban work rules designed to protect jobs by legal action have been almost uniformly unsuccessful,[25] perhaps because of the reluctance of the courts to engage in the kind of conflict to the bitter end likely to result where a direct threat to employees' livelihoods is posed by legal action.[26] Where options such as attrition, retraining, severance payments or job creation by trust funds are offered, however, and the alternative is loss of work to competitors, or a net expansion of jobs for the employee group is insured, resistance to abolition of even historically rooted work rules may be overcome

---

[20] Note 12 *supra; see generally* American Assembly, Automation and Technological Change (1962); Industrial Relations Research Assn, Collective Bargaining and Productivity (1965); Somers, Adjusting to Technological Change (1963); Kirshy, Automation and Collective Bargaining (1964).

[21] See Wayne, *Joys of Fleeing the Corporate Stable,* NY Times, Nov 15, 1981, §3, at 1.

[22] *See* 29 USC §10(k); Goldstein, *Electronic Journalism and Union Rivalry,* 29 Labor LJ 137 (1978); Note, *Work Assignment Disputes Under the National Labor Relations Act,* 73 Harv L Rev 1150 (1960); Note, 26 UCLA L Rev 805 (1979).

[23] Goldstein, *supra* note 22.

[24] See note 4 *supra;* on inchoate law restraining the parties in such cases, *compare* Berle, Power Without Property (1959); Kennedy, The Burden and the Glory 194-95 (Nevins ed 1964)(steel price controversy).

[25] See §§2.01, 2.02.

[26] *See* Dulles, Labor in America (1949); Perlman, A Theory of The Labor Movement (1928); Zausner, Unvarnished (1941); Barnett, Machinery and Labor (1926); *see, e.g.,* Bedford Cut Stone Co v Journeyman Stone Cutters Assn, 274 US 37 (1927); Coronado Coal Co v United Mine Workers, 268 US 295 (1925); Duplex Printing Co v Deering, 254 US 443 (1921); Hitchman Coal & Coke Co v Mitchell, 245 US 229 (1917); Lawlor v Loewe, 235 US 522 (1914); Lowe v Lawlor, 208 US 274 (1908); Handler, Cases and Materials on Labor Law 1-14, 101-11, 304-06 (1st ed 1944); Handler, *Labor and Antitrust: A Bit of History,* 40 ABA Antitrust LJ 233 (1971). *See* discussion in Atkinson v Sinclair Ref Co, 370 US 238, 274-79 (1962).

with union acquiescence or even agreement.[27]

Legal issues in this area are often generated by secondary boycotts, i.e., union pressure on third parties not to deal with employers who have violated union work rules or to use only items produced by others who have complied with work rules.[28] These issues arise under both labor law and antitrust provisions. Throughout the first third of the twentieth century, secondary pressures were considered unlawful restraints of trade under the Sherman Act of 1890[29] and prior sources of law.[30] The Clayton Act, in 1914,[31] declaring labor not to be a commodity, initially failed to reverse this judicial approach.[32] This contributed to pressure for passage of the Norris-LaGuardia Act in 1932.[33] This act, as an exercise of congressional power over lower federal court jurisdiction, banned many types of injunction in labor disputes. It was held, together with the Clayton Act, to generate a statutory antitrust exemption for unilateral union activity.[34] At about the same time, it was also held that such unilateral labor activity was outside the scope of the Sherman Act itself, which was found primarily aimed at commercial restraints in the marketplace.[35]

A nonstatutory exemption protecting activities necessary to collective bargaining—as contemplated by the National Labor Relations Act as passed in

---

[27] See §§2.01, 2.02.

[28] *E.g.,* NLRB v ILA, 447 US 490 (1980); National Woodwork Mfrs Assn v NLRB, 386 US 612 (1967); cases cited in Handler & Zifchak, *Collective Bargaining and the Antitrust Laws,* 81 Colum Law Rev 495 (1981).

[29] 15 USC §1; see cases cited note 26 *supra.*

[30] See Handler, Cases and Materials, *supra* note 26, at 1-14; Handler, *Labor and Antitrust, supra* note 26; Dulles, *supra* note 26. *Compare In re* Debs, 158 US 564 (1895); Frankfurter & Greene, The Labor Injunction (1930).

[31] §6, 38 Stat 731 (1914) (codified at 15 USC §17). Even today, this provision is rarely construed to bar antitrust involvement where collective bargaining is not affected. Radovich v National Football League, 352 US 445 (1957); Anderson v Shipowners Assn, 272 US 359 (1926); California State Council of Carpenters v Associated Gen Contractors, 648 F2d 527 (9th Cir 1980). *Compare* NLRB v Truck Drivers Local 449, 353 US 87, (1957); 57 Colum L Rev 1172 (1957); Industrial Assn v United States, 268 US 64 (1925); 60 Colum L Rev 205 (1960).

[32] *E.g.,* Bedford Cut Stone Co v Journeyman Stone Cutters Assn, 274 US 37 (1927); Duplex Printing Co v Deering, 254 US 443 (1921); *see also* Hitchman Coal & Coke Co v Mitchell, 245 US 229 (1917); Handler, Cases and Materials, *supra* note 26, at 1-14, 101-11, 304-06; Handler, *Labor and Antitrust, supra* note 26; Frankfurter & Greene, *supra* note 30.

[33] 47 Stat 70 (1932) (codified at 29 USC §101 *et seq,* upheld in Lauf v EG Shinner & Co, 303 US 323 (1938); *see* Senn v Tile Layers Union, 301 US 468 (1937); Truax v Corrigan, 257 US 312 (1921).

[34] United States v Hutcheson, 312 US 219 (1941); *see also* Milk Wagon Drivers Union v Lake Valley Farm Prods, 311 US 91 (1940); Handler & Zifchak, *Collective Bargaining, supra* note 28. For applications, *see, e.g.,* American Fedn of Musicians v Carroll, 391 US 99 (1968); Hunt v Crumboch, 325 US 821 (1945).

[35] Apex Hosiery Co v Leader, 310 US 469 (1940).

1935[36] and amended by the Taft-Hartley Act in 1947[37]—has also been developed by the courts.[38] The primary test under the nonstatutory exemption is whether any restraint is closely related to core areas of employer-employee relations, such as but not necessarily limited to wages, hours and working conditions,[39] work practices of the employer involved in regard to the employees concerned or in the same labor pool[40] or practices affecting work normally done by those employees.[41]

Work rules involving combinations with employers,[42] rather than mere unilateral union activity enforced against third parties, in a manner leading to significant restraint upon the marketplace may, however, remain subject to antitrust challenge if found not intended to be protected or left unregulated by the national labor laws.[43] This might be the case if a bottleneck[44] position of some employees were used to control an entire industry[45] in what the courts consider to be an unreasonable manner;[46] for example, restrictions on third parties freezing long-term barriers to innovation after rejection of other less

---

36 49 Stat 449 (1935) (codified at 15 USC § 151 *et seq*), upheld in NLRB v Jones & Laughlin Steel Corp, 301 US 1 (1937).

37 61 Stat 136 (1947) (codified at 29 USC §§141-197).

38 *See* Local 189, Amalgamated Meat Cutters v Jewel Tea Co, 381 US 676 (1965); United States v Hutcheson, 312 US 219 (1941).

39 Local 189, Amalgamated Meat Cutters v Jewel Tea Co, 381 US 676 (1965); *see generally* cases cited note 40 *infra;* Editorial Note, *Interface of National Labor and Antitrust Policies,* 23 Labor LJ 115 (1982).

40 *See* Home Box Office v Directors Guild of America, 531 F Supp 578 (EDNY 1982); *see also* Larry V Muko Inc v Southwestern Pa Bldg & Constr Trades Council, 670 F2d 421 (3d Cir 1982).

41 *See, e.g.,* Milk Wagon Drivers Union v Lake Valley Farm Prods, 311 US 91 (1940); Los Angeles Meat & Provision Union v United States, 371 US 94 (1962); Columbia River Packers Assn v Hinton, 315 US 143 (1942); United States v Brims, 272 US 549 (1926).

42 Connell Constr Co v Plumbers Local 100, 421 US 616 (1975); UMW v Pennington, 381 US 657 (1965); Allen Bradley Co v Local 3, IBEW, 325 US 797 (1945).

43 *See* Connell Constr Co v Plumbers & Steamfitters Local 100, 421 US 616, 625-30 (1975); Frito-Lay, Inc v Retail Clerks, 629 F2d 653 (10th Cir 1980); Larry V Muko, Inc v Southwestern Pa Constr Trades Council, 670 F2d 421 (3d Cir 1982).

44 *See e.g.,* Otter Tail Power Co v United States, 410 US 366 (1973); Times Picayune Pub Co v United States, 345 US 594 (1953); Associated Press v United States, 326 US 1 (1945); United States v Lehigh Valley RR, 254 US 255 (1920); United States v Reading Co, 253 US 26 (1920); United States v Reading Co, 226 US 324 (1912); United States v Terminal RR Assn, 224 US 383 (1912); United States v AT&T, 524 F Supp 1336, 1351-53 (DDC 1981).

45 *See* Hartford-Empire Co v United States, 323 US 386 (1945); Sullivan, Antitrust §48 (1977).

46 *See* North Am Soccer League v National Football League, 670 F2d 1249 (2d Cir 1982); Eiberger v SONY Corp of America, 459 F Supp 1276 (SDNY 1978), *affd in part,* 622 F2d 1068 (2d Cir 1980); Note, 58 Colum L Rev 673 (1958). On applicability of the rule of reason, see cases cited, note 40 *supra;* Berman Enters Inc v Local 333, 644 F2d 930 (2d Cir 1981) (alternative holding).

restrictive alternatives.[47]

The chief battleground in litigation over work rules to date, however, has been the secondary boycott provisions of the National Labor Relations Act added by the Taft-Hartley Act[48] and amended by the Landrum-Griffin Act in 1959 to include *hot cargo*[49] agreements between employers and unions not to handle goods made or sold by third parties. The courts have interpreted these provisions as not prohibiting work rules designed to protect or recapture work traditionally performed by the employee group involved.[50] The distinctions between permissible primary boycotts of participants in industrial conflict and prohibited secondary pressure on third parties has spawned numerous areas of conflicting interpretations.[51]

National Labor Relations Board and private suits for injunctions and private suits for damages are authorized for violations of the secondary boycott provisions.[52] Exemptions for the garment and construction industries, the former of wider scope, apply.[53]

§2.04 Jurisdictional Disputes Involving Union Representation

Jurisdictional conflicts as to which union will represent a given group of employees are often triggered by innovation. These frequently produce legal and industrial disputes among both the contending groups and the employer.[54]

Among the legal provisions which may be brought into play are the second-

---

[47] While *the least* restrictive alternative is not as such required, *see* Continental TV, Inc v GTE Sylvania Inc, 433 US 36 (1977), unnecessarily drastic restrictions are by definition not reasonably necessary and may be stricken down.

[48] 61 Stat 136 (1947), *as amended* (codified at 29 USC §§158(b)(4), 160(j)(1), 303(b)); see Lesnick, *Gravamen of the Secondary Boycott,* 62 Colum L Rev 1363 (1962).

[49] 73 Stat 519 (1959) (codified at 29 USC §158(b)(4), (e)).

[50] NLRB v International Longshoremen's Assn, 447 US 490 (1980); National Woodwork Mfrs Assn v NLRB, 386 US 612 (1967); see Lesnick, *Job Security and Secondary Boycotts: The Reach of NLRA §§8(b)(4) and 8(e),* 113 U Pa L Rev 1000 (1965); Previant, *Boycotts Under the 1959 Amendments: The Full Circle From Duplex v. Deering?,* Proc 13th Ann NYU Conf on Labor 141 (1960); Cox, *The Landrum-Griffin Amendments to the National Labor Relations Act,* 44 Minn L Rev 257 (1959). *See also* Handler & Zifchak, *supra* note 28.

[51] *See, e.g.,* NLRB v Pipefitters, 429 US 507 (1977); NLRB v Servette, Inc, 377 US 46 (1964). *Compare* NLRB v Retail Store Employees Union, 447 US 607 (1980) *with* NLRB v Fruit & Vegetable Packers Local 760, 377 US 58 (1964); see Castle & Pegnetter, *Secondary Picketing; The Supreme Court Limits the Fruit Tree Exception,* 33 Labor LJ 3 (1982).

[52] 29 USC §§160(k)(l), 187.

[53] Connell Constr Co v Plumbers & Steamfitters Local 100, 421 US 616, 628-33 & n 13 (1975); *see* Jou-Jou Designs, Inc v ILGWU, 643 F2d 905 (2d Cir 1981); Danielson v Joint Board, 494 F2d 1230 (2d Cir 1974) (garment industry); *compare* Note, 81 Colum L Rev 1702 (1981) (construction).

[54] See **§2.02.**

ary boycott provisions discussed in the preceding section, provisions barring a union from using coercion to demand bargaining rights where it does not actually represent the employees involved, §10(k) of the National Labor Relations Act,[55] which empowers the National Labor Relations Board to decide jurisdictional disputes once an illegal strike or strike threat to coerce work assignments is charged,[56] and §301 of the Taft-Hartley Act,[57] conferring jurisdiction on the federal courts in suits for breach of agreements between unions,[58] as well as for breach of agreements between unions and employers. Section 301 has been held to generate a federal law of collective agreements based upon the policies of the national labor laws.[59] It has also been held to support an inference of a no-strike pledge from the presence of an arbitration clause,[60] which is enforceable in suits under the section, including those where jurisdictional provisions of contracts are involved.[61]

Litigation under these various sources of legal intervention, however, has failed to lead to prompt or lasting resolution of controversies unless accommodations between the parties themselves have been found. Experience indicates that when faced with jurisdictional rivalry, management should choose the group of employees with the best ability to do the work involved, in which

---

[55] 29 USC §160(k). On the preceding phrase, *see* 29 USC §158(b)(1), (2).

[56] *See* 29 USC §158(b)(4)(D).

[57] 61 Stat 136 (1947) (codified at 29 USC §185).

[58] United Assn of Journeymen of the Plumbing & Pipefitting Industry v Local 334, — US —, 69 L Ed 2d 280 (1981).

[59] Textile Workers v Lincoln Mills, 353 US 448 (1957); *see* General Drivers v Riss & Co, 372 US 517, 520 (1963); Charles Dowd Box Co v Courtney, 368 US 502 (1962); Lesnick, *State Court Injunctions and the Federal Common Law of Labor Contracts,* 79 Harv L Rev 757 (1966); Givens, *The Validity of Delayed Awards Under Section 301, Taft Hartley Act,* 9 Arb J 161 (1961); Note, 75 Yale LJ 877 (1966).

[60] Local 174, Teamsters v Lucas Flour Co, 369 US 95 (1962); Isaacson, *The Grand Equation: Labor Arbitration and the No-Strike Clause,* 48 ABAJ 914 (1962).

[61] Carey v Westinghouse Elec Corp, 375 US 261 (1964); UMW v Consolidation Coal Co, 666 F2d 806 (3d Cir 1981). On enforcement of arbitration provisions generally, *see* United Steelworkers v American Mfg Co, 363 US 564 (1960); United Steelworkers v Warrior & Gulf Nav Co, 363 US 574 (1960); United Steelworkers v Enterprise Wheel & Car Co, 363 US 593 (1960) ("Steelworkers Trilogy"); Committee on Labor and Social Security Legislation, *Report Concerning (1) The Role of Judge and Arbitrator in Labor Arbitration and (2) Injunctions Against Strikers in Breach of Contract,* 20 Rec AB City NY 37 (1965).

See also Boys Markets, Inc v Retail Clerks, 398 US 235 (1970); Brotherhood of RR Trainmen v Chicago River & Ind R Co, 353 US 30 (1957); Note, 85 Harv L Rev 636 (1972); Note, 90 Harv L Rev 790 (1977). For an alternate approach to no-strike clause enforcement, *see* Ruppert v Egelhofer, 3 NY2d 576, 148 NE2d 129 (1958), discussed in Givens, *Section 301, Arbitration and the No-Strike Clause,* 11 Labor LJ 1005 (1960) and *Injunctive Enforcement of Arbitration Awards Prohibiting Strikes,* 17 Labor LJ 292 (1966).

On local arbitration laws, *see* Local Div 589 v Commonwealth, 666 F2d 618 (1st Cir 1981).

event its choice will influence whatever tribunals are invoked.[62]

Where one contending group is sufficiently powerful, of course, this may not be possible unless an effective counter-strategy is available, and unless any legal rulings obtainable can in fact be enforced, often the most difficult element. Accommodation at times may be attained through division of the elements of the work into subelements, but this may lead to inefficiency, friction and failure to do the job properly due to difficulties in communication or controversies during performance of the work.[63] Another approach to accommodation which is rarely possible but which can be dramatic is to find separate job opportunities for each contending group as part of a larger pie generated by a large-scale innovation.[64]

§2.05 The Free Election Model Contrasted With Specialized Groups of Employees: The Construction and Garment Industries

The National Labor Relations Act, both in its original 1935[65] version and as amended in 1947,[66] is based on the premise that the employees working for an employer select their representative, if necessary by secret ballot,[67] which in turn bargains for all of the employees.[68] The representative is then bound by implication to represent all fairly.[69]

Jurisdictional disputes, however, generally involve efforts to force an employer to hire different *employees* rather than merely to deal with a different union.[70] Ownership of work by a group of employees is a traditional industrial reality where jobs are short-term in nature so that workers have more connection with the industry itself than with a particular employer, or where employers are often thinly financed contractors for others who are the truly responsible parties in the relationship.[71] Such conditions may compel control over subcontractors if effective bargaining is to occur at all.

Based on the prevalence of these realities in the construction and garment

---

[62] Goldstein, *Electronic Journalism and Union Rivalry,* 29 Labor LJ 137 (1978).

[63] Egan, *To Err is Human Factors,* 85 Tech Rev 23 (Feb 1982).

[64] See **§§2.01, 2.02.**

[65] 49 Stat 449 (codified at 15 USC §151 *et seq*).

[66] 61 Stat 136 (codified at 29 USC §§141-197).

[67] 29 USC §159.

[68] JI Case Co v NLRB, 321 US 332 (1944); Weyand, *Majority Rule in Collective Bargaining,* 45 Colum L Rev 556 (1945); Finkin, *Limits of Majority Rule in Collective Bargaining,* 64 Minn L Rev 183 (1980).

[69] Steele v Louisville & N R, 323 US 192 (1944), and numerous cases following it, *e.g.,* those cited in American Postal Workers v American Postal Workers, 665 F2d 1096, 1105 (DDC 1981). *See also* Givens, *Federal Protection of Employee Rights Within Trade Unions,* 29 Fordham L Rev 259 (1960).

[70] *See, e.g.,* Goldstein, *Electronic Journalism and Union Rivalry,* 29 Labor LJ 137 (1978).

[71] *See, e.g.,* Jou-Jou Designs, Inc v ILGWU, 643 F2d 905 (2d Cir 1981).

industries, Congress exempted them—the garment field more completely than construction—from secondary boycott restrictions and also by implication from antitrust scrutiny of secondary boycott combinations with employers.[72]

Types of pressures which may be inevitable where this model applies have at times led to treatment of work as itself a franchise subject to trade without regard to the interests of the employees involved.[73] This tends to be perpetuated, even where these conditions do not obtain, by legal rules which give importance to the existence of a contract despite employee wishes to change their representative. This occurs under the contract bar rule which limits employee opportunity to change representatives to a narrow time window at contract expiration time (the very time when it is likely to be most counterproductive),[74] and by availability of state court injunctions against pressure by noncontracting union which may in fact speak for the employees.[75] The contract bar rule is based on the need to preserve contract stability throughout the contract period, but in nonlabor contexts a change of the representatives of a contracting party or third-party beneficiary obviously does not void the contract. Limitations on the bar have thus been proposed.[76]

§2.06 Initial Considerations When Innovation Creates Jobs

When innovation creates new jobs, the innovator has an opportunity to set the groundwork for dealing with future labor relations issues, and faces the challenge of seeking to avoid numerous legal pitfalls which surround the employment relationship. Among areas which should be considered at the earliest possible stage in structuring new jobs created by innovation are illegal

---

[72] **§2.03.**

[73] Comments, *see, e.g.,* 59 Colum L Rev 810 (1959); 69 Yale LJ 1393, 1405-06 (1960); NY Times, Feb 16, 1959, at 1, col 8. Laws against labor bribery may treat only symptoms of lack of adequate representation of true employee interests. *See, e.g.,* United States v Ryan, 350 US 299 (1956); United States v Ricciardi, 357 F2d 91 (2d Cir), *cert denied,* 384 US 942 (1966); United States v Gard, 344 F2d 120 (2d Cir 1965), decided under 29 USC §186.

[74] See Editorial Note, *When Has Substantial Agreement Been Reached For Application of the Contract Bar Rule,* 23 Labor LJ 121 (1982) and cases cited therein; Naumoff, *An Analysis of the Contract Bar Doctrine,* 7 Labor LJ 197 (1956).

[75] *See* J Radley Metzger Co v Fay, 4 AD2d 436, 166 NYS2d 87 (1st Dept 1957); Jarvis Surgical Co v Davis, 15 Misc 2d 1035, 186 NYS2d 357 (Sup Ct 1958); 59 Colum L Rev 810 (1959). For contrast, *see* Kaynard v Mego Corp, 33 F2d 1026 (2d Cir 1980); Wilkes-Barre Pub Co v Newspaper Guild, 504 F Supp 54 (MD Pa 1980). On employer-assisted organizations, *see* American President Line v NLRB, 340 F2d 490 (9th Cir 1965).

[76] Committee on Labor Law, Federal Bar Council, *Sweetheart Contracts,* 23 Indus & Lab Rel Rev 105 (1969), *reprinted in* 115 Cong Rec S6318 (daily ed June 12, 1969). On a related problem of short time limits to challenge lack of majority status, *see* Pick-Mt Laurel Corp, 1981-2 NLRB Dec (CCH) ¶18,505 (Nov 17, 1981).

discrimination, job safety, employee choice as to union representation, and trade secrets.

Under federal and numerous state and local provisions, it is illegal to discriminate on a wide variety of grounds defined as invidious, such as race, creed, color, national origin, sex,[77] age[78] and, frequently, handicap.[79] This may require recruitment from potential minority or disadvantaged applicants who may be qualified. It may also mean that any tests or other criteria used in hiring must be job-related if they disfavor any protected disadvantaged group.[80] Showing that tests or job qualifications that militate against particular groups are in fact job-related, or bona fide occupational qualifications, is often difficult and uncertain, especially if studies are undertaken after the fact. Job descriptions, if accurate, may become useful in these areas, but are dangerous if too detailed so as to become straitjackets.[81] Additional issues may be raised if existing employees are given preferred status in obtaining new jobs (often desirable in dealing with dislocation) and if the composition of the work force reflects any past discrimination.[82]

Job safety is another area which warrants advance consideration. Federal, state and local laws relating to occupational health and safety must be taken into account, both in technological choices and in work force arrangements.[83] An additional consideration in this area is that charges of discrimination may occur where selection of employees is based on perceived ability of employees to meet difficulties.[84]

The innovator should also remember in the advance planning stages that

---

[77] Civil Rights Act of 1964, tit VII, 42 USC §2000-e. *See also* Wood v Southwestern Bell Tel Co, 637 F2d 1188, 1191 (8th Cir 1981), *cert denied,* 101 S Ct 142 (1982) (harassment due to filing complaint); Givens, *Manhart in the Light of Bakke,* 4 Employee Rel LJ 330 (1978).

[78] Age Discrimination in Employment Act, 81 Stat 602 (1967), *as amended* (codified at 29 USC §621 *et seq*); *see, e.g.,* Haring v CPC Intl, Inc, 664 F2d 1234 (5th Cir 1981).

[79] *See* 29 USC §706; EE Black, Ltd v Marshall, 497 F Supp 1088 (D Hawaii 1980); S Rep No 316, 96th Cong, 1st Sess (1979).

[80] *See* Pothard v Rawlinson, 433 US 321 (1977); Firefighers Inst v City of St Louis, 616 F2d 350 (8th Cir 1980); Dickerson v United States Steel Corp, 472 F Supp 1304 (ED Pa 1978). Vulcan Socy v Civil Serv Commn, 360 F Supp 1265 (SDNY), *modified,* 490 F2d 387 (2d Cir 1973).

[81] Giving employees more challenging work not within too-specific and mundane job descriptions may be challenged as (1) underpayment of the employee, or, paradoxically, also (2) unfair preselection of the employee for promotion by affording better training not given to others. Thus does our legal system chase its own tail. *See* Stephens, *Supervisor Performs Bargaining Unit Work,* 31 Labor LJ 683 (1980).

[82] This may arise due to statistical tests relevant to prima facie showings of discrimination. *See* McDonnell Douglas Corp v Green, 411 US 792 (1973).

[83] See **§2.02**

[84] This may be rebutted by showing that a challenged requirement is a bona fide occupational qualification, see note 82 *supra. See generally,* Severo, *Should Firms Screen The Workplace or The Worker?,* NY Times, Sept 28, 1980. On another issue, see Sobel, *Testing the Psyche Can Strain Belief,* NY Times, Sept 28, 1980.

federal and state labor laws ban employer coercion or discrimination directed at collective employee activity or choice of union representation or lack thereof, as well as at establishment of or assistance to a labor organization.[85]

Finally, as to trade secrets, the innovator should be advised that if employees are to be privy to secret data, appropriate restrictions on disclosure should be formally contracted for,[86] but excessively sweeping bans on competition avoided.[87]

§2.07 The Importance of Employee Involvement in Innovation

Both conflict and cooperation are inherent in all organizations. The balance between them and how it is structured has great importance in promoting or retarding innovation, and in generating or avoiding legal and industrial confrontations.

The importance of involvement of higher-level employees in developing innovative methods is obvious;[88] involvement of working-level personnel is sometimes overlooked. It is critical not to overlook them because they know in detail how an operation is working and how it can be made to work (or, if they are so minded, not work).[89]

Conflict is, of course, inevitable whenever ways of doing things are to be changed. How such conflict is handled may determine whether it results in deadlock, accommodation, victory for one side or, at times, options that bene-

---

[85] *E.g.,* 29 USC §§157-159. For recent developments, *see, e.g.,* Lederer, *Wright Line or Spur Track?,* 334 Labor LJ 67 (1982); Statler Indus, Inc v NLRB, 644 F2d 902 (1st Cir 1981); Whittlesen, *Protection Against Discharge Without Cause,* NYSt BJ 101 (Feb 1981) (collecting cases); Note, 93 Harv L Rev 1816 (1980); Savodnik v Korvettes', Inc, 488 F Supp 822, 825 (EDNY 1980).

[86] *See, e.g.,* Lear Siegler, Inc v Ark-Ell Springs, Inc, 569 F2d 286 (5th Cir 1978); Kodekey Elecs, Inc v Mechanex Corp, 486 F2d 449 (10th Cir 1973).

[87] Hutter, *Drafting Enforceable Employee Non-Competition Agreements,* 45 Alb L Rev 311 (1981); *see* Business Foods Serv, Inc v Food Concepts Corp, 533 F Supp 992 (EDNY 1982); Goldschmid, *Antitrust's Neglected Stepchild: A Proposal for Dealing With Restrictive Covenants Under Federal Law,* 73 Colum L Rev 1193 (1973).

[88] This does not mean it is always encouraged. See Parkinson, Parkinson's Law (1957); MacKenzie, The Time Trap ch 7 (1972); Jay, Management and Machiavelli (1968); March & Simon, Organizations (1967). *Compare* Argyris, Organization and Innovation (1965); Caplow, Principles of Organization (1964).

[89] See Drucker, Concept of the Corporation (1953); *compare* Jacobs, The Economy of Cities (1969); Hall, The Competence Process (1980); Marrow, Management by Participation (1967); Pascale & Athos, The Art of Japanese Management (1981); Ouchi, Theory Z: How American Business Can Meet the Japanese Challenge (1981); Ewing, The Human Side of Planning (1969); Hayes & Abernathy, *Managing our Way to Economic Decline,* 58 Harv Bus Rev 67 (July-Aug 1980) (reprint no 80405); Lewis, *The Profits in Common Sense,* NY Times, Sept 26, 1980, at 35; Walsh, *Industrial Productivity and the Soft Sciences,* 209 Sci 476 (July 25, 1980); Miller, *Better Ideas Need More Watts,* NY Times, Nov 2, 1980, at E9. For congressional recognition, *see* 15 USC §2401 (enacted by 89 Stat 733 (1975)).

fit the variegated interests involved even more than the originally chosen approach did.[90]

The obstacles to affirmative employee involvement in innovation are not readily removed. Most important is fear of potential loss of jobs through facilitating the introduction of labor-saving methods. This fear includes peer pressure against any employee who breaks ranks, or who, as a "beaver," is seen as endangering the position of others. This pressure exists whether or not trade unions formalize it in the shape of bargaining for work restrictions.[91]

The attitudes of trade union organizations can be helpful or harmful depending on history and atmosphere of the relationship. Joint union–management productivity promotion on a firm or industry-wide level is developing in response to threats of job erosion. Use of formalized employee quality circle or similar structures may, however, at times tend to be seen as a threat to union representation, and generate hostility.

Management's own conduct can form an equally effective barrier to employee involvement in innovation. This can take the form of one-way communication downward, unwillingness to listen seriously to employee suggestions and a "do what you are told and don't ask questions" attitude, obsolete where sophisticated types of work are involved.

Another obstacle is created by multiple layers of review of innovative proposals emanating from the working level, middle management or even senior management. Deep layering of review processes may avoid some costly mistakes, but at the perhaps even greater cost of killing most new ideas be they good or bad, and apart from the usually unmeasured cost of operation of the bureaucratic procedures themselves.[92]

Where multiple levels of approval are required, it is almost inevitable that someone in the pathway will feel threatened by the potential success of an innovation sponsored by someone else in the organization. Some may seek to justify the importance of the review role by looking for objections. Even if senior executives make the final decision without requiring affirmative approvals at intermediate levels, the delay generated may be fatal to a project. It may be modified in harmful ways to meet politically motivated objections; the sponsors may also have incurred the hostility of intermediate executives who can later sabotage the innovation. Innovative employees may opt for silence,

---

[90] See Wertheimer, Productive Thinking (1959); DeBono, Lateral Thinking (1970); *see also* Koestler, Insight and Outlook (1949).

[91] On the power of peer pressure, see Asch, *Effects of Group Pressure Upon the Modification of Judgments,* in Readings in Social Psychology 2 (rev ed 1952), Lewin, *Group Decision and Social Change,* in Social Psychology 459 (rev ed 1952); sources cited in Note, 95 Harv L Rev 597, 612 n 44 (1982).

[92] "If too many approvals are required before an idea can be born, the probabilities are overwhelming against it." Tribus, *Applying Science to Industry,* 70 US News & World Rep 35 (Jan 8, 1971); see Givens, Advocacy, ch 14 (Shepard's/McGraw-Hill 1980); *compare also* H Rep No 575, 95th Cong, 1st Sess (1977); Winston Churchill, Address to the House of Commons, Jan 17, 1942, *quoted in* Never Give In! 22 (1965).

or simply leave if stifled by the weight of internal bureaucracy.[93] Such bureaucratic procedures can offset many of the advantages of scale otherwise enjoyed by large organizations.[94]

It is an illusion, however, to think that bureaucratic layering is necessary, inevitable or inherent in large organizations. External pressures for results can force delegation or decentralization within large structures, preserving overall direction while allowing local initiative, often with dramatic results.[95]

A deterrent to delegation can, however, be created by the legal system, to the extent that penalties are imposed on an entire organization because of errors committed within one segment, thus tending to cause higher executives to wish to review individual actions that might cause such penalties.[96] Realism counsels against serious organization-wide sanctions if the organization as a whole is not implicated. A wise response will isolate for central review decisions likely to result in the most serious legal consequences, without subjecting the entire organization to internal second-guessing of individual decisions.

---

[93] They may also, of course, protest externally. *See* Consolidated Edison Co v Donovan, 673 F2d 61 (2d Cir 1982); Davis v Williams, 598 F2d 916 (5th Cir 1979) *cert denied,* 449 US 937 (1980); The Whistleblowers: Report Prepared for the Senate Committee on Governmental Affairs, 95th Cong, 2d Sess (Comm Print 1978); Holden, *Scientist with Unpopular Data Loses Job,* 210 Sci 749 (Nov 14, 1980); Lublin, *Watchdog Has Hard Time Hearing Whistles,* Wall St J, Oct 17, 1980, at 30; Westin, Whistle Blowing: Loyalty and Dissent in the Corporation (1981), *reviewed,* Bus Wk 8 (Nov 24, 1980).

[94] "Increasingly, however, blame for the laggard performance of many large corporations is focusing on their structures and entrenched ways of doing things. A growing body of opinion has it that the 'economies of scale' made possible by bigness often are more than nullified by organizational rigidities and bottlenecks." Klein, *Some Firms Fight Ills of Bigness by Keeping Employee Units Small,* Wall St J, Feb 5, 1982, at 1, col 1. *See also* Carson, *Workers Aren't the Problem . . . Bosses Are,* Wall St J, May 8, 1981; Falvey, *The Benefits of Working With the Best Workers,* Wall St J, May 11, 1981; Lenmark-Ellis, *The Professional Troublemakers,* NY Times, Aug 22, 1981, at 23; Glaser, *Participatory Management,* 210 Sci 962 (Nov 28, 1980).

[95] *E.g., compare* Drucker, Concept of the Corporation (1953) *with* Speer, Inside the Third Reich (1970) in the industrial sphere; in the military field, *compare* Tuchman, Stilwell and The American Experience in China (1970) *with* Cookridge, They Came from the Sky (1965); in the law enforcement field, *compare* Special Committee on Consumer Affairs, *Creation of Consumer Fraud Units in United States Attorneys' Offices in Major Metropolitan Areas,* 32 Rec AB City NY 371 (1977) *with* Federal Trade Commission, Regional Office Operations, Hearings before a Subcommittee of the House Committee on Government Operations, 95th Cong, 1st Sess 66-81 (1977); HR Rep No 575, *supra* note 92; Givens, Advocacy, ch 14 (Shepard's/McGraw-Hill 1980).

[96] *See, e.g.,* 1 ABA Antitrust Sec, The FTC as an Antitrust Enforcement Agency; The Role of Section 5 of the FTC Act in Antitrust ch II(D)(2), at 50 (Monograph No 5, 1981); Kanwit, Federal Trade Commission §11.05 (Shepard's/McGraw-Hill 1979).

§2.08 —Overcoming Obstacles to Affirmative Involvement in Innovation

Means are available to deal with each of the major obstacles to affirmative participation in innovative efforts, although some may be difficult to implement.

The fear of adverse job consequences may be lessened if the economy as a whole or the industry is expanding rapidly, so that anyone displaced can expect to find other employment quickly. The firm itself can also offer job security arrangements through individual action, collective bargaining or customary practices. This is, of course, more realistic for both sides if the economy or industry is expanding.

The importance of conscious shaping of managerial structure and style so as to encourage innovative ideas percolating from below, and affirmative participation in implementing innovations sponsored from above is receiving increasing recognition. Among the means being given increased attention are:

1. Limiting the number of layers of review
2. Delegating increased authority to middle managers and working-level employees
3. Focusing central executive guidance upon generalized objectives and policies and on selection of effective middle managers, rather than upon detailed guidelines
4. Understanding of the value of taking appropriate risks

Replacement of detailed second-guessing of individual decisions or enforcement of detailed instructions by greater reliance on middle management can lead to a far more invigorating organizational climate *if* the managers are effective. The sanction of replacement of managers, if that confidence is found to have been misplaced, rather than imposition of detailed review or detailed written directives from above, is essential if this strategy is to succeed.

Greater reliance on middle-level management means that their views, if persuasive, must be allowed to prevail over data generated by quantitative management information systems. These necessarily only partially reflect actual progress toward organizational goals, and may at times give a short-term bias to evaluation of such progress.

One way of promoting communication without imposing bureaucratic delays as an automatic obstacle to innovation is the negative option procedure.[97] Potentially affected units of an organization are notified of impending action, but are *assumed to acquiesce* unless comments are received within a specified time. Comments may be required to be directed in the first instance to the initiating unit, not to a reviewing body, or be required to be made orally before

---

[97] *See* Federal Trade Commission, Trade Regulation Rule on Negative Option Sales in Commerce, 16 CFR pt 425.

being put in the form of memoranda. Such memoranda often start an internal paper war.

Guidance from central executive levels under this approach consists in development of overall objectives, general policy and major decisions, supported by technical assistance from central sources, not involvement in large numbers of detailed individual decisions at lower levels. Freedom from such involvement frees higher executives for tasks which are otherwise ignored or assigned to separate policy planners who lack responsibility for operations. Separate planners, lacking such responsibility, may themselves become an obstacle to action if they must be consulted on individual decisions.

In many organizations, a palace guard always has persuasive arguments why increased delegation of individual decisions to operating levels will lead to catastrophe. Those with knowledge about obstacles to innovation may be reluctant to communicate with executive levels about blockages at intermediate levels for fear of retaliation. To overcome such information barriers, assurances of anonymity may have to be given by executives or outside teams looking into organizational problems.

Another tool is preparation of a chart showing the steps required to accomplish common organizational tasks.[98] In such an effort, each step at which a piece of paper is generated or a signature required is to be set forth as a separate step. Each step can then be analyzed to determine its value, necessity and cost in the form of expense and delay. Enhanced error due to increased distance from the point of decision to the level of firsthand knowledge of the facts can also be taken into account.

Beyond these means of removing mechanical obstacles is the stimulation of enthusiasm based on the importance of the effort itself.[99] Increasing numbers of innovative organizations are stressing this incentive in recruitment.[100] At higher levels of sophistication in efforts required, monetary rewards are only partially sufficient incentives. The opportunity to make history is a vital ele-

---

[98] See HR Rep No 575, 95th Cong, 1st Sess (1977).

[99] See Hersey, Zest for Work (1955), Herzberg, Mausner & Snyderman, The Motivation to Work (1959); Work in America: Report of a Special Task Force to the Secretary of Health, Education and Welfare (1972); Maslow, Toward a Psychology of Being (1962); McGregor, The Human Side of Enterprise (1960); Givens, *Toward Fuller Use of Human Resources,* 17 Labor LJ 621 (1966); Allport, Becoming: Basic Considerations for a Psychology of Personality, chs 14-16 (1955); Pope John XXIII, Mater et Magistra, pt II, ¶¶82-83 (1961); Teilhard de Chardin, The Future of Man 16-18, 118 (1964); Gibran, *On Work* in The Prophet 27-31 ((Pocket ed 1955); Levenstein, Why People Work (1964); Houseman, Run-Through ch II, at 108 (1972). "Happiness lies not in the mere possession of money; it lies in the joy of achievement, the thrill of creative effort." Franklin D Roosevelt, First Inaugural Address, Mar 4, 1933, NY Times, Mar 5, 1933, *reprinted in* 2 Hofstadter, Great Issues in American History 353 (1958).

[100] This is evident from any issue of the classified section of, *e.g.*, the New York Times or Wall Street Journal. For one of the most dramatic, see *In technology, simplicity is the ultimate sophistication,* NY Times, Oct 10, 1977, at 29, *quoted in* Practising Law Institute, Drafting Documents in Plain Language 87 n 53 (1979).

ment in motivation in these instances.[101] While such motivation is more readily stimulated in glamorous fields which are obviously at the cutting edge of rapidly developing areas, every enterprise is seeking to deal with some kind of boundary situation. This affords the opportunity for drama if creative innovation is being risked. No area of effort need be mundane if its implications are evident.

Opening an internal frontier by unlocking dormant resources within organizations may permit dramatic innovation and productivity gains. Wartime experiences of opposing belligerents and other situations involving external pressures are one guide suggesting this.[102] Since a very large percentage of higher-level organizational energy is frequently consumed internally, even a small reduction in this proportion may yield startling results.

The consequences of unleashing human resources within organizations will powerfully affect legal and other conflicts over innovation. Outright opponents or suspicious allies can become enthusiastic partners *if* they see this to be in their interests.

Affirmative involvement within an organization will not, of course, remove barriers interposed by competitors or external labor groups who may believe they are threatened. Such external opponents must be won over by bargaining, circumvented or, if necessary, sought to be overcome frontally. In such conflicts, antitrust issues[103] or labor law and labor-antitrust issues will often come into play.

---

[101] For abuse of this motive, see Speer, Inside the Third Reich 385-87 (1970).

[102] See note 95 *supra.*

[103] **Ch 3.**

Antitrust and Innovation

Richard A. Givens

3

Antitrust Risks for Innovators

§3.01 Overview

Paradoxically, innovation launches its participants into situations involving antitrust risk even though innovation is virtually always competitive in its effects—and the preservation of opportunities for innovation is one of the objectives of antitrust law. The purpose of the discussion in the first part of this chapter is to suggest how this paradox has developed and to outline strategies and options for dealing with this challenge.

Innovation itself both creates new competition, and indeed sometimes entirely new industries, and creates unique advantages for the innovator. This uniqueness inherent in innovation itself may, of course, be claimed by others to hurt them competitively, and they may turn to the antitrust laws to seek relief. If granted, such relief in turn reacts against one of the original purposes of antitrust law—to stimulate aggressive competition and innovation.

To stimulate innovation, the law also provides affirmative protection of exclusiveness in certain cases in the form of patents, trademarks and similar protection, and protection of trade secrets against industrial espionage or improper disclosure by former employees. These legal protections of uniqueness both modify, and are modified by, antitrust concepts.

Beyond legally provided protections, the innovating enterprise may also seek to preserve or enhance the advantages of uniqueness by further measures of its own or by agreements of various kinds with others. These measures may be encouraged, permitted or prohibited by the antitrust laws and by other legal provisions which are interpreted in the light of antitrust principles.

In the present climate of recognition of the need for enhanced productivity and innovation in the economy, the chances of legal toleration or encouragement of measures necessary to innovation are increased. With this, however, there is also an increased risk that some innovators will overinterpret the shift in climate and read it as permitting additional restrictions which the decision makers may find not essential to innovation, and indeed to violate antitrust requirements designed to safeguard an open economy.

Antitrust issues are likely to become relevant both at the outset and in the implementation of any innovative activity. For example, assume that company A has developed a means of desalinization of sea water using a combination of temperature gradients in the water itself in the immediate continental shelf and solar energy tapped by a new type of process in dry and relatively hot and cloudless areas where irrigation by the water in question is needed. The innovation is radical and offers low-cost irrigation water at costs far below those otherwise encountered by users of publicly or privately supplied water or energy. Among the probable practical questions confronted might be:

1. Can company A enter a joint venture with energy producers, pipeline firms, electric utilities or others to develop the further hardware and software needed to make the innovation practicable?
2. If so, can the joint venture actually market the total product, i.e., offer water to farmers, or must it end its work with the development phase?
3. Can capital be obtained from firms likely to be involved in the affected industries?
4. What will happen when competitors seek legislation or interpretations to block the innovation and lobbyists for company A work, in combination with industry allies, for decisions by executive, legislative, regulato-

ry and judicial bodies that may result in success for the innovation but could put these competitors out of business?

5. If the innovation is dramatically successful and gives company A 90 percent of the irrigation market in a large area, can this itself be attacked as illegal?
6. Can company A agree with distributors of the water that they will carry only A's water, or not redirect the water to nonapproved distributors? Does it matter if company A is seeking a toehold or is already successful, or that ions in the water may affect untreated kinds of pipes?
7. If company A also provides fertilizer, can the two be sold as a package at a big discount?
8. Suppose company A sells water directly, but also sells it at a lower internal price to a fertilizer subsidiary which (a) operates its own farms, or (b) sells the package of water/fertilizer in partial competition with A? Does this affect A's ability to suggest water prices to water distributors and is it illegal in and of itself?
9. Can company A amass patents on eight different desalinization methods, use only some, and keep the others as "blocking" patents? Does it matter whether A is already dominant at the time? Does it matter whether A developed the inventions or bought the patents, and if so, whether it was dominant at the time?
10. If state X adopts a rule that bars (a) desalinized water, (b) desalinized water piped in for over 100 miles, or (c) more blatantly, water piped in from an ocean not bordering the state, can this be attacked under the antitrust laws or the federal interstate commerce clause affected by the spirit of the antitrust laws?

There are no simple answers to any of these questions. Indeed, there are no rules which will supply definitively reliable answers to most of them even if the facts are unearthed to the extent such unearthing is possible prior to litigation.

It is possible, however, to discern the legal elements which influence decision makers, making some answers more likely than others under different circumstances. At the most elementary level, the Adam Smith model of the open economy in which each participant can win by making a better mousetrap should be kept in mind. Anything consistent with that model starts the game ahead. Anything contrary to the model has at least one strike against it, though it may nevertheless be upheld because of the circumstances or the precise contours of legal authorities.

§3.02 Principal Antitrust Provisions

Several interlacing antitrust provisions are usually involved directly or indirectly in each factual situation addressed by the courts. Most frequently invoked is §1 of the Sherman Act of 1890, which prohibits: "[e]very contract, combination in the form of trust or otherwise, or conspiracy, in restraint of trade or commerce among the several States, or with foreign nations. . . ."[1] Section 1 is complemented by §2 of the act, which penalizes every person "who shall monopolize, or attempt to monopolize . . . any part of such trade. . . ."[2] Section 2 is the principal antitrust law applicable to purely unilateral conduct involving no combination.

The Sherman Act, as the basic United States antitrust law,[3] embodies a national policy favoring reliance on the marketplace to produce optimal economic results,[4] except where this policy is altered by more specific differing laws.[5] The Sherman Act is thus frequently used as a basis for interpreting other antitrust and related statutes.[6]

Perhaps most often debated in sales negotiations but less often successfully invoked in court is §2 of the Clayton Act of 1914, as amended by the Robinson-Patman Price Discrimination Act in 1936.[7] The core provision of the act, §2, bars discrimination in price in sales of commodities in commerce for resale within the United States in general where the effect may be substantially to lessen competition. The prohibition is subject to defenses where the discrimination is either given in good faith to meet competition, or makes only due allowances for differences in cost resulting from differing methods or quantities of sale or delivery.

Section 3 of the Clayton Act to a large extent overlaps a subject covered under §1 of the Sherman Act. It prohibits certain *tying* agreements conditioning sale of an item on not purchasing another item from competitors.[8]

The antimerger provision of the Clayton Act, §7, prohibits most corporate stock or asset acquisitions where "the effect . . . may be substantially to lessen competition; or to tend to create a monopoly" in any line of commerce in any section of the country.[9]

---

[1] 15 USC §1.

[2] Id §2.

[3] *See* United States v United States Gypsum Co, 438 US 422 (1978).

[4] *See* Catalano, Inc v Target Sales, Inc, 446 US 643 (1980); California Retail Liquor Dealers Assn v Midcal Aluminum, Inc, 445 US 97 (1980); National Socy of Professional Engrs v United States, 435 US 679 (1978); United States v Trenton Potteries Co, 273 US 392 (1927).

[5] *See generally* Report of the National Commission for Review of Antitrust Laws and Procedures, pt II (1979); Note, 58 Colum L Rev 673 (1958). *See* Silver v New York Stock Exch, 373 US 341 (1963).

[6] *E.g.*, United States v United States Gypsum Co, 438 US 422 (1978).

[7] 15 USC §13.

[8] *Id* §14. §§1 and 2 of the Sherman Act may also apply. *See generally* United States Steel Corp v Fortner Enters, Inc (Fortner II), 429 US 610 (1977). As to industry-specific regulatory statutes, *see, e.g.*, Black v Magnolia Liquor Co, 355 US 24 (1957); Federal Maritime Commn v Aktiebolaget Svenksa Amerika Linien, 390 US 238 (1968).

[9] 15 USC §18. Traditionally, technological gains were not only not a defense in merger cases, but could be cited as a reason for claiming a merger to be anticompetitive. *See* FTC v Procter & Gamble Co, 386 US 563, 580-87 (1967); Sullivan, Antitrust §214

In a far broader provision deemed to embrace all of the above and their spirit as well, §5 of the Federal Trade Commission Act of 1914, as amended, prohibits "unfair methods of competition" and "unfair or deceptive acts or practices" in or affecting interstate commerce.[10]

Foreign antitrust laws, including those of the European Economic Community,[11] and state antitrust laws are also often important,[12] as are patent and other nonantitrust laws.[13]

Finally, the interstate commerce clause of the federal Constitution[14] both measures the outer limits of federal antitrust authority,[15] and by its own force bars state or local action unduly burdening commerce[16] such as discriminatory

---

(1977). The current trend is to give more favorable consideration to types of transactions likely to promote efficiency without undertaking a case-by-case inquiry. Heublein, Inc, FTC Compalints & Orders (CCH) ¶21,783, Antitrust & Trade Reg Rep (BNA) F-1, F-10 n 67 (Oct 23, 1980) (FTC Oct 7, 1980); see Blake & Jones, *Toward a Three-Dimensional Antitrust Policy,* 65 Colum L Rev 422, 453-57 (1965); Givens, *Affirmative Benefits of Industrial Mergers and Section 7 of the Clayton Act,* 36 Ind LJ 51 (1960); *see also* Emhart v USM Corp, 527 F2d 177, 182 (1st Cir 1975). Merger law as such is beyond the scope of this chapter since the issues involved are only tangentially related to innovation in most instances.

[10] 15 USC §45.

[11] Atwood & Brewster, Antitrust and American Business Abroad (Shepard's/McGraw-Hill 2d ed 1981); Roberts, The Competition Law of Canada and Antitrust Law of the United States (1980); Blake & Rahl, Business Regulation in the Common Market Nations (1969); Kaye, Plaia & Hertzberg, International Trade Practice (Shepard's/McGraw-Hill 1981); Smit, The Law of the European Economic Community (looseleaf, updated); Von Kalinowski, World Law of Competition (same); Schapiro, *The German Law Against Restraints of Competition—Comparative and International Aspects,* 62 Colum L Rev 201 (1962); Buxbaum, *Antitrust Regulation Within the European Economic Community,* 61 Colum L Rev 402 (1961); *UK Commission Will Investigate Legality of Manufacturer's Refusal to Supply Discounters,* Antitrust & Trade Reg Rep (BNA) A-19 (May 21, 1981); Johnson, *Antitrust in France,* 22 Rec AB City NY 455 (1967); Kaplan, *Israeli Antitrust Law and Practice,* 26 Rec AB City NY 389 (1971); Hawk, United States, Common Market and International Antitrust (1979); Campbell, European Community Competition Law (1980); Wasserstein, *British Merger Policy From an American Perspective,* 82 Yale LJ 656 (1973). *See also* Fulda & Schwartz, Regulation of International Trade and Investment (1970).

[12] These now exist in most states. On limitations on state antitrust laws, *see* Flood v Kuhn, 407 US 258 (1972).

[13] See **§3.08.**

[14] Art I, §8, cl 3.

[15] *See* McLain v Real Estate Bd, 444 US 232 (1980); Scarborough v United States, 431 US 563 (1977); Hospital Bldg Co v Trustees of Rex Hosp, 425 US 738 (1976); Goldfarb v Virginia State Bar, 421 US 773 (1975); Burke v Ford, 389 US 320 (1967); United States v Womens Sportswear Mfrs Assn, 336 US 460, 464 (1949). *But compare* Gulf Oil Corp v Copp Paving Co, 419 US 186 (1974); National League of Cities v Usery, 426 US 833 (1976).

[16] Raymond Motor Transp, Inc v Rice, 434 US 429 (1978); Pike v Bruce Church, Inc, 397 US 137 (1970); Bibb v Navajo Freight Lines, 359 US 520 (1959); Dean Milk Co v Madison, 340 US 349 (1951); HP Hood & Sons v DuMond, 336 US 525 (1949); Southern Pac Co v Arizona, 325 US 761 (1945); Baldwin v GAF Seelig, Inc, 294 US 511

or protectionist regulation.[17]

§3.03 Judicial Interpretation of Relevant Antitrust Provisions

Section 1 of the Sherman Act,[18] read literally, bars *every* contract or combination that restrains trade at all. Finding this category too broad, the Supreme Court beginning in 1911 held that the section prohibits only *unreasonable* restraints.[19] To determine if a restraint is unreasonable the courts look to the purpose of the act—promotion of an open economy—and undertake a balancing process to determine whether the challenged conduct is more procompetitive or anticompetitive.[20] Contentions that competition itself is more harmful than helpful in some instances are, however, deemed for legislative consideration rather than for judicial evaluation under this rule of reason approach.[21]

Certain kinds of combinations are considered so inherently anticompetitive, and to have redeeming value so rarely, as to be deemed unlawful without application of this balancing process in each individual case. These *per se* categories, as defined by the courts, which expand and contract based upon experience in adjudicated cases and known from the business context,[22] generally include:

1. Horizontal price collusion or divisions of customers among competitors,[23] which in turn may also create illegal arrangements which facilitate

---

(1935); Buck v Kuykendall, 267 US 307 (1925); see Dowling, *Interstate Commerce and State Power: Revised Version,* 47 Colum L Rev 547 (1947); Givens, *The Commerce Clause and the Interdependent Economy,* 53 ABAJ 719 (1967). *Compare also* language in McLeod v JE Dilworth Co, 322 US 327, 330 (1944).

[17] Lewis v BT Inv Managers, Inc, 447 US 27 (1980); City of Philadelphia v New Jersey, 437 US 617 (1978); Hunt v Washington Apple Advertising Commn, 432 US 333 (1977); Nippert v Richmond, 327 US 416 (1946).

[18] 15 USC §1. See **§3.02.**

[19] Standard Oil Co v United States, 221 US 1 (1911); *see* Chicago Bd of Trade v United States, 246 US 231 (1918). *Compare* Appalachian Coals, Inc v United States, 288 US 344 (1933).

[20] Broadcast Music, Inc v CBS, 441 US 1 (1979); Continental TV, Inc v GTE Sylvania Inc, 433 US 36 (1977).

[21] National Socy of Professional Engrs v United States, 435 US 679 (1978); *see also* Catalano, Inc v Target Sales, Inc, 446 US 643 (1980).

[22] Broadcast Music, Inc v CBS, 441 US 1 (1979); United States v Topco Assocs, 405 US 596, 607-608 (1972); White Motor Co v United States, 372 US 253, 263 (1963). Compare also California Retail Liquor Dealers Assn v Midcal Aluminum, Inc, 445 US 97 (1980).

[23] As to price collusion, *see* Catalano, Inc v Target Sales, Inc, 446 US 643 (1980); United States v Socony-Vacuum Oil Co, 310 US 150 (1940); United States v Trenton Potteries Co, 273 US 392 (1927). *But compare* Broadcast Music, Inc v CBS, 441 US 1 (1979).

As to division of customers or territories, *see* United States v Topco Assocs, 405 US

these kinds of tacit or overt combinations[24]

2. Resale price agreeements between sellers and purchasers buying for resale[25]
3. Collective refusals to deal with others having actual or potential anticompetitive effects[26]
4. Use of coercive leverage based on power in one market to obtain advantages in another[27]

596, 608 (1972); Burke v Ford, 389 US 320 (1967); United States v Sealy, Inc, 388 US 350, 357 n 5 (1967); Timken Roller Bearing Co v United States, 341 US 593 (1951); Addyston Pipe & Steel Co v United States, 175 US 211 (1899), *affg* 85 F 271 (6th Cir 1898) (Taft, J); United States v Koppers Co, 652 F2d 290 (2d Cir 1981); Engine Specialties, Inc v Bombardier Ltd, 605 F2d 1, 8 (1st Cir 1979); United States v Consolidated Laundries Corp, 291 F2d 563, 574 (2d Cir 1961); Norfolk S Bus Co v Virginia Dare Co, 159 F2d 306 (4th Cir, *cert denied,* 331 US 827 (1947). For insufficient evidence of division of customers among competitors, *see, e.g.,* United States v Borden Co, 111 F Supp 562 (ND Ill 1953), *affd in part and remanded in part,* 347 US 514 (1954).

[24] United States v United States Gypsum Co, 438 US 422 (1978); United States v Container Corp of America, 393 US 333 (1969);Sugar Inst, Inc v United States, 297 US 553 (1936); American Column & Lumber Co v United States, 257 US 377 (1921); *see* Great A&P Tea Co v FTC, 440 US 69 (1979); Sullivan, Antitrust, ch 3(G) (1977). *See also* Zenith Radio Corp v Matsushita Elec Indus Co, 513 F Supp 1100, 1150-53 (ED Pa 1981). This area, however, is not defined as a *per se* category.

[25] California Retail Liquor Dealers Assn v Midcal Aluminum, Inc, 445 US 97 (1980); Simpson v Union Oil Co, 377 US 13 (1964); United States v Parke, Davis & Co, 362 US 29 (1960); FTC v Beech-Nut Packing Co, 257 US 441 (1922); Dr Miles Medical Co v John D Park & Sons, 220 US 373 (1911); see Levi, *The Parke, Davis-Colgate Doctrine: The Ban on Resale Price Maintenance,* 1960 Sup Ct Rev 258; Maslow, *Colgate Updated: Resale Price Maintenance and Antitrust Liability,* 48 NY St BJ 626 (1976); *see also* Continental TV, Inc v GTE Sylvania Inc, 433 US 36, 51 n 18 (1977). As pointed out by Assistant Attorney General William F. Baxter, "a pattern of vertical arrangements can have horizontal consequences," thus triggering enforcement activity. Remarks before the Senate Judiciary Committee, *reported in* Antitrust & Trade Reg Rep (BNA) at A-10 (Mar 2, 1981). This fact is also the basis for the rule of illegality laid down in California Retail Liquor Dealers Assn v Midcal Aluminum, Inc, 445 US 97 (1980). *See* Givens & Worsinger, *Vertical Restraints After Repeal of Fair Trade,* 45 Fordham L Rev 1093 (1977); Mannington Mills, Inc v Congoleum Corp, 610 F2d 1059 (3d Cir 1979).

[26] Radiant Burners v Peoples Gas Light & Coke Co, 364 US 656 (1961); Klor's, Inc v Broadway-Hall Stores, Inc, 359 US 207 (1959); Fashion Originators Guild v FTC, 312 US 457 (1941); *compare also* St Paul Fire & Marine Ins Co v Barry, 438 US 531 (1978); Eastern States Retail Lumber Dealers Assn v United States, 234 US 600 (1914); Montague & Co v Lowry, 193 US 38 (1904); Sullivan, Antitrust 260-61 (1977). If there is no threshold risk of anticompetitive effect, the rule may not apply. *E.g.,* EA McQuade Tours, Inc v Consolidated Air Tour Manual Comm, 467 F2d 178 (5th Cir 1972), *cert denied,* 409 US 1109 (1973); *see* Cartrade, Inc v Ford Dealers Assn, 446 F2d 289, 293-94 (9th Cir 1971), *cert denied,* 405 US 997 (1972); Joseph E Seagram & Sons v Hawaiian Oke & Liquors, Ltd, 416 F2d 71, 76-77 (9th Cir 1969), *cert denied,* 396 US 1062 (1970).

[27] United States Steel Corp v Fortner Enters, Inc (Fornter II), 429 US 610 (1977); Otter Tail Power Co v United States, 410 US 366 (1973); Northern Pac R v United States, 356 US 1 (1958); United States v Griffith, 334 US 100 (1948); International Salt Co v United States, 332 US 392 (1947); Yentsch v Texaco, Inc, 630 F2d 46 (2d Cir

Where conduct for some specific reason does not offend the objective of one of these judicially developed per se rules, it may be found to fall outside the scope of the rule.[28]

Monopolization under §2 of the Sherman Act is held to require both (1) substantial market dominance, and (2) its creation, maintenance or exercise by anticompetitive means.[29] Although formulations differ, attempt to monopolize generally requires specific intent, anticompetitive acts and dangerous probability of success.[30] For example, if firm X, with 50 percent of a relevant market during the prior year, is increasing its share toward 60 per cent largely by means of agreements with major industry-wide channels of distribution requiring them to carry X's products exclusively, there would be a high probability that an attempt to monopolize could be found.

The interpretation of antitrust provisions, because of their generality, is necessarily influenced by contemporary views of the problems of the national economy. The antitrust tradition also contains within itself a spectrum of views as to its goals and the means to attain those goals. Emphasis upon these differing strands changes with differing perceptions of the workings and difficulties of the economy at the time a decision is made, whether or not this is articulated.

Varying strands of antitrust tradition hold that bigness is inherently evil;[31] that the marketplace itself can destroy monopolistic practices with minimal or no political intervention;[32] and that overt or tacit horizontal pricing collusion[33] among competitors having the tendency to raise prices or limit output, wheth-

---

1980); SmithKline Corp v Eli Lilly & Co, 575 F2d 1056 (3d Cir), *cert denied,* 439 US 838 (1978); Spartan Grain & Mill Co v Ayers, 581 F2d 419 (5th Cir 1978), *cert denied,* 444 US 831 (1979); Sullivan, Antitrust §171 (1977).

[28] *See* Broadcast Music, Inc v CBS, 441 US 1 (1978).

[29] *See generally* United States v Grinnell Corp, 384 US 563 (1966); United States v EI duPont de Nemours & Co (cellophane), 351 US 377 (1956); United States v Griffith, 334 US 100 (1948); SCM Corp v Xerox Corp, 645 F2d 1195 (2d Cir 1981), *cert denied,* — US — (1982); Broadway Delivery Corp v United Parcel Service, 651 F2d 122 (2d Cir 1981); Berkey Photo, Inc v Eastman Kodak Co, 603 F2d 263 (2d Cir 1979), *cert denied,* 444 US 1093 (1980); United States v United Shoe Mach Corp, 110 F Supp 295 (D Mass 1953), affd, 347 US 521 (1954).

[30] See Handler & Steuer, *Attempts to Monopolize and No-Fault Monopolization,* 129 U Pa L Rev 125 (1980); Kintner & Goldstein, *Section 2 of the Sherman Act: The Attempt to Monopolize Offense,* 24 NYL Sch L Rev 573 (1979); Note, 73 Colum L Rev 1451 (1973).

[31] *See* Louis K Leggett Co v Lee, 288 US 517, 580 (1933) (Brandeis, J, dissenting); *compare* traditions discussed in Fox, *Monopoly and Competition: Tilting the Law Towards a More Competitive Economy,* 37 Wash & Lee L Rev 49, 56 (1980); Cary, *Legal Problems of Economic Power,* 60 Colum L Rev 4, 11 (1960).

[32] This view is sometimes attributed to Justice Holmes, who dissented in many early antitrust cases, *e.g.,* Northern Sec Co v United States, 193 US 197, 403-04 (1904). *See generally* Letwin, Law and Economic Policy, ch 7 (1965).

[33] Interstate Circuit, Inc v United States, 306 US 208 (1939); American Tobacco Co v United States, 328 US 781, 809-10 (1946); Bogosian v Gulf Oil Corp, 561 F2d 434 (3d Cir 1977), *cert denied,* 434 US 1086 (1978); C-O-Two Fire Equip Co v United States, 197 F2d 489 (9th Cir), *cert denied,* 344 US 892 (1952); Blechman, *Conscious Parallelism,*

er produced directly or as a consequence of other types of conduct, is the principal evil to be discouraged.[34] The third approach is currently in the ascendency,[35] perhaps in part due to recognition of the implications of the economy of scarcity which emerged into public consciousness in the mid-1970s.[36]

In this context, innovation is recognized as deserving of encouragement,[37] although substantial risks remain. For example, a ruling permitting business enterprises to proceed with caution in an area[38] may be misinterpreted as a green light implying the per se legality of formerly forbidden conduct.[39]

Antitrust concepts may also, of course, be used to attack regulatory or other barriers to technological innovation if they can be construed as inimical to an open marketplace.[40] Similarly, promotion of socially desirable innovation may

---

Signalling and Facilitating Devices: The Problem of Tacit Collusion Under the Antitrust Laws, 24 NYL Sch L Rev 881 (1979); Nye, *Can Conduct Oriented Enforcement Inhibit Conscious Parallelism?,* 44 Antitrust LJ 206 (1975); Givens, *Parallel Business Conduct Under the Sherman Act,* 5 Antitrust Bull 273 (1960); *but see* Theatre Enters, Inc v Paramount Film Distrib Corp, 346 US 537 (1954), and numerous cases following it, holding conscious parallelism alone insufficient.

[34] *See* Bork, The Antitrust Paradox (1978); Catalano, Inc v Target Sales, Inc, 446 US 463 (1980). On indirect factors causing the functional equivalent of overt collusion, *see* California Retail Liquor Dealers Assn v Midcal Aluminum, Inc, 445 US 97 (1980); William F Baxter, Assistant Attorney General, *quoted in* Antitrust & Trade Reg Rep (BNA), at A-10 (Mar 2, 1981); Practising Law Institute, Marketing and Distribution: Regulations and Restraints 17 (1978); Givens & Worsinger, *Vertical Restraints After Repeal of Fair Trade,* 45 Fordham L Rev 1093 (1977).

[35] National Socy of Professional Engrs v United States, 435 US 679 (1978); United States v Trenton Potteries Co, 273 US 392 (1927).

[36] *See, e.g.,* Baker, *Cheese & Macaroni,* NY Times, June 19, 1979, at A21.

[37] For recognition of the importance of innovation, see Terleckyj, Effects of R&D on the Productivity Growth of Industries (National Planning Assn 1974); Mansfield, *Basic Research and Productivity Increase in Manufacturing,* 70 Am Econ Rev 863 (Dec 1980); Starr & Rudman, *Parameters of Technological Growth,* 182 Sci 358 (Oct 26, 1973); Mansfield, *Research and Development, Productivity, and Inflation,* 209 Sci 1091 (Mar 2, 1979); Hutten, *On the Importance of Productivity Change,* 69 Am Econ Rev 126 (1979).

[38] *E.g.,* Continental TV, Inc v GTE Sylvania Inc, 433 US 36 (1977) (vertical territorial and customer resale restrictions subject to rule of reason and no longer illegal per se), *overruling* United States v Arnold, Schwinn & Co, 388 US 365 (1967).

[39] *See* Eiberger v SONY Corp of America, 622 F2d 1068 (2d Cir 1980), *affg in part* 459 F Supp 1276 (SDNY 1978).

[40] Deregulation to remove excessive or unnecessary barriers to legitimate economic activity was significantly advanced as a prominent part of antitrust concepts through the catalytic efforts of Lewis A. Engman while Chairman of the Federal Trade Commission from 1973 to 1975. *See generally* 1 ABA Antitrust Sec. The FTC as an Antitrust Enforcement Agency: The Role of Section 5 of the FTC Act in Antitrust Law 80-81 (Monograph No 5, 1981); Federal Trade Commission, Trade Regulation Rule Concerning Advertising of Ophthalmic Goods and Services, 43 Fed Reg 23992 (1978), 16 CFR pt 456, *remanded,* American Optometric Assn v FTC, 626 F2d 896 (DC Cir 1980).

Attacks on advertising bans in various fields during this period contributed to, *e.g.,* Virginia State Bd of Pharmacy v Virginia Citizens Consumer Council, 425 US 748 (1976); Bates v State Bar, 433 US 350 (1977). For description of issues, see Reich, *Consumer Protection and the First Amendment,* 61 Minn L Rev 705 (1977).

be a nonenforcement means of advancing antitrust objectives, since new technologies and new industries obviously offer the greatest source of new competition.[41]

§3.04 Sanctions for Antitrust Violations

Sanctions for antitrust violations are sufficiently severe to have an intimidating effect on many transactions or activities which can raise antitrust questions even if ultimately there might be no violation, or indeed the conduct might be found on balance to be procompetitive.

Private treble damage actions[42] and both private[43] and governmental[44] injunctive actions are authorized for violations of the Sherman and Clayton Acts. The Sherman Act likewise provides criminal penalties.[45] The antitrust provisions of the Federal Trade Commission Act[46] may be enforced by cease-and-desist orders. Whether Federal Trade Commission rule making in the antitrust field is authorized as well has not as yet been decisively determined.[47]

The deterrent effect of these remedies was somewhat attenuated for a long time by the long delays encountered in significant antitrust cases,[48] especially involving economic evidence concerning the market or under the rule of reason, tending to make enforcement both extremely expensive and uncertain for both plaintiffs and enforcement agencies. Since the 1979 report of the National Commission for Review of Antitrust Laws and Procedures, and because of

---

[41] Worsinger, *New Technologies and Antitrust,* 47 NY St BJ 651 (1975), *reprinted in* 81 Case & Com 33 (1976).

[42] 15 USC §15.

[43] *Id* §26.

[44] Id §4.

[45] *Id* §§1, 2.

[46] On the related Federal Trade Commission function of reporting on problems restraining the economy, see Hall, *The History of Section 6 of the Federal Trade Commission Act,* 27 Rec AB City NY 221 (1972).

[47] Section 5 of the act bars "unfair methods of competition" affecting interstate commerce. 15 USC §45. *See* FTC v Brown Shoe Co, 384 US 316, 321-22 (1966); FTC v Cement Inst, 333 US 683, 726 (1948); Fashion Originators Guild v FTC, 312 US 457 (1941); FTC v RF Keppel & Bro, 291 US 304, 310 (1934); Grand Union Co v FTC, 300 F2d 92 (2d Cir 1962). Rulemaking in antitrust matters has not been attempted and is specifically authorized by statute only if "unfair acts or practices," as distinct from "unfair methods of competition," are found. 15 USC §57a(a)(1). On the unfairness concept, *see* FTC v Sperry & Hutchinson Co, 405 US 233 (1972). On pre-1975 general rulemaking authority, *see* National Petroleum Ref Assn v FTC, 482 F2d 672 (DC Cir 1973), *cert denied,* 415 US 951 (1974), preserved in its original status as to possible general FTC rulemaking authority regarding unfair methods of competition by 15 USC §57a(a)(2). On the interpretation of §5 that it fosters innovative efforts even by leading firms, *see* EI duPont de Nemours & Co, Trade Reg Rep (CCH) ¶21,770 (FTC Oct 20, 1980).

[48] Documented in the Report of the National Commission for Review of Antitrust Laws and Procedures (1979).

increased judicial concern with expedition, however, a new climate exists. At the same time that defendants have a better chance to raise claims that challenged conduct is in fact procompetitive, the likelihood that challenges will be defeated by attrition alone has significantly decreased.

§3.05 Innovation as Monopolization

Successful innovation often leads to unique market advantages, which in turn sometimes lead to charges of monopolization in violation of §2 of the Sherman Act.[49] However, both a large market share and anticompetitive conduct must be shown for a violation of §2 to be found.[50]

Recent cases have uniformly rejected contentions that aggressive innovation constitutes monopolistic behavior upon which liability can be based.[51] In particular, suits based upon failure to inform competitors of projected product changes,[52] design to discourage use of competitors' related products where no regulatory monopoly is involved[53] and the obtaining of important patents

---

[49] 15 USC §2, set out in **§3.02.** SCM v Xerox Corp, 645 F2d 1195 (2d Cir 1981) *affg* 463 F Supp 983 (D Conn 1978) *cert denied,* — US — (1982); Berkey Photo, Inc v Eastman Kodak Co, 603 F2d 263 (2d Cir 1979), *cert denied,* 444 US 1093 (1980); California Computer Prods v IBM, 613 F2d 727 (9th Cir 1979); Telex Corp v IBM Peripheral EDP Devices Litig, 481 F Supp 965 (ND Cal 1979); Van Dyk Research Corp v Xerox Corp, 478 F Supp 1268 (DNJ 1979), *affd,* 631 F2d 251 (3d Cir 1980); ILC Peripherals Leasing Corp v IBM (Memorex), 458 F Supp 423 (ND Cal 1978), *affd,* 636 F2d 1188 (9th Cir 1981); Transamerica Computer Co v IBM, 481 F Supp 965, 1014-22 (ND Cal 1979); see Bock, The Innovator as an Antitrust Target (Conference Board Bull No 74 1980); Note, *Aggressive Innovation and Antitrust Liability,* 53 S Cal L Rev 1469 (1980); *compare also* Northeastern Tel Co v AT&T, 651 F2d 276 (2d Cir 1981) (but note regulatory monopoly consequences); Note, 28 Stan L Rev 285 (1976).

[50] See **§3.03.** *See, e.g.,* Broadway Delivery Corp v United Parcel Serv, 651 F2d 276 (2d Cir 1981); SCM v Xerox Corp, 645 F2d 1195 (2d Cir 1981), *affg* 463 F Supp 983 (D Conn 1978); Berkey Photo, Inc v Eastman Kodak Co, 603 F2d (2d Cir 1979), *cert denied,* 444 US 1093 (1980) and cases cited therein.

[51] See note 49 *supra; but see* GAF Corp v Eastman Kodak Co, 519 F Supp 1203 (SDNY 1981) (denial of summary judgment on this point). See Gribbon, *Does the Law of Monopolization Inhibit Innovative Behavior?,* 49 ABA Antitrust LJ 925 (1980); Jentes, *Assessing Efforts to Challenge Aggressive Competition as an Attempt to Monopolize,* 49 ABA Antitrust LJ 937 (1980).

[52] Berkey Photo, Inc v Eastman Kodak Co, 603 F2d 263 (2d Cir 1979), *cert denied,* 444 US 1093 (1980); ILC Peripherals Leasing Corp v IBM (Memorex), 458 F Supp 423, 437 (ND Cal 1978), *affd,* 636 F2d 1188 (9th Cir 1981); Note, 93 Harv L Rev 408 (1979). *See also* Rapid Print, Inc v Minnesota Mining & Mfg Co, Trade Reg Rep (CCH) ¶63,787 (D Mass Jan 30, 1981).

[53] *Compare* California Computer Prods v IBM, 613 F2d 727, 744 (9th Cir 1979) (referring to cost savings due to the challenged innovation); Response of Carolina, Inc v Leasco Response, 537 F2d 1307, 1330 (5th Cir 1976) *and In re* IBM Peripheral EDP Devices Antitrust Litig, 481 F Supp 965, 1003-05 (ND Cal 1979) *with* Northeastern Tel Co v AT&T, 651 F2d 276 (2d Cir 1981) (regulatory monopoly); *see also* Rapid Print, Inc v Minnesota Mining & Mfg Co, Trade Reg Rep (CCH) ¶63,787 (D Mass Jan 30, 1981); ILC Peripherals Leasing Corp v IBM (Memorex), 458 F Supp 423, 440-41 (ND Cal

without licensing competitors to use them[54] have been unsuccessful.

The clear trend in case law protecting innovators from monopolization charges flows both from concern with encouraging innovation in an era of economic scarcity[55] and from recognition that otherwise the very robust competition which antitrust seeks to promote[56] would itself be chilled.[57]

Section 2 prohibits attempts to monopolize as well as actual monopolization.[58] This is generally held to require anticompetitive behavior, specific intent to monopolize and a dangerous probability of success.[59] The attacks on innovation mentioned above have generally been brought under this theory as well, and have met the same fate.

A reevaluation of antitrust policy designed to insure that its strictures were not self-defeating[60] has been developing for the past decade,[61] both contributing to protection for innovators from monopolization charges and eroding the previous widely prevalent view that "bigness is badness."[62] The more widespread view now is that bigness is probably essential to some kinds of modern technological endeavors[63] and that the *exercise* of economic power so generated should become the focus of attention.[64]

1978), *affd,* 636 F2d 1188 (9th Cir 1981); Note, 89 Yale LJ 769 (1980); Comment, 1975 U Ill LJ 224; Note, 53 S Cal L Rev 1469, 1487 (1980).

[54] SCM v Xerox Corp, 645 F2d 1195 (2d Cir 1981), *affg* 463 F Supp 983 (D Conn 1978), *cert denied,* — US — (1982).

[55] See David, *Science Futures: The Industrial Connection,* 203 Sci 837 (Mar 2, 1979); Starr & Rudman, *Parameters of Technological Growth,* 182 Sci 358 (Oct 26, 1973).

[56] National Socy of Professional Engrs v United States, 435 US 679 (1978); United States v United States Gypsum Co, 438 US 422 (1978); Standard Oil Co v United States, 221 US 1, 58 (1911); see Smith, Wealth of Nations ch VII, at 62 (1812 ed).

[57] *See* Berkey Photo, Inc v Eastman Kodak Co, 603 F2d 263 (2d Cir 1979), *cert denied,* 444 US 1093 (1980); Van Dyk Research Corp v Xerox Corp, 478 F Supp 1268, 1325 (DNJ 1979), *affd,* 631 F2d 251 (3d Cir 1980).

[58] 15 USC §2.

[59] *See* Hunt-Wesson Foods, Inc v Ragu Foods, Inc, 627 F2d 919, 925-26 (9th Cir 1980); Note, 73 Colum L Rev 1451 (1973) and authorities cited therein.

[60] *See* Broadcast Music, Inc v CBS, 441 US 1 (1979); Great A&P Tea Co v FTC, 440 US 69 (1979); Continental TV, Inc v GTE Sylvania Inc, 433 US 36 (1977).

[61] See cases cited note 56 *supra* and also, *e.g.,* United States Steel Corp v Fortner Enters (Fortner II), 429 US 610 (1977); United States v General Dynamics Corp, 415 US 486 (1974); *and see* Tampa Elec Co v Nashville Coal Co, 365 US 320 (1961).

[62] *See* United States v Aluminum Co of America, 148 F2d 416, 427 (2d Cir 1945); Johnson, American Heroes and Hero Worship 234-48 (1941).

[63] United States v EI duPont de Nemours & Co (cellophane), 351 US 377, 386 (1956); Industrial Concentration: The New Learning (Goldschmid, Mann & Weston eds 1974); Scherer, Industrial Market Structure and Economic Performance (1970); Schwartzman, Oligopoly in the Farm Machinery Industry (Royal Commn on Farm Machinery, Study No 12, Ottawa 1970); Kessler & Stern, *Competition, Contract and Vertical Integration,* 69 Yale LJ 1, 8-10 (1959); Note, 68 Yale LJ 1627, 1654-62 (1959); Note, 58 Colum L Rev 673, 683 n 74 (1958).

[64] *See, e.g.,* SmithKline Corp v Eli Lilly & Co, 575 F2d 1056 (3d Cir), *cert denied,* 439 US 828 (1978); Eiberger v SONY Corp of America, 622 F2d 1066 (2d Cir 1980), *affg*

The high-water mark of antitrust interpretation against overly successful competitors was *United States v Aluminum Co of America* in 1945,[65] in which the Court of Appeals for the Second Circuit held that expansion to meet anticipated demand, even if "honestly industrial," could be evidence of monopolization.[66] No relief was ever granted, in part because of growth of competition with governmental assistance during World War II.[67] Subsequent to *Alcoa,* no case has held conduct of this or a similar type to be illegal.[68]

Thus, if a firm develops and surprises the market with secretly developed innovations, patents them and refuses to license the patents, and thereafter produces an integrated box of patented and nonpatented components in a way which minimizes its production costs, there should be no violation even if the effect is to gain a high market share in the industry, unless other vulnerable conduct is involved.

If physical technological steps taken by a firm, as distinct from marketing strategies, are challenged, they are likely to be upheld if there is a reasonably plausible technological justification. In other words, if the justification has merit on its face, the courts are unlikely to second-guess the innovator and become embroiled in acting as engineering personnel. The reasons are twofold: (1) the fear of chilling innovation if technological judgments were subject to antitrust attack based on unpredictable hindsight on the part of others, and (2) the difficulty facing the courts in making judgments of this type.

If there were no plausible technological justification at all, and a product design offered no cost or other engineering benefits, but, for example, was clearly instead designed solely to prevent competitors from offering a portion of the package, the firm's action might be judged in the same manner as a nontechnological marketing restraint. The degree of market power would become relevant. If the conduct were unilateral, the ordinary tests under §2 of the Sherman Act might apply;[69] if a combination were involved, the rule of reason or any applicable per se rule under §1 might, for instance, come into play.[70] Such a pure case of technological action taken for purely nontechnological reasons is, however, most rare.

Successful innovation does bring risks, however, both because success

---

in part 459 F Supp 1276 (SDNY 1978); *compare* Berle, Power Without Property (1959); *compare also* Steele v Louisville & N R Co, 323 US 192 (1944).

[65] 148 F2d 416 (2d Cir 1945), *cited with approval in* American Tobacco Co v United States, 328 US 781, 811 (1946).

[66] 148 F2d at 431.

[67] *See* United States v Aluminum Co of America, 91 F Supp 333 (SDNY 1957); HR Rep No 2954, 84th Cong, 2d Sess (1956); HR Rep No 2716, 85th Cong, 2d Sess (1959).

[68] *See, e.g.,* Berkey Photo, Inc v Eastman Kodak Co, 603 F2d 263 (2d Cir 1979), *cert denied,* 444 US 1093 (1980); EI duPont de Nemours & Co, Trade Reg Rep (CCH) ¶21,770 (FTC Oct 20, 1980).

[69] See **§3.03.**

[70] 15 USC §1. See **§3.03.**

brings more intense scrutiny,[71] and because power which can be used in ways subject to challenge is created. A defense of technological or business necessity can sometimes shield an innovator, who can show that the need for a marketing restraint is real, or such a defense can affect enforcement discretion.[72] This is not ordinarily possible, however, in litigation where the practice involved is per se illegal,[73] as in the case of price collusion between competitors,[74] unless the practice falls outside the scope of the purpose of the per se rule.[75]

Marketing practices are most frequently challenged under §1 of the Sherman Act. Monopolization may be claimed based on marketing strategies which have a specific exclusionary effect, such as rental-only policies in some circumstances,[76] whether or not §1 would independently be violated.[77]

§3.06 Extension of Power from One Market to Another

Successful innovation may create uniqueness which can permit the innovator to influence purchases or sales in other markets by those who want to deal with the innovator.[78] Tying of one purchase or sale of one product or service to another is, however, a practice commonly held to violate both the Sherman

---

[71] This is true even if due to visibility alone. In addition, market power becomes relevant to numerous conduct issues. *See* United States Steel Corp v Fortner Enters, Inc (Fortner II), 429 US 610 (1977); Continental TV, Inc v GTE Sylvania Inc, 433 US 36, 52 n 19 (1977); Spartan Grain & Mill Co v Ayers, 581 F2d 419 (5th Cir 1978), *cert denied,* 444 US 831 (1979).

[72] *Compare* Continental TV, Inc v GTE Sylvania Inc, 433 US 36 (1977); United States v Jerrold Elecs Corp, 187 F Supp 545 (ED Pa 1960), *affd per curiam,* 365 US 567 (1961); *In re* Data Gen Corp Antitrust Litig, 490 F Supp 1089, 1101 (ND Cal 1980) (antitrust) *with* McDonnell Douglas Corp v Green, 411 US 792, 802 (1973); Hatton v Ford Motor Co, 508 F Supp 621 (ED Mich 1981) (evidence of employment discrimination); *see also* Pitofsky, *Joint Ventures Under the Antitrust Laws,* 82 Harv L Rev 1007 (1969); Flikher, *Newcomer Defenses,* 18 Stan L Rev 457 (1966); Dehydrating Process Co v AO Smith Corp, 292 F2d 653 (1st Cir), *cert denied,* 368 US 931 (1961).

[73] For an exception, *see* Broadcast Music, Inc v CBS, 441 US 1 (1979).

[74] Catalano, Inc v Target Sales, Inc, 446 US 643 (1980); United States v Socony-Vacuum Oil Co, 310 US 150 (1940); United States v Trenton Potteries Co, 273 US 392 (1927). Resale price maintenance agreements are also deemed illegal by the courts, although the Antitrust Division of the Department of Justice favors a rule of reason approach based on horizontal impact between competitors at one or more levels. See **§3.03.**

[75] Broadcast Music, Inc v CBS, 441 US 1 (1979). See **§3.03.**

[76] *See* United States v United States Shoe Mach Corp, 110 F Supp 295 (D Mass 1953), *affd per curiam,* 347 US 521 (1954).

[77] On the scope of §1, see Baker, *Combinations and Conspiracies—Is There A Difference?,* 14 Antitrust Bull 71 (1969); *see also* Edward J Sweeney & Sons v Texaco, 637 F2d 105, 111 (3d Cir 1980), and authorities cited therein.

[78] *See, e.g.,* International Salt Co v United States, 332 US 392 (1947); IBM Corp v United States, 298 US 131 (1936); United Shoe Mach Corp v United States, 258 US 451 (1922).

Act[79] and other antitrust laws,[80] where the party has power in one market and uses this to secure a competitive advantage in another. Examples of classic tying arrangements are requirements that one use a seller's paper on its mimeograph machine or its salt in its salt tablet making machines. Reciprocity, on the other hand, involves demands to buy product X from a firm in order to be able to sell product Y to it. This is illegal where market power in one area is used to coerce sellers in that field to buy other items from the coercing party, or where the overall effect is viewed as sufficiently anticompetitive.[81]

The operative elements of violation under any of the statutes involved are increasingly treated as identical, although their formulations vary; these elements are power in market A; its use to secure an advantage in market B; that the use was coercive; and that significant interstate commerce was affected.[82]

A defense is recognized to tying charges where what may be defined as only one product is involved,[83] so that specific economies of scale are created by selling the items as a unit.[84] Temporary tying practices by a new entrant if

---

[79] *See, e.g.,* United States Steel Corp v Fortner Enters, Inc (Fortner II), 429 US 610 (1977); Yentsch v Texaco, Inc, 630 F2d 46, 56-57 (2d Cir 1980); Hill v A-T-O, Inc, 535 F2d 1349 (2d Cir 1976); *In re* Data Gen Corp Antitrust Litig, 490 F Supp 1089 (ND Cal 1980); *see also* Northern Pac Co v United States, 356 US 1 (1958) (§1); Otter Tail Power Co v United States, 410 US 366 (1973); United States v Griffith, 344 US 100 (1948); United States v Crescent Amusement Co, 323 US 173 (1944) (§2).

[80] §3 of the Clayton Act, 15 USC §14 (applicable to commodities only and only to items for domestic use or resale); §5 of the Federal Trade Commission Act, 15 USC §45 (which incorporates other antitrust laws and their spirit as described in **§3.04**). Under other statutes *compare, e.g.,* Federal Maritime Commn v Aktiebolaget Svenska Amerika Linien, 390 US 238 (1968); Black v Magnolia Liquor Co, 355 US 24 (1957).

[81] On the use of buying power, see Spartan Grain & Mill Co v Ayers, 581 F2d 419 (5th Cir 1978), *cert denied,* 444 US 831 (1979); Sullivan, Antitrust §§170, 171 (1977); Hale & Hale, *Reciprocity Under the Antitrust Laws,* 113 U Pa L Rev 69 (1914); Phillips, *Observations on the Hales' Comment,* 113 U Pa L Rev 77 (1914). On reciprocity as a factor in merger evaluation, *compare* Consolidated Foods Corp v FTC, 380 US 592 (1965); Allis-Chalmers Mfg Co v White Consol Indus, Inc, 414 F2d 506 (3d Cir 1969), *cert denied,* 396 US 1009 (1970) *with, e.g.,* United States v IT&T, 324 F Supp 19 (D Conn 1970), *appeal dismissed,* 404 US 801 (1971); United States v Penick & Ford, Ltd, 242 F Supp 518 (DNJ 1965); United States v Northwest Indus, Inc, 301 F Supp 1066 (1969); 5 Areeda & Turner, Antitrust Law §1134 (1980). For a stringent view on reciprocity, *see* United States v General Dynamics Corp, 258 F Supp 36, 59 (SDNY 1966).

[82] *See* Berkey Photo, Inc v Eastman Kodak Co, 603 F2d 263, 275-76, 291 (2d Cir 1979), *cert denied,* 444 US 1093 (1980); cases cited note 79 *supra.* A patent or copyright monopoly on the tying item appears to establish the requisite power. *See* United States v Lowe's, Inc, 371 US 38 (1962); International Salt Co v United States, 332 US 392 (1947); IBM Corp v United States, 298 US 131 (1936); United Shoe Mach Co v United States, 258 US 451 (1922); Motion Picture Patents Co v Universal Film Corp, 243 US 502 (1917); *compare* Sullivan, Antitrust §154 (1977). Some cases also require that the party engaged in tying have an economic interest in the tied item. Yentsch v Texaco, Inc, 630 F2d 46, 57 n 15 (2d Cir 1980).

[83] *See In re* Data Gen Corp Antitrust Litig, 490 F Supp 1089, 1101 (ND Cal 1980) and cases cited therein; Sullivan, Antitrust §155 (1977).

[84] See note 83 *supra;* United States v Jerrold Elecs Corp, 187 F Supp 545 (ED Pa 1960), *affd per curiam,* 365 US 567 (1961).

needed to establish viability in an industry have also been upheld.[85]

Use of combination package pricing may be regarded as less anticompetitive than strict tying, but is not per se immune if it shifts the profit the innovator can command on one item to other items and has the effect of tying.[86] Similarly, where a product line is involved and a seller wishes a dealer to take all in order to buy any, a rule of reason has generally been applied.[87]

§3.07 Predatory Pricing

Innovation can sometimes permit dramatic price reductions, and on the other hand, for success, may require high volume attainable only by an increased market share or increases in overall industry sales.[88] This in turn may necessitate price reductions.

Monopolization in violation of §2 of the Sherman Act is frequently claimed to result from predatory pricing where prices are set so low as to drive competitors out of business.[89] Such pricing, however, may constitute the very robust competition the antitrust laws were designed to promote in the first place.[90] Because of the difficulty of making such a distinction, the courts have been reluctant to find liability based on alleged cut-throat price competition.[91] This

---

[85] United States v Jerrold Elec Corp, 187 F Supp 545 (ED Pa 1960), *affd per curiam,* 365 US 567 (1961); *see* Dehydrating Process Co v AO Smith Corp, 292 F2d 653 (1st Cir), *cert denied,* 368 US 931 (1961); Flicker, *Newcomer Defenses,* 18 Stan L Rev 457 (1966).

[86] See Sullivan, Antitrust §157 (1977). Compare National Foam Sys v Urquhart, 202 F2d 659 (3d Cir 1953); Dehydrators, Ltd v Petrolite Corp, 117 F2d 183 (9th Cir 1941); Barber Asphalt Corp v Latera Grecco Contracting Co, 116 F2d 211 (3d Cir 1940).

[87] *See* Bob Maxfield, Inc v American Motors Corp, 637 F2d 1033 (5th Cir 1981); Heatransfer Corp v Volkswagenwerk, AG, 553 F2d 964 (5th Cir 1977), *cert denied,* 434 US 1087 (1978); Burstein, *A Theory of Full-Line Forcing,* 55 Nw UL Rev 62 (1960); Sullivan, Antitrust §158 (1977). Both the rule of reason and per se approaches are discussed in **§3.03.**

[88] *Compare* Adams, *The Automobile—A Luxury Becomes A Necessity,* in Hamilton, Price and Price Policies 27 (1938).

[89] See Handler & Steuer, *Attempts to Monopolize and No-Fault Monopolization,* 129 U Pa L Rev 125 (1980).

[90] On this concern, *see* Buffalo Courier-Express, Inc v Buffalo Evening News, Inc, 601 F2d 48 (2d Cir 1979); Pitofsky, *Beyond Nader: Consumer Protection and the Regulation of Advertising,* 90 Harv L Rev 661, 687-689 (1977); Kanwit, Federal Trade Commission §16.07 (Shepard's/McGraw-Hill 1979); *see also* United States v United States Gypsum Co, 438 US 422 (1978); Gold Fuel Serv, Inc v Esso Standard Oil Co, 306 F2d 61, 64 (3d Cir 1962); Weber v Wynne, 431 F Supp 1048, 1060 (DNJ 1977); Southern Blowpipe & Roofing Co v Chattanooga Gas Co, 360 F2d 79, 80 (6th Cir 1966), *cert denied,* 393 US 844 (1968); Report of the ABA Commission to Study the Federal Trade Commission 67-68 (1969). *Compare also* Turner, *Conglomerate Mergers and Section 7 of the Clayton Act,* 78 Harv L Rev 1313, 1341 (1965).

[91] *E.g.,* O Hommel Co v Ferro Corp, 659 F2d 340 (3d Cir 1981); William Inglis & Sons Baking Co v ITT Continental Baking Co, 652 F2d 917 (9th Cir 1981); Northeastern Tel Co v AT&T, 651 F2d 276 (2d Cir 1981); Janich Bros v American Distilling Co, 570 F2d 848, 855-57 (9th Cir 1977), *cert denied,* 439 US 829 (1979); Hanson v Shell Oil

result has been reached through a variety of restrictive tests as to what constitutes illegal predation.

The best-known such test is that proposed by Professors Areeda and Turner which argues that prices which exceed the additional or marginal cost of producing the items sold are profitable to the seller even apart from any adverse effect on competitors, and are hence nonpredatory.[92] Critics of this test point out that prices below *average* cost (so that the transaction makes no contribution to fixed plant and equipment or other overhead expenses) lead to long-run losses. Such prices, they assert, can be predatory[93] unless adopted defensively to meet competition.[94]

The courts, while influenced by the various tests, have with some exceptions generally avoided fixed commitment to any single test, and often take evidence of the actor's intent into account as well.[95] It is clear, however, that it is difficult or next to impossible for a plaintiff to prevail on a predatory pricing claim, because of the similarity of predatory and robust competitive price-cutting.[96]

Even nonpredatory price discounting may be subject to attack under the

---

Co, 541 F2d 1352 (9th Cir 1976), *cert denied,* 429 US 1074 (1977); Pacific Engg & Prod Co v Kerr-McGee Corp, 551 F2d 790 (10th Cir), *cert denied,* 434 US 879 (1977); International Air Indus, Inc v American Excelsior Co, 517 F2d 714, 723-24 (5th Cir 1975), *cert denied,* 424 US 943 (1976); Murphy Tugboat Co v Crowley, 454 F Supp 847, 853-56 (ND Cal 1978); William Inglis & Sons Baking Co v ITT Continental Baking Co, 461 F. Supp 410 (ND Cal 1978); *see also In re* IBM Peripheral EDP Devices Antitrust Litig, 481 F Supp 965, 987-996 (ND Cal 1979) (prices above average cost are conclusively legal; those below are illegal if unreasonable); Borden, Inc, 92 FTC 669, 817-32 (1978) (Pitofsky, Commr concurring); Axinn, *Predatory Pricing: Chaos in California,* NYLJ 1 (Nov 20, 1979); see also Island Tobacco Co v RJ Reynolds Indus, Inc, 513 F Supp 726, 737 (D Hawaii 1981).

[92] Areeda & Turner, *Predatory Pricing and Related Practices Under Section 2 of the Sherman Act,* 88 Harv L Rev 697 (1975), refined in 3 Areeda & Turner, Antitrust Law 148-94 (1978).

[93] On this debate, *see generally* Josko & Klevorick, *A Framework for Analyzing Predatory Pricing Policy,* 89 Yale LJ 213 (1979); Scherer, *Predatory Pricing and the Sherman Act,* 89 Harv L Rev 869 (1976); Greer, *A Critique of Areeda & Turner's Standard for Predatory Practices,* 24 Antitrust Bull 233 (1979); Williamson, *Predatory Pricing: A Strategic and Welfare Analysis,* 87 Yale LJ 284 (1977); Report of the National Commission for Review of Antitrust Laws and Procedures 149-51, 165-66 (1979); Koller, *When is Pricing Predatory?,* 24 Antitrust Bull 283 (1979); Areeda, *Predatory Pricing,* 49 ABA Antitrust LJ 897 (1980); *In re* IBM Peripheral EDP Devices Antitrust Litig, 481 F Supp 965, 987-996 (ND Cal 1979).

[94] Compare the meeting competition defense under the Robinson-Patman Price Discrimination Act, 15 USC §13. *See generally* Great A&P Tea Co v FTC, 440 US 69 (1979); United States v United States Gypsum Co, 438 US 422 (1978); Kuenzel & Schiffres, *Making Sense of Robinson-Patman: The Need to Revitalize its Affirmative Defenses,* 62 Va L Rev 1211 (1976); Note, 90 Harv L Rev 1476 (1977).

[95] See cases cited note 91 *supra* with the exception of *Northeastern Tel Co v AT&T.*

[96] Standard Oil Co v United States, 221 US 1, 58 (1911); *see* Catalano, Inc v Target Sales, Inc, 446 US 643 (1980); California Retail Liquor Dealers Assn v Midcal Aluminum, Inc, 445 US 97 (1980); National Socy of Professional Engrs v United States, 435 US 679 (1978); Smith, Wealth of Nations ch VIII, at 62 (1812 ed).

Robinson-Patman Price Discrimination Act[97] if the discounts are given only to favored customers.[98] Some courts have, however, applied the Areeda-Turner and similar doctrines under the act pursuant to the proof requirement of the Robinson-Patman Price Discrimination Act that competition be injured in order for there to be a violation.[99] Because price uniformity if mandated by the price discrimination law can encourage price collusion,[100] the courts have also given expansive treatment to the meeting competition in good faith defense under the Robinson-Patman Price Discrimination Act.[101]

The cost-justification defense,[102] which allows savings due to particular means of manufacture or delivery to be passed on to buyers, may also afford important opportunities to innovators. Favorable treatment of this defense is appropriate under evolving antitrust concepts in spite of past restrictive interpretations requiring impracticably precise accounting proof to support claimed savings.[103] For example, an innovator may be able to negotiate long-term sales commitments which can result in savings that can in turn be passed on to the buyer.[104] Indeed, such commitments may be a necessary means of

---

[97] 15 USC §13. *See generally* Practising Law Institute, 19th Annual Advanced Antitrust Seminar: Price Discrimination (1979); Rowe, Price Discrimination Under the Robinson-Patman Act (1962); Edwards, The Price Discrimination Law (1959).

[98] *See* Utah Pie Co v Continental Baking Co, 386 US 685 (1967); United States v Borden Co, 370 US 460 (1962); FTC v Morton Salt Co, 334 US 37 (1948). Buyers may be held liable for inducing illegal discrimination under §2(f) of the act, 15 USC §13(f).

[99] See cases cited note 91 *supra.* On the requirement of injury to competition generally, *see* O Hommel Co v Ferro Corp, 659 F2d 340 (3d Cir 1981); Interstate Cigar Co v Sterling Drug Inc, 655 F2d 29 (2d Cir 1981); Edward J Sweeney & Sons v Texaco, 637 F2d 105, 120-21 (3d Cir 1980); Janich Bros, v American Distilling Co, 570 F2d 848, 855 (9th Cir 1977), *cert denied,* 439 US 829 (1978); FLM Collision Parts, Inc v Ford Motor Co, 543 F2d 1019 (2d Cir 1976).

[100] *See* Automatic Canteen Co v FTC, 346 US 61 (1953); *compare* Schelling, The Strategy of Conflict (1960); Pribram, Cartel Problems (1935); Givens & Worsinger, *Vertical Restraints After Repeal of Fair Trade,* 45 Fordham L Rev 1093 (1977).

[101] Great A&P Tea Co v FTC, 440 US 69 (1979); United States v United States Gypsum Co, 438 US 422 (1978); *see also* Note, 90 Harv L Rev 1476 (1977).

[102] "*Provided,* That nothing herein contained shall prevent differentials which make only due allowance for differences in the cost of manufacture, sale, or delivery resulting from the differing methods or quantities in which such commodities are to such purchasers sold or delivered." 15 USC §13(a). *See generally* Rowe, *Cost Justification of Price Differentials Under The Robinson-Patman Act,* 59 Colum L Rev 584 (1959); Taggart, Cost Justification (1959).

[103] *See* Kuenzel & Schiffres, *Making Sense of Robinson-Patman: The Need to Revitalize Its Affirmative Defenses,* 62 Va L Rev 1211 (1976); American Motors Corp v FTC, 384 F2d 247, 251-52 (6th Cir 1967), *cert denied,* 390 US 1012 (1968). For a legislative proposal for "good faith cost justification", see Report of Legislative Committee, Antitrust Section, NY State Bar Assn, *Good Faith Cost Justification,* Trade Reg Rep (CCH) No 484, at 7 (Mar 23, 1981), *reprinted in* Practising Law Institute, 22d Annual Antitrust Law Inst 647 (1981) *and* 53 NY St BJ 376 (Aug 1981).

[104] *See* FTC v Standard Motor Prods, 371 F2d 613, 619 n 5 (2d Cir 1967); Borden Co, 62 FTC 130 (1963); National Biscuit Co, 50 FTC 932, 935 (1954); HR Rep No 2287, 74th Cong, 2d Sess 9 (1936); House Judiciary Committee, *Hearings on Bills to*

enabling a new development to be implemented at all.[105] Even if long-term commitments have escape clauses permitting the buyer to cancel orders, such commitments may furnish a basis for cost justification if the contract provides an incentive for the buyer to stick to the commitments. This may be achieved through cancellation fees or discounts conditioned on fulfillment of a projected order schedule.[106]

An unresolved question arises if an innovator has both old, high-cost and new, more efficient, low-cost facilities, and wishes to price the output of the new facility at a lower point, offering it to some customers. This could result in price discrimination claims unless a separate entity were created which would be treated as a different enterprise.[107] On the other hand, unless the

---

Amend the Clayton Act, 74th Cong, 2d Sess 257 (1935); Vawter, Rosner & Wegener, *Principal Issues Arising under the Robinson-Patman Act,* 49 Antitrust LJ 675, 683-84 (1981); Rowe, *Cost Justification of Price Differentials under the Robinson-Patman Act,* 59 Colum L Rev 584, 594-96 & 594 n 52 (1959); Jury Instructions, Transcript at 16, 453-61, Berkey Photo, Inc v Eastman Kodak Co, *revd in part on other grounds,* 603 F2d 263 (2d Cir 1979), *cert denied,* 444 US 1093 (1980), in Rayhill, *The Robinson-Patman Act: A Revival of Defenses* in Practising Law Institute, Marketing and Distribution: Regulations and Restraints 195, 210-11 (1978); Givens, *Cost Justification: Dead Letter or Sleeping Giant?* in Practising Law Institute, 19th Annual Advanced Antitrust Seminar: Price Discrimination 71 (1979).

Another defense is that prices for future and for current deliveries are not comparable. Texas Gulf Sulphur Co v JR Simplot Co, 418 F2d 793, 805-06 (9th Cir 1969); Austin, Price Discrimination 25 (2d rev ed 1959); Rowe, Price Discrimination Under the Robinson-Patman Act §4.2, at 48-50 & n 16 (1962); Kintner, A Robinson-Patman Primer 46 (1975); Trade Reg Rep (CCH) ¶3280.44; *see* National Biscuit Co, 50 FTC 932, 935 (1954); Lombino & Sons v Standard Fruit & SS Co, 1975-2 Trade Cas (CCH) ¶60,527 (SDNY 1975).

It is difficult to challenge long-term contracts as anticompetitive. *See* Tampa Elec Co v Nashville Coal Co, 365 US 320 (1961); Kinnett Dairies, Inc v Dairymen, Inc, 512 F Supp 608, 634-37 (MD Ga 1981); Taylor, *Antitrust Nominee Says He Plans to Limit Agency's Role, Prosecute Price Fixers,* Wall St J, Mar 20, 1981, at 4 (favoring long-term supply agreements); *see also* Salpukas, *Coal Strategy: Long Contracts,* NY Times, May 13, 1981, at D1. *Compare also* United States v Dairymen, Inc, 1978-1 Trade Cas (CCH) ¶62,053 (WD Ky 1978), *affd in part, revd in part on other grounds,* — F2d — (6th Cir 1981).

[105] See Edwards, The Price Discrimination Law 613-15 (1959); Sharav, *Cost Justification Under the Robinson-Patman Act,* Mgmt Acct 15 (July 1978).

[106] Terms may vary as long as estimated compensation and not a penalty is the result. *See* Priebe & Sons v United States, 332 US 407 (1947); 5 Corbin, Contracts §1038 (1964). On innovative terms, *compare* Pepper's Steel & Alloys v Lissner Minerals, 494 F Supp 487 (SDNY 1979).

[107] On treatment of a parent and subsidiary as a unit for purposes of showing that differing prices charged by the two amounted to discrimination, *see* Purolator Prods, Inc v FTC, 352 F2d 874, 883 (7th Cir 1965), *cert denied,* 389 US 1045 (1968); Island Tobacco Co v RJ Reynolds Indus, 513 F Supp 726, 735 (D Hawaii 1981)

On treatment of subdivisions of a single enterprise as one entity so as to disallow favorable intra-enterprise sales as grounds for charging discrimination, *see* Security Tire & Rubber Co v Gates Rubber Co, 598 F2d 962 (5th Cir), *cert denied,* 444 US 912 (1979); Schaben v Samuel Moore & Co, 606 F2d 831 (8th Cir 1979); Brown v Hansen Publications, Inc, 556 F2d 969, 971-72 (9th Cir 1977); Island Tobacco Co v RJ Reynolds Indus, Inc, 513 F Supp 726, 733-36; (D Hawaii 1981); Carroll-McCreary Co v New Jersey Steel

entity was *totally* insulated, partial independence could conceivably lead to pricing collusion charges.[108] Encouragement of innovation might be furthered by permitting partial independence of a wholly owned entity[109] where new competition is thereby created, rather than otherwise extant competition being destroyed.[110]

§3.08 Patent-Antitrust Strategies

The patent-antitrust interface is complex because of the wide ramifications of both fields, and is constantly changing because any shift in either side of the interface affects both.[111]

The basic principles and the core strategies open for dealing with them are simple, even if their application to particular facts is almost endlessly complex. The exclusive right granted by a patent and its use or nonuse is as such protected from antitrust attack.[112] The presence, however, of a patent does not

---

Corp, 1980-2 Trade Cas (CCH) ¶63,436 (SDNY 1980); Synder v Howard Johnson's Motor Lodges, Inc, 412 F Supp 724, 731 (SD Ill 1976); Reines Distribs, Inc v Admiral Corp, 256 F Supp 581 (SDNY 1966); Baim & Blank, Inc v Philco Corp, 148 F Supp 541, 544 (EDNY 1957); Emil J Lauter Co, v Brunswick Corp — F Supp — (ND Ill 1982).

[108] Perma-Life Mufflers, Inc v International Parts Corp, 392 US 134, 141-42 (1968); United States v Yellow Cab Co, 332 US 218 (1947); Kiefer-Stewart Co v Joseph E Seagram & Sons, 340 US 211 (1951); United States v Timken Roller Bearing Co, 341 US 593 (1951); United States v Cotton Baking Co, 513 F Supp 223, 225-27 (ED Pa 1981) and cases cited therein.

[109] *See* Sunkist Growers, Inc v Winckler & Smith Citrus Prods Co, 370 US 19, 28-29, (1962); Harvey v Fearless Farris Wholesale, Inc, 589 F2d 451, 455-56 n 8 (9th Cir 1979); Nelson Radio & Supply Co v Motorola, 200 F2d 911, 914 (5th Cir 1952), *cert denied,* 345 US 925 (1953); Island Tobacco Co v RJ Reynolds Indus, Inc, 513 F Supp 726, 739-41 (D Hawaii 1981); *see also* Joseph E Seagram & Sons v Hawaiian Oke & Liquors, Ltd, 416 F2d 71 (9th Cir 1969), *cert denied,* 396 US 1062 (1970). Where a severe restraint is regarded as clear, the plurality requirement for a combination in violation of 15 USC §1 may be more readily satisfied. *See* IT&T v General Tel & Elecs Corp, 449 F Supp 1158 (D Hawaii 1978); *see generally* Willis & Pitofsky, *Antitrust Consequences of Using Corporate Subsidiaries,* 43 NYU L Rev 20 (1968); McQuade, *Conspiracy, Multi-corporate Enterprises, and Section 1 of the Sherman Act,* 41 Va L Rev 183 (1955); Barmat, *Two Trees or One: The Problem of Intra-Enterprise Conspiracy,* 23 Mont L Rev 158 (1962); Kaysen, *Collusion Under The Sherman Act,* 65 QJ Econ 263 (1951); see Handler, Blake, Pitofsky & Goldschmid, Trade Regulation: Cases and Materials 473-75 (1975); Comment, 14 Duq L Rev 63 (1975); Note, 75 Mich L Rev 717 (1975); Sullivan, Antitrust 324-28 (1977); Report of the Attorney General's National Committee to Study the Antitrust Laws 30-36 (1955). On intra-corporate conspiracy, *see* Tose v First Pa Bank, 648 F2d 879, 893-94 (3d Cir 1981).

[110] See US Dept of Justice, Antitrust Guide for International Operations pt II, Case A, at 11-14 (1977).

[111] See Rupert, *The Relationship of Patent Law to Antitrust Law,* 49 ABA Antitrust LJ 755 (1981); Baxter, Legal *Restrictions on Exploitation of the Patent Monopoly: An Economic Analysis,* 76 Yale LJ 267 (1966). For a discussion of patent law, see **ch 4.**

[112] *See* SCM v Xerox Corp, 645 F2d 1195 (2d Cir 1981), *affg* 463 F Supp 983 (D Conn 1978), *cert denied,* — US — (1982); *see also* Automatic Radio Mfg Co v Hazeltine Re-

necessarily justify other conduct which would be illegal absent a patent. Although the original right to exclude others would seem automatically to include the lesser right to grant restricted use of the patent, this latter right is subject to illegal restrictions.[113] Restrictions relating solely to when, where or how a patent may be used appear permissible unless they unreasonably restrict competition in the overall market involved.[114]

The mere accumulation of patents,[115] and similarly, mere nonuse of patents developed by an inventor, are not illegal.[116] However, the use of fraud against the Patent Office to secure an otherwise unavailable monopoly, or extension of a patent monopoly beyond its terms by such means as industry-wide price controls, division of markets or, unduly broad reciprocity provisions or requirements to grant back patents developed by licensees may violate the anti-

---

search, Inc, 339 US 827, 834 (1950). On criteria for enforcement of patents against activities which contribute to infringement, *see* Dawson Chem Co v Rohm & Hass Co, 448 US 176 (1980).

[113] *See* United States v Line Material Co, 333 US 287 (1948); *see also, e.g.,* United States v Singer Mfg Co, 374 US 174 (1963); Timken Roller Bearing Co v United States, 341 US 593 (1951); United States v Imperial Chem Indus, 105 F Supp 215 (SDNY 1954); United States v General Elec Co, 80 F Supp 989 (SDNY 1952). Indeed, use of a patent to obtain market power beyond the scope of the grant may be treated as misuse of the patent barring its enforcement even without antitrust violation. *See* Zenith Radio Corp v Hazeltine Research, Inc, 395 US 100, 140 (1969); Morton Salt Co v GS Suppiger Co, 314 US 488 (1942); Report of the Attorney General's National Committee to Study the Antitrust Laws 254 (1955). For limitations on this doctrine, *see* Dawson Chem Co v Rohm & Hass Co, 448 US 176 (1980).

[114] Such restrictions are clearly permissible if reasonable. General Talking Pictures Corp v Western Elec Co, 305 US 124, 126 (1938); Atlas Imperial Diesel Engine Corp. v Lanova Corp, 79 F Supp 1002 (D Del 1948). On tests for reasonableness, *see* Continental TV, Inc v GTE Sylvania, Inc, 433 US 36 (1977); *see generally* Rupert, *The Relationship of Patent Law to Antitrust Law,* 49 ABA Antitrust LJ 755, 764-65 (1981); Sullivan, Antitrust §184 (1977); Gibbons, *Field Restrictions in Patent Transactions,* 66 Colum L Rev 423 (1966); Note, 75 Harv L Rev 602 (1962); *compare* Denison Mattress Factory v Spring Air Co, 308 F2d 403 (5th Cir 1962). For a finding of illegality, *see* United States v Ciba Geigy Corp, 508 F Supp 1118 (DNJ 1976).

[115] Note 112 *supra;* Continental Paper Bag Co v Eastern Paper Bag Co, 210 US 405 (1908); Bement v National Harrow Co, 186 US 70, 88-90 (1902); Handler, Blake, Pitofsky & Goldschmid, Trade Regulation: Cases and Materials 207 n 8 (1975).

[116] SCM v Xerox Corp, 645 F2d 1195 (2d Cir 1981); *see* Special Equip Co v Coe, 324 US 370 (1945); Continental Paper Bag Co v Eastern Paper Bag Co, 210 US 405 (1908); Handler, Blake, Pitofsky & Goldschmid, Trade Regulation: Cases and Materials 207 n 8 (1975); Frost, *Legal Incidents of Non-Use of Patented Inventions,* 14 Geo Wash L Rev 273 (1946); Note, 58 Harv L Rev 726 (1945); Note, 93 U Pa L Rev 456 (1945); Note, 31 Va L Rev 68 (1945); Note, 45 Colum L Rev 422, 601 (1945); Note, 53 Yale LJ 579 (1944); Note, 49 Mich L Rev 738 (1951). It may be noted that the first Justice Harlan dissented from the holding in *Continental Paper Bag,* 210 US at 430, that non-use of a patent had no effect on its enforceability and that the Court stressed that "there was no question of a diminished supply or of increase of prices. . . ." *Id* 429. Justice Harlan's dissents have often been prophetic, *e.g.,* that in Plessy v Ferguson, 163 US 537, 559 (1896), *overruled by* Brown v Board of Educ, 347 US 483 (1954).

trust laws.[117] Careful advice, indeed, would frown on seeking to control prices charged by licensees or others at all.[118] Combinations with others creating joint patent pools excluding competitors may also become illegal under the Sherman Act.[119]

A possible strategy for dealing with these questions is to estimate the dollar value to others of various options and charge for them rather than imposing trade restraints. For example, a royalty based on quantity sold may influence a licensee's cost and, hence, the licensee's price without seeking to control price as such.[120] Adopting the least restrictive alternative will generally minimize risk, even though the law may not require that such an alternative be adopted (generally, the courts apply a rule of reason analysis, balancing the

---

[117] *See* United States v Singer Mfg Co, 374 US 174 (1963); United States v Line Material Co, 333 US 287 (1948); Hartford-Empire Co v United States, 323 US 386 (1945); Sullivan, Antitrust ch 6 (1977). Concerning grant back provisions, *see* Transparent-Wrap Mach Corp v Stokes & Smith Co, 329 US 637 (1947); Santa Fe Pomeroy, Inc v P&Z Co, 569 F2d 1084 (9th Cir 1978); Antitrust Division, US Dept of Justice, Guide for International Operations 42-45 (1977). Concerning fraud on the Patent Office, *see* Walker Process Equip, Inc v Food Mach & Chem Corp, 382 US 172 (1965); cases cited in Rupert, *The Relationship of Patent Law to Antitrust Law,* 49 ABA Antitrust LJ 755, 759 n 13 (1981); Kayton, Lynch & Stern, *Fraud in Patent Procurement,* 43 Geo Wash L Rev 1 (1974).

For examples of the inquiry involved, *see In re* Coordinated Pretrial Proceedings in Antibiotic Antitrust Actions, 498 F Supp 28 (ED Pa 1980); Eutectic Corp v Metco, Inc, 505 F Supp 73 (EDNY 1980). On excessive threats of enforcement, *see* Analytichem Intl, Inc v Har-Lens Assocs, 490 F Supp 271 (WD Pa 1980). *See generally* Sola Elec Co v Jefferson Elec Co, 317 US 173 (1942); Morton Salt Co v GS Suppiger Co, 314 US 488 (1942); United States v Ethyl Gasoline Co, 309 US 436 (1940).

A licensee may no longer be precluded by contract from challenging the patent. Lear Inc v Adkins, 395 US 653 (1969); Bendix Corp v Balex, Inc, 471 F2d 149 (7th Cir 1972), *cert denied,* 414 US 819 (1973), nor, of course, is such preclusion implied by the license itself. Indeed, presence of a preclusion provision may constitute patent misuse. Kearney & Trecker Corp v Giddings & Lewis, Inc, 452 F2d 579 (7th Cir 1971), *cert denied,* 405 US 1066 (1972). *See generally* Beckman Instruments, Inc v Technical Dev Co, 433 F2d 44 (7th Cir 1970), *cert denied,* 401 US 976 (1971).

[118] *See* United States v United States Gypsum Co, 340 US 76, 84 (1980); United States v Line Material Co, 333 US 287 (1948); Sullivan, Antitrust ch 6(C) (1977); *and see* United States v General Elec Co, 272 US 476 (1926) (allowing price resale fixing where one licensee only is involved). *See generally* United States v New Wrinkle Co, 342 US 371 (1952); *In re* Yarn Processing Patent Patent Validity Litig, 541 F2d 1127 (5th Cir 1976); Newberg Moire Co v Superior Moire Co, 237 F2d 283 (3d Cir 1956); Tinnerman Prods, Inc v George K Garrett Co, 185 F Supp 151 (ED Pa 1960), *affd,* 292 F2d 137 (3d Cir), *cert denied,* 368 US 833 (1961); United States v General Elec Co, 80 F Supp 989 (SDNY 1948). For recent examples of traps in this area, *see* Mannington Mills, Inc v Congoleum Indus, Inc, 610 F2d 1059 (3d Cir 1979); American Cyanamid Co, 63 FTC 1747, 1981 (1963), *vacated,* 363 F2d 757 (6th Cir 1966), *on rehearing,* 72 FTC 623 (1967), *affd sub nom* Charles Pfizer & Co v FTC, 401 F2d 574 (6th Cir 1968), *cert denied,* 374 US 920 (1969).

[119] Hartford-Empire Co v United States, 323 US 386 (1945).

[120] *See* Westgo Indus, Inc v WJ King Co, 1981-1 Trade Reg Rep (CCH) ¶63,984 (DND 1981).

competitive benefits and detriments involved,[121] unless the activity involved is illegal per se, such as price collusion between competitors). The Department of Justice also strongly favors such an approach to most situations.[122]

Acquisitions of patents from others are subject to antitrust attack if they may tend substantially to lessen competition in any relevant market at the time of the acquisition[123] or are part of a broader combination involving patents or otherwise which the courts find monopolizes or unreasonably restrains trade.[124]

A more difficult question concerns the use of patents which are not exploited, to block development in a major aspect of an industry, perhaps to protect investment in existing technologies.[125] While such conduct has not been held

---

[121] *See* Continental TV Inc v GTE Sylvania Inc, 433 Us 36, 58 n 29 (1977). Patents affect reasonableness positively because of the legitimacy conferred by the patent laws and negatively because of the power created and the notion of not allowing its extension beyond the limits of the patent grant. This suggests the desirability of use of the power in some cases to obtain money payments rather than restrictions in return for use of patents or patented items. The fate of agreements to grant back to the patentee improvement patents developed by licensees seems to differ in part according to non-patent market power, *see* Transparent-Wrap Mach Corp v Stokes & Smith Co, 329 US 637 (1947); United States v General Elec Co, 115 F Supp 835 (DNJ 1953); United States v General Elec Co, 80 F Supp 989 1005-06 (SDNY 1948); Report of the Attorney General's National Committee to Study The Antitrust Laws 227-28 (1955); Sullivan, Antitrust 464-65 (1977).

[122] For a discussion of the various judicial approaches to antitrust analysis, see **§3.03.** See generally Sullivan, Antitrust §§182-191 (1977); Remarks of Abbott B. Lipsky, Jr, Deputy Assistant Attorney General, *Current Antitrust Division Views of Patent Licensing Practices,* before ABA Antitrust Section, Washington, D.C. (Nov 5, 1981).

[123] SCM v Xerox Corp, 645 F2d 1195, 1210-1212 (2d Cir 1981); United States v Lever Bros, 216 F Supp 886 (SDNY 1963); Dole Valve Co v Perfection Bar Equip Co, 331 F Supp 459 (ND Ill 1970); Sullivan, Antitrust §180 (1977); Turner, *Patents, Antitrust and Innovation,* 12 Antitrust Bull 277, 282 (1967).

[124] *See* United States v Singer Mfg Co, 374 US 174 (1963); Hartford-Empire Co v United States, 323 US 386 (1945); United States v Univis Lens Co, 316 US 241 (1942); *see also* United States v New Wrinkle Co, 342 US 371 (1952).

On risks of mandatory requirements to take licenses on more than one patent at a time, *see* Zenith Radio Corp v Hazeltine Research, Inc, 395 US 100 (1969); Automatic Radio Mfg Co v Hazeltine Research, Inc, 339 US 827 (1950); Standard Oil Co v United States, 283 US 163 (1931); *compare* United States v Paramount Pictures, Inc, 334 US 131 (1948); Alden-Rochelle, Inc v ASCAP, 80 F Supp 888 (SDNY 1948) (copyrights); Schine Chain Theatres, Inc v United States, 334 US 110 (1948); United States v Griffith, 334 US 100 (1948); United States v Crescent Amusement Co, 323 US 173 (1944) (films where exhibitor had only theater in some towns and requested exclusivity in a broader territory); Sullivan, Antitrust §159 (1977).

On antitrust risks in settlements of patent litigation, *see* Mallison-Cleveland-Akron S Co v Golden State Adv Co, 444 F2d 425 (9th Cir), *cert denied,* 404 US 873 (1971); Jack Winter, Inc v Koratron Co, 375 F Supp 1 (ND Cal 1974); United States v Krasnov, 143 F Supp 183 (ED Pa 1956); *see also generally,* United States v Vehicular Parking Ltd, 54 F Supp 828 (D Del 1944).

[125] *See* Continental Paper Bag Co v Eastern Paper Bag Co, 210 US 405, 428-29 (1908); *compare Bank Holding Company Act Amendments, Hearings Before the House Committee on Bank-*

illegal despite market power the issue has been specifically reserved insofar as equitable relief is concerned.[126] The implications of a perceived economy of scarce resources[127] and public desire to promote innovation may lead to vulnerability of such conduct in a case where a significant anticompetitive impact can be shown.[128] The risk is obviously greater if the patents were acquired from others, especially during a period of market dominance, rather than internally developed.[129] If a patent is not being exploited, and there is no closely adjacent patent reaching a similar technological goal, and if a major industrial breakthrough is being blocked by this situation, it may be prudent for the holder to consider licensing based on a fully compensatory royalty.[130]

---

ing and Currency, 91st Cong, 1st Sess 372-74, 388, 615-16 (1969). *See also* Special Equip Co v Coe, 324 US 370, 381-86 (1945) (dissenting opinion); Berkey Photo, Inc v Eastman Kodak Co, 603 F2d 263, 302 (2d Cir 1979), *cert denied,* 444 US 1093 (1980).

[126] SCM v Xerox Corp, 463 F Supp 983, 1012-13 (D Conn 1978), *affd,* 645 F2d 1195 (2d Cir 1981), *cert denied,* — US — (1982); *see also* Van Dyk Research Corp v Xerox Corp, 478 F Supp 1268, 1325 (DNJ 1979), *affd,* 631 F2d 251 (3d Cir 1980); United States v Aluminum Co of America, 91 F Supp 333, 386-87 (SDNY 1950).

[127] For public perception, *see, e.g.,* Baker, *Cheese & Macaroni,* NY Times, June 19, 1979, at A21.

[128] Rupert, *The Relationship of Patent Law to Antitrust Law,* 49 ABA Antitrust LJ 755, 761 (1981).

[129] *See* SCM v Xerox Corp, 645 F2d 1195 (2d Cir 1981), *cert denied,* — US — (1982); Turner, *Patents, Antitrust and Innovation,* 12 Antitrust Bull 277, 282 (1967).

[130] If the royalty is in fact what the patent's use is worth, the patentee should suffer no economic loss. On pitfalls in royalty computation, see Rupert, *The Relationship of Patent Law to Antitrust Law,* 49 ABA Antitrust LJ 755, 761-64 (1981). On reasonable royalty requirements as an antitrust remedy, *see* United States v National Lead Co, 322 US 319 (1947). *Compare* Allied Research Prods v Heatbath Corp, 300 F Supp 656 (ND Ill 1969); 42 USC §§1857-h-6, 2182-2190.

Refusal of injunctive relief against infringement can amount to compulsory licensing in some cases. Wisconsin Alumni Research Found v Vitamin Technologists, Inc, 146 F2d 941, 954-56 (9th Cir 1944), *cert denied,* 325 US 876 (1945); Milwaukee v Activated Sludge, Inc, 69 F2d 577, 593 (7th Cir 1933), *cert denied,* 293 US 576 (1934) and cases cited therein; 35 USC §283. Of course nonuse and refusal to license combined with other conduct may amount to a violation. Standard Sanitary Mfg Co v United States, 226 US 20 (1912). The so-called bottleneck cases may also be raised as analogous, *e.g.,* Associated Press v United States, 325 US 1 (1945); United States v Terminal RR Assn, 224 US 383 (1912); Byrnes, Lowry & Bordinant, *Product Shortages, Allocation and the Antitrust Laws,* 20 Antitrust Bull 713 (1975); Note, 87 Harv L Rev 1720 (1974); *see also* Berkey Photo, Inc v Eastman Kodak Co, 603 F2d 263, 284 (2d Cir 1979), *cert denied,* 444 US 1093 (1980). This would not, however, support a claim for damages, under SCM v Xerox Corp, 645 F2d 1195 (2d Cir 1981), *cert denied,* — US — (1982).

For some problems in implementing legally required licensing based upon reasonable royalties, *see* United States v Western Elec Co, 1980-2 Trade Cas (CCH) ¶63,519 (DNJ 1980). For rejection of or claim of illegality based upon the charging of excessive royalties, *see* Warner-Jenkinson Co v Allied Chem Corp, 477 F Supp 371, 397-398 (SDNY 1979), *affd,* 633 F2d 208 (2d Cir 1980).

On the problem of discriminatory royalties, *see* FaPeyre v FTC, 366 F2d 117 (5th Cir 1966); Laitram Corp v King Crab, Inc, 244 F Supp 9 (D Alaska 1965), *modified,* 245 F Supp 1019 (D Alaska 1966); Peelers Co v Wendt, 260 F Supp 193 (D Wash 1966);

Use of invalid patents as a basis for agreements between competitors is also subject to challenge. However, the invalidity or expiration of a patent no longer in and of itself makes an agreement involving know-how licensing (permission to use trade secrets or technical information for a fee) unlawful, since such licensing is recognized as an independently permissible arrangement.[131] As in the case of patent licensing, any restrictions inposed must be reasonable and cannot be of a type which would be illegal per se, absent a know-how license.

§3.09 Marketing Restrictions

Marketing advantages which an industrially innovative enterprise may seek in contracts with customers, distributors or suppliers are frequently formulated in terms which in one way or another bar dealings with competitors or control prices. Such restrictions, even if outside categories which are illegal per se,[132] always carry antitrust risks. The extent of risk often depends on unknown facts[133] or subjective judgments. In some instances these are minimal and taking the risk if justifiable. In other cases, alternatives may be developed to achieve critical business objectives while avoiding potential antitrust pitfalls. Direct financial benefits may frequently be bargained for instead of restrictions. For example, quantitative royalties[134] may be substituted for pricing restraints or covenants not to compete; long-term purchase contracts with compensation for cancellations may be substituted for strict requirements contracts,[135] perhaps with side benefits under the Robinson-Patman Price Discrimination Act.[136]

Dealers or distributors may request protection from competition or price-

---

Rupert, *The Relationship of Patent Law to Antitrust Law,* 49 ABA Antitrust LJ 755, 762-63 (1981).

[131] Painton & Co v Bourns Inc, 442 F2d 216 (2d Cir 1971); *compare* Aronson v Quick Point Pencil Co, 440 US 257 (1979) *and* Kewanee Oil Co v Bicron Corp, 416 US 470 (1974) *with* Brulotte v Thys Co, 379 US 29 (1964); *see also* Shin Nippon Koki Co v Irvin Indus, Inc, 186 USPQ 296 (NY Sup Ct 1975). A provision barring a licensee from challenging a patent is, however, invalid. See cases cited in note 117 *supra.*

Although patent license royalties cannot validly continue beyond patent expiration, royalties for transfer of know-how, if it is actually involved, may. Pitney-Bowes, Inc v Mestre, F Supp 52 (SD Fla 1981).

[132] For a list of per se categories, see **§ 3.03.**

[133] On uncertainty in fact finding, whether in adjudication or counseling, *see generally* Cahn, *Fact Skepticism: An Unexpected Chapter,* 38 NYU L Rev 1025 (1963); Frank, Courts on Trial (1963); Frank & Frank, Not Guilty (1957); Givens, Advocacy (Shepard's/ McGraw-Hill (1980). *Compare* England v Louisiana Medical Examiners, 375 US 411, 417 (1964); Speiser v Randall, 357 US 513, 525 (1958).

[134] *See* Westgo Indus Inc v WJ King Co, 1981-1 Trade Reg Rep (CCH) ¶63,984 (DND 1981).

[135] See notes 103-104 *supra.*

[136] 15 USC §13. See **§3.07.**

cutting in the form of distributional restraints.[137] These requests may sometimes run counter to the manufacturer's interest in maximizing total sales, but may be hard to resist because pressure is applied by parties, personally close to executives of the innovator,[138] who may individually or in concert with others decide not to deal with the innovating firm unless their requests are met.[139]

The common business reasons for distributional restraints[140] are the needs to obtain and retain high-quality dealer advertising, point-of-sale promotional support and warranty service.[141] Less often, there are problems of product deterioration or misuse.[142] A number of means may be used to meet these needs which may be less likely to cause antitrust problems. For example, cooperative advertising allowances may be granted to dealers,[143] subject to Robinson-Patman requirements of proportionately equal availability.[144] Further, warranty service, initially ordinarily a responsibility of the manufacturer,

---

[137] *See* Girardi v Gates Rubber Co, 325 F2d 196, 202 (9th Cir 1963). *Compare* Mannington Mills, Inc v Congoleum Indus, Inc, 610 F2d 1059 (3d Cir 1979).

[138] *Compare* Telser, *Why Should Manufacturers Want Fair Trade,* 3 JL & Econ 86 (1960) *with* Jay, Management and Machiavelli (1968); Palamountain, The Politics of Distribution (1955).

[139] *See* United States v General Motors Corp, 384 US 127 (1966), *discussed in* Continental TV, Inc v GTE Sylvania Inc, 433 US 36, 58 n 58 (1977); Girardi v Gates Rubber Co, 325 F2d 196, 202 (9th Cir 1963); Pribram, Cartel Problems (1935).

[140] *See* Continental TV, Inc v GTE Sylvania Inc, 433 US 36 (1977); Givens & Worsinger, *Vertical Restraints After Repeal of Fair Trade,* 45 Fordham L Rev 1093, 1098-1104 (1977).

[141] On warranties, *see generally* Magnuson-Moss Warranty-FTC Improvement Act, 15 USC §2301 *et seq* and regulations in 16 CFR §701 *et seq;* Eddy, *Effects of the Magnuson-Moss Act Upon Consumer Product Warranties,* 55 NCL Rev 835 (1977); Symposium, 45 Antitrust LJ 72 (1976); Special Committee on Consumer Affairs, Assn of the Bar of the City of New York, *Consumer Products Warranty Fulfillment* NYLJ, Nov 28, 1980, at 1. State laws dealing with payment to dealers for warranty service should also be reviewed. On problems if warranty reimbursements are excessive, *see* Eiberger v SONY Corp of America, 459 F Supp 1276 (SDNY 1978), *affd in part,* 622 F2d 1068 (2d Cir 1980).

[142] See Clairol, Inc v Boston Discount Center of Berkley, Inc, 608 F2d 1114 (6th Cir 1979); Tripoli Co v Wella Corp, 425 F2d 932 (3d Cir), *cert denied,* 400 US 831 (1970); Note, 72 Yale LJ 1171 (1963); *but see* Adolph Coors Co v FTC, 497 F2d 1178 (10th Cir 1974), *cert denied,* 419 US 1105 (1975); *see generally* Continental TV, Inc v GTE Sylvania Inc, 433 US 36, 55 n 23 (1977). *Compare* Direct Sales Co v United States, 319 US 703 (1943); HR Rep No 1808, 91st Cong, 1st Sess (1971).

[143] For issues raised by alleged claims for nonexistent advertising, *see* Schwimmer v SONY Corp of America, 637 F2d 41 (2d Cir 1980); Sadowy v SONY Corp of America, 496 F Supp 1071, 1075, 1080 (SDNY 1980).

[144] *See* Federal Trade Commission, Guides for Advertising Allowances and Other Merchandising Payments and Services, 16 CFR §240 (1973); *see also* FTC v Fred Meyer, Inc, 390 US 341 (1968); Sullivan, Antitrust §227 (1977); on the availability of meeting competition as a defense, *compare* Exquisite Form Brassiere, Inc v FTC, 301 F2d 499 (DC Cir 1961), *cert denied,* 369 US 888 (1962) *with* Exquisite Form Brassiere, Inc v FTC, 360 F2d 492 (DC Cir 1965), *cert denied,* 384 US 959 (1966).

may be delegated[145] to dealers who may, of course, be paid for the work,[146] provided that chargebacks to selling dealers elsewhere are not used to enforce territorial restrictions which the courts may find unreasonable and thus, antitrust violations.[147]

In addition, manufacturers may agree not to sell to other dealers within exclusive territories,[148] and may replace one dealer with another[149] in a territory or nationally, absent contract or state law violations,[150] or other evidence of unreasonable restraint of trade.

Agreements barring dealers from selling from unauthorized locations,[151] or selling to unauthorized customers such as other dealers or into unauthorized territories[152] are judged under the rule of reason under § 1 of the Sherman Act,[153] which involves a balancing of their pro-and anticompetitive effects.[154]

The extent of interbrand competition is the most important element in such balancing.[155] The legality of a restriction may thus depend significantly on the

---

[145] *See* 15 USC §2307.

[146] *Id;* on potential attacks on state warranty reimbursement laws, *compare* New Motor Vehicle Bd v Fox, 439 US 96 (1978); Exxon v Governor of Maryland, 437 US 117 (1978); American Motors Sales v Division of Motor Vehicles, 592 F2d 219 (4th Cir), *cert denied,* 444 US 836 (1979) *with* Flood v Kuhn, 407 US 258 (1972); Raymond Motor Transp, Inc v Rice, 434 US 429 (1978); Givens, *The Commerce Clause and the Interdependent Economy,* 53 ABAJ 719 (1967).

[147] Eiberger v SONY Corp of America, 622 F2d 1068 (2d Cir 1980), *affg in part,* 459 F Supp 1276 (SDNY 1978).

[148] *See* discussion and cases cited in GTE Sylvania Inc v Continental TV, Inc, 537 F2d 980 (9th Cir 1976), *affd on other grounds,* 433 US 36 (1977).

[149] Oreck Corp v Whirlpool Corp, 579 F2d 126 (2d Cir) *(en banc), cert denied,* 439 US 946 (1978).

[150] On state laws, see State Business Franchise Disclosure and Relationship Laws (CCH 1978); Note, 74 Colum L Rev 1487 (1974).

[151] *See* GTE Sylvania Inc v Continental TV, Inc, 537 F2d 980 (9th Cir 1976), *affd* 433 US 36 (1977).

[152] Continental TV, Inc v GTE Sylvania Inc, 433 US 36 (1977); see generally Practising Law Institute, Marketing and Distribution: Regulations and Restraints (1978); Pitofsky, *The Sylvania Case: Antitrust Analysis of Non-Price Vertical Restrictions,* 78 Colum L Rev 1 (1978); Louis, *Vertical Distributional Restraints After Sylvania: A Postscript and Comment,* 76 Mich L Rev 266 (1978).

[153] *See* Chicago Board of Trade v United States, 246 US 231, 238-41 (1918); Standard Oil Co v United States, 221 US 1 (1911); *see also* Appalachian Coals, Inc v United States, 288 US 344 (1933); Handler, Antitrust in Perspective ch I (1957).

[154] "The central teaching of *Sylvania* is that the overall effect of a vertical restraint must be the enrichment of competition in the market as a whole." Handler, *Antitrust—1978,* 78 Colum L Rev 1363, 1370 (1978). For an instance of a finding of violation, *see* Eiberger v SONY Corp of America, 622 F2d 1068 (2d Cir 1980), *affg in part* 459 F Supp 1276 (SDNY 1978). For positive answers, *see, e.g.,* Muenster Butane, Inc v Stewart Co, 651 F2d 292 (5th Cir 1981); Abadir & Co v First Miss Corp, 651 F2d 422 (5th Cir 1981).

[155] Continental TV, Inc v GTE Sylvania Inc, 433 US 36, 52 n 19 (1977).

market share of the firm involved.[156] The need for the restriction may also vary inversely with market share and is also of great importance.[157] If the restriction seems more stringent than necessary, it may be vulnerable, although it is not always necessary to choose the least restrictive alternative.[158] Customer restrictions may also be upheld if product misuse is a problem.[159] Patent field-of-use restriction limiting types of applications permitted may likewise fall under this rule of reason approach.[160]

Efforts to influence dealer prices are at the most dangerous end of the spectrum. While manufacturer suggestion of resale prices and price advertising[161] and product preticketing[162] are not illegal as such, agreements with dealers on resale prices are unlawful.[163] This long-standing per se rule[164] has been explained by the United States Supreme Court on the ground that vertical resale price maintenance creates horizontal effects at both manufacturer and dealer levels.[165]

Governmental authorities may look to horizontal effects as a guide to enforcement action.[166] If competing manufacturers can be held tacitly responsi-

---

[156] See ABA Antitrust Sec, Vertical Restraints Limiting Intrabrand Competition 63-64 (monograph No 2, 1977). Effective interbrand competition is also defined as essential to validity of territorial restrictions under the Soft Drink Inter-brand Competition Act, Pub L No 96-308, 94 Stat 939 (1980), bringing about reversal of finding of violation in Coca-Cola Co, 91 FTC 517 (1978) by Coca-Cola Co, v FTC, 642 F2d 1387 (DC Cir 1981). See S Rep No 645, 96th Cong, 2d Sess (1980), *reprinted in* 1980 US Code Cong & Ad News 4391.

[157] *See* Eiberger v SONY Corp of America, 622 F2d 1068 (2d Cir 1980), *affg in part* 459 F Supp 1276 (SDNY 1978).

[158] Continental TV, Inc v GTE Sylvania Inc, 433 US 36, 58 n 29 (1977).

[159] See note 142, *supra.*

[160] See **§3.08.**

[161] Chisholm Bros Farm Equip Co v International Harvester Co, 498 F2d 1137 (9th Cir), *cert denied,* 419 US 1023 (1974); Gray v Shell Oil Co, 469 F2d 742 (9th Cir 1972), *cert denied,* 412 US 943 (1973); *compare* Susser v Carvel Corp, 206 F Supp 636 (SDNY 1962), *affd,* 332 F2d 505 (2d Cir 1964), *cert dismissed,* 381 US 125 (1965).

[162] See Automobile Information Disclosure Act, 15 USC §1232(f)(1); United States v OM Scott & Sons, 303 F Supp 141 (DDC 1969); United States v Serta Assoc, Inc, 296 F Supp 1121 (ND Ill 1968), *affd,* 393 US 534 (1969); Bailey's Bakery v Continental Baking Co, 235 F Supp 705, 721-22, (D Hawaii 1964), *affd,* 401 F2d 182 (9th Cir 1968), *cert denied,* 393 US 1086 (1969); Engbrecht v Dairy Queen Co, 203 F Supp 714, 719 (D Kan 1962); *see also* Elman, *"Petrified Opinions" and Competitive Realities,* 66 Colum L Rev 625, 631-32 (1966).

[163] See **§3.03.**

[164] *See* Continental TV, Inc v GTE Sylvania Inc, 433 US 36, 51 n 18 (1977); Dr Miles Medical Co v John D Park & Sons, 220 US 373 (1911).

[165] California Retail Liquor Dealers Assn v Midcal Aluminum, Inc, 445 US 97 (1980); see Givens & Worsinger, *Vertical Restraints After Repeal of Fair Trade,* 45 Fordham L Rev 1093 (1977).

[166] As stated by Assistant Attorney General William F. Baxter, "a pattern of vertical arrangements can have horizontal consequences." *Quoted in* 1981 Antitrust & Trade Reg Rep No 1007, at A-10 (Mar 2, 1981).

ble for prices charged by their dealers, this may generate horizontal interaction or coordinative behavior.[167] Similarly, dealers may seek to suppress competition between themselves[168] utilizing manufacturers as enforcement agencies.[169] If particular conduct can be shown not to tend to produce these results or not to be inimical to the workings of a competitive marketplace, however, the reason for the per se rule and hence the rule itself may not apply.[170]

If customer or territorial resale restrictions are implicated in resale price maintenance efforts, they may likewise become illegal per se.[171] Indeed, combining customer resale restrictions even with otherwise permissible noncoercive efforts to influence dealer prices must be considered dangerous.[172]

Maximum price-fixing by sellers, while covered by the per se ban as articulated,[173] may be more often favorable to the other side of the marketplace (unless in practice minimum price-fixing) and hence a possible candidate for future rule of reason treatment.[174]

While it does not in and of itself offend the Sherman Act for a manufacturer to refuse to sell to discounters[175]—or indeed to any undesired

---

[167] *See* Continental TV, Inc v GTE Sylvania Inc, 433 US 36, 51 n 18 (1977).

[168] *Id.* 58 n 28.

[169] *See* United States v General Motors Corp, 384 US 127 (1966), discussed in Continental TV, Inc v GTE Sylvania Inc, 433 US 36, 58 n 28 (1977).

[170] *See* Broadcast Music, Inc v CBS, 441 US 1 (1979); *compare also, generally* Zenith Radio Corp v Matsushita Elec Indus Co, 513 F Supp 1100 (ED Pa 1981).

[171] *E.g.,* United States v Bausch & Lomb Optical Co, 321 US 707 (1944); Copper Liquor Inc v Adolph Coors Co, 506 F2d 934, 947-48 (5th Cir 1975); Adolph Coors Co v FTC, 497 F2d 1178 (10th Cir 1974), *cert denied,* 419 US 1105 (1975); see Denger, *Vertical Restrictions: The Impact of Sylvania,* 46 ABA Antitrust LJ 908, 915-16 (1977) and cases cited therein.

[172] It could be claimed that resale restrictions cause resale price persuasion to become illegal coercion. *See* Simpson v Union Oil Co, 377 US 13, 17 (1964); S Rep No 466, 94th Cong, 1st Sess 3 (1975). Similarly, the price effect of the restriction under the circumstances might be claimed to cause it to violate the rule of reason. *Compare* Oreck Corp v Whirlpool Corp, 579 F2d 126, 131, 130 n 5 (2d Cir) *(en banc), cert denied,* 439 US 946 (1978). On risks of the combination of restrictions on sales and price persuasion, *see* United States v Bausch & Lomb Optical Co, 321 US 706 (1944); FTC v Beech-Nut Packing Co, 257 US 441 (1922); see also Martin, *Transshipping,* Dealerscope 8 (Mar 1979).

[173] Albrecht v The Herald Co, 390 US 145 (1968); Kiefer-Stewart Co v Seagram & Sons, 340 US 211 (1951).

[174] See *Vertical Restrictions Limiting Intra-Brand Competition,* 1977 ABA Sec Antitrust 29, 84-96 (Monograph No 2). *Compare* Quinn v Mobil Oil Co, 375 F2d 273 (1st Cir), *cert dismissed,* 389 US 801 (1967); Albrecht v The Herald Co, 390 US 145, 157-60 (1968)(Harlan, J dissenting).

[175] State franchisor-franchisee relationship protection laws and special industry statutes such as the Automobile Dealers Day in Court Act and petroleum dealers' protection laws may, however, apply. *See* 15 USC §§1221-1225, 2802; Clark v Mobil Oil Co, 496 F Supp 132 (ED Mo 1980); State Business Franchise Disclosure and Relationship Laws (CCH 1978); Brown, Franchising: Realities and Remedies (1978); on § 5 of the Federal Trade Commission Act, 15 USC § 45, concerning unfair methods of competition, see discussion in Official Airline Guides, Inc v FTC, 630 F2d 920 (2d Cir 1980).

dealers,[176]—cutoffs of discounters may become evidence of, or in practice lead to, agreements with dealers on prices to be charged.[177] It is unrealistic to expect that an effort to control or influence dealer prices involving such measures as cutoffs can proceed without informal agreements developing as a lesser sanction at the working level.[178]

Dealer cutoffs also quite naturally provoke litigation on all kinds of grounds not limited to the Sherman Act and hence outcomes are normally difficult to predict.[179]

§3.10 Joint Research

Joint research efforts may offer important advantages for innovation through pooling of resources and information,[180] provided a strong incentive to succeed exists,[181] and provided internal bureaucratic layering does not stifle initiative within the research organization.[182] Multiple lines of attack on the same problem carried out under separate independent leadership are often essential to success,[183] and indeed many discoveries are made in unforeseen

---

176 United States v Colgate & Co, 250 US 300 (1918), reaffirmed in Reeves, Inc v Stake, 447 US 429, 450 (1980). *See* discussion in Zenith Radio Corp v Matsushita Elec Indus Co, 513 F Supp 1100, 1166 (ED Pa 1981).

177 In order to rely sucessfully on the *Colgate* doctrine permitting unilateral refusal to deal with discounters, it has been said that the arrangements of a seller would have to be "of such Doric Simplicity as to be somewhat rare in this day of complex business enterprise." George W Warner & Co v Black & Decker Mfg Co, 277 F2d 787, 790 (2d Cir 1960). See Levi, *The Parke, Davis-Colgate Doctrine: The Ban on Resale Price Maintenance,* 1960 Sup Ct Rev 258; Maslow, *Colgate Updated: Resale Price Maintenance and Antitrust Liability,* 48 N4 St BJ 626 (1976). *Compare* Canadian Am Oil Co v Union Oil Co, 577 F2d 468 (9th Cir 1978); Andersen, *The Antitrust Consequences of Manufacturer-Suggested Retail Prices,* 54 Wash L Rev 763 (1979). For the view that *Colgate* cannot legally be used for an improper purpose even if literally satisfied, *see* AC Becken Co v Gemex Corp, 272 F2d 1 (7th Cir 1959). For a venerable indication of a contrary view, *see* Frey & Son v Cudahy Packing Co, 256 US 208 (1921).

178 *See generally* Practising Law Institute, The Repeal of the Fair Trade Laws: Impact on Product Distribution (1976).

179 *See, e.g.,* Cernuto, Inc v United Cabinet Corp 595 F2d 164 (3d Cir 1979). *But compare, e.g.,* Jack Kahn Music v Baldwin Piano & Organ Co, 604 F2d 755 (2d Cir 1979).

180 Compare wartime experiences of opposing belligerents, US War Production Board, Industrial Mobilization for War (1947); Speer, Inside the Third Reich (1970).

181 Dilution of competitive rivalry might lessen the urgency of research absent a sufficient incentive, such as resource scarcity, regulatory pressure or a major market awaiting development. For an attempt to deal with these conflicting possibilities in the context of a consent decree extension, *see* United States v Motor Vehicle Mfrs Assn, 1979-2 Trade Cas (CCH) and ¶62,759, at 78,379 (CD Cal 1979), *revd on other grounds,* 643 F2d 644 (9th Cir 1980).

182 See MacKenzie, The Time Trap ch 7 (1972); Stein, Organizations (1967).

183 See Mendelssohn, The Search for Absolute Zero (1966).

ways.[184]

Small enterprises often have major advantages in innovative development in these respects.[185] The advantages of combination and decentralized initiative can both be reaped if and only if specific attention is paid to this problem.[186] Use of expert subcontractors or independent subunits with effective leeway may be vital. "If too many approvals are required before an idea can be born, the probabilities are overwhelming against it."[187]

Joint research efforts have never been held illegal per se under the antitrust laws.[188] Under the rule of reason, they are permissible if on balance, they are pro-competitive.[189] The cases[190] and the Department of Justice's 1980 *Antitrust Guide Concerning Joint Research Ventures*[191] are in essential agreement on the relevant criteria.

If the joint effort permits economic activity which otherwise could not occur at all, because it is beyond the ability of the individual participants to achieve alone, the effort is almost certainly permissible.[192] If, however, the research could just as well be undertaken individually, and the combination may serve to chill the race to succeed in the research sweepstakes, a joint program may well be vulnerable.[193]

---

[184] See Katz, *Harvey and Medical Research,* 160 JAMA 1137 (1956).

[185] See Jacobs, The Economy of Cities (1969).

[186] *See* Givens, Advocacy ch 14 (Shephard's/McGraw-Hill 1980).

[187] Tribus, *Applying Science to Industry,* 70 US News & World Report 35 (Jan 8, 1971).

[188] See Note, *Joint Research Under the Antitrust Laws,* 39 Geo Wash L Rev 1112 (1971); Bakke, *Joint Efforts in Developing Standards,* 44 ABA Antitrust LJ 337 (1975).

[189] For a discussion of the illegal per se approach to antitrust analysis, see **§3.03.** *Compare* Continental TV, Inc v GTE Sylvania Inc, 433 US 36 (1977); Broadcast Music, Inc v CBS, 441 US 1 (1979). For a discussion of the rule of reason approach to antitrust analysis, see **§3.03.**

[190] *See* Standard Oil Co v United States, 283 US 136, 170-71 (1931); Apex Elec Mfg Co v Altorfer Bros, 238 F2d 867 (7th Cir 1956); International Mfg Co v Landon, Inc, 336 F2d 723 (9th Cir), *cert denied,* 379 US 988 (1964); Old Dominion Box Co v Continental Can Co, 273 F Supp 550, 572 (SDNY 1967); United States v National Malleable & Steel Castings Co, 1957 Trade Cas (CCH) ¶68,890 (ND Ohio 1957), *affd per curiam,* 358 US 38 (1958); *compare* United States v Line Material Co, 333 US 287, 310 (1948) (dictum supporting the concept of joint research); United States v EI dupont de Nemours & Co, 118 F Supp 41, 218-19 (D Del 1953), *affd on other grounds,* 351 US 377 (1956); United States v National Lead Co, 332 US 319, 359 (1947); Pitofsky, *Joint Ventures Under the Antitrust Laws,* 82 Harv L Rev 1007, 1049 (1969).

[191] *Reprinted* 1980 Antitrust & Trade Reg Rep (BNA) No 992 (special supp Dec 4, 1980). *See also* US Department of Justice, Antitrust Guide for International Operations 23-25 (1977), *reprinted in* 1977 Trade Reg Rep (CCH) No 226, II (Feb 1, 1977); S Rep No 1301, 89th Cong, 2d Sess (1966).

[192] See Pitofsky, *Joint Ventures Under the Antitrust Laws,* 82 Harv L Rev 1007 (1969).

[193] This belief was the basis of the government's lawsuit against technological information-sharing limited to certain competitors in United States v Automobile Mfrs Assn, 307 F Supp 617 (CD Cal 1969), *affd sub nom* City of New York v United States, 397 US 248 (1970); *see* 643 F2d 644 (9th Cir 1980); *see also In re* Multidistrict Vehicle Air Pollution, 538 F2d 231 (9th Cir 1976); *In re* Multidistrict Vehicle Air Pollution, 481 F2d

Where restrictive agreements or joint activity beyond the research itself are needed to support the research effort by providing incentives to participants, and are no broader than reasonably required for that purpose, they could be upheld under the rule of reason, as in the case of exclusive distributorships under which a manufacturer agrees not to sell to other dealers within a territory.[194]

Joint production or distribution on the part of competitors is likely to be supportable as necessary to a joint research effort, only if the joint research itself is of significant importance and unlikely to succeed absent combined effort, and if, in addition, joint development, production or marketing is found necessary to make the joint research effort workable. For example, if research and development for a solar reflector manufacturing effort in space depends on a guaranteed market for distribution of any power later transmitted, which could only be assured in advance by a joint venture, the joint venture might be found necessary and hence reasonable.

Of course, this indirect source of support for a joint production or marketing venture would not be required if a joint production or marketing venture were itself independently permissible under the antitrust laws. Here, a similar but separate rule of reason analysis would apply.

If the joint venture were the only practicable way for the marketing or production activity to go forward, because the participants alone could not handle it, and if any restrictions involved were no broader than reasonably necessary, there would be no violation unless per se illegal features such as horizontal price-fixing were involved.

The closer the joint effort proceeds toward the marketing phase, necessarily involving questions of pricing and distribution, the greater the antitrust objections become.[195] Joint production falls somewhere in a continuous spectrum

---

122 (9th Cir 1973). Where the agreement involves joint delay or suppression, it is illegal. Berkey Photo, Inc v Eastman Kodak Co, 603 F2d 263, 301-04 (2d Cir 1979), *cert denied,* 444 US 1093 (1980). See generally Winfield, *Joint Ventures in the Implementation of New Technology,* 2 Legal Notes & Viewpoints 59 (Feb 1982).

[194] *Compare* discussion in Oreck Corp v Whirlpool Corp, 579 F2d 126 (2d Cir) *(en banc), cert denied,* 439 US 946 (1978); GTE Sylvania Inc v Continental TV, Inc, 537 F2d 980 (9th Cir 1976), *affd,* 433 US 36 (1977).

[195] Similarly, joint applied research raises greater antitrust questions than basic research. See Hoffman, *Joint Industry Research and Other Cooperative Programs: Antitrust Hazards and Lawful Opportunities,* 32 Food Drug Cosm LJ 444, 454 n 44 (1977). For an example of a joint venture preliminarily enjoined, *see* United States v Columbia Pictures Indus, 507 F Supp 412 (SDNY 1980), *affd*— F2d — (2d Cir 1981). For a view favorable to joint ventures for innovation, see Note, 53 S Cal L Rev 1409, 1490-93 (1980); *see also* Ginsburg, *Antitrust, Uncertainty and Technological Innovation,* 24 Antitrust Bull 635 (1979). For a discussion of bases of liability, *see* Berkey Photo, Inc v Eastman Kodak Co, 603 F2d 263, 301-03 (2d Cir 1979), *cert denied,* 444 US 1093 (1980). The market power of the participants is a significant consideration, as is the extent of horizontal competition between them and whether any is a monopolist. See Turner, *Patents, Antitrust and Innovation,* 12 Antitrust Bull 227 (1967). On joint ventures for marketing, *see* Sullivan, Antitrust §104 (1977); RSE, Inc v Pennsy Supply Co, 523 F Supp 954 (MD Pa 1981). *Compare also* Zenith Radio Corp v Matsushita Elec Indus Co, 513 F Supp 1100, 1147 (ED Pa 1981).

between joint marketing and pure research in this respect. Joint determination of the quantity to be produced rather than mere joint investment in facilities would implicate marketing aspects, inasmuch as quantity and price are interdependent.[196]

The ownership and use of any patents resulting from joint research are, of course, critical to the incentives to engage in the joint effort, and also to the competitive impact. At one extreme, a joint research effort open to anyone who wishes to invest, with licensing of any patents and know-how at compensatory royalties to all, would, of course, be the most certain to survive antitrust scrutiny.[197] At the other end of the spectrum, sole ownership of patents and know-how by specified original participants should be sustainable if it could be shown that without limiting participation, the venture would not be viable.

§3.11 Governmental Approval or Participation

The business review procedure[198] of the Department of Justice and the advisory opinion procedure of the Federal Trade Commission[199] may be utilized to obtain advice on the points discussed in the preceding section. If the question is close, though, approval may be unlikely, whereas if not, it may not be necessary.[200] Professor Eleanor Fox of New York University Law School has proposed that the law be amended to preclude trebling of damages for conduct approved under such procedures, but unless such an amendment is adopted, private treble damage as well as injunctive suits would be unaffected by successful running of the gantlet of the governmental approval procedure.[201]

---

[196] See Stigler, The Theory of Price (3d ed 1966). Quantity restrictions are indeed often traditionally regarded as more onerous than duties affecting price in the field of international trade, and the first to be eliminated when freer trade is promoted. See President's Message to Congress, Feb 6, 1961, NY Times, Feb 7, 1961, at 16, col 3, *quoted in* Givens, *The Search for an Alternative to Protection,* 30 Fordham L Rev 17, 21 n 19 (1961).

[197] See US Department of Justice, Antitrust Guide, *supra* note 191. This would, indeed, seem to involve no anticompetitive aspects at all, neither requiring a balancing process nor offending any per se rule. *Compare* Eliason Corp v National Sanitation Found, 614 F2d 126 (6th Cir), *cert denied,* 449 US 826 (1980); United States v Johns-Manville Corp, 259 F Supp 440 (ED Pa 1966); Tag Mfrs Inst v FTC, 174 F2d 452 (1st Cir 1949) (product standardization where not having purpose or effect of raising prices or excluding competition).

[198] 28 CFR §506. Several examples are summarized in the US Department of Justice's Antitrust Guide to Joint Research Ventures (1980), *reprinted in* 1980 Antitrust & Trade Reg Rep (BNA) No 992 (special supp Dec 4, 1980). See Brebbia, *The Role of Advisory Opinions and the Business Reveiw Procedure,* 18 Antitrust Bull 191 (1973).

[199] 16 CFR §§1.1(a), 3.61.

[200] See 2 Kanwit, Federal Trade Commission ch 25 (Shepard's/McGraw-Hill 1980); Note, 80 Harv L Rev 1005, 1086-87 (1967).

[201] If it were regarded as important, the governmental agency might file an amicus brief opposing a finding of liability, as in Broadcast Music, Inc v CBS, 441 US 1 (1979). *See* Statement of Assistant Attorney General William F. Baxter, 1981 Antitrust & Trade

Other forms of governmental participation may be more effective in providing antitrust protection for joint research. For example, where a statutory exemption exists, or can be ottained, this obviously provides the best umbrella.[202] However, Congress is often reluctant to enact exemptions because of concern that they may lead to anticompetitive conduct nor originally contemplated.[203] Specific approval by an agency having authority over the kind of activity involved may also create an implied complete or partial exemption, if this is necessary to effectuate the purpose of the underlying statute.[204] The courts are, however, often reluctant to imply exceptions to the basic national policy set forth by the Sherman Act.[205]

Short of exemption, governmental approval of participation may be relevant to evaluation of challenged joint conduct under the rule of reason,[206] or even in determining whether the purpose of a per se rule, and hence the rule itself,

---

Reg Rep (BNA) AA-1, AA-3 (July 3, 1981). An analogous issue concerns the circumstances under which a governmental body may institute litigation to protect private parties where governmental interests the agency is charged with protecting are also involved but where there is no express statutory authority for the agency to enforce the rights of the private party. *See, e.g.,* United States v Brand Jewelers, 318 F Supp 1293 (SDNY 1970).

[202] "What Congress has forbidden by the Sherman Act it can modify." Sunshine Anthracite Coal Co v Adkins, 310 US 381, 395-96 (1940); *see generally* Note, 58 Colum L Rev 673 (1958).

[203] *See* United States v Socony-Vacuum Oil Co, 310 US 150 (1940). For a negative attitude toward exemptions, see Report of the National Commission for Review of Antitrust Laws and Procedures, pt II (1979); Baker, *Agenda for the Reagan Administration: It's Time to Rethink Assumptions,* National LJ, Feb 23, 1981, at 32.

[204] *See* Gordon v New York Stock Exch, 422 US 659 (1975); *compare* Far East Conference v United States, 342 US 570 (1952).

[205] *E.g.,* National Gerimedical Hosp v Blue Cross, 101 SCt 2415, 69 LEd 2d 89 (1981); Otter Tail Power Co v United States, 410 US 366, 372-75 (1973); Carnation Co v Pacific Westbound Conference, 383 US 213 (1966); United States v RCA, 358 US 334 (1959); United States v Alkali Export Assn, 325 US 196 (1945). A sweeping, blanket exemption may be rejected as in *National Gerimedical,* and yet an exemption implied for specific aspect of certain conduct, as in Gordon v New York Stock Exch, 422 US 659 (1975); Alabama Power Co v Alabama Elec Power Coop, Inc, 394 F2d 672, 676-77 (5th Cir), *cert denied,* 393 US 488 (1968).

[206] *See* Broadcast Music, Inc v CBS, 441 US 1 (1979)(governmental consent decrees relevant); National Gerimedical Hosp v Blue Cross, 101 SCt 2415, 69 L Ed 2d 89, 102 n 19 (1981) (by implication); Continental TV, Inc v GTE Sylvania Inc, 433 US 36, 55 n 23 (1977); United States v North Am Van Lines, Inc, 1957 Trade Cas (CCH) ¶68,761, at 73,108 (DNM 1957); United States v United States Steel Corp, 251 US 417, 446-47 (1920), *affg,* 223 F 55, 134-35, 149-50 (DNJ 1915); Fosburgh v California & Hawaiian Sugar Ref Co, 291 F 29 (9th Cir 1923); United States v Railway Express Agency, Inc, 89 F Supp 981, 983 (D Del 1950); United States v Railway Express Agency, Inc, 101 F Supp 1008 (D Del 1951); Note, 58 Colum L Rev 673, 681-88 (1958); Note, 58 Yale LJ 764, 765-66 (1949), discussing United States v Columbia Steel Co, 334 US 495, 505 (1948); *compare also* Alabama Power Co v Alabama Elec Power Coop, Inc, 394 F2d 672, 676-77 (5th Cir), *cert denied,* 393 US 488 (1968); Consolidated Farmers Mut Ins Co v Anchor Sav Assn, 480 F Supp 640, 648-56 (D Kan 1979). The rule of reason approach to antitrust analysis is discussed in **§3.03.**

bar the conduct.[207]

Further, if a governmental body affirmatively requests that something be done, and private parties agree to it, there may be only separate agreements with the governmental entity and no horizontal combination at all for antitrust purposes,[208] even if steps such as pooling of technological information are required. Such requests may be made, for example, incident to loans or loan guarantees under federal programs.

If state rather than federal participation is involved, the potential immunity for private action is narrower, because no indirect congressional authorization can be made out.[209] Rather, challenged action may have to meet far stricter tests of having been compelled by state action for purposes specifically sought by the state legislature and carried out under detailed state supervision.[210] State action may still, however, be relevant to rule of reason or purpose-interpretation questions even absent any implied exemption.

Because of the recognized importance of promoting innovation,[211] joint research is likely to continue to enjoy a favorable antitrust climate, unless what are found to be unduly restrictuve marketing restraints accompany it. If experience indicates that a particular joint effort, or even joint effort generally,

---

[207] Broadcast Music, Inc v CBS, 441 US 1 (1979); see Note, 58 Colum L Rev 673, 677-88 (1958). The illegal per se approach to antitrust analysis is discussed in **§3.03.**

[208] *See* Consolidated Farmers Mut Ins Co v Anchor Sav Assn, 480 F Supp 640, 648-56 (D Kan 1979)(compliance with standards of government corporation required for participation in program); United States v Eli Lilly & Co, 1959 Trade Cas (CCH) ¶69,536, at 76,152 (DNJ 1959), discussed in Givens, *Parallel Business Conduct Under the Sherman Act,* 5 Antitrust Bull 273, 288-89 (1960); Givens, Advocacy 309 (Shepard's/McGraw-Hill 1980); Worsinger, *New Technologies and Antitrust,* 47 NY St BJ 651, 691-92 (1975), *reprinted in* 81 Case & Com 33 (1976); White, *Participant Governmental Action Immunity from the Antitrust Laws: Fact or Fiction?,* 50 Tex L Rev 474 (1972). This concept obviously would not apply to private action where merely undertaken in order to fulfill the purposes of a governmental program, where no specific governmental request was involved. *See* National Gerimedical Hosp v Blue Cross, 101 S Ct 2415, 69 L Ed 2d 89, (1981), nor would it apply to private agreements merely approved or purportedly authorized by a governmental body where no exemption is applicable. California Retail Liquor Dealers Assn v Midcal Aluminum, Inc, 445 US 97 (1980); *compare* Consumers Union v Rogers, 352 F Supp 1319 (DDC 1973), *affd as modified sub nom* Consumers Union v Kissinger, 506 F2d 136 (DC Cir 1974), *cert denied,* 421 US 1004 (1975); Sneaker Circus, Inc v Carter, 566 F2d 396 (2d Cir 1977), *on remand,* 457 F Supp 771 (EDNY (1978).

[209] *See* California Retail Liquor Dealers Assn v Midcal Aluminum, Inc, 445 US 97 (1980) and cases cited therein.

[210] *Id.*

[211] *See, e.g.,* Mansfield, *Basic Research and Productivity Increase in Manufacturing,* 70 Am Econ Rev 863 (Dec 1980); Starr & Rudman, *Parameters of Technological Growth,* 182 Sci 358 (Oct 26, 1973); Mansfield, *Research and Development, Productivity, and Inflation,* 209 Sci 1091 (Sept 5, 1980); David, *Science Futures: The Industrial Connection,* 203 Sci 837 (Mar 2, 1979); Hutten, *On the Importance of Productivity Change,* 69 Am Econ Rev 126 (1979); Terleckyj, Effects of R&D on the Productivity Growth of Industries (National Planning Assn 1974); US Dept of Commerce, Reports of the Industrial Innovation Advisory Committee Established as Part of the Demestic Policy Review (1978)(available from National Technical Information Service, Springfield, Va 2216l).

retards rather than advances innovation, this favorable climate could, and doubtless would, change.[212]

Private efforts to bring about governmental action are, in general, immune from antitrust attack as being both outside the scope of the antitrust laws, and protected by the First Amendment.[213] Sham litigations designed solely to wear down an opponent through such means as opposing all license applications regardless of the merits, however, fall outside this exception.[214] Similarly, where a state or local body is acting in a commercial as distinct from a strictly regulatory capacity, or beyond clear state authorization the exemption may not apply.[215] The precise boundaries of the exemption are in the process of definition. Such conduct as fraudulent submissions to state or local administrative bodies can also create liability.[216]

§3.12 International Aspects

In many instances, research or development efforts involve multinational participation.[217] This, of course, may make the antitrust laws of more than one jurisdiction relevant.[218] Some United States laws, such as the Robinson-Patman Price Discrimination Act, do not apply to transactions involving intended resale outside the United States.[219] The Sherman Act, however, does apply to

---

[212] On the effect of experience, *see* Broadcast Music, Inc v CBS, 441 US 1 (1979); United States v Topco Assoc, Inc, 405 US 596, 607-08 (1972); White Motor Co v United States, 372 US 253, 263 (1963); Stone, Law and Its Administration 39, 45 (1915).

[213] Eastern RR Presidents Conference v Noerr Motor Freight Inc, 365 US 127 (1961); UMW v Pennington, 381 US 657 (1965); Sullivan, Antitrust §238 (1977).

[214] California Motor Transp Inc v Trucking Unlimited, 404 US 508 (1972); Vendo Co v Lektro-Vend Corp, 433 US 623 (1977); LG Balfour Co v FTC, 442 F2d 1 (7th Cir 1971).

[215] City of Lafayette v Louisiana Power & Light Co, 435 US 389 (1978) (several opinions filed on varying grounds); *In re* Airport Car Rental Antitrust Litig, 474 F Supp 1072 (ND Cal 1979); *compare* George R Whitten, Jr, Inc v Paddock Pool Builders, Inc, 424 F2d 25 (1st Cir), *cert denied,* 400 US 850 (1970); Rangen, Inc v Sterling Nelson & Sons, 351 F2d 851 (9th Cir 1965), *cert denied,* 383 US 936 (1966).

[216] Ernest W Hahn, Inc v Codding, 615 F2d 830, 837-843 (9th Cir 1980); Woods Exploration Co v Aluminum Co of America, 438 F2d 1286 (5th Cir 1971), *cert denied,* 404 US 1047 (1972)(alleged submission of false data); Duke & Co v Foerster, 521 F2d 1277 (3d Cir 1975); Colorado Petroleum Marketers Assn v Southland Corp, 476 F Supp 373 (D Colo 1979); Note, 86 Harv L Rev 715 (1973).

[217] *E.g.,* US Dept of Justice, Antitrust Guide for International Operations, Case D, at 23 (1977), *reprinted in* 1977 Trade Reg Rep (CCH) No 226 (special supp Feb 1, 1977).

[218] 15 USC § 13.

[219] *Id.* Products must be "for use, consumption or resale" in the United States for the act to apply. *See* Zenith Radio Corp v Matsushita Elec Indus Co, 402 F Supp 244 (ED Pa), *petition denied,* 521 F2d 1399 (3d Cir 1975); Fimex Corp v Barmatic Prods Co, 429 F Supp 978, 980 (EDNY), *affd without opinion,* 573 F2d 1289 (2d Cir 1977); *compare also* as to jurisdiction of the act, Gulf Oil Corp v Copp Paving Co, 419 US 186, 200 (1974).

trade "with foreign nations".[220]

Much controversy has been generated concerning the degree to which the United States can and should impose its laws on world-wide economic transactions. Jurisdiction is generally held to exist over international transactions when there is a significant impact on domestic United States commerce; but the courts normally will invoke the Sherman Act only where the conduct involved either significantly limits competition in sales to United States customers or significantly restricts United States exports, and where all the circumstances indicate that exercise of jurisdiction is on balance appropriate.[221]

There has also been controversy surrounding the degree to which foreign governmental action provides a defense to Sherman Act charges. Obviously, conduct of foreign governments as such in their governmental (not commercial) capacities cannot become the basis of antitrust action.[222] Private action to induce such action would appear likely to be immune to at least the same degree that it is where domestic governmental entities are involved.[223] Although not all governments adhere to freedom of petition which is one basis for immunizing efforts to influence governmental authorities in the United States,[224] even partial recognition of such a right is important,[225] and it would

---

[220] 15 USC § 1.

[221] *See* Continental Ore Co v Union Carbide & Carbon Corp, 370 US 690, 705 (1962); National Bank of Canada v Interbank Card Assn, — F2d — (2d Cir 1981); United States v Aluminum Co of America, 148 F2d 416, 443 (2d Cir 1945); numerous additional judgmental factors may influence whether the exercise of jurisdiction is wise and hence appropriate. Timberlane Lumber Co v Bank of America, 549 F2d 597 (9th Cir 1977); Mannington Mills, Inc v Congoleum Corp, 595 F2d 1287 (3d Cir 1979); IAM v OPEC, 477 F Supp 553 (CD Cal 1979), *affd,* 649 F2d 1354 (9th Cir 1981). *See generally* Symposium, 49 ABA Antitrust LJ 1185 (1980); Fugate, Foreign Commerce and the Antitrust Laws (2d ed 1973); Note, 70 Yale LJ 259 (1960); Ongman, *"Be No Longer A Chaos": Constructing A Normative Theory of the Sherman Act's Extraterritorial Jurisdictional Scope,* 71 NW UL Rev 733 (1977); Kintner & Griffin, *Jurisidction Over Foreign Commerce Under the Sherman Antitrust Act,* 18 BC Indus & Com L Rev 199 (1977); Brewster, Antitrust and American Business Abroad (1958). *See also* Webb-Pomerene Act, 15 USC §§61-65. For legislative proposals in this area, *see, e.g.,* S 734 and HR 2326, 97th Cong, 1st Sess (1981).

[222] IAM v OPEC, 649 F2d 1354 (9th Cir 1981); Hunt v Mobil Oil Co, 550 F2d 68 (2d Cir), *cert denied,* 434 US 984 (1977); Occidental Petroleum Corp v Buttes Gas & Oil Co, 331 F Supp 92, 110 (CD Cal 1971), *affd,* 461 F2d 1261 (9th Cir), *cert denied,* 409 US 950 (1972). Although not followed in other respects, American Banana Co v United Fruit Co, 213 US 347 (1909) is still pertinent on this point. *But see* United States v Sisal Sales Corp, 274 US 268 (1927); *see generally* Note, 70 Yale LJ 259, 269-72 (1960).

[223] *See* Interamerican Ref Corp v Texaco Maracaibo, Inc, 307 F Supp 1291 (D Del 1970); Hawk, *Act of State Doctrine, Noerr-Pennington Abroad and Foreign Government Compulsion Defense,* 47 ABA Antitrust LJ 987 (1978).

[224] *Compare* Nebraska Press Assn v Stuart, 427 US 539 (1976); United States v Carolene Prod Co, 304 US 144, 152 n 4 (1938); Dowling, *The Methods of Mr. Justice Stone in Constitutional Cases,* 41 Colum L Rev 1160 (1941). *See also* Givens, *Chief Justice Stone and the Developing Functions of Judicial Review,* 47 Va L Rev 1321 (1961).

[225] Thus, the right to be heard can be important even if the tribunal is implementing laws enacted without full participation, as is indeed evident from the history of the effort to secure such rights. See Goebel, Cases and Materials on the Development of Legal Institutions (1946); *compare* Levy, Origins of the Fifth Amendment (1968).

hardly make sense for the United States to seek to impose harsher restrictions in the international than in the domestic context.[226]

Foreign governmental approval of a restraint which would otherwise be illegal will not, however, automatically immunize it if it is otherwise violative of the Sherman Act,[227] although compliance with a mandatory foreign statute may do so,[228] subject to evaluation of its applicability under conflict of laws principles.[229]

The greatest dangers obviously arise if international research and development efforts involve international divisions of territories between competitors, which in turn restrict competition to sell to the United States market, as well as exports by United States firms outside their assigned territories.[230]

Antitrust Policy Seeking To Promote Innovation

§3.13 Direct Enforcement

Antitrust enforcement designed to remove private barriers to innovation may either help to achieve that goal or on the other hand, work in reverse if concern over such enforcement itself has a chilling effect on innovative activity.[231] In either event, areas where such governmental or private enforcement may occur include the following:

1. Attacks upon patent pooling arrangements which are unnecessarily

---

[226] *See* US Dept of Justice, Antitrust Guide, *supra* note 217, *reprinted in* 1977 Trade Reg Rep (CCH) No 226, pt II (Feb 1, 1977).

[227] *See* Continental Ore Co v Union Carbide & Carbon Corp, 370 US 690 (1962); United States v Sisal Sales Corp, 274 US 268 (1927); Timberlane Lumber Co v Bank of America, 549 F2d 597 (9th Cir 1977); *compare* Mannington Mills v Congoleum Corp, 595 F2d 1287 (3d Cir 1979)(allegedly obtaining foreign patents by fraud); Interamerican Ref Corp v Texaco Maracaibo, Inc, 307 F Supp 1291 (D Del 1970).

[228] *See* US Dept of Justice, Antitrust Guide, *Supra* note 217, Case L, at 52-56, *reprinted in* 1977 Trade Reg Rep (CCH) No 226, pt II (special supp Feb 1, 1977); **§3.12**.

[229] See Note, 70 Yale LJ 259 (1960).

[230] The status of the participants as prior competitors is important here. US Dept of Justice, Antitrust Guide, *supra* note 217, Case A, at 15-22, *reprinted in* 1977 Trade Reg Rep (CCH) No 226, pt II (special supp (Feb 1, 1977); on divisions of territories or customers among competitors as illegal per se, *see e.g.*, United States v Topco Assocs, 405 US 596 (1972); Burke v Ford, 389 US 320 (1967); United States v Sealy, Inc, 388 US 350 (1967); Addyston Pipe & Steel Co v United States, 85 F 271 (6th Cir 1898) (Taft, J), *affd*, 175 US 211 (1899); Norfolk S Bus Corp v Virginia Dare Co, 159 F2d 306 (4th Cir), *cert denied*, 331 US 827 (1947). In the international context, *see, e.g.*, Timken Roller Bearing Co v United States, 341 US 593 (1951); More, *US Antitrust Laws and Territorial Provisions in Licensing Foreign Patents and Know-How for Foreign Use*, 24 Rec AB City NY 512 (1969).

[231] *See* Continental TV, Inc v GTE Sylvania Inc, 433 US 36 (1977); Bork, The Antitrust Paradox (1978).

broad in scope, while excluding competitors from participation, unreasonably restraining trade as a result [232]

2. Nonexploitation, by use of licensing or of blocking patents employed to protect existing or other technologies from competition, where the result is to prevent a major innovative breakthrough in an industry, especially if the patents are acquired rather than developed, in the absence of countervailing justification[233]
3. Collusion among competitors to retard innovation.[234] This may be overt or tacit,[235] provided the facts actually show that collusion and not separate independent decision making was involved[236]
4. Use of any form of leverage, e.g., through influence on third parties such as suppliers, financing sources or channels of distribution[237] to retard innovation by a competitor[238]

The key to prosecution or defense in these areas is likely to be detailed study of the practical effects of the challenged conduct.[239] Prelitigation or early post-litigation reference to all available sources of information about the industry is thus extremely important,[240] as is questioning designed to reach

---

[232] *See* Hartford-Empire Co v United States, 323 US 386 (1945); *compare* United States v Singer Mfg Co, 374 US 174 (1963).

[233] See **§3.08.**

[234] See **§3.10.**

[235] *See* Interstate Circuit, Inc v United States, 306 US 208, 226 (1939); American Tobacco Co v United States, 328 US 781, 809-10 (1946); Bogosian v Gulf Oil Corp, 561 F2d 434 (3d Cir 1977), *cert denied,* 434 US 1086 (1978); C-O-Two Fire Equip Co v United States, 197 F2d 489 (9th Cir), *cert denied,* 344 US 892 (1952).

[236] Theatre Enters, Inc v Paramount Film Distrib Corp, 346 US 537 (1954); First Natl Bank v Cities Serv Co, 391 US 253 (1968); see Nye, *Can Conduct Oriented Enforcement Inhibit Conscious Parallellism?,* 44 ABA Antitrust LJ 206 (1975); Givens, *Parallel Business Conduct Under the Sherman Act,* 5 Antitrust Bull 273 (1960); Turner, *The Definition of Agreement Under the Sherman Act: Conscious Parallelism and Refusals to Deal,* 75 Harv L Rev 655 (1962); Rahl, *Conspiracy and the Antitrust Laws,* 44 Ill L Rev 743 (1950); Blechman, *Conscious Parallelism, Signalling and Facilitating Devices: The Problem of Tacit Collusion Under the Antitrust Laws,* 34 NYL Sch L Rev 881 (1979).

[237] Compare criteria set forth in Oreck Corp v Whirlpool Corp, 579 F2d 126 (2d Cir) *(en banc), cert denied,* 439 US 946 (1978). The per se rule against boycott combinations may also come into play. *See* Silver v New York Stock Exch, 373 US 341 (1963); Radiant Burners v People's Gas Light & Coke Co, 364 US 656 (1961); Klor's, Inc v Broadway-Hale Stores, Inc, 359 US 207 (1959); Associated Press v United States, 326 US 1 (1945); Fashion Originators Guild v FTC, 312 US 457 (1941); *compare* St Paul Fire & Marine Ins Co v Barry, 438 US 531 (1978).

[238] *Compare Bank Holding Company Act Amendments, Hearings Before the House Committee on Banking and Currency,* 91st Cong, 1st Sess 372-74, 388, 615-16 (1969).

[239] *E.g.,* SmithKline Corp v Eli Lilly & Co, 575 F2d 1056 (3d Cir), *cert denied,* 439 US 838 (1978); Eiberger v SONY Corp of America, 622 F2d 1068 (2d Cir 1980), *affg in part,* 459 F Supp 1276 (SDNY 1978).

[240] *E.g.,* trade and general press and other published sources, other litigation, governmental reports, financial analyst and other industry studies, economic literature and interviews with knowledgeable persons who know the folklore of the field.

beyond preconceived labels to substance.[241]

§3.14 Innovation as Competition Policy

Promotion of innovation can itself bring about new competition. This is evident from numerous examples such as; the elimination of the monopoly power attributed to the railroads in the transportation industry through development of alternatives[242] rather than primarily through extensive antitrust litigation;[243] the similar dilution of the power attributed to major members of the movie industry through the impact of television rather than primarily through numerous antitrust suits;[244] and the dilution of the power attributed to Alcoa, largely through promotion of alternative sources during World War II rather than as a result of the famous litigation.[245]

Affirmative promotion of innovation may take many forms, such as: tax benefits;[246] procedures for fast track processing of required approvals,[247] including consolidation of approval functions in a single body where appropriate;[248] specific antitrust exemptions if required to avoid a chilling effect which may retard joint efforts if required,[249] or exemptions from other legal requirements if justified and necessary;[250] and measures to assure availability of financing where high-risk, postponed return and large capital requirements combine to make investment in particular vital research and development

---

[241] See Givens, Advocacy, ch 4 (Shepard's/McGraw-Hill 1980).

[242] See Nelson, Railroad Transportation and the Public Interest (1959).

[243] *E.g.,* Northern Sec Co v United States, 193 US 197 (1904); United States v Terminal RR Assn, 224 US 383 (1912); United States v Lehigh Valley R Co, 254 US 255 (1920); United States v Reading Co, 253 US 26 (1920); United States v Southern Pac Co, 259 US 214 (1922).

[244] *E.g.,* Interstate Circuit, Inc v United States, 306 US 208 (1939); United States v Crescent Amusement Co, 323 US 173 (1944); United States v Griffith, 334 US 100 (1948); Schine Chain Theatres, Inc v United States, 334 US 110 (1948); Bigelow v RKO Radio Pictures, Inc, 327 US 251 (1946); United States v Paramount Pictures, Inc, 334 US 131 (1948); Theatre Enters, Inc v Paramount Film Distrib Corp, 346 US 537 (1954).

[245] United States v Aluminum Co of America, 148 F2d 416 (2d Cir 1945), *on remand,* 91 F Supp 333 (SDNY 1950); see HR Rep No 2954, 84th Cong, 2d Sess (1956); HR Rep No 2716, 85th Cong, 2d Sess (1959).

[246] See **ch 5.**

[247] *Compare, e.g.,* procedures contained in HR 1445, 96th Cong, 1st Sess §204(b), at 26-27 (1979).

[248] See *id* §216, at 33-36.

[249] *See,* for a narrow exclusion, United States Synthetic Fuels Corporation Act, §175(e), 42 USC §701 *et seq,* discussed in Pozen, *Synthetic Fuels Corporation—Investment Bank or Government Agency,* 36 Bus Law 953, 967 (1981). *See also, e.g.,* §252(j) of the Energy Policy and Conservation Act, 42 USC §6272(j), discussed in Statement of Ronald G. Carr, Deputy Assistant Attorney General, before the Senate Committee on Energy & Natural Resources, July 20, 1981; 50 USC app §§2158, 2158a.

[250] This may be achieved by specific provision or implied pre-emption. *See* Flood v Kuhn, 407 US 258 (1972).

unattractive even if perhaps on the average profitable.[251]

Such measures, discussed in greater detail in Chapter 12, may involve use of central banking functions to enhance private sector activity.[252]

§3.15 Limitations on Economic Regulation

During the period from 1803 to 1937, the Supreme Court held invalid many types of state and federal economic regulation, most frequently on the ground of interference with the liberty of contract held implicit in due process after adoption of the Fourteenth Amendment.[253] The interstate commerce clause of Article I of the Constitution was also held to void state and local regulation imposing an undue burden on interstate commerce,[254] subject to congressional power to allow such regulation.[255]

Subsequent to 1937, exacting judicial review under broadly phrased constitutional provisions, such as the commerce and due process clauses, has generally been imposed where other political or judicial restraints are (1) absent,[256] (2) endangered by the very conduct under review,[257] or (3) clearly insuf-

---

[251] See Terleckyj, Effects of R&D on the Productivity Growth of Industries (National Planning Assn 1974). Such measures, as those listed here, discussed in greater detail in **ch 12,** may involve use of central banking functions to enhance private sector activity.

[252] See Worsinger, *New Technologies and Antitrust,* 47 NY St BJ 651 (1975), *reprinted in* 81 Case & Com 33 (1976).

[253] *E.g.,* Lochner v New York, 198 US 45 (1905); Adair v United States, 208 US 161 (1908); Adkins v Children's Hosp, 261 US 525 (1923). For the supervening view, *see, e.g.,* West Coast Hotel Co v Parrish, 300 US 379 (1937); Nebbia v New York, 291 US 502 (1934); United States v Darby, 312 US 100 (1941). A greater willingness to impose constitutional limitations on economic regulation where based on more specific constitutional provisions may be indicated by, *e.g.,* United States Trust Co v New Jersey, 431 US 1 (1977). For a contemporary defense of the pre-1937 line of cases, see Siegan, Economic Liberties and the Constitution (1980); *compare* Struve, *The Less-Restrictive-Alternative Principle and Economic Due Process* 80 Harv L Rev 1463 (1967). For attacks on the approach often taken before 1937, *see* United States v Butler, 297 US 1, 78-79 (1936)(Stone, J, dissenting) and Morehead v New York *ex rel* Tipaldo, 298 US 587, 663 (1936); *see generally* Bickel, The Least Dangerous Branch (1962). For description of the earlier period, *see also* Howell, *The Judicial Conservatives of Three Decades Ago: Aristocratic Guardians of The Prerogatives of Property and the Judiciary* 49 Va L Rev 1447 (1963).

[254] **§3.02.**

[255] *E.g.,* Prudential Ins Co v Benjamin, 328 US 408 (1946); see Dowling, *Interstate Commerce and State Power—Revised Version,* 47 Colum L Rev 547 (1947).

[256] *E.g.,* where regulation affects out-of-state interests, as pointed out in Hunt v Washington Apple Advertising Commn, 432 US 333,351-53(1977); Nippert v Richmond, 327 US 416, 434 (1946); Southern Pac Co v Arizona, 325 US 761, 767 n 2 (1945); Edwards v California, 314 US 160, 174 (1941); McGoldrick v Berwind-White Coal Mining Co, 309 US 33, 46 n 2 (1940); South Carolina State Highway Dept v Barnwell Bros, 303 US 177, 184 n 2(1938); *see also* Raymond Motor Transp, Inc v Rice, 434 US 429, 444 n 18 (1978).

[257] *E.g.,* denial of equally effective voting rights as in Wesberry v Sanders, 376 US 1 (1964); access to the judicial process as discussed in Bates v State Bar, 433 US 350

ficient.[258] Under this approach, most kinds of economic regulation which have been challenged have been upheld.[259]

State or local regulation impeding interstate commerce, however, insofar as it affects interests not represented in local political processes, is not subject to local political restraints.[260] Consequently, to protect the "national common market"[261] envisioned by the framers of the Constitution, the Court has struck down protectionist local regulation which directly or covertly discriminates against interstate commerce,[262] or which substantially interferes with such commerce without adequate redeeming justification.[263]

Affirmative congressional regulation can reach beyond areas covered by this implied limitation on state action, and embrace matters affecting as well as directly involving commerce.[264] A state's own direct governmental functions

---

(1977); interference with freedom of expression as in Landmark Communications, Inc v Virginia, 435 US 829 (1978) or with freedom of conscience as discussed by Stone, J, dissenting in Minersville School Dist v Gobitis, 310 US 586, 601-07 (1940), overruled by West Virginia State Bd of Educ v Barnette, 319 US 624 (1943); *see* Dowling, *The Methods of Mr. Justice Stone in Constitutional Cases,* 41 Colum L Rev 1160 (1941); Givens, *Chief Justice Stone and the Developing Functions of Judicial Review,* 47 Va L Rev 1321 (1961).

[258] *E.g.,* invidious discrimination against discrete minorities as discussed in United States v Carolene Prods Co, 304 US 144, 152 n 4 (1938); Brown v Board of Edu, 347 US 483 (1954), discussed in, *e.g.,* Black, *The Lawfulness of the Segregation Decisions,* 69 Yale LJ 421 (1960); Givens, *The Impartial Constitutional Principles Supporting Brown v. Board of Education,* 6 How LJ 179 (1960); see Lusky, *Minority Rights and the Public Interest,* 52 Yale LJ 1 (1942).

[259] *See* Exxon v Governor of Maryland, 437 US 117 (1978); Williamson v Lee Optical, 348 US 483, 487-89 (1955); Lincoln Fed Labor Union v Northwestern I&M Co, 335 US 525, 536-37 (1949). Concerning criteria for this approach, *see* United States v Carolene Prods Co, 304 US 144, 152 n 4 (1938);

Massachusetts Bd of Retirement v Murgia, 427 US 307, 313 (1976); Foley v Connelie, 425 US 291, 294 (1978); Dowling, *The Methods of Mr Justice Stone in Constitutional Cases,* 41 Colum L Rev 1160 (1941); Givens, *Chief Justice Stone and the Developing Functions of Judicial Review,* 47 Va L Rev 1321 (1961); Wellington, *The Nature of Judicial Review,* 36 Rec AB City NY 360 (Oct 1981). *Carolene Products* does not discuss access to the judicial process, which is as essential to preservation of all other rights as is access to the process, as recognized in, *e.g.,* cases cited in note 272 *infra.*

[260] See note 256 *supra.*

[261] Hunt v Washington Apple Advertising Commn, 432 US 333, 350 (1977).

[262] *See* Lewis v BT Inv Mgrs, Inc, 447 US 27 (1980); City of Philadelphia v New Jersey, 437 US 617 (1978); Dean Milk Co v Madison, 340 US 349 (1951); Nippert v Richmond, 327 US 416 (1946).

[263] *See* Raymond Motor Transp, Inc v Rice, 434 US 429 (1978); Hunt v Washington Apple Advertising Commn, 432 US 333 (1977); Bibb v Navajo Freight Lines, Inc, 359 US 520 (1959); Southern Pac Co v Arizona, 325 US 761 (1945).

[264] *See* McLain v Real Estate Bd, 444 US 232 (1980); Scarborough v United States, 431 US 563 (1977); Hospital Bldg Co v Trustees of Rex Hosp, 425 US 738 (1976); Wickard v Filburn, 317 US 111 (1942); United States v Ricciardi, 357 F2d 91 (2d Cir), *cert denied,* 384 US 942 (1966).

are, however, exempt from such federal interference.[265]

Similarly, other specific clauses limiting economic regulation, such as that prohibiting retroactive impairment of the obligation of contracts,[266] are now given a fair reading. Perhaps the most sensitive developing area concerns the line of demarcation between permissible regulatory activity and unconstitutional confiscation which violates due process under the Fourteenth Amendment or amounts to uncompensated taking under the Fifth Amendment.[267] Some believe that even beyond this, interference with legitimate economic activity which is more restrictive than necessary for valid purposes may again come under challenge under substantive due process concepts;[268] the facts presented in particular cases will undoubtedly determine whether and how this may occur.[269]

The antitrust laws themselves are, of course, a form of regulation subject to constitutional limitations such as the requirement of fair notice of what is prohibited. Notice is required to be more clear, the more severe the sanctions for past conduct imposed.[270] The First Amendment's protection of freedom of expression also protects access to political[271] and judicial processes,[272] even when used for anticompetitive objectives. Sham litigation designed merely to oppress opponents as a competitive weapon[273] or conduct which is part of a broader antitrust violation is, however, not necessarily protected.[274]

The antitrust laws are also relevant to interpretation of the commerce clause as a limitation on state and local action. Since judicial review of state and local regulation under the commerce clause is essentially a reading of the signifi-

---

[265] National League of Cities v Usery, 426 US 833 (1976).

[266] *E.g.,* United States Trust Co v New Jersey, 431 US 1 (1977); Allied Structural Steel Co v Spannaus, 438 US 234 (1978).

[267] *See* Penn Cent Transp Co v City of New York, 438 US 104 (1978); United States v Central Eureka Mining Co, 357 US 155 (1951); Miller v Schoene, 276 US 272 (1928); *cf* Givens, *The Antiredlining Issue: Can Banks Be Forced to Lend?,* 95 Banking LJ 515 (1978); *see also* Agins v City of Tiburon, 447 US 255 (1980).

[268] See note 253 *supra.*

[269] See Stone, Law and Its Administration 39, 45 (1915).

[270] *See* United States v United States Gypsum Co, 438 US 422 (1978); Lambert v California, 355 US 225 (1957); *compare* Practising Law Institute, Drafting Documents in Plain Language (1979, 1981).

[271] First Natl Bank v Bellotti, 435 US 765 (1978); Buckley v Valeo, 424 US 1 (1976); Wesberry v Sanders, 376 US 1 (1964); *see* Abrams v United States, 250 US 616, 630 (1919)(Holmes, J, dissenting).

[272] *See* Bates v State Bar, 433 US 350 (1977) and, *e.g.,* United Transp Union v State Bar, 401 US 576 (1971); UMW v Illinois State Bar Assn, 389 US 217 (1967); Brotherhood of RR Trainmen v Virginia State Bar, 377 US 1 (1964); NAACP v Button, 371 US 415 (1963).

[273] See **§3.11.**

[274] *E.g.,* LG Balfour Co v FTC, 442 F2d 1 (7th Cir 1971); see Robinson, *Reconciling Antitrust and The First Amendment,* 48 ABA Antitrust LJ 1335 (1979).

cance of congressional silence on the issue involved,[275] congressional action may be relevant even if not literally applicable.[276] In effect, the courts are fashioning federal common law subject to congressional revision,[277] and hence statutes, like judicial decisions, may thus become pertinent sources of reasoning entirely apart from the effect of their express language.[278]

The national policy of an open economy enshrined in the Sherman Act is thus highly pertinent under the commerce clause, just as it is in interpreting various federal statutes.[279] This does not mean, of course, that every type of conduct shielded from antitrust attack under an antitrust defense is likewise protected from state regulation.[280] It does allow the courts to strike down state or local regulation which burdens commerce and which interferes with the objectives of the antitrust laws as interpreted by the courts.[281]

This power, although sparingly exercised in the past, may be employed more vigorously as elements of the economy becomes more interdependent,[282] making previously existing restrictions more constricting. However, use of the power does not involved the risks of determining with finality that action of a challenged type cannot be taken at all, since Congress may choose to sanction state regulation of commerce.[283] On the other hand, of course, there may be little or no local political restraint on some protectionist regulation under review.[284]

Antitrust concepts are also indirectly relevant to the constitutionality of

---

[275] Lewis v BT Inv Managers, Inc, 447 US 27 (1980); Philadelphia v New Jersey, 437 US 617 (1978); Hunt v Washington Apple Advertising Commn, 432 US 333 (1977); Nippert v Richmond, 327 US 416 (1946).

[276] *E.g.,* Southern Pac Co v Arizona, 325 US 761, 773 (1945); Parker v Brown, 317 US 341, 367 (1943); Huron Portland Cement Co v Detroit, 362 US 440, 446 (1960).

[277] See Dowling, *Interstate Commerce and State Power—Revised Version,* 47 Colum L Rev 547 (1947); Givens, *The Commerce Clause and the Interdependent Economy,* 53 ABAJ 719 (1967); *compare* Friendly, *In Praise of Erie—and of the New Federal Common Law,* 39 NYU L Rev 383 (1964), *reprinted in* 19 Rec AB City NY 64 (1964).

[278] See Stone, *The Common Law in the United States,* 50 Harv L Rev 4, 12-18 (1936); Note, 82 Yale LJ 258 (1972).

[279] *E.g.,* FCC v RCA Communications, Inc, 346 US 86, 94 (1953); McLean Trucking Co v United States, 321 US 67, 85-86 (1944); *see* United States v RCA, 358 US 334 (1959); Note, 58 Colum L Rev 673 (1958).

[280] Exxon v Governor of Maryland, 437 US 117 (1978).

[281] See Gerhart, *Constitutional Limits on State Regulatory and Protectionist Policies,* 48 ABA Antitrust LJ 1351, 1357-68 (1979).

[282] This results from continuously improved transportation and communication. For historical background, see Commons, *American Shoemakers, 1684-1895,* Labor and Administration 219 (Commons ed 1913). For judicial recognition, *see, e.g.,* Heart of Atlanta Motel, Inc v United States, 379 US 241 (1964); United States v Women's Sportswear Mfrs Assn, 336 US 460 (1949); Polish Natl Alliance v NLRB, 322 US 643 (1944); Wickard v Filburn, 317 US 111 (1942).

[283] Prudential Ins Co v Benjamin, 328 US 408 (1946); see Dowling, *Interstate Commerce and State Power—Revised Version,* 47 Colum L Rev 547 (1947).

[284] See note 256, *supra.*

restrictions on commercial speech under the First Amendment.[285] The First Amendment's protection of freedom of conscience[286] and expression[287] is at the core of a free society[288] and placed beyond legislative impairment.[289] Some cases, however, invoke a balancing process to determine whether a marginal impact on First Amendment values is justified by a compelling governmental interest,[290] and whether, if so, the least restrictive alternative has been chosen.[291]

In evaluating asserted justifications for any impingement upon First Amendment values, defense of regulation which is antithetical to the Sherman Act objective of an open economy is more difficult.[292] In this respect attacks based upon antitrust concepts by the deregulation movement against advertising prohibitions may have contributed to their limitation on First Amendment grounds,[293] even though the First Amendment itself obviously stands on a far higher footing than the anti-trust laws or any other legislative enactment.

§3.16 Industrial Strategy and Antitrust

Antitrust considerations may be relevant to the long-term industrial strategy of an innovator. Apart from avoiding direct violations and ability to attack illegal restrictions imposed by others, antitrust may be pertinent to larger policy options.

Modern economic theory indicates that where innovation succeeds, an experience curve develops, characterized by initial investment followed by a strong market position (perhaps within an entirely new field that did not exist before) if the innovation is successfully marketed, which in turn may permit

---

[285] *See* Bates v State Bar, 433 US 350 (1977); Virginia State Bd of Pharmacy v Virginia Citizens Consumer Council, 425 US 748 (1976); Reich, *Consumer Protection and the First Amendment,* 61 Minn L Rev 705 (1977); Gerhart, *Constitutional Limitations on State Regulatory and Protectionist Policies,* 48 ABA Antitrust LJ 1351, 1353-57 (1979); Elman, *The New Constitutional Right to Advertise,* 64 ABAJ 206 (1978); Note, 44 Chi L Rev 205 (1976). *Compare generally* Metromedia, Inc v San Diego, 453 US 490 (1981).

[286] See note 257 *supra.*

[287] *Id.*

[288] *See* First Natl Bank v Bellotti, 435 US 765 (1978); Kingsley Intl Pictures Corp v Regents, 360 US 684 (1959); Abrams v United States, 250 US 616, 630 (1919) (Holmes, J, dissenting).

[289] See Black, The People and the Court (1960); authorities cited in Givens, *Power to Define the Constitutional Rights of Defendants: Congress and the Federal Courts,* 46 Fordham L Rev 383, 395-400 (1977).

[290] *E.g.,* Buckley v Valeo, 424 US 1 (1976).

[291] *See, e.g., id;* Nebraska Press Assn v Stuart, 427 US 539 (1976); Talley v California, 362 US 60, 66 (1960) (concurring opinion); Bates v Little Rock, 361 US 516 (1960); NAACP v Alabama, 357 US 449 (1958); Butler v Michigan, 352 US 380 (1957); Minersville School Dist v Gobitis, 310 US 586, 603-04 (1940) (Stone, J, dissenting).

[292] See **§3.03.**

[293] See note 271, *supra.*

profitable recoupment of the investment.[294] In order to encourage innovation, antitrust thinking has largely abandoned[295] the once-dominant "bigness is badness" concept which, articulated or not, underlay many antitrust decisions spanning a number of decades.[296] On the other hand, there is no guarantee that such a climate will persist.[297] Even within that climate, it is obvious that strong market positions necessarily expose their holders to closer public and regulatory scrutiny, as well as creating power which may be used in ways that can lead to antitrust challenge.[298] Thus, the policy choices confronting such firms, like those confronting the legal system, are multidimensional.[299]

The choice which maximizes short-term benefit from prior successful innovation is to seek the largest market share during the exploitation phase, while pricing the product or service at a level which will compensate for risks taken without being high enough to provoke extreme hostility.[300] Within this mode there may well be innovation in marketing or in details of product or service design, normally protected from successful antitrust attack.[301] This option may also involve marketing restrictions of various types which may be challenged

---

[294] See Bock, *Antitrust Blueprint for the 1980s,* 55 NYU L Rev 574, 578 n 11 (1980); Strategic Planning and the Future of Antitrust: The Conference Board Antitrust Forum 1980; Hayes & Wheelwright, *The Dynamics of Process-Product Life Cycles,* 57 Harv Bus Rev 127, 130 (Mar-Apr 1979); Wilcox & Shepard, Public Policies Toward Business 31 (5th ed 1975); Grabowski & Mueller, *Life Cycle Effects on Corporate Returns on Retentions,* 57 Rev Econ Statistics 400 (1975); Hayes & Wheelwright, *Linked Manufacturing Process and Product Life Cycles,* Harv Bus Rev 133 (Jan-Feb 1979); Scherer, Industrial Market Structure and Economic Performance ch 15 (1970); Williamson, Markets and Hierarchies: Analysis and Antitrust Implications ch 10 (1975); but see Dhalla & Yuspeh, *Forget the Product Life Cycle Concept,* Harv Bus Rev 102 (Jan-Feb 1976).

[295] **§3.05.**

[296] **§3.05.**

[297] The continuity provided by the courts as a non-elected branch of the government, seeking to reflect the sober second thought of the community is a vital aspect of our society, as discussed in, *e.g.,* Bickel, The Least Dangerous Branch (1962) and Black, The People and the Court (1960), but obviously does not prevent dramatic changes in antitrust policy, e.g., the enactment of the Clayton and Robinson-Patman Price Discrimination Acts. For retroactive impact of legal interpretations on long-past transactions, *see, e.g.,* United States v EI duPont de Nemours & Co (GM), 353 US 586 (1957); United States v General Dynamics Corp, 415 US 486, 505 n 13 (1974); Orrick, *The Clayton Act: Then and Now,* 24 ABA Antitrust LJ 44 (1964); Subcommittee on Section 7, *Report,* 32 ABA Antitrust LJ 306 (1966).

[298] See Berle, Power Without Property (1959).

[299] *Compare* Schelling, The Strategy of Conflict (1960).

[300] Berle, *supra* note 298, at 91-93, 114-16; Kennedy, The Burden and the Glory, docs 37-38, at 194-97 (Nevins ed 1964).

[301] See **§3.03;** *but compare* Scherer, Industrial Market Structure and Economic Performance 341-45 (1970); Report of the White House Task Force on Antitrust Policy pt II (1968), *reprinted in* 2 Antitrust L & Econ Rev 11 (Winter 1968-69); Rostow, *The New Sherman Act,* 14 U Chi L Rev 567 (1974); Snell, *Annual Style Change in the Automobile Industry as an Unfair Method of Competition,* 80 Yale LJ 567 (1971), criticized in Selander, *Is Annual Style Change in the Automobile Industry an Unfair Method of Competition? A Rebuttal,* 82 Yale LJ 691 (1973).

with necessarily uncertain results.[302] The uncertainty derives in part from the nature of antitrust, which looks to results and "substance,"[303] as well as particular, specific overt behavior in defining illegality.[304]

The inevitable quantum of uncertainty is further increased by continuing changes in antitrust thinking which cannot be predicted in advance, since they are dependent on interpretation of provisions of high generality.[305] Such interpretation is influenced by many intangible factors such as the nature of the primary problems facing the national economy. During the Great Depression, fairness in pricing and avoidance of cutthroat competition were considered paramount,[306] leading to fair trade[307] and price discrimination[308] legislation, whereas during an era of perceived scarcity, the federal antitrust exemption for fair trade was repealed,[309] and defenses to price discrimination provisions were interpreted broadly.[310]

---

[302] *Compare* Continental TV, Inc v GTE Sylvania Inc, 433 US 36 (1977) *with* Eiberger v SONY Corp of America, 622 F2d 1068 (2d Cir 1980).

[303] Continental TV, Inc v GTE Sylvania Inc, 433 US 36, 47 (1977), *quoting* Appalachian Coals, Inc v United States, 288 US 344, 377 (1933).

[304] *See* the approach taken in, *e.g.*, United States v United States Gypsum Co, 438 US 422 (1978); National Socy of Professional Engrs v United States, 435 US 679 (1978); Broadcast Music, Inc v CBS, 441 US 1 (1979); California Retail Liquor Dealers Assn v Midcal Aluminum, Inc, 455 US 97 (1980); Catalano, Inc v Target Sales, Inc, 446 US 643 (1980).

[305] "As a charter of freedom, the [Sherman] Act has a generality and adaptability comparable to that found to be desirable in constitutional provisions." Appalachian Coals, Inc v United States, 288 US 344, 359-60 (1933) (Hughes, CJ).

[306] Freidel, FDR: Launching the New Deal chs VI, XVIII, XXIV (1973); Galloway, Industrial Planning Under Codes (1935); Mack, Controlling Retailers (1936), dealing with the National Industrial Recovery Act of June 16, 1933, ch 90, 48 Stat 195, held invalid in ALA Schechter Poultry Corp v United States, 295 US 495 (1935).

[307] Miller-Tydings Act of Aug 17, 1937, ch 690, 50 Stat 673, and McGuire Act of July 14, 1932, ch 745, 66 Stat 631 (both formerly codified in 15 USC §§1, 45(a)), repealed by Consumer Goods Pricing Act of Dec 12, 1975, Pub L No 94-145, 89 Stat 801. For history, *see* Schwegmann Bros v Calvert Distillers Corp, 341 US 384 (1951); United States v McKesson & Robbins, Inc, 351 US 305 (1956); 1975 US Code Cong & Ad News 1569.

[308] Robinson-Patman Price Discrimination Act, 15 USC §§13, 13a, discussed most extensively in Rowe, Price Discrimination Under the Robinson-Patman Act (1962); Edwards, The Price Discrimination Law (1979) ABA Antitrust Sec, The Robinson-Patman Act: Policy and Law (Monograph No 4, 1980).

[309] Consumer Goods Pricing Act of Dec 12, 1975, Pub L No 94-145, 89 Stat 801, discussed in California Retail Liquor Dealers Assn v Midcal Aluminum, Inc, 455 US 97 (1980); Continental TV, Inc v GTE Sylvania Inc, 433 US 36, 51 n 18 (1977); *see* S Rep No 466, 94th Cong, 1st Sess (1975); Practising Law Institute, Marketing and Distribution: Regulations and Restraints (1978); Practising Law Institute, The Repeal of the Fair Trade Laws: Impact on Product Distribution (1976); Givens & Worsinger, *Vertical Restraints After Repeal of Fair Trade,* 45 Fordham L Rev 1093 (1977).

[310] Great A&P Tea Co v FTC, 440 US 69 (1979); United States v United States Gypsum Co, 438 US 422 (1978); *see* Kuenzel & Schiffres, *Making Sense of Robinson-Patman: The Need to Revitalize Its Affirmative Defenses,* 62 Va L Rev 1211 (1976); Practising

The fruits of industrial success often tend to generate momentum toward continuing the currently profitable mode of activity.[311] When competitive or technological conditions change, as they inevitably do, the previously leading contender may find itself outflanked and left behind—even without adverse antitrust intervention.[312] There are many illustrations of once leading firms or industries which were beached in this manner.[313]

One alternative strategy is diversification. Apart from the changing contours of merger law or other public policies[314] and their potential further development in uncertain ways,[315] competition for acquisition candidates may make this form of investment for large amounts of capital less optimal.[316]

A further or supplemental alternative is to continue the innovation process and launch further experience curves while the earlier ones are at a profitable state. This may well be the safest as well as potentially the most profitable option over time, even though it involves an increase in risks.[317] It entails, however, a conscious decision to go against the normal gravitational tendencies inherent in an organization enjoying short-run success.[318] Increased dele-

---

Law Institute, Marketing and Distribution: Regulations and Restraints 151-215 (1978); Practising Law Institute, 19th Annual Adv Antitrust Seminar: Price Discrimination 71 (1979); Note, 90 Harv L Rev 1476 (1977).

[311] Naturally this is the path of least resistance. On the importance of the ability to undertake change before the absolute necessity arises, compare the comments of Winston Churchill, *quoted in* Never Give In! 10 (1967): "as danger comes nearer and grows [the British] become progressively less nervous; when it is imminent they are fierce, when it is mortal they are fearless. These habits have led them into some very narrow escapes."

[312] See Jacobs, The Economy of Cities (1969).

[313] *E.g.*, W. T. Grant Co. and the nation's railroads, which were undoubtedly not that much affected by antitrust strictures notwithstanding the cases cited at n 243 *supra. See, e.g.*, Reed-Bulwinkle Act, 62 Stat 472 (1948), discussed in Note, 46 Colum L Rev 990 (1946), now revised in 49 USC §10706 *as amended by* the Revised Interstate Commerce Act of 1978.

[314] *E.g.*, financing for mergers, discussed in, *e.g.*, Wall St J, July 14, 1981, at 2, cols 3-4; Wall St J., Dec 5, 1979, at 1, 36; Chavez, *The Agglomeration of America*, NY Times, July 12, 1981, §3, at 1, 12; Farnsworth, *House Panel Approves Foreign Takeover Curb*, NY Times, July 17, 1981, at D1, 4; S 1994, 97th Cong, 1st Sess (1981).

[315] Compare the retroactive effect of United States v EI duPont de Nemours & Co (GM), 353 US 586 (1957), discussed in, *e.g.*, Dirlam & Stelzer, *The DuPont-General Motors Decision: In the Antitrust Grain*, 58 Colum L Rev 24 (1958); Givens, *Affirmative Benefits of Industrial Mergers and Section 7 of the Clayton Act*, 36 Ind LJ (1960).

[316] This will occur if the purchase and sale of going concerns is regarded as a market with unlimited supply, an idea derived from the concept of an acquisition market proposed to the author by Dr. Betty Bock of The Conference Board.

[317] *See* Terleckyj, Effects of R&D on the Productivity Growth of Industries (National Planning Assn 1974); David, *Science Futures: The Industrial Connection*, 203 Sci 837 (Mar 2, 1979); Starr & Rudman, *Parameters of Technological Growth*, 182 Sci 358 (Oct 26, 1973).

[318] See Jay, Management and Machiavelli (1968); Stein, Organizations (1967); *compare* Parkinson, Parkinson's Law (1957); Hayes & Abernathy, *Managing Our Way to Economic Decline*, Harv Bus Rev 67 (July-Aug 1980).

gation of authority within the organization and tolerance of inevitable mistakes may be necessary to maintain initiative[319] and permit new ideas to be tried without being stymied by too many levels of approvals.[320] A climate favorable to attraction, retention and exploitation of talented business and scientific innovators may have to be deliberately created.[321] Delegation has the further advantage of permitting central managers to focus upon long-term strategies, otherwise made more diffucult by the demands of involvement in detailed review of individual decisions at lower levels of a large organization.[322]

Similarly, the strength attainable by continuous innovation may permit a less risky antitrust strategy, based upon examination of all available options. This is the opposite of the approach of merely seeking protection from competition to the maximum extent believed legally sustainable at a given time.

Less restrictive alternatives may have business as well as antitrust benefits. In some instances, the riskier option may, of course, be necessary and may be upheld,[323] but its adoption without full consideration of alternatives is likely to be a significant error.[324]

Problem	**Option A**	**Less Restrictive Option**
Licensee of patent may cut price to compete with patentee to the latter's detriment.	Impose restriction on resale prices (legally vulnerable).[325]	Establish royalty which does not drop if the price is cut, protecting revenues of patentee while building fixed cost into price structure of licensee, and exacting added financial return in

[319] See Drucker, Concept of the Corporation (1953); Mackenzie, The Time Trap ch 7 (1972); Townsend, Up the Organization (1970); Pascale & Athos, The Art of Japanese Management (1981); Ouchi, Theory Z: How American Business Can Meet the Japanese Challenge (1981); Lewis, *The Profits in Common Sense,* NY Times, (Sept 26, 1980, at 35; Glaser, *Participatory Management,* 210 Sci 962 (Nov 28, 1980); Walsh, *Industrial Productivity and the "Soft Sciences",* 209 Sci 476 (July 25, 1980); *compare also* Kamien & Schwartz, *Market Structure and Innovation,* 13 J Econ Lit 9 (1975).

[320] See Givens, Advocacy ch 14 (Shepard's/McGraw-Hill 1980).

[321] On changing motivations for work, *see, e.g.,* Ewing, The Human Side of Planning (1969); McGregor, The Human Side of Enterprise (1960); Pope John XXIII, Mater et Magistra part II, ¶¶82-83 (1961); Work in America: Report of a Special Task Force to the Secretary of Health, Education and Welfare (1972); Levenstein, Why People Work (1964).

[322] See Churchill, Memoirs of the Second World War (1978).

[323] *E.g.,* Great A&P Tea Co v FTC, 440 US 69 (1979); *cf* Continental TV, Inc v GTE Sylvania Inc, 433 US 36 (1977).

[324] *See* Eiberger v SONY Corp of America, 622 F2d 1068 (2d Cir 1980).

Problem	Option A	Less Restrictive Option
		exchange for licensee's ability to set prices.[326]
Desire of seller to assure continued sales to major customer.	Requirements contract (subject to challenge but may be upheld under rule of reason).[327]	Long-term quantity supply contract with terms to cover contingencies [328] (potential Robinson-Patman cost justification benefit[329] as well as probable greater likelihood of validity because other suppliers are not entirely excluded; greater business benefit of assured sales or compensation under contract).
Need for dealer advertising, warranty and similar services.	Resale price maintenance program (maximum legal risk[330] or restrictions on territorial or customer resale restrictions (subject to attack under rule of reason,[331] especially if firm has large market share).[332] Business side effects may include silencing some complaining dealers[333] while generating new disputes over fairness of policing of restrictions.[334]	Cooperative advertising program,[335] and reimbursement for reasonable actual cost[336] of warranty service out of funds secured from wholesale price.[337] Business benefit may include avoiding excessive markup beyond what is needed to obtain services, which can cause interbrand competitive detriment in some products.[338]

The antitrust laws do not necessarily require selection of the least restrictive option in all cases,[339] but the degree of restraint in relation to the need for

---

[325] See **§3.03.**

[326] See **§3.08.**

[327] See **§3.07.**

[328] *Id; compare* United States v Dairymen, Inc, 1978-1 Trade Cas (CCH) ¶62,053, at 74,543-545 (WD Ky 1978), *affd in part, revd in part on other grounds.* 660 F2d 192 (6th Cir 1981). In regard to bargaining for quantitative benefits to a party in lieu of qualitative controls on the behavior of the other, *see also* Westgo Indus, Inc v WJ King Co, 1981-1 Trade Reg Rep (CCH) ¶63,984 (DND 1981).

[329] See **§3.07.**

[330] See **§3.03.**

[331] Continental TV, Inc v GTE Sylvania Inc, 433 US 36 (1977). For a negative outcome, *see* Eiberger v SONY Corp of America, 622 F2d 1068 (2d Cir 1980). For a less restrictive alternative, *see* James B Lansing Sound, Inc, 46 Fed Reg 29928 (FTC June 4, 1981) (modification of order to allow manufacturer to ban resale to those other dealers who do not meet stated qualifications).

[332] Interbrand competition is the key element, Continental TV, Inc v GTE Sylvania Inc, 433 US 36, 52 n 19 (1977). See Pitofsky, *The Sylvania Case: Antitrust Analysis of Non-Price Vertical Restrictions,* 78 Colum L Rev 1 (1978); Denger, *Vertical Restrictions: The Impact of Sylvania,* 46 ABA Antitrust LJ 908 (1977).

[333] But these may concern prices, suggesting a dealer cartel of the type posited in Continental TV, Inc v GTE Sylvania Inc, 433 US 36, 58 n 28 (1977) (citing United States v General Motors Corp, 384 US 127 (1966); *see* Girardi v Gates Rubber Co, 325 F2d 196, 202 (9th Cir 1963). Vertical restraints may also turn out to be part of a dealer-inspired resale price maintenance combination with horizontal aspects. *Compare* Simpson v Union Oil Co, 377 US 13 (1964); United States v Parke, Davis & Co, 362 US 29 (1960); United States v Bausch & Lomb Optical Co, 321 US 707 (1944); FTC v Beech-Nut Packing Co, 257 US 441 (1922); Copper Liquor Inc v Adolph Coors Co, 506 F2d 934, 947-48 (5th Cir 1975); Adolph Coors Co v FTC, 497 F2d 1178 (10th Cir 1974), *cert denied,* 419 US 1105 (1975); Denger, *Vertical Restrictions: The Impact of Sylvania,* 46 ABA Antitrust LJ 908, 915-16 (1977).

[334] *Compare* Vidich, The New York Cab Driver and His Fare (1976); Bowman, *The Prerequisites and Effects of Resale Price Maintenance,* 22 U Chi L Rev 825 (1955); Note, 69 Yale LJ 168 (1959).

[335] Such programs must comply with the Robinson-Patman Price Discrimination Act, particularly 15 USC §13(e); *see* Federal Trade Commission, Guides for Advertising Allowances and Other Merchandising Payments and Services, 16 CFR §240 (1973); FTC v Fred Meyer, Inc, 390 US 341 (1968); FTC v Simplicity Pattern Co, 360 US 55 (1959); Exquisite Form Brassiere, Inc v FTC, 301 F2d 499, 501-06 (DC Cir 1961), *cert denied,* 396 US 888 (1962), Exquisite Form Brassiere, Inc v FTC, 360 F2d 492 (DC Cir 1965), *cert denied,* 384 US 959 (1966); Sullivan, Antitrust §227 (1977). On the question of actual use of co-op monies, *see* Schwimmer v SONY Corp of America, 637 F2d 41 (2d Cir 1980); Sadowy v SONY Corp of America, 496 F Supp 1071, 1075, 1080 (SDNY 1980). On deductions from the invoice, *see* LS Amster & Co v McNeil Laboratories, Inc, 504 F Supp 617 (SDNY 1980).

[336] *See* Eiberger v SONY Corp of America, 459 F Supp 1276, 1283 (SDNY 1978), *affd in part,* 622 F2d 1068 (2d Cir 1980).

[337] *See also* 16 CFR §702.4 under the Magnuson-Moss Warranty Act, 15 USC §2301 *et seq,* regarding warranty documentation.

[338] See Telser, *Why Should Manufacturers Want Fair Trade,* 3 JL & Econ 86(1960).

[339] Continental TV, Inc v GTE Sylvania Inc, 433 US 36, 58 n 29 (1977).

it[340] is relevant to the balancing of pro- and anticompetitive effect and purpose[341] under the rule of reason.[342] This is also important if an exception to a normally applicable per se rule is sought.[343] These illustrations suggest that just as a flexible, innovative internal policy can seek to maintain the momentum of innovation, flexible external policies can minimize antitrust risks, especially for large successful innovators certain to come under close scrutiny.

---

[340] See note 336 *supra.*

[341] On interaction of purpose and effect, *see, e.g.*, Broadcast Music, Inc v CBS, 441 US 1 (1979); United States v United States Gypsum Co, 438 US 422, 436 n 13 (1978); Chicago Bd of Trade v United States, 246 US 231, 238 (1918). For an approach that a technologically based purpose is valid whereas "artificial competitive advantages" are contrary to the purposes of the antitrust laws, *see* Emhart Corp v USM Corp, 527 F2d 177, 182 (1st Cir 1975). *Compare generally* on legitimate justifications for challenged conduct, cases cited in Wood v Southwestern Bell Co, 637 F2d 1188, 1191 (8th Cir 1981).

[342] *See* Continental TV, Inc v GTE Sylvania Inc, 433 US 36 (1977); Chicago Bd of Trade v United States, 246 US 231 (1918); Handler, Antitrust in Perspective ch I (1957); Note, 58 Colum L Rev 673, 681-88 (1958).

[343] *Compare* Broadcast Music, Inc v CBS, 441 US 1 (1979) *with* Catalano, Inc v Target Sales, Inc, 446 US 643 (1980).

Innovation, Patents and Trade Secrets

Lloyd McAulay

4

§4.01 —Basic Patent Requirements

This chapter is intended to be useful to lawyers and business managers who are not patent experts. It is also intended to raise strategic questions which may be of relevance to those familiar with the field. Some basic concepts in patent law are presented to provide a context within which to understand and apply the more specific comments and suggestions appearing in this chapter.

For the subject matter of an invention to be patented, it must meet the three basic preliminary criteria discussed below.

Utility

First, it must be useful.[1] It is not likely that a patent would be sought if the subject matter were not useful. Most things one would bother to patent or make are useful. Even a game is useful.[2]

The main significant commercial impact of this requirement occurs where someone seeks to patent a new chemical composition or chemical process. The proprietor senses it might someday be useful and seeks to patent it "just in case." It cannot be patented, however, unless some utility can be found.[3]

Novelty

The second requirement is novelty.[4] The product or process must have some element of novelty. It may be, and almost always is, a combination of old or known features.[5] To be patentable, though, it must be a new combination.[6] This requirement accords with almost everyone's sense of the basis of a patent system so that it is rare for anyone to turn to the patent system unless novelty is (rightly or wrongly) believed to exist.

Nonobviousness

The third requirement is nonobviousness.[7] It is this rquirement that raises most of the questions and arguable issues in applying for a patent and in litigating the validity of a patent. Broadly stated, the requirement is that at the time the invention is made, the novel combination which constitutes the invention must not be obvious to someone skilled in the art to which the invention pertains.[8]

What constitutes obviousness or nonobviousness is a large part of what patent law is all about. It is not practical here to go into all of the doctrines concerning application of the nonobviousness standard to particular situa-

---

[1] 35 USC §101.

[2] *E.g.,* Patent No 2,026,082 issued in 1935 on the Monopoly game.

[3] Brenner v Manson, 383 US 519 (1966); *In re* Brenner, 182 F2d 216, 86 USPQ 74 (CCPA 1950).

[4] 35 USC §§101, 102.

[5] BG Corp v Walter Kidde & Co, 79 F2d 20, 22, 26 USPQ 228 (2d Cir 1935); Reiner v I Leon Co, 285 F2d 501, 503, 128 USPQ 25 (2d Cir 1960).

[6] Smith v Hall, 301 US 216, 217, 233, 33 USPQ 249 (1936); Piccard v United Aircraft Corp, 128 F2d 632, 625, 53 USPQ 563 (2d Cir 1942).

[7] 35 USC §103.

[8] *Id. See* Graham v John Deere Co, 383 US 1, 13-17 (1966).

tions. For example, there is a presumption that certain structurally similar compounds are obvious in view of one another, but a showing of unexpected utility might rebut such a presumption.[9]

One of the more common commercial situations in which a novel and useful combination is not patentable is one in which some new component or product has recently become available, which can be combined in a known fashion with other known components to provide an improved combination.[10] Usually it is obvious that one can adapt the new product or component to replace a predecessor component. For example, when the transistor was developed, it was obvious how to redesign electronic tube circuits to use transistors. Because the transistor is a device that is best analyzed in terms of how it manipulates current and an electron tube is a device which is best analyzed in terms of how it manipulates voltage, circuits employing transistors and electron tubes look quite different. However, to a circuit designer of ordinary skill, it would be obvious how to adapt an electron tube circuit to employ transistors.[11]

Similarly, new materials are frequently developed that provide characteristics of improved strength, corrosion resistance, wear and the like, which when used in known devices make it possible to use the devices for purposes that were not previously commercially feasible. Such applications are normally, but not always, obvious and thus not patentable. For example, the development of ceramic technology may have been what made it possible or convenient over a hundred years ago, to provide a ceramic doorknob. The Supreme Court held a patent on a ceramic doorknob invalid because it was an obvious substitution of superior material (ceramic) which had never been used as a doorknob for an old material (metal).[12]

Nonobviousness is clearly the most subjective of the requirements for patentability.[13] Accordingly, it is the criterion most used by decision makers, such as Patent Office examiners or judges, to put into effect their beliefs about the scope and value of the patent system.[14] It is the area in which there is the

---

[9] *In re* Papesch, 315 F2d 381, 391, 137 USPQ 43 (CCPA 1963).

[10] Levitan Mfg Co v Slater Elec, Inc, 331 F Supp 395, 406, 169 USPQ 202 (EDNY 1970) (concerns application of silicon controlled rectifiers in a light dimming circuit). HK Porter Co v Black & Decker Mfg Co, 182 USPQ 401, 414 (ND Ill 1974), *affd,* 518 F2d 1177, 187 USPQ 476 (7th Cir 1975) (concerns use of a miniature motor and rechargeable battery in a cordless electric grass shear).

[11] Forbro Design Corp v Raytheon Co, 532 F2d 758, 190 USPQ 49 (1st Cir 1976).

[12] Hotchkiss v Greenwood, 11 How (52 US) 248, 266 (1851), *cited with approval in* Graham v John Deere Co, 383 US 1, 11 (1966).

[13] *See* Graham v John Deere Co, 383 US 1, 18 where the court stated: "[w]hat is obvious is not a question upon which there is likely to be uniformity of thought in every given factual situation."

[14] *Contrast* the discussion by Justice Douglas in his concurring opinion in Great A&P Tea Co v Supermarket Equip Corp, 340 US 147, 154 (1950) *with* Judge Frank's discussion in his concurring opinion in Picard v United Aircraft Corp, 128 F2d 632, 642-43 53 USPQ 563 (2d Cir 1942).

greatest uncertainty in giving advice concerning patentability or patent validity.

If a patent is viewed as an award for invention, the nonobviousness standard will be strictly applied with the result that fewer patents will issue and fewer will be determined valid.[15] If, on the other hand, a patent is viewed as a means to induce the investment of time, money and effort in developing a new product or a new business, a greater likelihood exists that a novel product or process will be found to be patentable.[16]

§4.02 —Patentable Subject Matter

The patent statute sets forth a wide variety of things and processes that are patentable types of subject matter. To get a patent, one must meet the requirements of utility, novelty and nonobviousness set forth in the preceding section. As long as those three requirements are met, the kinds of subject matter that can be patentable may be any new and useful process, machine, manufacture or composition of matter as well as any improvement thereof.[17]

The patent statute also provides that any asexually reproduced distinct variety of plant may be protected by a plant patent.[18] The Agriculture Department issues a certificate providing patent-like protection over distinct and new varieties of sexually reproduced plants.[19] Although new and useful bacteria were once considered to be outside of the patent statute, it is now clear that such organisms may be covered by a patent.[20] Thus, the results of genetic engineering are generally available for patent coverage even though the subject matter is a living organism. A range of inventions connected with the use of computers and computer programming was similarly once considered to be not the type of subject matter entitled to patent protection. The Supreme Court, however, has ruled otherwise and, thus, a substantial range of inventions relating to computer programming may now be covered by a patent.[21]

Under the design sections of the patent statute, a new original and ornamental design for an article of manufacture may be covered by a design patent.[22]

---

[15] *See* Cuno Engg Corp v Automatic Devices Corp, 314 US 84, 91 (1941) and Great A&P Tea Co v Supermarket Equip Co, 340 US 147, 157 (1950).

[16] *See* Judge Frank's discussion in Picard v United Aircraft Corp, 128 F2d 632, 642-43, 53 USPQ 563 (2d Cir 1942). This policy was recognized in Diamond v Chakrabarty, 447 US 303, 100 S Ct 2204, 2212; (1980), where the Court stated: "[w]hether respondent's claims are patentable may determine whether research efforts are accelerated by the hope of reward or slowed by want of incentive."

[17] 35 USC §101. Diamond v Chakrabarty, 447 US 303, 100 S Ct 2204, 2207 (1980) recited the §101 language as showing Congressional intent to give the patent laws a wide scope.

[18] 35 USC §161.

[19] 7 USC §2402(a).

[20] Diamond v Chakrabarty, 447 US 303 (1980).

[21] Diamond v Diehr & Lutton, 450 US 175 (1981), *affg* 602 F2d 982, 203 USPQ 44 (CCPA 1979).

[22] 35 USC §6171.

A design patent covers the ornamental aspects of the design; that is, the way it looks. A design patent does not cover the utility aspects of the product involved.[23]

The categories of things that cannot be patented are substantially as follows. Methods of doing business in the sense of organization, marketing and distribution are not patentable even though industrial processes and industrial methods are patentable.[24] Thus, an improved technique of franchising would not be patentable subject matter.

Printed matter is not patentable, although often subject to copyright protection. However, if printed matter is part of some other structure, it may constitute one of the elements of the combination claimed that defines the scope of the patent and may even be the basis on which patentability is predicated.[25]

Although a scientific principle or law of nature cannot be patented,[26] a structural manifestation of a scientific principle can be. This exclusion from the scope of patentable subject matter rarely if ever causes real problems.

A naturally occurring article or plant not requiring human modification cannot be patented. A compound in nature or a plant or animal which occurs in nature is not patentable to its discoverer.[27] Except for such *naturally* occurring articles and scientific principles, the fact that a new and useful product and process is discovered rather than invented will not defeat the right to a patent. The statute is explicit that anyone who discovers or invents is entitled to a patent.[28]

A mathematical procedure or concept is not patentable. Essentially, this means that mathematical techniques and indeed even a new mathematical system would not be subject to patent protection. However, similar to the scientific principle and printed matter exception, a structure which includes the mathematical formula or concept may be patentable even though the claim recites the mathematical concept.[29]

---

[23] *See* Gorham Co v White, 81 US 511, 524 (1871) and Lancaster Colony Corp v Alden Acessories, Ltd, 506 F2d 1197, 1200, 184 USPQ 193 (2d Cir 1974).

[24] Hotel Sec Checking Co v Lorraine Co, 160F 467 (2d Cir 1908); *In re* Wait, 73 F2d 982, 24 USPQ 88 (CCPA 1934). For a contrary view, see Haasman, *Method of Doing Business,* 50 J Pat Off Socy 503 (July 1968).

[25] *In re* Royka, 490 F2d 981, 985, 180 USPQ 580 (CCPA 1974); *In re* Miller, 418 F2d 1392, 164 USPQ 46 (CCPA 1969); *In re* Jones, 373 F2d 1007 (CCPA 1967).

[26] *See* comment in Diamond v Chakrabarty, 447 US 303, 100 S Ct 2204 2208 (1980), and Funk Bros Seed Co v Kalo Inoculant Co, 333 US 127, 130 (1948). A good discussion of the meaning of this exception and of its history can be found in *In re* Bergy, Coats & Malik, 596 F2d 952, 988-995, 201 USPQ 352, 384-389 (CCPA 1979).

[27] Funk Bros Seed Co v Kalo Inoculant Co, 333 US 127, 130 (1948). *Also see In re* Thuau, 135 F2d 344, 347, 57 USPQ 324 (CCPA 1943).

[28] 35 USC §§101, 161.

[29] A mathematical algorithm is not patentable. Parker v Flook, 437 US 584, 196 USPQ 864 (1978); Gottschalk v Benson, 409 US 63, 67, 17 USPQ 673 (1972). But as a step in a process it may contribute to patentability. Diamond v Diehr & Lutton, 450 US 175 (1981).

Artistic devices that have no substantial utility other than to be decorative are not patentable, although copying may be limited by other proprietary rights.[30] Neither a painting nor a sculpture may be patented, although articles of manufacture, such as furniture, which have an original ornamental or aesthetic aspect may have the ornamental feature covered by a design patent.[31]

The scope of the patent system has been broadly conceived by Congress to include substantially anything that is made by human intervention. At times the courts have not seemed to view the scope quite so broadly;[32] but two recent Supreme Court cases, relating to genetic technology and to methods that include the use of a programmed computer, indicate a current willingness to invoke a broad scope.[33]

§4.03 —Critical Definitions and Concepts

With the above framework in mind, it is helpful to focus on a few terms and concepts that are often confusingly used.

The terms *invention* and *patentability* are often confused with one another. Some attention to the meaning of these terms is important to avoid confusion. Confusion between these two terms is particularly rampant prior to 1952 when the nonobvious standard of patentable invention was first introduced into the patent statue at 35 USC §103. See the discussion under nonobviousness at **§4.15.**[34]

Invention

The word *invention* is used in the patent field in a broad sense. Basically, it means anything new. It is substantially identical in import to the statutory requirement that the subject matter be new.[35] Thus, under this definition anything copied from someone else is not an invention. Because of the statutory time bars and the nonobviousness requirement, there may be invention without patentability.[36]

---

[30] In particular, copyright and trade dress may protect nonutilitarian features. There is even some overlap between copyright and design patent protection. Mazer v Stein, 347 US 201, 207 (1954); *In re* Yardley, 493 F2d 1389, 1395, 181 USPQ 331 (CCPA 1974).

[31] 35 USC §171. Schnadig Corp v Gaines Mfg Co, 494 F2d 383, 181 USPQ 417 (6th Cir 1974) (held a design patent on a sectional sofa valid and infringed).

[32] Brenner v Manson, 383 US 519 (1966); *In re* Arzeberger, 112 F2d 834, 46 USPQ 32 (CCPA 1940).

[33] Diamond v Chakrabarty, 447 US 303 (1980) and Diamond v Diehr & Lutton, 450 US 175 (1981), *affg* 602 F2d 982, 203 USPQ 44 (CCPA 1979).

[34] For an excellent discussion of the concepts of invention, patentability and nonobviousness, see Rich, *Monopolies, "Invention" and Section 103,* in The Encyclopedia of Patent Practice And Invention Management 592-602 (Calvert ed 1964).

[35] 35 USC §101. See **§4.01.**

[36] See **§4.05** for discussion of time bars under 35 USC §102 and **§§4.11, 4.13** and **4.14** for the effect of foreign time bars. See **§4.01** for a discussion of the nonobviousness requirement under 35 USC §103.

Patentability or Patentable Invention

These terms generally refer to those inventions which constitute subject matter recognized by the patent statute and which are sufficiently nonobvious so as to justify and warrant the issuance of a patent. Many court decisions, particularly older decisions, confuse the terms *invention* and *patentability*.[37]

Nonobviousness

The patent statute requires that an invention of patentable subject matter be nonobvious to one skilled in the art at the time the invention is made in order for that invention to be patentable.[38] Thus, the term nonobviousness is used to refer to that circumstance, feature or subject matter which supports patentable invention. Nonobviousness is the usual key issue in patent litigation, although a patentable invention must also meet other criteria.[39]

Valid Patent

Patents are generally issued in an ex parte context. The inventor presents an argument for the issuance of a patent to a patent examiner. The patent examiner may exercise a degree of diligence and is likely to have expertise, but does not have quite the incentive that is present in a full-scale litigation. Because of this exercise of expertise, the issued patent has a presumption of validity.[40] Because of the limits of the patent prosecution process, that presumption is rebuttable.[41] Many patents are issued which in litigation are deemed invalid because the defendant or infringer makes a thorough search of the prior art to provide evidence that the issuance of the patent was not warranted. Because this occurs often, it is useful to use the phrase *valid patent* to describe the situation either where the patent being considered has passed muster in court or where, based on a thorough search of the prior art, one concludes it is highly likely to pass muster.

The Patent Grant

The patent does not give its holder the right to do anything. It gives the owner the right to exclude others from practicing the invention as claimed.[42] Thus, it is not quite what one usually considers a monopoly. It is for this reason that the subject matter of an improvement patent can infringe the claims of an earlier more basic patent.

In 1933, the United States Supreme Court provided a useful description, if not definition, of the significant characteristics of the *patent grant*. In *United*

---

[37] *See* Graham v John Deere, 383 US 1, 11-12, 14 (1966).

[38] 35 USC §103. See discussion in **§4.01.**

[39] The most significant criteria are set forth in **§§4.01, 4.02, 4.05.**

[40] 35 USC §282.

[41] *See* Woodstream Corp v Herter's, Inc, 446 F2d 1143, 1156-57, 170 USPQ 380 (8th Cir 1971).

[42] 35 USC §154.

States v Dubilier Condenser Corp,[43] the Court stated:

> Though often so characterized a patent is not, accurately speaking, a monopoly, for it is not created . . . at the expense and to the prejudice of all the community except the grantee of the patent . . . . The term "monopoly" connotes the giving of an exclusive privilege for buying, selling, working, or using a thing which the public freely enjoyed prior to the grant. Thus a monopoly takes something from the people. An inventior deprives the public of nothing which it enjoyed before his discovery, but gives something of value to the community by adding to the sum of human knowledge . . . . He may keep his invention secret and reap its fruits indefinitely. In consideration of its disclosure and the consequent benefit to the community, the patent is granted. An exclusive enjoyment is guaranteed him for seventeen years, but, upon the expiration of that period, the knowledge of the invention inures to the people, who are thus enabled without restriction to practice it and profit by its use . . . . To this end the law requires such disclosure to be made in the application for patent that others skilled in the art may understand the invention and how to put it to use.

§4.04 —Identification of the Inventor

In the United States, a patent application must be made in the name of the inventor or inventors, who normally sign a declaration or oath that the inventorship claimed is accurate.[44] False claims concerning inventorship will invalidate the patent.[45]

Unfortunately, there are many circumstances where the head of a research unit arrogates all inventorship and becomes the inventor of all output. One should look skeptically at such situations. People who are very inventive and have a great deal of drive may well see all developments as essentially deriving from their own activities. It is necessary, however, to insist upon the designation of correct inventorship or risk a loss of important proprietary rights. It is important that clear and unambiguous authority be given to the patent attorney to make the determination of inventorship, regardless of internal political considerations, seniority, supervisory role in the company and similar factors.[46]

Unfortunately, it is often difficult to determine the inventor or inventors in

---

43 289 US 178, 186 (1933).

44 35 USC §111. But application can be made in the name of recalcitrant, unavailable or dead inventor. *Id* §118.

45 Campbell v Spectrum Automation Co, 513 F2d 932, 185 USPQ 718 (6th Cir 1975); Kennedy v Hazelton, 128 US 667, 672 (1888).

46 For example, in Campbell v Spectrum Automation Co, 513 F2d 932, 934, 185 USPQ 718 (6th Cir 1975), it appears that Mr. Campbell's employee was the inventor, yet Campbell filed for the patent as the alleged inventor.

a laboratory situation where a number of people interact. The courts are likely to be forgiving if a good faith error is made.[47] Inventors can be added or subtracted during the course of prosecution or even after a patent is issued, providing there is a showing that the inventorship originally set forth was in good faith.[48] When there is doubt, it is prudent to include someone as co-inventor rather than to exclude the individual. However, the fact that in some circumstances there are doubts should not be used as a subterfuge to permit someone to arrogate inventorship improperly.

The identification of inventorship becomes particularly subject to ambiguity and thus hazardous manipulation where an improvement in the invention is involved. The standards of inventorship, though, are no different in such a situation and an improvement must stand on its own two feet. The fact that an invention by inventor B is an improvement on an invention by inventor A does not make A a co-inventor of the improvement.

Ownership of the invention or of the patent, however, is a very different thing and can be quite separate from inventorship. The inventors who file for the patent can, at the time of filing, assign all rights to any corporate or noncorporate entity and that assigment can and should be recorded at the Patent Office.[49]

§4.05 —Statutory Time Bar for Filing Patent Applications

The basic concept of the patent law is that a right for a limited term (17 years in the United States) to exclude others is granted in exchange for disclosure to the public of a patentable invention. As a consequence, when one has a patentable invention, one must choose between exploiting it under the aegis of a patent or keeping it as a trade secret.[50]

Most inventions cannot practically be kept as trade secrets, and the real choice is between obtaining a patent or depending solely on the advantage of being first on the market with the best. One cannot, however, commercially exploit an invention for a significant period of time and then apply for a patent. The patent statute bars filing if there has been public use or disclosure; this is know as a *statutory bar*.[51]

The United States patent law gives a one-year grace period after any public disclosure, public use or commercial use (whether public or not) within which to file a patent application. If the application is not filed within the one-year

---

[47] AF Stoddard & Co v Dann, 564 F2d 556, 195 USPQ 97 (DC Cir 1977). *See also In re* Shibata, 203 USPQ 280 (Commr of Pats 1979).

[48] 35 USC §256.

[49] *Id* §152.

[50] See discussion in **§4.20.**

[51] The various statutory bars are set out at 35 USC §102.

time period, then the bar applies.[52] As a consequence, it is important to establish a policy which controls publication in scientific journals, commercial use, sale or even the offering for sale of any device which embodies an invention until a determination has been made concerning the filing of a patent application.

Almost no other country provides this one-year grace period. Thus, if the option of filing abroad is desired, the one-year grace period established in the United States must be ignored and the patent application filed prior to any publication or commercial use.[53]

The conservative strategy, therefore, is to immediately follow development in the laboratory or factory with a patent application, and only after the patent application has been filed to engage in any form of public disclosure or commercial use, either directly or indirectly. Even a secret use—possible in certain industrial process—in the manufacture of a product which is offered for sale, constitutes a commercial use that activiates the one-year grace period in the United States.[54]

Reality frequently makes it impossible to take this most conservative approach. In those circumstances, certain actions can be taken to minimize the problem. Disclosures can be made in confidence.[55] An experimental use can be documented.[56] However, in order to establish an optimum strategy, the relationship between the disclosure or use bars in the United States and the comparable disclosure or use bars in other countries must be considered.[57]

§4.06 —Effect of an Offer for Sale

It is an established rule that an offer to sell an on-the-shelf item constitutes the kind of commercial use which activates the one-year statutory time bar. The sale does not have to be completed. An offer to sell is enough.[58] Even if the product is not on the shelf, but clearly can be manufactured in response to orders placed, this constitutes an offer for sale which is a commercial use.[59]

However, in many situations involving complex technology, a company will

---

[52] *Id.*

[53] For useful manual covering this and other foreign filing requirements, see Manual For the Handling of Applications For Patents, Design and Trade Marks Throughout the World (Octrooibureau Los en Steiger, Amsterdam, Holland).

[54] Metallizing Eng Co v Kenyon Bearing & Auto Parts Co, 153 F2d 516, 68 USPQ 54 (2d Cir 1946).

[55] See **§4.07.**

[56] See **§4.08.**

[57] See **§4.09.**

[58] Chromalloy Am Corp v Alloy Surfaces Co, 339 F Supp 859, 173 USPQ 295 (D Del 1972) and cases cited therein. *Also see* Egbert v Lippman, 104 US 888 (1881); and Smith & Griggs Mfg Co v Sprague, 123 US 249, 258 (1887).

[59] Amphenol Corp v General Time Corp, 275 F Supp 903, 905, 155 USPQ 246 (D Ill 1967).

make an offer to provide equipment which performs a particular function without having completed the design of that equipment. The purchase assumes or subsumes a development stage. The customer usually, but not necessarily, understands that the product must be developed under the contract. Under such circumstances, the offer to sell and even the contract for sale of the as yet undeveloped product is not the kind of commercial use which activates the time bar.[60] However, in such circumstances the tendency to slip over into a commercial use situation is substantial. A product may be developed and delivered before the contract is at an end and before the production design is completed. Such delivery and use, absent a contract to maintain confidentiality and to provide the results of experimental customer use, is likely to constitute a commercial use, which activates the time period, barring later patent filing.[61]

§4.07 —Confidentiality Exception

Customers or other entities are often given a developed product for testing purposes or may be consulted in connection with development of a product. If that is done on a confidential basis or if the purpose is experimental, the disclosure will not consititue a statutory bar.[62]

Confidentiality may be implied by the circumstances. The requirement for confidentiality is implied by law when employees are involved.[63] It is nevertheless prudent to have a specific written agreement with both employees and other parties who have access to information in connection with a development of a possible patentable nature.[64] It may be crucial to enter into such agreements if the development can be most successfully protected as a trade secret.

The important point is that the maintaining of secrecy avoids the public disclosure bar, but it can be a serious mistake to rely on confidentiality. Maintaining secrecy does not avoid the commercial use bar. For example, the secret use of an industrial process to manufacture a commercial product is a use that bars the filing of a patent application after the one year grace period.[65] A

---

[60] Koehring Co v National Automatic Tool Co, 362 F2d 100, 149 USPQ 887 (7th Cir 1966).

[61] Hobbs v United States Atomic Energy Comm, 451 F2d 849, 857-861, 171 USPQ 713 (5th Cir 1971).

[62] Maintaining confidentiality is particularly important under most foreign laws which are keyed to whether or not the invention has been divulged.

[63] Tlapek v Chevron Oil Co, 407 F2d 1129, 161 USPQ 193 (8th Cir 1969); Byrne v Barratt, 268 NY 199, 197 NE 217 (1935). *See* Pratt v Shell Petroleum Corp, 100 F2d 833 (10th Cir 1939), for a broader discussion of the employee's duty.

[64] *But see* Water Serv, Inc v Tesco Chems, Inc, 410 F2d 163, 162 USPQ 321 (5th Cir 1969), for the need to carefully draft any covenant not to compete where trade secrets are also involved.

[65] Metallizing Engg Co v Kenyon Bearing & Auto Parts Co, 153 F2d 516, 68 USPQ 54 (2d Cir 1946).

confidential use in an attempt to get orders may run afoul of the commercial use bar.

§4.08 —Experimental Use Exception

An invention is not considered to be in commercial use if the activity is experimental. Unfortunately, guidelines as to what constitutes experimental use are not at all clear even though there has been a great deal of litigation. The general principles are clear,[66] but applying those principles to fact situations is not always easy. Accordingly, a prudent rule to follow is that any testing necessary to establish the value of the product must be conducted with as much formal confidentiality as possible. If customers or outsiders are required for the testing program, confidential disclosure agreements should be obtained. Furthermore, and most important, regular reports of the testing activity should be made and, if at all possible, the experimental product should be returned at the end of the expermental testing period.[67]

Although an invention may be a development of one company, there are often circumstances under which the testing of the invention is a joint activity of the developer and a customer or potential customer. The customer may even be willing to pay for the cost of manufacturing the device and/or the cost of the development work. Under such circumstances, it is particularly important that the transaction avoid the appearance of a sale and purchase. An explicit written agreement on the part of the customer to maintain confidentiality with respect to the invention and to provide regular reports to the developer is particularly important under those conditions.[68]

The sale of the product, even in limited quantities, to customers to await the outcome of their field experience, does *not* constitute experimental use. This is true even if the purpose is to determine the extent to which the developer wishes to commit funds and resources to the product. Market testing is *not* experimental use. Determining how acceptable a product is to an intended market is *not* considered experimental testing.[69] Experimental testing means determining whether or not the product will serve the technical or technologi-

---

[66] For useful list of facts that can satisfy this exception, *see* Robbins Co v Lawrence Mfg Co, 482 F2d 426, 433, 178 USPQ 577 (9th Cir 1973). For a leading case, with a thorough discussion of the experimental use exception, *see In re* Yarn Processing Patent Validity Litig, 498 F2d 271, 277-78, 183 USPQ 65 (5th Cir 1974). *But see* American Mach & Hydraulic, Inc v Mercer, 585 F2d 404, 200 USPQ 262 (9th Cir 1978).

[67] *In re* Blaisdell, 242 F2d 779, 113 USPQ 289 (CCPA 1957), illustrates the risk of not maintaining control over the experimental devices once the experiment is completed.

[68] Minnesota Mining & Mfg Co v Kent Indus, Inc, 409 F2d 99, 101, 161 USPQ 321 (6th Cir 1969).

[69] Cataphote Corp v De Soto Chem Coatings, Inc, 235 F Supp 936, 938, 143 USPQ 229 (ND Del 1964); Interlego AG v FAO Schwarz, Inc, 187 USPQ 580, 583-85 (ND Ga 1975), 191 USPQ 129, 132-34 (ND Ga 1976). *See also* supra ns 55 and 63.

cal function to which it is addressed.[70]

Refinements and improvements in a product probably constitute the most seductive situation for inducing belief that the experimental use exception applies. In a sense, however, a commercial product is never complete. It can always be improved. It can always be refined. It can always be optimized.[71]

Overall, to ensure that the experimental use tolls the statutory bar, it must be carefully monitored and kept as confidential as possible, and its results continuously monitored.

The leading case on the experimental use exception is *City of Elizabeth v Nicholson Pavement Co.*[72] The invention at issue was an improved wooden pavement. A 75-foot strip was covered with the improved pavement and it was publicly used for six years. That period of time was deemed to be an experimental test even though it was public, because any testing of a road construction for durability had to be, as a practical matter, public, and had to be for a substantial period of time. However, this probably would not have been sufficient to make the experimental use exception applicable except for the further important fact that the pavement was regularly inspected by the inventor for the purpose of determining how well it held up. Although this is a United States Supreme Court decision and generally considered to be a leading case, a lackadaisical attitude should not be taken as to the controls necessary to maintain a use as experimental.

It is instructive in this regard to contrast *Koehring Co v National Automatic Tool Co.*[73] In *Koehring,* the Seventh Circuit in 1966 held a patent invalid on the ground of a premature public use. The patent was on a prefill valve mechanism that was part of a much larger molding machine. It was tested in a showroom facility and demonstrated to customers without a requirement for secrecy.

The prototype was shipped to a customer, again allegedly for testing by the company's personnel, one year and three weeks before filing the patent application. The court held that the use amounted to commercial activity in that it was an attempt to obtain orders and an advantage over competitors by a demonstration to customers. Many, if not all, who viewed the molding machine were not privy to the specifics of the prefill valve mechanism that constituted the invention. However, that factor did not appear to carry weight with the court. This is an example of a situation where what might have been a proper initial expermental use was unable to successfully maintain its status during commercial activity.

---

[70] *See* Koehring Co v National Automatic Tool Co, 362 F2d 100, 103-04, 149 USPQ 887 (7th Cir 1966); Minnesota Mining & Mfg Co v Kent Indus, Inc, 409 F2d 99, 101, 161 USPQ 321 (6th Cir 1969).

[71] Minnesota Mining & Mfg Co v Kent Indus, Inc, 409 F2d 99, 101, 161 USPQ 321 (6th Cir 1969). For good survey of the issue in a contest where the equities seemed to favor the losing applicant for a patent, *see In re* Theis, 610 F2d 768, 204 USPQ 188 (CCPA 1979).

[72] 97 US 126 (1877).

[73] See note 60 *supra.*

Another important issue that should be addressed is what happens to the early stage devices that serve the primary purpose of providing information to direct the rest of the development but which might continue to be used for whatever value they have. Obviously legal prudence, ignoring economic considerations, dictates that these devices be scrapped. These devices, however, may be expensive and have some continuing commercial value. It is to be hoped that a continued use of a device that once clearly served an experimental purpose will be recognized by the courts as a use incidental to the experimental exception to the commercial use bar.[74]

In a new enterprise, a difficult allocation of funds determination has to be made. Manifestly, the best use of funds at the early stage of development is for the technical development of work and not for paperwork. The courts should, it is hoped, recognize the legitimacy of this problem. Activities, experimental or commercial, that are part of a bona fide development effort, and not an attempt to exploit an invention in a significant commercial way prior to filing for patent protection, are to be encouraged to minimize multiple filing and to increase the likelihood that the patent will disclose the better and more useful embodiments. The courts should recognize the purpose of the experimental use exception in a broad and generous way so as to minimize waste and encourage cost-effective decision making.

§4.09 —Multinational Patent Strategy

A patent is territorial.[75] It only provides coverage or protection as to the territory of the country that grants the patent. Thus, it is often desirable to seek foreign patents. There are, however, substantial costs in obtaining and maintaining a multinational patent portfolio.[76] For this reason, it is important to avoid needless filing.

§4.10 —Limiting the Extent of Foreign Filing

In a high technology situation or a specialized technology field, it is often not necessary to have patent protection in a host of smaller or low technology countries. In many countries, anyone in the market for a patented product would have to look to one of the few high technology countries where the product is likely to be made. As long as the high technology countries are covered, there is often little risk of effective competition because of activities in other countries.

In other circumstances, the effective commercial exploitation of an invention

---

[74] See note 67 *supra.*

[75] *See* 35 USC §154.

[76] The costs include translation costs, filing fees, agent's fees, cost of prosecution before each country's patent office and annual taxes or maintenance fees on issued patents and even on pending applications.

requires an appreciable amount of in-plant know-how which is developed with experience in producing a product, particularly where large volume production is concerned. Again, it may only be necessary to obtain patent protection in a limited number of countries to effectively assure dominance in the product involved. These strategies tend to make sense primarily where the patent proprietor is in a position to be commercially active either directly or through licensing.

§4.11 —Public Disclosure Bar

Almost all the technologically advanced countries (Canada is the main exception) require that a patent application be filed prior to any public disclosure of the invention whether the disclosure be through publication, commercial sales or even a showing to potential customers.[77] The specific rule varies from country to country. There is a tendency, however, accelerated with the advent of the European Patent Convention,[78] for convergence in the direction of rules that limit what the inventor can do before filing. The tendency is to define almost any divulgence as a bar to filing. Accordingly, if foreign patenting is contemplated, it is prudent to effect filing prior to the first public or commercial use or disclosure of the invention. If an option to file abroad is desired, the one-year grace period granted in the United States is of no benefit. In most cases, filing must occur before the United States grace period starts.

Immediate filing abroad, however, is not required. There is an international convention signed by almost every industrially significant country (Taiwan is an exception) which provides a one-year grace period after home country filing to effect a filing in a foreign country.[79] As long as the foreign filing is completed within that one year post-home filing period, the foreign filing adopts the home filing date for priority purposes. The result of this is that one can file in the United States prior to the first public or commercial disclosure or use and then wait up to a year to file abroad. Doing so will allow use of the United States filing date as the effective filing date abroad, thereby also complying with the foreign requirement that the effective filing date abroad be prior to the first public or commercial use or disclosure. This gives an inventor a one-year time period in which to assess the importance of the invention and how worthwhile it is to make the investment in multinational patent filing. One year is usually not enough time to make that assessment intelligently, so risks must be taken one way or the other.

---

[77] See note 53 *supra*.

[78] The European Patent Convention established the European Patent Office which is discussed in **§4.12**.

[79] The International Convention for the Protection of Industrial Property (1883); the most recent text of this convention is known as the Paris Convention, adopted at the Conference of Stockholm on July 14, 1967. The text of the convention may be found in 3 Manual for the Handling of Applications for Patents, Design and Trade Marks Throughout the World (Octrooibureau Los en Steiger, Amsterdam, Holland).

A contribution to industrial efficiency could be made if all countries established a two-year grace period in which to effect the initial patent application filing[80] and then provided a further two-year grace period in which to effect a filing of corresponding foreign applications.[81] The patent system, like many other legitimate and necessary government controlled and regulated systems, has been designed without an eye to minimizing costs. Short of a fully international patent system (which may eventually develop in nations with similar economic systems, e.g., market-oriented economies), adoption of a longer grace period would be a worthwhile start.

§4.12 —Selection of Countries and Use of the European Patent Office

The most important countries in which to file, apart from the United States, are usually Japan, Canada, the United Kingdom, Germany, France and Italy. In the case of Canada, proximity and the absence of maintenance fees are a factor which increase the frequency of filing by United States based companies.

The European Patent Office (EPO) has been established as the agent of nearly a dozen European countries, including the above four, to examine patent applications.[82] Once the EPO has completed its examination and has allowed the application, then the applicant may obtain a separate patent in as many of the individual member countries as desired. The applicant has to pay the individual national issue fees and costs of translation, and will automatically receive a patent from the member country. Annual maintenance fees on the patent will be incurred in each country.

Recent experience suggests that the cost of proceeding through the EPO is less than the cost of any three separate filings in member countries and usually is less expensive than the cost of any two such filings. Furthermore, prosecution through the EPO involves a standard of patentability comparable to that applied in the United States, and is neither as strict as that applied in the Netherlands nor as lenient as that applied in the nonexamination countries such as Italy. It must be remembered that at this stage there is no European patent and the EPO does not issue a European patent. It only issues an allowance which can then be converted into a separate patent in any one or more of the dozen or so contracting member countries.

Because of the costs of filing, the costs of prosecution, the issuing fees and the annual taxes or maintenance fees, the cost of maintaining a multinational patent portfolio can be substantial. If this is undertaken without an actual commercial opportunity in the country, it should be recognized as a fairly

---

[80] The United States had a two-year grace period prior to 1940, and Canada still has a two-year grace period.

[81] Filing under the recently adopted Patent Cooperation Treaty provides an extra eight months (for a total of 20 months from home country filing date). See note 87 *infra*.

[82] The countries are: Austria, Belgium, France, Germany, Italy, Luxemburg, the Netherlands, Sweden, Switzerland and the United Kingdom.

expensive option. Once undertaken, it should be monitored on a periodic basis to see if the continuing expense is worth the potential in view of the latest circumstances.[83]

§4.13 —The Disclosure Consideration

The usual foreign rule is that one must file prior to any commercial use or other public disclosure. A complication arises where the commercial use does not divulge the invention to the public. Such a commercial use often occurs where the invention is in an industrial process for making a product and may also occur where there are high technology items. For example the innards of an electronic system may not be available to the public although the equipment is functioning. In a number of countries a use which does not divulge information to the public does not constitute a bar to filing.[84] In other countries unless the public use is within the country involved, it does not constitute a bar.[85]

In most countries, any publication, including publicly available instruction manuals and advertising, will constitute a bar to the filing of the patent application if the publication divulges the invention to the public.

Thus, it may be possible to file in some countries when barred from filing in other countries. Such a possiblity should be remembered where there has been a commercial use in the United States prior to establishing a filing date in the United States. If the commercial use in the United States prior to the filing in the United States has been a confidential or secret use, in that the invention is not divulged to the public, then it may well be possible to take advantage of this situation. The most prudent course is to file in one's home country prior to the first commercial use. But it may be possible where the commercial use in the United States does not divulge the invention to file in the United States during the one year time period after the first nondivulging commercial use and then to file abroad prior to any public divulgence of the invention. This strategy should be available when filing through the EPO and thus might be of appreciable value.

§4.14 —Timing and Planning

Because of the commercial use and publication bars, timing is more critical in establishing patent rights than in establishing other proprietary rights (trade secrets, trademarks and copyrights). Considering only these bars, the prudent rule is to file an application at as early a date as possible. In the United States, at least, one can then file successive so-called continuation-in-part applications

---

[83] For example, the cost of the annual tax in West Germany rises from under $100 in the third year of the patent to just under $1,500 in the twentieth (last) year. Not all countries are as extreme. But even small countries can and do impose substantial annual taxes on a patent.

[84] *E.g.*, Australia and Canada.

[85] *E.g.*, Japan and Australia.

to update the development.[86] However, cost considerations dictate filing as late as possible to minimize the number of continuation-in-part applications and also to simplify the complexity of foreign filing where a succession of continuation-in-part applications exists.

Furthermore, there is the complex problem in meshing the one-year bar rule in the United States; the zero time period bar rule of foreign countries; and the one-year convention period, which provides one year after filing in the United States within which a foreign application must be filed in order to get the benefit of the home country filing date.

If development is viewed as a single invention occurring substantially at one point in time, the interrelationship between these requirements creates a minor problem and results in only limited inefficiencies in forcing premature filing. As is often the case, however, where development involves a series of stages requiring field testing, feedback and experimentation, the situation can become quite cumbersome, and the costs involved, substantial. Prudent early filing in the United States of the parent application (which forms the basis of a series of continuation-in-part patent applications) runs into the considerable problem that a foreign filing can acquire the benefit of the home filing date no further back than one year.[87]

For example, assume a commercial use in the United States of an early design of a development and prudent filing in the United States before the commercial use is under way. Then assume that over a period of two or three years the development is regularly improved and put into commercial use in order to develop experience with the product. A series of continuation-in-part applications are filed to reflect each stage of the development prior to each stage's becoming a commercial use. When the product is final or close to final (from one point of view no product is ever finally developed) and the development has reached a stage which warrants a major market push, then an application is filed abroad. The case as filed abroad will carry a priority date that is no earlier than one year prior to the filing abroad.[88] Thus, early developmental stages might well be prior art to the later design. Depending on what the most critical inventive steps are (it is not always clear what they are ahead of time),

---

[86] The term *continuation application* refers to the refiling of the same case, usually with some change in the claim. A change in the disclosure, usually to update the material, results in a *continuation-in-part* application. *See* 35 USC §120 regarding requirements for maintaining continuity.

[87] Filing under the Patent Cooperation Treaty will provide an extra eight months to perfect the foreign filing and thus delay the costs of foreign filing. Invoking this option will cost under $1,000 prior to the twelfth month after filing. *See* Patent Cooperation Treaty Schedule of Fees under Rule 15, and 37 CFR §1.445. The text of the Patent Cooperation Treaty is contained in 3 Manual for the Handling of Applications for Patents, Design and Trade Marks Throughout the World (Octrooibureau Los in Steiger, Amsterdam, Holland) and is discussed in detail in ch 1800 of the Manual of Patent Examining Procedures (Patent and Trademark Office, US Dep of Commerce).

[88] The PCT filing does not change this requirement. PCT filing just delays for eight months the paying of most of the foreign filing fees.

that could turn out to be fatal to patentability abroad. Alternatively, one could circumvent the problem by making sure that a series of foreign filings are made corresponding to each continuation-in-part United States case. That can be very expensive. Furthermore, each application filed abroad would have to be inventive over the preceding case because the availability of a continuation-in-part approach that exists in the United States is generally not available elsewhere.

In the hurly-burly of development, commercial investigation and the inevitable ignorance of much that is contained in the prior art, it is often difficult to focus on the most significant patentable features.

What this means is that if there is any thought of obtaining patent protection abroad, it becomes particularly useful to plan a patent strategy as early as possible in a development. Where possible, the early commercial uses might be under a specific confidentiality agreement with the customer and include specific requirements for feeding back information to aid in the continuing development.[89] Where that is possible, one may well be able to avoid a lot of expensive intermediate foreign filing at minimal risk.

§4.15 —Risks of Invalidity under the Nonobviousness Requirement—Supreme Court Opinions

During the last 50 years, mechanical patents have fared ill in the United States Supreme Court, though earlier Courts were more favorable.[90] Chemical and electrical patents have fared better.[91]

As to mechanical patents, in *Cuno v Automatic Devices Corp,* the Supreme Court in 1941 held invalid a patent on an improved automibile cigarette lighter. The Court said that this new device, however useful it might be, must reveal the flash of creative genius, not merely the skill of the calling.[92]

In *Great Atlantic & Pacific Tea Co v Supermarket Equip Co,* the Supreme Court in 1950 held invalid a patent on a pusher frame mechanism for guiding and moving groceries on a checkout counter. The Court stated that at least in the

---

[89] **§§4.07** and **4.08.**

[90] *E.g.,* Hildreth v Mastoras, 257 US 27 (1921), concerned a candy-pulling machine. The Court through Justice Taft stated at 34:The history of the art shows that Dickinson took the important but long delayed and therefore not obvious step from the pulling of candy by two hands guided by a human mind and will, to the performance of the same function by machine. The ultimate effect of this step with the mechanical or patentable improvements of his device was to make candy pulling more sanitary, to reduce its cost to one-tenth of what it had been before him, and to enlarge the field of the art. He was, therefore, a pioneer.

[91] *See* Goodyear v Ray-O-Vac, 321 US 275, 279 (1944) (a leakproof drycell for a flashlight battery); United States v Adams, 383 US 39, 51-52 (1966) (electric battery using plain or salt water electrolyte); Graver Tank & Mfg Co v Linde Air Prod Co, 336 US 271 (1949) and 339 US 605 (1950) (a flux for electric welding).

[92] 314 US 84, 91 (1941).

area of mechanical matters it would be unusual to provide a combination of parts that produces the new quality or function required to result in a patentable invention.[93]

It was hoped by many that the explicit adoption by Congress of the nonobviousness test in 1952 would put to rest the doubts spawned by the *Cuno* case in 1941 and reinforced by the *Great Atlantic & Pacific Tea* case in 1950.[94] Although the nonobviousness test was generally interpreted as codifying the previous requirement for *patentable invention,* it was also intended as a signal from Congress to the judiciary to be more lenient toward validity. The United States Supreme Court did not take the hint. Rather, in three fairly recent cases, the Court has reversed the lower courts and held several mechanical patents invalid.

In the landmark case of *Graham v John Deere Co.,*[95] the Court in 1966 held invalid a patent on a chisel plow in which the novel feature was reversal of a shank and hinge plate. This reversal of components resulted in improved ability to take shock as the chisel is driven through rocky ground. *Graham* is a landmark case because it set down a series of subtests for nonobviousness and made a distinction between primary considerations for determining obviousness and secondary considerations.[96]

In *Anderson's Black Rock v Pavement Salvage Co,*[97] the Court in 1969 held invalid a patent on a road-paving technique involving a novel combination of an asphalt spreader and a radiant heater for preventing cold seams. This case is notable for producing the term *synergism* as a test of nonobviousness. This concept that the whole must in some way exceed the sum of its parts has introduced further debate, much like the "flash of creative genius" expression introduced in the *Cuno* case in 1941. It is not clear whether the Court was indicating (a) that synergism is a prerequisite to a determination of nonobvi-

---

[93] 340 US 147, 151, 152 (1950).

[94] *See* Graham v John Deere Co, 383 US 1, 15-17 (1966). The nonobviousness test is incorporated in 35 USC §103. See discussion in **§4.01.**

[95] 383 US 1 (1966).

[96] 383 US 17-18. As Judge Markey pointed out in *Some Patent Problems,* 80 FRD 203, 211-12 (1979), the five secondary considerations set forth by the *Graham* court are more properly considered evidentiary factors which are significant in determining the issue of obviousness. The five significant evidentiary factors are:

1. A long unfilled need that people were seeking to fill without success and which the invention filled.
2. Prompt and significant commercial success not attributable to a marketing or advertising push.
3. Licenses sought by others.
4. The infringer copied the invention after the infringer was unable to solve the problem.
5. The acclaim accorded to the invention, particularly by others in the same technical or commercial field.

[97] 396 US 57 (1969).

ousness, or (b) that if synergism is found, then clearly nonobviousness would be met. Under the latter interpretation it could be possible for the nonobviousness requirement to be met without the finding of synergism. If the concept of synergism is to be deemed congruent with the nonobviousness criteria of the patent statute,[98] it introduces confusion without enlightenment. If it refers to a subset of circumstances under the statute, it may have some value. However, it is often taken to require a more stringent standard than was intended by Congress in its enactment of the nonobviousness standard.[99]

The third case is *Sakraida v Ag Pro*[100] in which the Court held invalid a patent on a waterflush system for cleaning out a dairy barn. The system involved the abrupt release of a reservoir of water so as to cause a sheet of water to flow across the floor to provide a rolling and cleaning action. Obviously the Court was not impressed that this technique warranted a patent.

In all three cases, useful, novel technology was provided that seemed simple and obvious enough once the technology was presented. All three inventions involved a novel and commercially useful advance in very old and familiar arts. None of these inventions involved the situation where the invention was awaiting the advent of some new technology to make the invention possible. For example, if it were obvious that a reversal of shank and hinge plate would be desirable but for some reason could not be provided until improved materials became available, then the invention on the heels of such new material would fail to meet the nonobviousness criteria.[101] But there was no such reason as to why the invention in *Graham* had not been made earlier.

Accordingly, in the mechanical arts at least, one must recognize that patent filing, although clearly of real value, does not guarantee that actual protection will be afforded.

If a patent is issued on something which is not novel (that is, which was known in the art prior to the patentee's invention), then the patent is invalid.[102] That is a normal business risk. To have a patent, though, on a a genuine contribution to technology deemed invalid because it looks simple is a deterrent to the exploitation of innovative ideas. In an era of increased recognition of the importance of innovation, perhaps the courts will be more likely to find patentable invention in mechanical innovation.[103]

For example, in all three of the above cases, there was prior art technology available for everyone to use, which performed the same general function as did the patented invention and which was only slightly less desirable or only

---

[98] 35 USC §103. See **§4.01.**

[99] Brennan v Mr Hangar, Inc, 479 F Supp 1215, 203 USPQ 697 (SDNY 1979); Smith v Acme Gen Corp, 614 F2d 1087, 1093, 204 USPQ 1060 (6th Cir 1980). *But see* Republic Indus, Inc v Schlage Lock Co, 592 F2d 963, 967-72; 200 USPQ 769 (7th Cir 1979).

[100] 425 US 273 (1976).

[101] See **§4.01.**

[102] 35 USC §102; see **§4.01.**

[103] *See* Judge Frank's comments in Picard v United Aircraft Corp, 128 F2d 632, 638, 53 USPQ 563 (2d Cir 1942), especially at 642-43.

slightly less efficient. Manifestly, the grant of a patent did not create any burden on these technologies. The holding of each of these patents as valid would not have provided the patent owner with a monopoly over an entire industry. The patent owner would only have been able to extract a fraction of any benefit of improved efficiency.

A patent on an improved product or technology or method of manufacture provides only a limited advantage to the patent holder, an advantage limited by the extent to which the improvement is more efficient or effective.

When the Courts strike down patents, the opinion is frequently expressed that an invalid patent or a patent on a slight improvement is a drag on the economy.[104] One is left with the impression that there is concern that patents on what appear to be relatively slight improvements have a detrimental impact while patents on what may be deemed to be significant improvements have a beneficial impact. Yet the slighter the improvement, and the less the benefit provided, the less can the patentee extract as a reward for the invention because the public is always free to copy the art or technology as it existed prior to the invention.

§4.16 —Synergism

The launching of this synergism standard by the U. S. Supreme Court in these recent three cases over the past ten years has generated more controversy than any other recent development in patent law. When the statute was amended in 1952 to add the explicit nonobviousness standard for patentable invention, it was hoped that this standard and the meaning given to this standard by the development of case law under the standard would provide as much of a legal criteria as could be expected in assessing the level of invention as warranting the issuance of a patent. The standard of a reasonable man's action in the law of negligence and the standard of the intent of the parties in the law of contracts establish broad standards by which individual events or agreements are evaluated or construed. So too it was hoped that the nonobviousness standard would perform the same function in the patent law. The nonobviousness standard stands for the philosophy that something more than mere novelty is required to warrant the issuance of a patent. But that something more cannot be explicitly defined for all situations. However, the principle is that there should be something which would not have been obvious at the time the invention was made to those skilled in the art knowing the prior art as it existed at that time. It is probably not possible or desirable to be more concrete or explicit. The philosophy behind the nonobviousness standard is clear enough to be applied to each situation as it arises. It requires an exercise of judgment. Any attempts to make that standard somehow more explicit result in a confonding of that judgment.

---

[104] *See* Sears Roebuck & Co v Stiffel Co, 376 US 225, 230 (1964).

Accordingly it would seem unwise to launch a synergism standard or any other type of standard in an attempt to explicate non-obviousness.

The actual statement in *Anderson's-Black Rock* v *Pavement Salvage Co* is:

> The combination of putting the burner together with the other elements in one machine, though perhaps a matter of great convenience, did not produce a "new or different function" . . . within the test of validity of combination patents. A combination of elements may result in an effect greater than the sum of the several effects taken separately. No such synergistic result is argued here.[105]

On its face, that language does not state that a synergistic result must be found in order to meet the nonobviousness standard of the statute. It implies that if a synergistic result were found, then the standard would be met. This does not mean that a synergistic result must be found. Furthermore, on its face, the language indicates that there is identity between the phrases "synergistic result," "the sum of the several effects taken separately" and "new or different functions." As a matter of doctrine, all three phrases can be taken as different ways of articulating situations in which nonobviousness is to be found. It does not follow that they define the borderline of nonobviousness. Part of the problem with these three phrases is that they are employed in assessing the validity of combination patents as if there were some objective difference between a combination patent and some other kind of patent. Unless it be a novel cam or a new chemical element, it is hard to imagine what could not be designated as a combination patent.[106] Accordingly, any attempt to deal with combination patents more strictly than other types of patents leads to treating virtually all inventions in accordance with the stricter criteria and subverts the intent of Congress to subject all inventions to the nonobviousness standard. In *Hotchkiss v Greenwood,*[107] the leading case for formulating the standard that patentability requires something more than novelty, that standard was stated in terms of the following holding:

> for unless more ingenuity and skill in applying the old method of fastening the shank and the knob were required in the application of it to the clay or porcelain knob than were possessed by an ordinary mechanic acquainted with the business, there was an absence of that degree of

---

105 396 US 57, 60-61 (1969).

106 In BG Corp v Walter Kidde & Co, 79 F2d 20, 22, 26 USPQ 288, 289 (2d Cir 1935), Judge Learned Hand stated:

> All machines are made up of the same elements; rods, pawls, pitmans, journals, toggles, gears, cams, and the like, all acting their part as they always do and always must . . . . But the elements are capable of an infinity of permutations, and the selection of that group which proves serviceable to a given need may require a high degree of originality. It is that act of selection which is the invention; and it must be beyond the capacity of common-place imagination.

107 11 How (52 US) 248, 266 (1851).

> skill and ingenuity which constitute essential elements of every invention. In other words, the improvement is the work of the skilled mechanic, not that of the inventor.[108]

This holding is quoted with approval by the United States Supreme Court in its 1966 *Graham v John Deere* decision, the *Graham* decision being presented as essentially a clarification of the *Hotchkiss* test.[109] The *Graham* court stated that the major distinction between then and now is that since 1952, Congress has emphasized nonobviousness as the operative test in order to avoid the imprecision of the *invention* language.[110] The *Graham* court did recognize that the *Hotchkiss* formulation (or any other formulation) lies not in any label but in its functional approach to questions of patentability.[111] Thus, the Court identified the *Hotchkiss* standard as the statutory standard.

In *Anderson's-Black Rock,* the Court continued to identify the *Hotchkiss* standard with the nonobviousness standard[112] and never suggested that language such as "new or different function," "an effect greater than the sum of the several effects taken separately" or "synergistic result" meant a stricter standard than that set down in *Hotchkiss* and the statute.

The language in *Sakraida v Ag Pro* is perhaps the most troubling of all:

> We cannot agree that the combination of these old elements to produce an abrupt release of water directly on the barn floor from storage tank pools can properly be characterized as synergistic, that is "resulting in an effect greater than the sum of the several effects take separately." . . . [R]ather this patent simply arranges old elements with each performing the same function it had been known to perform, although perhaps producing a more striking result than in previous combinations.[113]

If the combination did indeed produce a more striking result than in previous combinations, that would certainly suggest, although perhaps not be determinative of, nonobviousness. In any case, this language seems to treat synergism or "effect greater than the sum of the parts" as the standard, although, strictly speaking, the paragraph simply negates the finding of "synergistic result" by the circuit court. Recent circuit court decisions suggest that synergism may not be used to install a standard stricter than nonobviousness.[114]

---

[108] *Id* 267.

[109] 383 US 1, 15, 17 (1966).

[110] *Id* 14, 16-17

[111] *Id* 12.

[112] 396 US at 62.

[113] 425 US 273, 282 (1976).

[114] Republic Indus, Inc v Schlage Lock Co, 592 F2d 963, 200 USPQ 769 (7th Cir 1979): *see,* in particular, the thorough discussion at 967-72, Champion Spark Plug Co v Gyromat Corp, 603 F2d 361, 372 202 USPQ 785 (2d Cir 1979); Plastic Container Corp v Contenental Plastics of Oklahoma, Inc, 607 F2d 885, 904-05, 203 USPQ 27 (10th Cir 1979); Rengo Co Ltd v Molins Mach Co, 657 F2d 535, 211 USPQ 303 (3d Cir 1981); Sarkesian v Winn-Proof Corp, 662 F2d 596, 213 USPQ 912 (9th Cir 1981).

Although criteria of patentability are always ultimately subjective, selection of the terminology used not only reflects the nature of the judgment, but also affects the nature of the judgment.

Where the criterion "flash of creative genius" defines patentability, one will deem few inventions patentable and few patents valid. On the other extreme, where novelty or substantial novelty is the criterion for patentability, a great range of inventions will be deemed patentable.

The criteria "flash of creative genius," "synergism," "nonobviousness" and "substantial novelty" can all be worked over to produce the same operative results. However, there is no objective reality that gives these terms different meanings. There is no objective measure or objective category which can be imported to provide objective distinction between these criteria. However, it is perfectly clear that there is a subjective difference. Those who select terms such as "flash of creative genius" or "synergism" in order to create mandatory criteria for patentability are saying that they do not want many patents. Those who select "substantial novelty" as the criterion for the granting of a patent are saying that they want a lot of patents to be granted. The criterion of nonobviousness, which is the criterion of the patent statute, is subjectively a middle ground. Those who are determined to kill off patents will find it no deterrent. Those who are determined to grant a patent on as many inventions as possible will also find it little or no deterrent.

This exegesis on synergism and the like is set forth because one must deal with these doctrines. One should recognize that the underlying reality is not one of doctrine but the extent of a court's understanding and sympathy or suspicion and fear of the effects of the patent grant.

§4.17 —Evaluation of Patentability and Identification of the Patentable Feature

Many observations have been made as to what constitutes nonobviousness.[115] The courts and commentators have attempted to provide guidelines

---

[115] For a discussion of the nonobviousness requirement in patentability, see **§4.01.** One of the best observations as to nonobviousness is that of Judge Hand in Reiner v I Leon Co, 285 F2d 501, 503-04, 128 USPQ 25 (2d Cir 1960), where he stated:

> The test laid down [for obviousness] is indeed misty enough. It directs us to surmise what was the range of ingenuity of a person "having ordinary skill" in an "art" with which we are totally unfamiliar; and we do not see how such a standard can be applied at all except by recourse to the earlier work in the art, and to the general history of the means available at the time. To judge on our own that this or that new assemblage of old factors was, or was not, "obvious" is to substitute our ignorance for the acquaintance with the subject of those who were familiar with it. There are indeed some sign posts; e.g. how long did the need exist; how many tried to find the way; how long did the surrounding and accessory arts disclose the means; how immediately was the invention recognized as an answer by those who used the new variant.

See also the discussions in Judge Rich's article, *Monopolies, "Invention" and Section 103,*

and subtests to aid in the analysis. The Supreme Court has set forth what it considers to be primary considerations and secondary considerations for addressing the issue.[116]

Much of this is of little or no value to aid an inventor or entrepreneur in determining which inventions are more likely to meet the criteria for obtaining a valid and useful patent. The fact that a structure is complex is not a good indication of patentability because complexity can be built through fairly straightforward addition of a large number of parts to perform a large number of functions without requiring more than ordinary engineering ability. In addition, the fact that a particular structure or arrangement or combination is unusual is generally not too reliable an indication that it is patentable, because all that is required to negative patentability is a showing that somewhere in the prior art such a structure was known or that it is an obvious development or derivation from some known structure.

Consideration of a structure or method in terms of the problem it solves is more likely to provide some feeling for its patentability. The nonobvious step taken by an inventor is more likely to be identified by considering: (a) the problem the inventor has addressed, how the inventor has solved it, and what the inventor has done to effect the solution; or (b) the source of the problem that was addressed, whether the inventor recognized the source of that problem and developed a technique for solving the problem.

Thus, it is in the problem solution rather than in the artifact which solved the problem that patentable invention is most likely to be found.[117] This suggested approach should not mask the fact that the patent, and in particular the patent claims which define the legal field of exclusivity, must relate to artifact, to structure or to steps of a method.

It is often thought there is some sort of rough correlation between degree of structural uniqueness of a device and the quality or level or height of invention involved. Thus, it is often thought that the more likely valid patent is one that represents the more unique structural embodiment. This misses the fact that the greater and most useful insights are often those which provide simple solutions or simple structures, or see to the heart of the problem.[118] The elegance or beauty of the solution, inherent in its grand simplicity, may enhance recognition of nonobviousness and hence, patentability.

Once one has understood the nature of the problem being addressed, the consequent solution is often pretty straightforward. There is patentable inven-

---

in The Encyclopedia of Patent Practice And Invention Management 592-602 (Calvert ed 1964); Galston, *Invention And The "Obvious"*, 13 FRD 463-70 (1953); and Markey, *Some Patent Problems,* 80 FRD 203-18 (1979).

[116] Graham v John Deere Co, 383 US 1, 17-18 (1966).

[117] Eibel Process Co v Minnesota & Ontario Paper Co, 261 US 45, 68 (1923).

[118] *In re* Osplack, 195 F2d 921, 923-24, 93 USPQ 306 (CCPA 1952); Goodyear Co v Ray-O-Vac Co, 321 US 275, 279 (1944); Reeves Instrument Corp v Beckman Instruments, Inc, 161 USPQ 450, 480 (CD 1968) *affd,* 444 F2d 263 (9th Cir 1971); Manville Boiler Co v Columbia Boiler Co, 269 F2d 600, 604, 122 USPQ 545 (4th Cir 1959).

tion, however, in discovering the source of a problem. For example, one reason for weakness in weld beads is the existence of minute gas pockets which cause local stress concentration. Two inventors discovered that a cause of the porosity is the occurrence of minute amounts of organic particles distributed in the base metal. Under the high heat of welding these organic particles decompose and form gas pockets. They then found that the inclusion of metals, such as titanium and zirconium, which have a great affinity for oxygen and nitrogen, tends to solve the problem. Accordingly, they claimed a welded stainless steel article having a weld which included minute amounts of these getter elements. The Patent Office examiner and the Patent Office board of appeals rejected their claim on the ground that (a) the article was known to include titanium and zirconium in stainless steel, and (b) these elements had been previously known as *getters* with an affinity for oxygen and nitrogen. The Patent Office argued that the prior art therefore suggested using titanium or zirconium as a getter for nitrogen and that it was obvious to incorporate small amounts of such elements in the weld material in order to absorb the generated gases and in particular in order to absorb the nitrogen.

The Court of Customs and Patent Appeals reversed and held that the critical factor is that none of the prior art references recognized the cause of the problem of porous welds faced by these two inventors and, thus, the prior art did not suggest the solution which they sought to patent.[119]

It may well have been that once the cause of the problem of weld porosity was understood, one skilled in the art would have found it obvious to develop the solution of adding getter elements. The court, however, took the approach that one has to look at the subject matter as a whole and determine what would have been obvious at the time the invention was made. One then has to include a recognition of the cause of the weld bead porosity problem as part of that subject matter. The court concluded that the essence of the invention was in the determination of the cause of weld porosity and that since this determination was an inseparable part of the invention as a whole, it made the invention nonobvious and thus patentable.

§4.18 —The Many Natures of Invention

The major changes in history have never been predicted beforehand. Likewise, in technology, future developments are impossible to foretell. Further, even where certain trends appear highly likely to continue, one can never be sure when an occurrence will happen. For example, it is clear that the cost of memory and the physical size of memory in computers will continue to dramatically decrease. It is hard to predict, though, with any reasonable certainty when, for example, the cost will drop to 20 per cent of present levels.

In technology, the invention of the massproduced automobile was not predicted, much less its timing. Once invented, the impact of the mass-produced

---

[119] *In re* Linnert, 309 F2d 498, 135 USPQ 307 (CCPA 1962).

automobile was also not predicted. There is a lot of looking backward and selection of those guesses as to the future which best fit what in fact happened. This selective history leaves the impression that there was a basis for knowing what was going to happen. In fact, since almost everyone is constantly speculating about the future, some speculations are bound to turn out to be right. It is a rare prognosticator, however, who hits right on more than one major prediction.

This is not to say that the near future cannot be predicted in certain respects and, perhaps more importantly, controlled. Once the nature of intentional control is known, moderately decent predictions can be made on a near-term basis.

Thus, very short-term trend lines can be projected, and as long as they are continuously recomputed, They can be relied upon for short-term purposes.[120] Relying upon short-term trend lines tends to make them come true in many, but not all, circumstances. For example, technological development is in many respects hard to project. Technological development depends upon coping with the constraints of the real world. Unlike human society, the real world is uncaring about our goals and plans to achieve them. If a law is passed that is generally popular, the vast majority of people will obey the law and a mode of social behavior, caused by the law, will result. This will occur because it is desired, and in a sense, because it is desired and steps are taken to make it occur, the prediction that it will occur will come true in many circumstances.

By contrast, nature does not follow our laws. We try to impose heuristic laws on nautre but it does not change the way in which reality operates. Anyone who has attempted to develop a new product learns that the most reasonable projections as to how materials and mechanisms will operate cannot be relied on. Every minute step must be tested out. This is a long and tedious process. After one has developed the new product, it is usually perfectly clear why it operates and perfectly clear that the combination of elements and material selected were the ones that should have been selected. It is, nonetheless, remarkable how difficult it is to see this perfectly clear solution ahead of time.

Explanations as to why an invention operates and how it operates always give the impression that the invention is a simple and straightforward application of known concepts. Before the invention has been made, though, that straightforward application is never clear. There are a number of reasons for this. The major reason is that the explanation after the fact only appears to be a complete explanation. It in fact is not. It is only an explanation of the gross efects. There is always much more to be solved than ever appears in an after the fact explanation.

A second reason is that there are usually close to an infinite number of choices one can make ahead of time in the solving of a particular problem. Once a useful or near optimum solution has been devised, its very existence reifies it above all of the hypothetical explanations. Its existence gives it sub-

---

[120] For splendid discussion of the nature, limitations and problems with predicting, see Schumacher, Small is Beautiful chapter entitled *A Machine To Tell The Future* (1973).

stance. Its existence gives it visibility. It appears so much more obvious than all of the other possible solutions simply because it now exists.

Doing something in a simpler fashion than before may be a nonobvious and thus a patentable step. Optimizing conflicting parameters is sometimes patentable subject matter. Invention, whether patentable or not, has many natures. Following is a list of concepts that may aid in indentifying invention and determining patentability. One should remember, however, that the identification of the problem and its solution is the key to thinking about patentability.[121] In that context the following can aid in more precisely identifying the inventive features. Because there are many natures of invention, invention can be found in:

1. Bucking the teaching in the art[122]
2. Ignorance (especially in not knowing and not blindly following tried and true ways)
3. An accidental discovery[123]
4. The tension between understanding that "there is nothing new under the sun" and understanding that "you can never step into the same river twice"
5. Recognizing that a small technological step or small structural change often provides a significant or large functional difference[124]
6. Simplicity[125]
7. Optimization[126]
8. New use of a known artifact[127]
9. New use of a known or obvious chemical[128]
10. Marked improvement in result[129]

---

[121] See **§4.17.**

[122] United States v Adams, 383 US 39, 51-52 (1966); White v Ma-Bel, Inc, 509 F2d 287, 185 USPQ 129 (5th Cir 1975); Ingersoll-Rand Co v Brunner Lay, Inc, 474 F2d 491, 496; 177 USPQ 112 (5th Cir 1973).

[123] The patent statute at 35 USC §§103, 161 specifies that the manner by which the invention came about shall not adversely affect patentability.

[124] Carnegie Steel Co v Cambria Wire Co, 185 US 403, 445-46 (1902); Rosen v Lawson-Hemphill, Inc, 399 F Supp 532, 187 USPQ 158 (DRI 1975).

[125] See note 118 *supra.*

[126] *In re* Legator, 352 F2d 377, 147 USPQ 322 (CCPA 1965); *Ex parte* Kahn, 132 USPQ 359 (POBA 1961); *In re* Pearson, 494 F2d 1399, 181 USPQ 641 (CCPA 1974).

[127] *In re* Cox, 342 F2d 1017, 145 USPQ 128 (CCPA 1965); *In re* Foster, 438 F2d 1011, 169 USPQ 99 (CCPA 1971).

[128] *In re* Papesch, 315 F2d 381, 137 USPQ 43 (CCPA 1963); Commissioner of Patents v Deutsch, 397 F2d 656, 157 USPQ 549 (DC Cir 1968).

[129] Carnegie Steel Co v Cambria Iron Co, 185 US 405, 445-46 (1902); Rosen v Lawson-Hemphill, Inc, 339 F Supp 532, 187 USPQ 158 (DRI 1975).

11. Determining critical proportions or size[130]
12. The small final step that creates the practical product; for example:
    a. barbed wire[131]
    b. the Bell telephone[132]
    c. the Rust brothers spindle[133]

§4.19 —Licenses

The licensing of a development is often important in order to get maximum return on the development investment. Licensing is treated in detail in a number of useful works which provide caution against misuse and antitrust violations as well as in works which provide a range of useful licensing clauses.[134]

---

[130] Beckett v Coe, 98 F2d 322, 336, 38 USPQ 26 (DC Cir 1938); *Ex parte* Moore, 87 USPQ 378 (POBA 1950); *In re* Pearson, 494 F2d 1399, 181 USPQ 641 (CCPA 1974).

[131] In The Barbed Wire Patent, 143 US 265, 282-83 (1892), the Court stated:

> It is true that the affixing of barbs to a fence wire does not apparently give a wide scope to the ingenuity of the inventor; but from the crude device of Hunt to the perfected wire of Glidden, each patent has marked a step in the progress in the art. The difference between the Kelly fence and the Glidden fence is not a radical one, but slight as it may seem to be, it was apparently this which made the barbed-wire fence a practical and commercial success . . . . Under such circumstances courts have not been reluctant to sustain a patent to the man who has taken the final step which has turned a failure into a success. In the law of patents it is the last step that wins. It may be strange that, considering the important results obtained by Kelly in his patent, it did not occur to him to substitute a coiled wire in place of the diamond shaped prong, but evidently it did not; and to the man to whom it did ought not be denied the quality of the inventor. There are many instances in the reported decisions of this court where a monopoly has been sustained in favor of the last of a series of inventors, all of whom were groping to obtain a certain result, which only the last one of the number seemed able to grasp.

[132] *See* The Telephone Cases, 126 US 1, 531, 538-44 (1887). It is remarkable how close the prior art was to the telephone patented by Alexander Graham Bell. Everyone understood in principle that speech could be transmitted by a properly fluctuating current (*id* 538). Philip Reis provided a product which came close to anticipating Bell's appartatus (*id* 539) and which reproduced musical tones. As the Court stated, "he could sing through this apparatus, but he could not talk." (*Id* 540.) Apparently, Reis employed a intermittent current whereas Bell employed an undulating unbroken current (*id* 544) and that was the step that made all the difference. The Court seemed to assume that if Bell's mode of treating the electronic current were applied to the Reis apparatus, the Reis apparatus, would transmit speech (*id* 539, 540). The Bell patent was deemed to apply to any apparatus in which this undulating current principle was employed (*id* 539).

[133] The use of a smooth, moist spindle was the step that created the first useful spindle cotton picker. Scientific American 55-57 (Aug 1967).

[134] For a particularly useful book, see Nordhaus, Patent License Agreements. *See also* Nordhaus, Patent-Antitrust Law.

In general, the kinds of clauses which a licensee would find unacceptable are the kinds of clauses which are likely to be a patent misuse or an antitrust violation. Misuse may invalidate the provision or cause a court to refuse to enforce the patent being misused. Antitrust can result in graver penalties. Much of what is improper comes under the rubric of "tie-in" or "tie-out." One cannot tie to a license agreement restrictions on subject matter that is not the subject matter being licensed. For example, restrictions which require the purchase of a non-patented product from the licensor are generally improper.[135] However, technological necessity may validate such restrictions in rare cases.[136] Restrictions which limit the field of activity of the licensee with respect to unpatented products are generally improper.[137] Furthermore, in general, restrictions which require consultation between the licensee and licensor to include or exclude a third party in the licensing arrangement are also improper.[138] By contrast, restrictions which limit the scope of the license granted by limiting the quantity of products licensed,[139] the geographical area in which the license is effective[140] or the commercial field of the license grant[141] are generally upheld.

Although it does not answer every question, it is useful to consider the scope of the patent grant as defining the maximum that can be granted to the licensee. Thus, any portion of those patent rights may normally be granted or restricted. However, if the licensor were to condition that grant on anything other than appropriate payment (and those conditions necessary to assure payment), patent misuse or antitrust violation is a real risk.[142] Even the seem-

---

135 BB Chem Co v Ellis, 314 US 495 (1942); Leitch Mfg Co v Barber Co, 302 US 458 (1938).

136 Dehydrating Process Co v AO Smith Corp, 292 F2d 653 (1st Cir), *cert denied,* 368 US 931 (1961); United States v Jerrold Elecs Corp, 187 F Supp 545 (ED Pa 1960), *affd per curiam,* 365 US 567 (1961). *See also* Pick Mfg Co v General Motors Corp, 80 F2d 641 (7th Cir 1935).

137 See 135 *supra.* For an interesting case illustrating the fact that the limits of the patent right automatically limit the scope of the grant and of the control that can be exercised through the grant, *see* Munters Corp v Burgess Indus, Inc, 194 USPQ 146 (SDNY 1977) and 201 USPQ 756 (SDNY 1978).

138 United States v Krasnov, 143 F Supp 184, 110 USPQ 411 (ED Pa 1956), *affd per curiam,* 355 US 5 (1957); United States v Besser Mfg Co, 96 F Supp 304 99 USPQ 421 (ED Mi 1951), *affd,* 343 US 444 (1952).

139 United States v EI DuPont de Nemours & Co (cellophane), 118 F Supp 41, 99 USPQ 462 (D Del 1953), *affd,* 351 US 377 (1956).

140 35 USC §261. Brownell v Ketchum Wire Mfg Co, 212 F2d 121, 100 USPQ 338 (9th Cir 1954); Ethyl Gasoline Corp v United States, 309 US 436, 456, 44 USPQ 614 (1940).

141 Munters Corp v Burgess Indus, Inc, 194 USPQ 146 (SDNY 1977) and 201 USPQ 756 (SDNY 1978) provides a useful comparison of field of use licensing involving manufacture and sale and field of use licensing involving resale and use.

142 For a less restrictive view, see the remarks of Abbott B. Lipsky, Jr., Deputy Assistant Attorney General, Antitrust Division, given before the ABA Antitrust Section in Washington, D.C. on Nov 5, 1981.

ingly innocuous condition of requiring the licensee to concede patent validity is unenforceable.[143]

Where commercial necessity or a legitimate desire to maintain a quality level in a licensed product can justify restrictions or constraints in the license that otherwise might be suspect, it is advisable to employ the least restrictive constraint that will achieve the purpose.[144]

For example, if technological necessity requires that the licensee employ a licensor's nonpatented product in association with the patented device, that requirement should be limited to the time necessary for the licensee to develop the know-how or ability to act without the nonpatented product. The least restrictive alternative approach often requires limiting the time of the restriction.

Many license constraints are acceptable to obtain a legitimate purpose. Too often, though, a licensor will use a legitimate purpose as an excuse for excessive restrictions upon the licensee. The best guide for developing a licensee agreement that will run the minimum risk of being found to have unlawful restrictions is to use the least restrictive alternative for obtaining a particular legitimate type of object.

A licensor with a network of patents in a number of foreign countries may wish to grant an exclusive license under the patent in each country to a separate licensee. This result can be achieved by granting an exclusive license which is geographically limited to one or more specified countries in which a patent has issued.[145] Indeed, a license may even be geographically limited within the United States.[146] It is inadvisable, however, to include a clause which requires the licensee to agree to refrain from practice of the invention in a geographical area outside of the area of the grant. Although such a clause may be comforting to the licensor, there is no need for such a clause. Any valid patent rights outside of the area of grant would be and should be effective to bar the licensee from practicing the invention where patent rights exist outside of the area of the grant. It would be improper to try to bar the licensee from practicing the invention where patent rights do not exist, have been held invalid or of limited scope so as not to cover the licensee's activities, or have expired.[147]

The basis for licensee payments to licensor can be more flexible than the restraints which a licensor can impose on a licensee. However, royalties cannot in general be measured by unrelated or total income of the licensee.[148] Often, the patented item is a minor but essential component of a much larger commercial product which is sold as an entity. The license payment can be calculat-

---

[143] Lear Inc v Adkins, 395 US 653 (1969).

[144] See note 136 *supra.*

[145] Dunlop Co Ltd v Kelsey-Hayes Co, 484 F2d 407, 417-18, 179 USPQ 129 (6th Cir 1973).

[146] See note 140 *supra.*

[147] *See, e.g.,* Brulotte v Thys Co, 379 US 29 (1964); Timely Prods Inc v Costanzo, 465 F Supp 91, 201 USPQ 567 (D Conn 1979).

[148] Zenith Radio Corp v Hazeltine Research Inc, 395 US 100, 133-41 (1969).

ed as a negotiated fraction of the sales price of the entire equipment in which the patented product is employed.[149] There should be no difficulty in tying payment to whatever cost base is convenient or reasonable. However, such an action when designed to effectively force the licensee to purchase unpatented goods from the licensor would be improper.

One frequently used method which can readily lead to abuse is the requirement of a grantback from the licensee of all improvements. The risk of blocking patents or blocking technologies developed by a licensee provides a legitimate basis for permitting some grantback provisions in license agreements. It is legitimate to prevent a blocking situation in which the best technology cannot be employed by either party. All that is required is a nonexclusive grantback on those improvements which directly relate to the licensed subject matter. More extensive grantback clauses may have the appearnace of an attempt to use an initial patent position to force related technological development into the hands of a single entity.[150]

§4.20 —Trade Secrets

There are basically two ways to obtain proprietary rights over technological information. One is to disclose it and obtain a patent. The other is to use it in a fashion that keeps its secrecy without a patent, and thus establish a trade secret. Except for the one-year grace period prior to filing a patent application and for the time when an application is pending, these two approaches are mutually exclusive.[151] The choice of which approach to take with respect to a particular development usually has to be made at an early stage. Of course, there are many developments which are not subject to trade secret protection because the commercialization of the development will automatically disclose it to the public. There are many other developments, however, involving industrial know-how which are unlikely to result in a valid patent and thus, as a practical matter, can only be held as a trade secret. Where a choice can be made, the balance has to be struck between those factors favoring the trade secret approach and those factors favoring the patent approach.

In brief, the considerations favoring a trade secret approach include: (1) the avoidance of cost in preparing and prosecuting a patent, (2) the avoidance of risk that any patent might be held invalid in litigation, and (3) the possibility that substantial cost may be involved in enforcing the patent against an infringer.

The considerations in favor of the patent approach include: (1) the difficulty of keeping the trade secret a true secret, and (2) the likelihood of a com-

---

[149] American Optical Co v New Jersey Optical Co, 58 F Supp 601, 65 USPQ 114 (D Mass 1944).

[150] *See* Transpatent-Wrap Mach Corp v Stokes & Smith Co, 329 US 637 (1947).

[151] For a discussion of the relationship between patents and trade secrets and the option forced upon the inventor, see Milgram, Trade Secrets §8.02. The effect of the patent grant and patent application upon trade secrets is discussed therein at §2.06.

petitor's developing the know-how independently. Where the nature of the technology permits a trade secret to be maintained as an effective secret for any substantial period of time, the balance is usually in favor of maintaining the trade secret rather than obtaining a patent.

The decision can sometimes be postponed by filing a patent application and prosecuting it to allowance. If the patent application is abandoned prior to allowance by, for example, not paying the issue fee, then it is maintained as a confidential file in the Patent Office and the applicant can continue to use the know-how as a trade secret. However, if the option to keep the trade secret is desired, applications cannot be made for foreign patents because in most countries the application is automatically published within a short time after application is made.

In selecting the option for a trade secret, it must be remembered that a trade secret is meaningful only so long as it remains a secret. It can be licensed and the licensee by contract required to maintain secrecy.

A large bundle of detailed information which is generally referred to as know-how can only be dealt with as a form of trade secret. Even though no single feature of the know-how is significantly novel matter, the total combination of features often constitutes a very valuable set of information which has been culled from experience and from a vast body of more general information. These features often include dimensions and dimensional relationships, controls on timing, temperature and speed of processing, selection of materials, identification of the sources of supply and experience with what are the most likely problems to arise in an industrial process and how to avoid them. This type of information can be broadly considered trade secret information. It is often part of a technology license agreement and is often contained as part of a patent license agreement. Thus, it is often possible to obtain a patent on the invention and also have a bundle of trade secrets on the know-how developed out of experience in manufacturing the product. Significantly, however, one cannot hold back on relevant trade secrets known at the time of filing a patent application. The patent law is quite explicit that a patentee must both (1) reveal enough in the patent to enable one skilled in the art to make and use the invention, and also (2) set forth what the inventor considers the preferred embodiment of the invention or what is the best mode for carrying out the invention.[152] This latter, best mode-preferred embodiment requirement refers to the situation at the time of filing. An attempt to hide the essentials while applying for a patent results in an invalid patent.[153]

A good discussion of the present status of trade secrets under the law and a recognition of their value may be found in the Second Circuit opinion in

---

[152] 35 USC §112.

[153] Reynolds Metals Co v Acorn Bldg Components, Inc, 548 F2d 155, 192 USPQ 737 (6th Cir 1977); Engelhard Indus, Inc v Sel-Rex Corp, 253 F Supp 832, 149 USPQ 607 (DNJ 1966).

Painton & Co v Bourns Inc.[154] *Painton* settled the question that had been raised about the validity and enforceability of trade secret agreements. The Second Circuit effectively held that: (1) the United States patent laws do not preempt legal recognition and enforcement of private agreements for the sale and licensing of trade secrets; and (2) it is not an extension of the patent monopoly to recognize and enforce trade secret agreements covering a period of time prior to the issuance of a patent.

§4.21 —Fostering Innovation

A set of detailed constraints on development personnel may tend to inhibit innovation. Unfortunately, funds generally are not unlimited. Thus, given the constraints of available funds, some direction must be given and some limitations placed on research activities. However, it is important to remember that most inventions cannot be strictly programmed.

§4.22 —The Use of Generalists

Occam's Rule—The better theory is the one which explains the most with the least—is a statement of economy in scientific theory and reflects an aesthetic if not psychological desire or need for as simple a statement as possible of scientific theory. More particularly, it means that if two theories are proposed to explain a given set of phenomenona, that theory is to be preferred which employs the least number of factors or variables; that theory is to be preferred which commends itself as simplest.[155]

In spite of unanimous concurrence with Occam's Rule, once we have a theory and are comfortable with it, we do tend to a Tinker-Toy approach in making it fit new and additional observations. In spite of Occam's Rule, the usual thing to do (and thus we might say the obvious thing to do) when faced with new data that do not quite fit the theory is to craft an exception or an amendment to the theory. What is not the usual thing to do (and thus what we may assume is the nonobvious thing to do) is to revise or restate the whole theory to better fit the observed data. That is hard to do for may reasons. One of the reasons is that it requires changing some theory with which we have become comfortable.

Thus, it may be seen that a new and simple solution to a problem is a high level of invention and eminently patentable. Perhaps one of the reasons why those who know so much about a field often miss the invention is that they would have to throw over too much with which they are comfortable. It may often make sense to bring a problem that has confounded the experts to a generalist not already too steeped in prevailing views within the field involved.

---

[154] 442 F2d 216, 222-26, 169 USPQ 528 (2d Cir 1971).

[155] William of Occam's original pithy statement is that multiplicity ought not to be posited without necessity.

§4.23 —The Tension between Planning and the Unpredictable

From one point of view, almost every new discovery can be viewed as accidental. There is always discontinuity between what is already known and any discovery or invention. The jump or discontinuity may be considered in some cases to be slight and in other cases great. However, discontinuity and assessment or agreement that one jump is greater than another jump is not an objective matter; it is subjective or a matter of feel.

From another point of view, most discoveries or inventions occur because the inventor is addressing a problem and is looking for a solution. In that sense, invention is a directed, intentional, planned process.

These two seemingly contradictory viewpoints present a fair approximation of the reality. It is extremely rare that an invention occurs without direction or intent. It is also extremely rare that invention occurs without some unplannable, or at least unprogrammable, discrete break with prior understandings or viewpoints.

From the point of view of overall planning for development and invention, it is desirable to avoid being trapped by some variant of either of these viewpoints and to recognize that one must rely upon, expect and foster unplanned and unprogrammed approaches, attitudes and experiments. There may be rare cases where there are formulas for invention, for example, testing out all the homologs or isomers of an herbicide to determine if one of them has a substantially greater potency. Formulas for invention rarely apply, though, and when they do, they are expensive to follow.

§4.24 —Explaining the Invention Does Not Explain the Inventive Process

The patent law[156] requires that the patent disclosure convey to one skilled in the art the nature of the invention. However, the disclosure of the invention is *not* the disclosure of the inventive process. Once an invention has been made, it can be set forth in terms of logical thought patterns. It can be related in a logical fashion to the existing knowledge within the art involved. By contrast, the inventor's thought process in developing the invention is difficult to describe and perhaps cannot even be logically explained.[157] Once the invention, however, exists and is explained or described in the rational fashion which scientific explanation demands, it may not thereafter even be possible to imagine the thought process which led to the invention.

The context within which an invention is made by an inventor is entirely different from the nature of the explanation which conveys that invention in

---

[156] 35 USC §112.

[157] Judge Hand suggested that the test for obviousness is "misty enough." Reiner v I Leon Co, 285 F2d 501, 503, 128 USPQ 25 (2d Cir 1960). The process of invention is likely to be even murkier.

a unambiguous and generally understood fashion to others. In a very real sense scientific discovery is not congruent nor even similar to scientific explanation. It has been stated, "The path of discovery runs through a series of inferences which are deeply veiled in the darkness of instinctive guessing."[158]

We have a great tendency to look at the way things are as the natural order of things. It required a wrench of imagination to challenge the divine right of kings. Yet, today, it is hard to understand how people could have had such a concept. This phenomenon applies to the making of an invention. Once the invention has been made, it seems as if, in a sense, it has always been there at least inherently in the knowledge of human beings. It becomes an accepted fact, a given in the matrix of existence. It becomes hard to envision why the invention was ever not known. That it could ever have been considered anything other than an obvious application of known facts is beyond belief.

When assessing patentability, this tendency to merge the new with the old as a seamless web or continuum leads to the error of reconstructing the invention by hindsight, thereby making something appear obvious which was not at all obvious at the time the invention was made.[159] When planning a development program, the logical nature of the post-invention explanation of inventive development leads to the error of believing that the development of invention can be programmed in detail. In fact, what can be programmed is merely the direction of effort and at least some of the preconditions for success.

---

[158] The quote is based on a phrase once used by Schrodinger. See Reichenbach, Philosophic Foundations of Quantum Mechanics 67 (1946).

[159] *See* the comments by Justice Frankfurter in Marconi Wireless Tel Co v United States, 320 US 1, 62, 63 (1943), concerning the "mirage of hindsight."

Tax Planning for Innovation

*Robert S. Taft**

5

§5.01 Options for Consideration

Taxation inevitably affects all business decisions and, therefore, powerfully affects innovation. By the same token, innovators must pay close attention to

*The author wishes to acknowledge the assistance of Steven J. Sklar and Alan I. Appel.

the tax consequences of their business choices. Research, development and innovation can yield important tax benefits, which can be maximized through careful advance tax planning. Some of the options available which affect the taxation of innovators include the following:

1. The use of a parent, subsidiary, joint venture or offshore corporation to conduct the business activity in question[1]
2. The use of domestic or offshore entities[2]
3. An election to report research and development expenses as current costs or to capitalize these costs and amortize them over a period of 60 months[3]
4. The approach taken to deferred compensation for any key employees hired temporarily for specific research and development projects[4]
5. The use of internal or external financing, and debt-equity ratios[5]
6. The use of industrial revenue bonds or other tax-favored financing devices[6]
7. The construction, purchase or lease of equipment used in research or development[7]
8. Investment tax credits and other similar credits[8]
9. The taxation of the sale and exchange of patents, trademarks and similar property[9]

§5.02 Basic Elements of Federal Corporate Income Taxation

Since most research and development is carried on today in the corporate form, corporate income taxation has great relevance to tax planning for innovation. The basic structure of federal corporate income taxation is therefore highly relevant.

A corporation is an artificial person for tax purposes, subject to federal income taxation since 1913, an accumulated earnings tax since 1921 and a personal holding company tax since 1934.

A corporation has a well-defined, particular meaning in the law as a creation

---

[1] See **§5.10.**
[2] See **§5.20.**
[3] See **§§5.11-5.13.**
[4] See **§§5.14-5.15.**
[5] See **§5.16.**
[6] See **§5.17.**
[7] See **§§5.11-5.13.**
[8] See **§§5.12** and **5.19.**
[9] See **§5.19.**

of a state or federal statute. The 1954 Internal Revenue Code expanded that meaning beyond its conventional connotation for tax purposes. The Code defined the term *corporation* to include "associations, joint stock companies and insurance companies."[10] Federal law prevails over state law with respect to how an entity or transaction is characterized for federal tax purposes. Thus, it is federal law which will determine whether a corporate entity should be recognized or disregarded. Thus, organizations which are not corporations in the conventional sense (under state law) may be considered corporations for federal income tax purposes.

Although the Sixteenth Amendment was required to enable Congress to pass a valid general income tax on individuals, the power to tax a corporation as an entity separate from its shareholders was settled before that.[11] The Supreme Court has applied this doctrine of corporate entity to transfers between corporation and stockholder,[12] and it has been held that as a general rule a corporation is a "nonconductor that cuts off [any] connection between dividends to its stockholders and the corporate funds from which the dividends are paid."[13] This doctrine has been applied with respect to the determination of gain or loss arising from a sale[14] or exchange of property,[15] and so forth.

Inasmuch as a corporation and its stockholders are separate and distinct legal entities, it has been held that a change of stock ownership does not result in a new corporation for tax purposes;[16] that a change in name, business status or type of business does not create a new corporation;[17] that a corporation continues as a taxable entity even though its charter has been canceled;[18] and that the distinct entity should be recognized even though all of the stock of the corporation is owned by one individual or by one corporation.[19]

On the other hand, there have been instances where the courts have disregarded the corporate entity and looked to the stockholders themselves as the real parties in interest.[20] The Supreme Court has disregarded the corporate entity where the corporation combines its affairs with those of its shareholders to such an extent as to be a mere agent for its stockholders or where their

---

[10] Internal Revenue Code of 1954, as amended, §7701(a)(3); Treas Reg §301.7701-1(c).

[11] *See* Flint v Stone Tracey Co, 220 US 107 (1911).

[12] Burnet v Commonwealth Improvement Co, 268 US 541 (1932).

[13] Miller v Milwaukee, 272 US 713 (1927).

[14] Foster v Commissioner, 96 F2d 130 (2d Cir 1938).

[15] Rose v Trust Co, 77 F2d 355 (5th Cir 1935).

[16] Alprosa Watch Corp, 11 TC 240 (1948).

[17] *Id.*

[18] Hvidsten v United States, 185 F Supp 856 (DND 1960).

[19] Webber v Knox, 97 F2d 921 (8th Cir 1938); Benjamin F Haas, TC Memo 1956-165, *revd and remanded on other grounds,* 248 F2d 487 (2d Cir 1957), *on remand,* TC Memo 1959-86 (a corporate entity will not be disregarded simply because the sole stockholder of the corporation dominated and controlled it).

[20] Pauline W. Ach, 42 TC 114 (1964), *affd,* 358 F2d 342 (6th Cir 1966).

businesses are so commingled as to constitute a single business entreprise for all practical purposes.[21] In this regard, in determining whether profits should be taxed to a corporation or to its stockholders, the following are considered: the representations made to the public or the business world as to who conducts the business; to whom credit is extended; whether the income is derived from an activity commonly engaged in by the corporation and whether or not the corporation holds directors' meetings, declares dividends and holds stockholders' meetings.[22]

§5.03— Tax Computation

The ordinary business corporation is subject to tax on a graduated five-step tax rate structure. It may also be liable for an accumulated earnings tax and a minimum tax. Generally, corporate taxable income is computed in the same manner as an individual's, although some items peculiar to corporaions receive special tax treatment.

The graduated five-step tax is imposed on a corporation's taxable income. Taxable income is gross income less itemized and special deductions. Corporate tax rates are reflected in the following chart.

Corporate Tax Rates for Taxable Income

For tax years beginning in	Amount (Dollars)	Tax Rate (Per Cent)
1981	0-25,000	17
	Over 25,000-50,000	20
	Over 50,000-75,000	30
	Over 75,000-100,000	40
	Over 100,000	46
1982	0-25,000	16
	Over 25,000-50,000	19
	Over 50,000-75,000	30

[21] Southern Pac Co v Lowe, 247 US 330 (1918).

[22] *See* American Sav Bank, 56 TC 828 (1971). *See also* Keller v Commissioner, 77 TC No 70 (1981), *appeal filed* (10th Cir Apr 5, 1982).

For tax years beginning in	Amount (Dollars)	Tax Rate (Per Cent)
	Over 75,000-100,000	40
	Over 100,000	46
1983	0-25,000	15
	Over 25,000-50,000	18
	Over 50,000-75,000	30
	Over 75,000-100,000	40
	Over 100,000	46

The minimum tax on corporations is 15 per cent of the amount by which the tax preference items[23] exceed the greater of (1) $10,000, or (2) the regular income tax imposed for the tax year, less credits for foreign taxes, investment in qualified property, work incentive program expenses, targeted jobs credit and the adjustment for timber income. The regular income tax for the purpose of the deductions does not include the accumulated earnings or personal holding company taxes.

§5.04 —Existence of the Corporate Entity

In cases of tax avoidance, fraud, sham and other types of non-bona fide transactions and in cases involving no real business purpose, the courts have not hesitated to disregard the corporate entity by looking through the form to the substance of the transaction.[24] While it has been held that the taxpayer has the right to decrease taxes or to avoid them by methods permitted by law,[25] it has also been held that where the avoidance or evasion of taxes has been the intent and purpose of the transaction, the means and forms used for its accomplishment will be subject to careful scrutiny.[26] On the other hand, with today's tax structure, one of the first considerations in any business deal is the tax implications. The taxpayer will not be penalized for electing the legal or

---

[23] Tax preference items, in the case of a corporation, include deductions for accelerated depreciation, excess intangible drilling costs, and various other items. For a complete list, *see* IRC §57(a).

[24] National Investors Corp v Hoey, 144 F2d 466 (2d Cir 1944), *revg* 52 F Supp 556 (SDNY 1943).

[25] Helvering v Lazarus Co, 308 US 252 (1939).

[26] Gregory v Helvering, 293 US 465 (1935).

legitimate procedure that carries the least tax liability with it.[27] As a rule, the determination of the validity or invalidity of a corporate entity is a factual issue necessitating inquiry into whether the entity served a business purpose and earned income, or whether the income at issue was in reality produced and earned by an individual or other entity.[28]

Even in cases involving tax avoidance, the courts have often held that where there was no independent nontax business purpose, such as the mere holding of title to property for the convenience of the stockholders, the corporate entity should be disregarded.[29] On the other hand, even though a corporation may be a mere "dummy" for most purposes, if it engages in any business activity, such as the making of a loan, this scintilla of activity may be sufficient to balance the scales in favor or recognition of the corporate entity.[30]

Most recently, because of increasing use of tax shelters, there has been a closer look at whether certain types of partnerships are in reality corporations because of the limited nature of the partner's liability. Many such associations have been found taxable as corporations. The Supreme Court, in *Morrissey v Commissioner,*[31] delineated the major characteristics which are to be used to determine whether an organization is to be treated as a partnership or as an association. The major characteristics of a corporation were stated to be shareholders; an objective to carry on business as a common enterprise; continuity of life; centralized management; free transferability of beneficial interests; and limited liability. Treasury regulations include the *Morrissey* test of whether an organization *resembles* a corporation.[32]

Whether an organization has more corporate characteristics than noncorporate characteristics will generally determine, according to the regulations, whether the organization qualifies as a limited partnership under law or as an association taxable as a corporation. In *Larson v Commissioner*[33] the Tax Court held that it is sufficient if a limited partnership has two of the last four previously mentioned corporate characteristics. The Internal Revenue Service recently stated that it will follow *Larson.*[34]

Foreign corporations are subject to scrutiny. However, as long as the corporation is actually engaged in business, it does not matter that it is incorporated in one foreign country and conducting business in another, or that tax motivations were involved in its formation. It is also not controlling that the corporation's ownership of its business property is not revealed to the foreign country. Furthermore, a close relationship may be maintained with a domestic sister

---

[27] Consolidated Premium Iron Ores, Ltd, 27 TC 127 (1957), *affd,* 265 F2d 320 (6th Cir 1959).

[28] Lloyd F Noonan, 52 TC 907 (1969), *affd,* 451 F2d 992 (9th Cir 1971).

[29] United States v Jelenko, 23 F2d 511 (4th Cir 1927).

[30] Langdon P. Skarda, 27 TC 137 (1956), *affd,* 250 F2d 429 (10th Cir 1957).

[31] 296 US 344 (1935).

[32] Treas Reg §301.770(2)(a)(1).

[33] 66 TC 159 (1976).

[34] Rev Rul 79-106, 1979-1 CB 448.

corporation which is the source of its funds and the majority shareholder may be actively interested in the business.[35]

§5.05 Independent Business Purpose Requirement

It is a basic tax concept that transactions normally must have at least some nontax business purpose to be recognized for tax purposes. This principle like most others has its exceptions but it is a good general guide. For example, §269 of the Internal Revenue Code provides that "if any person or persons acquire, or acquired on or after October 8, 1940, directly or indirectly, . . . property of another corporation, not controlled, directly or indirectly, immediately before such acquisition, by such acquiring corporation or its stockholders, the basis of which property, in the hands of the acquiring corporation, is determined by reference to the basis in the hands of the transferor corporation and the principal purpose for which such acquisition was made is evasion or avoidance of Federal income tax by securing the benefit of a deduction, credit, or other allowance which such person or corporation would not otherwise enjoy, then the Secretary [of the Treasury] may disallow such deduction, credit, or other allowance."

Thus, if a corporation is to use innovative tax planning, it must also have a bona fide nontax business purpose and function. In addition to ignoring the corporate entity, where a transaction involves "related parties" (as described below), the commissioner may allocate income and deductions among entities owned or controlled directly or indirectly by the same interests under Internal Revenue Code §482.

§5.06 Tax Consequences of Corporate Status—Double Taxation

An initial determination should be made as to whether to organize the business activity as a corporation. Basically, the tax disadvantage in using a corporation is the element of double taxation. A corporation is taxed on its income and its shareholders are taxed again on dividends received from the corporation. If the corporation does not expect to pay any dividends for a significant period or at all, there is an advantage resulting from the lower tax rates on corporations. The highest rate on individuals is 50 per cent, while the top corporate rate is 46 per cent.[36] When funds are eventually taken out, the shareholder hopes to be taxed at a lower capital gains rate (e.g., upon liquidation or sale).

However, corporations paying no dividends run the risk of incurring an

---

[35] Ross Glove Co, 60 TC 569 (1973).

[36] IRC §1(d), *as amended by* the Economic Recovery Tax Act of 1981, §101(a), Pub L No 97-34, 95 Stat 176 (codified at 26 USC §1 *et seq*).

accumulated earnings tax.[37] Furthermore, the taxpayer may wish to use some of the funds generated by the corporation (over and above anything drawn as a reasonable salary) and not plow all the funds back into the corporation.

Apart from doing business as a partnership or sole proprietorship, double taxation may be avoided by using a Subchapter S corporation.[38] Where a qualified corporation elects Subchapter S status,[39] the income and losses then pass directly to the shareholders with no tax liability at the corporate level. This election is often advantageous when a corporation is first organized and incurs start-up losses. Losses then pass directly to the shareholders who are often individuals in a high tax bracket. If and when the corporation begins to make money, the Subchapter S election can be terminated.

A corporation must meet several requirements to qualify for Subchapter S status. It must have no more than 25 shareholders, all individuals, and none of whom are nonresident aliens. In addition, it can have only one class of stock.[40] If a corporation fails one of these tests during the year, its Subchapter S status is terminated.

In order to avoid triple taxation, corporations are allowed a deduction of at least 85 per cent of dividends received from a domestic corporation.[41] Triple taxation would result because the corporation paying, the corporation receiving, and eventually the shareholder, would all be taxed. In addition, when any corporation receives an in-kind property distribution from another corporation, double taxation of any appreciation is avoided by a special rule. The amount of the distribution for tax purposes is the lesser of: (1) the fair market value of the property received, or (2) the adjusted basis of the property to the distributing corporation immediately before the distribution, increased by any gain recognized to the distributing corporation on distribution. This treatment allows a step-up in basis to the extent of gain recognized by the distributor corporation in all situations where gain is recognized.[42]

Fair market value (instead of the lesser of fair market value or adjusted basis) must be used for distributions by a foreign corporation, unless the shareholder gets a deduction for dividends received.

Fair market value must also be used by foreign corporate shareholders receiving property distributions unless the distribution is effectively connected income.[43] The amount of the distribution received by any individual or corporation is reduced (but not below zero) by any liability of the distributing corporation assumed by the recipient shareholder in connection with the distribution, and any liability to which the property is subject immediately before

---

[37] See **§5.08.**

[38] IRC §1371 *et seq.*

[39] The election is made under §1372(a). IRC §1371(b).

[40] IRC §1371, *as amended by* the Economic Recovery Tax Act of 1981, §233.

[41] IRC §243(a).

[42] *Id* §301(b)(1)(B).

[43] *Id* §301(b)(1)(C)-(D).

and after the distribution.[44] The basis of the property received is the same as the amount of the distribution, as figured above without reduction for liabilities.[45]

As previously mentioned, corporations can generally deduct 85 per cent of dividends received from taxable domestic corporations. A 100 per cent deduction is allowed to small business investment companies and members of an affiliated group. A dividend from a foreign corporation paid out of earnings and profits accumulated while it was a domestic corporation subject to United States tax is treated as a dividend from a taxable domestic corporation.[46]

A corporation generally realizes no gain or loss when it distributes its stock or property to shareholders.[47] A corporation can distribute a part interest in a corporate asset without tax liability, unless the distribution is solely to avoid corporate tax on the latest sale of the property. Retirement of a shareholder's interest in the corporation is good reason to distribute a part interest. If the property is later sold, the corporation and the retired shareholder each transfer their own part interests.

No gain or loss is recognized to a corporation if it sells or exchanges its property and distributes its assets in a complete liquidation within a 12-month period. A corporation has no recognized gain or loss if this is done within 12 months from the time the plan of liquidation is adopted. Assets sufficient to meet claims can be retained.[48]

§5.07 —Deductions

Corporations generally can take the same itemized deductions as individuals except purely personal deductions, such as medical expenses, alimony payments and so forth.[49] A corporation may deduct charitable contributions.[50] Gifts or donations that are deductible by individuals qualify for the corporate deduction, except those made to fraternal societies and those made to trusts, community chests, funds or foundations for use outside the United States or its possessions. The deduction is ordinarily allowed for the tax year in which the contribution is actually paid. There is a special rule when inventory type property is given by a corporation to a public charity or private operating foundation. This rule limits the deduction to the sum of: (1) the corporation's basis in the property, plus (2) up to one-half of the unrealized depreciation. However, no deduction exceeding twice the basis of the property is allowed. No deduction is allowed for any part of the unrealized depreciation which

---

44 *Id* §301(b)(2).
45 *Id* §301(d)(2).
46 *Id* §243.
47 *Id* §311.
48 *Id* §337.
49 IRC §§211-216.
50 *Id* §170.

would be ordinary income resulting form recapture of certain items if the property were sold. The corporation's deduction for contributions cannot exceed 5 per cent of its taxable income, computed without regard to the charitable contribution deduction, special deductions, any net operating loss carryback to the tax year and any capital loss carryback to the tax year.[51] The same general rules apply to corporation contributions of appreciated property. Contributions over the 5 per cent limit are carried over five succeeding tax years. However, contributions actually made during a later year plus the carryover must fall within the 5 per cent limit. Also a carryover must be reduced to the extent that the excess contribution carried over reduced taxable income and increased a net operating loss carryover.[52]

If a corporation's deductions exceed its income, the corporation will have a net operating loss.[53] All corporations are entitled to a net operating loss deduction in computing their tax with certain exceptions. The general rules governing net operating loss are similar to those for individuals, but the adjustments for figuring net operating lossess and net operating loss carryovers are different. However, a net operating loss limits and defers a corporation's minimum tax liability in much the same way it does an individual's. A net operating loss may be carried back to each of the three preceding years and carried over to each of the fifteen following years. The loss is first carried to the earliest year, then to the next earliest year, and so forth. There are several exceptions to this rule, whose complexity is beyond the scope of this chapter.[54]

§5.08 —Accumulated Earnings Tax

Every taxable corporation (other than a domestic or foreign personal holding company) may be subject to the accumulated earnings tax when it is formed or used to avoid income taxes that would otherwise be paid by its shareholders.[55] The tax is imposed in addition to other corporate taxes at the rate of 27 1/2 per cent on the first $100,000 and then 38 1/2 per cent of the balance of all unreasonably accumulated taxable income.[56] Liability for the accumulated earnings tax is incurred if one of the purposes for an accumulation beyond the reasonable needs of the business is to prevent the imposition of income tax on the shareholders. If the accumulation is reasonable, the finding of the tax avoidance purpose and the liability for the penalty surtax is unlikely. The most important question is whether or not the surplus is beyond the reasonable needs of the business.[57] These needs include those that are reasonably

---

51 *Id* §170(b)(2).

52 *Id* §170.

53 *Id.*

54 *Id* §172(b).

55 IRC §532.

56 *Id* §531.

57 *See* Treas Reg §1.533-1(a).

anticipated. In other words, accumulated earnings and profits do not have to be invested immediately, so long as there is an indication that future needs of the business require the accumulation. If there is a definite plan for investment of earnings and profits, it is not necessary to complete the plan immediately after the close of the tax year; but when the future use of the funds is uncertain or vague, the accumulated earnings tax applies. Also, for tax years starting after September 30, 1979, corporations can accumulate earnings for payment of reasonably anticipated products liability losses.[58]

§5.09 —Capital Gains and Losses

The rules for capital gains and losses of individuals also apply, with some exceptions, to corporations. Important differences include: (1) corporations do not get long-term capital gain deductions (although long-term capital gains are taxed at a lower rate);[59] and (2) they cannot deduct a net capital loss from ordinary income in the year it is sustained.[60] A net capital loss may be carried back and carried over as a short-term loss and offset against the net capital gains.[61]

A corporation must pay a minimum tax equal to 15 per cent of the excess of its tax preference items over $10,000 or its regular income tax liability, whichever is greater.[62] A portion of the corporation's net capital gain is a tax preference. It is figured by applying 39.13 per cent to the net long-term capital gain over net short-term capital loss.[63] Although a corporation cannot take a deduction in the current year for a net capital loss, it can carry the loss back to each of the three years before the loss year.[64] If the carrybacks do not absorb the entire loss, the rest can be carried over to each of the five years after the loss year. Regardless of origin, all carrybacks and carryovers are treated as short-term capital losses for carryback and carryover purposes. Furthermore, capital loss carrybacks are deductible only if they do not increase or create a net operating loss for the tax year in which applied. The net capital loss is first carried back to the earliest tax year to which it is allowed and any excess thereof is carried to the succeeding tax year.

If the net long-term capital gain exceeds the net short-term capital loss, the tax is computed under two methods. The method ultimately used is that which produces the smaller tax.[65] In the regular method, the tax is figured on taxable income which includes net capital gain. In the alternative method, the tax is

---

58 IRC §537. See generally **ch 10.**

59 IRC §§1201(a), 1202.

60 *Id* §1211.

61 IRC §1212.

62 See **§5.03.**

63 IRC §57.

64 *Id* §121.

65 *Id* §1201.

figured on taxable income less the excess of net long-term capital gain over net capital gain. Twenty eight per cent of the net capital gain is added to the tax figured on the taxable income without regard to the net capital gain.

If the corporation had a net short-term capital gain instead of a net short-term capital loss, or if the corporation had no net short-term capital gain or loss but did have a net long-term capital gain, the alternative tax would be the tax figured on taxable income, less the net long-term capital gain, plus 28 per cent of the net long-term capital gain.

§5.10 —Affiliated Corporations

In determing whether to form a subsidiary, an independent business purpose is not a factor in tax treatment because members of a controlled group of corporations are treated for tax purposes as if they were one corporation, regardless of the business purpose of each corporation.

A group of affiliated corporations may file consolidated tax returns for the period that they are affiliated, but only if all the corporations that were members of the affiliated group at any time during the year consent before the last day for filing the return. An affiliated group is formed when (1) at least 80 per cent of all classes of voting stock and at least 80 per cent of each class of nonvoting stock of each includable corporation (except the common parent corporation) is owned directly by one or more of the other includable corporations, and (2) the common parent owns directly 80 per cent of all classes of the voting stock and at least 80 per cent of each class of nonvoting stock of at least one of the other includable corporations. Stock does not include nonvoting preferred stock, employers' securities under a tax credit employees' stock ownership plan or other special qualifying employers' securities.[66]

Whether or not a group of affiliated corporations elects to file a consolidated return, it will be a "controlled group", treated as one corporation for tax purposes.[67] Controlled groups are classified as either (1) parent corporations and their 80 per cent-owned subsidiaries, or (2) brother-sister corporations. A controlled group is two or more corporations at least 80 per cent-owned by five or fewer individuals, estates or trusts, who must also own more than 50 per cent of the stock, taking into account as stock owned by each only his or her least per cent of ownership in each corporation. In determining the 80 per cent control test, the five or fewer shareholders do not have to own stock in each corporation.[68]

In computing their tax, members of a controlled group of corporations are limited to a total of $25,000 in each of the graduated tax rate brackets below 46 per cent, which is divided equally among them or shared as they may elect.[69]

---

[66] IRC §1504.

[67] All affiliated groups are controlled groups, although all controlled groups are not affiliated groups, since the definition of the latter is broader.

[68] *Id* §1563.

Thus, a corporation cannot shift any income to lower tax rate brackets by simply forming subsidiaries. In addition, members of a controlled group of corporations are limited to one accumulated earnings credit totaling $250,000 which must be divided equally among the component members.

Finally, the $10,000 tax preference exclusion in computing a corporation's minimum tax is divided among the members of a controlled group in proportion to their respective regular tax deductions for the year.

§5.11 Tax Treatment of Research and Development Expenditures—IRC §174

An area to examine in connection with corporate innovation involving tax law is research and development expenditures. Section 174 of the Internal Revenue Code allows the taxpayer the option of currently deducting, or amortizing over a period of 60 or more months, most expenditures for research and experimentation. It does not define the term *research and experimental expenditures.* The Internal Revenue Service by Treasury regulation interprets the definition to include "costs incident to the development of an experimental or pilot model, a plant process, a product, a formula, an invention or similar property and the improvement of already existing property of the type mentioned" but to exclude "expenditures paid or incurred for research in connection with literary, historical, or similar projects."[70]

In 1980, the Service ruled that environmental impact study expenses are not deductible as research.[71] That ruling concerned a power company which incurred expenses in connection with environmental studies relating to proposed construction of a power plant and an ultra high voltage electricity transmission line. The Service ruled in technical advice that "expenses contributable to specific investigations of projected environmental impacts . . . are not research and experimental expenditures" and that payments to outside experts will qualify only if they "perform substantial original research with respect to the taxpayer's particular project. For example, payments to others primarily for appearing as expert witnesses before the state licensing agencies are not research and experimental expenditures."

The Service has implied that costs of developing "computer software" are not research and experimental expenditures under §174, but it has also announced in an audit procedure that costs of that character are so akin to expenditures of the type falling under §174 that for purposes of administrative convenience it will not disturb the taxpayer's treatment of such costs if the principles of §174 were consistently applied.[72]

Costs incident to the development of a market for new products would not

---

[69] *Id* §1561.

[70] Treas Reg §1.174-2(a)(1).

[71] Letter Rul 8028008.

[72] Rev Proc 69-21, 1969-2 CB 303.

seem to be within the intended scope of §174. In most cases, however, market development costs and the like will qualify as currently deductible *ordinary and necessary* business expenses under the more general provisions of §162 of the Code.

Unsuccessful development efforts can nevertheless yield deductions. In *Snow v Commissioner*,[73] the Supreme Court defined the scope of §174 to include deductions not only for expenditures of ongoing businesses, but also those of the investor in, and presumably for inventory of, a product which is ultimately neither patented nor marketed. The opinion, which was unanimous, apparently overruled prior cases which disallowed such deductions for nascent businesses. The Court held that the purpose of §174 was to encourage taxpayers to carry on research and experimentation. This opinion conveys implicit judicial recognition of the importance of deductions such as this to the national economy. The deduction is available to companies that are "upcoming and about to reach the market" as well as those with established business activities. The context and language of the decision indicate that there must be a bona fide purpose of seeking to realize a profit, not merely to create tax benefits.

A taxpayer may, of course, utilize the beneifts of §174 regardless of whether only a single project is undertaken, or whether a full-scale research program involving a multiplicity of projects is involved.[74] Where several separable projects are undertaken by (or for the benefit of) the taxpayer, it may be possible to get the consent of the Commissioner to treat some projects under the current deduction method and others under the deferred expense method.[75] This would be innovation at its best in this area.

Section 174 specifically states that it will not have application in the case of expenditures to ascertain "the existence, location, extent or quality of any deposit of ore or other mineral (including oil and gas)," undoubtedly because of the existence of other tax benefits in those fields. Thus, exploration costs apparently are not permitted under §174 as a deduction. On the other hand, Revenue Ruling 74-67[76] holds that a mining corporation is entitled to treat under §174 expenses in connection with developing new methods of extracting ore and minerals from the ground.

This section also specifically eliminates from treatment thereunder all expenditures for the acquisition or improvement of land and/or depreciable (i.e., equipment) or depletable property (normally, minerals) even if "to be used in connection with research and experimentation," and beneficial title to which is taken and held in the taxpayer's name.[77]

---

[73] 414 US 1128 (1974).

[74] Treas Reg §1.174-1.

[75] See **§5.13.**

[76] 1974-1 CB 63.

[77] IRC §174(c).

§5.12 —Impact of the Economic Recovery Tax Act on Research and Development

The Economic Recovery Tax Act of 1981[78] provides for a 25 per cent credit for incremental research expenditures paid or incurred after June 30, 1981. This credit is in addition to the deductions allowed by §174 for research and development.[79] The credit is based on the excess of qualified research expenditures for the year over the average yearly amount of such expenditures during the base period. The base period for the next year is the two preceding years. A new business will also qualify for the credit. A year in which no research expenses were incurred, because the taxpayer was not in existence, still counts (as zero) in the base period, but the base period average cannot be less than 50 per cent of current year expenses under Internal Revenue Code §44F(c)(3). However, the limit of incremental research expenses cannot be evaded by forming a subsidiary. Where research expenses are paid or incurred by a controlled group or by trades or business under common control, special apportionment rules govern allotment of the credit.[80]

Research expenditures qualifying for the new incremental credit consist of

1. In-house expenditures for research wages and supplies, plus certain lease or other charges for research use of computers, laboratory equipment and so forth
2. 65 per cent of amounts paid for contract research
3. 65 per cent of corporate grants for basic research to be performed by universities or certain scientific research organizations.

§5.13 —Current Expense Deduction or Deferred Expense Treatment

To the extent that research and experimental expenditures are paid or incurred in a trade or business, they can be deducted as current expenses. Where the current expense method is in effect, all expenditures qualifying thereunder must be deducted in the proper year. This applies to all qualified research and experimental expenditures of the taxpayer and cannot be altered, in whole or in part, as to any taxable year without the permission of the Commissioner. Section 174 provides that the current expense method of treating research and experimental expenditures may be adopted without the consent of the Commissioner where the adoption is made effective for the first taxable year (beginning after December 31, 1953 and ending after August 16, 1954) during which such expenditures are paid or incurred by the taxpayer. In the event that the taxpayer does not adopt the current expense method in

---

[78] Pub L No 97-34, 95 Stat 176 (codified at 26 USC §1 *et seq*).

[79] IRC §44F, as added by §221 of the Economic Recovery Tax Act.

[80] IRC §44F(f).

respect of the first taxable year (after 1953) during which research and experimental expenditures are incurred, adoption of this method in any subsequent year requires the consent of the Commissioner.

Deferred expense treatment may be elected by the taxpayer as to research and experimental expenditures which are paid or incurred in a trade or business but not treated as current expenses under §174, and are thus charged to capital account (but not with respect to depreciable or depletable property). As a consequence of such an election, the taxpayer may deduct deferred research and experimental expenditures ratably over the elected period (not less than 60 months), commencing with the month in which the taxpayer first realizes benefits therefrom. This permits offsetting these expenses against later gains. Where the deferred expense method is in effect, it applies to all qualified research and experimental expenditures (within the scope of the election made) paid or incurred by the taxpayer, and cannot be altered, either in whole or in part, as to any taxable year without the permission of the Commissioner. The election to defer may be made without the consent of the Commissioner (assuming that the current expenses method has not been adopted previously), but when made will have no retroactive effect.

The innovative tax planner must analyze the advantages of both methods of taking a research and development deduction. The current expense method involves bookkeeping cost savings, current deductions, avoidance of any abandonment loss problems, avoidance of realization of benefits problem and, most importantly, ability to offset against other current income. Where the current expense method has been adopted, the deductions which it permits will have the effect of offsetting other income earned during the taxable year, thereby reducing tax liability and increasing cash flow. This increased positive net cash flow results in savings because of the present value of money especially during periods of high inflation and interest rates. Current expensing is important during periods of high interest rates or high inflation since the current deduction increases in relative value.

There is one additional advantage of current expensing of research and development costs which deserves mention. Normally, where a capital asset is developed as a result of research and experimentation and is thereafter sold, and gain is realized on the sale, it will not matter which method has been utilized. Where a loss is realized, however, the current expense method completely avoids the problem which might be created by disposition of property at a loss, since all the expenditures involved will have been written off and the taxpayer will hold the property with a zero basis. If, on the other hand, the deferred expense method had been elected (and depreciable or amortizable property was not produced), the tax basis of the property would be represented by the net balance in the deferred expense account attributable to that property. A sale of that property at an amount less than such basis would give rise to a loss which would appear to be a capital loss and, as such, would be limited by the capital loss carryover provisions of §1211 of the Code.[81]

---

[81] Corporations may generally carry capital losses back three taxable years and forward five taxable years. IRC §1212(a)(1).

Looking at the advantages of the deferred expense method, where research and experimentation is being conducted by a new corporation with no other operations, other income as an offset may not be available. Thus, there would be a benefit in deferral of the deduction.

Innovative tax planning in this area thus requires a careful analysis of projected income and expenditures for future years of the corporation. If the corporation is not in a profitable position but anticipates large future growth, wise tax planning might suggest the deferred expense method rather than the annual deduction. However, there aften are alternatives more favorable than the deferred expense method. For example, Subchapter S status may be elected, allowing losses to be used by the shareholders.[82] In addition, the corporation itself, if not a Subchapter S corporation, may be able to use these losses as a net operating loss (NOL) carryforward. Then, if gains are realized in less than five years, these loss carryforwards (generated by the prior research and development deductions) may be utilized immediately. If more than one project is involved, different periods for deductibility may be elected.

Tax losses may also be used by profitable corporations to offset taxable income. A profitable manufacturing corporation may well be on the lookout for a research and development corporation in the same product area. A merger of the two corporations may produce beneficial tax consequences for the former company, provided there are no fatal antitrust objections.[83]

Careful tax planning is crucial in such a merger, or the NOL benefits may be lost. Current losses of the research and development corporation can be offset against current earnings of the profitable manufacturing corporation, even if the research and development corporation is held as a subsidiary, simply by filing a consolidated return. However, on a consolidated return, the NOLs of the research and development corporation can only be offset against profits of that corporation. Thus, it may be advisable to merge the manufacturing corporation into the research and development corporation. The research and development corporation can purchase the assets of the manufacturing corporation by issuing sufficient stock to the manufacturing corporation so that the former corporation controls (80 per cent or more) the entire merged entity. These shares can then be distributed tax-free (in a "D" reorganization) to the old shareholders of the manufacturing corporation, who will control the new merged entity. This new merged entity will still have the NOL of the research and development corporation to be offset against future earnings of the merged entity (the operations of both corporations). Whether this technique is worthwhile will depend upon the size of the NOL of the research and development corporation and the feasibility of the loss corporation's purchasing the profitable corporation. Other methods of utilizing NOLs of acquired corporations have been severely limited by amendments of §382, which were

---

[82] See **§5.06.**

[83] See **ch 3.**

to become effective July 1, 1982.[84]

§5.14 Deferred Compensation Issues

Innovative tax planning can be accomplished in the employee benefit area. This may be important if research and development efforts require recruitment of key personnel at high compensation rates possibly for a limited period, and where the employees desire to defer taxation of part of their compensation.

With the advent of the Employee Retirement Income Security Act of 1974 (ERISA),[85] it became more difficult to provide large tax-deferred compensation for key employees of a corporation. In order to qualify under ERISA's provisions, nondiscriminatory benefits under profit-sharing and pension programs was required. As a result, compensating the key employee with greater deferred benefits falls outside the scope of §401 of the Internal Revenue Code, the section that sets forth the requirements for qualified plans under ERISA. One of the methods available to do so is the deferred compensation arrangement. This arrangement does not have to meet the minimum funding, employee coverage and other ERISA-qualified plan requirements. Accordingly, a quite flexible plan can be tailored to permit selective recognition of the special services which may be rendered by a key employee.

Deferred compensation arrangements, whether included in group plans or in individual employment or other agreements, principally involve an agreement by an employer to make payments to an employee at a future date for present services.

The employee's tax objective in participating in such an arrangement is to insure taxation (generally, at ordinary income tax rates subject to the maxi-tax limitation)[86] only when payments are received under the agreement. The underlying assumption is that the employee's effective tax rate at the time of payment will be lower than the current tax rate, and that the employee will be able to retain a larger portion of the total amount to be paid. The key revenue ruling in this area is Revenue Ruling 60-31.[87] Basically, that ruling held that prior to the rendition of any services, an employee can enter into an agreement with the employer to defer compensation to be earned by that employee to a later date. The individual would be taxable only upon receipt of the deferred compensation payments. The ruling held that "a mere promise to pay, not represented by notes or secured in any way, is not regarded as a receipt of

---

[84] These amendments have been postponed twice and will probably be repealed. NOL's are discussed in IRC §172 as well as §382.

[85] Pub L No 93-406, 88 Stat 129 (codified at 29 USC §1001 *et seq*).

[86] Probably 50 per cent. The maxi-tax limitation is no longer important for federal taxes (under the Economic Recovery Tax Act of 1981), but may still be important for certain state and local income taxes (*e.g.*, New York).

[87] 1960-1 CB 174, *modified by* Rev Rul 64-279, 1964-2 CB 121.

income within the intendment of the cash receipts and disbursements method."

In the appropriate case, the Service may be willing to accept the position that compensation may be deferred even after services are performed. In *James F. Oates*,[88] a life insurance agent agreed with the insurance company that his renewal commissions should be paid at the maximum monthly rate of $1,000 for 15 years. The purpose of the agreement was to equalize the agent's income from these commissions over a 15-year period. The agent had performed all services necessary to earn the commissions at the time the agreement was signed, but the amount of the renewal commissions to be paid was uncertain. The court held that the agent was to be taxed on the renewal commissions only when they were paid. At first the Service nonacquiesced, but subsequently, it acquiesced in the decision. However, it has recently re-examined its position on the question and has stated that it will not issue advance rulings on the tax consequence of an elective type of deferred compensation agreement unless the election is made before the services are performed.[89]

Section 83 of the Internal Revenue Code, which was added by the Tax Reform Act of 1969,[90] states that if property is transferred to a person as compensation for services, such person will generally be taxed at the time of receipt if the property is transferable or is not subject to a substantial risk of forfeiture. The amount required to be included in income as compensation is the excess of fair market value of such property (determined without regard to any restrictions, except restrictions which by their terms will never lapse) over the amount, if any, that is paid for such property. Restrictions which by their terms will never lapse are taken into account in determining the fair market value of the property.

If the property is subject to a "substantial risk of forfeiture," as where the employee's right to retain the property is conditioned upon future performance of substantial services, no tax is incurred until such restrictions lapse or, if earlier, the time the property becomes transferable. However, at such time the employee must include in income the excess of the current market value of the property over the amount, if any, paid for such property. Any further appreciation in the value of the property is usually taxed as capital gain when the property is sold.[91]

The employee has an alternative to this tax treatment method. An election may be made to pay the tax at the time property is received, despite the fact that the property is forfeitable. If this course is chosen, the amount includable in the employee's gross income is the fair market value of the property at the

---

[88] 18 TC 570 (1952), *acq* 1960-1 CB 5, *affd,* 207 F2d 711 (7th Cir 1953).

[89] Rev Proc 71-19, 1971-1 CB 698.

[90] Pub L No 91-172, 83 Stat 487 (codified at 26 USC §1 *et seq*).

[91] Under the Economic Recovery Tax Act of 1981, Pub L No 97-34, 95 Stat 176 (codified at 26 USC § 1 *et seq*), employees receiving "qualified" options to buy their company's stock will not be taxed until they sell the stock and will receive long-term capital gains treatment if they have held the stock for more than 12 months.

time of receipt, determined without regard to the restriction (except a restriction which never lapses), and §83(b) of the Code provides that no deduction is allowed to the employee if the property is subsequently forfeited.

A major new tool for employee compensation was provided by the Economic Recovery Tax Act of 1981.[92] Under the act, employees receiving options to buy their company's stock would not be taxed until they sold this stock, and if it were held for more than 12 months, long-term capital gains rates would apply. The option could be exercisable for a 10 year period at a price not less than the fair market value of the stock at the time such option was granted. This type of plan is of additional value when the increased value of the stock may be a direct result of the employee's services. The employer is generally allowed the deduction for the amount included in the employee's income.

A similar tax planning device is an Employee Stock Ownership Plan (ESOP), as defined in IRC §4975(e)(7). An immediate deduction can be obtained by the employer for the fair market value of its stock contributed to the plan, while the employee beneficiary does not recognize income until receipt of the shares. In the alternative, the employer receives an immediate deduction upon the contribution of money to the plan, and this money can then be utilized to purchase the employer's stock resulting in a net effect of a deduction to the corporation of the fair market value of its own stock acquired by the trust. In addition, dividends paid on the stock are accumulated in the trust, in effect benefiting the employee without the recognition of income prior to actual receipt of the proceeds. In general, an ESOP must meet the same nondiscriminatory rules of ERISA that apply to all qualified employee benefit plans, which may make an ESOP undesirable to the employer in certain situations. Also, although an investment tax credit is allowed for contribution to certain ESOPs, such plans must satisfy more extensive requirements that those generally applicable to qualified employee benefit plans. *See* IRC §409A.

Another element in determining when an employee is taxed is that of constructive receipt. In Revenue Ruling 60-31, the Service stated that "under the doctrine of constructive receipt a taxpayer may not deliberately turn his back on income and thereby select the year for which he will report it . . . nor may a taxpayer, by a private agreement postpone receipt of income from one taxable year to another. . . ." Once someone has an absolute unrestricted right to the income, one cannot then postpone taking it and thus postpone tax on it.[93] In Revenue Ruling 60-31, the taxpayer was employed under a contract which provided that he was entitled to receive in 1941 a percentage of his employer's net profits for 1940. On November 1, 1940, the parties agreed that the taxpayer's share of the 1940 profits should be paid in 1942. The court held that the taxpayer was not in constructive receipt of income in 1941 and stated that the agreement to defer was a bona fide business transaction which should be respected. It should be noted that although the agreement to defer was

---

92 *Id.*

93 But note Howard Veit, TC Memo Docket No 5862, 8 TCM 809 (1947), *acq* 1947-2 CB 4.

executed after most of the services were performed, the amount due at such time was not definitely determinable since it was based upon 1940 profits.

The Tax Court in *Howard Veit*[94] considered the following fact pattern. On December 26, 1941, after employee services were performed and 1940 profits were determined, but before payment was due under the revised agreement, the parties agreed that the deferred amount should not be paid in 1942 but should be paid in five annual installments beginning in 1943. The Tax Court held that the taxpayer was not in constructuve receipt of income in 1942, reasoning that a cash basis taxpayer can effectively agree to defer income prior to the time payment of the income is due, provided that the agreement is not a sham. The Internal Revenue Service has been very unhappy with this decision.

§5.15 —Innovative Tax Planning Devices

In inflationary times, it is unrealistic for a corporation and its employees merely to defer income to a future date in the hopes of reducing tax brackets. Accordingly, innovative tax planning requires some sort of arrangement utilizing the employer's funds to increase the value of the amount deferred. For example, the employer could enter into an agreement with the employee that it would take an amount of its own funds each year during the period of deferral and invest that amount either in a diversified portfolio or in the employer's stock. At the end of each year, the employer would pay tax on the taxable gains in the portfolio and at the end of the deferral period the net proceeds would be paid to the employee. As long as the employee's rights under this type of program are those of a general creditor of the employer, with respect to the portfolio, the deferral of tax on the entire amount will be permitted.

Under some deferred compensation plans, the employer credits interest to the employee's account each year in order to alleviate the adverse effects of inflation.[95] The tax treatment of this type of arrangement would be identical to that afforded the investment type plans.

The employer may wish to establish a revocable trust to provide the deferred compensation payments to the employee. If this trust is created as merely another pocket of the employer, the employee should not be taxed currently on the promise of the employer to make deferred compensation payments.[96]

Similarly, an employer should be able to fund its promise to pay deferred compensation payments by the purchase of an insurance or annuity contract, provided that the employer is the absolute owner of the contract and the employee has no rights in and to such contract.[97]

---

94 8 TCM 919 (1949).

95 Rev Rul 71-419, 1971-2 CB 220.

96 Rev Rul 67-289, 1967-2 CB 163.

97 Rev Rul 68-99, 1968-1 CB 193.

A popular innovative tax planning device is known as the *phantom stock plan.* This type of program generally provides the employee with the financial benefits of stock ownership without the attendant cost of risk and the actual ownership. Under such a plan, the employer corporation awards bonuses to an employee in the form of hypothetical shares of its sotck. This stock, known as phantom stock, is credited to the employee's account as are all cash and stock dividends and stock splits which are attributable to the phantom shares. No tax is payable by the employee at the time of such crediting. Upon termination of employment (due to retirement, death or other cause), or at a specified future date, the employee becomes entitled to receive an amount equal to the excess (if any) of the market value of the phantom shares on the date of such termination over the value of such shares on the date (or dates) on which such shares were awarded. This amount can be paid out in installments (either in cash or common stock of the company) over a period of years. The deferral is acknowledged and recognized.[98]

A variation of this program would pay the employee at the deferred date the number of shares credited to his or her account or the equivalent amount of cash without an offset for the value thereof on the date the plan began. Thus, the employee is assured of receiving something for any phantom shares held even if the market value of the stock at the specified future date is less than what it was at the time the stock was awarded.

Under another type of program, the employer transfers its stock to an employee for an amount equal to its book value and agrees to buy back such stock when the employee leaves the company for an amount equal to the then book value of the stock. The gain on such retransfer of stock is treated as a long-term capital gain, provided the employee has held the stock for the requisite one year.

In other plans, the employer may transfer common stock to an employee without consideration, subject to the restriction that it will repurchase an amount equal to the appreciation in book value of the stock during the term of the employee's service, or the total cash dividends on such stock during such term. Here, the employee is deemed to have acquired an interest in the stock for the sole purpose of securing the payment of deferred compensation.

Under all these plans the employee runs a risk of current taxation. A tax is immediately imposed under a deferred compensation arrangement if an economic benefit has been conferred and such benefit is the equivalent of cash.[99] In example 4 of Revenue Ruling 60-31, the taxpayer, a football player, entered into an agreement which provided for the payment of salary and bonus. At the request of the taxpayer the bonus was paid to a bank, as escrow agent, pursuant to an agreement approved by the bank, the player and his employer. The escrow agreement provided that the amount of the bonus, together with accrued interest, was to be paid to the taxpayer in installments over five years, and the taxpayer's rights in such agreement were nonforfeitable. The Service

---

[98] Letter Rul 7914076 (Jan 8, 1979) and material cited therein.

[99] Commissioner v Smith, 324 US 177 (1945).

ruled that an economic benefit had been conferred on the taxpayer, and that he was taxable currently on the amount of the bonus.

In a subsequent ruling the Service held that where a taxable prize was placed in a noninterest-bearing escrow account for the benefit of the taxpayer, the discounted value of the amount held in the escrow account was currently taxable.[100]

In *E.T. Sproull,*[101] an employer deposited $10,500 in December of 1945 in an irrevocable trust for the benefit of its employee. One-half of the principal of such trust was to be paid to the employee in December, 1946 and the balance was to be paid in 1947. The court held that an economic benefit had been conferred on the employee and that the $10,500 was taxable to the employee in 1945. The taxability of the trust account was applied because the employer had given up its rights to the funds and thus the employee, in effect, had an absolute control over the funds.

Under the economic benefit theory, if an employee has nonforfeitable rights in an insurance or annuity contract, or has ownership rights in such contract, the income is generally taxed currently on the basis of the value of the contract.[102]

Most recently, innovative tax planning for key employees has led the employer to develop the concept of salary reduction agreements. The Internal Revenue Service has approved, in private rulings, qualified master plans pursuant to which employees can elect annually, prior to the rendition of services, to have their salary reduced and to have the amount of such reduction paid to a qualified trust adopted as part of the plan. The taxpayer is taxed only when the distributions are made to him or her from the qualified trust. In addition, the amount of such reduction is apparently treated as a contribution to the plan by the employer which may be deducted by it currently, subject to limitations set forth in the Internal Revenue Code with respect to qualified plans.

In line with this, many employers permit employees to individually elect irrevocably to receive pay in the form of deferred compensation. This type of election was not thought to lead to constructive receipt of income provided that it was exercised before payment was due. However, in a change of position, the Service for advance ruling purposes now imposes the additional requirement that such election be made before the services are rendered.[103]

One reason for deferring compensation is to try to choose the period when the taxpayer's overall income makes a lower tax bracket applicable. Because of the progressive tax rate under our income tax system, effective tax planning over the work life of an employee is important in maintaining the employee's standard of living after retirement or termination of service.

---

[100] Rev Rul 62-74, 1962-1 CB 68.

[101] 16 TC 244 (1951), *affd,* 194 F2d 541 (6th Cir 1962).

[102] Richard R. Deupree, 1 TC 113 (1942).

[103] Rev Proc 71-19, 1971-25 CB 698.

§5.16 Tax Aspects of Choice of Funding Sources—Debt or Equity

Tax planning for innovation should involve consideration of corporate financing. Whenever one is setting up a corporation, an initial determination must be made as to the amount of debt and equity to be taken back by the shareholder for contributions of money and/or property. There are several tax advantages to debt. Amounts paid out as interest as opposed to dividends are deductible to the corporation. The repayment of debt is simply a return of capital to the shareholder, while the redemption of only a portion of a sole shareholder's stock is a dividend to that shareholder to the extent of the corporation's earnings and profits.[104] In addition, a corporate lender is entitled to an ordinary loss or short-term capital loss (noncorporate lender) on bad debts from loans, as opposed to long-term capital loss treatment if stock becomes worthless.[105] Finally, a corporation may reasonably accumulate money to repay debt and thereby avoid an accumulated earnings and profits tax.[106]

There are, however, corporate disadvantages to debt, since high debt-equity ratio makes a company riskier and outsider creditors may be less willing to lend the company money. In addition, too high a debt-equity ratio could result in reclassification of the debt as stock by the Internal Revenue Service. The Service has issued complex proposed regulations for determining whether various interests are to be treated as debt or equity.[107]

Under these regulations, a safe debt-equity ratio is 10 to 1 when considering all debt, and 3 to 1 when considering only shareholder debt. In addition, a reasonable interest rate must be charged to avoid partial reclassification. If the interest rate is too low, the face value of the note will exceed the note's fair market value and any excess paid above that value will be considered a contribution to capital (equity). If the interest rate is too high, the note's value will exceed its face amount and, to that extent, a dividend will be deemed distributed to the shareholder. While a reasonable interest rate is an uncertain quantity, if the debt-equity ratio (including all debt) is less than or equal to 3 to 1, any rate will be considered reasonable if it is equal to or less than the difference between the §6621 rate[108] and (1) the prime rate in effect in any local commercial bank, (2) the prime rate plus 2 per cent, (3) the average yield or outstand-

---

[104] IRC §302(b).

[105] Losses on stock are normally capital losses (long-term if the stock was held more than one year). One exception is for losses on §1244 "small business stock." Losses up to $50,000 ($100,000 on a joint return) may be deducted as ordinary losses on the sale or exchange of such stock. The stock must be common stock of a domestic corporation which had $1 million or less of capital at the time the §1244 stock was issued. IRC §1244.

[106] See **§5.08**.

[107] Proposed Treas Reg §1.385.1 *et seq.*

[108] The §6621 rate is the interest rate established annually by the Secretary on underpayments and overpayments of taxes.

ing marketable United States obligations of comparable maturity, and (4) certain rates under the §482 regulations.[109]

Until these regulations become effective,[110] whether an equity interest will be reclassified as debt is determined under common law, and while there is no certain formula, several factors are taken into account in addition to the debt-equity ratio. Favorable factors to recognition of debt include a fixed maturity date, preferably in the near future, an unconditional right to repayment of principal and interest with normal creditor default remedies, a reasonable interest rate not related to the corporation's earnings, free transferability, no subordination, some security arrangement, an instrument documented and labeled as debt, nonprorata loans (a large amount of debt held by a small shareholder and viceversa) and so forth. In addition, the shareholder should act like a creditor, enforcing collection of principal and interest and acting as an arm's length creditor would under the circumstances.

§5.17 —Industrial Development and Aribtrage Bonds

The tax benefits of external debt may be greatly enhanced by §103(b) of the Internal Revenue Code which permits the use of the industrial development bond. Interest on this type of bond is tax-exempt. This permits a corporation to float its debt at more favorable rates than if the owner of the debt had to pay taxes on the interest. Issues of $10 million or less may also qualify for nontaxable status if a local government elects the small issue exemption.[111] Also exempt is interest on such bonds issued to finance sports, conventions, trade shows, transportation, public utilities, industrial parks, residential real property for family units, energy facilities, air and water pollution control facilities, as well as water facilities available on reasonable demand to the general public, and interest on bonds committed for sale before May 1, 1968 and issued before January 1, 1969.

In floating an industrial development bond, one must be careful not to have such debenture deemed an arbitrage obligation. Funds from tax-exempt issues are sometimes used to purchase higher yielding federal obligations the interest on which is not taxable in the hands of the state or local agency issuing the tax exempts. The tax-exempt obligations generally specify that the interest on the federal bonds is to be used to service the tax-exempts. Consequently, the

---

[109] Under §482, the Secretary may allocate income and deductions between and among related taxpayers. The regulations under §482 prescribe the interest rate which may be charged for tax purposes on loans to related taxpayers.

[110] As of the date this book went to press, the §385 regulations were scheduled to go into effect on July 1, 1982. However, according to Deputy Assistant Treasury Secretary for Tax Legislation David Glickman, the Treasury planned administrative action to pull back the regulations for an unspecified period of time. Daily Tax Report (BNA) No 112, at G-8 (June 10, 1982).

[111] IRC §103(b)(6)(D).

investor has the security of a federal bond without paying tax on the interest. This type of state or local obligation is known as an arbitrage bond.

An *arbitrage bond* is defined as an obligation issued where all or a major part of the proceeds can be reasonably expected to be used (directly or indirectly) to acquire securities or obligations which may reasonably be expected, at the time of issue, to produce a materially higher yield than the yield on the state local bond issue, or to replace funds used directly or indirectly to acquire such obligations.[112] Generally, a *major portion* of the bond proceeds is an amount exceeding 15 per cent of the original face amount of the issue.[113] An arbitrage obligation issued by a state or local government is not treated as a tax-exempt state or local government bond and for all such issues after October 9, 1969, interest on arbitrage bonds is not exempt from tax.[114]

The reasonable expectations of the issuer as of the date of issuance regarding the amount and use of the bond proceeds determine whether an obligation is an arbitrage bond. Bond holders may rely upon reasonable expectations of future events established by certification of the state or local government unit, or, in the case of an issue with a fact amount of $2.5 million or less, an unqualified opinion of counsel that such obligations are not arbitrage bonds, if such opinion is reasonable and given in good faith.[115]

The yield produced by acquired obligations is not *materially higher* than the yield produced by an issue of governmental obligations if the yield produced by the acquired obligations exceeds the yield produced by the issue of governmental obligations by more than (a) 1/8 of a percentage point, or (b) at the election of the issuer, 1/2 of a percentage point if the issuer waives certain temporary period provisions. The term *yield* refers to that yield which, when used in computing the present worth of all payments of principal and interest to be paid on the obligation, produces an amount equal to the purchase price.[116]

Under Internal Revenue Code §103(c), obligations will not be treated as arbitrage bonds solely because the proceeds of the issue are invested in securities or other obligations for a temporary period until the proceeds are needed for the purpose for which the state or local government bonds were issued.[117] Thus, the common practice of investing funds raised in tax-exempt and short-term Treasury bills while a building project is under way should not necessarily have an adverse effect on the tax-exempt status of the state or local bonds.

Generally, original and investment proceeds of an issue of governmental obligations invested in acquired obligations during a three year period beginning on the date of issue are deemed invested for a temporary period provided the following requirements are satisfied: (1) an amount equaling 85 per cent

---

[112] *Id* §103(c)(2).

[113] Treas Reg §1.103-13(b)(1)(ii).

[114] IRC §103(c)(1).

[115] Treas Reg §1.103-13(a)(2).

[116] *Id* §1.103-13(b)(5).

[117] *Id* §1.103-14(a)(1).

expendable proceeds are expended on the project by the end of such a period; (2) the state or local government unit incurs substantial binding obligations to commence or acquire the project to be financed by the issue within six months after the date of issue of the governmental obligations; and (3) due diligence is exercised regarding completion of the project.[118]

Investment proceeds invested within the above period during a one-year period beginning on the date of the receipt of such investment proceeds are considered as invested for a temporary period.[119]

The temporary period, for amounts treated as proceeds of an issue because they are accumulated in a sinking fund for the issue, is 30 days beginning on the date of accumulation. However, a one-year temporary period applies to amounts received from investments of a sinking fund and a 13-month temporary period is available if the sinking fund constitutes a bona fide debt service fund.[120]

Accordingly, an industrial development bond which is not deemed an arbitrage bond offers financial advantages under the tax law for corporations in their expansion programs. Recent tax rulings open the door to even more innovative tax planning using industrial development bonds. The Internal Revenue Service has indeed ruled that the purchase with proceeds of industrial development bonds of a covenant not to compete that is limited to a definite time period is an acquisition of property that qualifies under §103(b)(6)(A) of the Code. It further states that if a nonexempt person[121] purchases such a covenant with its own funds, together with facilities it will use in connection with its existing facilities in the same city that were financed the previous year by industrial development bonds, the purchase is a capital expenditure within the meaning of §103(b)(6)(D) counted toward the $10 million small issue limitation.[122] Thus, the interest on the bonds used to purchase the covenant not to compete will be excludable from gross income of the bondholder under both situations.

The Service has ruled, however, that the purchase with proceeds of industrial development bonds of good will as part of the assets of a business is not an acquisition of property that qualifies under §103(b)(6)(A) of the code, but that if a nonexempt person purchases good will with its own funds, together with facilities it will use in connection with its existing facilities in the same city that were financed the previous year by industrial development bonds, the purchase is a capital expenditure within the meaning of §103(b)(6)(D) counted toward the $10 million small issue limitation.[123] In this situation, the interest on the bonds will not be excludable from gross income. Accordingly, the

---

[118] *Id* §1.103-14(b).

[119] *Id* §1.103-14(b)(6).

[120] *Id* §1.103-14(b)(10).

[121] A nonexempt person is anyone except a governmental unit or a tax-exempt organization under IRC §501(c)(3). IRC §103(b)(3).

[122] Rev Rul 81-55, 1981-1 CB 52.

[123] Rev Rul 81-56, 1981-1 CB 53.

innovative tax planner will use the industrial development bond not just for the purchase of the facility, but for the purchase of a covenant not to compete if appropriate. He or she will not structure any part of the purchase as a purchase of good will.

§5.18 Corporate Spinoffs, Splitoffs and Splitups

Internal Revenue Code §355 provides that if a corporation distributes to a shareholder with respect to its stock, or to a securityholder in exchange for its securities, solely stock or securities of another corporation which it controls immediately before the distribution, no gain or loss will be recognized to the shareholder or securityholder by reason of the receipt of the stock or securities, and no amount will be included in income by reason of such receipt, if certain requirements are met.

First, immediately after the distribution, both the distributing corporation and the controlled corporation must be engaged in the active conduct of a trade or business, which must have been actively conducted throughout the five-year period ending on the date of distribution.[124]

Second, the distribution must consist of either (1) all of the stock and securities of the controlled corporation, or (2) an amount of stock in the controlled corporation sufficient to constitute *control* (that is, stock possessing at least 80 per cent of the total combined voting power of all classes of stock entitled to vote and at least 80 per cent of the shares of all other classes of stock of the corporation). If less than all of the stock and securities of the controlled corporation are distributed, it must be established to the satisfaction of the Commissioner that the retention of stock or stock and securities was not part of a plan having as one of its principal purposes the avoidance of federal income tax.[125]

Third, the transaction must not have been used principally as a device for the distribution of earnings and profits of the distributing corporation, the controlled corporation or both.[126]

Fourth, the purpose of the distribution must be germane to the business of the corporation.[127]

Fifth, there must be a continuity of the entire business enterprise under modified corporate forms and a continuity of interest in all or part of such business enterprise on the part of those persons who, directly or indirectly, were the owners of the enterprise before the distribution or exchange.[128]

Section 355 includes spin-offs, split-offs and split-ups. A parent-subsidiary relationship may be dissolved and replaced by direct ownership of the parent

---

[124] IRC §355(b).

[125] *Id* §355(a)(1)(D).

[126] *Id* §355(a)(1)(B).

[127] Treas Reg §1.355-2(c).

[128] *Id* §1.355-1(b).

and the subsidiary by stockholders who formerly held only stock of the parent. One corporation can divide itself into two or more corporations. By definition, a spin-off occurs when some of the assets of a corporation are transferred to a corporation controlled by the transferor, and the stock or securities in the transferee corporation are distributed to shareholders or securityholders of the transferor corporation without surrender by the shareholders or security-holders of stock or securities in the distributing corporation. A split-off is similar except that the sharehodlers surrender a part of their stock or securities in the distributing corporation for stock or securities in the controlled corporation. A split-up occurs when the distributing corporation transfers its assets to two or more newly created corporations in exchange for their stock or securities and then, effecting a complete liquidation, passes such stock or securities up to its former shareholders.

It should be noted that the stock that may be distributed under §355 without tax to the distributees may be either common or preferred stock. However, if preferred stock is used, it may be treated as §306 stock[129] and carry with it the taint thereunder.

If the above conditions are met, §355 will apply whether or not a distribution of stock or securities is pro rata among the shareholders or securityholders. Thus, a business owned by two shareholders can be divided into two distinct corporations and each shareholder may end up owning all the shares of one of the corporations. It should be noted that the regulations caution that a transaction which is in substance a mere exchange of stock in one corporation for stock in another corporation will not qualify under §355.

Section 355 does not apply to a distribution that includes *boot* (property other than stock or securities of the controlled corporation); such a distribution may be partially tax-free. Only the boot is taxable and such tax treatment is governed by §356. Boot includes not only cash and property other than stock or securities of the controlled corporation, but also the amount by which the principal amount of securities of the controlled corporation received exceeds the principal amount of such securities surrendered. Furthermore, it includes any stock of the controlled corporation received in a taxable transaction within the five-year period preceding the date of distribution. It also includes stock rights and stock warrants.

When a spin-off fails to meet the requirements of §355, distributions of all property (stock and securities as well as boot) are taxable as dividends under §301 to the extent of earnings and profits. In the case of a nonqualifying split-off, the distributions would be treated as in redemption of stock. They could be taxed either as dividends or as the proceeds of the sale or exchange, depending upon the application of §§302 and 346. In the case of a nonqualifying split-up, distributions would be a complete liquidation of the distributing corporation and should be treated as proceeds from a sale or exchange under

---

[129] Classification as §306 stock may result in ordinary income (instead of capital gain) treatment upon the eventual redemption or sale of the stock).

§331. There is a transfer of basis and holding period under a qualified §355 transaction.

There is therefore a method by which businesses under a corporate structure can be divided up among the shareholders without the immediate application of tax. In this age of conglomerates, §355 can be very effective when incorporated in the provisions of a shareholder's agreement so that if there is a dispute among groups of shareholders, businesses can be broken up in a tax-free transaction to resolve such disputes. The innovative tax planner, in representing a corporation which is being courted for a merger or for acquisition purposes, can often look to §355 as a way of bailing out for the client without adverse tax consequences. Of course, bailing out to convert ordinary income into capital gain is not available under §355.[130]

§5.19 Investment Credits and Taxation of Intangibles

Investment tax credits, often important to an innovator, are generally allowed on depreciable property other than real estate (i.e., buildings). There are special rules regarding, for example, self-constructed property, leased property and movie and television films. A credit is also allowed for certain employee stock ownership plans. In addition, there are energy credits for producing fuel from a nonconventional source[131] and alcohol used as fuel,[132] and depreciable property includes solar, ocean, thermal and other innovative energy-producing equipment. Finally, as previously mentioned, the 1981 Economic Recovery Tax Act added credit for increases in research and development expenditures.[133] A credit of 25 per cent is allowed for research and development expenditures which exceed the average of such expenditures over the base period.[134]

The taxation of patents, franchises, trademarks and trade names in general is also an important consideration.[135] Transfers of all the substantial rights in these items will generally be considered the sale or exchange of a capital asset held for more than one year (i.e., a long-term capital gain), although the rules regarding patents are much more liberal.

Proceeds from the transfer of all substantial rights in a patent, or an undivided interest therein which includes a part of all such rights, will be considered

---

[130] Rafferty v Commissioner, 452 F2d 767 (1st Cir 1971), *cert denied,* 408 US 922 (1972).

[131] IRC §44D.

[132] *Id* §44E.

[133] See **§5.12.**

[134] IRC §44F.

[135] Copyrights might also be relevant. In general, a copyright or similar property held by a taxpayer, whose personal efforts created such property, will not be considered a capital asset. IRC §1221(3). Thus, a sale by the author or creator of the copyright will give rise to ordinary income.

a long-term capital gain to any individual except the employer of the creator or someone related to the creator, including a corporation, trust, and so forth where the transferor has a 25 per cent or more interest, and any spouse, ancestors, and lineal descendants. The payments in consideration for the patent may be contingent on its productivity, use or disposition.[136] If the consideration for a franchise, trademark or trade name includes as a substantial element a right to payment, contingent on its productivity, use or disposition, the transfer will not be treated as the sale or exchange of a capital asset.[137] If the transferor retains any significant power, right, or continuing interest with respect to the subject matter of a franchise, trademark or trade name, the transfer will not be treated as a sale of a capital asset.[138]

§5.20 Use of Offshore Entities

One of the bases of United States taxation is source. The United States taxes income which is from sources within the United States. With some innovative tax planning, some otherwise applicable United States taxation can be avoided by the use of offshore entities. Foreign tax planning cannot be fully explained within the confines of this chapter. However, one exceedingly interesting issue involving the use of offshore entities to buy and sell real estate in the United States and avoid United States taxation will be discussed as an example of the kinds of questions which may have to be considered.

Prior to the enactment of §897 of the Internal Revenue Code, a nonresident alien not engaged in a trade or business within the United States and not present in the United States for more than 183 days could sell real property in the United States and avoid all United States tax on the sale of a capital asset.[139] Moreover, if the nonresident alien were in a treaty jurisdiction which taxed real estate in the country within which it was located, there would be no tax at all.

This situation was changed with the enactment of §897 on December 5, 1980.[140] It applies to the disposition of all interests in United States real estate after June 18, 1980. This section provides that any gains or losses realized by a foreign person or corporation from the disposition of a United States real property interest will be taken into account as if such person or corporation were engaged in trade or business during the taxable year and as if such gains or losses were effectively connected with that trade or business.

Until January 1, 1985, however, §897 does not adversely affect transactions which take place within countries with income tax treaties. For example, Article

---

136 *See generally* IRC §1235.

137 *Id* §1235(a), (b)(2)(E).

138 *Id* §1235(a).

139 *Id* §871(a)(2).

140 The Foreign Investment in Real Property Tax Act of 1980, Pub L No 96-449, 94 Stat 2599.

XI of the Income Tax Treaty between the United States and the Netherlands provides that gains derived by a resident or corporation of the Netherlands from the alienation of a capital asset (other than real estate) shall be exempt from tax in the United States. A structure known as the "Dutch Sandwich" can be utilized to avoid the effects of §897. A Netherlands Antilles corporation is formed by a foreign person. The Netherlands Antilles corporation forms a Dutch subsidiary, which in turn forms a United States corporation to own the real estate.

As structured, the principal profit in a venture may be the sale (alienation) of the stock of the United States corporation by a Dutch corporation. Under the treaty between the United States and the Netherlands, there should be no United States capital gains tax payable on this transaction.

It is necessary that the Netherlands Antilles corporation be engaged in the real estate business. This is so because the Netherlands imposes a 48 per cent tax on the income of a Dutch holding company where the income is from passive real estate investment. Otherwise there is no capital gains tax. It would appear that the income of the Netherlands Antilles corporation would not be regarded as passive if the management of the venture were handled by persons who participated in the management of both the Netherlands Antilles Company and the development corporation. Precisely, what else, if anything, will be required for favorable tax treatment will have to be clarified by Dutch counsel. In any event, a ruling from the Dutch authorities granting the lower Dutch tax rate would have to be obtained.

The Netherlands Antilles, for its part, would impose a tax at the rate of 2.4 to 3 per cent on dividends received by the Netherlands Antilles corporation from the Dutch corporation.

By the use of the method described above, the tax on the principal profits of the venture would be reduced from 28 per cent to a maximum tax of 3 per cent.

It should be noted that, though all the authorities[141] who have commented on the subject have concurred that the sale by a Netherlands corporation of stock in a United States real estate holding corporation should be exempt from United States capital gains tax under the Netherlands treaty, there are not yet any Internal Revenue rulings dealing with this subject.

This transaction exemplifies innovative tax planning by showing how foreign tax laws may be used to reduce United States taxation when nonresident aliens buy and sell real estate.

---

[141] *See, e.g.,* Round Table Conference on the Foreign Investment in Real Property Tax Act 1980, Tax Mgmt Intl J, Special Supplement (Apr 1981).

Innovation and Securities Law 6

Richard H. Rowe

Introductory Materials

§6.01 Generally

Any consideration of the various methods of external nongovernmental financing of innovation must include an analysis of the applicability of the federal securities laws.[1] While business, antitrust and tax considerations and other applicable regulatory schemes are clearly more significant to the overall success of an enterprise, failure to comply with the securities laws can have serious consequences. Failure to comply with those laws could close the doors to the financial markets and prevent an enterprise from raising the capital it needs to survive and prosper.

At the firm level, the securities laws, federal and state, and, to an increasing degree, international, may be perceived as an impediment to the financing of innovation. In theory those laws assist capital formation by assuring efficient, informed securities markets in which investors have confidence. Not all economists, however, would agree with that theory and the mandatory disclosure system established under United States federal securities laws has been subject to considerable criticism in recent years for imposing costs in excess of any benefits to be derived from the system.[2] In response to that criticism, regulators are making an effort to remove what may be viewed as unnecessary burdens imposed by the system, while preserving its basic features. Even the American Law Institute's proposed *Federal Securities Code* (1978) would retain the basic theoretical framework of the federal securities laws.

For the newer and smaller firm, the securities laws may impose costs and

---

[1] Principally: The Securities Act of 1933, *as amended,* 48 Stat 74 (codified at 15 USC §§77a-77aa); the Securities Exchange Act of 1934, *as amended,* 48 Stat 881 (codified at 15 USC §§78a-78jj); the Trust Indenture Act of 1939, *as amended,* 53 Stat 1149 (codified at 15 USC §§77aaa-77bbbb); the Investment Company Act of 1940, *as amended,* 54 Stat 789 (codified at 15 USC §§80a-1 to 80a-64); and the Investment Advisers Act of 1940, *as amended,* 54 Stat 847 (codified at 15 USC §§80b-1 to 80b-21).

[2] *See, e.g.,* Advisory Committee on Corporate Disclosure, 95th Cong, 1st Sess, Report to the Securities and Exchange Commission ch XX (Comm Print No 29, House Comm on Interstate and Foreign Commerce 1977).

substantive regulations, including regulations affecting potential investors, which preclude adequate financing. Generally, however, the federal and state securities laws impose costs and delay on a start-up company with an innovative idea, but will not be an absolute barrier to financing.

Investor acceptance of a financing arrangement also can be affected by the securities laws. Those laws may regulate the relationship between the investor and investee or the rights of the investor in the investment. The extent and nature of that regulation may be determinative of a decision to invest.

For the larger, more established company, in addition to regulating its financing activity, the securities laws regulate the nature and timing of the disclosure required about significant developments in the company's business. For these established firms, fear of liability or competitive effects of disclosure may inhibit risk taking.

While these impediments may never be completely eliminated, their impact can be reduced and the distinct trend is in that direction, particularly with respect to smaller businesses.[3]

§6.02 Sources and Types of Financing—In General

While the principal source of financing for United States business is internally generated funds, that is not the case for the newer company in need of start-up or seed capital or bridge financing. A new company engaged in an innovative project may incur substantial research and development expenses before it reaches the stage of commercial success. Even a larger company may not wish to carry the overhead for a research project, and a very large project with long-range potential may result in a substantial drain on earnings. Thus, both the newer, smaller company and the larger company may be required to seek external sources of financing. The type of financing desired by the company or available to it will depend upon many factors, including the company's circumstances; prevailing interest rates; the availability of credit; the state of various credit and securities markets; and the cost of compliance with applicable legal requirements.

Private capital in various forms is a typical source of financing for a start-up business. Private capital may be provided by a variety of means—loans; equity investment; or various forms of contractual arrangements. All involve securities laws considerations.

---

[3] *See, e.g.,* Q&A: Small Business and the SEC (SEC March 1980); Wertheimer, *Small Offerings: Recent SEC and Legislative Initiatives,* in Twelfth Annual Institute on Securities Regulation 507 (PLI Corporate Law & Practice, Course Handbook Series No 348, Fleischer, Lipton & Mundheim eds 1980); and System of Classification for Purposes of Exempting Smaller Issuers from Certain Reporting and Other Requirements, Securities Exchange Act of 1934 Release No 18189 (Apr 15, 1982).

§6.03 —Loans from Employees, Friends and Relatives

A new company may obtain loans or investment from enployees or from friends and relatives of the founders.

For a new company, its employees, who may also be its founders, are an important source of capital. Employees can lend money to the company or make an equity investment. They also can forgo the level of compensation they could command elsewhere. Investment by employees in a new venture not only provides necessary capital, but provides incentives to employees to work even harder to further the interest of the enterprise.

§6.04 —Commercial Loans

Commercial loans are another source of capital for the newer company. However, loans may be difficult to obtain. The company, in its earlier stages, may not be a good credit risk. It may have little in the way of assets to pledge as security for a loan. Personal guarantees of the debt by the founders of the business may help, but not always.

The company may not be able to generate sufficient cash flow to service the debt. Interest rates may be too high to be acceptable. Lenders may demand more of an equity participation than the owners are willing to relinquish. Lenders may also require a substantial equity investment by others before making a loan. The nature of the negative control over operations inherent in the typical loan agreement may not be acceptable to the company's management.

For other companies, commercial loans may well be the way to finance their operations, for tax or other considerations, particularly since ownership interests in the company will not be diluted.

§6.05 —Venture Capital

Venture capital firms are an increasingly important source of capital.[4] Membership in the National Venture Capital Association, the principal voluntary association of venture capitalists, increased from 69 in 1977 to 107 in 1981. During 1980, approximately $900 million was invested in venture capital firms, compared to only $10 million in 1975. These firms in turn invested about $1 billion in 1980 compared to $250 million in 1975. Revision of the capital gains tax rate in 1978 was an important factor in the growth of these venture capital firms. Further revision in 1981 and 1982 may result in even more growth.

Venture capitalists are willing to risk their capital in an innovative project if the chances for reward are reasonably assured or the possible rewards are great. They may, however, demand a large equity participation in the company.

---

[4] *See, e.g.*, Pollack, *Funds Outgrow Investment Opportunities*, NY Times, June 17, 1981, at D-1.

This equity participation, or the "kicker" which is the condition of a loan, may be in the form of warrants or other rights exercisable at a minimal cost to the investor.

These private venture capitalists also may insist upon the same personal guarantees and other concessions that a commercial lender would obtain. On the other hand, commercial banks may not be in a position to take the risks that a venture capitalist would be prepared to take.

Venture capitalists may provide valuable managerial and financial advisory assistance to the company. This can be of great importance if the innovator is a brilliant scientist but lacks experience in business and financial matters. Professional venture capitalists are also able to monitor their investments and provide additional financing if needed.[5]

Major corporations and their subsidiaries, General Electric Credit Corporation, for example, may also provide venture capital financing. For another example, Standard Oil of California, Standard Oil Company (Indiana) and National Distillers and Chemical Corporation are substantial investors in the Cetus Corporation, a genetic engineering firm which in turn has research contracts with those corporations. The innovative company may need substantial backing such as this to weather years of losses, before its operations become profitable.

§6.06 —Partnerships

Various forms of partnership can be used to finance a new venture. Partnerships may provide the partners with certain tax advantages, particularly if the partners are in a position to benefit from tax deductions for losses of the partnership in the start-up stage.

Offerings of a general partnership interest do not generally involve securities laws issues,[6] but subject the personal assets of the partners to substantial risks of liability for the partnership's debts. The limited partnership form permits the limited partners to limit their liabilities to their investment in the partnership. That form of organization is, however, more attractive to the passive investor who does not desire to participate in the active management of the company.

An interesting recent development has been the formation of limited partnerships for research and development projects.[7] These partnerships enter into contractual arrangements with other entities who perform the actual research and are licensed to market the products developed. The licensee (or

---

[5] See **§6.45**.

[6] *See, e.g.,* Frazier v Manson, 651 F2d 1078 (5th Cir. 1981); Williamson v Tucker, 645 F2d 404 (5th Cir), *cert denied,* — US —, 50 USLW 3278 (Oct 13, 1981); and Hirsch v Dupont, 396 F Supp 1214 (SDNY 1975).

[7] *See, e.g., R&D and Gene Amdahl form the foundation for a $55 million tax shelter,* Wall St J, July 28, 1981, at 1, col 5.

licensor) may be a not for profit organization, such as a state university.[8] The investors in the partnerships may benefit from tax deductions for the losses of the partnership in the early years of the project and later from royalties, if the project is commercially successful. Storage Technology reportedly raised $50 million privately with such a device, and Trilogy, Ltd. reportedly plans to raise money from the public through a tax shelter partnership to fund a $55 million portion of a $150 million venture capital investment. However, requirements for accounting and financial reporting by public companies involved in these types of arrangements is under study by the accounting profession. Until the results of that study are available, the accounting and financial reporting requirements imposed by the Securities and Exchange Commission, the agency responsible for administering the federal securities laws, may inhibit widespread use of these arrangements by companies subject to its jurisdiction.

§6.07 —Contractual Arrangements

Various types of contractual arrangements can be utilized to provide funds for innovation. These arrangements often will be combined with other types of financing arrangements.

In the genetic engineering field, the Cetus Corporation reportedly has been financed in part through research contracts with large corporations, including some of its major stockholders. The contracting out of research is also advantageous to the other contracting party, since it need not carry the overhead for research facilities and staff. These research services may also be performed by private nonprofit research institutes and foundations, such as Battelle Memorial Institute and smaller organizations springing up all over the country.

§6.08 —Leasing

Leasing is a financing alternative. Leveraged leases can be used for big ticket items, such as a satellite communications system or an expensive computer. It has been estimated that the original value of all outstanding equipment leases in 1979 amounted to $150 billion and that leasing may finance some 20 per cent of all capital equipment acquisitions.[9]

Due in part to tax implications, leasing may be an attractive method of financing for both lessees and lessors. Institutional investors, independent leasing companies and wealthy individual investors all engage in providing financing to companies with continuing capital equipment needs. In addition

---

[8] *See, e.g.,* Feder, *Gene Splicing Licensed,* NY Times, Aug 19, 1981, at D-1, col 4; and *College, Company Sign Research Pact on Artificial Cells,* Washington Post, Sept 13, 1981, at 2.

[9] *See generally,* The Conference Board, Report No 791, Leasing: Experiences and Expectations (1980); and *A Flood of Red Ink Swamps the Airlines,* Intl Bus Week 89 (Nov 16, 1981).

to buildings and space, computers and sophisticated office equipment are frequently leased.

Individual lease arrangements fall into two basic categories—financial (or capital) and operating. The term of a financial lease generally approximates the useful life of the asset, and the lessor generally recoups the initial investment and certain costs, and makes a profit from the lease. Sale and leaseback transactions and leveraged leases are variations of financial leases.

In sale and leaseback transactions, the owner of the asset sells the asset to an investor who leases it back to the seller. The seller thus improves its working capital position without loss of the use of the asset. Leveraged leasing involves a third party who finances a portion of the lessor's investment on the leased asset.

Leasing may be worthwhile for a reasonably new company whose operations have not become profitable because it cannot use the full benefits of depreciation or the investment tax credit associated with ownership. Any decision to engage in lease financing should involve extensive financial and tax analysis.

§6.09 —Profit Sharing

Profit sharing arrangements represent another method for an investor to obtain a return on an investment or for partners to share the profits of the enterprise. Such arrangements also may be used to provide incentive compensation to important executives and employees. On occasion, employees of a division or a subsidiary engaged in an innovative project will buy the division or subsidiary or the rights to a product or process when the parent no longer desires to continue to incur the drain on its earnings. The parent may also determine to distribute the division or subsidiary to its stockholders to avoid accounting for its operations.

§6.10 —Joint Ventures

Various types of joint ventures can be used to share the risk of a new enterprise. Even large corporations may not wish to expend millions of dollars to develop a new project. A joint venture investment with one or more other corporations will spread that risk. This may involve a partnership or a new corporation in which the venturers own all of the capital stock. That jointly held corporation may itself be financed from external sources other than the original venturers.

§6.11 —Franchising and Licensing

Franchising and licensing arrangements are means of financing innovation where the innovator may not have the facilities, capital or business experience to successfully develop, manufacture or market a product. What control the innovator would maintain would have to be provided by agreement and the return would be provided through royalties, fees and profit sharing arrangements.

§6.12 —Merger, Exchange or Sale of Assets and Other Buy-Out Arrangements

Some companies may find it appropriate to sell out to another company through a merger, exchange or sale of assets or other combination and receive equity securities of that company in exchange, in order to be able to benefit from any successful development of an innovation. This type of transaction may be particularly attractive if the acquiring company is in a position to benefit from the innovator's tax losses.

§6.13 —Public Offerings

During the hot issue securities markets of the late 1960s and earlier 1970s, it was often remarked that if all other means of financing were unavailable, the public was always ready to invest in an exotic new venture.[10] There is a germ of truth in these observations, since professional venture capitalists and institutional investors, due to regulatory or business considerations, may not be prepared to take certain risks where public investors would be prepared to gamble.

There are, of course, other reasons than a need for capital for making a public offering. If a public offering results in a market for the company's securities, the market will provide a measure of the value of the original investors' and venture capitalists' investments. This may facilitate the valuation of those investments for estate planning, tax and other purposes. The availability of a public market for the company's securities also would open up new opportunities for employee incentive compensation. It would also provide liquidity for investors, particularly venture capitalists who may desire to turn over their investments.

Going public also results in accepting responsibilities and obligations as to a new group of owners—the public stockholders—and operating in the fish bowl created by the periodic disclosure requirements of the federal securities laws. As evidenced by the number of public companies which have gone private

---

[10] *See* Public Investigation in the Matter of the Hot Issues Securities Markets, SEC Administrative Proceeding File No 4-148. Hot issue activity has increased in the 1980s. *See, e.g.,* SEC Monthly Statistical Review, Release PR-32 (July 1981). There are a number of available sources which provide an overview of the legal issues involved in going public. *See, e.g.,* Robinson & Eppler, Going Public (2d ed rev 1978). Also, various financial printers publish helpful reprints on the subject. *See, e.g.,* Schneider, Manko & Kant, Going Public, Practice, Procedure and Consequences (Packard Press reprint 1981); Halloran, Going Public (3d ed Sorg Printing Co Aug 1979).

recently, and the regulatory reaction,[11] operating as a public company may not be appropriate for every business.

Public offerings are also an expensive method of raising capital with costs, not including underwriters' compensation, often in the hundreds of thousands of dollars.

Accordingly, a company should carefully analyze all other alternatives before going public. Also, many companies should not go public before they are properly seasoned. These companies should have some reasonable history of operations and management experience. The two most publicized offerings by genetic engineering firms, Cetus Corporation and Genentech, Inc., which were made only after a number of years of seasoning, 10 years in the case of Cetus, are examples of this scenario. More seasoning may assist in avoiding the disasters which struck many of the companies which went public prematurely in the late 1960s and early 1970s, and have done so now in the 1980s.

§6.14 —Industrial Development Bonds

Public or private sale of industrial development bonds (IDBs) is an increasingly popular method of financing facilities for start-up ventures. It has been estimated that $8.4 billion was raised in 1981 by this method.[12] An IDB is issued by a municipality in accordance with the law of the state in which it is located, but it is not a general obligation of the municipality. The investors must look to the revenues from the facility which is privately owned or leased to fund payments of principal and interest. The interest on certain issues which comply with the provisions of §103 of the Internal Revenue Code of 1954, as amended, and related regulations, is exempt from federal income taxation, and the bonds and any separate but related security are exempt from registration under the 1933 Act.[13]

This type of financing is often used in conjunction with other sources. For example, a limited partnership could be formed to own the facilities and lease them to the innovator, and operating funds could be provided by venture capital equity investors.

The extent of IDB financing and its use to finance facilities for business such as fast food franchises has caused something of a political furor and its availability could be curtailed by Congress or the states.

---

[11] *See, e.g.*, Public Fact-Finding Investigation and Rulemaking Proceeding In The Matter of "Going Private" Transactions by Public Companies or Their Affiliates, SEC Administrative Proceeding File No 4-178. These proceedings ultimately resulted in the adoption of Rule 13e-3, 17 CFR §240.13e-3 and related Schedule 13E-3, 17 CFR §240.13e-100, which regulate going private transactions.

[12] *See, e.g.*, Sec Reg & L Rep (BNA) No 606, at A-10 (July 31, 1981).

[13] See **§6.20.**

§6.15 —Offshore Financing

There are investors in many parts of the world eager to invest in the United States because of our relatively stable economy and our stable political climate. These investments may take many different and often imaginative forms. For example, financing lease arrangements may be entered into which take advantage of foreign tax laws applicable to the foreign lessor, and which could reduce interest costs as much as 1 or 2 per cent.[14]

Applicability of the Securities Laws

§6.16 Overview

If a financing transaction involves an offer to sell or sale of a security, the federal and state securities laws will apply. The security will be required to be registered under the 1933 Act,[15] unless an exemption from registration is available. For the start-up venture, registration of a security under that act generally requires the preparation of a comprehensive disclosure document which is filed with the Securities and Exchange Commission (SEC) and furnished to prospective investors.[16] The registrant and its officers and directors and controlling persons are subject to civil liability if the registration statement contains materially misleading statements.[17]

Once a company has become public, it is subject to the full panoply of federal securities regulations. The company will be required to file an annual report with the SEC which contains disclosure comparable to that in the company's registration statement. The company will also be required to file current and other reports with the SEC, and the company's communications with its stockholders will be subject to the provisions of the federal securities laws.

The federal securities laws and, to a lesser degree, state securities laws may add to the cost of external financing of innovation and, to some degree, discourage the risk taking necessary for innovation. The direct expenses involved in complying with those laws, including professional fees, may be so high that some smaller companies will be prevented from raising capital. Distribution costs reduce the amount available for investment and investors will pay less if there are restrictions on the liquidity of their investments. The legal risks involved also impose a cost.

The existing system of dual state and federal regulation of securities transactions, although politically entrenched, has little theoretical justification. There have been efforts to reduce the burdens of dual regulation. The Small Business

---

[14] *See, e.g.,* NY Times, Apr 7, 1981, at D-1, col 3.

[15] Securities Act of 1933, *as amended,* 48 Stat 74 (1933) (codified at 15 USC §§77a-77aa).

[16] *Id* §, 15 USC §77e.

[17] *Id* §§11, 12, 15, 15 USC §§77k, l, o.

Investment Incentive Act of 1980[18] added §19(c) to the 1933[19] Act which provides for greater coordination between the SEC and state securities officials.

The American Law Institute's proposed *Federal Securities Code,* which is supported by the SEC, would effect a limited pre-emption of state blue sky laws in specific areas.[20] It is doubtful that the code will be enacted in its present form or that anything but the most limited pre-emption of state blue sky laws would be otherwise politically feasible. However, some further reform could be accomplished under existing law, particularly at the federal level. For example, the SEC has the authority to defer completely to the states with respect to any issue of securities which aggregates less than $5 million, if it can find that the states will provide adequate protection to investors.[21] Section 19(c) of the 1933 Act also provides the SEC with broad authority to adopt uniform state exemptions from registration for federal purposes. This type of reverse administrative pre-emption is worthy of serious consideration, particularly in view of the increasing scarcity of federal regulatory resources.[22]

In addition to increasing the cost of capital, the disclosure requirements of the federal securities laws could cause an innovator to pursue a course of conduct he or she might not pursue under a different set of rules. Those laws require public disclosure about important developments in a company's business in order to keep stockholders and the securities market informed about the company. However, there is considerable concern about the unintended effects of that disclosure. There may be competitive effects from disclosure about innovative developments. Liability may be imposed by the securities laws for unintentionally misleading disclosure. The information required to be disclosed may also be of value to persons with whom the company is engaged in some dispute. Moreover, existing financial reporting requirements, and executive compensation arrangements based on the resulting reports, tend to emphasize short-run performances at the expense of developments which may not produce results for longer periods.

Because of the degree of uncertainty as to the precise standards for compliance with the disclosure requirements of the securities laws and the fear of liability, as well as the emphasis on short-term results, the willingness to take the risks necessary to develop an innovative product or processes may be reduced. Liability concerns may be greater for a company engaged in innovation since there may be greater risk of failure and resultant loss of investment.

---

[18] Pub L No 96-477, 94 Stat 2275 (Oct 21, 1980).

[19] 15 USC §77s(c).

[20] §1904 (1978).

[21] Securities Act of 1933, §3(b), 15 USC §77c(b).

[22] The SEC has adopted a limited exemption from registration of securities under the 1933 Act, based, in part, upon state regulation. Regulation D—Revision of Certain Exemptions from the Registration Provisions of the Securities Act of 1933 for Transaction Involving Limited Offers and Sales, Securities Act of 1933 Release No 6389 (Mar 8, 1982).

In that event, investors might seek to recoup their losses through an action under the securities laws.

Efforts are continuously being made to reduce uncertainties and to develop disclosure requirements which are more future-oriented and sophisticated. Other efforts are being devoted to easing the burden of the securities laws on smaller companies and facilitating the raising of capital by new ventures.[23] This chapter is primarily devoted to federal securities regulation. One should never overlook the fact that the blue sky laws of most states also have registration provisions which may impact more directly on smaller companies.[24]

The securities laws should never prevent an innovative development. These laws should represent only another cost factor, and are perceived to have considerably less impact on innovation than do the federal income tax laws or other regulatory systems.[25]

§6.17 Is a Security Involved?

The securities laws will apply to a transaction engaged in to finance innovation, if the transaction involves the purchase or sale of a *security* as that term is defined under those laws. However, there are many types of financing transactions which do not involve sale of a security. Therefore, a financing transaction should be analyzed to determine whether it is subject to the securities laws at all. Unfortunately, that question may not always be easily answered.[26] Moreover, a public company may have an obligation to disclose material facts about a financing transaction, even if it is not a securities transaction.[27]

In §2(1) of the 1933 Act, the term *security* is defined.

> Unless the context otherwise requires—(1) The term "security" means any note, stock, treasury stock, bond, debenture, evidence of indebtedness, certificate of interest or participation in any profit-sharing agreement, collateral-trust certificate, pre-organization certificate or

---

[23] *See, e.g.,* Summary of Comments Relating to: Small Business Hearings and Proposed Form S-18, SEC File No S7-734. The SEC also has reduced periodic reporting requirements under the 1934 Act for smaller companies. System of Classification for Purposes of Exempting Smaller Issuers from Certain Reporting and Other Requirements, Securities Exchange Act of 1934 Release No 18647 (Apr 15, 1982). The Financial Accounting Standards Board, the private sector organization which sets accounting standards which are recognized by the SEC as authoritative, has requested public comment on the impact of financial reporting rules on smaller companies. 25 Sec Reg Guide (P-H) ¶25.3 (Dec 16, 1981).

[24] Commerce Clearing House publishes a looseleaf service covering state blue sky laws. Blue Sky Law Reporter (CCH).

[25] *See, e.g.,* SEC File No S7-734, *supra* note 23.

[26] For an excellent summary of the issues, see Schneider, *The Elusive Definition of a Security,* 14 Rev Sec Reg 981 (Jan 23, 1981).

[27] See **§6.56.**

> subscription, transferable share, investment contract, voting-trust certificate, certificate of deposit for a security, fractional undivided interest in oil, gas, or other mineral rights, or, in general, any interest or instrument commonly known as a "security", or any certificate of interest or participation in, temporary or interim certificate for, receipt for, guarantee of, or warrant or right to subscribe to or purchase, any of the foregoing.[28]

The definitions of that term in the other federal securities statutes are substantially similar.[29] Under the laws of some states, the term may be defined somewhat differently than it is under federal law.[30]

One cannot make a determination whether a particular arrangement or instrument is a security based solely on the words of the statute. Both the federal and state courts and the Securities and Exchange Commission (SEC) have placed an extensive gloss on the statutory definition. The federal courts tend to follow the rule of economic reality laid down by the Supreme Court in *United Housing Foundation, Inc v Forman,*[31] in determining whether a particular transaction involves a security. Although this results in some uncertainty, it is basically a sensible rule.

In *Forman,* the Court analyzed a transaction involving the lease of an apartment in a nonprofit housing cooperative where shares of stock in the cooperative had to be purchased as a condition to entering into the lease. The court found that the name given the instrument is not dispositive, unless the characterization of the instrument creates a reasonable expectation in the purchaser that the protections of the federal securities laws apply to the transaction. A person who is seeking residential housing for personal use is unlikely to believe he or she is making an investment in a security. The required purchase of stock was therefore only incidental to the primary purpose of the transaction, and the federal securities laws did not apply.[32]

It will be fairly clear in most situations involving the issuance of stock, debentures or bonds that a security is involved. However, there are at least

---

[28] 15 USC §77(b)(1).

[29] *See, e.g.,* Securities Exchange Act of 1934, §3(a)(10), 15 USC §78c(a)(10); Investment Company Act of 1940, §2(a)(36), 15 USC §80a-2(a)(36); Investment Advisers Act of 1940, §202(a)(18), 15 USC §80b-2(a)(18).

[30] *See, e.g.,* Silver Hills Country Club v Sobieski, 55 Cal 2d 811, 361 P2d 906, 13 Cal Rptr 186 (1961).

[31] 421 US 837 (1975).

[32] The nonprofit nature of the real estate investment may not be determinative. Grenader v Spitz, 537 F2d 612 (2d Cir 1976), *cert denied,* 429 US 1009 (1976). On the other hand, if the opportunity for profit is emphasized in the sales literature, it may be evidence that a security is involved in the transaction. SEC v United Benefit Life Ins Co, 387 US 202, 211 (1967), *quoting* SEC v CM Joiner Leasing Corp, 320 US 344, 352-53 (1943); and *see, e.g.,* Guidelines As To The Applicability of the Federal Securities Laws to Offers and Sales of Condominiums or Units in a Real Estate Development, Securities Act of 1933 Release No 5347 (Jan 4, 1973). The SEC has declined to take a position on the so-called "time sharing" arrangements for vacation housing. *See, e.g.,* Securities Act of 1933 Release No 6253 (Oct 28, 1980).

three types of transactions, relevant to the financing of innovation where some uncertainty exists. These are commercial loans from banks or other lenders; sale of a business; and investment contracts.

§6.18 —Commercial Loans

Commercial loans generally involve the issuance of a note by the borrower or the creation of some other evidence of the indebtedness. On their face, these instruments are securities. Some courts have gone behind the characterization of the instrument to examine the expectation of the parties to the transaction, e.g., was the loan made as an investment or was it of a commercial or mercantile character.[33] These courts appear to be becoming increasingly reluctant to permit the securities laws to be used as a remedy for commercial loan transactions gone awry. While such an analysis is consistent with the spirit of the *Forman* rule, the courts have not developed precise standards for resolving the issue and the Supreme Court has not addressed the issue directly.

Even if the loan transaction involves a security, the transaction is probably exempt from registration.[34] The issue, then, only becomes significant if litigation over the debt arises and one of the parties sees an advantage to be gained by asserting a claim under the antifraud provisions of the securities laws.

A careful lawyer should always assume that if any issue may be raised, it will be raised, and should take steps to protect the client's interest to the extent practicable. Drafting the loan agreement to reflect the intent of the parties to enter into a commercial loan arrangement may help as an evidentiary matter. It will not preclude the issue from arising, since under the federal securities laws, contractual waivers of rights created under those laws are voidable.[35] Therefore, care should be taken to assure that the other parties to the transaction are provided with disclosure about the borrower which is adequate under the circumstances should a securities law issue later arise.

---

[33] *See, e.g.,* Union Planters Natl Bank v Commercial Credit Business Loans, Inc, 651 F2d 1174, [1981-82 Transfer Binder] Fed Sec L Rep (CCH) ¶97,996 (6th Cir), *cert denied,* — US —, 50 USLW 3486 (Dec 15, 1981). Robins v First Am Bank, 514 F Supp 1183 (ND Ill 1981); and Bellah v First Natl Bank, 495 F2d 1108 (5th Cir 1974). *Cf* Marine Bank v Weaver, — US —, 50 USLW 4285 (Mar 8, 1982) (certificate of deposit and business agreement not securities). Meason v Bank of Miami, 652 F2d 542 (5th Cir 1981) (certificate of deposit a security), *cert denied,* — US —, 50 USLW 3668 (Feb 23, 1982), and McGovern Plaza Joint Venture v First of Denver Mortgage Investors, 562 F2d 645 (10th Cir 1977) (loan commitment is not a security). *See generally* Sonnenschein, *Federal Securities Law Coverage of Note Transactions: The Antifraud Provisions,* 35 Bus Law 1567 (July 1980).

[34] See **§6.23.**

[35] *See, e.g.,* Securities Act of 1933, §14, 15 USC §77n, and Securities Exchange Act of 1934, §29, 15 USC §78cc.

§6.19 —Sale of a Business

When an entire closely owned business is sold to another person by an individual or a closely related group, although the sale may be accomplished by the transfer of stock, the federal courts may utilize a *Forman* analysis to find that the stock is not a security for purposes of the federal securities laws.[36] This type of analysis may logically be extended to the sale of part of a business, a subsidiary or division, for example, or even to the sale of a controlling interest in a business.[37] In any of these situations, the purchasers of the business must look to their own efforts to produce profits, not to those of the seller. Application of the *Forman* analysis in this context has not, however, been universally accepted and has been opposed by the SEC as *amicus curiae*.[38]

The issue becomes significant if a dispute arises subsequent to the sale, since any securities sold in the original transaction probably would be exempt from registration under the 1933 Act. Again, careful drafting of the sales agreement may provide evidence of the intent of the parties to enter into an agreement for purchase and sale of an entire business which, for tax considerations or other reasons, may be evidenced by a transfer of stock. It is also important that disclosure adequate under the circumstances be provided, in the event that it is later found that the transaction did involve the sale of a security.

§6.20 —Investment Contracts

There are investment transactions in which the investment may not be evidenced by an instrument such as a stock certificate or note, or indeed, by any instrument. Nevertheless, it is well accepted that an investment of money (or other valuable consideration, including services) in a common enterprise, with an expectation of profits from the managerial efforts of others, involves the acquisition of a security, i.e., an investment contract. This is, in fact, part of the *Forman* economic reality analysis. There still may be issues such as how extensive the investors' own efforts to produce profits need to be to preclude finding an investment contract, but the investment contract concept is a rea-

---

[36] *See, e.g.,* Fredericksen v Poloway, 637 F2d 1147 (7th Cir 1981), *cert denied,* — US —, 49 USLW 3882 (May 26, 1981); Anchor-Darling Indus, Inc v Suzzo, 510 F Supp 659 (ED Pa 1981); and Canfield v Rapp & Son, 654 F2d 459 (7th Cir 1981).

[37] Reprosystem BV v SCM Corp, 522 F Supp 1257, [1981-82 Transfer Binder] Fed Sec L Rep (CCH) ¶98,207 (SDNY 1981); Oakhill Cemetery of Hammond, Inc v Tri-State Bank, 513 F Supp 885, [1981-82 Transfer Binder] Fed Sec L Rep (CCH) ¶98,275 (ND Ill 1981) (controlling interest); Somogyi v Butler, 518 F Supp 970 (DNJ 1981) (controlling interest). In addition to the *Forman* economic reality analysis, these cases utilize the investment contract analysis of SEC v CM Joiner Leasing Co, 320 US 344 (1943) and SEC v WJ Howey Co, 328 US 293 (1946), regarding an investment of money in an enterprise with the expectation of profit from the efforts of others. *See generally* Seldim, *When Stock Is Not a Security: The Sale of Business Doctrine Under the Federal Securities Laws,* 37 Bus Law 637 (Jan 1982). *But see* Golden v Garafalo, Fed Sec L Rep (CCH) ¶98,656 (2d Cir May 5, 1982).

[38] Golden v Garafalo, Fed Sec L Rep (CCH) ¶98,656 (2d Cir May 5, 1982).

sonably well-settled area of the law with a large body of reported cases at both the federal and state levels.[39]

Franchise, license, distributorship and similar agreements should be subjected to a *Forman* analysis. Communications relating to these arrangements should be carefully scrutinized. The arrangements should not be characterized as investments, which might be found to be evidence of a security.

The SEC, by adopting its Rule 131[40] has taken the position that an industrial development bond financing arrangement involves the issuance of a separate security by the owner or lessee of the facilities being financed. This security is in the nature of an investment contract. The SEC's theory is that the investors in the bonds, which are not backed by the full faith and credit of the municipal issuer, look to the efforts of the user of the financed facility to produce the revenue necessary to service the bonds. The SEC has, however, exempted separate securities related to certain industrial development bond financings from registration under the 1933 Act in Rule 131 under that act. The bonds generally are exempt under §3(a)(2) of the 1933 Act.[41] Notwithstanding these exemptions, the SEC has sought to raise the consciousness of participants in industrial development bond offerings with regard to their disclosure obligations through use of its enforcement powers.[42] For this and other reasons, investors in these offerings are generally furnished information comparable to that furnished in connection with an offering of securities registered with the SEC.

§6.21 Is a Sale Involved?

In addition to a determination of whether a security is involved, another approach for determining whether the securities laws apply would be to analyze the transaction to determine whether it involves a *sale* of a security for value. If it does not, the securities laws do not apply to the transaction. The analysis sometimes is blurred with the analysis of what constitutes a security, and arises most frequently in connection with employee benefit plans and bonuses in the form of securities, for example.

The sale issue can arise, however, where a company distributes the stock of a subsidiary to its stockholders as part of a corporate restructuring, i.e., a

---

[39] *See, e.g.,* SEC v CM Joiner Leasing Corp, 320 US 344 (1943); SEC v WJ Howey Co, 338 US 293 (1946); and United Housing Found, Inc v Forman, 421 US 837 (1975).

[40] 17 CFR §230.131.

[41] 15 USC §77c(a)(2).

[42] *See, e.g.,* SEC v Calhoun County Medical Facility, Inc, [1981-82 Transfer Binder] Fed Sec L Rep (CCH) ¶98,243 (ND Miss May 28, 1981) (law firm consented to adopt procedures for compliance with securities laws in connection with IDB offerings); and SEC Staff Report on Transactions in the Marine Protein Corporation Industrial Development Revenue Bonds, Securities Exchange Act of 1934 Release No 15719 (Apr 11, 1979).

spin-off. In that case, registration of the stock may be required.[43]

Exemptions under the Securities Laws

§6.22 Exempt Securities

There are two types of securities transactions under the Securities Act of 1933, as amended.[44] Those involving securities which are required to be registered with the Securities and Exchange Commission (SEC), and those involving securities or transactions exempt from registration. Every security which is sold by means of the facilities of interstate or foreign commerce or the United States mails must be registered with the SEC, unless an exemption under the 1933 Act is available.[45] Most securities transactions are exempt. There are only a few thousand each year which require registration. In connection with many thousands of other transactions, considerable effort and expense is required to perfect an exemption. In the past, the tendency of the SEC and, to a lesser degree, the courts has been to narrowly construe exemptions to further a perceived policy of encouraging registration. There is less of this tendency currently since many of the benefits of registration can now be provided in an exempt transaction.

Under the 1933 Act, securities may be exempt or the transaction in which securities are sold may be exempt. There are no specific exemptions for companies involved in innovative activities, although efforts are being made to provide exemptions which would be available for transactions to finance new ventures of all types.

The securities exemptions generally would not be directly relevant to the financing of innovation. There are exceptions. One is the exemption for certain industrial development bonds which could be used to finance facilities.[46]

While industrial development bonds are exempt from registration, underwriters of public issues generally insist on preparation and distribution of an offering circular for public offerings which contains much of the same information found in a 1933 Act prospectus. The SEC has actively enforced the antifraud provisions of the securities laws in connection with these types of offerings.[47] Accordingly, the same degree of care which goes into preparation of a 1933 Act prospectus should be used in the preparation of an offering circular for an exempt offering.

---

[43] *See, e.g.*, Application of the Securities Act of 1933 and the Securities Exchange Act of 1934 to Spinoffs of Securities and Trading in Securities of Inactive or Shell Corporations, Securities Act of 1933 Release No 4982 (July 2, 1969).

[44] 48 Stat 74 (1933) (codified at 15 USC §§77a-77aa).

[45] *Id* §5, 15 USC §77e.

[46] See **§6.14.**

[47] *See, e.g.*, SEC Staff Report on Transactions in the Marine Protein Industrial Development Revenue Bonds, Securities Act of 1934 Release No 15719 (Apr 11, 1979).

Another possible exception would be a project financed by an eleemosynary organization operated "exclusively for religious, educational, benevolent, fraternal, charitable, or reformatory purposes and not for pecuniary profit. . . ."[48] Innovative research externally funded could be performed by such an organization. However, funding would normally be accomplished by gifts, grants or contractual arrangements rather than by sale of securities exempt under §3(a)(4) of the 1933 Act.[49]

§6.23 Exempt Transactions—Generally

There are a myriad of transactional exemptions from registration provided in the 1933 Act and in the rules and interpretations the Securities and Exchange Commission (SEC) has adopted under that act.[50] Each should be considered to determine which best fits the client's circumstances and needs. Often, one of these transactional exemptions may not be used in combination with another transactional exemption.[51]

A company can privately place its securities with a small group of investors.[52] A local business can sell its securities to residents of the state in which it is incorporated or organized and does a substantial portion of its business.[53] An issue of securities in an aggregate amount of up to $5 million may be sold to accredited investors.[54] Several administrative exemptions are available for small offerings in aggregate amounts of from up to $500,000[55] to up to $5 million.[56] Also, a company can make an offshore offering solely to non-United States investors without registering the securities in the United States.

Federal law does not provide an automatic exemption under state law and the terms and conditions of the various exemptions may vary from state to state, although the Uniform Securities Act[57] has been enacted by 34 states and the District of Columbia. Thus, an analysis of laws of the states in which the offer will be made is required.

It also should be recognized that the exemptions provide relief only from the registration provisions of the securities laws. The antifraud[58] and certain

---

[48] Securities Act of 1933, §3(a)(4), 15 USC §77c(a)(4).

[49] *Id.*

[50] A chart comparing various exemptions appears in app 6-1.

[51] See **§6.24.**

[52] Securities Act of 1933, §4(2), 15 USC §77d(2). *See generally* Schneider, *The Statutory Law of Private Placements,* 14 Rev Sec Reg 869 (Aug 26, 1981).

[53] Securities Act of 1933, §3(a)(11), 15 USC §77c(a)(11).

[54] *Id* §4(6), 15 USC §77d(6).

[55] 17 CFR §230.504.

[56] *Id* §230.505.

[57] 1 Blue Sky L Rep (CCH) ¶5501.

[58] Principally Securities Act of 1933, 17(a), 15 USC §77q(a); Securities Exchange Act of 1934, §10(b), 15 USC §78j(b), and 17 CFR §240.10b-5.

express civil liability provisions will still apply.[59]

§6.24 Definition and Significance of an Offering

The term *offer* is defined in §2(3) of the 1933 Act to "include every attempt to offer, to dispose of, or solicitation of an offer to buy, a security or interest in a security, for value." The term *offering,* however, is not defined in the securities statutes or by Securities and Exchange Commission (SEC) rule. It is important to determine what activities constitute an offering, because of the SEC's *integration doctrine.* If offers and sales of securities are part of an offering (i.e., integrated), then the conditions to the availability of any exemption must be satisfied with respect to each offer and sale. Moreover, exemptions generally may not be used in combination for parts of a single offering. For example, the intrastate offering exemption may not be used as an exemption for offers and sales to the public in one state and the private placement exemption used for offers and sales in another state, if the offerings in both states are integrated.

The SEC has sought to provide some guidance by provision of five factors which should be considered in determining whether offers and sales should be integrated:

1. Whether the offers and sales are part of a single plan of financing
2. Whether the offers and sales involve the issuance of the same class of security
3. Whether the offers and sales are made at or about the same time
4. Whether the same type of consideration is to be received for the securities
5. Whether the offers and sales are made for the same general purpose[60]

The SEC has also provided objective standards in exemptive rules and regulations, specifically, Regulation D[61] and Rule 147[62] under the 1933 Act. Under those rules, offers and sales which are separated in time from other offers and sales by a period of six months during which, with certain limited exceptions, there were no offers or sales of any securities of the same issuer, will not be integrated. The clean six-month period is a safe harbor. If there are offers or sales within the six months, registration is not automatically required; rather, the five-factor analysis must be utilized to determine whether the transactions should be integrated.

The integration doctrine may be of particular concern to a company engaged in a new innovative venture which may need to raise additional capital

---

[59] Principally, Securities Act of 1933, §12(2), 15 USC §77l(2).

[60] 17 CFR §230.502(a), note.

[61] *Id* §§230.501-506.

[62] *Id* §230.147.

by various methods periodically. The problems are not, however, insurmountable.

Apple Computer, Inc., one of the more widely publicized new ventures to go public in the 1980s, reportedly raised capital through a number of different methods. It was initially financed through private placements of stock to its officers and directors and to venture capital investors, and through various forms of debt financing. Then it made a public offering of shares registered with the SEC in December of 1980. In January of 1981, it made an intrastate offering in California and, later in 1981, registered securities for a secondary offering by private investors. Various versions of this scenario may be typical for a successful venture.

§6.25 Private Placement Exemption §4(2))

The *private placement* is a typical method used to finance a new business. The private placement exemption under §4(2) of the 1933 Act[63] may be used to exempt sales of securities to a few close associates of an inventor with a laboratory in his or her garage or placement of a multimillion-dollar issue of debt securities with a group of financial institutions.

Section 4(2) of the 1933 Act exempts securities transactions by an issuer which do not involve any public offering. What may seem like a straightforward exemption is fraught with uncertainty. The two principal uncertainties involve: (1) the determination of what is a public offering; and (2) when the transaction can be considered complete, in order to permit the investors to resell their securities and realize on their investment.[64]

Determination of the public nature of the offering is not a numbers game. Under the rule established by the Supreme Court in *SEC v Ralston Purina Co*,[65] an offering will be public if an offer is made to a single individual who needs the protections that registration under the 1933 Act would provide (an *unsophisticated* investor). This would be the case even if no sale were made to that offeree.

The protections provided by registration under the 1933 Act consist primarily of the disclosure in the prospectus required to be provided to the investor and the civil liability provisions in §11 of the 1933 Act,[66] which permit the investor to recover damages if that disclosure was materially misleading and which cause the sellers to use greater care in preparation of that disclosure.

Some would argue that the SEC's review of the disclosure in a registration statement is an added protection for the investors. While that may well be the case, the SEC admittedly is not in a position to review every registration statement filed with it and cannot assure that those registration statements it

---

[63] 15 USC §77d(2).

[64] See **§6.49**.

[65] 346 US 119 (1953).

[66] 15 USC §77k.

does review do not contain materially misleading information. Indeed, §23 of the 1933 Act provides that:

> Neither the fact that the registration statement for a security has been filed or is in effect nor the fact that a stop order is not in effect with respect thereto shall be deemed a finding by the Commission that the registration statement is true and accurate on its face or that it does not contain an untrue statement of fact or omit to state a material fact, or be held to mean that the Commission has in any way passed upon the merits of, or given approval to, such security. It shall be unlawful to make, or cause to be made, to any prospective purchaser any representation contrary to the foregoing provisions of this section.[67]

Unsophisticated investors are sometimes lightly described as widows and orphans. However, some widows may be highly sophisticated in the securities law sense and some professionals, doctors and dentists, for example, may be unsophisticated in that sense. What, then, are the characteristics of an investor who does not need the protection of the 1933 Act?

The U.S. Supreme Court has held that an investor does not need the protection of the registration provisions of the 1933 Act, if the investor, through a relationship with the issuer or otherwise, has access to the information registration would provide and has the knowledge and experience to evaluate the risks of the investment (a sophisticated investor).[68] Satisfying this subjective standard is not much of a problem if the investor is a large financial institution. Otherwise, it is. Some states have substituted or added suitability standards based on the investor's net worth or income on the theory that the investor must also be able to bear the economic risk of the investment.

The risks are great if an offer is made to an unsophisticated investor, since loss of the exemption subjects the issuer to claims for rescission from all purchasers, even the most sophisticated.[69] The issue is further complicated, since the burden of establishing the exemption is placed on the person seeking the exemption. The Court of Appeals for the Fifth Circuit, which since *Ralston Purina* has been the de facto Supreme Court in this type of case, has narrowly defined the §4(2) exemption and gone as far as requiring defendants to prove that they have not made any offers to unsophisticated investors.[70]

Because of these risks, a private placement must be a carefully documented transaction. Care should be taken to establish criteria for the type of person to whom offers can be made and to keep track of offerees. Copies of any

---

[67] *Id* §77w.

[68] SEC v Ralston Purina Co, 346 US 119 (1953).

[69] *See, e.g.*, Henderson v Hayden, Stone, Inc, 461 F2d 1069 (5th Cir 1972).

[70] *Id. See also* Svenson v Engelstad, 626 F2d 421, 425 (5th Cir 1980) (four factors in determining availability of the §4(2) exemption are: the number of offerees and their relationship to the issuer; the number of units offered; the size of the offering in dollars; and the manner of the offering); and Mary S Krech Trust v The Lakes Apartments, 642 F2d 98 (5th Cir 1981).

offering material which is used should be numbered and their distribution monitored. Written representations should be obtained from offerees as to their investment intent, knowledge, experience and financial ability to bear the risk of the investment and as to the information furnished or access to information provided to them. Certificates for securities issued to these investors should bear a legend stating that the securities are not registered under the 1933 Act and are restricted as to resale.

§6.26 —Regulatory Interpretation (Rule 506)

In an attempt to alleviate some of the uncertainty about the availability of the private placement exemption, the SEC adopted Rule 506,[71] interpreting §4(2), which, if all of its terms and conditions are complied with, provides a safe harbor exemption from registration. The rule is a nonexclusive safe harbor, since failure to satisfy its terms and conditions is not unlawful and does not render the §4(2) exemption unavailable.

Under Rule 506, any amount of securities may be sold to up to 35 persons, and, in addition, to anyone who is an *accredited investor*.[72] The offering cannot be made by any means of general solicitation or general advertising.[73] General solicitation or general advertising include, but are not limited to, advertisements, articles, notices or other communications published in print media or broadcast, as well as meetings or seminars attended by persons invited by general solicitation or general advertising.

The purchasers must be sophisticated or have a sophisticated purchaser representative.[74] capable of advising them on the merits and risks of the investment. There is no limit on the number of offerees. The standards for sophistication under Rule 506 may be more subjective than those under §4(2). This may be of particular concern to those attempting to finance highly technical innovation since, in the view of the SEC, the purchaser or purchaser representative may have to have the sophistication to evaluate the particular investment, not just investments in general.

The purchasers must be provided with the kind of information they would receive if the securities were registered, including, in many instances, certified financial statements and an opportunity to ask questions to verify the information.[75] This is a reasonably simple procedure for a public company which may utilize the reports it files with the SEC under the 1934 Act.[76] It is more a

---

[71] 17 CFR §230.506.

[72] *Id* §230.501(a). Principally, enumerated financial institutions or investors who satisfy specified income or net worth standards.

[73] *Id* §230.502(c).

[74] *Id* §230.501(h).

[75] *Id* §230.502(b).

[76] The Securities Exchange Act of 1934, *as amended,* 48 Stat 881 (1934), (codified at 15 USC §§78a-78jj).

burden on a closely held company which must prepare an offering document which resembles a 1933 Act prospectus.

The issuer also must exercise reasonable care to assure that the purchasers are not underwriters engaging in distributing the securities to the public. Reasonable care includes inquiry of the purchasers to determine if they are purchasing for their own accounts, and placing a legend on stock certificates referring to restrictions on their transfer.[77] These restrictions on resales obviously make the investment less attractive.[78]

Filing of a notification with the SEC is necessary to perfect the Rule 506 exemption.[79] In addition to the specific requirements of the rule, a careful lawyer will elaborately document a Rule 506 transaction, in order to provide evidence of compliance. The documentation will include representations from investors similar to those obtained in §4(2) transactions.[80]

Rule 506 can still be criticized at several levels. Its terms and conditions are complex and some of its provisions are too subjective or ambiguous. Its informational requirements are too burdensome. An inadvertent failure to comply with any provisions of the rule in an immaterial respect may result in loss of the safe harbor. The necessity of professional assistance and elaborate documentation probably make compliance with Rule 506 too costly for smaller offerings.

There also may be concern, which the SEC has sought to dispel in connection with the predecessor to Rule 506, Rule 146,[81] that the courts will apply by analogy some of the more exacting standards of Rule 506 in cases involving only the §4(2) exemption. The Fifth Circuit gave some indication of leaning in that direction with respect to Rule 146.[82] However, attorneys should advise their clients to comply with the provisions of Rule 506 as far as is reasonably possible, but to rely on the statutory exemption which is considered to be subject to more flexible interpretation.

Neither §4(2) nor Rule 506 provides any exemption from the antifraud provisions of the federal securities laws or from any similar state law. Furthermore, if an issuer supplies an investor with materially misleading information in connection with the offering or omits to provide material information, even if only through negligence, the issuer may be liable for rescission or damages under §12(2) of the 1933 Act.

While the uncertainties inherent in the private placement exemption have not prevented its use in raising billions of dollars in capital, the resulting risks

---

[77] 17 CFR §230.502(d).

[78] The standards for determining when and under what circumstances the investors may resell are discussed in **§6.49.**

[79] 17 CFR §§230.503, 239.500.

[80] See **§6.25.**

[81] Securities Act of 1933 Release No 5913 (Mar 6, 1978).

[82] *See, e.g.*, Doran v Petroleum Management Corp, 545 F2d 893 (5th Cir 1977); and Securities Act of 1933 Release No 5913 (Mar 6, 1978).

and restrictions on resales[83] have raised the costs of that capital. Moreover, the costs are relatively greater for the smaller company which needs venture capital to develop an innovation.

Part of the problem is that the private placement exemption is forced to serve too many disparate functions. It must provide an exemption for a placement of debt securities with a group of financial institutions which can fend for themselves and the major concern of which is restriction on the liquidity of their investment. It must also provide an exemption for placement of an equity interest with a group of sophisticated venture capital investors whose interests are not unlike those of a financial institution. On the other hand, it must provide an exemption for sales of securities to friends, relatives, business associates or employees with various levels of sophistication. This is probably too much baggage for one exemption to carry and some efforts are being made to create a more logical set of exemptions.

Notwithstanding these drawbacks, the §4(2) exemption and the Rule 506 safe harbor are of utility in raising capital in a number of situations. Even with the need to extensively document the transaction, it takes considerably less time to complete a private placement—a few days to a few weeks—than to prepare and review a registration statement, which may take months. As a practical matter, some companies may not be seasoned enough for a public offering or a company may want to avoid the burdens of being a public company. Other companies may have difficulty complying with the registration requirements, including requirements for certified financial statements, in many instances. For certain types of transactions, offerings to executive officers, for example, there may be very little benefit to be derived from registration and almost no burden in effectuating a private placement under §4(2).

In other circumstances, even where the risks of the unavailability of the §4(2) exemption are minimal, the company or institutional investors may want the added comfort of the Rule 506 safe harbor because of the severe consequences of loss of the exemption. The greater the amount of money to be raised, the greater the degree of comfort which may be sought.

§6.27 Intrastate Offerings Exemption §3(a)(11) and Rule 147)

In §3(a)(11) of the 1933 Act,[84] Congress has provided an exemption from registration for securities sold to local investors to finance an essentially local business. This is a somewhat artificial construct, relying as it does on the confinement of the offering within the boundaries of a single state and requiring the issuer to be incorporated or organized in that state as opposed to just conducting a substantial portion of its operations in the state. The exemption is therefore unavailable to the many companies which incorporate in Delaware

---

[83] See **§6.49.**

[84] 15 USC §77c(a)(11).

to obtain the advantages of Delaware corporate law. The intrastate exemption also involves significant risks, since, theoretically, a single offer to a person who is not a resident of the state, even if present in the state and even if no sale is consummated, could invalidate the exemption and subject the issuer to claims of rescission from all purchasers.

Nevertheless, the exemption has some value, since it can even be complied with in some circumstances by sheer inadvertence. For example, the inventor who raises a few thousand dollars from friends and neighbors in her home state, assuming she can comply with state law, probably complies with federal law. It is not generally, however, a viable exemption for a multistate enterprise or a widespread offering.

Under §3(a)(11), unaffiliated investors are not subject to limitations on resales when a reasonable period of time has elapsed after the completion of the offering. Resales in the state may be made at any time. Under the SEC's intrastate offering safe harbor rule, Rule 147,[85] resales may be made nine months after the last sale in the initial offering. Under §3(a)(11), resales may be made after the initial offering has come to rest. There are no precise standards for determining when this occurs, although the period of time elapsed after the last sale may be considerably shorter than the period of human gestation.

Those using the intrastate exemption remain subject to state blue sky filing and other requirements which can be almost as extensive as federal registration requirements, including requirements to provide certified financial statements. Federal antifraud requirements and §12(2) of the 1933 Act[86] also apply to intrastate offerings.

The SEC has attempted to reduce some of the uncertainty under §3(a)(11) by adopting an interpretive safe harbor rule, Rule 147, which provides more objective standards for determining the availability of the exemption. The price of certainty was a degree of rigidity, but the rule has been of some utility. The courts may interpret the §3(a)(11) exemption with greater flexibility than the SEC does. For example. §3(a)(11) only requires that a corporation be doing business in the state. The courts have interpreted this to mean a substantial portion of the company's business.[87] Rule 147 requires that 80 per cent of the issuer's consolidated gross revenue be derived from operations within the state; that 80 per cent of its consolidated assets be located within the state; and that the issuer intend to use 80 per cent of the net proceeds of the offering in connection with the operation of a business in the state.

The American Law Institute's proposed *Federal Securities Code* (ALI Code) would provide a more flexible regional exemption[88] which might be a more viable method of raising capital, since many metropolitan areas where investors may be found overlap the boundaries of one or more states. The price

---

[85] 17 CFR §230.147.

[86] 15 USC §77l(2).

[87] *See, e.g.*, Chapman v Dunn, 414 F2d 153 (6th Cir 1969).

[88] §514 (1978).

would be state blue sky regulation, which may not be a bargain, particularly in those states which impose merit regulations. *Merit* regulations could be particularly troublesome to an innovative company. If a state determines the investment to be too risky, it might impose limitations on the financing activities of the company.

Theoretically, under the federal securities laws, once the required disclosure has been made, the investor evaluates the merits and risks of the investment and makes a decision. Under state blue sky laws which provide for merit regulation (California[89] for example), state securities officials may judge the merits and risks or fairness of the investment and either prevent the offer from being made or impose substantive conditions. This may be of greater concern to a company which desires to finance an innovation than to the promoter of some more prosaic and less risky type of business.

If there is to be an exemption from the federal securities laws based on the degree of federal interest, it perhaps should be based on the standards, such as the amount of money to be raised or the number of investors involved, rather than on the degree of state interest. The ability of a state to protect the interests of investors varies from state to state and is an imperfect standard on which to base an exemption from federal law. The SEC's Regulation D[90] incorporates certain of these concepts.

The proposed ALI Code would simplify matters by providing for an exempt limited offering which results in sales to financial institutions and to up to 35 other persons.[91] Those limitations and certain restrictions on the buyer's resales would be the only conditions to be satisfied. Also, various exemptions could more readily be used in combination under the ALI Code's more limited integration provisions.[92]

Notwithstanding the difficulties involved in complying with the intrastate offering exemption, it could be of some utility to the company engaged in an innovative project. A start-up company often may wish to sell a small amount of stock to technical or scientific employees, for example. The intrastate offering exemption may provide the exemption for such a sale.

§6.28 Sales to Accredited Investors Exemption (§4(6))

In order to assist in the raising of venture capital, Congress, in the Small Business Investment Incentive Act of 1980,[93], provided a simplified exemption

---

[89] *See, e.g.*, Cal Admin Code, tit 10, ch 3, R 260.140.05, Securities of speculative businesses.

[90] 17 CFR §§230.501-.506.

[91] §202(88) (1978).

[92] §202 (110) (1978).

[93] Pub L No 96-477, 94 Stat 2275 (Oct 21, 1980) (codified in scattered sections of 15 USC).

from registration by adding §4(6) to the 1933 Act.[94] Section 4(6) permits the issuer to offer an issue of securities in the aggregate amount of up to $5 million to accredited investors[95] as long as there is no advertising or public solicitation in connection with the offer. No definitions are provided for the terms *advertising* or *public solicitation.* Presumably, the Rule 506 standards could be used as a guideline.[96]

The exemption must be perfected by filing a reasonably simple notification with the Securities and Exchange Commission (SEC).[97] The term *accredited investor* is defined essentially as it is under Regulation D.[98]

There is no definition in the 1933 Act of the phrase *an issue of securities.* Presumably the criteria for judging what constitutes an issue could be similar to those used to determine what is an offering.[99]

Issuers using the §4(6) exemption are still subject to the antifraud provisions of the federal securities laws, §12(2) of the 1933 Act and state law. The SEC takes the position that investors in §4(6) transactions are subject to the same restrictions on resale as investors in §4(2) private placement transactions.[100]

There has yet to be sufficient experience with §4(6) to determine its utility, but it should simplify private placements to institutions.

§6.29 Small Offerings Exemptions—Regulation A

The Securities and Exchange Commission (SEC) has increasingly looked to §3(b) of the 1933 Act[101] as the basis for its authority to exempt transactions from the registration provisions of the 1933 Act. There are several §3(b) exemptions which could be available to persons involved in financing innovation. Regulation A,[102] which, in many respects, is not a true exemption, but a modified version of registration, permits offering of an issue of securities in an aggregate amount of up to $1.5 million every 12 months. Rule 504[103] permits relatively closely held companies to offer up to $500,000 in securities every 12 months. Rule 505[104] permits an issue of securities of up to $5 million every 12 months to 35 purchasers, not counting an unlimited number of

---

[94] 15 USC §77d(6).

[95] As defined in §2(15) of the Securities Act of 1933, 15 USC §77b(15) and 17 CFR §230.215 thereunder.

[96] See **§6.26.**

[97] 17 CFR §239.500.

[98] 17 CFR §§230.501-.506.

[99] Discussed in **§6.24.** *See also* 17 CFR §230.502(a).

[100] *See, e.g.,* Securities Act of 1933 Release No 6274 (Dec 23, 1980). See **§6.49.**

[101] 15 USC §77c(b).

[102] 17 CFR §§230.251-.264.

[103] *Id* §230.504.

[104] *Id* §230.505.

accredited investors.[105]

Regulation A, the oldest of the SEC's small offering exemptions, as a practical matter establishes a simplified registration procedure. A Canadian or United States issuer may sell securities with an aggregate offering price of up to a ceiling of $1.5 million during any period of 12 months. The limitation on the aggregate offering price of securities which may be sold during a period of time resolves many of the questions involved in determining what is an issue.

The statutory ceiling on the amount of the aggregate offering price of securities which may be sold pursuant to rules adopted under §3(b) of the 1933 Act is $5 million during any period the SEC establishes by rule. In part, to facilitate raising of capital by smaller businesses, the SEC has commenced rulemaking proceedings in order to consider raising the ceiling or shortening the period during which sales must be aggregated for purposes of Regulation A and other §3(b) exemptions.[106]

To comply with Regulation A, an issuer must file an informational notification with the SEC and provide an offering circular to prospective purchasers. The notification, among other things, contains information about the issuer and any underwriter and persons associated or affiliated with the issuer or the underwriter, in order for the SEC to compare that information with its electronic data bank. If any of these persons have been involved in specified violations of law, Regulation A may be unavailable to the issuer under the "Bad Guy" provisions of that regulation.[107] The SEC, pursuant to Rule 252(f)[108] under the 1933 Act, may relieve an issuer of such a disability.

The SEC also may temporarily suspend the exemption administratively for failure to comply with the terms and conditions of Regulation A or for specified violations of law.[109] After notice and opportunity for hearing, the SEC may make the suspension permanent.

The SEC recently has adopted amendments to Regulation A which codify present informal information practices and cause a Regulation A offering circular to resemble more closely the prospectus used in connection with a registration statement for certain smaller offerings, Form S-18.[110] Either to avoid misleading omissions or as the result of SEC staff comments after review of the offering circular, offering circulars generally contain more than the specified information.

Under Regulation A, the required financial statements need not be audited by an independent accountant, if the issuer is not required to provide audited financial statements in reports filed under the 1934 Act and if audited financial

---

105 *Id* §230.501(a).

106 Securities Act of 1933 Release No 6275 (Dec 23, 1980). The Regulation A ceiling has not been raised and the SEC is considering rescission of the entire regulation. Securities Act of 1933 Release No 6339 (Aug 7, 1981).

107 17 CFR §230.252(b)-(e).

108 *Id* §230.252(f).

109 *Id* §230.261.

110 *Id* §239.28.

statements are not otherwise available.[111] This could result in significant savings, or permit an offering to be made which could not otherwise be made because the company is unable to obtain a clean accountant's opinion on its financial statements. However, many states require audited financial statements as a condition for qualification of Regulation A offerings. An underwriter also may insist on an audit for its own protection and to make the offering more acceptable to investors. Requirements for audits of the financial statements required by Regulation A are presently under consideration by the SEC, and could be imposed as a condition to use of Regulation A for offerings in excess of $1.5 million if the Regulation A ceiling is raised above that amount.

Regulation A notifications and offerings circulars are filed with and reviewed by the SEC regional office for the region in which the issuer conducts its principal business operations. Regional office review, as an alternative to review in Washington, D.C., can result in the saving of time and expense and can be more convenient for a local company. The notification must be filed at least 10 days prior to the date the initial offers or sales are made, and this period may be extended informally to allow time for SEC staff review and comments. If the offering is underwritten, a preliminary offering circular may be circulated to prospective investors during the 10-day waiting period and any extensions. This represents a recent revision of Regulation A intended to facilitate financing of smaller businesses.

§6.30 —Circumstances under which the Regulation A Exemption Should Be Used

Why would Regulation A be used, rather than some other exemption or registration of securities on Form S-18, which could permit a greater amount of capital to be raised? There are several reasons why Regulation A may remain an important exemption for transactions to finance innovation, if not rescinded by the SEC or combined with Form S-18. Certain of the present advantages of Regulation A may become less important if Regulation D[112] is utilized extensively, since Regulation D permits up to $5 million or more to be raised every 12 months.

1. Some businesses may find it impracticable to obtain the audited financial statements required by Form S-18[113] or other SEC registration forms
2. Under Rule 257,[114] for offerings of less than $100,000, no offering circular is required, if advertising is limited as specified in that rule. Since

---

[111] Reg A, Form 1-A, Schedule I, Item 11, 17 CFR §239.90.

[112] 17 CFR §§230.501-.506.

[113] *Id* §239.28.

[114] *Id* §230.257.

general advertising may not be permitted under Rule 504,[115] and Rule 504 may not be utilized by a reporting company, in some circumstances, Rule 257 may be an alternative to Rule 504, for an offering to employees to provide them with incentive through ownership of an equity interest in the business, for example

3. Regulation A offerings may be made to an unlimited number of unsophisticated persons; an offering under §4(2), Rule 506 or §4(6) may only be made to sophisticated or accredited investors and offerings under Rule 505 may not be made to more than 35 unsophisticated persons
4. Some states may qualify Regulation A offerings by less burdensome procedures than those used in connection with intrastate offerings. There is also no risk of offers to out of state residents, as there would be under §3(a)(11) or related Rule 147
5. Issuers filing under Regulation A are not automatically subject to reporting under the 1934 Act and the recordkeeping and internal accounting controls provisions of that act,[116] as they would be if they filed a registration statement under the 1933 Act[117]
6. Regulation A filings are not subject to the civil liability provisions of §11 of the 1933 act.[118] Under some circumstances, the general antifraud provisions of the securities laws, and the civil liability provisions of §12(2) of the 1933 Act[119] which are applicable, may be less onerous than §11
7. Professional fees associated with a Regulation A offering may be lower than those associated with a registered offering

The principal advantages of using Regulation A may be lower professional fees and, where appropriate, avoiding audited financial statements and 1934 Act reporting requirements, which can become expensive.

§6.31 —Rule 504

Rule 504[120] would be useful for exempting transactions to provide the initial capital for a small venture; to provide for investments by employees in a new enterprise; or to provide an immediate infusion of capital into a smaller company in financial difficulty.

Under the rule, an issuer which is not subject to the periodic reporting

---

[115] If the issuer involved in a Rule 504 offering files with a state and complies with the state's blue sky law requiring delivery of information to purchasers, the issuer would be permitted to advertise and solicit generally.

[116] Securities Exchange Act of 1934, §13(b)(2), 15 USC §78m(b)(2).

[117] *Id* §15(d), 15 USC §78o(d).

[118] 15 USC §77k.

[119] *Id* §77l(2).

[120] 17 CFR §230.504.

provisions of the 1934 Act, may issue securities for an aggregate sales price of up to $500,000 during any 12-month period, if it does not generally advertise or solicit buyers and does not pay anyone for soliciting prospective buyers. If a single sale fails to satisfy the conditions of the rule, the other sales could still be exempt, unlike other exemptions where one violation causes the exemption to be lost for the entire offering.

The exemption must be perfected by filing of a notice of sales with the SEC.[121]

Securities acquired in a transaction exempt under Rule 504 are restricted as to resale to the same extent as they would be if acquired in a private placement, and stock certificates must bear a legend to that effect.

Rule 504 does not provide an exemption from state securities laws or from the antifraud and certain civil liability provisions of the federal securities laws.

§6.32—Rule 505

Rule 505, one of the newest SEC exemptive rules, should be of great utility in financing innovative operations. The rule permits an issuer to sell up to $5 million in securities during any period of 12 months. Sales may not be generally advertised or solicited, but may be made to any number of accredited investors and up to 35 other persons. If offers are made to nonaccredited persons, the information required by Part I of the SEC's Form S-18 registration statement, other available registration forms or 1934 act reports must be provided to all purchasers in writing.[122]

A notice must be filed with the SEC to perfect the exemption[123] and the securities are restricted as to resale to the same extent they would be if they were acquired in a private placement.[124]

Rule 505 does not provide an exemption from state securities registration and other requirements or from the antifraud or certain civil liability provisions of the federal securities laws. Rule 505 also has "Bad Guy" rules which prohibit its use by persons who have been involved in specified violations of law.

§6.33 Offshore Offerings

There is no specific exemption in the federal securities laws for offshore financing. Presumably, a United States company financing abroad could not avoid using the facilities of United States interstate or foreign commerce or the United States mails in connection with that financing. However, as a practical matter, the Securities and Exchange Commission (SEC) has taken the position

---

[121] 17 CFR §230.503.

[122] 17 CFR §230.502(b).

[123] *Id* §230.503.

[124] *Id* §230.502(d).

that the securities registration provisions of the 1933 Act and the broker-dealer registration provisions of the 1934 Act are designed to protect United States citizens and residents. Where a domestic company makes an offering solely to residents of a foreign country, the SEC took the position in 1964[125] that registration was not required. Although this position was based in part on the belief that the United States interest equalization tax then in effect would inhibit flowback of those securities into the United States, the SEC has adhered to this position subsequent to the expiration of that tax in the 1970s. Steps must still be taken to prevent securities flowing back into the United States securities markets until the offering has "come to rest" abroad. While the SEC has issued no action letters[126] relating to offshore offerings, it has consistently declined to address the flowback issue or the efficacy of the various measures used to prevent or impede flowback. It is possible that the standards to be applied to determine whether an intrastate offering has come to rest could be applied by analogy to offshore offerings. Thus, the nine-month standards in Rule 147 could be used. Some believe that trading in the United States may commence 90 days after completion of the offering abroad, since 90 days is the period under §4(3) of the 1933 Act[127] during which securities dealers in the United States would be required to deliver a prospectus after an initial offering.

If contacts can be made with foreign investors, an offshore offering may be a viable method of financing innovation for either a small or a very large company. Foreign financial intermediaries could be used and would not be required to register as broker-dealers under the 1934 Act, if the SEC guidelines[128] are followed. Foreign financial institutions may have a greater flexibility in the types of investments that they may make. Hundred of millions of dollars were raised by United States companies through Eurodollar offerings in the 1960s and 1970s. In addition, the SEC will not object if a domestic private placement is made in conjunction with a foreign public offering.

While the registration provisions of the federal securities laws (and presumably state law) would not apply to an offshore offering, the SEC and federal judges frown upon the exportation of securities frauds. The antifraud provisions could apply if there are sufficient United States contracts to satisfy a requirement for a jurisdictional nexus.[129] For this reason, because it is sound business practice and because of increasingly sophisticated disclosure requirements in European countries, underwriters generally insist on the preparation of offering circulars for offshore offerings which bear a resemblance to a prospectus prepared under SEC regulations.

---

[125] Registration of Foreign Offerings by Securities Issuers; Registration of Underwriters of Foreign Offerings as Broker-Dealers, Securities Act of 1933 Release No 4708 (July 9, 1964).

[126] See **§6.34.**

[127] 15 USC §77d(3).

[128] Securities Act of 1933 Release No 4708, *supra* note 125.

[129] *See, e.g.*, SEC v United Fin Group, Inc, 474 F2d 354 (9th Cir 1973).

Compliance with the Securities Laws

§6.34 Assistance of the Securities and Exchange Commission

Once a determination is made that a financing transaction is subject to federal and state securities laws, the steps necessary to comply with such laws must be determined. The Securities and Exchange Commission (SEC) staff may provide assistance in this regard.[130] The Office of the Chief Counsel, SEC's Division of Corporation Finance, provides informal administrative interpretations of the provisions of the Securities Act of 1933, as amended[128] and the rules thereunder to interested persons. The staffs of the other SEC offices provide similar informal advice concerning the laws which they administer. In many instances, rather than providing an informal interpretation, the staff will indicate whether or not they would recommend enforcement action to the commission, if a transaction proceeds as described by the person requesting the advice (a *no action* letter).

The division issues hundreds of interpretive or no action letters every year concerning the definition of the term *security* and the exemptive provisions of the 1933 Act and the rules thereunder. These interpretations and no action letters together with interpretive releases issued by the commission, although they do not have the force of law, are relied on in practice as if they did. The SEC staff views are sometimes the only source of guidance on a particular issue. There are also issues which the staff will not address, including some investment contract issues.[129]

The incoming letters and the staff's responses become public 30 days after the date of the staff's responsive letter. The SEC periodically issues a list in its News Digest of what it believes to be significant letters, and some letters are listed, summarized or included in various periodical publications on securities laws and included in research and information retrieval systems such as LEXIS.

Companies should exercise judgment and seek the SEC's advice only when necessary. In many instances, an opinion of counsel may be relied upon. This may save time, since it generally takes the SEC about three weeks to respond to a request and several days to prepare the initial request. However, in circumstances where the SEC will be reviewing a filing, it may be more expeditious to resolve the issues with the SEC in advance. Broker-dealers and investment bankers may also insist on the comfort of an SEC no action letter for certain

---

[130] *See, e.g.*, 17 CFR §§200.81 and 202.2; and Securities Act of 1933 Release No 6253 (Oct 28, 1980).

[128] 48 Stat 74 (1933) (codified at 15 USC §§77a-77aa).

[129] Securities Act of 1933 Release No 6253 (Oct 28, 1980); and Regulation D—Revision of Certain Exemptions from the Registration Provisions of the Securities Act of 1933 for Transaction Involving Limited Offers and Sales, Securities Act of 1933 Release No 6389 (Mar 8, 1982).

transactions, since they have licenses at stake.[130]

§6.35 Registration of Securities

If an exemption is not available, the securities to be offered must be registered with the Securities and Exchange Commission (SEC).[131] Under the 1933 Act, unlike the 1940 Act,[132] securities are registered, not companies. In order to register securities with the SEC, a registration statement must be prepared and filed with the SEC. The registration statement is required to contain information concerning the terms of the offering, the securities to be offered, and the business property and management of the company as well as certified financial statements.[133] The SEC has broad authority to vary the information required in the registration statement[134] and the prospectus which is required to be delivered to investors.[135]

Offers of any sort must be avoided once the company has determined to file the registration statement.[136] *Offer* is broadly defined under the securities laws,[137] and "gun jumping"[138] before filing can result in delay in the effectiveness of the registration statement to permit a "cooling off" period. Written offers may only be made by means of a preliminary prospectus from the date of the filing of the registration to the effective date[139] (the *waiting period*). Oral offers are permitted during that period.

The registration statement may become effective automatically 20 days after filing;[140] the SEC may accelerate the effective date;[141] or, as is often the case, the SEC staff will review the registration statement, comment upon it in writing and one or more pre-effective amendments will be filed in response to those comments before the registration statement is ordered effective.[142] If the re-

---

[130] See **§6.47.**

[131] For a general introduction, see Robinson & Eppler, Going Public (2d rev ed 1978), and Schneider, Manko & Kant, Going Public, Practice, Procedure and Consequences (Packard Press reprint 1981).

[132] The Investment Company Act of 1940, *as amended,* 54 Stat 789 (1940) (codified at 15 USC 15 USC §§80a-1 to 80a-52).

[133] Securities Act of 1933, Schedule A, 15 USC §77aa.

[134] *Id* §7, 15 USC §77g.

[135] *Id* §10, 15 USC §77j.

[136] *Id* §5(b), 15 USC §77e(b).

[137] *Id* §2(3) and (10), 15 USC §§77b(3) and (10).

[138] *See, e.g.,* Securities Act of 1933 Release No 5009 (Oct 7, 1969).

[139] Securities Act of 1933, §5, 15 USC §77e.

[140] *Id* §8(a), 15 USC §77h(a).

[141] *Id.*

[142] In order to provide time for staff review, Rule 473, 17 CFR §230.473, requires a statement in the registration statement to the effect that the effective date of a registration statement is automatically delayed until declared effective by SEC order or the statement is removed. Failure to include such a statement risks an enforcement proceeding or the threat of such a proceeding.

view process takes too long, the financial statements will be required to be updated.[143] The registration process may take months to consummate and cost hundreds of thousands of dollars.

§6.36 Securities and Exchange Commission (SEC) Administration and Enforcement

The SEC's informal review and comment procedures[144] have been restructured in recent years. The Branches of Corporate Analysis and Examination or the SEC's Division of Corporation Finance now specialize by industry. A company in the communications field, for example, would have its filings reviewed by professionals who, if not experts in that industry, are familiar with it. However, due to budgetary restraints and turnover in personnel, many of the SEC's professional staff are relatively inexperienced. This can lead to delay in review of a registration statement while the staff goes through a learning process.

Counsel may be able to alleviate some of this delay by requesting a prefiling conference[145] to discuss important issues which are complex or otherwise troublesome. Although no longer required, submission with the registration statement of an informative transmittal letter which points out some of the more important issues and anticipated staff comments may also assist in avoiding delay. This is even more important when transmitting an amendment responsive to SEC staff comments.

SEC staff review probably can be expected for first-time registrants, which would include many companies seeking to finance an innovative project. The review process may take 90 days or more to complete and this should be factored into any financing schedule. The company should also plan for standby sources of funds to weather the storm while waiting for the SEC to complete its review.

Larger companies which file periodic reports under the 1934 Act[146] or which filed registration statements under the 1933 Act in the past may have their registration statements reviewed in a few weeks or not reviewed at all. The SEC has generally declined to announce its internal guidelines for determining when to review or not review a registration statement. Informally, the staff has indicated that a registration statement for securities with novel or complex terms, even if by a reporting company, may be reviewed at least to the extent of the description of the securities.[147]

A registration statement which is materially misleading when filed (or after

---

[143] Reg S-X, Rule 3-12, 17 CFR §210.3-12.

[144] 17 CFR §202.3.

[145] *Id* §202.2.

[146] The Securities Exchange Act of 1934, as amended, 48 Stat 881 (1934) (codified at 15 USC §§78a-78jj).

[147] *See, e.g.*, Sec Reg & L Rep No 625, at (BNA) F-3 to F-4 (Oct 21, 1981).

it becomes effective) may not be subject to the review and comment procedure, if the SEC determines that enforcement procedures are more appropriate. The SEC may investigate informally[148] or may order a private investigation[149] and examination[150] relating to the registration statement or other matters relating to the registrant. A formal order of investigation permits SEC officials designated in the order to subpoena documents and compel attendance of witnesses from anywhere in the United States and to take testimony under oath.[151]

An investigation could result in a stop order proceeding[152] or civil injunctive action,[153] or in a criminal reference to the Department of Justice.[154] A stop order suspends the effectiveness of the registration statement and prevents the covered securities from being offered or sold. A stop order must be lifted if the disclosure deficiencies found by the commission are cured by amendment.

In a civil injunctive action, the SEC may obtain injunctive and other equitable relief, including impoundment of proceeds of an illegal offering and disgorgement from the company and other persons involved in violations of the securities laws. The defendants may also be collaterally estopped from contesting in private litigation facts found by the court in a contested SEC action.[155]

SEC enforcement proceedings should be avoided if at all possible. Defending such a proceeding could cost hundreds of thousands of dollars and could even bankrupt a company. Careful preparation of the registration statement and full and fair disclosure should avoid a situation resulting in an investigation. This requires a delicate balance between adequate disclosure of the risks of the investment and avoiding such a negative presentation that the securities cannot be sold. This may be particularly difficult for the newer company which has experienced losses because of the requirement to write off research and development expenditures, for example.

§6.37 Which Registration Form?

The determination as to which registration form of the Securities and Exchange Commission (SEC) is available for registration of the company's securities is often important.[156] The SEC has virtually completed a decade-long project of integrating the disclosure requirements of the 1933 and 1934 Acts.[157] The 1933 Act requires ad hoc disclosure when securities are regis-

---

[148] 17 CFR §202.5.

[149] *Id* §202.5 and pt 203.

[150] Securities Act of 1933, §8(e), 15 USC §77h(e).

[151] *See, e.g., id* §19(b), 15 USC §77s(b).

[152] *Id* §8(d), 15 USC §77h(d).

[153] *See, e.g., id* §20(b), 15 USC §77t(b).

[154] *Id.*

[155] Parklane Hosiery Co v Shore, 439 US 322 (1979).

[156] A chart comparing various forms appears in **app 6-2.**

[157] *See generally,* Securities Act of 1933 Release No 6383 (Mar 3, 1982).

tered. This may result in considerable duplication for a company required to file annual, quarterly and current reports under the 1934 Act and to send proxy statements and annual reports prepared in accordance with SEC rules[158] to stockholders. The SEC has developed a three-tier structure for its registration forms to avoid some of this duplication.

A company which has been reporting under the 1934 Act for three years, and which satisfies certain other conditions, may register securities on Form S-3[159] for secondary offerings and many types of primary offerings, including offerings to raise capital from the public, if those securities are underwritten. The form is only available for securities of companies with a substantial market following by securities analysts. That following will be assumed, if the company has a class of equity securities outstanding and the aggregate market value of the shares of that class held by nonaffiliates of the company is at least $150 million or $100 million with a trading volume of 3 million or more shares.

A company which satisfied the standards of Form S-3 would be able to incorporate most of the disclosure required in the registration statement and prospectus for its 1934 Act reports. While this involves some important liability issues, particularly for underwriters,[160] use of this form results in a significant saving in time and cost. Moreover, Form S-3 may be reviewed by the SEC in a week to 10 days or not at all which means that market windows can be taken advantage of more readily.

Issuers who may use Form S-3 may also be able to register large amounts of securities for the shelf which may be sold when the funds are needed and market conditions are appropriate.[161]

Since Form S-3 permits a relatively speedy and inexpensive registration process, it is important for the company to remain in compliance with the 1934 Act's periodic reporting provisions so that the form will be available.

Companies eligible to use Form S-3, may also use Form S-2[162] which is available for certain types of offerings not covered by Form S-3. Form S-2 relies on incorporation by reference of information from 1934 Act reports, and would permit delivery of annual and quarterly reports to securityholders with the prospectus in lieu of including the information in those reports in the prospectus.

Since prospectus delivery requirements for reporting companies are generally less burdensome than for nonreporting companies, this requirement should result in considerable cost saving. The 1933 Act requires delivery of

---

[158] 17 CFR §240.14a-3(b).

[159] *Id* §239.13.

[160] *See generally,* Securities Act of 1933 Release No 6335 (Aug 6, 1981), and 17 CFR §230.176.

[161] 17 CFR §230.415. This is a temporary rule which will expire on December 10, 1982, unless renewed.

[162] *Id* §239.12.

a final prospectus to a purchaser prior to sale.[163] For offerings of securities listed on a stock exchange at the market over the exchange, this requirement may be satisfied by delivery of a relatively small quantity of the prospectus to the exchange.[164]

The 1933 Act also requires securities dealers to deliver a final prospectus for a period of 40 days (or 90 days in the case of a first-time registrant).[165] Dealers are relieved of this requirement if the issuer of the securities is a reporting company.[166]

Form S-2 is not available for exchange offers, although the SEC has adopted a form, Form S-15, with similar disclosure requirements, which is available for an exchange offer by a reporting company for securities of a company approximately one-tenth its size.[167] Form S-15 should be made available for a broader range of transactions.

For nonreporting companies and reporting companies unable to satisfy the standards of Form S-2 or S-3, the registration form in general use is Form S-1.[168] Form S-1 does not permit incorporation by reference of information from 1934 Act reports. Thus, most start-up companies would be required to provide extensive information about their business, properties, management and financial condition and operations in the registration statement and in the prospectus.

In order to relieve smaller businesses of some of the burdens of registration and facilitate their raising of capital, the SEC has adopted a somewhat simplified registration form, Form S-18[169] which may be used in lieu of Form S-1. That form is available to certain reporting companies for primary offerings of $5 million or less or secondary offerings of $1.5 million or less.

Form S-18 may be filed and reviewed in an SEC regional office which may be advantageous, particularly since the registration process often involves considerable personal contact with representatives of the agency. The disclosure requirements of Form S-18 are somewhat simpler than those of Form S-1, particularly the financial statement requirements. Certified financial statements for two years (three years for Form S-1) may be prepared in accordance with generally accepted accounting principles and are not subject to the SEC's accounting regulation.[170] Disclosure and financial statement requirements similar to those of Form S-18 also apply to a Form S-18 registrant's first annual

---

[163] Securities Act of 1933, §5(a), 15 USC §77e(a).

[164] 17 CFR §230.153.

[165] Securities Act of 1933, §4(3), 15 USC §77d(3).

[166] 17 CFR §230.174.

[167] *Id* §239.29.

[168] *Id* §239.11.

[169] *Id* §239.28. Many small companies have used Form S-18 to register securities. Form S-18, A Monitoring Report on the First 18 Months of its Use (SEC March 1981) (summarized Sec Reg & L Rep (BNA) No 595, at A-17 (Mar 18, 1981)).

[170] Reg S-X, 17 CFR pt 210.

report filed with the SEC pursuant to §15(d) of the 1934 Act.[171] (All companies which register securities under the 1933 Act are required to file reports under the 1934 Act for at least the remainder of the fiscal year in which the securities are registered.)[172]

The SEC's disclosure requirements have often been criticized as being too negative. In recent years, the SEC has taken steps to encourage disclosure of more favorable information. However, fear of liability still results in the drafting of prospectuses as insurance policies rather than selling documents.[173]

If disclosure of favorable information in the form of projections of future revenue or income, for example, would be helpful to a client in raising capital, the SEC's safe harbor rules for disclosure of projections[174] should be considered. If they can be complied with, the risk of liability can be significantly reduced.

§6.38 Offerings to Employees

Employees can be an important source of venture capital either directly or through various types of employee stock ownership plans. In addition, employees may be given additional incentive through equity participation in the enterprise.

Under the securities laws, as interpreted by *Ralston Purina,*[175] employees generally have the same status as any other investor. The Securities and Exchange Commission (SEC) has made it easier to register securities to be issued to employees pursuant to certain employee benefit plans; securities issued in connection with certain other types of plans are exempt from registration under §3(a)(2) of the 1933 Act;[176] and some plans do not involve the issuance or sale of securities at all.[177]

---

[171] Form S-8, General Instruction E.

[172] See **§6.51.**

[173] There are indications that the SEC may attempt to change this situation. The SEC has hired a consultant to develop approaches for "reducing the complexity and improving the effectiveness of corporate disclosure documents required to be filed with the Commission." SEC News Digest, Issue 81-46, July 30, 1981, at 1. *See also,* Gerth, *Shad of SEC Favors Bright Corporate Image,* NY Times, Aug 3, 1981, at D-1, col 1 (SEC Chairman in interview reportedly stated that, in order to assist raising capital, steps should be taken so that companies would not be required to disclose so much unfavorable information); and Sec Reg & L Rep (BNA) No 619, at B-1 (Sept 2, 1981) (summary of remarks of former SEC Commissioner Stephen J. Friedman—primary job is to "get out of the way: to remove barriers to capital formation").

[174] *E.g.,* 17 CFR §230.175.

[175] SEC v Ralston Purina Co, 346 US 119 (1953).

[176] 15 USC §77c(a)(2).

[177] *See, e.g.,* Securities Act of 1933 Release No 6281 (Jan 15, 1981); and Securities Act of 1933 Release No 6188 (Feb 1, 1980).

Resales of shares issued to affiliate employees is also simplified.[178] Any incentive plan for officers and directors involving the issuance of the securities of the employer raises concerns about the short swing profit recapture provisions of §16(b) of the 1934 Act.[179] Under the draconian provisions of that section, any profits from the purchase and sale, or sale and purchase within a six-month period, by an officer or director of the issuer of an equity security of an issuer which has a class of equity securities registered under the 1934 Act are recoverable by the issuer. (Any such profits made by the beneficial owner of more than 10 per cent of a class of registered securities would also be recoverable). These provisions are particularly troublesome with respect to stock option and other incentive plans. However, the SEC has adopted a safe harbor rule, Rule 16b-3, which alleviates some of the problems, if the plan complies with the provisions of that rule, including stockholder approval of the plan.[180]

Form S-8[181] is a simplified registration form for offerings of securities issued in connection with employee benefit plans by companies subject to the reporting provisions of the 1934 Act. A registration statement on Form S-8 generally is not reviewed by the SEC staff and becomes effective automatically 20 days after filing. Post-effective amendments on Form S-8 are generally effected automatically by the filing of the company's annual report with the SEC.

§6.39 Business Combinations and Other Restructurings

One method of financing innovation for the financially troubled business is to enter into an agreement with a financially stronger company to combine the business through a merger, an exchange of shares or a sale of assets in exchange for shares of the stronger company.

The company's assets may also be sold to a group of private investors in a *leveraged buy-out.*[182] These agreements will often be accompanied by arrangements for key individuals to continue to operate the acquired business, with perhaps a much improved compensation package which a larger company can make available as an incentive. Thus, the innovative project can be continued under new ownership.

Another possibility is the acquisition by employees of a piece of a business engaged in an innovative project, when the owners of the business no longer want to continue to incur the risks involved in the project.

All of these situations involve securities law considerations.

Under state corporation laws, mergers, sales of all or substantially all of the

---

[178] Form S-8, 17 CFR §239.16b.

[179] 15 USC §78p(b).

[180] 17 CFR §240.16b-3.

[181] 17 CFR §239.16b.

[182] *See, e.g.,* Securities Exchange Act of 1934 Release No 15572 (Feb 15, 1979).

assets of a corporation or the authorization of the issuance of stock require the approval of stockholders.[183] Listing agreements with national securities exchanges, such as the New York Stock Exchange, may require transactions involving the issuance of a significant amount of securities to be submitted to stockholders for their approval.[184] If the company seeking stockholder approval is a public company, it may be subject to the proxy rules of the Securities and Exchange Commission (SEC),[185] or if proxies are not required to be solicited under the applicable state law, information substantially equivalent to a proxy statement prepared in accordance with SEC rules may have to be provided to stockholders.[186] In addition, mergers, acquisitions of assets for stock and exchanges of stock all involve the offer and sale of securities for purposes of the federal securities laws, including the registration provisions of the 1933 Act.[187] This has not always been the case with mergers.

Until 1973, the SEC had provided by rule (Rule 133) that a merger did not involve a sale for purposes of the registration provisions of the 1933 Act, although a sale was involved for purposes of the antifraud provisions of the 1934 Act.[188] The SEC's conceptual theory was that there were no individual investment decisions involved in approving the merger, only corporate action. The practical basis for Rule 133 may have been the SEC's perception that registration in a merger context would be difficult to accomplish. In 1973, the SEC rescinded Rule 133 and adopted Rule 145[189] which provides that a merger in which securities are to be issued does involve a sale of those securities. It determined that a decision to grant a proxy or to vote on a proposed merger was conceptually an investment decision and that accommodations could be made to facilitate registration of the securities to be issued, if required.

Business combination transactions may be exempt from registration under one of the transactional exemptions, such as the private placement exemption.[190]

If the terms and conditions of the issuance and exchange of the securities are approved by any court or specified governmental authority, after a hearing upon the fairness of such terms and conditions at which all persons to whom the securities are to be issued have the right to appear, the transaction may be exempt under §3(a)(10) of the 1933 Act.[191] This exemption is often overlooked, but should not be. Counsel should carefully consider any applicable

---

[183] *See, e.g.*, Del General Corporation Law, §271.

[184] *See, e.g.*, New York Stock Exchange Company Manual A-283, 284.

[185] Reg 14A, 17 CFR §§240.14a-1 through .14a-101.

[186] Securities Exchange Act of 1934 §14(c), 15 USC §78n(c); and Reg 14C, 17 CFR §§240.14c-1 through .14c-101.

[187] 17 CFR §230.145.

[188] SEC v National Sec, Inc, 393 US 453 (1969).

[189] 17 CFR §230.145.

[190] Discussed in **§§6.25, 6.26.**

[191] 15 USC §77c(a)(10).

state law to determine whether a fairness hearing might be a practical means of avoiding some of the time and expense involved in registration.

If the transaction is not exempt from registration, or even if it is exempt, but is subject to the SEC's proxy rules, compliance with the federal securities laws may be more burdensome than in other situations.

While the SEC's rules permit the proxy statement to serve as the prospectus for the securities offering, it is not uncommon for a merger proxy statement to exceed 200 pages, because of the need to describe the businesses and provide financial information for two or more companies.[192] Prospectuses for exchange offers suffer from comparable problems. The registration fees, professional fees and printing costs incurred can amount to hundreds of thousands of dollars and in very large transactions can exceed a million dollars.

The risks of a disgruntled securityholder commencing litigation to challenge the transaction may also be greater than in the typical financing transactions. The Supreme Court in recent decisions has narrowed permissible scope of these challenges, however.[193]

The old owners of the business may be restricted in their ability to resell the securities they obtain in exchange for their business without registration of those securities.[194] They also may be restricted because of provisions for recapture of any profit they may realize on resale[195] or because of possible adverse accounting consequences.[196]

The SEC has made two abortive attempts to simplify disclosure requirements relating to business combination transactions[197] and recently provided some relief for combination of a smaller company with a company 10 times its size. The simplified registration procedures made available through use of its new Form S-15 provide relief from the merger proxy rules, if either company is subject to the SEC's proxy rules and must obtain stockholder approval of the transaction.

There are indications that the SEC administratively is making Form S-15 available for transactions which exceed the size tests in the instructions to the form. The SEC is also continuing its efforts to simplify the merger proxy

---

[192] Schedule 14A, 17 CFR §240.14a-101.

[193] *See, e.g.,* Santa Fe Indus, Inc v Green, 430 US 462 (1977).

[194] See **§6.49.**

[195] See **§6.38.**

[196] Generally accepted accounting principles permit pooling of interest accounting for a business combination only where there is a continuity of interest between the equity interests in the combined businesses. This results in certain limitations on resales of securities of the new owners received by the old owners. Accounting Series Release No 130 (Sept 29, 1972); and Accounting Series Release No 135 (Jan 5, 1973). *See also In re* CGA Computer Assocs, Inc, Securities Act of 1933 Release No 6328 (July 16, 1981) (stop order proceeding involving allegations of improper pooling of interest).

[197] Securities Act of 1933 Release No 5744 (Sept 27, 1976); and Securities Act of 1933 Release No 5806 (Feb 16, 1977).

statement required by Form S-14[198] and Schedule 14A[199] of the proxy rules.[200]

Exchange offers, sales of assets for stock, leveraged buy-outs and spin-offs of part of a business all may involve compliance with the registration provisions of the 1933 Act or the SEC's proxy rules and lengthy, complex and expensive disclosure documents which are time-consuming to prepare. Efforts to reduce the cost and other burdens of compliance in this area have so far accomplished very little.

Assuming the business deal is right, for the privately held company to sell out to a large public company, these rules may provide little initial problem. Some large companies register securities for the shelf for offering to smaller companies. What counsel for the sellers should be particularly concerned about are the restrictions on the ability of the owners of the business being sold to sell the securities they receive as consideration.[201]

§6.40 Trust Indentures

If the securities to be offered are debt securities, compliance with the Trust Indenture Act of 1939, as amended,[202] may be required. If the aggregate principal amount of the securities is less than $10 million, the Securities and Exchange Commission may exempt them from the 1939 Act.[203] An issue of less than $2 million need not be issued under an indenture.[204] An issue under an indenture of under $5 million is exempt from the Trust Indenture Act.[205]

If an exemption is not available, the indenture must be qualified and an independent trustee appointed to protect the interest of the securityholders. The qualification process is coordinated with the process for registration of the securities under the 1933 Act and is generally routine.

Debt securities issued in certain transactions exempt from registration under the 1933 Act are subject to the Trust Indenture Act;[206] securities issued after a fairness hearing by a state authority,[207] or securities exchanged for other securities held by existing securityholders,[208] for example.

---

198 17 CFR §239.23.

199 *Id* §240.14a-101.

200 *See, e.g.,* Freund & Greene, *Substance Over Form S-14: A Proposal to Reform SEC Regulation of Negotiated Acquisitions,* 37 Bus Law 1483 (July 1981).

201 See **§6.49.**

202 53 Stat 1149 (1939) (codified at 15 USC §§77aaa-777bbb).

203 *Id* §304(a)(9), 15 USC §77ddd(a)(9).

204 *Id* §304(a)(8), 15 USC §77ddd(a)(8), and Rule 4a-1, 17 CFR §260.4a-1.

205 17 CFR §260.4a-2.

206 *Id* §304(b), 15 USC §7ddd(b).

207 Securities Act of 1933, §3(a)(10), 15 USC §77c(a)(10).

208 *Id* §3(a)(9), 15 USC §77c(a)(9).

§6.41 Investment Company Act Issues

Most companies seeking to finance an innovative enterprise would be surprised to learn that there is some risk, that in doing so, they may inadvertently become investment companies subject to the Investment Company Act of 1940, as amended.[209] Regulation as a registered investment company may severely affect a company's ability to conduct an ordinary business enterprise and to provide adequate incentives to its employees.[210] Under §7 of the 1940 Act,[211] a company which is required to register under the act, but which operates as an unregistered investment company, is subject to having all of its transactions voided.

A company may become an *inadvertent investment company* if it has more than 100 stockholders and more than 40 per cent of its assets are invested in investment securities, as defined in the 1940 act, and it is not primarily engaged in a business other than investing or trading in securities.[212] Thus, a company which raises money from the public, a large portion of which it cannot immediately employ in its business should be concerned.[213] A company can also become an inadvertent investment company if it fails to obtain control of another company, but acquires a large block of its securities.[214]

If, however, the company's board of directors resolves to make a good faith effort to engage in a business other than that of investing or trading in securities within 12 months of the adoption of the resolution, the Securities and Exchange Commission has provided, in Rule 3a-2 under the 1940 act, a one-year safe harbor from registration under the act.[215] Longer periods may be permissible depending upon the nature of the company's investments and its efforts to engage in an operating business.[216]

Companies can also become investment companies as the result of leveraged buy-outs where management contracts with the buyers of the business to continue on as its managers while the stockholders of the seller are left with investments in money market instruments or municipal securities.[217] In these situations, provision is generally made for the company to register under the 1940 Act for the tax advantages that will provide, i.e., no tax at the corporate level if substantially all investment income is paid out as dividends.

A company which is in the business of buying up failing companies, turning

---

[209] 54 Stat 789 (codified at 15 USC §§80a-1 to 80a-64).

[210] See **§6.44.**

[211] Investment Company Act of 1940, 15 USC §80a-7.

[212] *Id* §3(a)(3), 15 USC §80a-3(a)(3).

[213] See Genentech, Inc, [1981] Fed Sec L Rep (CCH) ¶76,707 (available Sept 24, 1980).

[214] Investment Company Act of 1940 Release No 11552 (Jan 14, 1981).

[215] 17 CFR §270,3a-2.

[216] *See, e.g.,* Mallory Randall Corp, [1981] Fed Sec L Rep (CCH) ¶76,868 (available June 1, 1981).

[217] *See, e.g.,* Securities Exchange Act of 1934 Release No 15572 (Feb 15, 1979).

them around, and selling them may be a *special situation* investment company although that animal, like the unicorn, is often pictured but seldom seen in the flesh.[218]

Most companies can avoid 1940 act problems with a little careful business planning and appropriate legal advice.

§6.42 Liability

Violations of the provisions of the 1933 Act are subject to both civil and criminal penalties. In addition, unless the Supreme Court should rule otherwise, conduct which violates the 1933 Act may also violate provisions of the 1934 act.

The civil liability provisions of the 1933 Act are designed to reverse the old legal adage of "let the buyer beware," by imposing liability on the seller, if the buyer is not adequately informed about the securities he or she is buying. There are several gradations of liability. Controlling persons may be liable for the acts of persons they control, unless they can prove they did not know of and did not have reasonable grounds to believe the facts upon which liability is predicated.[219] Thus, a company which establishes a subsidiary to carry on its innovative projects may still be liable for violations of the securities laws committed by the subsidiary.

The courts have also imposed derivative liability on the basis of criminal law concepts of aiding and abetting and the common law doctrine of respondeat superior.[220]

The harshest liability is imposed for failure to register a security required to be registered.[221] In that case, even if the failure was through innocent inadvertence, the entire transaction can be rescinded and the investor's money returned. Liability is imposed, even if the seller provided the buyer with all of the information registration would have required. About the only mitigation of this remedy is that an action under §12(1)[222] must be commenced within one year of the violation.[223]

Under §12(2) of the 1933 Act,[224] a person who offers or sells securities by means of misstatements of material facts, or who omits to state material facts which, under the circumstances, were required to make other statements not misleading, is also liable to the buyer. The seller's only real legal defense, other

---

[218] *See, e.g.*, Bankers Sec Corp v SEC, 146 F2d 88 (3d Cir 1944).

[219] Securities Act of 1933, §15, 15 USC §77o.

[220] *See, e.g.*, Rolf v Blyth, Eastman Dillon & Co. 570 F2d 38 (2d Cir 1978), *cert denied*, 439 US 1039 (1978) (aiding and abetting); Coopers & Lybrand v Sharp, 649 F2d 175 (3d Cir 1981), *cert denied*, — US —, 50 USLW 3668 (Feb 23, 1982) (respondeat superior).

[221] Securities Act of 1933, §12(1), 15 USC §77l(1).

[222] *Id.*

[223] *Id* §13, 15 USC §77m.

[224] 15 USC §77l(2).

than the three-year statute of limitations (one year after discovery of facts underlying the violation),[225] is that he or she did not know, and in the exercise of reasonable care could not have known, the untruth or omission. In some circumstances, proof of a due diligence investigation similar to that required under §11 of the 1933 Act[226] may be required to establish the defense.[227] Section 12(2) applies to transactions in both registered and unregistered securities.

Section 11 of the 1933 Act is the no fault provision of the securities laws. It provides a civil remedy for persons who purchase registered securities and are misled by materially misleading statements or omissions in registration statements filed under that act. As a practical matter, other than the statute of limitations in §13,[228] the issuer of the registered securities may have no defense under §11[229] against claims of purchasers in the initial offering if the registration statement is materially misleading.

Underwriters, all the directors of the issuer and persons signing the registration statement may also be liable, but they may have certain defenses available. If they have no reason to believe otherwise, reliance on statements of experts made in the registration statement, accountants, actuaries, lawyers and engineers, for example, will relieve the person so relying from liability based on the expertised statements.[230] As to other statements in the registration statement, they must prove that they had, after a reasonable investigation, no reason to believe that the statements or omissions were misleading.[231]

The fear of liability, particularly since *Escott v BarChris Construction Corp,*[232] has led to extensive investigation of the facts underlying the disclosure in registration statements, particularly by underwriters and their counsel and counsel for the company. The resulting professional fees have contributed to the cost of registration which may run into hundreds of thousands of dollars for larger companies, excluding filing fees.

Fear of liability may have had an even more detrimental effect by turning the prospectus into an insurance policy (or what an insurance policy read like before the impact of plain English laws) rather than an informational document. There are indications that the Securities and Exchange Commissior (SEC) might wish to change this situation,[233] and those involved in the preparation of registration statements also bear a responsibility to make it a communicative document.

---

[225] Securities Act of 1933, §13, 15 USC §77m.

[226] 15 USC §77k.

[227] John Nuveen & Co v Sanders, 619 F2d 1222 (7th Cir 1980), *cert denied,* 450 US 1005 (1981). Mr. Justice Powell dissented from the denial of certiorari.

[228] 15 USC §77m.

[229] Id §77k.

[230] *Id.*

[231] *Id.*

[232] 283 F Supp 643 (SDNY 1968).

[233] *See, e.g.,* Rule 175, 17 CFR §230.175, and Rule 176, 17 CFR §230.176.

The Supreme Court has not definitely ruled on the availability of an implied private remedy under the anti-fraud provisions of the 1933 act, §17(a).[234] Some lower courts have refused to imply such a remedy, in part because it is reasonably well-established that there is such a remedy under the antifraud provisions of the 1934 act, principally §10(b)[235] and Rule 10b-5[236] thereunder.[237] Conversely (or perversely), other courts have implied a private remedy under §17(a) for the same reason.[238] That remedy may be available to persons acquiring registered securities or acquiring securities in exempt transactions.

In any event, the SEC can and does enforce §17. This is important, since the target of an SEC investigation or enforcement action may have to expend hundreds of thousands or even millions of dollars in defending and resolving the action and the outcome of the action may have collateral effects on private actions.[239] In some cases, the SEC may obtain equitable relief in the form of disgorgement to create a fund for investors who have suffered damages.[240]

The impact of the securities laws can be attributable in large measure to their rather stringent liability provisions, enforced by a government agency with investigative and enforcement powers more extensive than most regulatory agencies and by private plaintiffs with adequate economic incentives to pursue the litigation. Some believe that the system is tilted too much in favor of plaintiffs and that the liability to which companies, their managements and professional advisers may be exposed is too great. The American Law Institute's proposed *Federal Securities Code* (1978) would provide additional defenses, negligence standards rather than strict liability in some areas,[241] for example, and would limit monetary liability in some circumstances.[242] To the extent that fear of liability may inhibit innovative risk taking, the Code's provisions would be an improvement over the current law.

---

[234] 15 USC §77q.

[235] *Id* §78j(b).

[236] 17 CFR §240.10b-5.

[237] *See, e.g.,* Hill v Der, 521 F Supp 1370, [1981-82 Transfer Binder] Fed Sec L Rep (CCH) ¶98,311 (D Del 1981).

[238] *See, e.g.,* Roth v Bank of the Commonwealth, [1981-82 Transfer Binder] Fed Sec L Rep (CCH) ¶98,267 (WDNY Aug 17, 1981).

[239] Parklane Hoisery Co v Shore, 439 US 322 (1979).

[240] *See, e.g.,* SEC v Texas, Gulf Sulphur Co, 401 F2d 833 (2d Cir 1968), *cert denied,* 394 US 976 (1969).

[241] *Federal Securities Code* §1702(c).

[242] *Id.* §1708.

Regulation of Investors and Investments

§6.43 Direct Regulation of Investors—Overview

Some of the more significant impediments to the external nongovernmental financing of innovation result from the direct and indirect restrictions imposed by the federal securities laws on investors who provide funds for new ventures or otherwise assist in financing new ventures. Extension of credit to these investors for purposes of investing in or carrying securities are subject to the regulations adopted by the Board of Governors of the Federal Reserve System pursuant to §7 of the 1934 Act.[243] State securities or blue sky laws may also be an impediment, although to a lesser degree. This maze of overlapping, duplicative and complex regulations may discourage venture capital investment by certain types of financial institutions and affect private investment in innovation through the imposition of restrictions on the liquidity of those investments. While providing a degree of investor protection, this makes private investment riskier and less attractive and increases the cost of capital.

§6.44 —Investment Companies

The 1940 act[244] provides for regulation of public companies which are engaged in investing in securities, i.e., an *investment company* as that term is defined in §3 of the act.[245] An investment company must register with the Securities and Exchange Commission (SEC).[246] The 1940 act provides for regulation, among other things, of a registered investment company's management[247] and capital structure;[248] the relationship between the company and its owners (stockholders);[249] the compensation the company may pay to its management and investment advisers;[250] and transactions between the investment company and companies in which the investment company has significant investments or between the investment company and its affiliates.[251] The 1940 act also subjects those who violate its rather complex provisions to liability to those who suffer damage as the result of the violation.[252]

This regulatory structure, together with the obligations of a public company

---

[243] Securities Exchange Act of 1934, *as amended,* 48 Stat 881 (1934), 15 USC §78g.

[244] Investment Company Act of 1940, *as amended,* 54 Stat 789 (codified at 15 USC §§80a-1 to 80a-64).

[245] *Id,* 15 USC §80a-3.

[246] *Id* §§7 and 8, 15 USC §§80a-7 and 80a-8.

[247] *E.g., id* §§9, 10, 15 and 16, 15 USC §§80a-9, 80a-10, 80a-15 and 80a-16.

[248] *E.g., id* §§17 and 21, 15 USC §§80a-17 and 80a-21.

E.g.; id §§15 and 20; 15USC §§80a-15 and 80a-20.

[250] *E.g., id* §§15, 15 USC §80a-15.

[251] *E.g., id* §17, 15 USC §80a-17.

[252] *E.g., id* §5, 15 USC §80a-35.

to its stockholders under state corporate law, can be a disincentive to venture capital investments by registered investment companies. The experience of many small business investment companies illustrates the problem.

The ability of public small business investment companies to make successful investments, in accordance with the provisions of the Small Business Investment Act of 1958,[253] is affected by regulation under the 1940 act. Transactions between an investment company and companies in which it has a significant investment, or transactions in which an investment adviser or significant stockholder or control person of the investment company has an interest require SEC approval, after application and notice and opportunity for hearing.[254] The nature of venture capital investments may make it impractical to follow such procedures, and could discourage a second or third infusion of capital into a struggling venture.

In addition, restrictions on compensation which an investment company may pay to its investment advisers and management may reduce the incentives to engage in venture capital transactions.

These restrictions as well as restrictions on the liquidity of private investments in securities imposed by the federal securities laws,[255] may have prevented investment companies from becoming a more important source of venture capital. In addition, the perceived demand of public investors for evidence of short-run investment performance may tend to discourage publicly owned investment companies from making capital investments in projects which may take years to develop.

§6.45 —Business Development Companies

Recently, efforts have been made to stimulate more venture capital investment by permitting an investment company to elect to be regulated as a business development company.[256] A business development company is subject to a different scheme of regulation under the 1940 Act than other investment companies. A business development company is expected and indeed required to provide venture capital to new, emerging businesses and financially troubled ventures.[257] To elect and maintain business development company status, the company is required to invest a specified portion of its assets in securities of companies whose securities are not widely held, small business investment companies or financially troubled companies.[258] The business development company must also make available *significant managerial assistance* to

---

[253] 15 USC §§661-687, *See generally*, Miller, *Small Business Investment Companies: Licensing, Tax and Securities Considerations*, 36 Bus Law 1679 (1981); and Investment Company Act of 1940, §17, 15 USC §80a-17.

[254] Investment Company Act of 1940, §17, 15 USC §80a-17.

[255] See **§6.49.**

[256] Investment Company Act of 1940, §54, 15 USC §80a-53

[257] *Id* §55, 15 USC §80a-54.

[258] *Id.*

those companies.[259]

The 1940 Act's restrictions on the capital structure of investment companies do not apply business development companies,[260] so that those companies may issue debt securities with equity features to finance their own investments. A business development company may also grant options or other rights to acquire securities of the business development company to its directors, officers and employees as incentive compensation.[261]

The 1940 Act's provisions relating to transactions between a business development company, its affiliates and companies in which it has significant investments and others were also relaxed.[262] This should permit greater flexibility to a business development company which is considering an additional investment in a portfolio company, particularly where the portfolio company is experiencing financial problems and needs an immediate infusion of capital.

The amendments also provide for an added incentive to venture capital investment, in that investment advisers to business development companies may charge performance-related fees within certain specified limits.[263]

There has not been sufficient experience to determine whether the business development company experiment in deregulation will result in a significant infusion of capital into innovative ventures. Although the amendments may have made venture capital investments more attractive, many other types of investments are competing for the available supply of investment dollars.

§6.46 —Investment Advisers

There are a number of closely held companies which are a source of venture capital and which may not be regulated as investment companies, if they have fewer than 100 stockholders and are not offering their own stock to the public.[264] These venture capital companies are not to be confused with *hedge funds* which generally were speculative vehicles, more popular in the late 1960s and early 1970s.[265]

The incentive to organize, manage and advise these private venture capital companies is somewhat limited by the restrictions imposed by the Investment Advisers Act of 1940, as amended,[266] on advisory compensation based on performance in situations where the adviser is required to register under that

---

[259] *Id.*

[260] *Id* §61, 15 USC §80a-60.

[261] *Id.*

[262] *Id* §57, 15 USC §80a-56.

[263] Investment Advisers Act of 1940, §205, 15 USC §80b-5.

[264] Investment Company Act of 1940, §3(c)(1).

[265] *See, e.g.*, 3 Institutional Investor Study Report of the Securities and Exchange Commission 887, HR Doc No 64, 92d Cong, 1st Sess (Mar 10, 1971).

[266] 54 Stat 847 (codified at 15 USC §§80b-1 to 80b-21).

act.[267]

Since persons who manage these entities are generally engaged in the business of advising others as to the value of securities or the advisability of investment in, purchasing or selling securities for compensation, they would be investment advisers by definition under §202(a)(11) of the Advisers Act.[268] If an adviser holds himself or herself out generally to the public as an investment adviser or had 15 or more clients in the preceding 12 months, the adviser must register with the SEC[269] and may not be compensated on the basis of the capital appreciation of the assets managed.

One of the more troublesome aspects of these provisions relates to the determination of the number of clients of an investment adviser. It is unclear, for example, whether an adviser to a limited partnership may also be deemed to be the adviser to the limited partners.[270]

The standards for determining whether an adviser is holding out generally to the public as an investment adviser are also not sharply defined.

The most significant disincentive to venture capital investment imposed by the Adviser Act is, of course, the restriction on the manner of the adviser's compensation.[271] The Small Business Investment Incentive Act of 1980[272] relaxed these restrictions somewhat with respect to investment advisers to business development companies.[273]

§6.47 —Financial Intermediaries

Investment bankers can be of great assistance to a company in raising capital and to investors in realizing on their investment. For smaller innovative companies; the efforts of regional investment banking firms are of importance.[274] These financial intermediaries are directly regulated under the federal securities laws, principally the 1934 Act. State securities regulation of financial intermediaries, on the other hand, probably does not have as great an impact on capital formation.

Commercial banks also may assist in raising private capital, assuming com-

---

[267] *Id* §205, 15 USC §80b-5.

[268] *Id* §80b-2(a)(11).

[269] *Id* §203(b)(3), 15 USC §80b-3(b)(3).

[270] See, *e.g.*, Abrahamson v Fleschner, 568 F2d 862 (2d Cir 1977), *cert denied,* 436 US 913 (1978).

[271] Investment Advisers Act of 1940, 15 USC §80b-5.

[272] Pub L No 96-477, 94 Stat 2275 (Oct 21, 1980)(codified in scattered sections of 15 USC).

[273] Investment Advisers Act of 1940, §205, 15 USC §80b-5.

[274] *See* Initial Public Offerings of Common Stock: The Role of the Regional Broker-Dealer in the Capital Formation Process (US Small Business Admijnistration and Securities and Exchange Commission, Phase I Report, Mar 1980); and *see generally,* Wolfson, Phillips & Russo, Regulation of Brokers, Dealers and Securities Markets (1977).

pliance with the Glass-Steagall Act of 1933,[275] but they are not regulated by the SEC. This may put investment bankers at a competitive disadvantage.

Investment bankers generally must register with the SEC as brokers and dealers in order to do business. The SEC on various grounds may administratively revoke or suspend that registration and bar or suspend individuals from being associated with a broker or dealer.

Broker-dealers must maintain a specified level of net capital under a complex set of regulations[276] and their business conduct and capital structure is regulated by self-regulatory organizations, the stock exchanges and the National Association of Securities Dealers, Inc. acting under authority granted by the 1934 Act.[277]

While investment banking houses engage in all facets of venture capital investing, a disincentive to their involvement in private placements may result from the legal uncertainties involved in placing securities privately on behalf of the issuer.[278]

A securities firm risks both monetary liability under the 1933 Act[279] and regulatory action under the 1934 act which could result in limitations on the conduct of its business. The firm faces similar risks if it participates in resales of privately placed securities.

Also, anyone, including a registered broker-dealer, who engages in advising investors with respect to private investments could become subject to the Advisers Act, which would require registration with and reporting to the SEC and affect the amount of their compensation. Here again the applicable standards are not sharply defined.

In addition, recent judicial decisions, which the Supreme Court has thus far declined to review, have imposed an obligation on securities professionals to conduct a due diligence investigation in connection with an exempt securities transaction.[280] If that investigation is required to be as extensive as that required to establish a defense for an underwriter under §11 of the 1933 Act[281] against civil liability in connection with an offering of registered securities, it could be an impossible burden. That type of investigation is costly and time consuming and may not be economically feasible with respect to many exempt offerings.

The nature of the underwriter's due diligence obligation in connection with offerings of registered securities is currently being re-examined as a result of the increasing utilization of registration forms which rely on incorporation by reference of the issuer's periodic reports under the 1934 Act (disclosure inte-

---

[275] 48 Stat 162 (codified in scattered sections of 12 USC).

[276] 17 CFR §240.15c3-3.

[277] Securities Exchange Act of 1934, §§15A and 19, 15 USC §78o-3 and 78s.

[278] *See, e.g.,* Henderson v Hayden-Stone, Inc, 461 F2d 1069 (5th Cir 1972).

[279] Securities Act of 1933, *as amended,* 48 Stat 74 (codified at 15 USC §77a-77aa).

[280] *See, e.g.,* John Nuveen & Co v Sanders, 619 F2d 1222 (7th Cir. 1980), *cert denied,* 450 US 1005 (1981).

[281] 15 USC §77k.

gration).[282] Underwriters are generally not involved in preparation of those reports and are concerned about the nature and extent of their due diligence obligation with respect to such registration statements. The SEC has been willing to listen to these concerns and has taken the position that the nature and extent of the underwriters' investigation in an exempt transaction should not be as great as that required in connection with an underwritten offering of registered securities.[283]

§6.48 Restrictions on Liquidity of Investments

The restrictions imposed by the federal securities laws on the liquidity of private investments may be a greater disincentive to investment in innovation than direct regulation of institutional investors and investment intermediaries. These restrictions apply to all investors, whether directly regulated or not. In addition, these restrictions substantially increase the cost of capital, since invested funds may be tied up for two or three or more years.

The 1933 act is drafted in such a way as to prohibit the use of any facility of interstate or foreign commerce or the United States mails to offer or sell a security unless there is compliance with the registration provisions of the 1933 act.[284] Unless an exemption from registration is available, the security must be registered with the Securities and Exchange Commission (SEC) and a prospectus containing specified disclosure about the issuer and the security provided to the purchasers.

Private transactions between an issuer and a limited number of sophisticated investors or investors who otherwise do not need the protections that registration would afford are exempt under §4(2) of the 1933 act.[285] Transactions pursuant to §§4(6), 3(a)(9) and 3(a)(10) of the 1933 act[286] and Rules 504, 505 and 506[287] under that act also may also may result in the securities which were sold in these transactions being restricted as to resale.

The issue then becomes—when, how and to whom these investors can resell their securities and realize on their investments. If resales occur too soon after the private placement, the presumption may be that the investors were actually participants with the issuer in a distribution of the securities to the public, i.e., they were underwriters[288] and the securities should have been registered. On the other hand, if the investors have been at risk for an appropriate period of time, there is current information about the issuer available to the public

---

[282] *See generally* Securities Act of 1933 Release No 6383 (Mar 3, 1982).

[283] *See, e.g.*, SEC Brief Amicus Curiae in John Nuveen & Co v Sanders, 619 F2d 1222 (7th Cir 1980), *cert denied*, 450 US 1005 (1981).

[284] Securities Act of 1933, §5, 15 USC §77e.

[285] 15 USC §77k. See **§§6.25** and **6.26.**

[286] *Id* §§77d(6), c(a)(9) and (10).

[287] 17 CFR §§230.504-506.

[288] Securities Act of 1933, §2(11), 15 USC §77b(11).

similar to that which would be provided by a prospectus, and there is a market for the securities of sufficient depth and liquidity that an unusual effort and high selling costs are unnecessary to sell the securities, registration should not be and is not required. Section 4(1) of the 1933 act[289] provides an exemption for such transactions where the seller is neither the issuer, a securities dealer or an underwriter. However, the precise standards for determining when that exemption is available were never clearly anticulated and uncertainty resulted.[290]

§6.49 —Safe Harbor Rule (Rule 144)

The SEC has done a great deal to reduce the uncertainties in this area, principally through its nonexclusive safe harbor rule, Rule 144.[291] Even then, one must master literally hundreds of pages of gloss on the rule,[292] to be satisifed that a particular transaction is in compliance.

Compliance with Rule 144 requires an investment to be held subject to investment risks for a minimum of two years before it can be resold publicly. While securities can be re-placed privately,[293] this type of transaction has distinct disadvantages. First, under Rule 144, at least, the new purchaser would be required to hold the securities at risk for at least two years from the date of its purchase. Then, the law concerning replacements in non-Rule 144 resales (with one very limited exception)[294] is uncodified and uncertain.[295] This uncertainty also imposed risks on financial intermediaries who may assist in these transactions. Because of these restrictions and uncertainties, the investor cannot realize the same value from an investment as could be realized if it were freely salable.

Rule 144 also provides the standards for resales of securities acquired in transactions which are not strictly private. Transactions by affiliates (control persons) of the issuer are specifically provided for in Rule 144. Because a broker who sells securities on behalf of an affiliate of the issuer of the securities might be considered an *underwriter* of those securities, as that term is defined

---

[289] *Id* §77d(1).

[290] *See, e.g.*, Committee on Developments in Business Financing, *Resale of Debt Securities Acquired in Private Placements,* 34 Bus Law 1927 (July 1979); Study Group on §"4 (1 1/2)", *The Section "4 (1 1/2)" Phenomenon: Private Resales of "Restricted Securities",* 34 Bus Law 1961 (July 1979).

[291] 17 CFR §230.144.

[292] *See, e.g.*, Securities Act of 1933 Release No 6099 (Aug 2, 1979).

[293] *See, e.g.*, Committee on Developments in Business Financing, *Resale of Debt Securities Acquired in Private Placements,* 34 Bus Law 1927 (July 1979); and Study Group on §"4(1 1/2)", *The Section "4(1 1/2)" Phenomenon: Private Resales of "Restricted" Securities,* 34 Bus Law 1961 (July 1979).

[294] 17 CFR §230.237.

[295] *See generally* Schneider & Zall, *Section 12(1) and the Imperfect Exempt Transaction: The Proposed I & I Defense,* 28 Bus Law 1011 (1973).

in §2(11) of the 1933 Act,[296] an exemption such as Rule 144 is generally necessary for the transaction, although the §4(1) exemption might also be available.

Rule 144 also provides an exemption for restricted securities acquired in transactions exempt pursuant to Regualtion D[297] or §4(6) of the 1933 Act.[298] The standards of Rule 144 are also referenced in the exemptive provisions provided in Rule 145[299] for resales of securities acquired in business combination transactions. The SEC staff also refers to provisions of Rule 144 for guidance in responding to no action requests relating to resale of securities acquired in transactions exempt under other provisions of the 1933 Act, such as §3(a)(10).

The legal risks connected with resales of securities acquired in private transactions are compounded by the manner in which both the SEC and the courts have consistently interpreted the provisions of §12(1) of the 1933 Act.[300] These interpretations impose almost absolute liability for rescission or damages, if the security should have been registered. The good faith or the reasonableness of the seller's conduct are not defenses.[301]

There is, however, considerable support for imposing a standard which would at least excuse inadvertent and immaterial departures from the statutory provisions. The SEC has not taken steps to implement such a standard, except indirectly, by supporting the American Law Institute's *Federal Securities Code.*[302]

There is also a limited exception for private resales provided by Rule 237[303] under the 1933 Act. That exemption is only available after the securities have been owned, at risk, for five years. The rule also limits the proceeds from sale to the lesser of the gross proceeds from the sale of 1 per cent of the class of securities sold or $50,000. Further, the issuer must have been a going concern during the five years preceding the sale. Rule 237 may have some viability, however, as it relates to securities of nonpublic companies where securities are not traded in organized securities markets.

Under Rule 144, an unlimited amount of restricted securities of public companies may be resold by nonaffiliates of the company, if there is current information about the company available to the public and the securities have been owned at risk for a minimum of three years. The major uncertainty remaining is the lack of standards for determination of who is an affiliate.

If the three-year holding period has not been satisified, and the sale is to be made pursuant to Rule 144, there are limitations on the amount of securities which may be sold in any three-month period (the greater of 1 per cent of the

---

296 15 USC §77b(11).

297 17 CFR §§230.501-.506.

298 15 USC §77d(6).

299 17 CFR §230.145.

300 15 USC §77l(1).

301 See **§6.32.**

302 See §1702(c) (1978).

303 17 CFR §230.237.

outstanding shares of the class or a test based on the average trading volume in the securitiy), and the manner of sale (which requires a professional intermediary); and there must be current information (as defined in the rule) available about the company. Thus, a venture capitalist which makes a private investment that amounts to 5 per cent of the outstanding common stock of a company may have to wait longer than three years to turn over its investment. For example, a business development company which owns a substantial block of stock of an issuer and exerts some influence on its management, as required by the 1940 act,[304] might be considered an affiliate of the issuer and be required to sell within the volume limitation of the rule and in the manner required by the rule. This is an important factor in capital formation since a portion of the funds of a venture capitalist available for new investments is provided by turnover of old investments.

§6.50 —Other Methods of Resale

Other methods of offering the securities may have certain disadvantages. The investor may have contracted for rights to cause the issuer to register the securities with the SEC. Registration, however, is expensive and at times impracticable. There is a simplified registration form available in certain circumstances, Form S-18.[305] However, the amount of securities which may be sold by affiliates is limited to an aggregate of $1.5 million and sales by an affiliate may prevent the issuer from using the form to register securities for its own account. Moreover, the registration process, even for a Form S-18, could delay sales for months.

The small offering exemption under Regulation A[306] is a less expensive alternative[307] but may be just as impracticable at times and the amounts of securities which may be sold by affiliates are limited. Also, sales by affiliates may limit the use of any other 1933 Act §3(b)[308] exemptions by the issuer.

These issues are further complicated by the fact that the definition of affiliate is ambiguous and the SEC will not render informal interpretive or no action advice as to whether a person is an affiliate for purposes of a particular rule.[309] The term *affiliate* generally is defined to mean: "a person that directly, or indirectly through one or more intermediaries, controls, or is controlled by, or is under common control with, such issuer."[310] The term *control* has been defined to mean: "the possession direct or indirect of the power to direct or cause the direction of the management and policies of a person, whether

---

[304] Investment Company Act of 1940, §55, 15 USC §80a-64.

[305] 17 CFR §239.28.

[306] *Id* §§230.251-.264.

[307] See **§6.29.**

[308] 15 USC §77c(b).

[309] Securities Act of 1933 Release No 6253 (Oct 28, 1980).

[310] 17 CFR §230.405(a).

through ownership of voting securities, by contract, or otherwise."[311]

The control standards are obviously imprecise and are further burdened by concepts such as group control and related concepts. These concepts require that, for equity accounting purposes, a 20 per cent or greater equity owner be able to exert a degree of influence on the investee[312] and a business development company also must exert an influence on the management and operations of its portfolio companies.[313]

The SEC could reduce some of the legal uncertainties involved in raising capital by stating what its views are as to the standards for determining the availability of:

1. The statutory private placement exemption, §4(2) of the 1933 Act
2. An exemption for private resale of securities obtained from the issuer in a private placement
3. The exemption for public resales under §4(1) of the 1933 Act of securities acquired in private placements
4. Factors for determining who is an affiliate for purposes of those exemptions

All resale transactions are also subject to the general antifraud provisions of the federal securities laws and state securities laws principally §17(a) of the 1933 Act[314] and Rule 10b-5 under the 1934 Act.[315] The transactions may also be subject to the express civil liability provisions of §12(2) of the 1933 Act.[316]

If the investor is aware of material nonpublic information about the company, and a substantial venture capital investor generally would be, the investor may not resell securities of the company unless and until that information is disclosed to the public, or in a private transaction, to a purchaser.[317] This problem is of less significance as it relates to a public company, reporting to the SEC and announcing material events to the press and its stockholders. In those situations, only the timing of the sales may be affected. However it would be otherwise with respect to investments in closely held companies.

Another issue relates to the nature of the information which is required to be provided by the private purchaser on resale. In certain circumstances, it may be sufficient if, through a relationship with the issuer (contractual or otherwise) or economic clout, the purchaser from the private investor has access to all material information that registration under the 1933 Act would provide.

---

[311] *Id* §230.405(f).

[312] Financial Accounting Standards Board, Interpretation No 35, AICPA Professional Standards (CCH) Accounting. §5131-1 (May 1, 1981).

[313] Investment Company Act of 1940, §55, 15 USC §80a-54.

[314] 15 USC §77q(a).

[315] 17 §240.10b-5.

[316] 15 USC §771(2).

[317] See **§6.25.**

This was the theory behind the informational access requirements of old Rule 146. For example, the purchaser from the private investor may be a partner in the venture or an important officer of the company, or it may be the principal lender. Generally, however, some written information will be provided. The risk is that the information is false or misleading or other information has not been provided which is necessary to make that which is provided not misleading. The issues then become the state of mind of the person providing the information (was the person negligent or did the person act with scienter, depending upon the applicable standard?) and the materiality of the information. The Supreme Court has defined materiality by an objective standard:

> There is a substantial likelihood that, under all the circumstances, the omitted fact would have assumed actual significance in the deliberations of the reasonable shareholder. Put another way, there must be a substantial likelihood that the disclosure of the omitted fact would have been viewed by the reasonable investor as having signfiicantly altered the "total mix" of information made available.[318]

Under that standard, there is no litmus test as to what would be material under a particular set of facts and each situation requires analysis with regard to the surrounding facts and circumstnaces. Moreover, privately negotiated transactions, if face to face, may be tested by a subjective materiality standard, i.e., what was important to that particular purchaser. When financing innovative tèchnology, there may be a desire to preserve business confidences or trade secrets and thus to disclose as little as possible to the investors. This requires difficult judgements as to the materiality of the undisclosed information. Moreover, materiality is generally judged with the benefit of hindsight. Thus, the risk associated with raising capital for such a venture may be increased.

There are no easy solutions to these liability issues. However, the Supreme Court, concerned about the tremendous liability which could result, has attempted to restrict the situations in which liability under the federal securities laws can be imposed by various means. It has limited the availability of implied private rights of action;[319] limited those who have standing to bring actions;[320] required plaintiffs to prove defendants acted with scienter;[321] and refused to extend the coverage of the federal securities laws to acts of mismanagement or breach of fiduciary obligations which it believes should be the province of state statutory or common corporate law.[322] These efforts by the Supreme

---

[318] TSC Indus, Inc v Northway, Inc, 426 US 438, 449 (1976).

[319] Touche Ross & Co v Redington, 442 US 560 (1979); TransAmerica Mortgage Investors, Inc v Lewis, 444 US 11 (1979).

[320] Blue Chip Stamps v Manor Drug Stores, 421 US 723 (1975); Piper v Chris-Craft Indus, Inc, 430 US 1 (1977).

[321] Ernst & Ernst v Hochfelder, 425 US 185 (1976).

[322] Santa Fe Indus, Inc v Green, 430 US 462 (1977).

Court may eventually reduce some of the legal risks involved in raising capital to finance innovation.

The fraud and civil liability provisions of the American Law Institute's proposed *Securities Code* also represent an effort to reduce the uncertainties in this area.[323] Perhaps the most significant provisions are those which would set specific limits on the amount of damages for certain types of violations.[324] However, political support for the Code is scant and the chances for its enactment in the near future are slim.

Disclosure Requirements

§6.51 The Aftermath: Periodic Reporting—Overview

What consideration should the larger publicly owned company give to the federal securities laws when it decides to develop an innovative product or process?

Assume the company decides to finance the project with internally generated funds, either alone or through a joint venture with another company which will also use its own funds. This will limit the necessary considerations of the securities laws to three principal areas:

1. Timely disclosure of the development to the securities market
2. Insider trading or tipping about the development
3. Mandatory periodic disclosure requirements

§6.52 —What Must Be Reported and When?

An hypothetical example might assist in analyzing the disclosure considerations involved. Assume that I company decides to develop a method for generating electricity powered by a fusion process. I company is presently engaged in an unrelated manufacturing industry. Its common stock is listed on the New York Stock Exchange (NYSE). The company intends to invest $1 billion in the project over the next 10 years with an initial investment of $100 million. I Company's total assets are $1 billion. No site has been selected and no agreement executed, as yet.

There may be both tangible and intangible benefits from prompt voluntary disclosure of an innovative development or plan. The stock market's rational assessment of the probability of commercial success of the project, could result in an increase in the market value for the company's stock. The market is a fickle mistress, however, since it generally emphasizes short-term performance

---

[323] Pt XVII (1978).

[324] *Federal Securities Code* §1708.

at the expense of long-range development. Nevertheless, it is also attracted by glamour issues and industries which may involve more speculative investments. Depending upon the available managerial compensation arrangements, management could also derive benefits from these increased stock values. Moreover, the company's corporate image could be enhanced. Voluntary disclosure of favorable information may also be of value to a company which is concerned about being taken over.

There is another side to this coin. There are legal and competitive risks which can result from disclosure, to say nothing of the damage of the corporate and managerial image if the proudly publicized attempt at innovation fails.

At what stage should the benefits of disclosure be weighed against these always attendant risks? Generally, long before there is any mandatory requirement for disclosure. Regulators are reluctant to state so loudly and clearly, but there is no specific legal requirement for a company to disclose a significant development in its business, absent circumstances specified in the law, at least where there is a business reason for nondisclosure.[325] In our example, I company may well have good business reasons to avoid premature disclosure—to avoid premature activity by environmental groups in opposition to the project, for example. While there is an infinite variety of factual situations within each category, there are really only three basic types of circumstances which would trigger a disclosure requirement:

1. The company or its insiders are engaged in purchasing or selling the company's own securities or have provided nonpublic information to others who are purchasing or selling securities of the company[326]
2. The company is contractually obligated to a stock exchange to make disclosure[327]
3. The periodic disclosure provisions of the federal securities laws require disclosure in a report filed with the Securities and Exchange Commission (SEC)[328]

---

[325] SEC v Texas, Gulf Sulphur Co, 401 F2d 833 (2d Cir 1968), *cert denied,* 394 US 976 (1969); and State Teachers Retirement Fund v Fluor Corp, 654 F2d 843, 1981 Fed Sec L Rep (CCH) ¶98,005 (2d Cir 1981).

[326] Chiarella v United States, 445 US 222 (1980).

[327] There may be no right of action implied under §6 of the Securities Exchange Act of 1934, 15 USC §78f, for violation of a stock exchange's timely disclosure policies. *See, e.g.,* Pittsburgh Terminal Corp v Baltimore & ORR, 509 F Supp 1002 (WD Pa 1981) (plaintiff not a third-party beneficiary of listing agreement). A corporation may exercise its business judgment to delay disclosure of even the most material contract. *See, e.g.,* State Teachers Retirement Fund v Fluor, 654 F2d 843, 1981 Fed Sec L Rep (CCH) ¶98,005 (2d Cir 1981) ($1 billion contract for coal gasification plant; government of South Africa requested embargo on news, NYSE did not object; but did suspend trading; no duty to request trading suspension; unless source of rumors).

[328] Principally Securities Exchange Act of 1934, §13(a), 15 USC §78m(a). *See generally* Schneider & Shargel, *"Now That You Are Publicly Owned . . ."*, 36 Bus Law 1931 (July 1981).

§6.53 —Insider Trading

If the company is engaged in purchasing or selling its own securities, the securities registration[329] or issuer tender offer provisions[330] of the federal securities laws may apply, as well as the general antifraud provisions of these laws.[331] In these circumstances, under our hypothetical, some disclosure would be required. The considerations involved where an insider is purchasing or selling are similar.

Companies have a degree of control over the first type of circumstances, in that they may be able to refrain from buying or selling their own securities. There is an exception. Many public companies will have an employee stock option plan or some type of employee benefit plan which involves purchases or sales of the company's securities. In theory, the company's disclosure obligations to its employees who are buying its stock are the same as to members of the general public.

One of the objectives of a company's listing agreement with a major stock exchange is to provide timely disclosure to the exchange and to the public of information that may affect security values or influence investment decisions or reasonably be expected to materially affect the market for the company's securities.[332] At that point, some disclosure is required in the exchange's view.[333]

The major stock exchanges recognize that information concerning such matters as new contracts, products or discoveries may be confidential matters which do not require disclosure unless leaks occur.[334] Also, premature announcement of new products or discoveries whose commercial application cannot be reasonably evaluated should be avoided.[335]

The stock exchanges are of the view, however, that at some point in time, contract awards, expansion plans and major new products reach the stage where immediate disclosure is required.[336] Immediate disclosure would not be required in our I Company example.

Timely disclosure requirements of the major stock exchanges, which are strongly supported by the SEC,[337] do not specifically require disclosure where

---

[329] Securities Act of 1933, §5, 15 USC §77e.

[330] Securities Exchange Act of 1934, §13(e), 15 USC §78m(e), and Rule 13e-4 thereunder, 17 CFR §240.13e-4.

[331] Principally, Securities Exchange Act of 1934, §§10(b) and 14(e), 15 USC §§78j(b) and n(e).

[332] *See, e.g.,* New York Stock Exchange Company Manual A-18 to A-25.

[333] *See, e.g., id* A-20, A-21.

[334] *Id.*

[335] *See, e.g., id* A-23.

[336] *See, e.g., id* A-19.

[337] *See* Securities Act of 1933 Release No 5092 (Oct 15, 1970); *In re* Spartek, Inc and John A Cable, Securities Exchange Act of 1934 Release No 15567 (Feb 14, 1979); and Report of Investigation in the Matter of Sharon Steel Corporation, Securities Exchange Act of 1934 Release No 18271 (Nov 19, 1981).

there is a business purpose for nondisclsure, at least if there are no rumors rampant in the marketplace. However, once a stock exchange inquires of the company about a rumored development, disclosure generally will result, if there is a basis for the rumor. Responses to inquiries from stock exchange representatives should not be purposefully misleading.[338]

Although the disclosure which results from a stlock exchange's urging may be voluntary, in the sense that it is not compelled by the securities laws, like any other voluntary disclosure made by a public company, the disclosure may not be misleading.[339]

SEC periodic reporting requirements are concededly not timely disclosure requirements. Those requirements require filing of current reports with the SEC containing information about specified events, an acquisition of a significant amount of assets, for example.[340] A plan to develop a new project, as in our hypothetical example, is not among those specified events.

The SEC's current reporting requirements do provide for strictly voluntary reporting of information which the company believes to be relevant.[341] Otherwise, an innovative development would have to involve acquisition or disposition of a material amount of the company's assets to require current reporting. For example, establishing a joint venture by transfer of assets might trigger a current report requirement. However, a current report on Form 8-K[342] filed within 15 days of an event is not a timely report and would not satisfy an obligation to promptly report a material development.[343]

Persons associated with the company may not buy or sell securities of the company on the basis of material information about the innovative development which has not been made available to the public and may not "tip off" others about that information before it becomes public.[344] The extent of the company's legal liability for illegal trading by insiders is unclear. However, once the company learns of insider trading in its securities, it probably cannot remain silent.

The company may also have some degree of control over this type of situation. It can establish policies and procedures to reduce the possibility of insider trading. However, no procedures are airtight. Once the company learns of a leak, it may be under an obligation to make prompt disclosure.[345]

Insider trading is also controlled by short swing profit recovery provisions

---

338 *In re* Spartek, Inc and John A Cable, *supra* note 337.

339 *E.g.*, SEC v Texas, Gulf Sulphur Co, 401 F2d 833 (2d Cir 1968), *cert denied*, 394 US 976 (1969).

340 Form 8-K, 17 CFR §249.306, Item 2.

341 *Id* Item 5.

342 17 CFR §249.306.

343 *See, e.g.*, Securities Act of 1933 Release No 5092 (Oct 15, 1970).

344 Charella v United States, 445 US 222 (1980).

345 *Cf* 17 CFR §240.14e-3 (in connection with tender offers).

of §16(b) of the Securities Exchange Act of 1934, as amended.[346] Under §16(a) of that act,[347] officers and directors and persons who own beneficially more than 10 per cent of a class of the company's equity securities registered under the 1934 Act must report their ownership of all equity securities of the company and changes in ownership. Under §16(b), any profit by one of these persons on a sale of such a security within six months of purchase or vice versa is recoverable by the company or a stockholder suing derivatively on behalf of the company. The SEC also views any such profits as indebtedness to the company and requires disclosure of that indebtedness to stockholders in proxy statements.[348]

§6.54 —Periodic Disclosure

The periodic disclosure provisions of the 1934 act[349] require qarterly[350] and annual reports[351] containing specified information and current reports of material acquisitions or dispositions of assets.[352]

The SEC requires financial and other information to be reported for the first three quarters of a company's fiscal year. While the quarterly report form[353] provides for voluntary disclosure of information not called for by the form, it does not specifically require information about innovative developments. Therefore, companies are encouraged, but not required, to disclose a major innovative development in a quarterly report.

An exception would occur if the development involved the company entering into a material contract during the quarter. In that case, the contract would be required to be filed as an exhibit to the report for that quarter.

The SEC's annual reporting requirements[354] are considerably more extensive than its quarterly reporting requirements. Annual reporting requirements might trigger disclosure of an innovative development pursuant to provisions for describing developments in the company's business during the year; for discussing developments which might involve material capital expenditures or show that the company's historical financial statements were not indicative of future operations; or for filing material contracts. The possible environmental

---

[346] 15 USC §78p(b).

[347] *Id* §78p(a).

[348] Schedule 14A, 17 CFR §240.14a-101, Item 7; and Reg S-K, 17 CFR §229.20, Item 4. For an extensive discussion of the staff's views on the requirements of §16 and the rules thereunder, *see* Securities Exchange Act of 1934 Release No 18114 (Sept 23, 1981).

[349] Securities Exchange Act of 1934, §§13(a) and 15(d), 15 USC §§78m(a) and o(d).

[350] *E.g.,* 17 CFR §240.13a-13.

[351] *E.g., id* §240.13a-1.

[352] *E.g., id* §240.13a-11.

[353] Form 10-Q, 17 CFR §249.308a.

[354] Form 10-K, *id* §249.310.

impact of the project might be required to be described.[355]

In the I company hypothetical, the decision to expend a significant amount of capital in an entirely different business should be disclosed pursuant to these requirements.

§6.55 Disclosure of Material Information—Overview

Generally, only material information is required to be disclosed under the federal securities laws unless confidential treatment is afforded the information. The major exception is forward looking information, which, although often very material, is not generally required to be disclosed.[356] This is the greatest anomaly in the securities laws—that forward looking information which may be of the greatest importance to investors is not required to be disclosed, for a variety of unarticulated reasons.

One can, however, see reasons for not requiring disclosure of forward looking information, which all boil down to a balancing of the interests of the company and its owners in keeping the information in confidence against the benefits and disadvantages of disclosure.

Disclosure of an innovative development may have disadvantageous competitive effects. While the Securities and Exchange Commission (SEC) is required by law to consider the impact of its rules on competiion,[357] it is only required to balance the competitive impact of the rules against the benefits of the rule in light of the statutory purposes to be served. The SEC is ill-equipped to make that analysis, at least with respect to disclosure rules of general applicability. Accordingly, the exercise does not serve to protect a company faced with a specific factual situation.

There are other avenues of relief which are somewhat cumbersome and provide no absolute assurance against ultimate revelation of the information. The SEC can grant confidential treatment to information otherwise required to be disclosed.[358] The standards applied in considering applications for confidential treatment are those found in the SEC's rules under the Freedom of Information Act (FOIA).[359] By statute or under its own rules, the SEC may furnish the confidential information to Congress or other government agencies.[360] This could be of concern when a company is engaged in a dispute with another agency or because of fear of leaks.[361]

---

355 Reg S-K, 17 CFR §229.101(c)(1)(xii).

356 See **6.59.**

357 Securities Exchange Act of 1934, §23(a)(1), 15 USC §78w(a)(1).

358 *E.g., id* §24(b), 15 USC §78x(b), and 17 CFR §230.485.

359 Securities Exchange Act of 1934, §24(b), 15 USC §78x(b).

360 *Id.*

361 *See, e.g.,* Permian Corp v US, 665 F2d 1214, [1981-82 Transfer Binder] Fed Sec L Rep (CCH) ¶98,280 (DC Cir 1981).

Any grant of confidential treatment is subject to reconsideration when appropriate, including receipt of a request for the information under the FOIA. After reconsideration, which could be on the SEC's own motion, the grant of confidential treatment may be revoked.[362] The SEC will attempt to give the person requesting confidential treatment notice of the intention of revocation wherever possible.[363] Because of the weak notice provisions of those rules, a company's ability to challenge the SEC's determiantion to revoke a grant of confidential treatment could be severely curtailed and its standing to challenge such a determination in court is not clear.[364]

The SEC's rules also provide that it may disclose a record even if exempt from disclosure under the FOIA, if the application shows good cause for such disclosure.[365]

The SEC's rules under the FOIA provide that it will generally not make available records which disclose trade secrets and commercial or financial information obtained from a person and priviledged or confidential. . . ."[366] The judicial gloss on FOIA Exemption 4 makes it unclear whether potential competitive harm from release of the information is a factor to be considered in connection with assertion of the availability of the exception.[367]

Because of the nature of the SEC's procedural rules for seeking and maintaining confidential treatment and the uncertainties of the substantive standards to be applied, no assurance can be given that competitively harmful information may not be divulged. On the other hand, the SEC generally is sympathetic to requests for confidential treatment of trade secrets and commercial and financial information or provisions of contracts, disclosure of which "would impair the value of the contract and is not necessary for the protection of investors." An application for confidential treatment will protect the information from disclosure for some period of time.[368]

§6.56 —What Is Material?

If confidential treatment of the information cannot be obtained, the information must be disclosed if material. Although the term *material* is defined under the SEC's rules,[369] the prevailing definition is that provided by the Supreme Court in the *Northway* case:

---

[362] 17 CFR §240.24(b)-2.

[363] *Id.*

[364] Chrysler Corp v Brown, 441 US 281 (1979).

[365] 17 CFR §200.80(d)(6)(iv). *Cf* Securities Exchange Act of 1934, §21(a), 15 USC §78u(a) (SEC authorized to publish reports of investigations).

[366] 17 CFR §200.80(b)(4).

[367] *See, e.g.,* National Parks & Conservation Assn v Morton, 498 F2d 765, 771 (DC Cir 1974).

[368] *See, e.g.,* Securities Act of 1933, Schedule A(30), 15 USC §77aa(30), and Rule 406, 17 CFR §230.406.

[369] *E.g.,* 17 CFR §230.405.

> There is a substantial likelihood that, under all the circumstances, the omitted fact would have assumed actual significance in the deliberations of the reasonable shareholder. Put another way, there must be a substantial likelihood that the disclosure of the omitted fact would have been viewed by the reasonable investor as having significantly altered the "total mix" of information made available.[370]

Under this definition, a situational analysis of the factual circumstances is required and no hard and fast objective standards have been developed. It is reasonably clear, however, that forward looking information generally is not required to be disclosed. The exception could be a situation where an insider has traded on the basis of or "tipped off" someone as to forward looking information.[371]

The SEC has adopted safe harbor rules to protect those disclosing certain types of forward looking information in filings with the SEC from liability under the federal securities laws, if the information has a reasonable basis and is disclosed in good faith.[372] The precise coverage of the safe harbor rules is ambiguous and they have yet to be tested in court.[373]

The availability of those rules must be balanced against the risks a public company takes in disclosing forward looking information, if the market price for its stock rises upon the disclosure but subsequently declines.

The Supreme Court has sought to limit this exposure to almost open-ended liability by restricting standing to sue;[374] narrowing the definition of materiality;[375] imposing requirements to prove that defendants acted knowingly or recklessly;[376] and limiting the availability of private rights of action.[377] The ultimate answer may be statutory limitations on the amount of damages, as found in the American Law Institute's proposed *Federal Securities Code.*[378]

---

[370] TSC Indus, Inc v Northway, Inc, 426 US 438, 449 (1976).

[371] *Cf* 17 CFR §240.14e-3 (person who receivers material forward looking information in connection with a tender offer may be required to disclose that information or abstain from trading).

[372] *E.g.,* 17 CFR §§230.175 and 240.3b-6.

[373] *See generally* Rowe, *SEC Clarifies "Safe Harbor" Projection Rules,* Legal Times of Washington, July 16, 1979, at 9, col 1.

[374] *See, e.g.,* Blue Chip Stamps v Manor Drug Stores, 421 US 723 (1975); and Piper v Chris Craft Indus, Inc, 430 US 1 (1977).

[375] TSC Indus, Inc v Northway, Inc, 426 US 438 (1976). Although *Northway* was limited to the materiality standard under the Securities Exchange Act of 1934 Rule 14a-9, 17 CFR §240.14a-9, the lower federal courts have applied the same standard under other provisions of the securities laws.

[376] Ernst & Ernst v Hochfelder, 425 US 185 (1976).

[377] Touche Ross & Co v Redington, 442 US 560 (1979); TransAmerica Mortgage Investors, Inc v Lewis, 444 US 11 (1979).

[378] §1708 (1978).

§6.57 —Regulatory-Related Disclosure in General

Another troublesome aspect of materiality analysis is the extent to which disclosure is required about the impact of a proposed course of action (or ongoing developments) on aspects of the company's operations subject to other regulatory schemes—labor relations, including equal employment opportunity and occupational safety and health; product safety and liability; antitrust concerns; environmental regulation; and tax laws.

One of the areas of disclosure which should be of particular concern to a company engaged in an innovative project is that concerning disclosure about the impact of various regulatory systems on the company's operations. It is an area where traditional concepts of economic materiality tend to erode. Disclosure may be required not because of the direct impact of the regulatory system on the company's operations, but because the company's reaction to the system may reflect adversely on the company's management.[379]

Disclosure requirements under the securities laws may also be developed in part to serve the needs of another regulatory system,[380] and the disclosure requirements of the securities laws may be amended so as to no longer defer to the informational requirements of another regulatory system. This is the situation in the transportation and banking industries where reports filed with the Interstate Commerce Commission, Department of Transportation or federal bank regulatory agencies may no longer satisfy 1934 act reporting requirements.[381]

Many of the requirements in this area are administrative, not statutory or judge-made laws, and even administrative requirements may not be codified in SEC rules. Indeed, some SEC positions have been rejected in court actions when they were advanced by private plaintiffs.[382]

Some of these regulatory disclosure requirements arguably increase the cost of compliance with the securities laws without any clearly perceived benefits.

Some disclosure of the impact of other regulatory schemes is traditional. The most obvious is disclosure relating to federal income taxes.[383] Disclosure about the impact of federal antitrust laws is also accepted at least at the point where a claim has been asserted.[384] In some instances, disclosure of facts

---

[379] *See, e.g.,* SEC v Jos Schlitz Brewing Co, 452 F Supp 824 (ED Wis 1978).

[380] *See, e.g.,* Energy Policy and Conservation Act, Pub L No 94-163, 89 Stat 958, 959 (codified at 42 USC §6383).

[381] *See, e.g.,* Railroad Revitalization and Regulatory Reform Act of 1976, Pub L No 94-210, 90 Stat 31 (amending §13(b) of the act to permit the SEC to establish standards for filings under that act by issuers in certain regulated industries); and Securities Act Amendments of 1975, 89 Stat 97 (amending Section 12(i) of the 1934 Act to require federal bank regulatory agencies to pattern disclosure required for banks on SEC requirements).

[382] *See, e.g.,* Amalgamated Clothing & Textile Workers Union v JP Stevens & Co, 638 F2d 7 (2d Cir 1980).

[383] 17 CFR §210.4-08.

[384] *Id* §229.103.

relating to possible violations of law may be required before any claim is asserted.[385]

The first departure from traditional economics-related disclosure was with regard to environmental regulation. The SEC was embroiled in litigation for almost a decade over the issues involved[386] and some of the issues have yet to be clearly resolved.

At present, a company describing its business in a registration statement or an annual report is required to disclose any material effects governmental environmental regulations may have upon its capital expenditures, earnings and competitive position.[387] It is also required to disclose its estimated capital expenditures for environmental control facilities for the remainder of the current fiscal year and the next fiscal year and such further periods "as the registrant may deem material."[388] While this would seem to rely on the company's judgment as to what needs to be disclosed, there would have to be an exercise of reasonably based judgment. Moreover, under SEC catchall disclosure rules, Rule 408[389] under the Securiites Act of 1933, as amended,[390] and Rule 12b-20 under the 1934 Act[391] and general antifraud provisions[392] as well, a company cannot omit to state material facts which are necessary to make other statements not misleading. For example, if a company estimates that it will make minimal expenditures in the current year and in the next year, but will have to expend a billion dollars over the following three years, disclosure may well be required. This was apparently the SEC's position in connection with its administrative proceeding against United States Steel Corporation.[393]

§6.58 —Environmental Disclosure

In addition to the more traditional economically material information, a company may be required to disclose all governmental proceedings arising under environmental protection laws, regardless of the more traditional eco-

---

[385] *See, e.g.*, Securities Act of 1933 Release No 5466 (Mar 8, 1974); and Report of the Securities and Exchange Commission on Questionable and Illegal Corporate Payments (submitted to the Senate Banking, Housing and Urban Affairs Committee, May 12, 1976).

[386] Natural Resources Defense Council, Inc v SEC, 606 F2d 1031 (DC Cir 1979).

[387] 17 CFR §229.101.

[388] *Id.*

[389] *Id* §230.408.

[390] 48 Stat 74 (1933) (codified at 15 USC §§77a-77aa).

[391] 17 CFR §240.12b-20.

[392] *E.g.*, Securities Exchange Act of 1934, §10(b), 15 USC §78j(b), and 17 CFR §240.10b-5.

[393] *In re* United States Steel Corp, Securities Exchange Act of 1934 Release No 16223 (Sept 27, 1979). *See generally* Rowe, *"US Steel" Reflects Disclosure Governance Trends,* Legal Times of Washington, Nov 5, 1979, at 10, col 1.

nomic materiality of the proceeding.[394] The term *proceeding* was very broadly interpreted by the SEC in the *US Steel* proceeding and in the interpretive release which was issued contemporaneously with the *US Steel* findings.[395]

Nongovernmental environmental litigation must be disclosed, if it satisfies certain materiality thresholds.[396]

While the SEC has instituted rulemaking proceedings to reconsider the extent of disclosure required about governmental environmental proceedings,[397] it has recently brought administrative compliance proceedings against corporations for alleged inadequate disclosure about environmental matters. In *US Steel*, the SEC took the position that where a company voluntarily addresses environmental matters in disclosure documents, that voluntary disclosure must not be misleading.

In *In re Occidental Petroleum Corp*,[398] the SEC found that the respondent failed to provide required information about discharge of chemical and toxic wastes into the environment and about other environmental matters. The respondent neither admitted nor denied those findings, but undertook in settlement to have one of its directors prepare an environmental report on specific matters and to assure that potential liabilities relating to environmental matters were identified to the board of directors and that appropriate disclosure was made in filings with the SEC.

Earlier, the SEC had brought a civil injunctive action against Allied Chemical Corporation alleging failure to disclose certain of its activities which had a detrimental effect on the environment and the resulting material economic impact on that company.[399] The relief obtained in settlement of these and other types of cases generally results in some impact on the governance structure of the company. For example, new directors may be appointed to the board; new committees of the board may be created or existing committees reconstituted; new officers may be appointed; and new lines of authority constructed. Generally, it is this type of relief that is significant, rather than the usual injunction to go forth and sin no more.

The SEC recently had occasion to reconsider its environmental disclosure rules, but, other than limited modifications to the requirements for disclosure about environmental litigation, it has proposed no revisions to its rules.[400] Therefore, companies engaged in innovative projects which might have an impact on the environment, such as chemicals, nuclear energy or other activities generating nuclear radiation or toxic wastes, should consider the SEC's disclosure rules if they are raising capital or filing reports with the SEC. The disclosure required will often involve estimates of future expenditures or the

---

[394] 17 CFR §229.103.

[395] Securities Act of 1933 Release No 6130 (Sept 27, 1979).

[396] 17 CFR §229.103.

[397] Securities Act of 1933 Release No 6315 (May 4, 1981).

[398] Securities Exchange Act of 1934 Release No 16950 (July 2, 1980).

[399] SEC v Allied Chem Corp, No 77-373 (DDC March 4, 1977).

[400] Securities Act of 1933 Release No 6315 (May 4, 1981).

impact on operations of a particular compliance activity and litigation-related uncertainties.[401] Provisions of this kind of forward looking information in filings with the SEC is covered by SEC safe harbor rules, if the estimates are reasonably based and provided in good faith.[402]

One should be aware that these safe harbor rules are only applicable to statements in registration statements or reports filed with the SEC or in annual reports to securityholders prepared in accordance with the SEC's proxy rules which are technically furnished to the SEC rather than filed.[403] The safe harbor rules are also applicable to statements made elsewhere and reaffirmed in an SEC filing.

Accordingly, if forward looking statements are made in connection with exempt securities transactions, in a private placement memorandum or intrastate offering circular, for example, one could argue that the standards of the safe harbor rules apply by analogy, even if a filing is not made. The statements could also be in a filing. The latter can be accomplished by a public company through the filing of a current report on Form 8-K[404] responding to Item 5, Other Materially Important Events.

A company whose activities may have an environmental impact, therefore, must be familiar not only with SEC's environmental disclosure rules, but with its rules and guidelines relating to disclosure of soft or forward looking information.

§6.59 Disclosure of "Soft" Information

The historical information about a company which is required by the securities laws to be disclosed may be indicative of the company's future, but not always. This may be particularly true of a company engaged in a major new innovative venture. Moreover, investors are more interested in an assessment of the company's future prospects by its management, than in a rehash of its past. While it is conceded that such an assessment is material, the securities laws generally do not directly mandate projections of future prospects or other forward looking or soft information.

Historically, disclosure of soft information was discouraged by the SEC due to its perceived unreliability.[405] The SEC now encourages, but still does not require, disclosure of projections of future economic performance and other

---

[401] *See, e.g.,* Financial Accounting Standards Board, Statement of Financial Accounting Standards No 5: Accounting for Contingencies (1975); and SEC Staff Report on Corporate Accountability, Printed for the Senate Committee on Banking, Housing and Urban Affairs, 96th Cong, 2d Sess (Sept 4, 1980).

[402] *E.g.,* 17 CFR §230.175.

[403] *Id* §240.14a-3(b).

[404] 17 CFR §249.308.

[405] *See generally* Advisory Committee on Corporate Disclosure, 95th Cong, Report 1st Sess, to the Securities and Exchange Commission (Comm Print No 29, House Comm on Interstate and Foreign Commerce 1977).

forward looking information, and has authorized publication of SEC staff guidelines applicable to such disclosure.[406] The SEC has also adopted safe harbor rules which protect those making reasonably based projections in good faith from liability under the federal securities laws.[407] The accounting profession has also developed guidelines for limited reviews of financial forecasts (a somewhat narrower and more precise term than projections) by independent auditors.[408]

Despite this encouragement and available guidelines, the vast majority of companies do not publish projections, even though many of them prepare projections for internal use.[409] In some instances, projections may even be provided to outsiders such as independent accountants auditing the company's financial statements, lenders, investment bankers considering an underwriting of the company's securities or another company with which a friendly business combination is being considered.

The reluctance to publish projections stems from a variety of factors. The projections prepared for budgetary purposes or for use as an employee incentive may not be suitable for public dissemination. Projections may be seen as inherently unreliable, despite improved forecasting techniques and systems. There is a fear of liability despite the SEC's safe harbor rules. Even if the company is confident that it would prevail in any litigation and avoid liability, it may not desire to incur the risk or the expense of litigation. The risk of litigation over projections which prove to be inaccurate is high, since a projection of earnings may be more directly traced to a movement in stock prices than other information and market reaction to material variances between the projection and the acutal results may be more dramatic.

Companies may be concerned about the market reaction to projections, favorable or unfavorable or fear loss of credibility or just plain public embarrassment, if a projection does not prove to be accurate. Moreover, the demand for projections may be dampened somewhat by the availability of financial analyst projections for larger companies, although the value of these forecasts may not be as great as an inside management assessment of a company's future prospects.

Finally, companies tend to publish favorable projections and avoid publication if the projection is unfavorable.

Since there is no specific requirement under the securities laws to publish projections, the determination to do so or not is primarily a business decision. Most public companies apparently do not perceive advantages to be gained from projecting which would outweigh the risks. However, a company engaged

---

[406] 17 CFR §229.10(b).

[407] *E.g., id* §230.175.

[408] Guide for Review of a Financial Forecast (AICPA 1981).

[409] There may be no legal duty to disclose these projections. *See, e.g.,* Freeman v Decio, 584 F2d 186, 199-200 (7th Cir 1978). This may even be true when the company is engaged in the purchase of its own securities. *See, e.g.,* Vaughn v Teledyne, Inc, 628 F2d 1214, 1221 (9th Cir 1981).

in an innovative project could weigh the factors described above against advantages to be derived from making projections, and decide to publish projections. Some of the possible advantages might include enhanced marketability of the company's securities, if it has reported poor performance historically but expects improvement in the near future. For example, current generally accepted accounting principles require research and development expenditures to be expensed for financial reporting purposes.[410] Thus, a new venture heavily dependent on research and development is likely to report losses in the earlier years of its existence. If its management has reasonable grounds to believe that it will reach a turning point in the next year and be able to report profits, it may be in a reasonable position to project profits.

Projections could also affect management incentive compensation based on stock performance; although this cuts both ways, since the good faith of executives who stand to gain directly as the result of projections for which they are responsible could be open to challenge.

Publication of projections may also produce intangible benefits by improving the corporate image; improving shareholder relations; and improving relations with financial analysts or developing a following among financial analysts.

Publication of favorable forward looking information may also diminish the impact of bad news.[411]

If a company determines to publish projections, it should carefully consider the SEC's rules and policies relating to projections and other types of soft information as well as the related accounting literature.

§6.60 —Guidelines for Disclosure

The Securities and Exchange Commission policies for projections[412] do not mandate disclosure of projections of future economic performance. However, those policies might be viewed as an interpretive gloss on the reasonable basis and good faith standards in the SEC's safe harbor rules.[413] as they provide guidelines for the basis and presentation of projections. The SEC policies provide that:

1. Projection should be a good faith assessment by management of the company's future performance
2. Management must have a reasonable basis for the projection. (This means that a sound system for developing projections must be in place

---

[410] Financial Accounting Standards Board, Statement No 2, Accounting for Research and Development Costs (Oct 1974).

[411] *Cf* Lynch, *Polaroid Plans "Entirely New" System for Another Market; Details Are Secret,* Wall St J, Apr 22, 1971, at 3, col 1 (announced at annual meeting; also announced that first quarter earnings down 45% and sales down 10%).

[412] 17 CFR §229.10(b).

[413] *E.g.,* 17 CFR §230.175.

and documented to serve as an evidentiary base to assert the availability of the safe harbor rule)

3. Although important factors, a history of operations and experience in projections are not always necessary to formulate a reasonable projection. (Thus, a company with a new idea in need of financing could conceivably publish projections)
4. Outside review may provide additional support for the reasonableness of the projection. (Reports of these outside reviews are protected by the safe harbor rules)
5. If a report of an outside review of a projection is included in a filing, there should be disclosure of the reviewer's qualifications; the extent of the review; any relationship between the company and the reviewer; and other material concerning the obtaining of the review. The reviewer would be considered an expert for purposes of the 1933 Act liability provisions and his or her consent to use of the report would be required to be filed with a registration statement under that act, if the report is referred to in a registration statement[414]
6. Projections of revenues, net income and earnings per share are considered of primary importance, but projections need not be limited to those items
7. Selective projection of favorable items, such as increased sales, but not of unfavorable items, such as decrease in net income, should be avoided
8. The period covered by the projection would depend upon the circumstances of the company
9. Projectons may be expressed in terms of the most probable specific amount or a reasonable range
10. Several projections, based on varying assumptions, could be presented
11. Management should disclose its intentions with regard to updating of projections. (Updating is not specifically required)
12. The most significant assumption on which the projections are based should be disclosed. (Selection of the assumptions to be disclosed will be no easy task)
13. Consideration should be given to disclosure of the company's track record for projecting; and variances from projections and actual results analyzed
14. Caution against attributing too great a certainty to the projections should accompany the projections

It should be emphasized that projection is voluntary and management has a great deal of flexibility.

---

[414] 17 CFR §230.436.

§6.61 —Duty to Update Information

One of the more difficult legal issues relates to the company's responsibility to update projections it has previously made. SEC rules do not specifically mandate updating or continuing to make projections. Therefore, a company which publishes a projection need not continue to publish projections. However, in general, if a public company has published a projection and it has reason to believe that investors are still relying on the projection, the company may have a responsibility under the general antifraud provisions of the securities laws to publish a correction, if the company has reason to believe the projection has become materially inaccurate.[415]

If a company chooses to include a good faith reasonable projection in a filing with the SEC, or confirms in a filing a projection initially published elsewhere, the company and those responsible for the projection would be protected from liability by the SEC's safe harbor rules for forward looking statements, Rule 175 under the 1933 Act,[416] Rule 3b-6 under the 1934 Act,[417] and substantially identical rules under the other federal securities statutes. Those rules basically would require that a plaintiff prove that a projection made in an SEC filing which turned out to be inaccurate was not made in good faith and did not have a reasonable basis.

§6.62 —Other Considerations

The safe harbor rules also protect other types of forward looking information, such as those in the Management Discussion and Analysis of Financial Condition and Results of Operations required in registration statements and reports filed with the SEC.[418] Thus, a new venture which has been reporting losses due to start-up costs and significant research and development expenditures, in discussing its liquidity (i.e., cash needs) and financing needs, could provide its management's future plans and their good faith, reasonable assessment of future prospects with considerably less fear of liability.

The American Institute of Certified Public Accountants (AICPA), the professional association which sets the auditing standards for the accounting profession, has established guidelines for review of financial forecasts by its members. However, under the AICPA's Code of Professional Ethics, Rule 201E (1978), an auditor may not vouch for the forecast. While the SEC's safe harbor rules cover reviewer's reports on forecasts, the fear of liability and damage to professional reputations remains.

It is unlikely that the SEC will mandate disclosure of forward looking information in the predictable future. It is likely, however, that the SEC and the accounting profession will continue to encourage disclosure of forward look-

---

[415] *See generally* ALI *Federal Securities Code* §1602, commentary at 651-54 (1978).

[416] 17 CFR §230.175.

[417] 17 CFR §240.3b-6.

[418] 17 CFR §229.303.

ing information as a matter of good practice, with the further incentive of expanded safe harbor protection from liability.

§6.63 Accounting

The Securities and Exchange Commission (SEC) has broad authority to establish the requirements applicable to the financial statements required in registration statements and reports filed under the federal securities laws.[419] While the SEC requires those financial statements to be prepared in accordance with generally accepted accounting principles (GAAP),[420] it permits the private sector to establish those principles.[421] The SEC does, however, regulate certain aspects of the form, content and timing of the financial statements and related disclosure it requires in its accounting regulation, Regulation S-X.[422]

The private sector now establishes standards through a formal process. The Financial Accounting Standards Board, a privately funded organization, establishes GAAP[423] which are accepted as authoritative by the SEC.[424] However, compliance with GAAP may only be evidence that financial statements are not misleading.[425]

Auditing standards are established by the American Institute of Certified Public Accountants.[426] It is not clear that the SEC's authority extends to auditing standards,[427] but it asserts authority to discipline accountants who practice before it for failure to adhere to generally accepted auditing standards.[428]

Despite criticism of the SEC's deference to the private sector in establishing auditing standards,[429] the process has worked reasonably well. The SEC has

---

[419] Securities Act of 1933, §19, 15 USC §77s, and Schedule A(25), (26), 15 USC §§77aa(25) and (26); Securities Exchange Act of 1934, §13(b), 15 USC §78m(b) and 17 CFR §210.3-13.

[420] 17 CFR §210.4-01(a)(1).

[421] *Id* §211.150.

[422] *Id* pt 210.

[423] Official pronouncements of the accounting profession are published in AICPA Professional Standards (CCH).

[424] 17 CFR §211.150.

[425] *See, e.g.,* United States v Simon, 425 F2d 796 (2d Cir 1969), *cert denied,* 400 US 827 (1970).

[426] Auditing standards are published in AICPA Professional Standards (CCH).

[427] *See, e.g.,* Securities Act of 1933 Release No 6242 (Sept 18, 1980).

[428] SEC Rules of Practice Rule 2(e), 17 CFR §201.2(e); Reg S-X, Rule 2-02, 17 CFR §210.202; and Release No 6242, *supra* note 324; Touche Ross & Co v SEC, 609 F2d 576 (2d Cir 1979).

[429] *See, e.g.,* Kripke, *Where Are We on Securities Disclosure After the Advisory Committee Report?,* 6 Sec Reg LJ 99 (Summer 1978); Kripke, *Required Periodic Disclosure Under the Securities Acts and the Proposed Federal Securities Code,* 33 U Miami L Rev 1425-648 (Summer 1979); and Briloff & Engler, *Accountancy and the Merger Movement: A Symbiotic Relationship,* 5 J Corp L 81 (Fall 1979).

not been reluctant to exercise its authority to require supplemental disclosure when it believes that the accounting profession is not proceeding rapidly enough in a particular area.[430]

SEC certified financial statement requirements under the 1933 Act are enforced strictly with few exceptions.[431] These are perhaps the only SEC requirements which can operate as an absolute bar to a company registering securities under the 1933 Act. The requirements are particularly troublesome and complex as they relate to financial statements of predecessor business[432] and financial statements of businesses recently acquired or to be acquired.[433]

Perhaps the initial thing that a first-time registrant should determine when considering whether to go public is whether it can satisfy requirements as to certified financial statements. The company should also recognize that obtaining certified financial statements may be one of the larger expense items it will incur in order to comply with the federal securities laws.

While the SEC cannot take away the keys to the marketplace if financial statements in 1934 Act reports are not certified, unless the registrant desires to incorporate those reports in a 1933 Act registration statement, the SEC may still bring an enforcement action to compel compliance with its requirements.[434]

The SEC tends to exercise its discretion and not object to financial statement deficiencies to a much greater degree under the 1934 Act than under the 1933 Act. This is another area where a careful lawyer will utilize prefiling conferences and other informal procedures to keep the SEC informed of issues which are bound to arise. This type of up front presentation is generally superior to the head in the sand approach of those who believe the SEC may never uncover the possible deficiency.

The historical financial statements produced by application of GAAP do not produce the type of forward looking information which investors seek. SEC efforts to produce this type of information, such as encouragement of projec-

---

[430] For example, the SEC has withdrawn accounting disclosure rules relating to leasing and replacement costs and has proposed to rescind other rules relating to disclosure about nonaudit services upon becoming satisfied that standards adopted by the accounting profession were adequate. On the other hand, the SEC has taken a rigid position on accounting for research and development funded by limited partnerships.

[431] In very unusual and limited instances, such as those involving Curtis Publishing Company and Chrysler Corporation, requirements for unqualified accountants' opinions on financial statements may be waived by the SEC. These generally involve situations where disclosure of the risks is adequate and a seriously financially troubled company must make the securities offering in question to survive.

[432] 17 CFR §210.3-07.

[433] *Id* §§210.3-07 and 3-08.

[434] *See generally,* Rowe, *SEC Shifts to Administrative Enforcement, Remedies,* Legal Times of Washington, Sept 14, 1981, at 23, col 1; Rowe, *SEC Enforcing Through Administrative Proceedings,* Legal Times of Washington, June 2, 1980, at 15, col 1; and Rowe, *Administration and Enforcement of the Periodic Reporting Provisions of the Securities Exchange Act of 1934,* 25 U Okla L Rev 157 (May 1972).

tions[435] and requiring a management discussion and analysis of a company's financial condition and results of operations,[436] have not been overly successful. There is some indication, however, that the SEC may be moving in the direction of requiring more meaningful cash flow presentations.[437]

§6.64 Foreign Corrupt Practices Act

Companies subject to the periodic reporting provisions of §§13 or 15(d) of the 1934 act[438] are also subject to the antibribery,[439] recordkeeping[440] and internal accounting control[441] provisions of the Foreign Corrupt Practices Act of 1977.[442]

While the antibribery provisions are troublesome for a company with foreign business, it is the accounting provisions which have had the greatest impact on public companies.

Under §13(b)(2)(A)[443] of the 1934 Act, an issuer must make and keep books and records which "in reasonable detail, accurately and fairly" reflect the issuer's transactions. Under the 1934 act, Rule 13b2-1[444] provides that no person shall falsify such books and records.

Section 13(b)(2)(B)[445] requires the issuer to devise and maintain a system of internal accounting controls sufficient to provide "reasonable assurance" that transactions are executed in accordance with management's directions; accountability for assets is maintained; and financial statements in accordance with generally accepted accounting principles can be prepared.

These provisions have been subject to considerable criticism and bills have been introduced in Congress or legislative suggestions made to amend or eliminate some or all of these provisions.[446]

---

[435] See **§6.59.**

[436] 17 CFR §229.303.

[437] Securities Act of 1933 Release No 6316 (May 11, 1981).

[438] 15 USC §§78m, o(d).

[439] Securities Exchange Act of 1934, §§30A and 32(c), 15 USC §78dd-2 and ff(c)—United States nonreporting companies are *domestic concerns* and are subject to antibribery provisions under §104 of the Foreign Corrupt Practices Act of 1977, Pub L No 95-213, 15 USC §78dd-2.

[440] Securities Exchange Act of 1934, §13(b)(2)(A), 15 USC §78m(b)(2)(A).

[441] *Id* §13(b)(2)(B), 15 USC §78m(b)(2)(B).

[442] Pub L No 95-213, 15 USC §§78m, ddd-1, ddd-2 and ff.

[443] 15 USC §78m(b)(2)(A).

[444] 17 CFR §§240.13b2-1 and -2 prohibit officers and directors from misleading accountants in connection with an audit or examination of financial statements or preparation of a report filed under than 1934 Act.

[445] 15 USC §78m(b)(2)(b).

[446] See proposed amendments to the Federal Corrupt Practices Act of 1977, Business Accounting and Foreign Trade Simplification Act, S 708, 97th Cong, 1st Sess, 174 Cong Rec S13983 (daily ed Nov 23, 1981).

§6.65 Stockholder Involvement

After raising money from the public, a company must be prepared to deal with the issues involved in having public stockholders. While these issues generally arise under state corporate law governing relationships with stockholders,[447] the federal securities laws do apply in certain circumstances. They do not, however, directly govern the conduct of internal corporate affairs.[448]

The proxy rules of the Securities and Exchange Commission (SEC)[449] proxies from holders of securities registered under §12 of the 1934 Act. Those rules require that those solicited be furnished a proxy statement containing information about any matter to be acted upon and background information about directors and nominees for election of directors, as well as information about management remuneration and transactions.[450] Similar information is required to be furnished to stockholders in an information statement if proxies are not solicited in connection with a meeting of stockholders.[451] Thus, even where the insiders hold sufficient stock to carry the vote, stockholders will receive information about the matters to be acted upon.[452]

Proxy and information statements must be accompanied or preceded by an annual report containing specified business and financial information about the company.[453]

SEC rules also require proposals by stockholders satisfying certain standards to be included in proxy or information statements.[454] This is a somewhat controversial provision and can result in major corporate decisions, such as a determination to develop a new product or process, being submitted to something akin to a stockholder referendum.

While stockholder proposals can be troublesome they seldom receive significant voting support. It is often more expedient to negotiate with a stockholder proponent, and arrive at a resolution which obviates the need for inclusion of the proposal in the proxy statement. This will avoid unnecessary publicity of the issue or the expense involved in resting conclusion of the proposal before the SEC or in the courts.

---

[447] *See, e.g.*, Del Code Ann tit 8, §211 (1981).

[448] Santa Fe Indus, Inc v Green, 430 US 462 (1977).

[449] 17 CFR §§240.14a-1 through .14a-101.

[450] 17 CFR §240.14a-101.

[451] Securities Exchange Act of 1934, §14(c), 15 USC §78n(c).

[452] Stock exchange rules may require solicitation of proxies in such a situation, in any event. *See, e.g.*, New York Stock Exchange Company Manual A-134.

[453] 17 CFR §240.14a-3(b).

[454] *Id* §240.14a-8.

§6.66 Securities and Exchange Commission (SEC) Administration and Enforcement—1934 Act

Except for proxy statements, SEC administration of 1934 periodic reporting provisions differs markedly from its administration of the 1933 Act registration provisions.[455] Under the 1934 Act, the staff does not have the clout of the acceleration power it has under the 1933 Act[456] to encourage amendments in response to staff comments. Accordingly, in many instances futures comments are given on 1934 act reports. For example, the staff will suggest that its comments on disclosure in an annual report on Form 10-K[457] be responded to in the report for the next year or in the next quarterly report on Form 10-Q.[458]

Proxy statements are an exception. Preliminary proxy material is required to be submitted to the SEC 10 days in advance of mailing of the definitive statement to stockholders.[459] If the staff fails to comment within 10 days, the definitive statement may be mailed.[460] If the staff does comment, the mailing may have to be delayed while a responsive amendment is prepared, filed and reviewed by the staff.

Another factor in the difference in administration of the 1934 act periodic reporting provisions results from the sheer volume of the reports and proxy statements filed. There are approximately 9,000 reporting companies, approximately 7,000 of which file proxy statements. This results in tens of thousands of documents being filed with the SEC every year. With a staff of about 100 professionals available for reviewing filings, nowhere near all of these filings can be reviewed.

If there is serious concern about the disclosure in a 1934 Act filing, the SEC may commence enforcement proceedings instead of informally commenting on the filing.[461]

The SEC could commence an informal investigation.[462] Depending upon the results of an investigation, the SEC may institute administrative proceedings

---

[455] See **§6.36.**

[456] Securities Act of 1933, §8(a), 15 USC §77h(a).

[457] 17 CFR §249.310. The staff seldom comments on annual reports to stockholders which are furnished to the SEC for its information, but not filed. *Id* §240.14a-3(b). The principal reason for this practice may be to avoid the expense (and possible liability exposure) involved in recirculation of an amended annual report to stockholders. Recirculation of the entire report could be avoided by making a separate mailing or providing the disclosure in the next quarterly report to stockholders or preparing the next year's annual report in accordance with the comment.

[458] *Id* §249.308a.

[459] *Id* §240.14a-6.

[460] SEC News Digest, Nov 17, 1980, at 1.

[461] 17 CFR §202.3(a). *See generally* Rowe, *SEC Shifts to Administrative Enforcement Remedies,* Legal Times of Washington, Sept 14, 1981, at 23, col 1; and Rowe, *SEC Enforcing through Administrative Proceedings,* Legal Times of Washington, June 2, 1980, at 15, col 1.

[462] 17 CFR §200.5(a).

to compel compliance with the 1934 Act reporting provisions[463] or commence a civil injunctive action seeking to enforce compliance.[464] The SEC could also issue a public report of investigation or require a filing by the company of a statement about the matters under investigation.[465]

In extreme situations, the SEC is authorized to refer a matter to the Department of Justice for consideration of criminal prosecution.[466]

Most SEC enforcement proceedings are settled without trial or hearing and without defendants or respondents admitting or denying any facts. The SEC obtains various forms of ancillary relief in connection with those settlements. That relief may take the form of an undertaking to establish procedures to assure future compliance and/or the appointment of new directors to oversee these procedures.

Negotiating settlements of SEC enforcement proceedings avoids the expense of litigation, adverse publicity[467] and admissions or estoppels[468] which could be used by private litigants.

§6.67 Liability

Liability under the 1934 Act is a subject for a multivolume treatise[469] and cannot be dealt with at any length in this chapter, although a few practical suggestions can be made.

The express civil liability provisions[470] under the 1934 Act, unlike those under the 1933 Act,[471] have not been a source of great concern to public companies. There has never been a reported recovery on the merits in an action under those provisons.

However, the courts have recognized implied private rights of action under various antifraud provisions of the 1934 Act, principally Securities and Exchange Commission (SEC) Rule 10b-5[472] and Rule 14a-9,[473] the antifraud

---

463 Securities Exchange Act of 1934, §§12(j) and 15(c)(4), 15 USC §§78l(j) and o(c)(4).

464 *Id* §21, 15 USC §78u. Civil injunctive actions are most frequently used to remedy failures to file reports or a pattern of late filing.

465 *Id* §21(a), 15 USC §78a(a), and Securities Exchange Act of 1934 Release No 15664 (Mar 21, 1979).

466 Securities Exchange Act of 1934, §21, 15 USC §78u(a).

467 The SEC publishes litigation releases describing settlements of civil litigation and its orders accepting settlements in administrative proceedings. However, this one-shot publicity is probably preferable to the continuing publicity given to a contested proceeding as it progresses through its various stages.

468 Parklane Hosiery Co v Shore, 439 US 322 (1979).

469 *See, e.g.*, Bromberg, Securities Law: Fraud, SEC Rule 10b-5 (Shepard's/McGraw-Hill 1967).

470 Principally, Securities Exchange Act of 1934, §18, 15 USC §78r.

471 See **§6.42.**

472 Superintendent of Ins v Bankers Life & Casualty Co, 404 US 6 (1971); and Affiliated Ute Citizens v United States, 406 US 128 (1972).

provisions of the SEC's proxy rules. Under these provisions, liability could be based on materially misleading statements in 1934 Act periodic reports or proxy statements, press releases or communications with shareholders. Reports filed under the 1934 Act which are incorporated by reference in a 1933 act registration statement may be subject to liability under that act, as well as the 1934 Act.

In order for a company to avoid liability under the 1934 act for misleading statements, it should establish a system for controlling all communications which could be subject to such liability. The nature and size of the company will determine the nature and extent of the system, but it is important that the system be documented and information about the system and related policies communicated to officers, directors, employees and others involved in communicating information about the company to persons outside the company. The company may wish to limit the persons who have such contacts, particularly contacts with securities analysts, securities brokers and other securities professionals.

Directors and officers should be advised of their responsibilities under the securities laws and given an opportunity to query counsel on issues they do not understand.

Reports and other documents filed pursuant to the 1934 Act should be reviewed by counsel. Press releases and communications to securityholders should also be reviewed by counsel, or issued in compliance with established written guidelines.

Dissemination of information is often important. In this connection, a company may wish to consider retaining a public relations firm or inside public relations staff with experience in the mechanics of distributing information to the news media.

A system for controlling public communications cannot provide absolute assurance that a claim will not be asserted under the securities laws in connection with such a communication. However, it will reduce the possibility of such a claim and assist in demonstrating the good faith and reasonableness of the company's actions should such a claim be asserted. Being able to demonstrate the good faith and reasonablemeness of the company's action may negate claims that the company knowingly or recklessly violated the securities laws[474] or provide a defense to the entry of an injunction in a civil action where scienter is not an element of the violation.[475]

---

[473] JI Case v Borak, 377 US 426 (1964); and Mills v Electric Auto-Lite Co, 396 US 375 (1970).

[474] Ernst & Ernst v Hochfelder, 425 US 185 (1976).

[475] *Cf* Aaron v SEC, 446 US 680 (1980) (scienter not an element of violation of §17(a)(2) of the Securities Act of 1933, 15 USC §77q(a)(2), in an SEC injunctive action).

Conclusion

§6.68 Professional Responsibility

Professionals play indispensable roles in assisting a company in complying with the securities laws. However, the professionals themselves may be subject to civil liability under the securities laws;[476] Securities and Exchange Commission (SEC) enforcement action;[477] or SEC administrative disciplinary proceedings.[478]

The securities laws explicity provide for liability of professionals who render materially misleading expert opinions for use in certain securities transactions.[479]

It is also evident that a professional who participates directly in a securities transaction and violates the securities laws or aids or abets such a violation[480] should be liable. There is considerable controversy, however, as to whether a lawyer acting in an advisory capacity should be subject to SEC disciplinary proceedings which could result in loss of the opportunity to practice before the SEC. Indeed, the SEC interprets what constitutes practice before the SEC so broadly, that a securities lawyer's practice would be severely limited if he or she were barred from practices before the commission. In the SEC's view, practice before the SEC includes not only appearing at the agency in a representative capacity, but assisting in the preparation of documents to be filed with, or communications to, the agency.[481]

The SEC has also taken the position that it may discipline lawyers for conduct, which does not occur in a proceeding before the agency[482] and for violation of applicable ethical standards as well as violations of law.[483] The commission has stated:

> A lawyer engages in "unethical or improper professional conduct" under

---

[476] *See, e.g.,* SEC v National Student Marketing Corp, 457 F Supp 682 (DDC 1978); SEC v Calhoun County Medical Facility, Inc, [1981-82 Transfer Binder] Fed Sec L Rep (CCH) ¶98,243 (ND Miss 1981); SEC Staff Report on Transactions in the Marine Protein Corporation Industrial Development Revenue Bonds, Securities Exchange Act of 1934 Release No 15719 (Apr 11, 1979); *In re* North Am Acceptance Corp, 513 F Supp 608 (ND Ga 1981) (law firm liable where it had knowledge investors would rely upon misleding advertisements upon which it advised issuer); and Stokes v Lokken, 644 F2d 779 (8th Cir 1981) (attorney not liable where he qualified his opinion; relied on facts provided by others; and his opinion was not used directly in offering literature).

[477] *E.g.,* SEC v National Student Marketing Corp, 457 F Supp 682 (DDC 1978).

[478] 17 CFR §201.2(e).

[479] Securities Act of 1933, §11, 15 USC §77k.

[480] *E.g.,* Rolf v Blyth, Eastman Dillon Co, 637 F2d 77 (2d Cir 1980).

[481] 17 CFR §201.2(g).

[482] *E.g., In re* William R Carter & Charles J Johnson, Jr, Securities Exchange Act Release No 17597 (Feb 28, 1981).

[483] *Id.*

> the following circumstances: When a lawyer with significant responsibilities in the effectuation of a company's compliance with the disclosure requirements of the federal securities laws becomes aware that his client is engaged in a substantial and continuing failure to satisfy those disclosure requirements, his continued participation violates professional standards unless he takes prompt steps to end the client's noncompliance.[484]

The SEC has further observed:

> We do not imply that a lawyer is obliged, at the risk of being held to have violated Rule 2(e), to seek to correct every isolated disclosure action or inaction which he believes to be at variance with applicable disclosure standards, although there may be isolated disclosure failures that are so serious that their correction becomes a matter of primary professional concern. . . .
>
> Initially, counselling accurate disclosure is sufficient, even if his advice is not accepted. But there comes a point at which a reasonable lawyer must conclude that his advice is not being followed, or even sought in good faith, and that his client is involved in a continuing course of violating the securities laws. At this critical juncture, the lawyer must take further, more affirmative steps in order to avoid the inference that he has been coopted, willingly or unwillingly, into the scheme of non-disclosure.
>
> The lawyer is in the best position to choose his next step. Resignation is one option, although we recognize that other considerations, including the protection of the client against foreseeable prejudice, must be taken into account in the case of withdrawal. A direct approach to the board of directors or one or more individual directors or officers may be appropriate; or he may choose to try to enlist the aid of other members of the firm's management. What is required, in short, is some prompt action that leads to the conclusion that the lawyer is engaged in efforts to correct the underlying problem, rather than having capitulated to the desires of a strong-willed, but misguided client.
>
> Some have argued that resignation is the only permissible course when a client chooses not to comply with disclosure advice. We do not agree. Premature resignation serves neither the end of an effective lawyer-client relationship nor, in most cases, the effective administration of the securities laws. The lawyer's continued interaction with his client will ordinarily hold the greatest promise of corrective action. So long as a lawyer is acting in good faith and exerting reasonable efforts to prevent violations of the law by his client his professional obligations have been met. In general, the best result is that which promotes the continued, strong-minded and independent participation by the lawyer.
>
> We recognize, however, that the "best result" is not always obtainable, and that there may occur situations where the lawyer must conclude that

---

[484] *Id* 79-80.

> the misconduct is so extreme or irretrievable, or the involvement of his client's management and board of directors in the misconduct is so thoroughgoing and pervasive that any action short of resignation would be futile. We would anticipate that cases where a lawyer has no choice but to resign would be rare and of an egregious nature.[485]

This position of the SEC places a tremendous burden on the securities lawyer and highlights the complexities involved in compliance with the federal securities laws.[486]

§6.69 Commentary—Suggested Improvements in the Securities Laws

If a substantial revision of the federal securities laws is not feasible, there are several amendments to those laws which could be enacted to facilitate the raising of capital by removing burdens not significantly related to investor protection.

1. State blue sky laws, which are essentially duplicative of federal requirements, should be pre-empted
2. A person who reasonably believes the conditions for an exemption from registration under the 1933 Act have been satisfied should not be subject to liability for violation of the registration provisions of that act
3. Liability of individuals and firms under the federal securities laws should be limited to amounts sufficient to assure compliance with those laws
4. Sales of securities to accredited investors in compliance with §4(6) of the 1933 Act[487] should not be limited to $5 million, but should be unlimited in amount
5. Section 4(6) of the 1933 Act should be amended to permit secondary transactions in compliance with the provisions of that section

---

[485] *Id* 80-82.

[486] The SEC has requested public comments on its *Carter-Johnson* position. Securities Act of 1933 Release No 6344 (Sept 21, 1981).

[487] 15 USC §77d(6).

APPENDIX 6-1

Comparison of 1933 Act Statutory Exemptions, Exemptive Rules and Small Issuer Registration Form

	Sec. 3(a) (11)[1]	Rule 147[2]	Sec. 4(2)[3]	Rule 506[4]
1. Availability to Issuers	Incorporated or organized and doing business in state	Incorporated or organized and doing business in state as defined in rule	Available to any issuer	Available to any issuer
2. Limitations or General Solicitation or Advertising	No solicitation outside state	No solicitation outside state	Cannot make a public offering	Yes
3. Nature of Offerees or Purchasers	Residents of state	Residents of state	Sophisticated (*SEC v Ralston Purina Co.*[5])	Accredited or sophistication standards in rule
4. Number of Purchasers	No limit	No limit	Cannot make public offering	35 plus accredited investors
5. Amount Which Can Be Sold	No limit	No limit	No limit	No limit
6. Informational Requirements	Subject only to antifraud provisions and state law	Subject only to antifraud provisions and state law	Subject only to antifraud provisions and state law	Yes
7. Financial Statement Requirements	None specified	None specified	None specified	Yes, audited
8. Restrictions on Resales of Securities	Offering must come to rest in state	No sales outside state for 9 months	Yes	Yes
9. Sales Compensation Prohibited	No	No	No	No
10. Filing Requirement	Only under state law	Only under state law	No, unless state requirement	Yes

1 15 USC §77c(a)(11). See **§6.27.**
2 17 CFR §230.147. See **§6.27.**
3 15 USC §77d(2). See **§6.25.**
4 17 CFR §230.506. See **§6.26.**
5 346 US 119 (1953).

Comparison of 1933 Act Statutory Exemptions, Exemptive Rules and Small Issuer Registration Form

	Sec. 4(6)[6]	Rule 504[7]	Rule 505[8]
1. Availability to Issuers	Available to any issuer	Not available to investment company or reporting company	Not available to investment companies
2. Limitations on General solicitation or Advertising	Yes	Yes, unless subject to state registration	Yes
3. Nature of Offerees or Purchasers	Accredited[9]	No standards	No standards for non-accredited
4. Number of Purchasers	No limit	No limit	35 plus accredited persons
5. Amount Which Can Be Sold	$5,000,000 an *issue*	$500,000 every 12 months	$5,000,000 every 12 months
6. Informational Requirements	Subject only to antifraud provisions and state law	Subject only to antifraud provisions and state law	Yes
7. Financial Statement Requirements	None specified	None specified	Yes, audited
8. Restrictions on Resales of Securities	Yes	Yes, unless state registration required	Yes
9. Sales Compensation Prohibited	No	No	No
10. Filing Requirement	Yes	Yes	Yes

[6] 15 USC §77d(6). See **§6.28.**
[7] 17 CFR §230.504. See **§6.31.**
[8] 17 CFR §230.505. See **§6.32.**
[9] 1933 Act, §2(15), 15 USC §77b(15); Rule 215, 17 CFR §230.215.

Comparison of 1933 Act Statutory Exemptions, Exemptive Rules and Small Issuer Registration Form

	Regulation A[10]	Rule 257[11]	Form S-18[12]
1. Availability to Issuers	Only available to domestic and Canadian companies	Same as Regulation A with certain limitations	Nonreporting domestic or Canadian companies
2. Limitations on General Solicitation or Advertising	Not after offering circular delivered	Yes	Not after prospectus delivered
3. Nature of Offerees or Purchasers	No standards	Yes	No standards
4. Number of Purchasers	No limit	No limit	No limit
5. Amount Which Can Be Sold	$1,500,000 every 12 months	$100,000 every 12 months	$5,000,000 reduced by specified sales
6. Informational Requirements	Yes	Notification, but no offering circular	Yes
7. Financial Statement Requirements	Yes, audited if available	None specified	Yes, audited
8. Restrictions on Resales of Securities	No	No	No
9. Sales Compensation Prohibited	No	No	No
10. Filing Requirement	Yes	Yes	Yes

10 17 CFR §§230.251–.264. See §§**6.29–6.30.**

11 17 CFR §230.257. See §§**6.29–6.30.**

12 17 CFR §239.28. See §§**6.30** and **6.37.**

Appendix 6-2

Comparison of 1933 Act Registration Forms

	S-1[1]	S-2[2]
1. Availability to issuers	Available to almost all types of issuers except investment companies	Reporting companies meeting certain standards
2. Types of offerings covered	All	All except exchange offers
3. Amount which can be registered	No limit	No limit
4. Information requirements	Based on Schedule A of 1933 Act[3]	Based on Schedule A, except reports to stockholders may be incorporated by reference and delivered to investors

[1] 17 CFR §239.11. See §**6.37.**
[2] 17 CFR §239.12. See §**6.37.**
[3] 15 USC §77aa. See §**6.35.**

Comparison of 1933 Act Registration Forms

	S-3[4]	S-8[5]	S-18[6]	S-14[7]	S-15[8]
1. Availability to issuers	Certain reporting companies	Reporting Companies	Non-reporting companies	Reporting companies subject to SEC proxy rules[9]	Reporting companies subject to SEC proxy rules
2. Types of offerings covered	Laundry list	Offerings to employees pursuant to employee plans; no secondaries	All; secondaries limited to $1,500,000	Business combinations	Business combination where company is 10× the size of acquired company
3. Amount which can be registered	No limit	No limit	$5,000,000	No limit	Not to exceed 10% of total assets
4. Information requirements	Basically the same as old Form S-16	Except for description of plan, relies on incorporation by reference of 1934 Act reports	Based on Schedule A of 1933 Act;[10] simplified version of Form S-1	Based on SEC proxy rules	Relies on reports to stockholders in part

4 17 CFR §239.13. See §**6.37.**
5 17 CFR §239.16b. See §**6.38.**
6 17 CFR §239.28. See §§**6.30** and **6.37.**
7 17 CFR §239.23. See §**6.39.**
8 17 CFR §239.29. See §**6.39.**
9 17 CFR §§240.14a-1 to .14a-101.
10 15 USC §77aa.

Innovation and Environmental Law

*Frank F. Skillern**

7

* The author wishes to acknowledge with gratitude the capable assistance of Rosanna B. Ciupek, Class of 1982, The University of Tulsa, College of Law, in preparing and reviewing the citations in this chapter.

General Introduction

§7.01 Demands and Expectations for Innovation

Presently society demands from innovators and scientists an inordinate amount of new and creative technology. We are in a period largely dominated by science and technology. This dominance, in turn, generates expectations that very complicated problems such as the water shortage in the arid Southwest not only can, but will, be solved by new scientific discoveries or technological breakthroughs, for example, in desalinization of ocean water. Furthermore, solutions are expected as soon as a problem is uncovered or very soon thereafter.

In this technological age, Americans in particular consider that any problem can be solved if sufficient money and time are expended. We assume that science and technology will answer each and every problem regardless of its complexity. This includes everything from interplanetary travel to eliminating cancer to instantaneous cleanup of hazardous waste disposal sites that have been in use over several years.

With technological advances people also want unlimited economic and industrial growth. Technology and science are expected to provide the means to feed the exponentially growing population of the world, to provide the energy needed for desired growth, and to provide the jobs and housing individuals have come to expect.

Another effect of technological progress is a change in the nature of research. As technology and science have been successful in developing solutions, they have become more practical. Pure research is being replaced by ever-increasing research and development that is result-oriented. This also contributes to society's expectations from technological innovation.

§7.02 Conflicts between Scientific and Technological Advances and the Environment

Scientific and technological progress has produced a serious threat to the human and natural environment. The demand for peaceful use of nuclear power for energy and medical purposes, for example, could be technologically met, but the problem of disposing of radioactive waste—some of which has a half-life of hundreds of thousands of years—remains unresolved. Similarly, the demand on science for miracle drugs to cure ailments and diseases and extend life expectancy or for pesticides and fertilizers to increase food production results in products whose environmental side-effects are not known.

The societal demand for solutions to complex problems and for unlimited growth has caused other environmental impacts. Many of those impacts have been discovered by the persons who developed the environment-threatening technology in the first place. For example, scientific studies of the atmosphere and oceans demonstrated the threats to the ozone layer and to aquatic habitats and coastal regions from fluorocarbons and other air pollutants and from oil spills, respectively. Another example is the danger to groundwater and drinking water supplies caused by improper disposal or leaching from waste disposal sites. Scientific studies are confirming not only the existence but the severity of this problem in several areas of the country.

§7.03 Interaction between Science and Technology and the Environment

One ramification of identifying with greater precision the impacts of scientific and technological advances on the environment has been a demand that the source of the problem also solve it. That is, society wants the environmental problems solved scientifically and technologically. In addition, the expectation is that the solution will be forthcoming. Hence, even more innovation is anticipated and awaited.

Also, as people become more aware of environmental problems, they are more concerned about perpetuating them. However, at the same time they want to take advantage of the scientific and technological advances of the past four decades. Thus, as consumers, the public may use disposable or energy-intensive products, but does not want the resulting waste put in rivers, oceans or other sites posing contamination problems or the air and water pollution from power plants. The public also wants to use new medicines and chemicals, but without any dangerous waste or by-products. Furthermore, the public requires oil transportation by water, but expects that no oil spill will occur, or if it does, that it will be cleaned up completely and quickly, without environmental harm.

Needless to say, this situation places scientists and others developing or considering innovation in a quandary. The dilemma for an innovator is how to make a technological breakthrough without any adverse environmental effects, or at least minimal or controllable impacts. When the results of the innovation are not known or reasonably predictable or foreseeable, the innovator has a difficult decision whether to undertake the expense and time to develop new technology or chemicals. Particularly when the success of the venture is uncertain, these risks may deter innovation. Environmental law has tried to respond to these issues in some areas through planning and encouraging innovative pollution control technology.

§7.04 —Planning Stage

Today an innovator can no longer be concerned merely with technological and economic factors of a project. Environmental factors also must be considered. The use of innovative technology which is environmentally unsound or which creates a hazard to human health by pollution or other means may be restricted, notwithstanding its economic or technological advantages. The restrictions come from environmental regulations that may impose planning, licensing, performance or other requirements on the use of the technology.

An innovator has to anticipate the environmental ramifications of an experimental or test project at each stage of its development. Initially, the innovator should identify the environmental impacts of the project. In a desalinization project, for example, potential emission of air or water pollutants, disposal of salt, chemicals or other by-products and the location of the desalinization facility would have to be considered. As the work progresses, changes in the quantity or quality of pollutants as the project expands toward commercial operation have to be anticipated. Likewise, the impacts on the environment from the innovative technology if it proves feasible are subject to planning before the innovation is applied on a large-scale basis.

Environmental regulation also may have a coercive effect on innovation. For example, a plant or operation that is causing pollution of the air or water may be required to comply with air or water quality standards. Those standards may force the polluting source to reduce or eliminate the discharge of any or of particular pollutants. To comply, the source may have to apply improved means of pollution controls or cut back its operation. In order to continue the same level of operation the source has to develop the new technology itself or purchase it from another firm or innovator.

Similarly, innovation may be affected by licensing or permit requirements. Some innovations have to be approved by a governmental agency before use. New chemicals or different uses of existing chemicals are subject to permitting or testing requirements before they can be marketed in the United States. The developer of new chemicals or different uses of chemicals must anticipate the permit requirements and conduct the necessary tests to establish the safe use of the chemical. During development of the chemical, the developer should watch and test for adverse effects, such as a risk to human health, that might prevent later approval. If such adverse effects are identified, the developer should eliminate or otherwise control them so that the product can be approved for manufacture and use.

§7.05 —Enforcement Stage

A recent decision by the Fifth Circuit Court of Appeals illustrates well the dilemma of a pollution source in using new technology to comply with regulatory standards. The source may want to use new technology but may run the risk of not complying with the standard if the untested technology proves unsuccessful.

In *Alabama Air Pollution Control Commn v Republic Steel Corp,*[1] the company was charged with violating the Alabama state implementation plan (SIP) emission limitations. The company developed a technique to capture pollutants that were not regulated by the state Air Pollution Control Commission. The new

---

[1] 646 F2d 210 (5th Cir 1981), discussed in 12 Envt Rep (BNA) 223 (June 12, 1981).

technology captured particulate emissions at each of three phases of the steelmaking process. However, the emission limitation for the first phase for the entire operation was violated. The trial court accepted the company's argument that the steelmaking process had three separate phases. By testing each phase separately, the company proved that the SIP standard was satisfied.

The Fifth Circuit disagreed with the trial court. It held that steelmaking was a single process under the applicable SIP regulations. Hence, its emissions could not be measured at separate phases. Nonetheless, the appellate court did not allow the commission to fine the company for its violation. In the words of the court, a penalty would be "a travesty of justice" where the company voluntarily applied experimental technology to control unregulated pollutants. Thus, the company was not subject to a fine where the violation was based on an unsuccessful effort to control pollution by innovative technology. The failure, though, must be beyond the control of the source.

The *Republic Steel* opinion represents a practical judicial response to a very real problem for innovators. If a penalty ensues from use of experimental or untested technology, new techniques cannot be demonstrated. Further, a polluting source will have no incentive to improve or change existing pollution control techniques. The Fifth Circuit recognized this dilemma for a source and reasonably responded to it.

§7.06 —Marketing Stage

The impact of environmental regulation also can be analyzed economically according to its effect on market innovation and social innovation.[2] Market innovation is improvement in existing or development of new products or processes that provide the innovator with market benefits which can be realized through increased sales.[3] The return to the innovator from process innovations can come through reduced production costs or simplified production techniques.

Social innovation deals with social, as opposed to market, benefits from process or product improvement or original development.[4] Social benefits

---

[2] *E.g.*, Stewart, *Regulation, Innovation, and Administrative Law: A Conceptual Framework,* 69 Cal L Rev 1256 (1981); Symposium, *Regulation and Innovation,* 43 Law & Contemp Prob 1-149 (1979); 88 Yale LJ 1713 (1979). For an excellent discussion of the issues of externalities, see Coase, *The Problem of Social Cost,* 3 JL Econ 1 (1960).

[3] Market innovation refers to product or process innovations that create benefits that firms can capture through the sale of goods and services in the market. These innovations include new or improved products with greater appeal to consumers and process changes that reduce manufacturing costs or facilitate the manufacture of new or improved products.

Stewart, *supra* note 2, at 1279.

[4] Social innovation refers to product or process innovations that create social benefits, such as cleaner air, that firms cannot directly capture through market sales. Such innovations may, however, confer private economic benefits on the firms if they reduce the costs of responding to government regulatory requirements or incentive systems.

Id.

include enhancement of environmental quality and similar factors that cannot be recovered in the market. Social innovation may provide benefits to the firm using it, if the innovation reduces the firm's costs of complying with governmental regulation.

Significantly, social and market innovation are not mutually exclusive. For example, an innovative pollution control device that enables a factory to burn its fuel more efficiently can result in a cleaner environment and may reduce a firm's production costs. Similarly, the firm that develops and sells that innovative device may have a competitive advantage in marketing the product. A firm using the innovative control technology also may get a market advantage over similar producers if they all have to satisfy industry-wide standards.[5]

§7.07 Environmental Law

Environmental law has an impact on innovation and research and development by imposing planning requirements on particular projects. For example, a federal agency funding, licensing or sponsoring innovative technology or an R&D project may have to prepare an environmental impact statement under the National Environmental Policy Act of 1969.[6] Advance planning may also be needed under regulatory programs. The effect of new technology in reducing or eliminating emissions has to be considered in granting or planning a new plant or a modification to an existing one under the Clean Air Act.[7]

The Clean Air and Clean Water Acts illustrate regulatory programs that affect use of innovative technology.[8] These regulatory programs are intended to be technology-forcing, i.e., they establish standards that require development and use of new technology. Innovation, by improving the state of the art, is an objective of technology-based standards.

Other regulatory programs can affect innovation and R&D by testing and licensing requirements. New chemicals and substances must undergo testing prior to use.[9] Pesticides must be tested and registered for general, restrictive or experimental use.[10] Restrictions also apply to how pesticides are used. In addition, the government may impose restrictions on the use of technology through licensing or permit requirements. A license or permit may be condi-

---

[5] *Id. But see* Magat, *The Effects of Environmental Regulation on Innovation,* 43 Law & Contemp Prob 4, 18 (1979).

[6] 42 USC §4321 *et seq.*

[7] See **§7.26.**

[8] See **§7.23-.42.**

[9] *See,* Toxic Substances Control Act of 1978, 15 USC §2604 *et seq,* discussed at **§7.45.**

[10] *See* Federal Insecticide, Fungicide, and Rodenticide Act, 7 USC §§136-136y *et seq, as amended by* Federal Environmental Pesticide Control Act of 1972, Pub L No 92-516, 86 Stat 973, and Federal Pesticide Control Act of 1978, Pub L No 95-396, 92 Stat 819, discussed at **§7.44.**

tioned on the activity meeting particular standards or using specific technology after the activity is operational.

Other federal programs provide incentives for new technology. Under the Resource Conservation and Recovery Act,[11] for example, the government is authorized to financially assist state programs that develop new methods for recovering or reusing waste. The act encourages communities and states to develop programs for the recovery and recycling of solid waste or using it to generate energy.

§7.08 Effects of Environmental Regulation on Innovation Generally

Environmental regulation affects innovation and research and development in basically three ways: performance standards, licensing requirements and planning processes. Some environmental regulatory programs impose performance standards or other requirements on operations to reduce or eliminate pollution. Standards requiring new or improved technology may stimulate innovation or R&D.

A direct result of this regulation may be increased production costs and an increased price to the consumer. The innovator or the regulated company must initially sustain the cost of developing and applying new technology that satisfies the regulation. Those costs probably are less for a new plant than for retrofitting an existing one. In either case, the costs usually will be passed on to the consumer of the product.

A second impact is time delay. Time may be required to develop the required new technology and that will delay construction of a plant or production of goods. Additional time also may be needed to apply and test new technology and to show compliance with applicable statutes.

Some environmental programs operate by screening a product or the location of a business before allowing it in the market. This type of regulation is seen in the registration of pesticides and land use programs that require a permit before an activity may be conducted in a particular area.

Environmental regulation also entails early planning of innovative or R&D projects to avoid identifiable problems. Planning may be expressly required, as under the National Environmental Policy Act through its environmental impact statement.[12] Alternatively, it may be necessary to get prior registration or approval of a new product before marketing it. Early planning can identify the legal requirements both procedurally and substantively that must be met in conducting innovation or R&D. Careful planning will assure that the work will be legally sufficient when completed.

Thus, not only can government regulation curtail innovation by making it

---

[11] 42 USC §6901 *et seq, as amended by* Pub L No 95-609, 92 Stat 3081 (Nov 8, 1978). This statute is discussed at **§7.52.**

[12] 42 USC §4332(2)(C).

too expensive, it can also delay approval of innovation, sometimes indefinitely. That delay may result in a reluctance to spend funds on innovative technology. At the same time, technology-forcing standards are intended to encourage innovative technology. Finding the point at which innovation is stimulated and not unreasonably deterred has not been an easy task for either the legislature or administrative agencies.

§7.09 —Costs

Often, innovative technology is expensive to produce and to apply. Regulation that requires development and application of the technology may have negative results if the costs are too high. Small firms may be forced out of the market if they cannot afford the technology. If development costs are excessive compared to the return, firms may decide not to do the research and development necessary for more efficient control technology.

There are other, more positive, ramifications than the cost of new technology. Use of technology-based standards can open a new market for firms producing innovative processes. However, if the regulations require technology that is prohibitively expensive, no innovator will enter the market and develop improved techniques.

§7.10 —Uncertainty of Standards

Environmental regulation poses another problem for innovation—uncertainty of standards. Innovation needs a relatively stable regulatory environment for maximum development. If standards or requirements are regularly in flux, the innovator will be deterred from experimental or new work. This problem may arise under some programs of environmental regulation.

Under several regulatory programs the applicable standard is based on currently available technology. These state of the art standards are regularly changing as technology improves. Hence, innovators are not sure that what is developed today will meet the standard or, if so, for how long. This uncertainty is not conducive to the large economic investment often required for industrial innovation, particularly by a small firm.

Uncertainty in standards because they are regularly changing may also discourage compliance. A firm obligated to meet a particular standard may take the chance of nonenforcement if the only option is expensive technology that must constantly be improved. This uncertainty could also discourage firms from beginning a new business to develop and manufacture innovative technology for other companies. Their success may be short-lived.

§7.11 —Delay

Another effect of environmental regulation is delay in getting the product on the market or getting approval from a governmental licensing or registration agency. Delay serves to increase the price of the product to the consumer. It also deters innovation because of the uncertainty that, when developed, the technology will be marketable or usable. Delay in this instance ties into the uncertainty that attaches to changing or evolving standards.

The National Environmental Policy Act

§7.12 Overview

The National Environmental Policy Act of 1969 (NEPA), although basically a planning statute,[13] can affect development of new technology or federally sponsored research and development programs. NEPA requires every federal agency to prepare an environmental impact statement (EIS) on any proposed project or legislation "significantly affecting the quality of the human environment."[14]

The EIS must describe the proposed project and its environmental impact, state the project's adverse environmental effects, including those that cannot be avoided if the action is undertaken and discuss alternatives to the proposed action and their environmental consequences. Moreover, the EIS must identify the short-term benefits and long-term ramifications of the proposed project as well as any irreversible or irretrievable commitment of resources.[15]

The draft EIS then must be circulated to appropriate state and federal agencies and the public for comment. The preparing agency must promulgate the final EIS and submit it to the Council on Environmental Quality and the public.[16]

§7.13 —Duty to Protect the Environment

Under NEPA, a federal agency must strive to minimize the adverse environmental effects of its actions.[17] In approving projects, the agency must balance and consider environmental matters, along with economic, technological, foreign policy and other matters relevant to the proposed action.[18] This objective can be achieved by a properly prepared EIS that is considered by the agency

---

[13] 42 USC §4321 *et seq. See also* Skillern, Environmental Protection: The Legal Framework, ch 2 (Shepard's/McGraw-Hill 1981), for a more detailed discussion of NEPA.

[14] 42 USC §4332(2)(C).

[15] *Id.*

[16] *Id.*

[17] *See* Council on Environmental Quality, Final Regulations on Implementing NEPA, 40 CFR §§1505.1-.3 (1981).

[18] Calvert Cliffs' Coordinating Comm v Atomic Energy Commn, 449 F2d 1109 (DC Cir 1971). *See* Scientists' Inst for Pub Information, Inc v Atomic Energy Commn, 481 F2d 1079 (DC Cir 1973) (liquid metal fast breeder reactor research and development program). *See also* 42 USC §4332(B).

in making its decision. That decision should include any terms or conditions necessary to assure maximum environmental protection and minimum harm. The agency also can impose post-decision requirements such as inspections, monitoring or other review necessary to assure compliance with conditions.[19]

§7.14 —Federal Actions

An agency must be proposing a *major federal* action before an EIS is required.[20] As applied to research and development or innovative technology programs, this requirement means that a federal agency be involved in some way with the innovation project. Federal involvement can arise in various ways. Sometimes the federal government is funding a pilot program to determine if a particular innovation or technology is feasible.[21] Alternatively, the government may directly sponsor and fund a program to have new technology used by industry.[22] Federal action also may occur through agency licensing or granting a permit for the initial use of a newly developed chemical or substance.[23]

A proposed action by a federal agency is not immune from NEPA requirements simply because the action involves research and development or unproven technology. NEPA requires that the environmental effects of government projects be examined and considered before the decision to undertake the activity is made. It avoids a premature commitment to innovative technology or an R&D project without full awarenesss of the environmental aspects of the proposed action.

§7.15 —Comprehensive and Supplemental Environmental Impact Statements

Federal research and development projects present other unique problems under NEPA. Research and development projects, particularly large-scale ones, often are done in phases or stages. At different points in the work, decisions may be required to determine if the entire project should be completed. This phase development raises questions about the scope of the original EIS and whether at different points of the project a new or supplemental

---

19 *See* CEQ Final Regulations, *supra.* note 17, at §1505.2.

20 "(2) agencies of the Federal Government shall . . . (C) include in every recommendation or report on proposals for legislation and *other major Federal actions* significantly affecting the quality of the human environment, a detailed statement by the responsible official. . . ." 42 USC §4332(2)(C) (emphasis added).

21 *See* Scientists' Inst for Pub Information, Inc v Atomic Energy Commn, 481 F2d 1079 (DC Cir 1973) (liquid metal fast breeder reactor program). *See also* Council on Environmental Quality, Final Regulations on Implementing NEPA, 40 CFR §§1502.4 and .5.

22 *See* 33 USC §1285(i) (Clean Water Act); 42 USC §7411(j) (Clean Air Act).

23 *See* 7 USC §136a (pesticides registration); 15 USC §2603 (toxic substances premanufacture testing). See also **§§7.44** and **7.45** for discussion of these acts.

EIS may have to be prepared.[24]

The issue of the scope of an initial EIS concerns whether a program or a site-specific EIS will be needed. A program EIS is required on projects that affect broad policies or programs of agencies. It examines generally the broader aspects of the project, without examining the individual or site-specific aspects of the project. A site-specific EIS focuses on the individual aspects of a project. It localizes the project and discusses the particular environmental effects.

The responsible agency and innovator should anticipate identifiable changes in the project at different stages or different phases of development of the program. If environmental impacts, alternatives or mitigation measures as originally described and discussed in the initial EIS do not adequately project the progress or status of the work at various stages, the agency should consider requiring a supplemental EIS at later stages of the project. Alternatively, as further information becomes available or environmental impacts are identified, a new or site-specific EIS should be required on the project.

The research and development proponent and the responsible agency should also anticipate the environmental impacts when the project finally is completed. Given that work on a product may continue for several years, the research and development project may establish the feasibility of technology and lead to proposals for expanded use of the anticipated product. In that event, another EIS may be necessary before authorizing expanded use of the technology that the project has now proven feasible. In short, an EIS may be required at the completion of the project because of long-range effects that were anticipated (and in fact even hoped for) when the project began.

§7.16 —Purposes and Contents of the Environmental Impact Statement

In providing information to the responsible agency, the innovator should remember the purposes of the EIS. One objective has been described judicially as environmental full disclosure.[25] This means that the EIS must be as complete and thorough in its treatment of environmental factors as available data permit. Withholding or not making available information relevant to the environmental effects, unavoidable impacts or alternatives and their environmental effects may only operate to later delay commencement of a project. Moreover, if the information leads to the agency's deciding not to undertake the project, that decision is best made in the planning stages rather than after the project has begun.

The EIS also is intended to provide the decision maker with the basis to make an informed decision after full consideration of all factors, including

---

[24] *See* Skillern, Environmental Protection: The Legal Framework §§2.34-.36 and 2.48 (Shepard's/McGraw-Hill 1981).

[25] Environmental Defense Fund, Inc v Corps of Engrs, 325 F Supp 728 (ED Ark 1971) (Gillham Dam).

environmental ones.[26] The decision must balance various factors. The EIS should enable the decision maker to consider available options and compare them on environmental grounds. Failure to include known alternatives or to discuss the environmental effects of the alternatives will render an EIS inadequate. Then the government cannot go ahead with the project until an adequate EIS is prepared.

If an innovator has done initial or preparatory work before seeking governmental participation, if pertinent, that work should be given to the agency for incorporation into the EIS.[27]

§7.17 —Responsibility for Environmental Impact Statement Preparation

The responsibility for preparing an EIS lies with the agency proposing major federal action. Moreover, it cannot be delegated to an applicant or other person.[28] Information not only can but must be gathered from other federal and state governmental authorities and the public. Ultimately, however, the responsible agency must incorporate those data as appropriate into the EIS for the project.

This does not, however, mean that the nonfederal participant cannot assist

---

[26] *See* Council on Environmental Quality, Final Regulations on Implementing NEPA, 40 CFR §1500.1 and .2.

[27] See **§7.21.**

[28] *E.g.*, Greene County Planning Bd v Federal Power Commn, 455 F2d 412 (2d Cir), *cert denied*, 409 US 849 (1972).

In 1975, NEPA was amended to allow delegation under limited circumstances. As amended, NEPA provides:

> (D) Any detailed statement required under subparagraph (C) after January 1, 1970, for any major Federal action funded under a program of grants to States shall not be deemed to be legally insufficient solely by reason of having been prepared by a State agency or official, if:
>
> (i) the State agency or official has statewide jurisdiction and has the responsibility for such action,
>
> (ii) the responsible Federal official furnishes guidance and participates in such preparation,
>
> (iii) the responsible Federal official independently evaluates such statement prior to its approval and adoption, and
>
> (iv) after January 1, 1976, the responsible Federal official provides early notification to, and solicits the views of, any other State or any Federal land management entity of any action or any alternative thereto which may have significant impacts upon such State or affected Federal land management entity and, if there is any disagreement on such impacts, prepares a written assessment of such impacts and views for incorporation into such detailed statement.
>
> The procedures in this subparagraph shall not relieve the Federal official of his responsibilities for the scope, objectivity, and content of the entire statement or of any other responsibility under this chapter; and further, this subparagraph does not affect the legal sufficiency of statements prepared by State agencies with less than statewide jurisdiction.

42 USC §4332(2)(D).

the federal agency.[29] The participant can provide data and information to the responsible agency. It also can undertake studies and tests to provide the agency with necessary background or baseline work for the EIS. The agency, though, must write the EIS and determine how, if at all, the work from outside sources, including the applicant or participant under the proposed project, should be used in the EIS.

§7.18 Preparation of an Environmental Impact Statement on R&D Projects

An environmental impact statement (EIS) on R&D or developing technology presents special difficulties for federal agencies. Typically the scope of the project may be uncertain or hard to describe. The potential effects may be unknown, especially if innovative technology or research is being undertaken for the first time or is being done on an expanded scale. Alternatives to the proposal, in particular, may be hard to ascertain. A frequent complaint of federal agencies preparing an EIS on an innovative R&D project is that the alternatives and effects are too remote and speculative.[30]

The Council on Environmental Quality (CEQ) in its Final Regulations on the Implementation of the National Environmental Policy Act (NEPA)[31] addresses how these problems can be handled in an EIS. Those regulations do require an EIS on R&D projects when their end product, implementation or work in progress would significantly affect the quality of the human environment.[32] The CEQ final NEPA regulations require that the environmental factors of proposed research and technological development be examined prior to a commitment by an agency to the project.

§7.19 —Alternatives to the Proposed Action

Under the CEQ final regulations, an agency must consider all reasonable alternatives.[33] Some guidance is provided by the regulations to aid in determining what constitutes reasonable alternatives. Reasonable alternatives include ones that are available within the time frame of the proposed action and

---

[29] The proponent of the proposed action is to be invited by the lead agency to participate in the scoping process. 40 CFR §1501.7(a)(1) (1981). During the scoping process the scope and content of the EIS and schedule for its preparation are established. *Id.* The draft and final EIS must be circulated to an applicant. *Id.* §1502.19. The preparing agency must also solicit from an applicant comments on a draft EIS before preparing the final one. *Id* §1503.1.

[30] *E.g.*, Scientists' Inst for Pub Information, Inc v Atomic Energy Commn, 481 F2d 1079 (DC Cir 1973).

[31] 40 CFR §1500 *et seq.*

[32] *Id* §1502.4. *See also id* §§1502.5(a) (timing preparation of EIS) and .9(c)(1)-(2) (supplemental EIS).

[33] 40 CFR §1502.14.

even options beyond the jurisdiction of the preparing agency. The objective is to make the ultimate decision maker aware of environmentally more desirable options, including avoidance and mitigation measures.

With innovative R&D projects, particularly if they deal with untested technology, the identification and discussion of alternatives and their adverse effects can be difficult for an agency. The agency may be confronted with engagement in sheer speculation, on the one hand, or inclusion of every feasible alternative, however remote, on the other. The CEQ final NEPA regulations urge agencies to use reasonable alternatives, based on available or reasonably available data, and not to engage in mere guesswork.

§7.20 —Incomplete or Unavailable Information

Especially troublesome for agencies preparing an EIS on R&D or innovative projects is when data are not available or complete. The CEQ final regulations address this problem by requiring the preparing agency to make the lack of relevant information known to the ultimate decision maker. Then the decision maker will be aware of the problems the incomplete or uncertain data create and will consider them in deciding whether to proceed with the project.[34]

The final regulations go further. They require the preparing agency to obtain the information if it is essential to a reasoned decision among alternatives and the cost of obtaining the data is not exorbitant. If essential data are lacking and the cost of getting those data is not known, the agency must decide whether to proceed in light of that uncertainty. In deciding to go ahead without essential information, the agency has to balance the "need for the action" against "the risk and severity of possible adverse impacts" if the action is undertaken without the information. If the agency decides to proceed without the information, it must prepare a "worst case analysis" and indicate the probability of its occurrence.[35]

§7.21 —Participation by the Project Proponent or Innovator

The proponent or innovator can assist in preparing an EIS for a project that involves sufficient federal involvement to require an impact statement. The agency responsible for preparing the EIS on an R&D or innovative project has a difficult task because by its nature the project may be novel. In that case, the agency does not have prior experience or information readily available to complete an EIS.

The proponent or innovator of the project can help the agency by providing it with information or data that relate to the EIS. The proponent or innovator may have studies or baseline information that will aid in determining the scope

---

[34] 40 CFR §1502.22(a).

[35] *Id* §1502.22(b).

or extent of the project and its effect on the environment. In addition, the proponent or innovator may have preliminary environmental assessments or studies that identify possible adverse environmental effects. In deciding to pursue the proposed project, the proponent or innovator may have considered possible alternative strategies to the one being examined. That information can be useful to the agency in identifying reasonable alternatives that must be studied in depth in an EIS. It also might help in assessing the environmental effects of various alternatives.

By working with the responsible agency by supplying available data and studies related to the project, proponents and innovators will be able to help expedite preparation of the EIS and a decision on their proposed project. It should be noted that the agency cannot delegate or assign the task of preparing the EIS to others, and particularly not to an applicant for permission or financial assistance, because of likely bias and vested interests.[36] However, the agency can use reliable materials and studies, regardless of their source, in writing the EIS and evaluating the environmental aspects of a project. Moreover, the agency is required to solicit comments on the draft EIS from affected individuals, including the applicant.[37]

A proponent or innovator is not helped by an inadequate or incomplete EIS. The EIS, if challenged in court, would be subject to questions of sufficiency, especially if data known and available were not used by the agency. If the EIS for a project is challenged, the project's starting date will be delayed, and cost projections may be increased accordingly. In short, the agency will be required to prepare an adequate EIS before the program can be undertaken. It is in the best interests of the proponent or innovator that the EIS be as thorough and complete as practicable early in the planning process to avoid subsequent litigation and unnecessary delays.[38]

Litigation not only delays the use of the innovation but also increases its cost to the consumer. This delay only disserves the public interest which the innovation is intended to advance.

---

[36] See **§7.17**.

[37] *See generally* CEQ final regulations, 40 CFR §1500.1 *et seq;* Skillern, Environmental Protection: The Legal Framework ch 2 (Shepard's/McGraw-Hill 1981).

[38] The CEQ final regulations try to minimize litigation delay from lawsuits alleging an agency failed to comply with the regulations:

> It is the Council's intention that judicial review of agency compliance with these regulations not occur before an agency has filed the final environmental impact statement, or has made a final finding of no significant impact (when such a finding will result in action affecting the environment), or takes action that will result in irreparable injury. Furthermore, it is the Council's intention that any trivial violation of these regulations not give rise to any independent cause of action.

CEQ Final Regulations, 40 CFR §1500.3 (1981).

§7.22 —Avoiding Bias

The nonfederal participant who provides information should present all available relevant and reliable data in a nonadversarial and impartial manner. An unbiased presentation will be more useful to the responsible agency that has to evaluate and determine how, if at all, to use the material in the EIS. The agency preparing the EIS runs the risk of being challenged in court for lack of objectivity in preparation of the EIS if the EIS is conclusionary and argumentative or skewed toward the proposed action.[39] The supplier of information can help the agency avoid that charge of bias by presenting material as factually and neutrally as possible.

The Clean Air Act

§7.23 Technology-Forcing Standards

The principal way that the federal Clean Air Act (CAA)[40] affects innovation is by use of technology-forcing standards to control air pollution. Under the CAA, the Environmental Protection Agency (EPA) promulgates national standards for ambient air quality,[41] emission limitations for hazardous pollutants[42] and performance standards for stationary sources of pollution.[43]

The ambient air quality standards are based on health factors. They are intended to protect the public from harmful levels or concentrations of pollutants in the ambient air.[44] The other standards are technology-based. They require the application of a certain degree of technology to reduce or eliminate the emission of pollutants into the air.[45] Individually, or in combination, these standards form the technology-forcing features of the CAA.

The standards under the CAA are carried out by implementation plans.[46] Each state has to submit a plan to implement the national standards. The plan, commonly called a state implementation plan (SIP), must be submitted to the EPA for approval. If the SIP meets the statutory requirements for approval,[47] the EPA must approve the plan. If the SIP does not, it must be revised in order

---

[39] *E.g.*, Environmental Defense Fund, Inc v Corps of Engrs, 470 F2d 289 (8th Cir 1972) ("good faith objectivity" standard of review); Silva v Lynn, 482 F2d 1282 (1st Cir 1973).

[40] 42 USC §7401 *et seq.* For a more detailed discussion of the Clean Air Act see Skillern Environmental Protection: The Legal Framework, ch 3 (Shepard's/McGraw-Hill 1981).

[41] 42 USC §7409.

[42] *Id* §7412.

[43] *Id* §7411.

[44] *Id* §7409.

[45] *Id* §§7411, 7412.

[46] *Id* §7410.

[47] *Id. See* Train v Natural Resources Defense Council, Inc, 421 US 60 (1975).

to be approved. If a state chooses not to submit an implementation plan or submits an inadequate one which the state refuses to modify, the EPA will promulgate a federal implementation plan for the noncomplying state. Thus, in order to avoid federal administration and enforcement of the CAA, a state must adopt an SIP meeting minimum federal requirements. The SIP standards have to be at least as stringent as, and may be stricter than, the federal ones.

§7.24 —National Ambient Air Quality Standards: *Union Electric*

The technology-forcing nature of the national ambient air quality standards (NAAQS) was illustrated in *Union Electric Co v Environmental Protection Agency.*[48] In that case, the company tried to challenge the EPA's approval of the Missouri implementation plan which contained ambient air standards stricter than the federal ones. The company argued that the state standards were economically and technologically impossible to meet. Hence, according to the company, the EPA improperly approved the Missouri implementation plan.

The United States Supreme Court in *Union Electric* disagreed with the company's technological and economic feasibility argument.[49] According to the Court, economics and technology were not factors the EPA could consider under the CAA when approving or disapproving an SIP. Moreover, if an SIP meets the eight (now eleven) statutory criteria,[50] then the EPA is obligated to approve the plan. Under the CAA, the EPA has no discretion not to approve a state submitted program that contains the statutory elements.[51]

The Court in *Union Electric* went on to hold that the CAA authorizes states to adopt more stringent NAAQS than the federal ones, if they so choose. The Court emphasized the technology-forcing aspects of the CAA. It concluded that Congress intended to force development and application of better technology to solve the air pollution problem. A state's adoption of more stringent standards than the federal ones is consistent with this technology-forcing strategy.

§7.25 —Emission Limitations

The EPA must promulgate emission limitations for hazardous air pollutants and new stationary sources of pollution. The standards for hazardous air

---

[48] 427 US 246 (1976).

[49] The Court also rejected procedural and jurisdictional arguments of the company based on its interpretation of the CAA. *Id.*

[50] The original criteria for approving a state plan included eight requirements. 42 USC §1857 c-5(a)(2)(A)to(H). The 1977 amendments to the Clean Air Act added three new requirements. The SIP now has to provide for nonattainment areas, prevention of significant deterioration (clean air) areas and noncompliance penalties. *Id* §7410(a)(2)(I)-(K).

[51] *See* Train v Natural Resources Defense Council, Inc, 421 US 60 (1975).

pollutants require that they not be released into the air in amounts that would be detrimental to public health.[52] That standard must allow an adequate margin of safety, which is determined by the EPA administrator. If the emission limitation is more stringent than the current state of the art technology, then new technology must be developed to reduce the emissions.

Similarly, the new source performance standards (NSPS) are based on emission limitations and are technology-forcing.[53] Those standards are set by the EPA and must be based on the state of the art, i.e., control technology available to the source to reduce or control the emission. The technology required also must be adequately demonstrated and economically achievable. As the state of the art for controlling emissions for the categories of stationary sources improves, the improved technology must be applied by the new or modified source. In this manner NSPS compel development and use of new technology.

§7.26 —Clean Air Areas

The standards for new sources planning to locate in clean air areas directly require application of innovative or improved technology for reducing or controlling emissions. Clean air areas are those in which the air has not been polluted, i.e., it is pristine, and the ambient air quality standards are being met. For those areas, the EPA, under the 1970 amendments to the CAA, developed the prevention of significant deterioration (PSD) program.[54] The EPA's initial PSD program was basically adopted by Congress, with some modifications, in the 1977 amendments to the CAA.[55] Following extensive and comprehensive litigation in *Alabama Power Co v Environmental Protection Agency*[56] the EPA promulgated its current PSD program by regulations issued on August 7, 1980.[57]

Under the PSD program a new source locating in a PSD area must get a preconstruction permit. The source also must apply the best available control technology (BACT) to control emissions of every pollutant regulated under the act. This includes criteria pollutants and hazardous pollutants, as well as any subject to NSPS.[58]

---

[52] 42 USC §7412(b)(1)(B).

[53] *Id* §7411. *See* Alabama Power Co v Costle, 636 F2d 323 (DC Cir 1979).

[54] The PSD program was originally called the nondegradation policy. That policy was required by the per curiam decision in Sierra Club v Ruckelshaus, 4 Envt Rep Cas (BNA) 1815 (DC Cir 1972), *affd by an equally divided court sub nom* Fri v Sierra Club, 412 US 541 (1973). The EPA then issued its final rule promulgating the nondegradation or prevention of significant deterioration program. 39 Fed Reg 42510 (Dec 5, 1974).

[55] 42 USC §§7470-79.

[56] 636 F2d 323 (DC Cir 1979).

[57] 45 Fed Reg 52676 (Aug 7, 1980).

[58] 636 F2d 323.

§7.27 —Nonattainment Areas

The 1977 amendments added another new program for nonattainment areas. Nonattainment areas are those regions that had not met the NAAQS by the deadline in the 1970 amendments. In 1977, Congress extended the time for attaining new source performance standards because most air quality control regions had not yet achieved the NAAQS. It also adopted, with minor modifications, the EPA's emission offset policy (EOP) as the program to handle pollution in those nonattainment areas.[59] Under the EOP, construction or modification of a major source can only be undertaken in a nonattainment area if emissions of regulated pollutants are decreased in the area after the construction or modification.

A new or modified major stationary source wanting to locate in a nonattainment area must get a preconstruction permit. The permit must require the source to apply the lowest achievable emission rate (LAER).[60] The state nonattainment plan must provide for "reasonable further progress" in reducing pollution from existing sources by requiring "at a minimum, reasonably available control technology" (RACT).[61]

§7.28 Need for Innovative Technology

Because of the congressional objective to force development and application of more efficient pollution control technology, new and improved means of reducing or eliminating emissions are needed by industrial polluters. Most major sources of pollution are bound to strict technology-based standards that require them to reduce the amount of pollutants emitted. Recognizing this need, Congress in 1977 provided incentives to innovators to develop improved means of pollution control. These included additional time to meet applicable standards for sources using innovative, but unproven, technology and a waiver of the standards for sources using innovative technology.

§7.29 —Time Extension for Innovative Technology

The 1977 amendments included a new provision that gives a source additional time to meet otherwise applicable new source performance standards (NSPS).[62] To qualify for the extra time the source must be using new techniques of emission reduction that have not been adequately demonstrated, but

---

[59] The nonattainment program is found at 42 USC §§7501-08. *See also* Emission Offset Interpretive Ruling, 40 CFR pt 51, app S (1981), *amended as necessary* to comport with the EPA's 1980 PSD Regulations, 45 Fed Reg 52676 (Aug 7, 1980).

[60] 42 USC §7503(2).

[61] *Id* §7502(b)(3).

[62] 42 USC §7413(d)(4).

are likely to be demonstrated by the final compliance date, or the new technology would not be used without a compliance order.

Before issuing a delayed compliance order, the Environmental Protection Agency (EPA) administrator must make two findings. The first finding is that the new technology is likely to achieve a greater emission reduction level than would otherwise be achieved, or if the same level is being achieved, it will be done at lower energy, economic or nonair quality costs than by other means. Second, the administrator must find that compliance with the requirements of an applicable state implementation plan (SIP) is impracticable while the new technology for emission reduction is used.[63]

Thus, by allowing additional time to comply with emission limitations, Congress sought to encourage sources to develop and use new means of emission limitation. This, in turn, would spur innovation in developing the technology that would be in compliance while it was being applied or adequately demonstrated.

§7.30 —Innovative Technology Waiver

In 1977, Congress also added another provision to the Clean Air Act (CAA) to encourage or provide an incentive to industry to develop new technology. That provision allows a waiver from applicable standards for a source using innovative technology.[64] This is the innovative technology waiver (ITW). The ITW is intended to encourage sources to develop and use new technology for emission control.

Under the 1977 amendments, a source may receive a waiver if it has been applying innovative technology for pollution reduction. The CAA provides:

> Any person proposing to own or operate a new source may request the Administrator for one or more waivers from the requirements of this section for such source or any portion thereof with respect to any air pollutant to encourage the use of an innovative technological system or systems of continuous emission reduction.[65]

The administrator may grant an ITW after statutory procedural requirements have been met and the requisite findings have been made by the administrator.

Before granting an ITW, the administrator must provide notice and an opportunity for public hearing on the waiver request. In addition, the administrator must find

---

[63] *Id* §7413(d)(4)(C). The delayed compliance order must include a timetable and schedule for compliance and be preceded by statutory procedural requirements. In addition, a delayed compliance order for a major stationary source must require it to pay the §7420 noncompliance penalty. *Id* §7413(d)(1)(A)-(E).

[64] 42 USC §7411(j)(1).

[65] *Id.*

1. That the new innovative system has not been adequately demonstrated
2. That the new system will operate effectively and achieve better or at least the same continuous emission reduction than otherwise would be required. but at lower energy, economic or nonair quality environmental costs
3. That the new system will be safe and not create an unreasonable risk to public health, welfare or safety from its operation[66]

A source that has received an ITW does not have to meet otherwise applicable standards while the innovative system is being used and tested. The time for compliance is extended.[67] However, if the technology is unsuccessful, then the source must meet the original standard within the minimum time necessary to comply.[68]

By the ITW, Congress acknowledged the need to develop improved pollution control technology and that new technology may require more time to be tested and proven successful. The ITW encourages companies to consider more efficient and ideally less costly control devices by not penalizing those who do so with noncompliance penalties or enforcement actions.[69] A plant operating under an ITW is deemed in compliance for purposes of enforcement and noncompliance penalties.[70]

The ITW, however, is not intended to be a simple expedient to delay compliance with NSPS. The burden is on the applicant for the ITW to prove that the innovative technology qualifies and the waiver is otherwise needed. In addition, consent to the ITW must be obtained from the governor of the state in which the source is located.[71]

§7.31 Bubble Concept

The *bubble concept* is a technique adopted by the Environmental Protection Agency (EPA) to measure emissions from a source.[72] Under it, a plant is treated as an entity for determining total emissions. Thus, if a facility or part of a plant achieves increased emission reduction of a pollutant by use of new technology, emission reductions of that pollutant can be set off against emis-

---

[66] *Id* §7411(j)(1)(i)-(iii).

[67] The time extension runs until the earlier of seven years after the source receives an ITW or four years after the source commences operation. *Id* §7411(j)(1)(E).

[68] *Id* §7411(j)(2)(A).

[69] *See* Alabama Air Pollution Control Commn v Republic Steel Corp, 646 F2d 210 (5th Cir 1981), discussed in 12 Envt Rep Cas (BNA) 223 (June 12, 1981). See **§7.05.**

[70] 42 USC §7411(j)(2)(B). For the enforcement and noncompliance penalty provisions, see *id* §7413.

[71] *Id* §7411(j)(1)(A).

[72] See Landau, *Alabama Power Co v Costle: An End to a Decade of Controversy over the Prevention of Significant Deterioration of Air Quality?*, 10 Envtl L 585, 603.07 and 623.28 (1980).

sions of it in other parts of the plant. Under the bubble concept, the total emissions may not change, but the source is encouraged to apply proven technology in limited areas. The source is not penalized for a modification by not receiving credit for the reduction obtained from the use of innovative technology and is encouraged to use new technology.

Thus, the bubble concept can be an incentive to innovation. Industrial sources can use innovative technology to reduce emissions in part of a facility without having to modify, retrofit or replace an entire plant which may be using dated equipment.

§7.32 —Prevention of Significant Deterioration

Under the prevention of significant deterioration (PSD) program, a source in a clean air area must get a permit before construction of a new source or before undertaking a major modification of an existing source.[73] Each state must designate the area where the national standards are not yet violated. If a source in a PSD area wants to modify its facility, certain requirements must be met.

The source must obtain a permit before commencing a major modification in a PSD area. However, if the source can show that after modification it will not cause a net increase in the amount of pollution or emit a new pollutant, it is not subject to PSD review. Avoiding PSD preconstruction review for a permit is important because the source can undertake the modification without time delays and increased costs associated with the PSD permit application process.

In determining whether the proposed modification is major and hence subject to PSD review, the source can apply the bubble concept. If a source, through application of innovative technology, can achieve a decrease in pollution emissions in one part of the plant, the emission reduction may be credited against emissions from other parts of the plant to determine if a net increase in pollution occurs. If not, the modification is not major, and not subject to PSD review and permit requirements.

The District of Columbia Circuit Court of Appeals approved the use of the bubble concept for the PSD program in *Alabama Power Co v Environmental Protection Agency*,[74] It distinguished its earlier decision in *Asarco Inc v Environmental Protection Agency*[75] in which the bubble concept was not allowed for new source performance standards (NSPS). The *Alabama Power* court concluded the PSD program is designed to allow development under strict conditions in clean air areas. According to the court, the bubble concept is appropriate for that objective because the concept considers net, not merely any, increase in emissions.

---

[73] The PSD program requirements for SIPs are set forth at 40 CFR §51.24 (1981).

[74] 636 F2d 323 (DC Cir 1979).

[75] 578 F2d 319 (DC Cir 1978).

§7.33 —New Source Performance Standards

In *Asarco v Environmental Protection Agency*,[76] the District of Columbia Court of Appeals interpreted the Clean Air Act (CAA)[77] as requiring reduction in pollution from new sources or from major modifications to existing sources.[78] Because the emphasis is on emission reduction, according to the court, the bubble concept is ill suited to attain NSPS. The bubble concept is not appropriate because it maintains the status quo through its trade-off, rather than achieving reductions in the emission of pollutants.

§7.34 —Nonattainment Areas

After *Alabama Power Co v Costle*,[79] the EPA did not apply the bubble concept to nonattainment areas. It instead defined a source subject to review under two alternative tests. The regulations included either an entire plant *or* an individual piece of process equipment within a plant.[80] Hence, a source could not offset emissions to determine whether a modification was major in areas where federal air quality standards were not met.

In October, 1981, the EPA amended its nonattainment area plan requirement.[81] As amended, the dual definition was eliminated, and the bubble concept was applied. The regulations now apply a plant-wide definition for nonattainment purposes.[82] That definition is the same as the one applied for PSD.

§7.35 —Current Environmental Protection Agency Regulations

The EPA has amended its PSD[83] and nonattainment area regulations.[84] to comport with the judicial interpretation of the CAA and use of the bubble concept. The concept is available to determine whether a major modification has been proposed under the PSD program. It cannot be used to establish compliance with NSPS or to avoid NSPS on modifications to an existing plant. In line with the opinion in *Alabama Power*, the bubble concept has been continued in the emission offset policy for nonattainment areas. The plant-wide

---

[76] 578 F2d 319 (DC Cir 1978).

[77] 42 USC §7411.

[78] See **§7.32.**

[79] 636 F2d 323 (DC Cir 1979).

[80] 45 Fed Reg 52676 (Aug 7, 1980).

[81] 46 Fed Reg 50766 (Oct 14, 1981).

[82] *Id.*

[83] 40 CFR §51.24 (1981).

[84] *Id* §51.18 and the interpretive ruling for that section in app S, *as amended by* 46 Fed Reg 50766 (Oct 14, 1981).

definition is now used for both nonattainment and PSD areas to determine if new source review will be necessary.

The Clean Water Act

§7.36 Technology-Forcing Aspects

As under the Clean Air Act,[85] innovation is effected under the Clean Water Act (CWA)[86] primarily by technology-forcing standards to control discharge of effluents into the nation's waters. The 1972 amendments to the Federal Water Pollution Control Act (FWPCA) declared a goal of no discharge of pollutants into the nation's waters by 1985.[87] This no-discharge goal was restated in the 1977 amendments to the FWPCA.[88] Clearly in 1972, and probably in 1977,[89] the technology did not exist to meet this no-discharge goal. The goal is to be reached by requiring state of the art technology which reduces and ultimately eliminates the discharge of industrial pollution into the nation's waters.

The no-discharge goal is to be attained by use of several programs, including water quality standards,[90] effluent limitations[91] and the National Pollutant Discharge Elimination System (NPDES).[92] Other standards also are imposed on publicly owned treatment works (POTWs).[93] These programs are intended to reduce, over time, the pollution discharged into the nation's waters by application of new technology for treatment or control of pollutants.

Industrial sources are affected primarily by the effluent limitations which are implemented by a federal- or state-administered NPDES permit program. Under the CWA, a source must have an NPDES permit before it can legally discharge effluents into the nation's waters. The NPDES permit program can be administered by a state, if it has had its program approved by the Environmental Protection Agency (EPA); if not, then the EPA will administer the

---

[85] 42 USC §7401 *et seq.* See **§§7.23-7.35.**

[86] 33 USC §1251 *et seq.* The 1972 and 1977 amendments to the Federal Water Pollution Control Act are commonly referred to as the Clean Water Act or CWA. For a more detailed discussion, see, Skillern, Environmental Protection: The Legal Framework ch 4 (Shepard's/McGraw-Hill 1981).

[87] 33 USC §1251(a)(1).

[88] *Id.*

[89] *See* S Rep No 370, 95th Cong, 2d Sess, *reprinted in* 1977 US Code Cong & Ad News 4341-42.

[90] 33 USC §1313.

[91] *Id* §1311.

[92] *Id* §1342.

[93] *Id* §1311 (1)(A)-(B) and (2) (B). *See also id* §1281(g)(2)(A).

program.[94]

The permitting authority must assure that the applicable water quality, effluent limitations and other standards under the CWA will be met by the permittee. Hence, an industry or source must apply state of the art technology to control its effluent discharges. Because the standards are based on currently available technology, they can be regularly changing as technology improves and require regular upgrading by sources to the current level of technology.

The effluent limitation standards are technology-based. They require a source within an industry to apply the best available technology to achieve emission reductions. In promulgating effluent limitations the EPA has to consider availability of technology and its cost. After promulgation, the industry or source is allowed a specified time to comply with the technological requirements. In this manner sources continuously need innovative technology.

§7.37 —Technological Factors

The industrial point source effluent limitations are based on best available technology economically achievable (BAT).[95] These standards are meant to force industry to develop and apply improved technology as it becomes available. In establishing BAT, the EPA must consider current technology and the availability of new technology. Moreover, the EPA must establish that the technology required is feasible and effective.

Effluent limitations also may differ according to the nature of the pollutant. Under the 1977 amendments, conventional pollutants are treated differently from nonconventional ones. The conventional pollutants must apply the best available control technology (BACT), but are given a longer time to achieve it.[96] Nonconventional pollutants must be controlled by use of BAT by 1984 or three years after the limitation is set, and in no case later than July 1, 1987.[97] BAT are required to be achieving "reasonable further progress" toward the no-discharge goal.

Under the 1977 amendments, a new time schedule for compliance was set. By 1977, the deadline of the 1972 amendments had not been met and, hence, most sources were in noncompliance. The 1977 amendments allowed additional time, but were stricter in requiring technology for increased pollution reduction.

§7.38 —Economic Factors

Innovation also may be affected by other factors that the EPA must consider in establishing the standards under the CWA. The innovative technology must

---

[94] *Id* §1342(b).

[95] 33 USC §1311(2)(A).

[96] *Id* §1311(2)(E). *See also id* §1314(a)(4).

[97] *Id* §1311(b)(2)(F).

not only be feasible, it must be cost-effective. The EPA must consider the cost of requiring an industry to apply innovative technology as a factor of its being *economically achievable*.[98]

As water quality is improved and fewer effluents are discharged by individual sources, the cost of new technology must be considered again. Innovation for means to control conventional pollutants may be deterred because the improvement in effluent limitation has to be balanced against the benefit of cleaner water. The EPA's effluent limitation guidelines for conventional pollutants have to consider, among other things, "the reasonableness of the relationship between the costs of attaining a reduction in effluents and the effluent reduction benefits derived."[99] A similar comparison is required by the guidelines for POTWs. The EPA guidelines of POTWs have to compare their costs and levels of reduction to those for industrial sources.[100]

Economics is also a factor in issuing an NPDES permit to an individual source. The effluent limitation required under the permit has to consider the age of the source, its costs, process used and others available, the cost of achieving the effluent reduction and other factors relevant to the source's ability to meet the national effluent limitation.[101]

§7.39 —Variances

Industry has challenged standards under the CWA on the issue of whether effluent limitations promulgated by the EPA must provide for variances.[102] An additional question is, if this is the case, what criteria may be used to allow exceptions from the national effluent limitations. Industry typically seeks to have economic and technological as well as water quality factors used to grant a variance.

The courts have generally held that the EPA effluent limitations are not invalid for their failure to include a variance procedure. This is especially true in situations where the EPA has examined all relevant factors in setting standards that are strictly technology-based.[103] Moreover, variances do not have to consider the quality of the receiving waters.[104]

The EPA has established a general variance procedure for individual sources. Under this procedure the EPA will allow a variance to a source only

---

[98] 33 USC §1311(b)(2)(A).

[99] *Id* §1314(b)(4)(B).

[100] *Id.*

[101] *Id* §1314(b)(2)(B) (effluent limitation guidelines for BAT).

[102] *See* Appalachian Power Co v Train, 620 F2d 1040 (4th Cir 1980); Weyerhaeuser Co v Costle, 590 F2d 1011 (DC Cir 1978); EPA v National Crushed Stone Assn, 449 US 64 (1981).

[103] *See* EI duPont de Nemours & Co v Train, 430 US 112 (1977).

[104] EPA v National Crushed Stone Assn, 449 US 64 (1981).

if the source alleges fundamentally different factors[105] or conditions from those originally considered in setting the applicable industry effluent limitations. Hence, if a source wants a variance based on economics or technology availability factors, it must show how those issues differ for the source because of location, processes, age of operation or other reasons before a variance can be allowed. Although the variance provisions have been frequently litigated, they have generally been upheld.[106]

§7.40 —Confirmation of Technology-Forcing Aspects: *National Crushed Stone*

In 1981, the United States Supreme Court addressed the variance issue and, in its analysis, confirmed the technology-forcing purposes of the CWA. In *Environmental Protection Agency v National Crushed Stone Assn*[107] the coal industry's trade association argued that the EPA had to consider the same factors in granting a variance that the agency considered in establishing the effluent limitation guidelines.

Specifically, the petitioner wanted the cost of implementation to each individual source considered, and if excessive, a variance granted. The EPA maintained that individual source costs were relevant only if they greatly exceeded normal implementation costs for the plants the EPA studied in setting the limitation. Under the EPA's approach, the financial inability of an individual plant to comply would not justify a variance.[108]

The Court noted that in *EI du Pont de Nemours & Co v Train,*[109] it had indicated variances would be appropriate for individual sources under effluent limitation guidelines for classes or categories of sources. However, the Court rejected the argument that economic inability of a source has to be a factor in granting a

---

[105] Consolidated Coal Co v Costle, 604 F2d 239 (4th Cir 1979), *revd on other grounds sub nom* EPA v National Crushed Stone Assn, 449 US 64 (1981).

[106] The EPA variance procedures provide in part:

> An individual discharger or other interested person may submit evidence to the Regional Administrator (or to the State, if the State has the authority to issue NPDES permits) that factors relating to the equipment or facilities involved, the process applied, or other such factors related to such discharge are *fundamentally different* from the factors considered in the establishment of the guidelines. . . . If such *fundamentally different* factors are found to exist, the Regional Administrator or the State shall establish for the discharger effluent limitations in the NPDES permit either more or less stringent than the limitations established herein, to the extent dictated by such *fundamentally different* factors. [emphasis added]

Any variance also must be approved or disapproved by the administrator. 40 CFR §434.22 (1981).

[107] 449 US 64 (1981).

[108] The *National Crushed Stone* Court did note that the act contains a provision for individual variance from BAT to be met by 1987. That variance can be based on the economic inability of the source to comply. 33 USC §1311(c) (Supp II 1978).

[109] 430 US 112 (1977).

variance. The "economic capability of an individual discharger to afford the costs of the BPT [best practicable control technology currently available] limitation" does not have to be considered in granting a variance.[110]

The *National Crushed Stone* Court first reviewed the statute. Under the act, the first level of technology required by 1977 was the BPT.[111] The Court concluded that the variance provision[112] provided no exception from the 1977 BPT. According to the Court, that provision applies only to BAT. Next the Court emphasized the technology-forcing aspects of the act. BPT is based on balancing the benefits of effluent reduction with the costs of pollution control. BPT standards, which are set for a class or category of sources, are based on normal costs for an individual discharger. A source is not required under BPT to apply the highest degree of technology that it can afford.

The second level of technology is the best available technology currently achievable (BAT).[113] BAT limitations are to be met over a longer period of time. They require the maximum commitment of technology and money to achieve the statutory no-discharge goal. BAT limitations assume compliance by a source with BPT. The variance provision requires that a source show it is making reasonable further progress beyond BPT toward BAT and that the source is using the maximum technology it can afford to attain BAT limitations.[114]

Hence, according to the Court, variances from BPT would undercut the technology-forcing aspects of these limitations. Allowing variances based on economic capability would not get the source to the level of BPT. In addition, variances based on economic capability would allow noncompliance because of individual source impacts, contrary to the finding the administrator must make that BPT technology is economically achievable for a category or class of sources. In the words of the Court:

> Because the 1977 limitations were intended to reduce the total pollution produced by an industry, requiring compliance with BPT standards necessarily imposed additional costs on the segment of the industry with the least effective technology. If the statutory goal is to be achieved, these costs must be borne or the point source eliminated. In our view, requiring variances from otherwise valid regulations where dischargers cannot afford normal costs of compliance would undermine the purpose and the intended operative effect of the 1977 regulations.[115]

---

[110] EPA v National Crushed Stone Assn, 449 US 64 (1981).

[111] 33 USC §1311(b)(1)(A).

[112] *Id* §1311(c).

[113] *Id* §1311(b)(2)(A).

[114] *Id* §1311(c).

[115] 449 US at 64.

Moreover, the Court found the administrator's interpretation was supported by legislative history.[116]

§7.41 —Publicly Owned Treatment Works

The CWA also affects technological innovation by requiring the upgrading and improvement of municipal water treatment plants. The CWA requires POTWs to build new facilities that achieve effluent reduction at least equal to that attained by secondary treatment.[117] If industrial sources are using the POTW, then additional pretreatment requirements are placed on them before entering the POTW facility.[118] In effect, in many areas—especially industrial metropolitan ones—this may require upgrading of the POTW or pretreatment of the water by the industrial source before going to the POTW.

Upgrading for new facilities is encouraged by the act. The pertinent provision addresses waste treatment management plans and the construction grants program.[119] That program is designed to establish and implement the best practicable waste treatment technology on a regional basis.[120] Applicants for grants must propose facilities that meet the requirements of 33 USC §§1281, 1288 and 1313.[121]

§7.42 —Innovative Technology

The CWA authorizes the administrator or state after consultation with the administrator to set a date for compliance with BAT for nonconventional pollutants on or before July 1, 1987.[122] This delayed compliance is incorporated into the NPDES permit.[123] The point source must be replacing existing production capacity with an innovative production process in order to seek the compliance delay.

In order to qualify for the delayed compliance date, the applicant's innovative technology must obtain or have a substantial likelihood of obtaining greater effluent reduction than would otherwise be necessary while progressing toward a no-discharge goal, or it must be found that the innovative techniques will attain the required reduction but at a significantly lower cost than the systems the administrator has determined are economically achievable. The only other finding the administrator must make is that the innovative system

---

[116] *Id* 80 (quoting from congressional hearings about closing marginal operations).
[117] 33 USC §1311(b)(1)(B).
[118] *Id* §§1311(b)(1)(A), 1317(c).
[119] *Id* §1281.
[120] *Id* §1281-1297.
[121] *Id* §1284.
[122] 33 USC §1311(k).
[123] *Id* §1342.

has the potential for industrywide application.[124] If that finding is made and the applicant shows the innovative process fits into one of these three situations, the administrator may set a later compliance date.

Regulation of Chemicals and Wastes

§7.43 Chemical Innovation

Chemistry may be the scientific field with the greatest potential for innovation. In areas of medicine and agriculture, for example, development of new chemicals or new uses of existing ones is helping to cure diseases and to increase food production or lessen crop damage. The environmental difficulty with new chemicals and substances is determining which are safe for human or restricted use and under what circumstances. Annually, many new chemicals, compounds or other substances are developed to alleviate a particular problem. Secondary, long-term and other effects, however, are not always known when the chemical or substance is first developed.

Two basic approaches are applied to regulate new or modified uses of chemicals or substances that may be hazardous, toxic or carcinogenic to humans. First, the Federal Insecticide, Fungicide, and Rodenticide Act[125] as amended by the Environmental Pesticide Control Act of 1976,[126] requires that pesticides, herbicides and similar substances be registered for particular uses. The other important statute is the federal Toxic Substances Control Act of 1976[127] which requires testing of toxic substances (as defined by the act) before their manufacture or production.

§7.44 Registration of Pesticides

The Federal Insecticide, Fungicide, and Rodenticide Act, as amended,[128] establishes a registration program for different chemicals which are to be used as pesticides or herbicides. The proponent of a chemical for a different or new use first must have the product registered by the Environmental Protection Agency (EPA). Several requirements have to be met to get a product regis-

---

[124] *Id* §1311(k). See **§§7.29** and **7.30** for discussion of similar provisions in the Clean Air Act.

[125] 7 USC §136 *et seq.*

[126] *Id* §136y *et seq. See* the Federal Pesticide Control Act of 1978, Pub L No 95-396, 92 Stat 842. *See also* Skillern, Environmental Protection: The Legal Framework, §§5.37-.49 (Shepard's/McGraw-Hill 1981) for more detailed discussion of the federal pesticides program.

[127] 15 USC §2601 *et seq. See also* Skillern, *supra* note 126, at §§5.23-.36 for more detailed discussion of the federal act.

[128] 7 USC §136 *et seq.*

tered.[129]

The innovator must show that the chemical or substance can be used without causing unreasonable adverse effects to the environment. Any danger in use must be noted by appropriate labeling to notify users by conspicuous markings on the container. The container label also must note any restrictions on use and clear instructions about safe application.

A registered pesticide also must be classified as to its use.[130] The EPA may register pesticides for several different uses. A permit may be issued allowing experimental use while further information necessary for the product's registration is being gathered.[131] Registration for general use means that no restrictions other than proper labeling are imposed on use of the product. The third type of registration is for restrictive use. This permit applies to pesticides that cause unreasonably adverse effects to the environment.

The act also commenced a program of training and certifying persons who qualify as applicators of restricted use chemicals.[132] The EPA or state may conduct the certification program. For restrictive use registration, only certified applicators may apply the product and then only under safe conditions. Under the Federal Pesticides Act of 1978,[133] the federal government is to develop a certification program which is to be generally available as a model to the states.

§7.45 Regulation of Toxic or Dangerous Substances: The Toxic Substances Control Act

Under the Toxic Substances Control Act (TSCA),[134] Congress imposed more stringent safeguards on the use of toxic or other substances that create an unreasonable risk of danger to health and the environment. The TSCA restricts use of chemical substances or mixtures that may be dangerous to human health or the environment. It does so by requiring the developer of a chemical or mixture to prove that it is safe for the use intended.

Under the TSCA, a developer of a toxic substance must apply to the Environmental Protection Agency (EPA) before manufacturing or processing it, or applying it to a new use.[135] The EPA may require information on the nature of the substance, its proposed use, the manner of use and potential harmful effects on human health. If the EPA finds the existing test information is

---

[129] *See* 7 USC §135b (registration procedures for economic poisons); *id* §136a (registration procedures for pesticides).

[130] *Id* §136a(d).

[131] *Id* §136c.

[132] *Id* §136b (certification procedures).

[133] Pub L No 95-396, 92 Stat 842.

[134] 15 USC §2601 *et seq.*

[135] *Id* §2604.

inadequate to assess accurately the harmful effects, the agency may deny permission to market the product pending additional testing and further information.[136]

§7.46 —Testing Requirement Rule

If the EPA finds (1) that the manufacture, use, processing or disposal of a chemical substance or mixture is creating or may create an unreasonable risk of harm to human health or the environment, or (2) that a chemical substance or mixture is entering or may enter the environment in substantial quantities or may have significant human exposure, the agency may issue a testing requirement rule.[137] That rule must identify the substance or mixture to be tested and include standards for testing. The rule must establish a time schedule for submitting the test data to the EPA. The schedule may consider the relative costs of test procedures and the personnel and facilities.

A testing requirement rule must require a substance to be evaluated for harmful effects, including risks of cancer, birth defects, transmittable genetic damage and behavioral disorders. The characteristics of the product also must be examined. These include the product's persistence, acute, subacute and chronic toxicity, and cumulative or synergistic effects.

§7.47 —Testing Exemption and Reimbursement of Testing Costs

The TSCA has a procedure for exemption from the testing requirement rule.[138] If a manufacturer or processor is preparing a product equivalent to one for which test data have previously been supplied, an exemption is then allowed to avoid duplication. The TSCA provides for voluntary reimbursement of testing costs among the manufacturer providing the data and those getting the exemption. If reimbursement cannot be arranged voluntarily, the TSCA authorizes mandatory reimbursement.

This reimbursement provision may be highly beneficial to the innovator who develops the product. Without it, the cost of mandatory testing would be paid by the original discoverer, but the benefit of expensive testing would inure, without cost, to all users. The exemption and reimbursement are reasonable means to reduce unnecessary paperwork and costs. They also allocate those costs among all potential users, not merely the person developing the chemical substance or mixture. Without reimbursement, the high cost of testing and the potential of a windfall to another user could deter innovative research and development or confine it to large companies.

---

[136] *Id* §§2604(e) and 2605.

[137] Testing also may be required if available data are insufficient to determine the health effects of the product. 15 USC §2603.

[138] 15 USC §2603(c).

§7.48 —Priority Testing List

Under the TSCA, a committee is established to study chemical substances and mixtures and to determine which are the more dangerous ones.[139] Those are then recommended to the EPA for priority testing. The committee first sets priority for testing of those chemicals listed. The list cannot number more than 100 chemicals or mixtures. The EPA must then expedite issuing testing requirement regulations for the listed products. Those products have to be tested first.

§7.49 —Premanufacture Notice

The TSCA imposes a heavy burden of proof on a person seeking to use new chemical substances or mixtures. The act requires a proponent to file a premanufacture notice (PMN) that indicates what product is being tested and for what purpose.[140] This PMN must be published in the *Federal Register* and is subject to public review.

The EPA after receiving a PMN must evaluate the test results. The PMN must show that the substance has been adequately tested, no harmful health effects were discovered and the proposed use is safe. The EPA has 90 days to confirm or agree with the PMN results or to delay manufacture pending additional tests. Those tests may be done by either the manufacturer or the EPA. In any event, the substance cannot be manufactured or marketed for the proposed use until approved by the EPA.

The EPA by rule may prohibit or otherwise limit or condition the manufacture, processing or distribution of listed substances. Any such rule must be preceded by a finding supported by a reasonable basis by the administrator that use, manufacture, processing, commercial distribution or disposal of the substance presents "an unreasonable risk of injury to health or the environment."[141]

§7.50 Regulation of Carcinogens

Innovators developing food additives or substances intended for human consumption or entry into the human food chain have to be aware of the potential carcinogenic effect of their products. The Delaney Amendment[142] prohibits the Department of Agriculture from approving any food additive that may cause cancer in humans or animals. The Delaney Amendment and the

---

[139] 15 USC §2603(e).

[140] 15 USC §2604.

[141] *Id* §2605.

[142] The Delaney Amendment added §348c, to the Food and Drug Act, 21 USC §§301 *et seq.*

Food Color Additives Act[143] establish the only federal regulation of food on strictly carcinogenic grounds.

Thus, it is important for producers of new food products to conduct the appropriate tests to assure that those products have no carcinogenic effects from food additives. If at any time it is shown that a product may cause cancer, the product cannot be approved for human use or must be removed from the market. The carcinogenic test is absolute; the appropriate agency has no discretion to permit approval of carcinogenic products for human use.[144]

§7.51 Regulation of Hazardous and Solid Waste

The field of hazardous and solid waste will be one of the most demanding areas of law in the 1980s. The explosive growth of industrial innovation and scientific knowledge has challenged and will continue to challenge legal efforts to contain or control that growth when it threatens human health or the natural environment. As this new learning and technology expand, there is even greater growth but that growth, in turn, often produces hazardous or dangerous by-products that must be dealt with. For example, increasingly plants are able to remove hazardous waste from manufacturing processes as a result of using innovative pollution control technology required by the Clean Air Act[145] and Clean Water Act.[146]

Many areas relating to hazardous and dangerous wastes remain open for development of innovative technology or other solutions to the waste disposal problem. Hazardous waste, perhaps, is the most obvious example. First, new and safe ways of on-site storage, transportation and disposal of hazardous waste are needed. The technological advances also should include ways to handle hazardous waste that are safe from sabotage or vandalism. Safety may be a more serious problem for radioactive waste disposal.

Second, a new means of permanent storage must be found. The cases occurring regularly[147] involving disposal directly into surface waters, containers into lakes, lagoons or wetlands, or burial in nonpermanent containers all illustrate the problems that must be solved. These situations have resulted in dangerous chemicals leaching into public water supplies or ground waters. They also have

---

[143] *Id* §376.

[144] Exceptions have been made for some food additives, e.g., saccharin. *See* 21 USC §343(o)(1), which conditions saccharin use on appropriate labeling and notice of possible harmful effects.

[145] 42 USC §7401 *et seq.*

[146] 33 USC §1251 *et seq.* See Canter, *Hazardous Waste Disposal and the New State Siting Programs,* 14 Nat Resources Law 421 (1982).

[147] One of the more publicized examples involves the contamination and environmental pollution at Love Canal. *See* United States v Hooker Chem & Plastics Corp, 16 Envt Rep Cas (BNA) 1268 (SDNY 1981) (denying company's motion for blanket protective order). For an excellent discussion of Love Canal and the hazardous waste problem, see Brown, Laying Waste: The Poisoning of America by Toxic Chemicals (1981).

created gaseous and other emissions over time that present serious threats to human health and the natural environment.

Third, new criteria to evaluate sites for hazardous waste disposal have to be established. Increased on-site storage and diminishing availability of existing sites illustrate the need for new ones. Those sites should meet minimum standards based on proven engineering practices which have to be developed. In the future, only sites using the engineering practices or other technology necessary to protect the public health and natural environment should be approved as disposal sites.

Means of detecting and eliminating the risks from hazardous and other dangerous substances must be discovered. Situations arise all too often where unsafe disposal presents an immediate and long-range threat to health. In one recent case, the Illinois Supreme Court upheld a ruling that a hazardous waste disposal site constituted a public nuisance, and, because of the imminent threat to health, ordered the toxic waste exhumed and removed. The company then had to restore and reclaim the site.[148] Technology to handle hazardous and dangerous wastes must be developed before the problems are so acute that drastic—and expensive—remedial action is the only solution.

§7.52 Resource Conservation and Recovery Act

The Resource Conservation and Recovery Act (RCRA)[149] is intended to promote technological advances and innovation in solving the problems of hazardous and dangerous wastes. Under the act, hazardous waste is dealt with differently from solid and liquid waste. Hazardous waste is subject to close regulation from the moment of generation to its ultimate storage or disposal.

§7.53 —Hazardous Waste

The Environmental Protection Agency (EPA) must promulgate standards regulating how persons handle hazardous waste at different stages of its disposition process. Generator, transporter and disposer standards must be set.[150] Those standards will require improved technology and greater care at each stage of disposal to assure protection of the public health. The hazardous waste standards also require the use of a manifest system.[151] The manifest must

---

[148] Village of Wilsonville v SCA Servs, Inc. 16 Envt Rep Cas 1105 (Ill 1981).

[149] 42 USC §6901 *et seq, as amended by* Pub L No 95-609, 92 Stat 3081 (1978). For more detailed discussion of RCRA, see Skillern, Environmental Protection: The Legal Framework, §5.01 (Shepard's/McGraw-Hill 1981).

[150] 42 USC §§6922 (generator standards), 6923 (transporter standards) and 6924 (disposer standards).

[151] *Id* §§6922(5), 6923(a)(3) and 6924(2). *Manifest* is defined as "the form used for identifying the quantity, composition, and the origin, routing, and destination of hazardous waste during its transportation from the point of generation to the point of disposal, treatment, or storage." *Id* §6903(12).

be prepared initially by the waste generator and placed on the storage or transportation containers. The manifest must identify the nature of the waste, its composition, quantity and source. In addition, the manifest must identify the transporter and ultimate disposer or storer of the waste.

The RCRA creates a permit system for hazardous waste treatment, disposal or storage facilities.[152] The permit system can be administered by either the state or the federal government. Since passage of RCRA in 1976 more than 30 states have adopted some type of hazardous waste management statutes.[153] The state may issue permits if it adopts a hazardous waste program approved by the EPA. The EPA has developed a Model State Hazardous Waste Management Act to assist states in designing a hazardous waste program.[154]

The permit or certification will be given only to owners or operators of qualified facilities that meet the minimum RCRA requirements.[155] Under the RCRA, a state must assure that all persons dealing with hazardous waste use safe means to avoid unnecessary exposure or risk to human health. This means that the generator, transporter and storer must use the best technology to avoid or eliminate undue risk at different stages of dealing with hazardous waste.

§7.54 —Solid Waste and Resource Recovery

The RCRA uses a different approach to deal with solid waste problems. Nonhazardous waste is not subject to a mandatory federal permit program. Rather, the federal government provides financial and technical assistance to assist states in the improvement of waste disposal techniques.[156]

The RCRA requires the EPA to review landfills and dumps throughout the country and determine which ones do not meet minimum statutory standards. The act requires a site to qualify as a sanitary landfill.[157] This inventory is national in scope and intended to inform states which sites must be upgraded.

An objective of the RCRA is to eliminate waste disposal at open dumps and to upgrade existing dumps to the level of sanitary landfills. The EPA is to promulgate nonbinding guidelines for states concerning the state of the art and available technology for waste disposal and resource recovery.

---

[152] *Id* §6925.

[153] *See,* Canter, *Hazardous Waste Disposal and the New State Siting Programs,* 14 Nat Resources Law 421 (1982).

[154] *Id* 434-37, where the model act is discussed and evaluated. For identifying suitable locations for treatment, storage or disposal facilities for hazardous waste, the model act includes "engineered" as well as natural sites.

[155] 42 USC §6925(c).

[156] 42 USC §§6947-6948.

[157] *Id* §6944.

§7.55 — Incentive to Innovation

Resource recovery is the other aspect of the RCRA that offers new areas to innovation. The RCRA encourages the development of innovative technology to reduce the hazards of waste disposal.[158] A major thrust is to encourage recycling, reuse or recovery of the usable materials from waste for later use. The subsequent use can be in varied forms such as recycling or as power from a municipal generating plant.

Similarly, the RCRA encourages innovation in developing ways to handle municipal waste. New ways to separate and dispose of different types of waste are needed. By financial and technical assistance and dissemination of information, the RCRA encourages development of new techniques to handle municipal solid waste. Innovation and R&D are needed to develop improved means of recycling or reusing materials suited for recovery. Other means of disposing of nonrecoverable solid waste are also sought. The act permits use of landfills only for solid or liquid waste suitable for that disposition.

Resource conservation and recovery is particularly important because of the competing demands for the nation's lands. New uses are regularly arising that threaten the availability of land for waste disposal in the near future.[159] Resource recovery and recycling also alleviate the energy problem by reducing the demand for natural resources for power generation. The city of Tulsa, Oklahoma, for example, is reviewing plans for a municipal waste recovery plant because of the lack of available landfill sites and increasing energy costs.[160]

Miscellaneous Considerations

§7.56 Need for Innovation in Testing, Monitoring and Other Compliance and Standard-Setting Techniques

Environmental regulation has created other demands for innovation or R&D. Those regulated need to know if their operation is complying with applicable standards. For large operations, compliance can require regular monitoring of various components of a plant. Innovation in the areas of checking and reviewing an operation under varying conditions may be developing.[161]

---

[158] 42 USC §§6981-6987.

[159] *Cf* City of Philadelphia v New Jersey, 437 US 617 (1978) (state justified ban on importing waste for disposal in New Jersey on grounds of limited land for disposal and need in other areas).

[160] Tulsa World, Jan 27, 1982, at B-8, col 4.

[161] For example, Sun Company has developed an environmental compliance system designed to monitor a plant and its different operations. The system was developed by Sun Information Services Company, a Sun Company subsidiary. One objective is to reduce compliance and reporting costs. How to Prove You're Clean (advertising brochure 1980).

The government as regulator also needs innovation in ways to measure compliance or to determine appropriate standards. Visibility standards, for example, for certain air pollutants have often been challenged.[162] Their potential lack of uniform application and subjective nature may suggest that more precise scientific means of measurement need to be found.

In setting standards, an agency often has to make certain assumptions or projections as the starting point for a standard. For example, in setting prevention of significant deterioration standards under the Clean Air Act,[163] the Environmental Protection Agency (EPA) has first to determine the baseline concentration of pollutants in the area. Actual emissions are not available for the proposed new source. Actual conditions vary dependent on climatic, meteorological, geographical and other factors. To determine a standard, the EPA uses modeling,[164] which has been challenged as being too imprecise a scientific method on which to set pollution levels.[165]

These problems carry over to testing and monitoring for enforcement purposes. This problem of more precise means to determine standards and establish violations needs additional scientific work and R&D.[166] Further investigation may provide more accurate methods of measurement which would be welcomed by both the regulated and regulator.

§7.57 Institutional Changes

As innovation is needed to improve technology to enhance the quality of the environment, reform in the method of regulating is also desirable. The basic programs for environmental regulation are in place and apply traditional planning, licensing and permitting techniques. With the high cost and time delay that regulation can cause,[167] these programs could be examined to see if they can be simplified or streamlined without sacrificing any of their environmental protection objectives.

Although each would need additional study, the following are ways by which the regulatory process might be simplified. Duplication of studies, application data and other paperwork could be minimized by putting environmental review in one agency. For example, the Council on Environmental Quality or the

---

[162] *E.g.*, Portland Cement Assn v Ruckelshaus, 486 F2d 375 (DC Cir 1973); Alabama Power Co v Costle, 636 F2d 239 (DC Cir 1979).

[163] *See also* 42 USC §7475. For further discussion of PSD standards, see **§7.32.**

[164] See Kramer, *Air Quality Modeling: Judicial, Legislative and Administrative Reactions,* 5 Colum J Envtl L 236 (1979); Sullivan & Shanoff, *Air Quality Assessments: Dispersion Modeling,* 17 Trial 50 (Dec 1981).

[165] *See* Texas v EPA, 499 F2d 289 (5th Cir 1974); Portland Cement Assn v Ruckelshaus, 486 F2d 375 (DC Cir 1973).

[166] The Clean Air Act requires further research into modeling techniques. 42 USC §7620 (standardized air quality modeling).

[167] See Friedman, *Environmental Checklist and Outline of Impacts of CAA Amendments of 1977,* 14 Real Prop Prob & Tr J 873 (1979).

Environmental Protection Agency could be authorized to examine major energy or other environment-threatening projects and resolve any environmental issues. That decision, either way, would then be binding on all other agencies. This centralizing also might be used for large projects under the National Environmental Policy Act.[168] The Council on Environmental Quality might be designated as the screening agency to determine and resolve the environmental issues on particular projects like federal energy programs.[169] Once the agency approved or disapproved a project, that decision could be binding on all other agencies as to environmental problems.

A specialized environmental or technical court has also been proposed by Congress in several bills. This court could hear all appeals on environmental matters or regulations, including standard-setting. These proceedings could provide expedited judicial review of agency action in environmental matters and minimize litigation delay. Ideally, better decisions would ensue because of the expertise and qualifications of the judges in scientific, technological and environmental matters.

Stricter and more precise regulations are also needed and can result in product and process innovation. Those being regulated often are responding to the regulation. Vague, confusing or ambiguous regulations do not stimulate innovation or new technologies. "The regulatory agencies should be aware of the fact that it is possible to design regulations to stimulate the development of new technologies whose performance exceeds the expectations of both industry and government."[170]

Any reform in environmental regulation should be preceded by a study evaluating the effectiveness of a particular program.[171] Each environmental program typically is designed to address one or a limited number of environmental problems. Modifications should seek to lessen the burden of regulation on those subject to it, but should not do so at the expense of progress the program has made in providing a better environment.[172] Maintaining a quality environment should remain the objective because a quality environment is

---

[168] 42 USC §4321 *et seq.*

[169] A similar proposal that was never implemented was the Energy Mobilization Board. See, Christian & Kovall, *Simplifying the Federal Permitting Process: the Proposed Energy Mobilization Board Legislation,* 12 Nat Resources Law 721 (1979).

[170] Ashford & Heaton, *Regulation and Technical Innovation,* 5 EPA J 32, 34 (1979) (discussing an MIT study of the effects of environmental, health and safety regulations on technological change in the chemical industry).

[171] For example, while energy issues and environmental protection often seem in conflict, it has been found that innovation induced by environmental regulation in some cases has produced an ancillary benefit of reduced energy costs. See *id* 33 (1979).

[172] The Clean Air Act regulations have helped reduce hydrocarbon and carbon monoxide emissions from automobiles. Now nitrogen oxide emissions need to be addressed. See LaPierre, *Technology-forcing and Federal Environmental Protection Statutes,* 62 Iowa L Rev 771, 795 (1977). When the act is reviewed by Congress in 1982, any legislative changes should maintain the progress achieved on some of the automotive pollutants while devising means to reduce emissions of the others.

obviously in everyone's best interest.[173]

§7.58 Innovation and State Environmental Regulation

Innovation can be affected by environmental regulation at the state level. As previously indicated,[174] many state procedures were developed under federal cooperative federalism programs such as the Clean Air[175] and Clean Water Acts.[176] Under most of the federal programs the state is not pre-empted from applying its own regulations. Usually, the federal program merely requires that any state regulation be at least as strict as federal regulation.[177] Hence, the state standard may be more stringent than the federal one.

Several states have enacted statutes modeled after the federal National Environmental Policy Act.[178] The state laws usually require an environmental impact report that must include the same basic information as the federal environmental impact statement. Some states impose this requirement on all agencies. Others limit it by naming specific agencies or particular projects, such as interstate highways, regional shopping centers, airports or projects for a minimum size (e.g., over 20 acres). Other states have adopted state-wide or regional planning for particular projects.[179] For example, a development permit may be required from a state agency for a project in a wetlands or coastal zone, in scenic mountain areas or other environmentally sensitive regions. These statutes may impose conditions and restrictions on development in the potential areas. At a minimum, they require early planning and approval by an appropriate board.

States have also adopted statutes to expedite the permit process.[180] Fast-track legislation has been used to centralize permit applications and data in one agency. After review is concluded, the agency's decision on the environmental aspects of a project will be final. Other agencies may review the record for necessary conditions relevant to their legislative mandate, but are limited concerning matters resolved by the central agency.

---

[173] However, maintaining a quality environment often requires scientific studies and data. The EPA, for example, is considering allocating federal funds for scientific research on the effect of acid rain on aquatic ecosystems, crops, soils and to study other environmental impacts. 417 Pollution Cont Guide (CCH) 481 (Oct 29, 1981).

[174] See **§§7.23** and **7.36.**

[175] 42 USC §7401 *et seq.*

[176] 33 USC §1251 *et seq.*

[177] *See* Union Elec Co v EPA, 427 US 246 (1976), discussed at **§7.24;** Homestake Mining Co v EPA, 477 F Supp 1279 (DSD 1979) (stricter state water quality standards under the Clean Water Act).

[178] 42 USC §4321 *et seq,* discussed in **§§7.12-7.22.** See Skillern, Environmental Protection: The Legal Framework §§7.03-.05 (Shepard's/McGraw-Hill 1981).

[179] Mich Comp Stat Ann §§691.1201 *et seq.* See Skillern, *supra* note 178, at §§7.12-.21.

[180] *Id* §§9.12-.13.

Michigan has enacted a citizen suit act[181] which arguably expedites judicial review. No longer is review delayed until the administrative process is concluded or action begun. The Michigan act allows any citizen to seek to abate or enjoin any activity that threatens the natural environment or violates the public trust therein. Thus, judicial scrutiny can begin when action is commenced or as standards are being determined, and litigation delays may be lessened.

Many states recently have adopted hazardous waste management statutes. Typically these statutes address problems of waste disposal. They establish procedures for locating hazardous waste disposal facilities. Although these programs usually deal with evaluating natural conditions for site selection, some states are developing engineering and other technological standards that new facilities would have to meet.[182]

---

[181] *Id* §§7.06-.10.

[182] These state programs are discussed in detail in Canter, *Hazardous Waste Disposal and the New State Siting Programs,* 14 Nat Resources Law 421 (1982).

State and Local Regulation of Innovation

8

Bruce M. Kramer

§8.01 Introduction

The purpose of this chapter is not to present a detailed account of how each state and local governmental unit in the United States affects or can affect the use of innovative technologies or products. Given the fact that there are 50 states and some 80,000 substate units of local government,[1] such an attempt would be doomed to failure. The objective of this chapter is to provide a framework for comprehending the nature of powers exercised by state and local governments as they could potentially and realistically affect innovative new products and technologies.

The chapter includes an analysis of how new technologies have in certain cases been able to get the support of state and local governmental units to eliminate barriers and create incentives for the marketing and distribution of innovative new technology and products. Examples are provided of existing state and local regulations which have had an impact on innovation and how in certain circumstances the deleterious impacts have been eliminated. The chapter also addresses the taxing structure at the state and local level in order to illustrate how the tax structure may inhibit or foster the use of innovative products. State and local governmental promotion of innovation is also mentioned, although briefly.[2]

To provide a more realistic approach to these problems, a model will be used to show how a particularly broad innovative technology with many products and subtechnologies has dealt with the issue of state and local taxation and regulation. This model will be the solar energy revolution of the late 1970s, more particularly the use of solar energy at the residential level. Although not meant to be totally exhaustive of the potential problems of the innovator, the issues facing the solar energy innovators will provide a good illustration of how one can work with state and local governments to eliminate barriers and establish incentives for the use of innovative products and technologies. Lessons learned from this experience may be transferable to other new products and technologies so as to guide decision-makers in dealing with the impact of state and local governmental regulation and taxation on their commercialization and marketing process.

The lessons to be learned from this chapter are relatively simple. State and local regulatory and taxing programs can have a deleterious impact on the marketing of innovative products and technologies. State and local governments can, on the other hand, be a positive force supporting innovation if the

---

[1] US Bureau of the Census, 1977 Census of Governments.

[2] For an in-depth discussion, see **ch 9.**

innovator will do its homework properly. The support of state and local governmental officials and employees may be critical to the elimination of barriers and the erection of incentives. State and local interest groups such as unions, chambers of commerce, consumer organizations and so forth can be of significant assistance to those seeking to work within the state and local governmental system. Careful attention must be paid to the persons and process involved if delay, higher costs or total prohibitions are to be avoided.

§8.02 Overview of State and Local Governmental Powers—Police Power and Home Rule

Under our federal system, the states are the basic repository for the police power.[3] Unlike the federal government, which is a government of enumerated powers,[4] each state, and through the state, each substate unit, possesses without specific constitutional authority the inherent ability to legislate and regulate for the protection of the public health, safety, morals and general welfare.[5] The police power cannot be easily defined.[6] It clearly changes over time and with the exigencies of each individual circumstance.[7] It is the broadest of all powers and is only limited by other specific federal or state constitutional provisions.[8]

The state as holder of the police power can transfer some or all of that power to substate units. In the absence of a specific state constitutional or legislative grant of some form of local autonomy or home rule as it is called, the local governments are treated as the creatures of the state.[9] Local governmental units are called *creatures* because they cannot exercise any governmental powers, including the police power, without an express grant of authority by the

---

[3] US Const amend X.

[4] Gibbons v Ogden, 22 US (9 Wheat) 1 (1824).

[5] Hamilton v Kentucky Distilleries & Warehouse Co, 251 US 138 (1919). *See also* Clements v McCabe, 210 Mich 207, 177 NW 722 (1920); People v Wood, 151 Misc 66, 272 NYS 258 (Broome County Ct 1934).

[6] *See, e.g.*, Stone v Mississippi, 101 US 814 (1880); Munn v Illinois, 94 US 113 (1876). *See generally* E Freund, Police Power (1904).

[7] *See, e.g.*, Slaughter House Cases, 83 US (16 Wall) 35 (1873); Charles River Bridge v Warren Bridge Co, 36 US (11 Pet) 420 (1837).

[8] *See, e.g.*, Pennsylvania Coal Co v Mahon, 260 US 393 (1922) (conflict between taking clause of the federal Constitution and state exercise of police power); Moore v City of East Cleveland, 431 US 494 (1977) (conflict between family rights protected under various provisions of the federal Constitution and municipal exercise of zoning powers); United States Trust Co v New Jersey, 431 US 1 (1977) (conflict between federal prohibition against state impairment of obligation of contract and state exercise of police power to deal with urban mass transit crisis).

[9] *See, e.g.*, Hunter v City of Pittsburgh, 207 US 161 (1907); City of Trenton v New Jersey, 262 US 182 (1923).

state.[10] Each local government must look to the state legislature for specific enabling statutes before it can exercise any regulatory powers.

Local governmental exercise of police powers under a creature theory was further hampered by the application of the canon of judicial construction known as *Dillon's Rule.*[11] Dillon's Rule required a strict or narrow interpretation of state enabling statutes, which weakened local governmental efforts to react to new problems such as tenement housing, the automobile and other conditions brought on by the urbanization of America. To a certain degree, the deleterious impact of Dillon's Rule has been recently repealed by judicial and legislative actions which broadly grant and then construe general grants of police power to local governments.[12] Prior to this trend the principal reaction to Dillon's Rule was the enactment of home rule provisions.

The home rule movement was spawned by several different and converging trends including the rural to urban movement first begun in the late 1800s. The first real constitutional provision was enacted as part of the 1875 Missouri Constitution.[13] The movement toward local autonomy or home rule has continued to the present day so that today a majority of states authorize at least some of their municipalities to enjoy home rule autonomy.[14]

The granting of home rule authority to local governments in effect allows them to engage in substantial police power regulation without having to secure specific state authorization.[15] In a limited number of states the home rule unit is given plenary authority over matters relating solely to local concern.[16] In other states the grant of local autonomy is made subject to ultimate control by the state legislature.[17] In either case the home rule unit is usually free to act in a regulatory fashion unless the state has a similar, competing or conflicting regulatory scheme. A good example of state preemption of home rule authority is the situation relating to commercial business transactions covered by the Uniform Commercial Code.[18] No home rule city could be able to adopt its own

---

[10] See cases cited in 2 McQuillin, The Law of Municipal Corporations 12-13 (1979). *See also* City of Philadelphia v Fox, 64 Pa 169 (1870); City of Fargo v Sathre, 76 ND 341, 36 NW2d 39 (1949).

[11] Dillon, Municipal Corporations 89 (5th ed 1911). *See, e.g.,* Nance v Mayflower Tavern, Inc, 150 F2d 773 (1944).

[12] *See, e.g.,* State v Hutchinson, 624 P2d 1116 (Utah 1980); Wilson v City of Council Bluffs, 253 Iowa 162, 100 NW2d 569 (1969).

[13] Mo Const 1875 art IX, §§16, 17, 20-25.

[14] *See, e.g.,* Mont Const art VI, §6; Ohio Const art XVIII; Wash Const art XI. *See generally* Vanlandingham, *Constitutional Municipal Home Rules Save the AMA (ULI) Model,* 17 Wm & Mary L Rev 1 (1975); Vanlandingham, *More Home Rule in the United States,* 10 Wm & Mary L Rev 289 (1968).

[15] Ohio Const art XVIII; ND Const art VI.

[16] Colo Const art XX; Cal Const art XI. *See, e.g.,* Sato, *Municipal Affairs in California,* 60 Calif L Rev 1055 (1972); Raich v Meyers, 27 Cal2d 773, 167 P2d 198 (1946).

[17] Iowa Const art 3, Mass Const amend art II.

[18] The Uniform Commercial Code was developed by the National Conference of Commissioner on Uniform State Laws in 1962 and adopted in almost all of the states. It affects most commercial and secured transactions.

rules relating to commercial transactions taking place within the municipal boundaries because the matter has either been preempted by the state or is a matter of state-wide, not local, concern.[19]

As with the police power, no substate unit inherently possesses taxing or revenue-raising authority. In fact, while most states now allow for home rule over police power matters, little general local authority exists in the taxing area. Only a few states give cities a free hand in local taxing authority.[20] Most severely restrict or give qualified home rule taxing authority to home rule units.[21] Thus, most local governments must receive specific revenue-raising authority from the state prior to utilizing a particular form of taxation. It must be noted, however, that user fees and police power fees designed to recoup the costs of providing municipal services are not taxes and therefore would not have to be state-authorized for a home rule unit.[22]

In the revenue arena several general conclusions may be reached regarding the impact of the taxation scheme on innovation. States are moving away from the ad valorem real and personal property tax as the main source of revenue. The principal source of state revenue is sales, gross receipts and excise taxes.[23] A substantial secondary source of revenue is individual and corporate income taxes.

On the other hand, local governments predominantly rely on the ad valorem tax on real or personal property for their revenue.[24] Whether personal property is subject to the ad valorem tax is usually dependent on the state constitution. In most states, individual property not held for business purposes is not subject to the ad valorem property tax. Several states allow their cities to utilize an income tax although the amount raised is somewhat small. Localities supplement the property tax through some type of revenue sharing or independent sales tax, user fees and occupation and licensing taxes.

Thus, depending on the product or technology involved, the state or local taxation scheme may provide a barrier to the entry of such technologies or products. In many cases, particularly the solar energy field, the enactment of tax incentives for innovative technologies has been considered very important for their initial success or failure.

The range of possible regulatory and taxation programs at the state and local level which can affect innovative technologies and products is great. For example, such diverse subject areas as fertilizer,[25] barber shops,[26] banks,[27] solar

---

[19] *In re* Lane, 58 Cal2d 99 372 P2d 897, 22 Cal Rptr 857 (1962). *See e.g.*, Jefferson v State, 527 P2d 37, 45-46 (Alaska 1974) (Connor, J, concurring).

[20] Colo Const art XX, §6(g).

[21] Ill Const art 7, §6(e); Iowa Const art 3, §40.

[22] *See, e.g.*, Post v City of Grand Junction, 118 Colo 434, 195 P2d 958 (1948).

[23] Bureau of the Census, 1978 Statistical Abstract of the United States 293.

[24] *Id.*

[25] *See, e.g.*, Minn Stat §17.701 *et seq;* SC Code §3-601 *et seq.*

[26] Tex Rev Civ Stat Ann art 8402 *et seq* (Vernon).

[27] *Id* art 342-901 *et seq.*

energy devices,[28] and plastic pipe[29] are just a few of the many that may be regulated by the state or local government. In most cases, the regulatory program will be in place because of some social problem that the program was at one time perceived to correct. Unfortunately, as years pass the program remains although the problems may change and the use of innovative technology or products will be impeded.

A simple example of this problem will be sufficient to illustrate the basic problem. Building codes have been used in the United States since the seventeenth century in order to provide safe, fireproof and well-built structures for persons to live and work in.[30] Unfortunately, as building practices have improved and workmanship has become more professional, the building code has ceased to achieve its original objective and has developed into a fairly expensive regulatory program which has as one of its principal functions the maintenance of the status quo in the building industry.[31] This scenario can probably be repeated in many states and cities with only the products or technologies different. In many cases, only persistent efforts by innovators or user groups or other influential lobbying groups can seek to remove the regulatory and taxing impediments, and in addition, replace them with real incentives for the introduction of innovative technologies and products into the marketplace.

§8.03 —How State and Local Regulations and Taxes Affect Innovators

State and local government regulatory and taxing programs can affect innovators in four general ways. They can provide research and development aid or assistance to an area undergoing technological change.[32] This direct assistance may also be in the form of state preferential purchasing for certain types of innovative technologies or products.[33] Similarly, however, one must be aware of state restrictions on state purchasing which may provide a barrier to innovation. One example would include a buy-state provision which gives preference in state contracts to in-state manufacturers.[34]

A second form of regulation will usually involve direct governmental inter-

---

[28] NM Stat Ann 70-80-1.

[29] *See, e.g.*, Seidel, Housing Costs and Government Regulations 90-91 (1978); Field & Rivkin, The Building Code Burden (1975).

[30] Field, *Judicial Solutions to the Tensions Between Construction Codes and Innovation*, 4 Urb L & Poly107, 113 (1981). *See also* Oster & Quigley, *Regulatory Barriers to the Diffusion of Innovation: Some Evidence from Building Codes*, 8 Bell J Econ 361 (1977).

[31] See Field, *supra* note 30, Oster & Quigley, *supra* note 30; Seidel, *supra* note 29, at 71-99; Field & Rivkin, *supra* note 29, at 33-101.

[32] *See, e.g.*, Ariz Rev Stat Ann §41-571-574; Ohio Rev Code Ann §122.63 (Page). This topic is more fully discussed in **ch 9.**

[33] *See, e.g.*, Tex Rev Civ Stat Ann art 678i, §§1-3.

[34] American Yearbook Co v Askew, 339 F Supp 719, 723-25 (MD Fla), *affd mem*, 409 US 907 (1972).

vention in the marketplace through registration, licensing or permit requirements. Direct prohibitions against products or technologies would also fall under this rubric. This would involve such diverse matters as pesticide or commercial feed registration, licensing of products prior to use or prohibition against specified uses of certain materials and land use controls.

A third form of intervention will involve indirect impacts on the marketplace. One example would include the requirement that certain types of products sold to the consumer be warrantied in a specific fashion.[35] Another type of indirect regulation would occur where the product may be unregulated but the installers or sellers of the product or technology are regulated by either a licensing or a testing scheme.[36]

The fourth program which will impact on the innovator is the taxing policy of the state and local governments. Because many new innovative technologies and products are not initially price-competitive, the additional burdens placed on such innovations by taxes may provide the difference between success and failure. Again, the forms are many and varied, including income tax credits for industries,[37] property tax exemptions[38] and franchise tax exceptions for corporations,[39] sales and use tax exceptions[40] and tax benefits for small business corporations.[41]

Existing state and local taxing schemes initially may provide barriers to entry for innovative products or technologies. Products subject to the ad valorem real or personal property tax may need exemptions, partial or total, from the tax to make them price-competitive. Sales and excise tax exemptions may also provide an incentive for the new product. Income tax credits or deductions may also provide a benefit to the innovator who can convince the appropriate level of government that the product or technology deserves that type of preferential treatment. While the federal government may provide greater incentives or barriers in the revenue or taxation area, no innovator should

---

35 *See generally,* Lawrence & Minan, *The Role of Warranties and Product Standards in Solar Energy Development,* 34 Vand L Rev 537 (1981).

36 Many local governments require plumbers, electricians and other artisans to be locally certified before they can be utilized in a construction project. This is often part of the municipality's general building or housing code. *See, e.g.,* Hall v City of Taft, 47 Cal2d 177, 302 P2d 574 (1956), where the court struck down a city building and housing code as it pertained to a state junior college district's construction program including the licensing of workers to be employed on the job.

37 *See, e.g.,* Kan Stat Ann §§79-32, 79-168, 79-1118; ND Const rev §57-38-01.8; 68 Okla Rev Stat §§2357.1-.3.

38 *See, e.g.,* Tex Const art VIII, §2(a); Fla Const art 7, §4(b)(3); Fla Stat §193.622 (supp 1978).

39 Tex Tax Code Ann tit 122a, §12.03(1)(r) (Vernon).

40 *See, e.g.,* Ariz Rev Stat Ann §42-1312.01(A)(9); Ga Code Ann §92-3403a(c)(2)(2.1).

41 The Wall Street Journal reported that Indiana was providing a 30% tax credit plus tax-free dividends for investors in publicly owned corporations providing capital to small business corporations or small business investment corporations. Wall St J, Oct 21, 1981, at 1, col 4.

overlook the problems inherent in the state and local tax and revenue strategies which may have to be overcome to successfully market an innovative technology or product.

§8.04 Solar Energy as an Example

In discussing the impact of state and local governments on innovation, the recent experience of the solar energy industry in the development and marketing of its technologies and products provides an excellent model. While not encompassing all possible problems facing innovators, the various solutions and responses of state and local governments to the problem of supplementing fossil-fuel energy with alternative sources does provide a wide range of solutions and problems raised by state and local governmental regulatory and taxing practices.

In particular, the state and local governmental impact on innovation in the solar energy field has been extensively recorded, studied, analyzed and digested.[42] State and local governments have resorted to many different statutory strategies designed in most cases to foster the development of a viable solar energy industry as it relates to residential energy needs. In fact, in some cases innovation has been mandated so that solar energy devices are required in certain types of new construction in several cities.

The solar energy industry also provides a good cross-section of both innovative technologies and products. State and local governments have to be aware that supporting one particular technology or product may have a deleterious impact on future innovations in the field or competing products or technologies within the industry. For example, a local government might create incentives for the use of an active solar energy system, e.g., solar panels, which may create disincentives for the use of passive solar energy systems. The local taxation scheme may run counter to the regulatory scheme, or in certain circumstances, private law matters might have an impact which effectively blunts a legislative initiative designed to increase the availability of innovative products or technologies.

Another problem is that while a state or local government may embrace a new or innovative technology such as solar energy, there is normally no consensus on what particular technologies or products are part of the industry. This leads the state and local government into a trap that stifles innovation,

---

[42] The Solar Energy Research Institute, an arm of the Department of Energy, has recently published a series of papers solely on the impact of state and local laws and taxes on the solar energy industry. *See, e.g.,* Solar Energy Research Institute, A Survey of State Approaches to Solar Energy Incentives (1979); Solar Energy Research Institute, Land-use Barriers and Incentives to the Use of Solar Energy (1979); Solar Energy Research Institute, Problems in the Administration of State Solar Legislation (1979); Solar Energy Research Institute, Wind Energy: Legal Issues and Institutional Barriers (1979); *see also* Kraemer, Solar Law (Shepard's/McGraw-Hill 1978).

i.e., the legislative ratification of an existing innovative product, creating barriers to further technological innovation.

Overall, the innovator must be prepared for possible delays in the marketing of the improvement or technology due to state and local regulation. The impediments or barriers will usually not be insurmountable, although they may be expensive to overcome. The experience of innovators in the solar energy and other fields may make the barriers somewhat easier to scale for those who follow.

§8.05 Seeking Legislative Change in State and Local Income Taxation—In General

Most states have the authority to engage in taxation of the incomes of corporations, partnerships and individuals. There is no federal constitutional limitation affecting the taxation of the income of the residents of any particular state except possibly the very broad parameters of the due process or equal protection clauses.[43] Thus, unless a state engages in a blatantly discriminatory income tax program, there will be no federal constitutional remedy available. Today, approximately 45 states impose either a personal or corporate income tax or some combination of both.[44] In addition, states are free to authorize local governments to impose municipal income taxes on persons and corporations who are either employed or reside within the municipality. Although not as widespread as state income taxation, municipal income taxes are prevalent in several states including New Jersey, New York, Pennsylvania, Ohio and Michigan.[45]

In many situations, innovations are capital-intensive. That was particularly true for many solar energy devices. The problem with capital-intensive products is that there is an initial high cost to the consumer which is only recoupable if the consumer includes in the purchasing decision the operating costs of the product over its useful life.[46] The impact of the income tax incentive is to make the initial cost to the consumer lower and thus encourage purchase and installation of the innovative product.

While the federal government may possess the power to make the greatest income tax incentives for an innovative product or technology, one should not overlook the opportunity to seek state or local income tax credits or deductions for the proposed innovative product.

The chances for success appear to be greater at the state and local than at the federal level. For example, prior to 1973, there were no income tax credits

---

[43] New York *ex rel* Cohn v Graves, 300 US 308 (1937).

[44] *See generally* Commerce Clearing House, State Tax Guide 1001-04, 1501-10 (1979).

[45] *Id* 1032, 1514.

[46] See examples cited in Minan & Lawrence, *State Tax Incentives to Promote the Use of Solar Energy*, 56 Tex L Rev 835, 835-36 n 2 (1978).

or deductions for solar energy facilities.[47] Yet by 1980, some 20 states had enacted some form of state income tax credit or deduction designed to encourage the installation of solar energy devices.[48] Even with that success the number of proposed bills and legislative proposals that did not meet with success cannot be ignored.[49]

Before an innovator invests the time, effort and money needed to seek a legislative change in the state or local income tax laws, the potential problems should be studied carefully. One initial issue should be whether the innovator would be benefited most from an income tax deduction or a credit. A second issue is whether the income tax is a significant inhibitor of the innovative product or technology. Finally the innovator must determine whether the definitional problem inherent in limiting the application of the incentive to the targeted groups of innovative products or technologies is too difficult to overcome.

§8.06 —Deductions and Tax Credits

There are two basic types of income tax incentives, the deduction and the credit. The income tax deduction is normally considered less effective than the income tax credit.[50] Income tax deductions operate by allowing taxpayers to deduct some percentage of the cost of a solar energy facility from adjusted gross income for that year.[51] Several states have enacted income tax deductions for solar energy devices, but they were normally considered part of a larger package of solar energy tax incentives.[52]

There are, however, several limits on the effectiveness of the income tax deduction. Normally, only those taxpayers who itemize deductions will be able to take advantage of the tax deduction. Secondly, the tax deduction is usually most beneficial to those in the highest income tax brackets who may or may

---

[47] US Dept of Housing & Urban Development, HUD Solar Status 1 (1978). For a discussion of the general problem of innovation and the solar energy field, See **§8.04.**

[48] Solar Energy Research Institute, A Survey of State Approaches to Solar Energy Incentives 14-17 (1979). For recent changes in income tax incentives for solar energy devices, *see, e.g.*, ND Cent Code §57-39-01.8 (1980); 1981 Mont Laws ch 266. Each volume of the Solar Law Reporter, first begun in 1979, provides a scoresheet for the success or failure of state legislation affecting solar energy utilization. *See, e.g., Summary of State Legislation,* 3 Solar L Rep 294-355 (1981).

[49] *See, e.g., Summary of State Legislation, supra* note 48.

[50] Minan & Lawrence, *State Tax Incentives to Promote the Use of Solar Energy,* 56 Tex L Rev 835, 849-50 (1978); American Bar Foundation, Overcoming Legal Uncertainties About Use of Solar Energy Systems 80 (1978).

[51] Solar Energy Research Institute, Legal and Institutional Implications of Providing Financial Incentives to Encourage the Development of Solar Technologies 8 (1979). *See generally* Federal Energy Administration, II Solar Heating and Cooling of Buildings, Commercialization Report, Opinions and Strategies pt A, 101 (1977).

[52] *See, e.g.*, Ariz Rev Stat Ann §43-123.37 (West supp 1979); Idaho Code §63-3022C (cum supp 1980); NM Stat Ann §§72-15A011.3 to .4.

not need the added incentive to overcome the initial higher price for the innovative product or technology. In addition, some states do not use income tax forms which allow for itemized deductions.[53]

On the other hand, one of the leading reasons for utilizing the income tax deduction is its simplicity of administration once the definitional problems are overcome. In states in which itemized deductions are authorized, the taxpayer need merely add another deductible item in order to take advantage of the tax incentive.[54]

In addition to providing income tax deductions for purchasers, a state or local income tax could also provide income tax deductions for the producers of the innovative technology or product. Other energy industries such as the oil and gas industry have traditionally received preferential federal legislative tax deduction treatment.[55] These deductions could be geared to specific types of investments although a persistent definitional problem might discourage passage of such tax relief measures.

Another constraint on income tax deductions is a deduction's impact if it is restricted to individuals rather than corporations. In most cases relating to solar energy devices the states have limited the deduction to individual purchasers for residential purposes only.[56] Thus, apartment house developers would not receive this added incentive to install solar energy devices to meet the apartment house's energy needs. Another problem with corporate or business deductions is the usually deleterious impact on the depreciation of the equipment used.[57] A countervailing benefit occurs because the deduction normally allows the corporate purchaser a shortened amortization period which will create an incentive for the corporate purchaser to choose the innovative product or technology.[58]

The book is still open on the income tax deduction as a valuable incentive for the innovative product. Clearly, a number of states have adopted income tax deductions for the homeowner who purchases certain solar energy devices in the belief that it will encourage the development and use of solar energy technologies and products in the residential market.

While income tax deductions raise more questions than they resolve, some commentators feel that income tax credits are the most effective income tax

---

[53] American Bar Foundation, *supra* note 50, Solar Energy Research Institute, *supra* note 51, at 8.

[54] Solar Energy Research Institute, *supra* note 51, at 8.

[55] Arthur D Little, Inc, Residential Solar Heating and Cooling Constraints and Incentives, A Review of the Literature 57 (1976).

[56] Solar Energy Research Institute, A Survey of State Approaches to Solar Energy Incentives 20-21 (1979). For a general discussion of innovation and solar energy, see **§8.04.**

[57] *See, e.g.,* Mass Ann Laws ch 63, §38H(b)(b) (Michie/Law Co-op 1978).

[58] Minan & Lawrence, *supra* note 50, at 850-51. *See, e.g.,* Ariz Rev Stat Ann §43-123.37(a) (West supp 1979).

incentive available for the solar energy industry.[59] Income tax credits operate differently than income tax deductions. Normally, the credit allows the taxpayer to deduct from his or her total income tax liability for a given year some specified percentage of the costs of the targeted innovative product or technology. Unlike the situation of the deduction, the individual taxpayer's bracket will not affect the availability of the credit. All taxpayers, low-income, middle-income or high-income, will receive the same tax advantage, although in most cases it will be the upper and middle income groups which can take full advantage of it.[60] In most states, the income tax credits are based on a combination of a specified percentage of the total cost of installing the solar energy device with a maximum-dollar cap on the credit.[61]

§8.07 —Other Possible Forms of Tax Credits

In addition to the standardized form of tax credit, several commentators have suggested other forms which would be more effective. The first method is to allow a taxpayer to reduce annual tax liability by a fixed percentage of the mortgage or debt costs of the solar energy device. The advantage of this system is that it is more than a one-shot credit and covers the continuing costs of the facility, especially if it is incorporated into a residence and increases the total cost of the mortgage loan of the person who desires to have the solar energy device in his or her home.[62]

A second alternative for the tax credit, to make it more effective, is to allow it to reduce tax liability by a greater percentage or a higher maximum limit.[63] In addition, several states have a lower maximum credit or percentage of expenditures for nonresidential individual purchasers.[64] The removal of those limits for the business or multifamily unit purchaser would undoubtedly also spur greater use of the solar energy devices.

A fourth alternative to increase the effectiveness of the income tax credit is to alter the general rule that carryovers of credits from year to year are not allowed. Thus, where the credit may exceed the tax liability for the taxpayer some states would not allow the taxpayer to carry over that credit to future

---

[59] *Minan & Lawrence, supra* note 50, at 843-44.

[60] *Solar Energy Research Institute, supra* note 51, at 9.

[61] *See, e.g.,* Alaska Stat §43.20.039(d)(4); Ariz Rev Stat Ann §§43-128.03 to .04 (West supp 1979); Hawaii Rev Stat §235-12.

[62] Arthur D Little, Inc. Residential Solar Heating and Cooling Constraints and Incentives, A Review of the Literature 57 (1976); Bezdek, Margolin, Sparrow, Sponsler, Ezra, Spongberg, Miller, Meeker, Roseman & Misa, Analysis of Policy Options for Accelerating Commercialization of Solar Heating and Cooling Systems 50 (1977).

[63] Arthur D Little, Inc, *supra* note 62. *See, e.g.,* Alaska Stat §43.20.039(d)(4); Ariz Rev Stat Ann §§43-128.03 to .04 (West supp 1979); Hawaii Rev Stat §235-12. *See also* Solar Energy Research Institute, A Survey of State Approaches to Solar Energy Incentives 14-17 (1979).

[64] *See, e.g.,* Cal Rev & Tax Code §17052.5(a)(3) (West 1979); ND Cent Code §57-38-01.8 (supp 1980).

years where income would be available.[65] Only Hawaii allows unlimited carryover of the tax credit from year to year.[66]

A complicating factor that the innovator must consider in proposing income tax credits is the effect the credit will have on the investor's basis. In most cases, the purchaser of a residential solar energy device who receives a tax credit will have the basis of that device lowered by the amount of the credit which upon the sale of the house will be translated into a higher capital gain for the taxpayer.[67]

For the business purchaser, the income tax credit is treated as an investment tax credit. The results should be the same in the commercial sector as they are for the individual purchaser. The commercial purchaser will be provided significant tax advantages to purchase the targeted product or technology which should encourage its development.[68]

In addition to the business purchaser, investment tax credits may be available for the producers of the targeted innovative product. This credit could be on either research and development or certain depreciable machinery or equipment.[69] Again, the oil and gas industry has received substantial investment tax credits from the federal government governing machinery and equipment. Similar credits are also available at the state level.[70] The expected result from the investment tax credit for the producer is the lowering of the manufacturing cost of the product, which should result in a lower initial cost to the consumer for that innovative product.[71]

Another form of investment tax credit that may aid the smaller innovator has been adopted in Indiana.[72] This tax incentive provides both a 30 per cent credit and tax-free dividends to investors in publicly held corporations who provide capital to small businesses or small business investment corporations. There is a maximum tax credit of $5 million.

§8.08 —Proposing Legislation and the Problem of Definitions

In drafting income tax incentive legislation, a basic problem that must be overcome is definitional in nature. In certain cases, the definitional problem

---

[65] Ariz Rev Stat Ann §43-128.03(c) (West supp 1979); Kan Stat Ann §79-32.166 (1977).

[66] Hawaii Rev Stat §235-12(a) (1976).

[67] *See generally* Minan & Lawrence, *State Tax Incentives to Promote the Use of Solar Energy,* 56 Tex L Rev 835, 847 (1978).

[68] Solar Energy Research Institute, Legal and Institutional Implications of Providing Financial Incentives to Encourage the Development of Solar Technologies 8, 9 (1979).

[69] Federal Energy Administration, II Solar Heating and Cooling of Buildings, Commercialization Report, Opinions and Strategies pt A, 103, 119 (1977).

[70] Solar Energy Research Institute, *supra* note 68, at 9.

[71] Federal Energy Administration, *supra* note 69, at pt A, 119.

[72] Wall St J, Oct 21, 1981, at 1.

may be simple, but in most situations, such as solar energy where the innovative product and technology is really a multiplicity of sometimes similar and sometimes dissimilar products, how the legislation includes or excludes these products will be critical.

For example, in defining solar energy devices that might be subject to the income tax incentive there are at least seven different types of systems which may or may not be covered.[73] These would include active solar heating and cooling facilities for buildings, passive solar energy architectural designs, hot water systems, wind energy systems, biomass-related systems, photovoltaic cells or hydroelectric facilities.[74] In certain cases it might be advantageous to bring in other related innovative products and technologies in order to broaden the support for the legislation. Yet at some point too much will be added to the legislation so as to threaten its passage.

One approach that might be taken to alleviate the definitional problem at the legislative level is to delegate specific rule-making authority to the state or local taxing agency which can make a case-by-case decision as may be appropriate. The laundry list approach sometimes taken by state legislation is probably the least desirable in that there are inevitably innovative technologies or products that are omitted, either intentionally or unintentionally, which is counterproductive. In addition, the listing of specific products or technologies does not leave room for future innovations which may achieve the same social policy that caused the legislature to create the original tax incentive.

Perhaps the best thing the legislature can do is enact a statute which provides in clear terms the policy behind the provision and how it is to be interpreted by the agency responsible for administering the tax system.[75] A good example is provided by the Arizona definition of solar energy for purposes of tax incentives. It provides: " 'solar energy device' means a system or series of mechanisms designed primarily to provide heating, to provide cooling, to produce electrical power, to produce mechanical power, or any combination thereof. . . ."[76]

Definitional problems need to be worked out prior to the submission of the draft bill. Consensus-building among supporters of the tax bill is important to the ultimate success or failure of any legislative proposal. Legislatures are most successful in declaring basic policies. As many as possible of the definitional issues should be given to administrative agencies so that even newer and more innovative products and technologies can take advantage of the proposed tax incentives in the future.

Innovative products and technologies often need special tax advantages in order to be price-competitive with the products and technologies they seek to

---

[73] For a general discussion of innovation and solar energy, see **§8.04.**

[74] Solar Energy Research Institute, A Survey of State Approaches to Solar Energy Incentives 14-17 (1979).

[75] Minan & Lawrence, *State Tax Incentives to Promote the Use of Solar Energy*, 56 Tex L Rev 835, 838-43 (1978).

[76] Ariz Rev Stat Ann §43-123.37(e) (West supp 1979).

replace. In some instances, state and local income taxes may be a substantial barrier or hindrance to the widespread commercialization of these products. As shown with the solar energy model, legislative amendments to existing state and local income tax programs are possible where there is a confluence between the innovative product and some public policy, however defined.

§8.09 State and Local Property Taxation

Local governments rely on the ad valorem property tax for a substantial part of their revenue. State governments are gradually moving away from reliance on the property tax although it still is an important part of most state revenue schemes.

In most cases, the individual state constitutions provide the basic rubric for the ad valorem tax.[77] Most property tax systems fall into one of two basic categories, those in which all property is taxed unless specifically exempted, and those states in which only enumerated property is taxable.[78]

The lion's share of all ad valorem tax systems is the taxation of real property. All state and local governments which employ the ad valorem property tax provide for taxation of real property.[79] Where innovative products are offered and become part of the real property, they normally assume the value of that real property. That leads to a higher assessed valuation, which in turn leads to higher tax payments. The result in the solar energy field is a disincentive to install solar energy devices. At least one study has shown that some 25 per cent of the operating costs of solar heating and cooling devices could be attributed to the property tax.[80] Given this impediment it is not surprising that many states have provided some sort of property tax relief for solar energy devices.[81]

Similar problems arise in the case of property tax incentives as arise with income tax incentives. Although property tax credits are sometimes given,[82] the primary form of property tax incentive is to exempt the value added by the solar energy device from the property's assessed valuation.[83] In some cases, constitutional amendments are required because of the uniformity provisions

---

[77] *See, e.g.*, Tex Const art 8, §1.

[78] For a summary of state property tax laws, see Commerce Clearing House, State Tax Laws 2005 *et seq* (1979).

[79] Environmental Law Institute, Legal Barriers to Solar Heating and Cooling of Buildings 158 (1978).

[80] Solar Energy Research Institute, Legal and Institutional Implications of Providing Financial Incentives to Encourage the Development of Solar Technologies, 14 (1979). For a general discussion of innovation and solar energy, see **§8.04.**

[81] Solar Energy Research Institute, A Survey of State Approaches to Solar Energy Incentives, 8-11 (1979).

[82] *See, e.g.*, Kan Stat Ann §79-45a-1; SD Comp Laws Ann §§10-6-35.8 to .18.

[83] *See, e.g.*, Colo Rev Stat §§39-1-104, 39-5-105; Ind Code §6.1-1-12 to 26.

contained in most state constitutional property tax sections.[84] These uniformity provisions normally require all kinds of property to be taxed equally with no special treatment allowed for a loss of taxable property. The administration of the exemption can be simple if the solar energy device is simply not assessed rather than being assessed and then exempted.[85]

Definitional problems also arise where property tax incentives are used. Most states give the property tax break to both commercial and residential property.[86] Which of the solar technologies that are affixed to realty receive that incentive is a matter of legislative discretion. As with the income tax incentives, the state and local government is free to include or exclude any of the various forms of solar energy devices including active and passive heating and cooling devices, photovoltaic or wind systems.[87]

The most common form of exemption is for the entire value of the solar energy device.[88] Other states use various formulas that are not as advantageous to the property taxpayer.[89] Certain states provide further impediments to the use of the tax exemption by requiring the taxpayer to affirmatively apply for the exemption.[90] The more effective view is to make the exemption automatic upon installation of the qualifying product.[91]

Generally, property tax exemptions are viewed as administratively inexpensive, while not greatly decreasing the tax revenue of the state or local governmental body. These incentives are seen as a relatively easy way to provide further impetus to the development of solar energy devices.[92]

Certain types of innovative products or technologies may run into attempts to include such products in the state or local governmental property taxation scheme. That was the case with the computer software industry which arose in the 1970s.[93] Most states only apply the property tax to tangible personal property. When software was first introduced, most state and local property tax administrative agencies labeled it as tangible, and thus subject to the property tax.[94] Through judicial challenges to these administrative rulings, the software owners were able to overturn the findings that software was tangible

---

[84] *See, e.g.,* Tex Const art 8, §1.

[85] Thomas, Miller & Robbins, Overcoming Legal Uncertainties About Use of Solar Energy Systems 79 (1978).

[86] The only exceptions are Louisiana, La Rev Stat Ann §47-1706 (West), and Nevada, *Nev Rev Stat* §361.

[87] Compare Hawaii Rev Stat §246-34.7 with Ind Code §6.1-1-12-26.

[88] Solar Energy Research Institute, *supra* note 81, at 7.

[89] *See, e.g.,* Nev Rev Stat §361; Vt Stat Ann §53-15.

[90] *See, e.g.,* Mich Stat Ann §7.7(4e); NH Rev Stat Ann §72.61-64.

[91] *See, e.g.,* Minn Stat Ann §273-11 (West); NC Gen Stat §105-277(g).

[92] Solar Energy Research Institute, *supra* note 80, at 14-15.

[93] For a complete discussion of the taxation of software, see Comment, *Software Taxation: A Critical Reevaluation of the Notion of Intangibility,* 1980 BYU L Rev 859.

[94] *See* 2 Computer L Serv (Bigelow) apps 2-3.2e, .2d (1979); Ariz Rev Stat Ann §42-1301.

property.[95] Again, the exception from property taxation is a positive factor that affects the marketability of innovative technology. Following in the footsteps of the software industry, litigation may provide relief from overreaching state and local property tax administrators who seek to enrich governmental treasuries at the expense of the innovator.

§8.10 Sales, Use and Other Tax Incentives

State and local governments are presently the sole units capable of imposing sales or use taxes. A sales tax is normally imposed upon the sale of a covered product within the taxing jurisdiction. A use tax is a tax on the sale of a covered product which is going to be used within the taxing jurisdiction but is actually purchased outside the boundaries of that jurisdiction.[96] All states except Delaware impose some combination of sales and use taxes.[97] Local governments in most jurisdictions can piggyback onto the state sales tax by levying an additional percentage for sales within that municipality. In most cases, the state administers and collects sales and use taxes and remits the city's share on a regular basis.[98]

To the extent that sales and use taxes add to the initial purchase price of an innovative product which may already lack price competitiveness, sales and use tax exemptions may provide an incentive for the purchase of such products. Again, in the solar energy field several states have enacted sales and use tax exemptions for certain solar energy devices.[99] In most states, amendments to state statutes dealing with sales tax are easier to manage than property tax amendments because there are fewer constitutional requirements for uniformity regarding sales or use taxes than for property tax.[100] In most cases, sales and use taxes are not set high enough to make an exemption very effective in assisting the market for an innovative product with a large price tag. State sales taxes usually range in the 4 to 5 per cent range with cities normally being able to tack on another 1 to 2 per cent for local revenue.[101] Thus, at least with the solar industry, the sales and use tax exemptions have been seen as part of an overall tax package with the sales tax exemption playing a rather minor role in the larger picture.[102]

Where the sales tax exemption has been utilized, eligible products as defined

---

[95] See cases cited in *Comment, supra* note 93, at 361 n 3.

[96] *See generally* Due, *The Nature and Structure of Sales Taxation,* 9 Vand L Rev 123 (1956); Oldman & Schoettle, State and Local Taxes and Finance 444-549 (1974).

[97] Commerce Clearing House, State Tax Guide 6021 (1979).

[98] Oldham & Schoettle, *supra* note 96, at 463.

[99] Solar Energy Research Institute, Legal and Institutional Implications of Providing Financial Incentives to Encourage the Development of Solar Technologies 16 (1979).

[100] *Id.*

[101] Commerce Clearing House, State Tax Guide 6021 (1979).

[102] Solar Energy Research Institute, A Survey of State Approaches to Solar Energy Incentives 24 (1979).

by the statute are totally exempt from the payment of sales tax.[103] The same definitional problems arise here as they do for the property and income tax incentives although the states that have adopted sales tax exemptions have been somewhat more liberal in allowing exemptions from the sales tax than they are with the other tax incentives.[104]

Another possible tax exemption that might directly affect the producers of innovative technologies or products is an exemption from the state franchise tax. For example, Texas allows a corporation which is exclusively engaged in the manufacture, sale or installation of solar energy devices to be exempt from the state's rather minimal franchise tax.[105] Relief in such a limited form is not likely to create much of an incentive. Subsidiaries of larger corporations engaged in innovative research would not be eligible for the tax exemption nor would smaller companies which have a diversified product or service line.

A final type of tax exemption might occur if the innovative product is subject to one of several traditional areas of state excise taxes such as fuel or tobacco. In response to the energy crisis, some 25 states passed legislation exempting gasohol, a combination of ethanol and gasoline, from the state fuel tax that normally applied to the sale of gasoline.[106] Again, the purpose was to make the innovative product, in this case gasohol, price-competitive with gasoline. Gasohol needed the additional incentive because the biomass-created additive to the traditionally refined gasoline cost more to produce per gallon than the gasoline itself. While the results have been mixed, the removal of the state excise tax on fuels undoubtedly increased the sale of gasohol by making its price closer to gasoline although apparently the innovative product's price has never been able to match the standard gasoline prices.

§8.11 State and Local Prohibitions against Innovative Products

In certain instances, state and local governments may have constitutional provisions, statutes or ordinances which directly prohibit an activity, thereby having an impact on an innovative product or technology. In those cases, the only way for the innovator to market a product is to seek a change in the constitutional provision, state statute or municipal ordinance. Obviously it will be easier to change or alter a municipal ordinance than a state constitutional provision. The impact of a state-wide rather than a local change, however, will be greater in creating or opening up potential markets for the distribution of an innovative product or technology.

Changes at the state constitutional level are not impossible, although they

---

[103] *See, e.g.,* Ariz Rev Stat Ann §42-1312-01(A)(9); Mich Stat Ann §§7.525(8), 7.555(4e).

[104] Solar Energy Research Institute, *supra* note 102, at 25-26.

[105] Tex Tax Code Ann tit 122A, §12.03(1) (r) (Vernon).

[106] US National Alcohol Fuels Commission, Fuel Alcohol, An Energy Alternative for the 1980s 50 (1981).

may be time-consuming and expensive. Nonetheless, the time and money spent in removing these absolute barriers to an innovative product may be shared with others interested in the innovation or may be small when compared to the research and development costs of the product itself. A recent state constitutional change in Texas will illustrate how an innovative product can be marketed even where there is some substantial direct impediment in the way.

Prior to 1979, the Texas Constitution and statutes prohibited branch banking in any form.[107] This prohibition had been in force since 1904. During the 1970s, computer technology had been utilized to develop unmanned bank teller machines which would allow for 24-hour transactions by bank clients. In Texas, the use of these machines was directly prohibited by the Texas Constitution and state statutes because each unmanned teller machine would be considered a branch bank.

Because the user group for the innovative product was convinced that unmanned teller machines would be profitable, they sought to have the constitutional and statutory prohibitions against branch banking removed.[108] The effort began in late 1976 and early 1977 just before the start of the biennial session of the Texas legislature. Success was not achieved until late 1980.

The problems that were facing the marketing of the innovative product were substantial. Not only would the legislature have to amend a statute, it would also have to adopt a constitutional amendment which would then have to be approved by the voters in a special election. The user group in this instance was also not unified in support of the policy of eliminating the branch bank prohibition. Nonetheless, under the aegis of several of the user group trade associations, a bill and constitutional amendment were drafted and submitted to the legislature for approval in 1977. The state legislature then both enacted the statutory change and agreed to have the constitutional amendment placed on the ballot. The statutory changes were made conditional on the approval by the people of the constitutional amendment. There was no organized effort by any of the user groups or innovators to influence the electorate and the amendment was defeated in late 1977 by approximately a 2-1 margin.

The trade associations were not deterred and again held a series of meetings in order to hammer out a consensus among the disparate elements within the association. The result in 1978-1979 was a newly drafted constitutional amendment and statutory changes which were again submitted to the 1979 biennial Texas legislature. The legislature approved both the amendment and the statutory changes and authorized an election to vote on the constitutional

---

[107] Tex Const art 16, §16; Tex Rev Civ Stat Ann art 342-903 (Vernon), repealed in 1979.

[108] The following narrative is gleaned from a conversation between the author and Leonard Pasmore, Esq., Executive Director of the Texas Bankers Association, who was directly involved in the effort to have unmanned teller machines legalized in Texas.

amendment.[109]

Learning from their past failures, the trade associations created a public relations fund to inform the electorate about the virtues of unmanned teller machines. They had also, through consensus building, eliminated almost all of the internal opposition to the change in branch banking rules. Armed with approximately $200,000, plus locally spent funds, the trade associations were able to make the case for unmanned teller machines with the electorate. In 1980, the electorate voted to change the constitution and the statutory provisions were thus made effective.[110]

This example illustrates how a well-organized and financed lobbying and public relations effort can successfully change a constitutional provision which prohibits the introduction of an innovative product into the marketplace. In this case, it was not the innovators themselves who had to lead the fight, but the more powerful and influential user groups for the innovative product. Having created a product with the potential of providing better service and higher profits to a user group, the innovators could basically stand on the sidelines and allow the user group to eliminate those barriers to commercialization.

§8.12 Direct Governmental Intervention: Building Codes—In General

State and local governments engage in a varied set of direct regulatory schemes through permits, licenses and registration requirements. These regulatory schemes include such diverse programs as state registration and regulation of commercial feed,[111] state registration for commercial fertilizer,[112] state and local housing and building code licensing schemes[113] and state regulation of activities which produce air, water, solid waste or other forms of pollution.[114]

One problem with a permit or licensing program is its failure to readily accept new and innovative products or technologies. This may be caused by the use of specificity standards. Under these programs, in order to receive a

---

[109] 1979 Tex Sess Law Serv A-54; 1979 Tex Sess Law Serv 1187-1190. The legislation was adopted virtually without opposition. *Id.*

[110] Tex Const art 16, §16; Tex Rev Civ Stat art 342-903A.

[111] *See, e.g.,* Ark Stat Ann §78-701 *et seq;* Neb Rev Stat §54-820.01.

[112] *See, e.g.,* Ga Code §5-1001 *et seq;* Ohio Rev Code Ann §905.31 *et seq* (Page).

[113] Field & Ventre, *Local Regulation of Building, Agencies, Codes and Politics,* in Municipal Yearbook (International City Management Assn 1971).

[114] All 50 states have some form of environmental agency or agencies which enforce both the state segment of the federal pollution statutes and the state statutes which may be more stringent. *See generally State Air Laws, State Water Laws, State Solid Waste-Land Use Laws,* Envt Rep (BNA) (1981).

permit, only the enumerated products or technologies may be used.[115] More recently, however, regulatory programs have begun to utilize performance standards. These entail the creation of objective criteria, allowing the private party to determine how and with what materials and techniques those standards should be met.[116]

In addition to licensing and permit programs, state and local governments can regulate through a registration system. These require that a product or technology be registered prior to use, sale or distribution. The feed and fertilizer statutory schemes fit within this rubric. In these situations, there may or may not be minimum standards imposed prior to getting the registration. In certain situations the act of registration and disclosure is sufficient; no standards are imposed.

The building code regulatory system provides the innovator with a typical governmental regulatory system which has historically prevented innovation.[117] It has also been the subject of investigation and scrutiny even before it was directly involved in the solar energy boom of the past few years.[118] A review of the problems associated with building code regulations will provide an overview of how many state or local regulatory programs can have a deleterious impact on innovation. It also illustrates how the problems in dealing with state and local regulatory schemes can be overcome, or at least made minimally intrusive into the plans to market innovative products or technology.

Building codes are not of recent origin. They have been around since pre-Revolutionary days in urban areas.[119] Building codes were originally deemed to be matters solely of local interest. As a result almost every local government enacted some form of building code.[120]

This leads to problems in meeting different local requirements in order to have a sufficiently large market area for the innovation. There has been a recent trend, however, toward state-wide building code regulation, either covering a limited segment of the building industry, as with energy conservation

---

[115] See story on prescriptive nature of Chicago building code which was hindering the development of solar energy devices in the city. 2 Solar L Rep 885-87. *See also* Advisory Commission on Intergovernmental Relations, Building Codes: A Program for Intergovernmental Reform (1966).

[116] Gardiner & Lyman, Decisions for Sale—Corruption and Reform in Land-Use and Building Regulation 10 (1978).

[117] For a complete study of the burdens placed on the innovator by the building code system, see Field, *Judicial Solutions to the Tension Between Construction Codes and Innovation*, 4 Urb L & Poly 107 (1981) *See also* Oster & Quigley, *Regulatory Barriers to the Diffusion of Innovation: Some Evidence from Building Codes*, 8 Bell J Econ 361 (1977); President's Commission on Urban Problems, Building the American City (1968).

[118] See references cited in note 117 *supra*. For a general discussion of innovation and solar energy, see **§8.04.**

[119] *Field, supra* note 115, at 113.

[120] Field & Rivkin, The Building Code Burden 43 (1975).

standards, or total building code regulatory schemes.[121] While local control over building codes has been the general rule, there has developed in the field the use of several model codes, developed by private organizations.[122] While the adoption of model codes may have lessened the problem of meeting different standards in different cities, there is still not that much uniformity in the building code system.

There are several reasons for this lack of uniformity in building codes which makes it more difficult to market an innovative product or technology. The first is that neighboring cities or cities within the marketing area have the freedom to adopt any one of the four codes. There are also separate codes dealing with individual trades such as plumbing, electricity, mechanical or fire codes. In addition, the cities are free to adopt all or part of any of the codes and may omit or add additional material as they deem necessary. Furthermore, changes to the model codes adopted by the code-creating organization may or may not be adopted by the local government. All of these conditions still make for a highly balkanized building code system.[123] Where states have enacted building standards, a higher degree of uniformity may be achieved, although under most circumstances the state codes are designed to set only minimum standards. Thus, local governments have the right to impose more stringent standards.[124] Over all, local governments still retain substantial authority to choose from any of the model codes or develop their own codes using strictly local criteria.

One commentator has suggested that the building code regulatory scheme has four component parts, all of which may have a deleterious impact on the innovator.[125] These four components include the substantive provisions of the regulatory scheme, the provisions for modifying or updating the scheme, the enforcement of the scheme and the personnel who implement it. To that we may add a fifth item, the role of the courts in reviewing decisions made regarding the regulatory scheme.[126] These five components of the building code scheme are usually transferable in whole or in part to other direct regulatory schemes that involve permits, licenses or registration requirements imposed by the state or local governmental body.

---

[121] See statutes listed in National Institute of Building Sciences, States' Energy Conservation Standards for New Construction: State Legislative Analysis (1978).

[122] Field & Rivkin, *supra* note 120, at 34-35. The four major codes are the Uniform Building Code produced by the International Conference of Building Offices; the Basic Building Code, developed by the Building Officials and Code Administrators, International, Inc.; the Southern Standard Building Code developed by the Southern Building Codes Conference; and the National Building Code developed by the National Board of Fire Underwriters. *Id.*

[123] *Id* 35-36.

[124] Comment, *The Home Buyers' Protection Act: An Alternative to Building Codes for Single Family Homes,* 54 S Cal L Rev 529, 543-44 (1981).

[125] *Id* 531-36.

[126] *Field, supra* note 117.

§8.13 —Substantive Provisions

The nature of the substantive regulations may provide the greatest initial hurdle for an innovator. Normally, the standards can be of two varieties, specificity standards or performance standards. Specificity standards require that a particular product, technology or system be utilized with no exceptions. Examples would be requirements of metallic pipe, 2 x 4 wood studs, spacing of studs and so forth.[127] The second type of standard normally involves the setting of performance criteria in which the builder has the choice of any number of techniques and equipment to use as long as the minimal performance standards are met.[128]

Obviously, the innovator would prefer that the performance standard type of regulatory scheme be used since it would allow flexibility in designing and marketing the innovative product or technology. For example, instead of requiring that a wall contain a certain amount of wood studs at a specific distance, the performance standard would merely state that the wall must be built so as to meet certain load-bearing criteria.[129] Unfortunately, most local building codes and other regulatory schemes utilize specificity and not performance standards although the trend is toward the greater use of performance standards.[130] Where the innovative product is incompatible with the specificity standard or where the performance standard is not translatable given the innovation, the innovator will be stymied in its marketing efforts. Delay will be a factor that must be considered since the innovator will have to resort to the second leg of most regulatory schemes, namely the provisions for modifying or updating the scheme.

Most police power regulatory schemes which provide for licensing, permitting or registration normally contain a variance or exception policy. In addition to the variance policy, most regulatory schemes also can be updated, revised, amended or repealed by subsequent legislative or administrative actions. Local or state legislation can only be changed by subsequent legislative acts. On the other hand, administrative policy can be changed by either legislative or administrative acts. In either case, the innovator may be able to work with the administrative variance or legislative amendment process in order to market an innovative product or technology. Recent experiences with building codes and solar energy provide examples of how the system, while appearing to be flexible, operates so as to stifle innovation.[131]

---

[127] Oster & Quigley, *Regulatory Barriers to the Diffusion of Invention: Some Evidence from Building Codes,* 8 Bell J Econ 361, 368-69 (1977).

[128] Sanderson, Codes and Code Administration 15 (1969).

[129] Field, Building Regulatory Practices and the Courts 15 (1980).

[130] *Id. See also* Seidel, Housing Costs and Government Regulations 71-99 (1978).

[131] Comment, The Home Buyers' Protection Act: An Alternative to Building Codes for Single Family Homes, 54 S Cal L Rev 529, 532-33 (1981); Field, *Judicial Solutions to the Tension Between Construction Codes and Innovation,* 4 Urb L & Policy 107, 118-22 (1981). *But cf* Mara & Engel, *Institutional Barriers to Solar Energy: Early HUD Demonstration Experiences,* 1 Solar L Rep 1095, 1104-07 (1980).

In response to demands for flexibility in building codes, many local governments have adopted equivalency standards which allow the innovator to make a special showing that its technology is the equal of the performance standards or specificity standards embodied in the building code.[132] All four of the major model building codes have an equivalency provision.[133] The administrative agency that deals with building code matters is the normal decision-making body for the equivalency decision. Although this will undoubtedly aid innovators, there are impediments to its full use which destroy the flexibility inherent in this type of variance or exception provision.

An initial problem that the innovator runs into is what is equivalent. While the general standard is considered to include such factors as "suitability, strength, effectiveness, fire resistance, durability, safety and sanitation,"[134] the innovator is not given any specific guidelines. As such, the innovator is in a sea of uncertainty with a difficult burden of persuasion with the local building code official to show that it has met the equivalency provisions.[135] In addition, the performance standards which are set may be difficult to translate into equivalency provisions given an innovative product or technology. Another impediment to the use of equivalency provisions is the performance standard which may be set too high because of outdated information or industry pressure.[136]

§8.14 —Modification of the Regulatory Scheme

Besides utilizing the administrative variance or equivalency scheme, the innovator may seek to legislatively amend the statute, ordinance or administrative regulations embodying the substantive regulations. As with the prohibitory regulation, change is possible but may be difficult. In certain cases, the state legislature would have to be the target of a well-financed and documented lobbying effort in order to change the substantive rules. As we have seen with the banking statutes[137] and the recent spate of state statutes dealing with solar energy,[138] that can be accomplished although at some substantial delay in marketing the innovative product.

State legislative changes must be carefully planned. Affected interest groups, legislators and legislative staffs must be contacted and informed of the problems facing the innovator. Broadening the support for the innovative product and technology beyond those with a direct pecuniary interest in the innovation will undoubtedly be important, as has occurred with solar energy legislation.

---

[132] Field, *supra* note 131, at 120-21.

[133] See code provisions cited in *id* 115, 152-53 n 47.

[134] *See, e.g.,* Uniform Building Code 105 (International Conference of Building Officials 1979).

[135] Field, *supra* note 129, at 27-28.

[136] Comment, *supra* note 131, at 533.

[137] See **§8.11.**

[138] *See, e.g.,* Utah Code Ann 63-9-46; Nev Rev Stat §341-260(3).

Identification of potential opponents to the measure prior to the lobbying effort will also minimize delay and unnecessary conflict.

With local regulations, the chance for administrative or legislative change may be greater and less time and capital-intensive. An interesting problem with the building codes is that many local governments are hesitant to change the model codes unless the issuing organization first makes such changes. Therefore, the innovator has two potential targets for lobbying efforts, the model code-setting organization and the local government.[139] In this case, however, convincing the code-setting organization does not necessarily guarantee local adoption but it will undoubtedly carry a persuasive if not controlling weight among the local legislators.

An additional problem in seeking changes to an existing regulatory scheme is that the status quo situation will probably be supported by those receiving benefits under the present standards.[140] The battle over the case of plastic pipe illustrates the difficulty an innovator may incur in the attempt to introduce a new product which is a compatible substitute for an existing product.[141] Again, planning can diminish the impact of potential opponents by the development of countervailing strategies and arguments. The opposition will not go away, but at least two sides may be heard. Innovators are also well-advised to spend some time with local governmental officers and employees. Time spent with staff-level personnel will undoubtedly benefit the innovator in the long run. In addition, presentations and meetings with local chambers of commerce, labor unions, potential user groups and other community leaders will undoubtedly make the path to change smoother. While nothing can guarantee that the local government will make the requested legislative or administrative changes, the likelihood of success will increase with time spent dealing with the local people on a one-to-one basis. Studies on the building code industry suggest that local officials tend to disfavor new and innovative ideas. Making those officials feel more comfortable with the innovative product or technology should lead to more fruitful results.

§8.15 —Enforcement Mechanism

The third element of the basic regulatory program involving licenses or permits is the enforcement scheme. In most situations, the person seeking to

---

[139] Comment, *The Home Buyers' Protection Act: An Alternative to Building Codes for Single Family Homes,* 54 S Cal L Rev 529, 533 (1981); Field, *Judicial Solutions to the Tension Between Construction Codes and Innovation,* 4 Urb L & Poly 107, 120 (1981).

[140] *Comment, supra* note 139, at 539.

[141] The plastic pipe battle pitted the plastic industry against the Cast Iron Soil Pipe Institute and labor unions. See Seidel, Housing Costs & Government Regulations 90-91 (1978); Field & Ventre, *Local Regulation of Building Agencies, Code and Politics,* in Municipal Yearbook 139, 145 (International City Management Assn 1971); Oster & Quigley, *Regulatory Barriers to the Diffusion of Innovation: Some Evidence from Building Codes,* 8 Bell J Econ 361, 370-77 (1977).

market the innovative product or technology will have to apply for local or state governmental permission, either showing compliance with the substantive standards or showing that the applicant is qualified for an exception or variance under the terms of the regulatory scheme. It is here that the problem of local discretion, especially in the variance provision, raises its ugly head.

In the building code arena, several studies have shown that code enforcement officials are hesitant to approve variances or exceptions or to make equivalency findings.[142] The burden to show that the product or technology is equivalent is on the innovator. In many cases, as occurred with solar energy devices, the code official will not be aware of the system being offered and its technological or engineering components. A public relations strategy like the one described above might overcome some of this hesitancy and meet the burden placed on the person or organization seeking a change from the status quo.

§8.16 —Implementation

The enforcement element blends with the fourth element of any regulatory scheme, namely the personnel who are the implementers of the governmental policy. Again, while generalizations are hard to make regarding any type of civil servant, several studies in the building code field have drawn an unflattering view as to the competence and honesty of the personnel who implement the building code regulatory scheme.[143] Even assuming an honest and competent bureaucracy, the innovator faces substantial problems. Given the general tendency for public employees to be paid at a lower scale than their counterparts in private industry, some commentators have argued that there is great unevenness in the implementation of building codes and other municipal regulatory schemes.[144] Although not every municipal employee or municipal department will fit the general pattern found prevalent in the building code enforcement infrastructure, the innovator still faces some difficult hurdles to cross before it can successfully market a product or technology in the face of this strong bias in favor of not rocking the boat and staying with that which has already been approved and used by one's peers in the private sector.[145]

---

[142] Field & Rivkin, The Building Code Burden ch 3 (1975); Bunnell & Kahn, Removing Obstacles to Building Reuse and Community Conservation at the Local Level 51-52 (1979); Field, *Judicial Solutions to the Tension Between Construction Codes and Innovation,* 4 Urban L & Policy 107, 120-21 (1981).

[143] Field & Ventre, *Local Regulation of Building Agencies, Code & Politics,* in Municipal Yearbook 139 (International City Management Assn 1971).

[144] Administration of Building Regulations (Claiborne ed 1973); Gardiner & Lyman, Decisions for Sale—Corruption and Reform in Land-Use and Building Regulation (1978).

[145] Comment, *The Home Buyers' Protection Act: An Alternative to Building Codes for Single Family Homes,* 54 S Cal L Rev 529, 535-36 (1981).

§8.17 —Judicial Review

The discretion granted to the local governmental official in deciding whether to grant a variance or exception or otherwise administer the regulatory program has a great impact on the fifth component of the regulatory mechanism having a deleterious impact on the innovator, namely, the role of the courts in reviewing local legislative or administrative decisions. Judicial review of local legislative or administrative actions is generally available. The scope of judicial review for most state or local administrative decisions is that they will be upheld unless they were arbitrary, capricious or an abuse of discretion, in violation of state statutory or constitutional limits, or in violation of federal constitutional limits.[146]

In the building code field, there are a number of reported cases in which the courts have overturned rejection by local officials of an equivalency argument on the grounds that the rejection was arbitrary, capricious or an abuse of discretion.[147] While a number of innovators have met with success, the obstacles that must be overcome by an innovator in these cases are very difficult. Typically, local or state actions are clothed with a presumption of validity. This is especially true with respect to legislative decisions. The plaintiff must shoulder a heavy burden of proof to show that the local administrative action was arbitrary, capricious or an abuse of discretion. For legislative actions, substantive review on the merits of the regulatory program is very difficult. For local ordinances involving nonhome rule cities, one can occasionally argue that the local legislative act was ultra vires, meaning that it was not authorized by a state enabling statute.[148] In addition, litigation is a slow, time-consuming process. The end result might also be a pyrrhic victory, in that the court will invalidate the prior governmental action but still not grant the required permit, license or registration.[149]

§8.18 —Conclusion

The problem inherent in regulatory programs involving direct intervention in the marketplace through licensing, permit or registration requirements is that they tend to stifle innovation. Although most of the regulatory systems will

---

[146] *See generally* Bruff, *Judicial Review in Local Government Law: A Reappraisal,* 60 Minn L Rev 669 (1976). *See, e.g.,* Boise Cascade Corp v Gwinnett County, 272 F Supp 847 (ND Ga 1967); City of Hutchinson v Otto, 306 Minn 36, 235 NW2d 604 (1975); Bennet v Ca Board of Appeal, 268 Mass 419, 167 NE 659 (1929).

[147] *See, e.g.,* Kingsberry Homes v Gwinett County, 248 F Supp 765 (ND Ga 1965); City of Montgomery v Robbins, 277 Ala 29, 166 So 2d 869 (1964); Johnson Constr Co v Township of White Lake, 351 Mich 374, 88 NW2d 426 (1958).

[148] *See generally* Field, *Judicial Solutions to the Tension Between Construction Codes and Innovation,* 4 Urb L & Poly 107, 123-51 (1981).

[149] *See, e.g.,* Casey v Zoning Hearing Board, 459 Pa 219, 328 A2d 464 (1970). *See also* Pa Stat Ann tit 53, §§10609.1, 11004(1)(b). *But cf* Sinclair Pipeline Co v Village of Richton Park, 19 Ill 2d 370, 167 NE2d 406 (1960).

provide for variances or exceptions, the variance process is normally not an easy road to follow. Administration of these systems also tends to support the status quo to the detriment of new and different products and technologies. Again, while the barriers exist to the marketing of these innovative products, sufficient planning and careful attempts to work with local officials, as discussed above, whether they be legislative or administrative in nature, will undoubtedly diminish the delay in the successful market entry of the target innovative product or technology. Judicial review must be seen as a last resort to be taken when all other measures have failed.

§8.19 Direct Governmental Intervention: Land Use Controls—In General

Another pervasive form of direct governmental intervention which may have a deleterious impact on innovation is the system of local land use controls. Traditionally, local governments have engaged in land use regulation in the form of zoning and subdivision control ordinances which initially were fairly inflexible, specificity-type regulatory programs.[150] Zoning and subdivision controls are even more balkanized than building code regulations, in that each locality has traditionally developed its own land use controls pursuant to state enabling or home rule charter authority.[151] Unlike the building code scheme, there are no model zoning or subdivision ordinances that are widely applied.

The focal point of the zoning system is the control of the use, bulk and height of buildings within a city. The regulatory scheme involves the creation of zones or districts within the city where minimal requirements for use, bulk and height are individually regulated.[152] Zoning controls are regulated by the building permit process. Subdivision control, on the other hand, deals not with the building process per se, but with the development process whereby raw parcels of land are divided into separate lots for sale to potential users, whether they be residential, commercial or industrial purchasers.[153] Subdividers must normally have their plat, or map of their subdivision, approved by a local governmental agency before they can begin to actually sell the individual lots. Subdivision and zoning controls may overlap and, normally, permits and approval from both the zoning authority and the subdivision authority are required.[154] Subdivision ordinances normally require the developer to dedicate needed improvements such as streets, sidewalks, curbs, paving and lighting.

---

[150] For a general history of zoning, see Hagman, Urban Planning and Land Development Control Law 66-99 (1971).

[151] The Department of Commerce in 1926 published the Standard Zoning Enabling Act and Standard Planning Enabling Act which were later adopted by most states as the basic authority for cities to zone themselves. See Hagman, *supra* note 150, at 80-82.

[152] *Id* 102-15.

[153] *Id.* 245-53.

[154] *Id* 251-52.

In addition, the subdivision ordinance may have minimum lot size and alley requirements.

The combination of zoning and subdivision control can limit innovative technology that relates to building. This has occurred with solar energy devices. While land use barriers to solar energy have been deemed to be not as obstructive as building codes and the tax system, they can provide some problems to the innovator prior to completion of a successful marketing campaign for an innovative product or technology.[155] For example, height restrictions might limit special types of roof-top solar collectors, antennae, chimneys, water tanks and similar products which are attached to the roof of a house or commercial structure.[156] Zoning and subdivision ordinances cannot be ignored lest they impose substantial and unexpected costs for the marketing of an innovative product or technology.

§8.20 —Substantive Provisions

Land use controls, being a direct form of governmental intervention in the marketplace, may be analyzed using the same five component parts as was done with building codes.[157] Zoning controls over use, height and bulk, and subdivision controls regarding the platting of parcels can involve substantive regulations of either the specificity or performance variety. Traditionally, zoning and subdivision controls utilize specificity standards. Specified uses, limited height and bulk regulations, minimum lot size, minimum setback and sideyard requirements are all specificity-type substantive standards. As such they must be met or the innovative product or technology cannot be built or developed.

Early land use controls were seen as being rigid in nature and not very capable of dealing with new concepts and ideas.[158] Thus, as new development and building techniques were being created, the substantive regulations prevented their widespread use. Such diverse new ideas as planned unit developments, cluster-type housing, floating zones, earth-sheltered housing and zoning to guarantee access to sunlight have faced difficult paths to acceptance because of the substantive regulations of most local land use control programs. Limited flexibility has been incorporated into the land use control system in response to innovative products and technologies.[159]

---

[155] Spivak, Land-Use Barriers and Incentives to the Use of Solar Energy (Solar Energy Research Institute 1979).

[156] *See, e.g.*, Miller, *Solar Energy and Land Use in Colorado,* 6 Envtl L Rep (Envtl L Inst) 50039, 50042 (1977).

[157] See **§§8.12-8.18.**

[158] *See, e.g.*, Gillespie, *Industrial Zoning and Beyond: Compatibility Through Performance Standards,* 46 J Urb L 723 (1969); Rockhill v Chesterfield Township, 23 NJ 117, 128 A2d 473 (1957).

[159] Hagman, Urban Planning and Land Development Control Law 117-22 (1971). *See also* Comment, *Contract and Conditioning Zoning: A Tool for Zoning Flexibility,* 23 Hastings

While there is growing acceptance of the use of performance standards in certain types of zoning matters, the bulk of land use controls today is still clearly a matter of specificity standards which must be met.[160] Thus, innovators have sought to utilize the second component of direct regulatory systems, namely, the use of administrative variances or legislative amendments to achieve their goal of marketing their innovative product or technology.

While there is great diversity among municipalities as to the use of specificity standards for land use controls, there is a fairly uniform treatment of administrative variances common to most of the municipal zoning legislation. This is due in part to the use of the Standard State Zoning Enabling Act which provides for two different types of administrative exceptions, the first being a variance and the second a special exception.[161] A variance is the allowance of an otherwise prohibited use or bulk or height regulation because if the regulation would apply it would cause unnecessary hardship.[162] The variance is to be granted by an administrative agency, normally called the board of appeals or zoning board of adjustment. Typically, the administrative board will rely on the expertise of the municipal planning and zoning staff regarding the variance application.

It is important for innovators not to file a petition for a variance without first going through negotiations and meetings with the local governmental staff. Typically, rough edges may be worked out and compromises developed so that the path to a variance will be somewhat smoother. One would normally stand a better chance for approval of a variance if the city's staff either supported the variance or at least did not vigorously oppose it during the hearings on the petition.

Variances may be granted in two forms, one a use variance and the other a bulk variance. Typically, the use variance is the more difficult to receive because it involves the greater departure from the specificity standards contained in the ordinance.[163] The administrative history of variances is that they are fairly easily given even where the legislative standards for variances have not been met.[164]

In seeking a variance, the administrative body will normally consider such factors as the effect on the public, the effect on the neighborhood, the unique characteristics of the property in question and the plaintiff's showing of unnecessary hardship.[165] A great deal of discretion is afforded the zoning board of adjustment in granting a variance although there are a number of standard factors that are typically considered. In almost all cases, there will be a formal

---

LJ 825 (1972); Aloi, *Legal Problems in Planned Unit Development: Uniformity, Comprehensive Planning Conditions, and the Floating Zone,* 1 Real Est LJ 5 (1972).

[160] Gillespie, *supra* note 158; 3 Williams, American Land Planning Law 244-46 (1975).

[161] Hagman, *supra* note 159, at 194-95.

[162] *Id* 195.

[163] *Id* 196-97.

[164] *Id.*

[165] *Id* 197-205.

hearing before the administrative board prior to the decision in which any interested party may comment on the proposed variance.

The other form of administrative relief available from a zoning ordinance is called a special exception, or special use permit, or special permit. The special exception differs from a variance in that the zoning ordinance itself will contain the specific standards that will allow a party to seek a different treatment.[166] For example, a commercial use zone may allow certain uses as of right, meaning that a permit will be issued without question if the developer states that the building will be put to that use. Alternatively, the zoning ordinance may provide that other types of uses can only be permitted in a commercial zone after the administrative body has had a chance to review that particular permit application and determine that the proposed use will not have a deleterious impact on the neighborhood in question.[167] Typically, the special exception will be accompanied by negotiated conditions which will be designed to minimize the deleterious external effects of the proposed use. Again, negotiation with the local planning staff will be important in the ultimate success or failure in the attempt to have a special exception issued.

§8.21 —Modification of the Regulatory Scheme

Zoning and subdivision ordinances are also subject to legislative changes through the local city council. Traditionally, city councils have been amenable to changes in the zoning district lines where a developer has requested a change in order to construct some beneficial improvement. In addition to this traditional form of rezoning, the zoning or subdivision ordinances can be changed to amend the substantive rules, allow for different types of special exceptions or create new zoning use, height or bulk districts. It is important in dealing with the local legislature to have the confidence of the local planning staff upon whom the legislators rely for their expertise in land use matters. Thoughtful planning and negotiations with local business persons, persons in the affected neighborhood and potential users of the innovative product or technology will again make the chances of success much greater.

§8.22—Enforcement and Implementation

The third and fourth components of a direct regulatory system relating to personnel and enforcement can be analyzed together. In most cases, municipalities have independent planning staffs composed of planning professionals who provide the city with expertise in the land use control business.[168] In most cases, the administrative bodies are staffed by local citizens without

---

166 *Id* 206-07.

167 *Id* 208. *See also* Kotrich v County of DuPage, 19 Ill2d 181, 166 NE2d 601 (1960).

168 *See, e.g.*, Mandelker, *Deregulation of Power and Function in Zoning Administration*, 1963 Wash ULQ 60; Dukeminier & Stapleton, *The Zoning Board of Adjustment: A Case Study in Misrule*, 50 Ky L Rev 273 (1962).

land use expertise. Both the administrative bodies and the local legislature rely on the professional planning staff for counsel and advice. As such, it is important for persons seeking to work within the system to become acquainted with the professional staff. In addition, the staff is usually responsible for the imposition of conditions on certain types of permits, variances or exceptions. As such, having a good working relationship with the staff will inure to the benefit of the innovator. Good public relations efforts among the city's business people including the chamber of commerce will also assist the innovator in achieving an objective, removing impediments to the marketing of an innovative product or technology. This would be especially true for legislative changes, where the local chambers of commerce will have more influence than over the appointed administrative bodies.

§8.23 —Judicial Review

There is a well-defined system for judicial review of local land use decisions. While each state has its own system and its own scope of judicial review, it is not an overgeneralization to conclude that the decision of the local legislature or administrative body will be difficult to overturn.[169] As with the judicial review of building codes, a presumption of validity will attach to the municipal order.[170] The party attacking the judgment of the legislative, or administrative body normally has the difficult burden of proof to show that the decision was unreasonable or arbitrary, capricious or an abuse of discretion.[171]

The chances for success are relatively slim and in most cases, the result will not be the granting of the required permit or variance but merely the invalidation of the legislative or administrative decision. The developer or builder then must persuade the municipality to give him or her what was originally requested although the municipality is not under a court order to grant the specific request of the winning party.[172]

---

[169] In Cow Hollow Improvement Club v DiBene, 245 Cal App2d 160, 53 Cal Rptr 610 (Ct App 1966), neither the court nor the attorneys could find one appellate decision in the state which had overturned the granting of a variance.

[170] The landmark case is Village of Euclid v Ambler Realty Co, 272 US 365 (1926), which upheld zoning as not violative of any federal constitutional protection. It also set the stage for a general hands-off attitude towards judicial review of local zoning decisions. *See, e.g.*, Levy, *Judicial Review of Zoning Cases—New Rules,* 6 Tulsa LJ 1 (1969).

[171] *See, e.g.*, Puritan-Greenfield Improvement Assn v Leo, 7 Mich App 659, 153 NW2d 162 (1967); Fritts v City of Ashland, 348 SW2d 712 (Ky Ct App 1961). *But cf* Fasano v Board of County Commrs, 264 Or 574, 507 P2d 23 (1973).

[172] *See, e.g.*, Casey v Zoning Hearing Bd, 459 Pa 219, 328 A2d 464 (1974); Ellick v Board of Supervisors, 17 Pa Commw 404, 333 A2d 239 (1975); Hyson, *The Problem of Relief in Developer-Initiated Zoning Litigation,* 12 Urb L Ann 21 (1976).

§8.24 —Conclusion

The need for a premarketing public relations campaign to acquaint local officials and staff of the need for changes to accommodate the innovative product and technology is obvious. Creating antagonistic local officials or staff will only cause delay if not total rejection of any changes that are required for the successful marketing of the target product. The more work done before the actual marketing, the better off the innovator will be in overcoming state and local regulatory barriers.

§8.25 Indirect Governmental Intervention—Warranties

Certain types of state or local regulatory programs will have an indirect impact on the successful marketing of innovative products or technologies. The impacts are indirect in that there is normally not a preclearance barrier to the marketing, such as a license or a permit, but merely the creation of some rules which the product or technology must comply with that may lead to personal liability or difficulties in the marketing arena.

The most pervasive form of indirect governmental intervention which the innovator must face is the problem of product warranties.[173] Every state but Louisiana had enacted, in complete or substantially complete form, the Uniform Commercial Code (UCC).[174] The UCC has within it two implied warranties which affect the sales of goods by merchants. The first is called the implied warranty of merchantability,[175] while the second is the implied warranty of fitness for a particular use.[176] The implied warranty of merchantability merely states that the goods must be of a merchantable quality when sold by a merchant.[177] The UCC defines *merchants* as persons who deal in the kind of goods involved and persons who hold themselves out as being knowledgeable about the goods being sold.[178]

Normally, the determination of whether a good or product meets the implied warranty of merchantability is made on an ad hoc basis. Several factors are relevant in any finding that goods are merchantable. These would include

---

[173] For a complete discussion, see Lawrence & Minan, *The Role of Warranties and Product Standards in Solar Energy Development,* 34 Vand L Rev 537 (1981). There are several federal warranty statutes which will apply to the marketing of any consumer product. *See, e.g.,* Magnuson-Moss Warranty-FTC Improvement Act, 15 USC §2301 *et seq.* These federal standards will not be discussed here.

[174] Lamm, Photovoltaic Commercialization: An Analysis of Legal Issues Affecting a Government-Accelerated Solar Industry 16 (Solar Energy Research Institute 1979); Noun, *Product Liability and Wind System Manufacturers: Proposals for Protection,* 2 Solar L Rep 73, 85-88 (1980).

[175] UCC §2-314.

[176] *Id* §2-315.

[177] *See generally* Nordstrom, Law of Sales 231-37 (1971).

[178] UCC §2-104(1). See the discussion in Nordstrom, *supra* note 177, at 82-84.

that such goods pass without objection in the trade, are of average quality, are fit for the ordinary purposes for which the goods are used and are adequately packaged and labeled.[179] The sale of any innovative product would normally carry with it the implied warranty of merchantability under state law unless it has been properly excluded under the provisions of §2-316 of the UCC.

The implied warranty of fitness for a particular purpose normally arises where the purchaser has good reason to rely on the seller's skill or judgment regarding a specific good.[180] The two key elements in showing that there is a breach of the implied warranty of fitness for a particular purpose are that the seller knows that the buyer has a particular purpose in mind for the goods and that the seller knows the buyer is relying on the seller's skill and judgment.[181]

The seller can attempt to exclude or modify either of these implied warranties. This can be accomplished by course of dealing or course of performance, or by using language such as *as is* which makes it plain that the goods are not warranted. In addition, if the buyer inspects the goods and should have discovered a defect but did not, there will be no warranty as to that discoverable defect.[182]

In addition to the implied warranties, the parties may consensually agree to any express warranties covering the quality or performance criteria of the goods being sold. In those situations, the buyer and seller are naturally bound by their agreement and if the product fails to live up to any express warranties, the buyer is entitled to any one of the UCC remedies available to a buyer.[183]

The impact of the UCC warranty provisions on the sale of innovative products is to add to the costs of the developer of innovative technology. Undoubtedly, the innovator will have to consider the potential actions by purchasers of the product for breach of warranty, whether implied or express. The existence of a warranty, however, may encourage purchasers of innovative products because they will be assured that if the product does not live up to its warranty the buyer will be entitled to relief. Obviously, no innovative product that does not meet minimum performance or merchantability levels will be successful. The warranty provisions of the UCC seek to protect the purchaser under certain circumstances from goods which do not live up to minimum standards under the agreement.

In addition to these general warranties applicable to the sale of most goods, several states have enacted legislation dealing with warranties for a particular industry such as solar energy devices[184] or new homes.[185] The purpose of these

---

[179] Nordstrom, *supra* note 177, at 232-33.

[180] *Id* 242-43.

[181] *Id* 243-44.

[182] UCC §2-316. Lamm, *supra* note 174, at 16; Nordstrom, *supra* note 177, at 269-74.

[183] UCC §2-711.

[184] *See generally* Solar Energy Research Institute, National Conference of State Legislatures, Solar Warranties Workshop: A Summary (1978); Me Rev Stat Ann tit 10, §§1491-1494.

[185] Minn Stat Ann §327A.01-.07 (West); NJ Stat Ann §§46.3B-1 to 3B-2(West).

specific warranties is to promote consumer confidence as well as protect the customer from shoddy equipment which may be prevalent in innovative product areas.[186] In several states, express warranties for solar energy products are required, and failure to include them will eliminate the tax incentives otherwise available for the purchase or installation of the innovative product.[187] In other states, sellers are subject to monetary penalties for failing to provide the required warranties.[188] Other approaches to specific warranties include merely requiring disclosures of certain information rather than mandating a warranty.[189] At least one commentator feels that mandatory warranties will retard solar energy development because the warranties will do little to instill buyer confidence or protect the buyer's investment.[190]

Warranties will usually be imposed at the state level since the state had traditionally dealt with areas of commercial transactions to the exclusion of municipalities, including home rule entities. However, a home rule entity might be able to enact warranty provisions for specific products or technologies in the absence of state legislation if it were deemed to be a matter of local concern. Thus, the innovator must be aware of any state or local attempts to impose warranties or disclosure requirements on their products or technologies because failing to do so will have a negative effect on the marketing of those innovations.

§8.26 —Product Standards

Another form of governmental regulation affecting the marketing of innovative products and technologies is the imposition of minimum product standards.[191] In this context, these standards may be a more direct intervention than the warranty because they can require that the product or technology receive approval prior to its sale or distribution.[192] Product standards achieve the same objective as product warranties, namely, the instilling of consumer confidence by providing that the product must meet some minimum objective characteristics. The problem with product standards is that they tend to be uniform and thus do not consider the newest innovation in a particular product or technology area which may not meet the prior standard. This problem is similar to the building code issue even though in most cases the product standards will be based more on performance standards than on the more

---

[186] Lawrence & Minan, *supra* note 173, at 542-43.

[187] *Id* 543 citing Cal Rev & Tax Code §23601(a)(2) and Cal Admin Code tit 20, §2601.

[188] Me Rev Stat Ann tit 10, §§1491-1494.

[189] NY Energy Law §§12-101 to-104 (Consol).

[190] Lawrence & Minan, *supra* note 173, at 544-48.

[191] *See generally* Lawrence & Minan, *The Role of Warranties and Product Standards in Solar Energy Development,* 34 Vand L Rev 537 (1981).

[192] Ashworth, Green, Pollack, Odland, Saltonstall & Perelman, The Implementation of State Solar Incentives: A Preliminary Assessment (1979).

inflexible specificity standards.[193]

As with building codes, product standards have been developed by national organizations, with a principal purpose of developing standards within their field of experience. For example, the American Society of Heating, Refrigerating and Air-Conditioning Engineers, Inc. and the American Society for Testing and Minerals have had long histories in the development of product standards.[194] However, unlike the model building codes, the private product standards, unless incorporated into building codes, have normally not been adopted at either the state or local level.

In the solar energy arena, only two states have enacted product standards legislation involving state certification and standard-setting powers.[195] In Florida, the statute created a Florida Solar Energy Center which was delegated the responsibility for developing not only the product standards but the test methods and a certification program.[196] The Florida provision is mandatory on all solar energy device vendors. Without certification, the product cannot be sold.[197]

The California program, on the other hand, is voluntary. Again, the powers to develop standards for testing, inspection and certification were delegated to a state administrative body, in this case the California Energy Commission.[198] Because of its voluntary nature, the principal purpose of the program has been one of providing public information and assurance to the consumer that at least some solar energy vendors have agreed to have their products tested and certified. Some other states have also developed regulatory programs designed to inform the public of innovative products through either public disclosure or requirements of product disclosures at the time of sale.[199]

Product standards legislation may be crucial to ultimate consumer acceptance of innovative technology. While some might conclude that product standards interfere with the operation of the free market system and tend to impede further innovation, there is some evidence that the principal impediment to the success of innovative solar products was consumer fear and lack of knowledge.[200] While product standard legislation or use of private standard-setting organizations may create some barriers to further innovation, if they

---

[193] Lawrence & Minan, *supra* note 191, at 168-70.

[194] ASHRAE was one of the first organizations to develop minimum product standards for solar energy devices, *ASHRAE Standards* 94, 95, 96, in Second National Conference on Standards for Solar Energy Use 19 (1979).

[195] Fla Stat Ann 377.705 (West); Cal Pub Res Code §25605 (West). See full discussion in Lawrence & Minan, *supra* note 191, at 573-75.

[196] Florida Solar Energy Center, Operation of the Collector Certification Program (1978).

[197] Fla Stat Ann §377.705 (West).

[198] Cal Pub Res Code §25605 (West).

[199] Minn Stat Ann §116H.127 (West).

[200] US Dept of Energy, Psycho-Economic Factors Affecting the Decision Making of Consumers and the Technology Delivery System (1978).

utilize performance rather than specificity standards they will encourage the use of existing innovative products while minimizing their deleterious impact on future innovative products.[201]

§8.27 —Worker Certification

A final indirect form of governmental regulation that may affect innovative products or technologies arises through employee certification programs. These can occur at the state or local level, although once a state program has been enacted, unless it provides for more stringent local programs, the local governments will usually be prohibited from having a duplicative licensing ordinance.

A good example of how employee or worker certification programs can inhibit the use of innovative products and technologies was recently raised in the case of *Heisse v State of Vermont.*[202] In *Heisse,* the plaintiffs were challenging the Vermont Polygraph Examiner's Act[203] which restricted the granting of licenses to engage in the truth and deception profession to those individuals trained and expert in the use of the polygraph machine.[204] The plaintiff was a medical doctor and a board-certified otolanygologist who was an expert in the innovative technique known as the Psychological Stress Evaluator (PSE). PSE was a system by which physiological events, including the voice, were studied for the determination of whether the individual being tested was telling the truth.[205]

Under the Vermont statute, Dr. Heisse was barred from engaging in the use of PSE to determine the truth or deception of a particular individual. Under the licensing scheme, only the use of a polygraph machine by a licensed technician could be used in the truth and deception profession.[206] In this case the plaintiff was arguing, among other grounds, that the statute was a denial of equal protection of the laws in that certain individuals could practice in the truth and deception profession while others could not. The court found that the regulation of professional conduct by the state was clearly within its police power. It then went on to state that the prohibition of the use of PSE in the truth and deception profession was supported by a rational basis and therefore there would be no constitutional violation.[207] The rational basis in this case was the long-standing use and knowledge about the polygraph machine versus the innovative and as yet untested knowledge and use of the PSE techniques.

Thus, it appears that a state can discriminate against the use of innovative

---

[201] Lawrence & Minan, *supra* note 191, at 584-95.

[202] 519 F Supp 36 (D Vt 1980).

[203] 26 Vt Stat Ann §2901 *et seq.*

[204] 519 F Supp at 39-40.

[205] *Id.*

[206] *Id* 45-46.

[207] *Id* 46-47.

technologies through a licensing scheme by basically requiring that only an individual expert in the settled technology receive licenses. This is a clear barrier to the advancement of innovation which must be overcome by legislative acts eliminating the employment licensing requirement.

Employee certification programs are a matter of long standing in the building trades.[208] These are traditionally run by the city, although state-wide certification is normally required for certain professions. Again, depending on the nature of the certification process, these local acts can inhibit the marketing of innovation by placing requirements on the installers of such innovative products as to make it difficult if not impossible for such installation to occur.

One of the problems with innovative products and technologies may be the lack of qualified installers. This may lead to shoddy installation, consumer fraud and dissatisfaction. In the solar energy area, these problems have been dealt with by some state and local governments through an employee certification program. These programs normally require that an installer receive a state or local permit prior to installation of any covered product.[209] Thus, before a mass marketing program for an innovative product begins, the innovator should be cognizant of potential shortages in competent installers of the product which might lead to state or local certification programs. In addition, if the innovative product is to be installed by an already regulated profession, the innovator should check to see if the certification program includes testing and competence in the installation procedures to be used on the innovative product.

§8.28 Some Limits on the Exercise of State and Local Regulatory and Taxation Authority—Federal Preemption

Because of the nature of our federal system, the states are not free to act without constraints in the regulatory and taxation field notwithstanding the language of the Tenth Amendment.[210] Although the Tenth Amendment seemingly reserves to the states all authority not otherwise delegated to Congress, it must be read in context of both the commerce clause[211] and the supremacy clause.[212] Although other constitutional constraints such as the equal protection and due process clauses also apply to state action, their impact on economic, regulatory or taxation problems is minimal unless a state has set up a blatantly discriminatory program.

---

[208] *See, e.g., In re* Means, 14 Cal2d 254, 93 P2d 105 (1939); City of Milwaukee v McGregor, 140 Wis 35, 121 NW 642 (1969).

[209] *See, e.g.,* Me Rev Stat Ann tit 5, §4004; Broward County solar heater installers certification program as reported in 3 Solar L Rep 201-02 (1981).

[210] US Const amend X.

[211] *Id* art I, §8, cl 3.

[212] *Id* art VI, §2.

Under the supremacy clause, federal statutes are considered to be the supreme law of the land prevailing over any conflicting state law.[213] Congress has, under the commerce clause, been granted "the Power . . . to regulate Commerce with foreign Nations, and among the several States. . . ."[214]

This grant of authority to Congress over interstate and foreign commerce is considered plenary in scope. It is not considered a limitation on congressional authority.[215] Thus, Congress had wide-ranging powers to regulate economic affairs. Similarly, Congress has broad powers under several different provisions of the Constitution regarding taxation.[216] If Congress has acted under either the commerce clause or the taxing clauses it may preclude state legislation concerning the same subject matter covered by state law.

This concept of preemption may occur in three different situations. The first occurs where the federal statute comprehensively deals with a subject matter and the purposes would be undermined if state regulation would be allowed.[217] The second occurs if the federal and state legislation are in direct conflict so the regulated party would be unable to comply with both schemes.[218] The third situation, which is more difficult to predict, is where preemption will be implied even in the absence of federal legislation if uniformity of regulation at the national level is considered vital.[219]

Since preemption can occur across a wide spectrum of activities, from alien registration laws[220] to truck weight limits,[221] there is no general test that can be applied to determine when and where preemption may occur. In all three of the above situations, the courts normally make an ad hoc decision based on the pervasiveness of the federal regulatory scheme, federal occupation of the field and danger of conflict between state laws and the objectives of the federal program.[222] While there has been preemption in the regulatory field, there is little effort at preemption in taxation. In the past, Congress has not intended to preempt local taxation even when the federal government has imposed a taxation program. For example, states are free to impose their own income and inheritance taxes even though there are federal income and inheritance taxes, because Congress has not intentionally chosen to preempt their taxing authority.

---

213 *See, e.g.,* Northern States Power Co v Minnesota, 447 F2d 1143 (8th Cir 1971), *affd mem,* 405 US 1035 (1972).

214 US Const art I, §8, cl 3.

215 Prudential Ins Co v Benjamin, 328 US 408 (1946).

216 US Const art I, §2, cl 3; *id* art I, §8, cl 1.

217 *See, e.g.,* Hines v Davidowitz, 312 US 52 (1941).

218 *See, e.g.,* Florida Lime & Avocado Growers, Inc v Paul, 373 US 332 (1963).

219 *See, e.g.,* Southern Pac Co v Jensen, 244 US 205 (1917).

220 Hines v Davidowitz, 312 US 52 (1941).

221 Castle v Hayes Freight Lines, 328 US 61 (1954).

222 Pennsylvania v Nelson, 350 US 497, 502-05 (1956).

§8.29 —Dormant Commerce Clause

While the commerce clause is an affirmative grant of federal authority, the courts have developed a dormant or negative side of the commerce clause. This dormant commerce clause limits the state's ability to regulate interstate commerce.[223] The dormant commerce clause only arises in the face of state action and congressional inaction. It was originally developed to fill the interstices of the Constitution to prevent the economic balkanization that was prevalent in the period between the adoption of the Articles of Confederation and the Constitution.[224] Dormant commerce clause cases only arise when states exercising their reserved police powers affect an area of interstate commerce where Congress has remained either neutral or silent.

A good example of how the dormant commerce clause operates is illustrated in the case of *City of Philadelphia v New Jersey*.[225] In this case, New Jersey had attempted to prohibit the importation of certain types of waste material into its borders.[226] The city of Philadelphia, which had been using New Jersey landfills to deposit its wastes, sued to prevent the New Jersey ban from taking effect. The Supreme Court initially disposed of a preemption argument made by the city under the Resource Conservation and Recovery Act.[227] Finding no express intent to preempt state laws and no need to occupy the field, the Court rejected the preemption argument.[228]

In dealing with dormant commerce clause attacks on state regulatory schemes the Supreme Court has developed two basic tests. In situations where the principal purpose of the state regulation is economic protectionism the Court has applied a *virtually per se* rule of invalidity.[229] In these cases, the issue which is resolved is whether the state regulatory enactment is basically a protectionist measure or is merely a scheme with incidental burdens on interstate commerce while implementing a primarily local purpose.

In situations where the per se invalidity test does not apply, the United States Supreme Court has adopted a much more flexible test to determine the validity of the state statute. The leading modern case which articulates that test is *Pike v Bruce Church, Inc.*[230] In *Pike,* the issue surrounded the validity of an Arizona statute which mandated that certain fruits grown in the state be packed inside that state. Justice Stewart speaking for the Court enunciated the test:

---

[223] *See, e.g.,* Southern Pac Co v Arizona, 325 US 761 (1945); Pike v Bruce Church, Inc, 397 US 137 (1970).

[224] Nowak, Rotunda & Young, Constitutional Law 244 (1977). *See also* Tribe, American Constitutional Law 319-30 (1978).

[225] 437 US 617 (1978).

[226] NJ Stat Ann 13-1I-10 (West).

[227] 42 USC §6901 *et seq.*

[228] 437 US at 620-21 n 4 (1978).

[229] *Id* 624.

[230] 397 US 137 (1970). *See also* Dowling, *Interstate Commerce & State Power,* 27 Va L Rev 1 (1940).

> Although the criteria for determining the validity of state statutes affecting interstate commerce has been variously stated, the general rule that emerges can be phrased as follows: Where the statute regulates even-handedly to effectuate a legitimate local public interest, and its effects on interstate commerce are only incidental, it will be upheld unless the burden imposed on such commerce is clearly excessive in relation to the putative local benefits. If a legitimate local purpose is found then the question becomes one of degree. And the extent of the burden that will be tolerated will of course depend on the nature of the local interest involved, and on whether it could be promoted as well with a lesser impact on interstate activities.[231]

This is basically a balancing test with a study of less onerous or burdensome alternatives. The court will first balance the local interests with its burden on interstate commerce. If the burdens are small, the state regulatory scheme will be upheld. If the burdens pass a certain undefined threshold, then the court will determine if less onerous means were available to the state to achieve the same objective. If the state could have achieved the same objective with a program that was less burdensome on interstate commerce the state program must fall.

In the state taxation arena the federal system provides a series of checks on state taxation of items in interstate commerce or located outside of the taxing jurisdictions. No state can apply its ad valorem tax to real property located outside its jurisdiction.[232] As for personal property, the same rule applies for tangible personal property as real property.[233] However, the Supreme Court has allowed potential multiple taxation of intangible personal property by holding that it may be taxed in either the domicile or residence of the owner or its situs.[234] In addition, there are some limitations on the taxation of items used in interstate or foreign commerce.[235]

State income taxes are not as limited by various constitutional limitations as is the state or local property tax. The Supreme Court has upheld the concept of taxing nonresidents who either work or earn other types of income within the income taxing jurisdiction.[236] The real hornet's nest comes in the imposition of state and local taxes on unitary interstate businesses. While state and local governments must apportion their income or gross receipts or franchise

---

[231] 397 US at 142 (case citations omitted).

[232] Union Refrigerator Transit Co v Kentucky, 199 US 194 (1905).

[233] *Id.*

[234] *Id* 205-06.

[235] *See, e.g.*, Pullman's Palace Car Co v Pennsylvania, 141 US 18 (1891); Japan Line Ltd v County of Los Angeles, 441 US 434 (1979). *But cf* Michelin Tire Corp v Wages, 423 US 276 (1976) (upholding a nondiscriminatory local ad valorem tax on imported goods).

[236] New York *ex rel* Cohn v Graves, 300 US 308 (1937); Shaffer v Carter, 252 US 37 (1920).

taxes to take into account the interstate nature of the businesses being taxed, there is some confusion as to what apportionment tests will pass constitutional muster.[237] One must be aware of these potential limits, although because of the recent upheaval in the traditional rules regarding taxation of interstate commerce, the area is in a complete state of flux.[238]

The Supreme Court has also recently upheld substantial state severance taxes on coal[239] and nonresident fees for hunting[240] as not violative of either the dormant commerce clause or the privileges and immunities clause. However, in another case, *Maryland v Louisiana,*[241] the Court, on dormant commerce clause grounds, struck down a state tax primarily imposed on gas sold in interstate commerce as opposed to that sold and used within the state.[242] State taxation of interstate commerce is an area undergoing change which must be followed carefully in the future before the limits of state and local taxation will be readily defined.

§8.30 Conclusion

This chapter's purpose was to address certain limitations on the state's power to impose regulatory and taxation schemes that may have an impact on innovation. These limitations can only be argued after a decision has been made to litigate the validity of a statutory or regulatory scheme. As such it might be somewhat of a last-gap alternative since litigation should probably be viewed only as a last resort by the innovator. Nonetheless, the dormant commerce clause and supremacy clause as well as the limits on state taxation might provide sufficient ammunition to either succeed in the litigation or cause a reflection and change in the regulatory or taxation scheme which is deleteriously affecting the innovator.

---

[237] *See, e.g.,* Complete Auto Transit, Inc v Brady, 430 US 274 (1977); Moorman Mfg Co v Bair, 437 US 267 (1978); Department of Revenue v Association of Wash Stevedoring Cos, 435 US 734 (1978). *See generally* Lockhart, *A Revolution in State Taxation of Commerce,* 65 Minn L Rev 1025 (1981).

[238] See Lockhart, *supra* note 237; Hellerstein, *State Income Taxation of Multijurisdictional Corporations: Reflections on Mobile, Exxon and HR 5076,* 79 Mich L Rev 113 (1980).

[239] Commonwealth Edison Co v Montana, 453 US 609 (1981).

[240] Baldwin v Montana Fish & Game Commn, 436 US 371 (1978).

[241] 451 US 725 (1981).

[242] *Id* 730-31.

State and Local Promotion of Innovation 9

Frank J. Mauro

State Government and Industrial Innovation: Origins of the Relationship

§9.01 State Mercantilism and Respect for the Contributions of Science

A growing number of states and localities are pursuing efforts to promote industrial innovation and others are likely to join them in the near future. While state and local governments have only recently begun undertaking specific programs for the explicit purpose of promoting industrial innovation, such efforts have many precursors in American economic history.

Since the earliest days of this nation, state and local governments have been involved in the promotion in industrial innovation. These early efforts, while not initiated in the name of innovation, are suggestive of the kinds of initiatives that state and local governments can undertake to promote innovation. Moreover, many of the industrial innovation programs begun by individual states in the nineteenth century continue to assist many individual businesses and certain entire industries in the development and utilization of new technologies.

During the first 50 to 60 years of our nation's history, state government played a very active role in the economy. The best known examples of this are the roles played by state governments in the financing of major internal improvements, such as highways, canals and railroads, which were undertaken in order to facilitate commerce and industry. Less well-known are early efforts to promote the growth of emerging new enterprises, including those based on innovative products and processes. These efforts were part of a state mercantilism which, during the eighteenth century and the first half of the nineteenth century, dominated American thinking about the role that government should play in the economy.[1] As part of this approach, for example, "many states undertook subsidy programs: New York paid bounties to salt producers, Louisiana and Alabama to ship building firms, and many other states to farmers who were experimenting with new techniques such as silk culture."[2]

---

[1] *See, e.g.,* Friedman, A History of American Law 157-178 (1979):

> By reputation, the 19th century was the high noon of *laissez-faire*. Government, by habit and design, kept its heavy hand off the economy. But recent historical research and explication have shown 19th-century developments in a different light. Certainly, during much of the century, there was a great deal of public sympathy for business, productivity, growth. Particularly, the first half of the century was a period of promotion of enterprise, of the release of creative energy; government, reflecting its powerful constituencies, wanted business and the economy to grow; where this required subsidy or intervention, no overriding theory held government back.

Id 157.

[2] Scheiber, Vatter & Faulkner, American Economic History 103 (9th ed 1976). See Hartz, Economic Policy and Democratic Thought (1948), for a discussion of one state's efforts to promote silk production. The records of the Pennsylvania auditor-general,

As the nineteenth century progressed, there developed in the United States a widespread appreciation of the role that scientific advances and applied research and development could play in industrial development, and government became involved in supporting education and research efforts which were important to industrial growth but which individuals and corporations could not afford to underwrite.[3]

§9.02 Scientific Agriculture and Dissemination of Innovation

America's predominant industry during the nineteenth century was agriculture, and both the federal government and most state governments undertook to foster research and education in what was then being called scientific agriculture. The states were at the forefront of this movement and, prior to 1862 when the federal government first acted, Florida, Georgia, Massachusetts, Michigan, Missouri, New Jersey and Pennsylvania had already established colleges of agriculture. Furthermore, despite the fact that an American version of the laissez-faire philosophy was then emerging to challenge, and eventually to eclipse, the initially dominant federalist-mercantilist philosophy, the federal

---

as cited by Hartz, indicate that the following payments were made under an 1838 act authorizing bounties for silk production:

1839	$ 568.44
1840	2,010.89
1841	4,418.55
1842	6,716.77
1843	3,425.76
1844	62.71
1845	18.00

Id 333.

[3] Rasmussen, *The Impact of Technological Change on American Agriculture, 1862-1962,* Journal of Economic History 22 (1972). *See generally* Mansfield, Technological Change (1971). Mansfield's discussion of the need and rationale for programs of federal support for research and development applies to state programs as well. He argues that such programs frequently

> are designed to offset imperfections in the system that would otherwise lead to an under-investment in R&D. For example, the value of an improvement due to R&D may be less to the individual citizen than to the community as a whole. . . . Industrial firms almost certainly will invest less than is socially optimal in basic research. This is because the results of such research are unpredictable and usually of little direct value to the firm supporting the research, although potentially of great value to the society as a whole. With regard to agricultural research, federal support could be defended originally on the grounds that the smallness of firms in farming would result in smaller R&D expenditures than would be socially desirable.

Id 120-21.

and state governments continued to expand their support for agricultural education and research. As the need for research in this field became more widely understood, states began to establish agricultural experiment stations. Eighteen states had set up such stations by 1887 when the Hatch Act[4] established a program of federal support for experiment stations and other agricultural research. Most of the research was conducted at state universities with additional support from state treasuries being used to augment the Hatch Act funds. It has been noted that "the resultant research in the state land-grant colleges was marked by a rich diversity that matched the diversity of agriculture itself."[5]

Government also became a facilitator of industrial innovation both through efforts to disseminate the findings of the research it was sponsoring in such areas as pest control and fertilizer techniques and to introduce new ideas such as the importation and distribution of new plant varieties.[6] As early as 1843, the New York Assembly authorized the State Agricultural Society "to employ a practical and scientific farmer to give public lectures throughout the State."[7] As more scientific research was undertaken as a result of the Hatch Act, the role of dissemination of innovation efforts grew in importance as a means of carrying the findings of this research to the people.[8] Much of this dissemination function was accomplished through the fieldwork of county agents or cooperative extension agents. The agricultural experiment stations had conducted such activities from the mid-1870s until 1912, when the agricultural extension services of various states began to organize.

In 1908, a committee of the Association of American Agricultural Colleges and Experiment Stations requested federal legislation to foster extension work. It was not until 1914, however, that a bill, which came to be known as the Smith-Lever Act,[9] was finally signed into law. This act was intended "to foster agricultural extension in the states and to coordinate the extension work of the land-grant colleges and the USDA [United States Department of Agriculture], both of which had been active in certain parts of the country."[10] The Cooperative Extension Service, which was organized pursuant to this act, received financial support from the USDA on a national basis, state support through the land-grant colleges and universities (which received both federal and state support pursuant to the Morrill Act of 1862[11]) and local support through county governments.

While such education, research and extension programs have a long history,

---

[4] Act of Mar 2, 1887, ch 314, 24 Stat 440 (codified at 7 USC §§361a-i).

[5] Scheiber, Vatter & Faulkner, American Economic History 218 (9th ed 1976).

[6] *Id.*

[7] Works & Morgan, The Land Grant Colleges 47 (US Department of Agriculture, Staff Study No 10, 1939).

[8] *Id* 48.

[9] Act of May 8, 1914, ch 79, 38 Stat 372–74 (codified at 7 USC §§346-348).

[10] Works & Morgan, *supra* note 7, at 50.

[11] Act of July 2, 1862, ch 130, 12 Stat 503-05 (codified at 7 USC §§301-308).

they are still an important resource today and should not be overlooked by public policymakers or by business people. The Geneva Agricultural Experiment Station, for example, which was initially authorized by the New York state legislature in 1881, is still in operation today with an annual budget of approximately $8 million and a staff of about 250 scientists, other professionals and laboratory and technical support personnel involved in both basic and applied research. Significantly, the state of New York, from its own resources, provides approximately three-fourths of the total funding for the Geneva program. The state contribution was $5.75 million in 1979. Also significant is the relationship between the station's research program and the private sector. During the 1977-1978 fiscal year, the station received $328,155 in research support grants from 56 corporations and industrial consortia for such projects as "evaluation of certain new organic fungicides for fruit disease control in the greenhouse," "elaboration of biochemical, microbiological and engineering knowledge in the field of enology," and "research and development for uses of processing waste from the shellfish processing industry to be examined for use as an industrial and technological enzyme source."[12] The station is also involved in research and development leading to the introduction of new products and processes. For example, the station has worked with growers to develop new products such as new grape varieties and productivity-enhancing processes such as the mechanical pruning of grapevines.[13]

§9.03 Early Promotion of Innovation in Other Key Industries: The Cases of Mining and Textiles

While state support of industrial innovation has traditionally been most prominent in agriculture, similar efforts have been undertaken in other fields over the years. In states where other industries were as central to local economies as agriculture was in most states, institutions focusing on research and training in those fields were established. The Michigan Mining School provides a good example of the close relationship between such efforts and innovation in a locally important industry. The rush to Michigan's copper country in 1844 was the first major mining boom in the United States. The area flourished, cities grew and foundries and factories producing mining machinery were established. There also developed an awareness of the advantages that came from applying modern technology to the activities of the area's mining and mining-related industries. Through the efforts of a local leader, the state legislature was moved to establish the Michigan Mining School in 1885. By 1887, the school had become the Michigan College of Mines and 12 years after that it had 30 faculty members and 266 students and was an important

---

[12] New York State Agricultural Experiment Station, Ninety-Seventh Annual Report 29, 31, 32 (1979).

[13] *Id* 1-36.

asset to the economy of this mining area.[14] Now called the Michigan Technological University, this university still works very closely with industry, but in a broader spectrum of areas than in its early years. Industry contracts with the university for research and problem-solving work and the university also proposes projects on industrial problems which faculty members detect in the course of other research. For example, the M-1 tanks had welding-related problems which Chrysler Corporation asked the university to solve and, in response to a university initiative related to the presence of squeaky brakes in certain cars, Chrysler funded the work to develop solutions to the problems. Dissemination of research findings is accomplished through an annual distribution of research reports to a mailing list of interested businesses and through the publication of findings and other articles in professional journals.[15]

In other mining states, schools specializing in mining technology were also established. The Colorado School of Mines, which now produces 25 per cent of all of the bachelor of science degrees in mining engineering awarded each year in the United States, was established in 1874.[16] The New Mexico School of Mines (now the New Mexico Institute of Mining and Technology) was founded in 1889,[17] and four years later, the Montana School of Mines (now the Montana College of Mineral Sciences and Technology) was established.[18]

All of these schools are still involved very intimately with the mining and related industries. The Colorado School of Mines, for example, works to disseminate information on new technologies and new processes through an extensive schedule of short courses and special programs in cutting-edge subjects. One such special program of this type is an oil shale symposium which has been gaining in popularity each year.

Another example of this technology transfer process is the School's Research Institute (CSMRI) which was created as a research arm of the college and exists today as an autonomous distinct organization. The president of the Colorado School of Mines is also the Chief Executive of the CSMRI, with funding for the institute derived from fees for services. The CSMRI is very involved with industry as a consultant and contractor and usually concentrates on quick turnaround projects, whereas the school deals with long-term, less intense research efforts. Fifty per cent of the work done at CSMRI involves developing processes to recover mineral values from ore (for example, recovering cobalt, nickel or other minerals from ocean modules), with the other 50 per cent of its work divided between geological research for the mineral re-

---

[14] Dunbar, Michigan: A History of the Wolverine State 498-99 (1965).

[15] For further information, contact Michigan Technological University, Houghton, Mich 49931.

[16] Colo Rev Stat §§23-41-102 through 23-41-122.

[17] 1889 NM Laws ch 138 (codified at NM Stat Ann §§21-11-1 through 21-11-27).

[18] Mont Rev Code Ann §75-8407.

source industry and environmental technology.[19]

Another key industry in the history of America's economy was the textile industry. Not only was textile production important to the self-sufficiency of the new nation, but it was the first industry in America to introduce modern methods as part of the industrial revolution. The first school to provide instruction in branches of the textile business other than design was established in Philadelphia in 1884. This was followed shortly thereafter by the Lowell Textile Institute located in Lowell, Massachusetts which eventually became the state-supported Lowell Technological Institute and which combined with Lowell State Teachers College in 1975 to form the University of Lowell, at which time the textile department was phased out of existence for several reasons. The textile industry in Massachusetts has declined substantially, total enrollment has decreased and the enrollment of foreign students has increased significantly, making it difficult for the state to rationalize continuing support for a program that was, in a sense, educating the nation's competition with taxpayers' dollars.[20]

Two other institutes established in Massachusetts to assist the textile industry were the New Bedford Textile School which began in 1899 and the Bradford Dunphy Textile School which opened in 1904 in Fall River. These institutions were consolidated in 1964 forming the Southeastern Massachusetts Technological Institute (since then renamed Southeastern Massachusetts University) which now conducts educational, research and extension programs in a number of technological and engineering fields, including textiles.

The textile department at Southeastern Massachusetts University is primarily a teaching institution and the limited work that it does do directly with industry is on a small, local scale and generally deals with analysis and problem solving. Research at the textile department centers around chemicals in the textile industry (dying, finishing, printing, etc.) and also involves conversion of fibers to various yarns and yarns to fabrics. Papers are given and published in professional journals in order to disseminate the research information.[21]

Textile education in the South originated through the efforts of Daniel Augustus Tompkins. Tompkins first planned to establish a private textile school, but as this project soon appeared impractical, he turned to promoting the establishment of textile schools under the control and support of state governments. He began with his native state of South Carolina, rallying public support for a textile department at Clemson College. He was soon invited by the trustees of the college to present his views concerning the proposed textile department. During his presentation, Tompkins stated that

---

[19] For further information, contact Office of Research, Colorado School of Mines, PO Box 112 Golden, Colo 80402.

[20] For further information, contact University of Lowell, One University Avenue, Lowell, Mass 01854.

[21] For further information, contact Textile Department, Southeastern Massachusetts University, North Dartmouth, Mass 02747.

> Such a department at Clemson College would undoubtedly contribute to diversifying the manufacture of cotton goods in the state, and would probably at an early date bring about the establishment of one or more bleacheries and finishing works. Besides being of advantage to the youth of the state in qualifying them for profitable employment a textile school should be of great value in stimulating the development of the resources of the state in the line of textile manufacturers.[22]

The trustees decided to establish the Clemson Textile School in 1898. The school cost about $30,000, of which $15,000 was given by the state.[23] Clemson College has evolved into Clemson University and remains a state-supported school. The university still maintains a textile department within which research into the manufacturing and marketing of textiles is conducted. Research and development at the textile school involves spinning yarn, knitting, weaving, formation of fabric (using nonwoven technology), work on composite materials and wet processing. Trade publications and school journals are the major vehicles of information dissemination.[24]

Tompkins next turned his energies toward the establishment of a textile school in North Carolina. With rising public support, a bill was introduced to the North Carolina legislature for a state textile school. During the debate over the bill, it was noted that "with astonishing frequency mill owners are hunting men who know enough to supervise a mill, and not finding them at home have to send to New England or Old England."[25]

Tompkins went before the legislature stating that, "no money could be better spent than in training specialists to take charge of North Carolina's greatest manufacturing interest, the cotton mills."[26] Tompkins finished his presentation by saying, "The question is not—'Can we afford to build a school' —but, 'Can we afford not to build it?' "[27] Although the committee reported favorably on the bill, a deficit in the state treasury and a low price for cotton led to the bill's defeat. Two years later, however, the legislature yielded to strong popular demand and provided for a textile department at the State College of Agriculture and Mechanic Arts in Raleigh, which exists today as North Carolina State University. Research in preparatory processing and manufacturing of textiles is conducted at the university's David Clark Lab in conjunction with many of the major firms in this industry. The lab receives funding from the following sources:

---

[22] Winston, A Builder of the New South 194 (1920).

[23] *Id* 206.

[24] Agricultural Services Division, Food and Agricultural Organization, United Nations, Worldwide List of Textile Research Institutes 44 (1974). For further information, contact School of Textiles, Clemson University, Clemson, SC 29631.

[25] Winston, *supra* note 22, at 204.

[26] *Id* 206.

[27] *Id.*

1. Its own research and development budget
2. Research grants from industry
3. Contracts for specific research
4. Service agreements, which are short-term research projects ranging from one day to six months

Information gained through each of these activities is dispersed differently. The various means of dissemination of research information include distribution of research and development abstracts and status reports on projects in process to industry based on a current mailing list and on request, short courses, journal articles, presentation of papers, graduate student seminars, plant site lectures and consulting work.

The research that is performed at the David Clark Lab covers new polymers, treatments for fibers, the manufacture of fibers, manufacture of yarns from fibers, fabric formation, knitting and weaving and air-jet weaving to increase weaving speed. This last effort allows old looms to be converted to new air-jet looms. In this particular case there was a good deal of collaboration between the school and industry experts. With the lab acting as a catalyst, the industry (specifically two firms, to date) commercialized this technology.[28]

The Georgia state legislature originally appropriated $10,000 in 1897 for a textile department at the Georgia School of Technology, provided friends of the institution would also contribute $10,000. Thirty-five thousand dollars were donated in 1898 for this purpose,[29] and the textile program was established the following year. It exists today as a part of the School of Industrial and Systems Engineering of the state-supported Georgia Institute of Technology.[30]

In the early 1900s, two other states, Texas and Alabama, established textile schools or departments which still exist today—Texas A&M University's Wool and Mohair Laboratory and Auburn's Textile School.[31]

The early state actions which fostered industrial innovation—whether in the form of bounties for silk production, or in the financing of major internal improvements, or in the establishment and support of experimental stations and research-oriented universities—were not characterized by any explicit attention to the idea of promoting industrial innovation per se. Nevertheless, these actions, to the extent that they were successful, did result in the origination, dissemination and/or utilization of technological innovations.

---

[28] For further information, contact Department of Textile Chemistry, David Clark Lab, Box 5666, North Carolina State University, Raleigh, NC 27650.

[29] Feiker, The Training of Men for the Textile Industry 21 (1934).

[30] Agricultural Services Division, *supra* note 24, at 40. For further information, contact School of Industrial and Systems Engineering, Georgia Institute of Technology, Atlanta, Ga 30332.

[31] *Id* 45. For further information, contact Kleberg Center—Wool and Mohair Lab, Texas Agricultural and Mechanical University, College Station, Tex 77843; and Department of Textile Engineering, Auburn University, Auburn, Ala 36830.

The Science and Technology Path to Economic Growth

§9.04 State Responses to the Growth of Organized Research and Development

In the last two decades, there have been two waves of state actions based on the relationship between technology and economic development, with the first taking place in the 1960s and the second, now in progress, having started in the mid-1970s. As opposed to the earlier state efforts, the initiatives of the 1960s focused on science and technology as a path to economic prosperity, and those of the 1970s and 1980s have paid explicit attention to the idea of promoting industrial innovation. These initiatives are similar to the earlier efforts, however, in that they have as their basic objective the development of the economy of the state involved.

The programs established in the 1960s and, to a slightly lesser extent, those established in the 1970s, were a reaction by the states to the emergence and growth of organized research and development and to the growth of the federal government as the patron of much of the research and development that is undertaken in this country, particularly in terms of large-scale projects. Prior to these developments, individual inventors, working alone or in small groups and not generating much economic activity while they developed their ideas, were responsible for most major technological breakthroughs, new products and new industrial processes.

While "each individual inventor built upon the work of innumerable other inventors," industrial innovations were developed "largely by hit-or-miss, trial-and-error methods" rather than in "a cooperative or orderly fashion."[32] All this began to change in 1900 with the establishment of the first American research laboratory by the General Electric Company.

By the beginning of World War I, the number of industrial laboratories was still limited to only about 100. By 1920, there were an estimated 300 industrial labs, employing about 7,000 scientists and engineers and spending a little less than $50 million on research and development.[33] Before World War II, the annual amount industry spent on research and development was up to nearly a third of a billion dollars. By 1953, the total of research and development expenditures was up to $5.2 billion, with industry performing research and development valued at $3.6 billion. Over the course of the next 10 years, each of these figures had more than trebled to $17.4 billion and $12.6 billion, respectively.[34] This tremendous increase was principally the result of a dramat-

---

[32] Krooss, American Economic Development 343-344 (1966).

[33] *Id* 362.

[34] National Science Foundation, NSF Publication No 78-313, National Patterns of R&D Resources, Funds and Personnel in the United States, 1953–1978-79, at 28 (1978).

ic increase in federal financial support for research and development. Federal support for research and development increased twice as fast as support from industrial sources during this period—307 per cent compared to 143 per cent.[35] The primary component of this expansion of federal support came in the area of defense and space exploration.

Research and development expenditures increased from about one-tenth of 1 per cent of the gross national product (GNP) in 1920, to a little less than 1 per cent at the beginning of World War II, to 2.67 per cent in 1960, peaking at 2.97 per cent in 1963, making it clear that research and development was a big business.[36]

While all research and development activities are, in a sense, part of the workings of one or another particular industrial sector, research and development came to be seen by many people, including state government officials, as an increasingly important type of economic activity in and of itself. State governments were also quick to recognize the significant impact which federal outlays for research and development in the defense and space fields were having on the economies of those places in which they were being concentrated through the establishment of major government-administered, university-administered, industry-administered or independent research installations or through various forms of contracting out. It became apparent to state and local officials and business organizations that the efforts to meet major national technological challenges, primarily in the defense and space areas at the time, had some very important local effects. The direct economic effects of a research and development installation—constructing it and servicing it—and of the payroll of its employees were observable and significant. Besides the direct effects, many state leaders (in both the public and private sectors) believed that such centers of research and development activity "became magnets for the nation's top scientific and engineering brain power, attracted major industrial R&D laboratories to their areas and frequently created new products and companies providing many thousands of new local jobs."[37]

State officials also saw research and development as a particularly desirable type of economic activity for other reasons. First, research and development activities were seen as relatively labor-intensive, with the jobs involved being higher income, professional positions which would contribute more to the locality economically and civically than the kinds of jobs associated with most other types of economic activity. Second, research and development was seen as a type of economic activity which did not generate the undesirable side effects of heavy manufacturing, such as air and water pollution and toxic wastes.

During the early 1960s, a number of states moved to take advantage of the

---

[35] *Id* 36.

[36] *Id* 2; Krooss, *supra* note 32, at 362.

[37] Colosimo, Wilkof & Duga, The State Development Foundation: Meeting Social Needs Through Centers of Innovation (Ohio Dept of Economic and Community Development 1974).

relationship between technology and economic development. This wave of state involvement, which peaked in the middle of the decade, was characterized by a desire to promote research and development as a type of economic activity, with most of the states focusing on improving the quality of their state's university research and graduate education programs in science and engineering as a means of attracting industrial and federal research and development activities. All of the state science foundations established during this period[38] were premised, to one degree or another, on this theory and in states both with and without such agencies there were numerous other state efforts during this period to upgrade the quality of these programs.[39]

The relationship between academic strength in scientific areas and the growth of research and development activity may appear to be a very indirect and loose one, based more on a wish and a hope than on an actual direct linkage. In reality, however, it was based on very pragmatic considerations. State officials could observe that major federal research and development installations were being located in areas close to strong research universities. If their observations were not sufficient, the federal government helped to make the connection for them. A June 1962 Department of Defense report entitled *Changing Patterns of Defense Procurement* stated this very clearly:

> Management planners, in considering sites for new or expanding facilities, have found that the availability of trained minds overshadows even such factors as the labor market, water supply and power sources. The evidence is overwhelming: Route 128 encircling Boston; the industrial complex around San Francisco Bay; that related to the California Institute of Technology and UCLA in the Los Angeles area, and other similar situations are cogent examples of the clustering of industry around centers of learning. Such a migration arises from the need by industry for access to persons with advanced training who can translate the new science into vastly improved or wholly new products.[40]

§9.05 The State Science Foundations—Historical Establishment

In 1962, the Kentucky Science and Technology Commission was established and became the first of eight such agencies created for the purpose of promot-

---

[38] See **§9.05.**

[39] State support for higher education grew generally during this period and a good deal of support for this growth was generated by the desire to attract federal and industry research and development facilities. In Texas, for example, it is reported that Governor John Connally's "emphasis on higher education for Texas had its roots in his discovery (as Secretary of the Navy in 1961) that defense contracts went where the knowledge was." Burka, *The Truth About John Connally*, 7 Texas Monthly 156, 158 (Nov 1979).

[40] As quoted in Shapp & Jurkat, New Growth . . . New Jobs for Pennsylvania 43 (The Shapp Foundation, 1962).

ing research and development activities in their particular states. All of the other state science foundations, as they are frequently called, were established during the next three years, except for the agencies established in Pennsylvania in 1967 and in Massachusetts in 1969. In the year after the enactment of the Kentucky statute, Kansas, New York and North Carolina acted, and Louisiana and Connecticut established similar agencies in 1964 and 1965, respectively.[41]

The burst of enthusiasm with which the state science foundation concept was originally put forth was not sustained over time. Three of the foundations (Kansas, Kentucky and Massachusetts) never received enough funding to undertake the grant programs they were created to implement[42] and, along with the Connecticut and Louisiana agencies, have since been abolished.[43] When the Connecticut and Massachusetts foundations were abolished, they were replaced by agencies focused on helping small, technologically innovative firms to meet their capital requirements. Two of the foundations which are still in existence (New York and Pennsylvania) have been close to extinction on several occasions and are now at different stages of revitalization.[44]

---

[41] Kentucky Science and Technology Commission, 1962 Ky Acts ch 100 (repealed 1976; see note 43); Kansas Research Foundation, 1963 Kan Sess Laws ch 416 (repealed 1974; see note 43); New York Science and Technology Foundation, 1963 NY Laws Ch 432 (repealed by 1981 NY Laws Ch 197, which also formally established the foundation within the context of the state's Public Authorities Law, *see* N Y Pub Auth Law §§3100-3109 (McKinney); North Carolina Board of Science and Technology, 1963 NC Sess Laws ch 846 (current version at NC Gen Stat §143B-441); Louisiana State Science Foundation, 1964 La Acts ch 10, Extraordinary Session (agency abolished and functions transferred in 1977; see note 43); Connecticut Research Commission, 1965 Conn Pub Acts 459 (repealed 1974; see note 43); Pennsylvania Science and Engineering Foundation, 1967 Pa Laws 922 (current version at Pa Stat Ann tit 71, §670 (Purdon); and Massachusetts Science and Technology Commission, 1969 Mass Acts ch 843 (repealed 1978; see note 43).

[42] Colosimo, Wilkof & Duga, The State Development Foundation: Meeting Societal Needs Through Centers of Innovation (Ohio Dept of Economic and Community Development 1974).

[43] Kansas and Kentucky stopped funding their foundations in 1971 and Connecticut stopped funding its in 1972. Colosimo, Wilkof & Duga, *supra* note 42, at 15. The statutes establishing the Connecticut Research Commission and the Kansas Research Foundation were repealed by 1974 Conn Pub Acts 338 and 1974 Kan Sess Laws ch 413, respectively. The statute creating the Kentucky Science and Technology Commission was repealed by 1976 Ky Acts ch 299. The Louisiana State Science Foundation was abolished effective June 22, 1977, and its powers, duties, functions and responsibilities were transferred to the Secretary of the Department of Commerce by 1976 La Acts ch 518 and 1977 La Acts ch 83. The Massachusetts Science and Technology Foundation was "abolished and dissolved" by 1978 Mass Acts ch 497, which also established the Massachusetts Technology Development Corporation. (See **§9.14** for further information regarding the Massachusetts Technology Development Corporation and similar initiatives in other states.)

[44] The New York State Science and Technology Foundation did not receive funds in 1969, 1971 and 1976-1978, and no appointments were made to its board from 1975—1978. The Pennsylvania Science and Engineering Foundation did not receive any funds from 1976—1979.

While the state science foundation boom was short-lived, the experiences of these agencies represent an important part of the history of state involvement in the promotion of industrial innovation and should be taken into consideration as future policy directions are explored.

All of the state foundations shared one basic operational approach—the idea of using state funds to make grants for various research and education purposes with the expectation that this would lead to an expansion of research and development activity specifically, and to economic development in general. However, the different states emphasized different specific ways of employing this grant-making authority and the individual foundations were charged with a number of other responsibilities for promotion and coordination which varied from state to state.

Another common characteristic of all of the foundations was that, to one degree or another, they were premised on the belief that the enhancement of the graduate education and research programs at a state's colleges and universities would have a beneficial effect on its economy.[45] The theoretical and pragmatic considerations which led to the widespread acceptance of this belief also led some states to use their foundations primarily as vehicles for building up the capabilities and reputations of both public and private universities in various scientific disciplines. The New York Foundation was the foremost proponent of this approach with all of its funds, during its first incarnation (1964 to 1975), going to purchase equipment or otherwise improve university programs. This, however, was not the universal practice.

The Louisiana State Science Foundation served basically as a vehicle for channeling financial support to develop and fund the Gulf South Research Institute.[46] The Connecticut Research Commission made a large number of relatively small, one-shot research grants (averaging $34,000) in 18 program

---

[45] The statute creating the Kentucky Science and Technology Commission, for example, contained a declaration of public policy recognizing that the ability of the commonwealth to meet the objectives of providing new opportunities for gainful employment through the promotion of science and technology-oriented industry "depends on the qualitative and quantitative enhancement of the excellence of education, research and training at all levels." Ky Rev Stat §152.510(3) (repealed 1976; see note 43). The record surrounding the enactment of the 1963 statute creating the New York Science and Technology Foundation also includes several examples of this orientation. The sponsor of that legislation, for example, wrote, in a memorandum in support of that bill, that "the improvement of scientific research and education is of real benefit to the state. It would make the state an even more attractive location for research-oriented industry, which would contribute much to the economic growth of New York State." Letter from Edward J. Speno, to Honorable Sol Neil Corbin, Counsel to the Governor (Apr 2, 1963). The New York State Department of Commerce was even more specific in its memorandum in support of the bill. It referred to "the excellence of the education and training given in science and technology" as "the base upon which pre-eminence in research and development depends," and said the state "should do everything in [its] power to strengthen [that] base." Memorandum in support of S 3226 (introduction number), 4155 (print number) (Apr 10, 1963).

[46] Colosimo, Wilkof & Duga, *supra* note 42, at 15.

areas which it identified as having the greatest potential.[47]

The North Carolina Board of Science and Technology appears to have been the most effective of the foundations in two respects: first, in meeting the tests of the political marketplace by being continued and refunded by the North Carolina legislature (it is the only one of the foundations to have operated continuously from its establishment until the present time); and second, in leveraging outside research and development funds from federal and private sources. Programmatically, it has employed a combination of the approaches used in the other states: the funding of research projects such as in Connecticut; the funding of research and equipment to enhance the research capabilities of existing institutions such as in New York; and the development of new nonprofit centers of innovation as in Louisiana.

A number of the enabling statutes establishing foundations contained authorizations for them to get involved in "bricks and mortar" approaches to fostering research and development. The Louisiana statute provided that the foundation could acquire or contract for the construction of buildings, facilities, equipment and furnishing to be used for scientific research.[48] The Kentucky foundation was authorized to acquire land and facilities and to convey or lease them to private enterprise for such purposes as scientific and technological research, a "satellite facility" and nuclear storage. It was given revenue bonding authority (which was repealed in 1964, two years after enactment) and the power of condemnation.[49] The North Carolina Board of Science and Technology, in addition to its early, critical support for the Research Triangle Park, was and still is authorized to use its available funds for the construction and modification of facilities, as well as for other appropriate purposes.[50]

The foundations, in several cases, were also charged with performing a diffusion of innovation function. This responsibility was emphasized in the Connecticut Research Commission statute, which required that agency to propose ways to translate published research findings into more jobs in Connecticut industry.[51] The New York foundation was given the general authority "to sponsor and conduct conferences and studies, collect and disseminate information, and issue periodic reports relating to scientific and technological research, development and education in the state."[52] The North Carolina statute was the most specific in this regard, with that state's foundation being charged with encouraging "the cooperation of the state's industrial community to the end that industry shall assist in screening and identifying research results for possible industrial applications," and arranging for "seminars, short courses,

---

[47] *Id.*

[48] La Rev Stat Ann §51:1301 (West) (agency abolished and functions transferred in 1977; see note 43).

[49] Ky Rev Stat §152.510 (repealed 1976; see note 43).

[50] NC Gen Stat §143B-440.

[51] Conn Gen Stat Ann tit 32, §§32-24 to 32-31 (West) (repealed 1974; see note 43).

[52] Originally enacted by 1963 NY Laws ch 432, and recodified as NY Pub Auth Law §3102(2) by 1981 NY Laws ch 197.

visits and practical demonstrations . . . to foster interest in the results of research as a means of fostering industrial progress."[53]

These responsibilities, particularly the more specific Connecticut and North Carolina provisions, are indicative of the fact that while the predominant objective of the 1960s wave of state involvement may have been to enhance local economies by increasing the amount of research and development as a specific type of job-creating economic activity, there also was a broader view of the relationship between innovation and economic development.

§9.06 —Present Status

The three foundations which remain in existence—those of North Carolina, Pennsylvania and New York—have all been revitalized or redirected in recent years.

The North Carolina Board of Science and Technology[54] has provided a large share of its funds ($800,000 as of 1974) to the Research Triangle Institute, a not for profit, independent research center with broad capabilities in physical, biological, engineering and social sciences.[55] In fact, one of the seats on the board is assigned to a nominee of the executive committee of the Research Triangle Institute,[56] and its enabling statute initially provided that its "activities shall be centered in the Research Triangle and will be closely allied with the Research Triangle Institute."[57]

The North Carolina foundation "also provided funds for a regional nuclear laboratory and the Triangle Universities' Computation Center, as well as a series of small grants to university research groups."[58] These grants to university groups, as in the New York experience, were used to purchase equipment and specialized facilities. In North Carolina, however, the ability of these

---

[53] This provision, which was found at NC Gen Stat §143B-331, was dropped from the North Carolina law when the North Carolina foundation was reorganized by 1979 NC Sess Laws ch 668.

[54] Originally established in 1963 as the North Carolina Board of Space and Technology, this agency's name was changed to the North Carolina Board of Science and Technology by 1967 NC Sess Laws ch 69, to the North Carolina Science and Technology Committee by 1963 NC Sess Laws ch 1263, and back to the North Carolina Board of Science and Technology by 1979 NC Sess Laws ch 668. A provision of the 1979 statute which is considered extremely significant by those involved with the board is the designation of the governor as the chairman of the board. Previously, the governor designated a member of the committee to serve as chairman and was not a member of the committee himself. For further information contact the North Carolina Board of Science and Technology, Office of the Governor, North Carolina State Government, Raleigh, NC 27611, 919-733-6500.

[55] Colosimo, Wilkof & Duga, The State Development Foundation: Meeting Societal Needs Through Centers of Innovation 15 (Ohio Department of Economic and Community Development 1974).

[56] NC Gen Stat §143B-440.

[57] *Id* §143B-441(2).

[58] Colosimo, Wilkof & Duga, *supra* note 55.

universities "to compete more effectively for federal R&D funds" was a more specific and direct objective of their program support grants.[59] In fact, the North Carolina statute was the only one of the eight to specifically target the primary federal R&D agencies: "The Committee shall encourage liaison between industry, educational interests, the Research Triangle of North Carolina, and federal agencies such as the National Aeronautics and Space Administration, Atomic Energy Commission, Department of Defense, National Science Foundation, and National Institute of Health."[60]

The Pennsylvania Science and Engineering Foundation, the last of the state science foundations to be established, has been most successful in helping "to launch and build a number of university-affiliated and independent non-profit research centers of innovation which then attracted other funding."[61] In some cases, the foundation was the primary source of the seed funding which led to the attraction of funding from other sources. In other cases, it is claimed that by being able to provide support at a particularly critical time, the foundation was able to ensure the continued existence and thus allow for the ultimate success of the center. Among the programs fostered by the foundation in these ways were the Monnell Chemical Senses Center at the University of Pennsylvania and the Knowledge Availability Systems Center at the University of Pittsburgh.[62]

---

[59] *Id.*

[60] NC Gen Stat §143B-441(3).

[61] Colosimo, Wilkof & Duga, *supra* note 55.

[62] The Monnell Chemical Senses Center, which received seed funding of $100,000 from the Pennsylvania Science and Engineering Foundation (PSEF), had attracted $8.6 million in funding from other sources by 1974. It conducts "basic research on chemical senses, relating to food industries, pollution and taste and smell identification." For further information, contact the Monnell Chemical Senses Center, University of Pennsylvania, 3500 Market Street, Philadelphia, Pa 19104, 215-243-5000. The Knowledge Availability Systems Center, which received seed funding of $106,000 from the PSEF, is now one of six regional NASA-funded Industrial Application Centers (IACs) and had attracted $1,208,700 in support from other sources by 1974. For further information, contact the Knowledge Availability Systems Center, University of Pittsburgh, LIS Building, 135 North Bellfield Avenue, Pittsburgh, Pa 15260, 412-624-4141. Both of these are successful centers of innovation whose existence has been credited in large part to state support initially in the case of the Knowledge Availability Systems Center and at a critical time in its history in the case of Monnell. Colosimo, Wilkof & Duga, *supra* note 55, at 16.

One of the other NASA-funded IACs is the North Carolina Science and Technology Research Center (NCSTRC) at the Research Triangle Park. NCSTRC was originally established as the North Carolina Space and Technology Center by 1963 NC Sess Laws Ch 846. Responsibility for the administration of NCSTRC was originally vested in the North Carolina Board of Science and Technology (See notes 41, 53 and 54). 1977 NC Sess Laws ch 198 transferred NCSTRC to the state Department of Commerce and 1979 NC Sess Laws ch 668 amended the relevant statutes to formally change the responsibility for administration of the NCSTRC from the committee to the state Department of Commerce. *See* NC Gen Stat §143B-442. For further information, contact North Carolina Science and Technology Research Center, PO Box 12235, Research Triangle Park, NC 27709, 909-549-0671.

While the Pennsylvania Science and Engineering Foundation (PSEF) had one of the most impressive records of all of the state science foundations, it did not receive any appropriation from the Pennsylvania legislature between 1976 and 1979, but in 1980-1981, received $350,000. In the past, PSEF has, on occasion, received National Science Foundation funds which it passed through to other entities. Since its refunding in 1980-1981, PSEF has focused on providing seed financing in the form of grants to fund research and development projects in the hope of creating jobs in the immediate and distant future. Although eligibility requirements are not restrictive (i.e., educational institutions, nonprofit organizations, business enterprises, individuals concerned with scientific and technological research and development, and state and federal agencies are all eligible to apply for grants), colleges and universities are the usual recipients of funds. Projects must have technical merit and aid in the creation of jobs. To ensure that education is served, the foundation reserves 20 to 25 per cent of its funds for the study of mathematics and the sciences at the secondary school level and for the scientific and technical training of personnel needed in innovative industries. The rest of the foundation's funds are used to fund basic and applied research in the following areas:

(1) the modernization of Pennsylvania's manufacturing and industrial plant, equipment and processes;

(2) the development of innovative consumer, commercial and industrial products;

(3) the development and application of innovative engineering systems, methods and practices in the solution of problems affecting the public welfare; and

(4) the innovative use of Pennsylvania's indigenous natural and cultivated resources.[63]

PSEF maintains a policy of not funding 100 per cent of any project, except in very unusual cases. During its first 10 years of operation the foundation's total expenditures of $8.2 million were used as seed funds and "attracted over $68.3 million in funds from industry, local governments and the federal government for applied science and engineering projects designed to meet public needs in energy, health, transportation, environment and education, as well as to strengthen Pennsylvania's economic position through technological innovation."[64] PSEF has also emphasized research projects which bring a particular

---

[63] Pennsylvania Science & Engineering Foundation, PSEF—Funding Strategies 3, (undated). For further information, contact the Pennsylvania Science and Engineering Foundation, Department of Commerce, 462 Forum Building, Harrisburg, Pa 17120, 717-787-4148.

[64] Brodsky, Kaufman & Tooker, University/Industry Cooperation: A Preliminary Analysis of Existing Mechanisms and Their Relationship to the Innovation Process 42 (Center for Science and Technology Policy, Graduate School of Public Administration, New York University, 1980).

university's research capabilities together with a particular business's research needs. For example, "a PSEF grant was used to facilitate cooperative research between the university (Pennsylvania State University's Materials Research Laboratory) and Erie Technical Ceramics on the development of a new capacitor material."[65]

The New York State Science and Technology Foundation had made numerous research and equipment grants between 1974 and 1975, but was defunded and became inactive in the 1976-1978 period. In January 1979, Governor Carey announced his intention to revitalize the foundation and to give it more of an economic development orientation. Since that time, a number of major steps have been taken. First, funding was restored and the commission reestablished a research grant program. Second, legislation was adopted in 1981 giving the foundation the following new duties:

1. To assess the technological development potential of the various regions of the state and to work with local and state groups and agencies to help realize the potential that exists
2. To disseminate to New York firms information on the availability of federal research grants and contracts and to assist small firms in applying for such grants and contracts
3. To provide information on where new, technologically oriented businesses can obtain management assistant services on a free or for-hire basis
4. To identify emerging technologies that have significant job-creating potential and to provide information on individuals and firms involved in those technologies to the New York State Job Development Authority, local industrial development agencies and other state, local and nonprofit agencies responsible for economic development
5. To facilitate interaction between inventors and entrepreneurs so as to foster the development and commercialization of new ideas.

This legislation also authorized a three-year test of a program under which the foundation can award up to three $25,000 matching grants each year to local technology development agencies for the purpose of promoting scientific and technologically oriented economic activity in their areas of the state.[66]

---

[65] *Id* 20.

[66] 1981 NY Laws ch 197. All of the provisions of this chapter except the promotional matching grants provision have been codified as NY Pub Auth Law §§3100-3109 (McKinney). The matching grant program was established by 1981 NY Laws ch 197, §4. For further information, contact the New York State Science and Technology Foundation, 99 Washington Avenue, Albany, NY 12245, 518-474-4348.

§9.07 Tax Incentives

Another approach which has emerged during the last two decades as a means of promoting industrial innovation is the use of state tax policy. By 1966, three states offered tax incentives to encourage research and development. By 1975, this number was up to seven and by 1979, the figure stood at twelve.[67] Starting in 1955, the state of Minnesota allowed a taxpayer when computing net income to treat research and experimental expenditures paid or incurred during the taxable year in connection with a trade or business as expenditures, deductible in that year or ratably over a period of not less than five years.[68]

In 1970, Massachusetts amended its tax law to allow any domestic research and development corporation to exempt from local taxation the machinery of the corporation. A domestic research and development corporation, as defined by Massachusetts, must derive more than two-thirds of its receipts from research and development activities and derive more than one-third of these receipts from the research and development of tangible personal property capable of being manufactured in the state.[69]

Pennsylvania began offering tax incentives for research and development in 1971. That year Pennsylvania enacted legislation providing that in determining franchise tax liability, the capital stock of a business invested purely in research and development should be relieved from state taxation.[70]

In New York State, the Advisory Council for the Advancement of Industrial Research and Development[71] recommended and the legislature adopted legislation providing for accelerated depreciation of investments in research and development facilities. This provision of New York state tax law, which is still in effect, allows a taxpayer to deduct capital investments in the state for research and development as current expenses in the same year the investments are made *and* if the costs of such investments exceed taxable income, the excess can be deducted in the following years. A similar Rhode Island statute, adopted in 1974, allows for a one-year writeoff of new research and development facilities in lieu of depreciation or an investment tax credit. In this case, excess deductions may be carried over for a period of up to three years.[72]

In North Dakota and Maryland, research and development equipment is classified as manufacturer's machinery and equipment and, as such, is eligible for tax exemptions.[73]

---

[67] *The Fifty Legislative Climates, 14th Annual Report,* Indus Dev 5, 7 (Jan/Feb 1980); *The Fifty Legislative Climates: What A Difference A Decade Makes,* Indus Dev 11, 13, 15 (Nov/Dec 1975).

[68] Minn Stat Ann §290.09(18) (West).

[69] Mass Gen Laws Ann ch 63, §38c.

[70] Pa Stat Ann tit 72, §7602 (Purdon).

[71] See **§9.08** for a discussion of the council.

[72] NY Tax Law §§210, 612; RI Gen Laws §44-32-1.

[73] Md Ann Code art 81, §9C.

Virginia (1) exempts tangible personal property purchased for use or consumption directly and exclusively for research and development in the experimental and laboratory sense from the state's sales and use tax; and (2) authorizes local governments to classify tangible personal property used or employed in a research and development business separately from other tangible personal property, and to tax such property at a lower rate.[74]

§9.08 Advisory Councils

In addition to the state science foundations,[75] a number of other state efforts were initiated during the first wave of state attention to the economy-technology linkage in the 1960s. One approach was the creation of a broad-based advisory body to advise the state governments on actions to be taken to expand science and technology-oriented industry in the state. In some states, such an advisory function was combined with other science advice functions. An example of this approach was the 50-member Georgia Commission on Science and Technology created by the Georgia legislature in 1964.[76] A purer version, i.e., without the other science advice to government responsibilities, was the Advisory Council for the Advancement of Industrial Research and Development (ACAIRD) which was established in New York state by Governor Nelson Rockefeller in 1960. This group's charge was "to examine the State's competitive appeal for attracting and holding research and development establishments associated with industry, universities and philanthropic organizations" and "to recommend steps which the State properly can take to expand its already substantial contributions to scientific and industrial progress." Among ACAIRD's recommendations was the creation of the New York State Science and Technology Foundation.[77]

In the late 1970s and 1980s, as interest grew in the relationships between industrial innovation and the economy, there emerged a renewed interest in the use of broad-based advisory mechanisms as a means of exploring and realizing a state's high technology potential. In 1981, for example, Governor William G. Milliken of Michigan created a high technology task force comprising business, government and education leaders to develop a strategy for achieving a greater high technology base for his state. That task force identified robotics and molecular biology as initial targets for Michigan's high technology development efforts, and the state has taken decisive action in the first

---

[74] Va Code §§58-441.6, 58-851.

[75] See **§9.05.**

[76] Moch, *The Role of a State Science Commission,* in Science for Society, Proceedings of the National Conference on Goals, Policies and Programs of Federal, State and Local Science Agencies, Atlanta, 1970, at 107-113 (Moch, ed, Georgia Science and Technology Commission 1971).

[77] *Statement by Governor Rockefeller Announcing the Appointment of the Advisory Council for the Advancement of Industrial Research and Development in New York State,* in Public Papers of Governor Nelson A. Rockefeller, 1960, at 927 (State of New York).

area[78] and has indicated its intention to begin similar efforts in selected areas of molecular biology. As a follow-up to the task force, a permanent Michigan High Technology Development Corporation was established as a high-level, policy-setting and coordinating organization composed of leading business, academic and governmental officials.

§9.09 Promotion of Innovation in Particular Industries and Technologies

During the 1960s, state efforts to promote research and development economic activity were frequently intertwined with efforts to foster innovation and development in particular sectors of the economy, primarily in the then-emerging domestic nuclear power industry, but occasionally in other industries as well. This was seen, for example, in the legislation enacted in Kentucky in 1962 establishing the Kentucky Science and Technology Commission.

The most common example of this approach during the 1960s was in regard to what was then being called "the peaceful uses of atomic energy."[79] Many states adopted legislation placing special emphasis on the promotion of atomic energy. As in the general programs described earlier, the emphasis was on economic development, the rhetoric of the time focusing on the development of atomic energy-related enterprises in the state.[80] There were examples in other areas, however, including Kentucky's program of contract research in regard to developing new and more effective uses for coal.[81] In California, in 1964, the state government entered into four $100,000 contracts with different aerospace companies "to direct some of their fabulous skills" to such "down-to-earth problems" as waste management and transportation.[82] The Kentucky

---

[78] See the discussion of Michigan's Industrial Technology Institute in **§9.16.**

[79] Remarks by Gov William G. Milliken to the Robots VI Conference and Exposition, Detroit, Mich (Mar 2, 1982).

[80] Other examples included New York Governor Nelson A. Rockefeller's continuing emphasis during the 1960s on improving opportunities for the development of atomic industry within New York; the legislation enacted in 1962 establishing the Kentucky Science and Technology Commission; and the enactment by most of the southern states of laws authorizing their participation in the Southern Interstate Nuclear Compact. An example of the atomic energy/economic development connection being made at the time is found in a December 1, 1959, report by the New York State Office of Atomic Development to Governor Rockefeller which noted that

> probably the most disturbing single fact is that the expanding non-weapons portion of the atomic industry, where the bulk of private investment in research and manufacturing facilities is being concentrated in anticipation of a large future market, has to date not been seeking New York as a principal site.

Quoted in New York State Atomic and Space Development Authority, Ten Years of Atomic Progress in New York State 7-8 (Aug 29, 1969).

[81] Ky Rev Stat §152.320.

[82] Bickner, Science at the Service of Government: California Tries to Exploit an Unnatural Resource 4-5 (Public Policy Research Organization, University of California at Irvine 1971).

program has been sustained and is highly regarded, while California's initiative proved to be a one-time effort with minimal results.

Since 1975, a number of important new state initiatives have occurred in regard to the promotion of particular innovative industries and the support of research and development in such areas. In New York, the state's Atomic and Space Development Authority was reconstituted as an Energy Research and Development Authority (NYSERDA) with an independent source of revenue (a levy on the state's electric and gas utilities) and a charge to contract out for the completion of projects involving research, development and demonstration of new energy production and energy conservation technologies. NYSERDA is currently supporting projects in a wide variety of areas including fluidized bed fermentation of cellulose, alcohol production from agricultural waste, development and testing of sterling engine concepts, hydrogen production storage and electrical energy recovery using solid polymer electrotype, and many solar and wind energy applications.[83] Kentucky has upgraded its coal research program into a more general aid program to develop new and more effective uses for fossil fuels, hydrocarbons and other natural resources and has increased the funding for the program.[84] North Carolina has established a State Energy Institute within its Department of Commerce "to fund projects that are to aid in developing efficient energy utilization and supplying appropriate energy technology and development for North Carolina."[85]

---

[83] For further information, contact the New York State Energy Research and Development Authority, Building 2, Empire State Plaza, Albany, NY 12223.

[84] Ky Rev Stat §§152.310-.390, establishing the state's "fossil fuel and hydrocarbon research" program; Ky Rev Stat §§152.750-.785, establishing a program to encourage and promote the development and demonstration of environmentally and commercially feasible coal conversion technologies; and Ky Rev Stat §§152A.001-.180, establishing a cabinet-level Department of Energy and within it a 15-member energy research board with responsibility for approving demonstration projects to be supported by the state's energy development and demonstration trust fund. For further information, contact the Kentucky Department of Energy, PO Box 11, 888 Ironworks Road, Lexington, Ky 40578-1916, 606-252-5535 and the North Carolina Science and Technology Research Center at Research Triangle Park, PO Box 12235, Research Triangle Park, NC 27709, 919-549-0671.

[85] For further information, contact the North Carolina Energy Institute, PO Box 12235, Research Triangle Park, NC 27709, 919-549-0671.

Alaska established a similar program in 1980. The Alaska Energy Center, Alaska Stat §§46.12.010-.500 as enacted by 1980 Alaska Sess Laws ch 148, is charged with "promot [ing] the most efficient and appropriate technologies for the use and conservation of the state's energy resources" and "bring[ing] existing and new technologies to a state of commercial feasibility." Alaska Stat §46.12.020(1) and (6). In accomplishing these overall purposes, the board of the center, subject to the availability of money, is required to "sponsor energy research projects," "conduct and sponsor applied research, development and demonstration projects of energy technologies" and "provide financial and other support to inventors and businesses engaged in the development, demonstration and commercialization of energy technologies." Alaska Stat §46.12.120(2). No funding was provided for the program in 1981-1982, but the statute may still be a useful model for others.

Some recent efforts to promote innovation and development in particular fields are closely tied to the movement in the late 1970s and early 1980s toward state involvement in the promotion of industrial innovation per se. The California Commission on Industrial Innovation, for example, administers a program called MICRO (Microelectronics Innovation and Computer Research Opportunities) under which the state and industry jointly fund university research projects in the microelectronics and computer fields which are such a significant part of California's economic base, but which are also integrally related to productivity improvement in all industries. An indication of the innovation orientation of small, high tech firms is in the fact that many small businesses are applying to participate in the program.[86]

§9.10 Research and Technology Parks

While there are countless research, technology and other industrial parks in the United States, and many examples of university-related parks, most state government interest stems from the highly successful Research Triangle Park (RTP) in North Carolina. Initiated by Governor Luther Hodges in 1959, the effort to establish RTP was clearly part of the state response to the growth of organized research and development and was oriented toward the attraction of jobs to the area.

With support from the state of North Carolina, the three universities that form the points of the triangle, the University of North Carolina at Chapel Hill, North Carolina State University at Raleigh and Duke University at Durham, formed a nonprofit consortium, the Research Triangle Institute, governed by a board made up of university, business and state government representatives to oversee the development of the park.

Over the years, the state has provided continuing leadership and financial support to RTP, both directly and through the location of many of its other research-related activities in the park, including the North Carolina State Energy Institute, the North Carolina Science and Technology Research Center and the state's new Microelectronics and Biotechnology Centers.

As of 1980, 40 research enterprises maintained facilities in the park. Thirty of these were private companies, with the remainder including professional associations (e.g., the Instrument Society of America) and industry-supported research institutes (e.g., the Chemical Industry Institute of Toxicology), feder-

---

[86] Banschich, *High Tech Fever Grabs States, Cities,* High Tech 18, 20 (Mar/Apr 1982). *See also* McDonald, *California Pushing Research in Microelectronics,* Chron Higher Educ. 11-12 (Apr 7, 1982). A unique feature of the program "allows faculty members to develop research projects with their peers in industry, instead of leaving the arrangements to administrators unfamiliar with technical developments." *Id* 12. In addition to the jointly funded research projects (31 in 1982-1983, involving six campuses of the University of California), the state is also building a new microelectronics research facility at the university's Berkeley campus which has the largest microelectronics program in the university system and which will do 40 per cent of the MICRO research. *Id* 11-12.

al agencies and university-related research activities.[87]

Despite the success of RTP and several other research parks in increasing the amount of technology related business activity in a particular area, it is questionable whether such parks can be presented as playing a role in promoting innovation or whether they are simply an approach to attracting research and development and other technology-related activities to an area. Does working in a research and development park setting, especially one with a university affiliation, provide a vehicle for interaction with other firms and with university researchers which will make a material contribution to the rate of innovation? Even if the answer to that question is no, it appears that the success of RTP has guaranteed that other states will undertake, or at least consider undertaking, efforts to develop research and development parks for economic development reasons.

Florida, for example, in 1978 enacted a statute, made much more rigorous by 1979 amendments, creating a state Research and Development Commission which could designate local research and development authorities which would be empowered to develop research and development parks in conjunction with one or more universities.[88] This law as amended provides that the state's public and private universities will be represented on the commission, which will also include two state government department heads (Commerce and Agriculture) and three individuals who are actively involved in the areas of research and development, industrial development or banking and finance. Research and development authorities would be created, on the local initiative of one or more counties and the approval of the state commission, for the purpose of financing and refinancing capital projects related to the establishment of a research and development park. In order to receive state approval as a research and development authority, a county or counties must present a statement of affiliation with one or more universities.

Not only is this Florida program indicative of the current importance being placed on university-industry linkages, but it is also similar to the "bricks and mortar" approach legislated for several of the state science foundations in the 1960s.

The state of Maryland pursuant to its Industrial Land Act[89] provides low-

---

[87] Brodsky, Kaufman & Tooker, University/Industry Cooperation: A Preliminary Analysis of Existing Mechanisms and Their Relationship to the Innovative Process 62 (The Center for Science and Technology Policy, Graduate School of Public Administration, New York University 1980). *See also* Lopata, *Research Triangle—A Far Out Concept That Worked,* Iron Age 105-08 (Nov 23, 1981) and *North Carolina's Research Triangle,* The Chapel Hill Newspaper, Feb 1980, particularly *The Research Triangle Park–A Dream Takes Shape Among the Scrawny Pine,* at 49, 51-52.

[88] 1978 Fla Laws ch 402, as revised by 1979 Fla Laws ch 101. *See also* O'Neal, *In Search of High-Brow Industry,* The South Magazine 33-35 (Oct 1979). Another state which recently created a public authority for the development of a research and technology park is Kentucky. *See* Ky Rev Stat §§152A.501-.543, establishing the chemical/energy park authority of Kentucky.

[89] Md Code Ann art 41, §§440-443.

cost loans to local governments in Maryland for the acquisition and development of industrial parks in instances where private development is not available. Localities can use these funds for the acquisition of land, the acquisition of options on sites, the planning and development of parks and shell building construction.[90]

§9.11 Dissemination of Innovation

As in nineteenth century efforts to foster innovation in agriculture and other locally important industries, there emerged during this first modern era of interest in scientific research a recognition of the importance of dissemination of innovation. During the 1960s, there was a heightened awareness of the importance of getting the results of government and university scientific research into the hands of businesses, particularly small businesses, which could use that information to solve problems, seize advantages and increase productivity. In 1965, the federal government established a program of assistance for states that established state technical assistance services for this purpose.[91] While the federal funding for this program was curtailed, several states continued their programs. One example of a state program that has continued and actually grown since that time is the Pennsylvania Technical Assistance Program, known as PENNTAP.[92]

---

[90] Maryland Office of Business and Industrial Development, 1748 Forest Drive, Annapolis, Md 21401, 301-269-3514. In addition, the University of Maryland Foundation is also considering a joint venture with two local entrepreneurs to establish a research park. Zoning plans are in and they hope to hire an international developer and start marketing in the near future. For further information, contact the University of Maryland Foundation, Metzerott Road, Central Administration Building, Adelphi, Md 20783, 301-454-4309.

The University of Utah Research Park, which was started by the university on an abandoned army base acquired from the federal government, was recently reported to have 25 tenants with employment approaching 2,000 and to be rapidly growing. The state of Utah is involved in this project not only through the university but also through the Utah State Department of Development, which has played a key role in the marketing of the park. For further information, contact the University of Utah Research Park, 505 Wakara Way, Salt Lake City, Utah 84108, 801-581-8133.

Several cities have also been involved in the development of research and technology parks. For example, the city of Philadelphia joined with a consortium of local universities and area businesses to form an independent nonprofit entity called the University City Science Center Corporation to develop what amounts to an inner city research park on a 16-acre tract within the city of Philadelphia. As of 1980, seven buildings had been completed at a cost of $37 million representing about 50% completion of the original 1964 conceptual plan. For further information, contact the University City Science Center Institute, 3624 Market Street, Philadelphia, Pa 19104, 215-387-2255.

[91] State Technical Services Act of 1965, Pub L No 89-182, 79 Stat 679.

[92] For further information contact PENNTAP, J. Orvis Keller Building, University Park, Pa 16802, 814-865-0427.

The High Tech Future: The 1970s, 1980s and Beyond

§9.12 Innovation as the Key to Economic Success

In the early to mid 1970s, a new wave of interest began to build in the relationship between technology and economic development. The disappointing experience of the state science foundations did not dampen a renewal of interest in "the technological prospect," generated in large part by the economic problems of the early 1970s. The recession that had such a severe impact on the Northeast during that period generated a good deal of thinking and analysis regarding the future of that region and of post-industrialism in general. The economic problems of that period also precipitated a greater degree of attention to the impact of federal spending on local economies with particular concern focusing on the regional impact of federal procurement practices.[93] Part of this attention resulted in a raising and popularizing of the discussions (that had never really stopped) about the geographic distribution of federal spending for research and development.[94]

During this period, for the first time, conscious, explicit attention on the part of state and local governments was given to the concept of innovation per se. Whether such attention can be chanelled into efficacious programs is another question. This wave of state involvement has also been characterized by an emphasis on several additional approaches, at least some of which seem to be more realistic than those of the earlier phase. The underlying concern with economic development which was characteristic of earlier state efforts continues and, if anything, is more generalized. The difference is that the key link underlying current efforts is the relationship between economic development and innovation through the dissemination of innovation as a means of increasing industrial productivity and the growth of new enterprises. In the 1960s, the emphasis was more on the expansion of research and development-type economic activity as an end in and of itself. The latter approach still exists, but it is not playing the dominant role it once did.

The approaches undertaken by state governments during the 1970s and 1980s fall into three major categories: (1) programs designed to directly or indirectly assist innovative enterprises in obtaining the capital necessary to become commercially successful;[95] (2) technical assistance to inventors and

---

[93] *See, e.g., Federal Spending: The North's Loss is the Sunbelt's Gain,* Nat J 878-91 (June 6, 1976).

[94] *See e.g.,* Silverman, *Trends in Research and Development: New York State and the Nation,* Chronica, 1-12 (Sept-Oct 1978); and the section on *Federal R&D Expenditures in Pennsylvania,* in Feller, Long & Nelson, Economic Aspects of the Relationship Between Research and Development Activities and State Economic Growth (Institute for Research on Human Resources, Pennsylvania State University 1972).

[95] See **§§9.12, 9.13.**

entrepreneurs;[96] and (3) beside these additional types of state involvement, renewed interest in all of the earlier approaches to promoting industrial innovation, with some of these established methods being pursued more ardently and more creatively. As discussed elsewhere,[97] several states have continued to use their state science foundations not only to foster research and development-type activity and to strengthen college and university programs, but also to more directly support industrial innovation. The number of states utilizing their tax laws to promote innovation through increased spending on research and development and in other ways has grown,[98] as has the use of advisory board mechanisms for a number of purposes related to the promotion of industrial innovation.[99] States have also continued

1. To provide support for research, development and demonstration of particular technologies and of innovative approaches in critical industries[100]
2. To develop and maintain research and technology park facilities as a means of attracting research and development activities and of encouraging the growth of high tech firms[101]
3. To support university-based education, research and dissemination of innovation programs.[102] This category has been the subject of considerable interest and attention during the last several years as state governments have undertaken to promote a number of types of university-industry cooperative arrangements and have endeavored to do so for a variety of purposes.

§9.13 Helping to Meet the Capital Requirements of Innovative Firms

While state and local governments had been involved for many years with programs designed to make capital more readily available to business enterprises and to reduce the effective cost of capital, it was not until the enactment in 1972 of a statute creating the Connecticut Product Development Corporation (CPDC) that a state or local government had specifically tried to promote industrial innovation through a capital investment or loan program.[103]

---

[96] See **§9.15.**

[97] See **§9.06.**

[98] See **§9.07.**

[99] See **§9.08.**

[100] See **§9.09.**

[101] See **§9.10.**

[102] See **§9.16.**

[103] An act creating the Connecticut Product Development Corporation, 1972 Conn Pub Act 248 (codified at Conn Gen Stat §§32-32 through 32-47). *See also* Jarrett & Jones, Creating Jobs: Connecticut's Product Development Corporation (Council of State Gov-

Earlier state efforts in this field (which included state-chartered business development, or development credit corporations, state loan guarantee programs, state and local industrial revenue bond programs, revenue bond insurance programs, bond-financed loan programs such as the Connecticut Development Authority and New York's Job Development Authority, tax revenue-financed loan programs such as the Pennsylvania Industrial Development Authority) were intended to make capital available to firms that were not able to secure adequate financing from conventional sources or to reduce the cost of capital to firms receiving such loans as a means of encouraging firms to locate or expand in the state.

While firms involved in the development of innovative products or in the utilization of innovative processes are not precluded from applying for assistance under such programs, the eligibility and selection criteria for these programs focus on other considerations. These criteria typically include limits on the types of businesses that are eligible (the usual focus is on manufacturing), the job creation or retention potential of the project and the creditworthiness of the firm.

The CPDC, on the other hand, represented the first of an increasing number of state programs specifically designed to assist technologically innovative firms in meeting their capital requirements. While these new financing assistance programs vary in terms of the point in the product or business development life cycle at which they are directed, they proceed from the same basic premises: (a) that many innovative ideas for new products and processes are developed by new and other small businesses; and (b) that such firms often have a difficult time obtaining the capital that is necessary for the development of their innovative ideas. Whereas the state science foundation and the other efforts associated with the 1960s wave of state interest in this field were concerned with increasing job opportunities in research and development itself, and the earlier state financing assistance programs were oriented toward influencing industrial location decisions, the focus of the CPDC and similar programs is on the jobs that will be created when an innovative firm establishes itself and begins to flourish or when a new product becomes viable and is manufactured and sold and on filling the capital shortages which can prevent firms from realizing their potential in such cases.

The CPDC was also the first state financing program to focus on the need for equity capital rather than or in addition to debt capital. All of the earlier state programs provided debt capital either directly or indirectly through insurance programs or at lower rates through explicit or hidden subsidies, whereas the CPDC was the first of a number of state programs to recognize the potential benefits of a state program to help meet the equity capital needs of certain promising, technologically oriented businesses. Two advocates of this type of state program have characterized the need that exists as follows:

---

ernments 1978). For further information, contact the Connecticut Product Development Corporation, 78 Oak Street, Hartford, Conn 06106, 203-566-2920.

> In contrast to the inexorable principal and interest repayments required of debt, equity represents capital supplied to a firm in return for some share of its uncertain future income. This type of financing gives brand new businesses or ones branching into new fields a critically necessary holiday from capital costs until they begin generating positive cash flow.[104]

CPDC focuses on new products rather than on the development of new industrial processes or on the development of firms with significant potential in technologically emerging fields. It provides risk capital for the expenses associated with the development and demonstration of a new product with the potential for commercialization and does so by investing in a particular project, not in the corporation as a whole.

While its enabling legislation was originally enacted in 1972, the corporation did not being operating until 1974. As of June 30, 1980, the end of its third year of operation, the CPDC had approved product development grants totaling $2,104,781 for 28 projects. Fifteen of those projects had received their grant commitments in full ($1,017,011), eight had utilized a part of their approved funds of $731,770 and five projects had not, as of June 30, 1980, drawn any of their approved grant funds ($356,000).

A time lag obviously exists between the awarding of a product development grant and the development of the product, its production, actual sales and, finally, the payment of royalties to the corporation. Despite this fact, the corporation's income from royalties amounted to $21,228, $62,246 and $44,-128, respectively, during its first three years of operation.

§9.14 —Recent Developments

In the last few years, several other states have followed Connecticut's lead and have established programs designed to help innovative firms in meeting their capital needs at one or more critical points in their development cycle. In 1978, Massachusetts created an entity called the Massachusetts Technology Development Corporation[105] and New Jersey established an Office for Promoting Technical Innovation (OPTI),[106] which has, to date, done much more

---

[104] Litwak & Daniels, Innovations in Development Finance 111 (one in a series of Studies in State Development Policy published by the Council of State Planning Agencies, 1979).

[105] 1978 Mass Acts 497 (codified at Mass Gen Laws Ann ch 406, §§1-9). This act also dissolved the Massachusetts Science and Technology Foundation. For further information, contact the Massachusetts Technology Development Corporation, 131 State Street, Boston, Mass 02109, 617-723-4920.

[106] An act to promote technical innovation and new enterprise in New Jersey, 1977 NJ Laws ch 429 (codified at NJ Stat Ann §§34:1A-64 to 69). 1981 NJ Laws ch 53 expanded the office's powers and duties to authorize it to make investments in as well as loans to firms proposing to develop and produce specific inventions or new products in New Jersey. This chapter was entitled the New Products, New Jobs Act of 1980. For

with some of its other authorized roles than it has with the financing assistance authority given it by its enabling statute.[107] In 1979, Maine and Alaska also entered the lists with the enactment, respectively, of statutes establishing the Maine Development Foundation[108] and the Alaska Renewable Resources Corporation.[109] In 1980, New York[110] and California[111] also moved to establish similar financing assistance programs in response to the availability of federal funds for that purpose, and in 1981, Indiana established a similar entity capitalized by the purchase of stock by individuals, businesses, banks and insurance companies who were offered tax credits for such purposes.[112]

The Massachusetts Technology Development Corporation (MTDC) was formed to provide a source of capital to new and expanding high technology firms that have significant potential for success but which are not able to secure capital necessary to realize that potential. The initial capitalization of the

---

further information, contact New Jersey Office for Promoting Technical Innovation, John Fitch Plaza, Room 1307, Trenton, NJ 08625, 609-292-7138.

[107] See **§9.15**.

[108] Me Rev Stat tit 10, §§915-954. The Maine Capital Corporation (*see* Me Rev Stat tit 10, §§950-954) was established as a state SBIC and capitalized by stock purchases encouraged by state tax incentives. The Maine Development Foundation (*see* Me Rev Stat tit 10, §§915-928) provides investment management and administrative services to the Maine Capital Corporation serving, in effect, as the Capital Corporation's "deal" packager. For further information, contact the Maine Development Foundation, One Memorial Circle, Augusta, Me 04330, 207-622-6345.

[109] Alaska Stat §§37.12.010 to .080. In addition to being authorized to "invest, in such form as it considers appropriate, in projects which are economically viable and income-producing," the corporation is also empowered to "provide grants for projects having broad public application which do not have direct income-producing potential." Alaska Stat §37.12.070(11) and (12). Among the purposes for which such investments and grants may be made are "research and development of technologies and innovations for the rehabilitation and enhancement of the state's renewable resources to achieve an appropriate use of those authorities," research and development of "new products, markets and technologies for renewable resource industries in the state," "the demonstration of their technical and economic feasibility" and "their introduction into commercial markets." Alaska Stat §37.12.015(2) and (3). For further information, contact the Alaska Renewable Resources Corporation, 319 Seward Street, Suite 203, Juneau, Alaska 99811, 907-586-1786. In 1979, a provision was added charging the corporation with providing funds to the Alaska Council on Science and Technology for grants to be awarded by the Council for the development and dissemination of low-cost and small-scale technologies appropriate to Alaska. *See* 1979 Alaska Sess Laws, ch 56, §1; Alaska Stat §37.12.070(16); Alaska Stat §§44.21.241 to .255.

[110] For further information, contact the New York State Corporation for Innovation Development, c/o New York State Science and Technology Foundation, 99 Washington Avenue, Albany, NY 12245, 518-474-4348.

[111] The California program is scheduled to become operational in the near future and will probably be run by the state's Department of Economic and Business Development. For further information, contact the Department of Economic and Business Development, 1030 13th Street, Suite 2000, Sacramento, Cal 95814, 916-322-1394.

[112] Ind Code §6-3.4. For further information, contact the Indiana Corporation for Innovation Development, 125 West Market Street, Indianapolis, Ind 46204.

corporation's development fund was by a $2 million grant from the Economic Development Administration (EDA) of the United States Department of Commerce. The corporation operates as a revolving loan fund with principal repayments and equity earnings credited to the fund for subsequent reinvestment. Under the terms of its grant from the EDA, loans may be made to Massachusetts-based businesses with operations involving a significant amount of technology located in EDA Title IV redevelopment areas. The corporation's policy is to use its funds for what it refers to as "co-venturing with traditional sources of private investment capital." During the corporation's 1980 fiscal year, its first full year of operation since the original capitalization of its investment funds, it made its first five investments. In the aggregate, MTDC invested $950,000 in these companies, and its private sector coventurers invested a total of $3,020,000 in the same companies contemporaneously with the MTDC investments.[113]

The MTDC's investment guidelines indicate that investments are most likely to be made in firms with strong potential for employment growth in Massachusetts, strong technological bases for their activities and strong management teams. In comparison with the Connecticut Product Development Corporation, which is intended to help firms with good ideas which need money for the development and demonstration of those ideas, MTDC is targeted at a later point in the product or business development process. It is looking for firms with a proven technology that need financing to begin or expand production.

The New Jersey OPTI can make grants or guaranteed low-interest loans to help finance the development and marketing of an innovative idea. In reality, however, its financing assistance role is a very limited one given the size of its revolving loan fund—$250,000.

New York's Corporation for Innovation Development (CID), initially capitalized by a $1 million grant from the federal government and $1 million in matching funds appropriated by the state legislature, has been established as a subsidiary of the state's Science and Technology Foundation. The CID will operate as a revolving loan fund, but is authorized to make equity investments as well as loans. The objective of the program is "to stimulate the formation and development of innovative, technology-based new business ventures that have a significant primary employment growth potential."[114]

The Foundation has established the following five key criteria for evaluating given investment proposals:

> The venture must have an innovative technology orientation.
>
> The venture must have a good probability for commercial success. This entails an evaluation of the company's management team, the product and the market for the product.

---

[113] Massachusetts Technology Development Corporation, Annual Report.

[114] New York State Science & Technology Foundation, CID Information Guide 4 (Oct 1981). See note 110 for contact information.

The venture must have made every effort to secure the necessary private capital prior to approaching the foundation.

The venture must have, or be prepared to have, collaborative relationships with other sources of capital. Generally, the ratio of foundation to other investments should be a minimum of 1:3.

After receiving the proposed investment funds, the venture must be able to provide substantial benefits to New York. This entails substantial primary employment and, where applicable, utilization of manpower training programs, existing industrial spaces and other infrastructure.[115]

The CID investments will usually be in the range of $50,000 to $100,000, but may be higher in exceptional cases.

In addition to the financing assistance programs established by state and local governments in the last 10 years, two recent state efforts to indirectly assist new firms in meeting their capital needs are worthy of note. While not limited to technologically oriented or innovative businesses, these provisions will give such firms a significant assist in attracting the capital they need for commercial success. In 1981, both New York and California adopted amendments to their state tax laws designed to encourage individuals to invest in small, independently owned, in-state businesses and thus, at the same time, to help such businesses attract equity capital. The California plan will abolish state capital gains taxes on stock held for three years or more in most California companies with fewer than 500 employees, maintain existing tax rates on investments and increase the taxes on gains from investments in nonproductive assets like metals, gems, art, antiques and stamps.[116] New York's new law provides for a reduction in the maximum tax on all capital gains from 7 per cent to 5.6 per cent and provides for a further reduction in the maximum tax on capital gains from investments in new, small businesses to 4.2 per cent if the investment is held for at least three years, 2.8 per cent if it is held for four years and 0 per cent if it is held for five years or longer.[117]

§9.15 Technical Assistance to Inventors and Entrepreneurs

While the New Jersey Office for Promoting Technical Innovation (OPTI), as discussed in the previous section, has a limited financing assistance role, it has established a substantial record in the area of technical assistance. OPTI

---

[115] *Id* 5.

[116] California modified its existing graduated capital gains tax, which bases the per cent of exclusion on the length of time the investment is held (Cal Rev & Tax Code §18162.5(a)) to include a preference for gains from investments in the stock of businesses with less than 500 employees. *See* 1981 Cal Stat ch 534. The definition of *small business stock* for this purpose may be found in a new §13161.5 added to Cal Rev & Tax Code by ch 534. The provisions establishing lower rates on gains from such investments and higher rates on gains from investments in nonproductive assets were included in Cal Rev & Tax Code §18162.5(b) and (c) *as amended by* ch 534.

[117] NY Tax Law §612(o), *as amended by* 1981 NY Laws ch 103, §3.

provides individual inventors, entrepreneurs and small businesses with assistance in the form of in-depth technical evaluations and marketing surveys to help them in realizing the full potential of their ideas.[118]

Other states have also become involved in providing similar assistance to individual inventors, entrepreneurs and small businesses through facilities at their state universities. A number of such *innovation centers* have been established with National Science Foundation (NSF) support at several state universities, including the University of Oregon at Eugene, the University of Utah and the University of California at Santa Cruz. The Oregon center, the Experimental Center for the Advancement of Invention and Innovation, for example, will provide a detailed report evaluating any invention, for a fee of $50.[119] NSF also funded similar centers at a number of private universities, including the Massachusetts Institute of Technology and Carnegie Mellon University,[120] and in 1980, NSF funded the first joint, public-private innovation center. Polytechnic Institute of New York, a private university, and the State University of New York at Stony Brook came together with the active assistance and financial support of the Action Committee for Long Island, an organization that represents a coalition of business and industry on Long Island which has as its goal the stimulation of enterprise in that region, to form the Center for Research and Education to Advance Technological Enterprise (CREATE) and to secure NSF support for the center.[121]

State support for these centers has to date been indirect through the funding of the universities involved. Because of the recent cutoff of federal funds for this and other NSF-supported programs, however, state governments may be playing a more direct role in the future.

§9.16 University-Based and University-Related Research and Education Centers

Another part of the new wave is a redirection of the earlier belief in the importance of university resources to the development of science and technology-oriented industries. The current orientation seems to be that the relationship between the university and industry can and should be more direct. Among the manifestations of this new orientation has been a recent concern with making university patent and licensing policies more attractive to industry

---

[118] Office for Promoting Technical Innovation, informational brochure published by New Jersey Office for Promoting Technical Innovation (undated).

[119] Burger, An Analysis of the National Science Foundation's Innovation Centers Experiment (National Science Foundation, July 1977).

[120] *Id.*

[121] For further information, contact Dr. James J. Conti, Director of CREATE and Director of the Long Island Campus of the Polytechnic Institute of New York, Route 110, Farmingdale, NY 11735, 516-454-5100, or Dr. Marign Visich, Jr., Co-Director of CREATE and Associate Dean of Engineering and Applied Science, SUNY at Stony Brook, Stony Brook, NY 11794, 516-246-6750.

so as to encourage the commercialization of new idea development in conjunction with university-based research—for both the economic development and social benefit that may result—and to promote greater industry support of university research.[122] Along this same line, state universities are pursuing the idea of ongoing liaison or affiliation arrangements with businesses.[123] Such arrangements are advocated on the basis that they would provide additional support for basic research, with the participating firms receiving briefings on research findings and interaction with research personnel on a priority basis. One example of an ongoing liaison between a state university and the business world in a particular technological area is the Center for Microelectronics and Information Services formed recently at the University of Minnesota in Minneapolis. As of April 1980, the center has two formally declared participants—the university's Institute of Technology and Control Data Corporation, which has granted $2 million toward the funding of the center. The center hopes to draw in more United States and international companies and at least three more universities.[124]

The University of Florida has established a Center for Intelligent Machines and Robotics Research as a focus for research and development of robotic technology for the manufacture of the next generation of robots. Specifically, the center's studies deal with machine technology where precision is important, sensors, man-machine relationships, machine dynamics and control hardware and software.

It is the top priority of the center to be of service to industry but to remain within the university structurally and functionally and not to become autonomous of it. Professors are appointed jointly to teaching positions and industrial or governmental research. Currently, the center has a $750,000 laboratory and during the last three years has received $2.5 million in funding. Eighty per cent of the center's funding originates outside the university (10 per cent from industry and 70 per cent from the federal government). The university itself has provided $150,000 to $200,000 annually for the past three years through its Center of Excellence funding program which was established to provide support for the initiation of projects such as the Robotics Center. Since the Center for Intelligent Machines and Robotics Research is now considered to be established, it may no longer receive Center of Excellence funding.[125] Research consortia have also been established at other universities, including

---

[122] *See, e.g., Patents at Colleges and Universities,* Service Supplement 2:4:1:1 to 2:4:1:12 (Mar 1978); *Survey of Institutional Patent Policies and Patent Administration,* Service Supplement 2:4:2:1 to 2:4:2:8 (Mar 1978).

[123] Roark, *Academic Scientists Eye the "Industry Connection,"* Chron Higher Educ 1, 8 (Apr 30, 1979).

[124] For further information, contact the Center for Microelectronics and Information Sciences, University of Minnesota, 235 Smith Hall, 207 Pleasant Street, S.E., Minneapolis, Minn 55455, 612-373-2236.

[125] For further information, contact the Center for Intelligent Machines and Robotics Research, University of Florida, Room 300A MEB, Gainesville, Fla 32611, 904-392-0844.

North Carolina State, Cornell, the University of Delaware and Carnegie-Mellon.[126]

The University of Delaware has established cooperative efforts with industry in the areas of catalytic science (known as the Catalysis Center), composite materials and solar energy. The university has established an industrial associates program which requires a $25,000 contribution from partners in catalytic science and $40,000 for partners in composite materials. The partners are allowed access to research, early notice of publications, positions on the board of directors of the particular program and consultation services. There are 23 partners in catalytic science and 17 in composite materials. There are no partners per se in solar energy, but industry provides 35 to 40 per cent of the support for this research program and the United States Department of Energy pays for 60 per cent of the program's $2 million budget. The research budgets for catalytic science and composite materials are $2 million and $1 million, respectively, and industry support in each is 50 per cent. Industrial contracts outside of the associate program are also accepted. Patents from partnership projects are owned by the university and ownership of patents from industrial contracts are negotiated. Partners are guaranteed access to licenses for patents from partnership projects.[127]

In 1980, North Carolina enacted legislation establishing the Microelectronics Center of North Carolina (MCNC) as a not for profit corporation. The center, which is located at Research Triangle Park,[128] was initiated with a $24.4 million state appropriation and expects to attract federal and industry funds as well. General Electric, which announced plans in August 1980 to establish a $100 million microelectronics center on a 90-acre site at Research Triangle Park, became the MCNC's first industrial affiliate with a $250,000 payment.[129]

Some industries are very involved with cooperative, industry-wide research programs at university-based institutes, and a recent study of university-industry cooperation by New York University's Center for Science and Technology Policy (CSTP) found that involvement with such institutes seems to be growing in all fields. The CSTP study cited the paper industry as an example of an

---

[126] *Education for Invention,* Indus Research/Dev 121-24 (Apr 1978).

[127] Untitled and undated informational brochures published by the Center for Catalytic Science and Technology and the Institute of Energy Conservation; Center for Catalytic Science and Technology, Summary of Research in Progress (undated); and Center for Composite Materials, 1981-82 Applications of Composite Materials to Industrial Products (undated). For further information, contact the Center for Catalytic Science and Technology, Department of Chemical Engineering, University of Delaware, Newark, Del 19711, 302-738-8056; the Institute of Energy Conversion, University of Delaware, One Pike Creek Center, Wilmington, Del 19808, 302-995-7155; and the Center for Composite Materials, College of Engineering, University of Delaware, Newark, Del 19711.

[128] See **§9.10.**

[129] For further information, contact the Microelectronics Center of North Carolina at Research Triangle Park, PO Box 12235, Research Triangle Park, NC 27709, 919-541-7286.

industry which "has succeeded in developing an effectively integrated program of institute-based research where specific spheres of expertise are exploited for the mutual benefit of participating companies."[130] Two institutes at state universities with which the paper industry is involved on an ongoing basis are the Empire State Paper Research Institute, part of the State University of New York's College of Environmental Science and Forestry, and North Carolina University's Southern Forestry Research Institute.

The paper industry's involvement with university-based research institutes is due to the nature of this industry as contrasted with other industries, which are characterized by strongly competitive market situations which demand heavy emphasis on proprietary, in-house research and development (pharmaceuticals was cited as an example) or which, like the chemical industry, are "dominated by a few highly profitable companies with a heavy emphasis in R&D" and, therefore, with less dependence on the basic research capability of such institutes.[131] In the chemicals field, CSTP identified the Coatings Research Institute at Lehigh University (a private institution), which works on multiclient projects in areas such as the nature of emulsion stability and principles of adhesion; the Catalysis Center at the University of Delaware, which has been discussed earlier in this section; and the Chemical Industry Institute of Toxicology, a nonprofit research institute located at the state-initiated and state-supported Research Triangle Park in North Carolina.[132] While this last institute, which at the time of the CSTP study had a staff of 80 to 85 and was receiving support from 34 companies, was not directly associated with a university, "it maintains close working relationships with the local academic institutions which jointly sponsor Research Triangle Park"—the University of North Carolina, Duke University (a private institution) and North Carolina State University. This institute was formed to provide services in testing new substances, and CSTP concludes that it provides an example of the kind of cooperative research which "may serve to remove regulatory barriers to innovation by clearing the way for new classes of materials."[133]

An interesting example of a state government cooperating with a private

---

[130] Brodsky, Kaufman & Tooker, University/Industry Cooperation: A Preliminary Analysis of Existing Mechanisms and Their Relationship to the Innovation Process 20 (The Center for Science and Technology Policy, Graduate School of Public Administration, New York University 1980). For further information, contact the Empire State Paper Research Institute, College of Environmental Science and Forestry, Syracuse, NY 13210, 315-470-6500, and the Southern Forestry Research Center, School of Forest Resources, North Carolina State University, Box 4588, Raleigh, NC 27650, 919-737-2883.

[131] *Id* at 19-20.

[132] See **§9.10.**

[133] *Id* 20-21. For further information, contact the Center for Surface and Coatings Research, Sinclair Lab, Lehigh University, Bethlehem, Pa 18015, 215-861-3600; the Center for Catalytic Science and Technology, Department of Chemical Engineering, University of Delaware, Newark, Del 19711; and the Chemical Industry Institute of Toxicology, PO Box 12137, Research Triangle Park, NC 27709, 919-541-2070.

university in this field is taking place in New York state, where the state government in 1982 authorized the financing on a 40-year no-interest basis of a $30 million facility for a New York State Center for Industrial Innovation (NYSCII) on the campus of the Rensselaer Polytechnic Institute (RPI), a private university in Troy, New York, with an additional $35 million to be provided by private industry and other non-state government sources, for equipping the facility. The NYSCII would house two university-based research institutes which RPI had already established, one in computer-aided design, another in manufacturing technology, and a third in integrated electronics which it was in the process of establishing in 1982. The first of the institutes, the Interactive Graphics Center, which was established in 1976, now has 26 associates, each of which provides a general contribution to operating support annually and also contracts individually with the center for particular problem-solving and training projects. The Center for Manufacturing Productivity and Technology Transfer has five founding partners (General Electric, General Motors, Boeing, Norton and United Technologies) and five affiliate members (Cincinnati Milacron, Kodak, Fairchild Republic, Fairchild Mechanical Division-Schlumberger and Al Tech Specialty Street) who collectively support the center's ongoing work in robotic/sensor integration, robotic safety/collision avoidance and robotic grippers/hand tool applications. In addition, the center contracts with its members and other individual companies to complete specific projects related to the improvement of their manufacturing processes.[134]

The state of Michigan, following the identification by the Governor's high technology task force of robotics as a prime target for that state's development, established an Industrial Technology Institute to focus the resources of Michigan's higher education system on meeting the needs for advanced robotics research, applied research and training and retraining programs that will integrate advanced robotics technology into the workplace. The institute is intended to work with private industry, to be supported by private corporations, foundations, individuals and government sources, and "to help transfer leading-edge robotics technology from the classroom and laboratory to the production process."[135]

§9.17 Establishing an Effective State Role in Promoting Innovation

Much of what has been done by state governments during the last 20 years in the name of promoting research and development or, more recently, in the

---

[134] As of February 4, 1982, RPI already had a commitment for more than $10 million of the $35 million to be raised from private industry and other non-state sources. *Joint Hearing of the Senate Finance Committee and the Assembly Ways and Means Committee on the State Operations and Capital Projects Budgets,* 205th Annual Legislative Sess (Feb 4, 1982, Albany, NY) (testimony of George M. Low, President, Rensselaer Polytechnic Institute).

[135] Remarks by Gov William G. Milliken to the Robots VI Conference and Exposition, Detroit, Mich (Mar 2, 1982).

name of promoting industrial innovation, has really been done for the purpose of encouraging the federal government or private firms to locate research and development and other technologically oriented installations or activities in the particular state. The efforts during the 1960s to strengthen university research and graduate programs in science and engineering, for example, were designed in large part to make particular areas more attractive locations for research and development facilities—to help make the XYZ metropolitan area the kind of place where the Department of Defense or a particular company would be likely to place its laboratory. Similarly, the laws authorizing significant tax incentives for the construction of research and development facilities and those authorizing state or local authorities to finance the construction of research and development facilities with tax-exempt industrial development bonds were (and are) designed to provide a comparative advantage to locating such facilities in the particular state. Such activities do not focus on expanding the total amount of innovation-related economic activity which is going on, but rather they are concerned with influencing the location of the jobs associated with such activity at its present level. In this sense, much of what has been done has been similar to those state economic development activities which have been called, pejoratively, "smokestack chasing" and the state can be seen as a player in a zero-sum game.

Some of the more recent state initiatives, however, offer the promise of a state role which will help to expand the total amount of innovation-related economic activity. If state governments are interested in playing a truly important and effective role in regard to industrial innovation, they should look in the direction of these positive-sum approaches. By doing so, they can have the dual effect of (1) benefiting the economy of their particular state by fostering the development of job-creating enterprises which may not have emerged or grown in the absence of one or another type of outside assistance, while (2) benefiting the overall national economy by fostering the development, adoption and production of innovative products and processes which lead to productivity improvements, overall employment growth and an improvement in the nation's competitive position in the world economy.

Interestingly enough, it seems that the various innovation promoting activities which state governments undertook during the nineteenth and early twentieth centuries were more in the nature of positive-sum approaches than many of the industrial development strategies which developed in the 1960s and 1970s. The support which state governments gave to research and development, extension services and educational programs in agriculture (and in several other industries in some states) served to increase the productivity and success of this industry in the particular state, not to attract to that state a larger share of a finite pie. The farmers in one state or another may have gained a relative advantage at various times because of the adoption of new methods or crop varities, but at the same time, these advances had the overall effect of increasing the national product. Similarly, state support for capital-intensive innovations such as canals and railroads facilitated the absolute growth of the nation's economy and the early efforts to subsidize risky ventures were designed to encourage economic activities that would not have taken place otherwise.

Some of the recent state initiatives together with several additional proposals currently being discussed provide a framework for establishing an effective state role in promoting industrial innovation. Part of this role[136] is based on the state's changing its policies and practices in certain areas, such as taxation and regulation, so as not to discourage innovation. Another part involves positive approaches in such areas as capital formation and research and development, and several catalyst, facilitator and consciousness raising responsibilities as described in this chapter.

---

[136] See **ch 8.**

Innovation and Products Liability

10

*Kenneth Ross**

* The opinions expressed herein are those of the author and not necessarily those of Emerson Electric Company or its Legal Department.

Introduction

§10.01 Products Liability in General

One of the most significant developments in the legal environment of the United States in the last two decades has been the use of new legal theories of liability to recover against manufacturers and others in the chain of distribution for personal injury, property damage and economic loss suffered from the use of products.[1] These theories of liability, which together make up the law of products liability, have had a significant impact on those concerned with the design, manufacture and marketing of products for sale in the United States and elsewhere.

One of the more significant effects is that many manufacturers have seen a tremendous increase in the amount of insurance premiums paid for products liability coverage. This has caused many manufacturers to operate bare, which is to operate with no insurance at all, or to increase greatly their deductible levels. In addition, some manufacturers have gone out of business entirely

---

[1] Products liability can be defined broadly to include any liability of manufacturers or others in the chain of distribution for personal injury, property damage or economic loss suffered by an individual or legal entity as a result of the sale or use of a product. This definition is broad enough to include a breach of contract action between two large commercial entities. However, this chapter will focus primarily on the more traditional types of products liability actions—lawsuits for injury or damage by an individual against a manufacturer or distributor.

because profits were not possible given the large amounts of money required to cover the products liability risk and resolve lawsuits. Some diversified manufacturers have ceased manufacturing the products which have encountered the greatest loss experience (e.g., football helmet manufacturers).

Manufacturers are now being forced to defend their design and manufacturing procedures and, in many situations, to show that the benefits of the existing product outweigh the risks inherent in a particular design. If the jury disagrees with the manufacturer's decision making process, it can even award punitive damages.

The products liability situation has generated a great deal of demand for reform. This has resulted in the passage of substantial state products liability reform legislation and the introduction of a number of proposed federal products liability laws which would supersede all state laws in this area and arguably make the law fairer to manufacturers. Products liability has also resulted in a great increase in attention to the process of design, manufacture and marketing, thus increasing the number and stature of manufacturing personnel concerned with safety, design reliability, quality assurance, safety engineering and products liability prevention.

After tracing the historical development of products liability and describing the major bases of liability, the remainder of this chapter will concentrate on the practical methods that can be utilizied by attorneys, either in-house or outside, to prevent or minimize the risk of products liability.

Critical Issues in Historical Context

§10.02 Abolition of the Privity Requirement

In order to further the growth and progress which were the foundation of the Industrial Revolution, the courts and government decided that it would be necessary for injured persons to prove that an employer or manufacturer was at fault before they could recover damages. It was felt that allowing recovery on such a basis would not impede industrial progress in that in many situations the burden of proof would be too great.

At the same time, the concept of privity of contract was formulated. In 1842, in the landmark case of *Winterbottom v Wright*,[2] the court decided that an injured party must have a direct contractual relationship with the other party being sued. Thus, if A and B have a contract, there is privity between them. If B sells the product to C, there is privity between B and C but not between A and C. In that situation, if C is injured by the product, C may not sue A because there is no privity of contract. The rationale for this decision was a fear of unlimited lawsuits which might impede the development of new industrial growth.

Thus, an injured party could not sue the person at fault unless there was

---

[2] 10 M&W 109, 152 Eng Rep 402 (Ex 1842).

some privity of contract. Courts recognized problems with literally following this doctrine and, as a result, the privity requirement slowly began to erode. In 1852, the court in *Thomas v Winchester*[3] enunciated an exception to this rule in a case of mislabeled poisons which were classified as inherently dangerous to life. In these limited cases, privity was not required and an injured party was free to sue the manufacturer for negligence. Another exception was created in 1903 for a manufacturer who delivered a product without giving notice of a known dangerous condition. In that situation, privity was not required for the injured party to sue the manufacturer.[4]

In 1916, the New York Court of Appeals issued a landmark opinion that established negligence as a readily available cause of action for a consumer injured by an inadequate product. In *MacPherson v Buick Motor Co,*[5] the plaintiff was injured by the collapse of an automobile wheel. Buick Motor Company, the defendant, sold the car to a retailer who resold it to the plaintiff. Under the concept of privity of contract, Mr. MacPherson's suit for negligence against Buick would not be allowed. Justice Benjamin Cardozo, in an opinion for the court, held that the manufacturer was liable for injuries resulting from the use of this product whether or not the product was inherently dangerous if there was evidence of negligence by the manufacturer in assembling the product. Furthermore, Justice Cardozo stated: "If to the element of danger there is added knowledge that the thing will be used by persons other than the purchaser, and used without new tests, then, irrespective of contract, the manufacturer of this thing of danger is under a duty to make it carefully."[6]

The *MacPherson* court substantially departed from prior opinions. The concept of inherent danger was eliminated and the court concluded that any negligently made product that is likely to be dangerous is inherently dangerous. The *MacPherson* case also established a duty of care to all foreseeable users and showed that an injured party could sue a final manufacturer for failure of a component part. Thus, an injured party could sue anyone within the chain of manufacture or distribution.

The *MacPherson* suit involved negligence, not contract or warranty. Privity of contract was still necessary to sue for breach of warranty or breach of contract. Erosion of this requirement took much longer.

In 1960, the New Jersey Supreme Court, in *Henningsen v Bloomfield Motors, Inc,*[7] eliminated the privity requirement for actions brought for breach of warranty. The injured plaintiff in the *Henningsen* case was driving an automobile manufactured by Chrysler. The steering mechanism failed and Mrs. Henningsen was injured. Suit was brought against the dealer and Chrysler, and both were held liable despite the lack of privity between the injured party, Mrs. Henningsen, who did not purchase the automobile, and the defendants. Thus,

---

[3] 6 NY 397 (1852).

[4] Huset v JI Case Threshing Mach Co, 120 F 865 (8th Cir 1903).

[5] 217 NY 382, 111 NE 1050 (1916).

[6] *Id* at 389, 111 NE at 1053.

[7] 32 NJ 358, 161 A2d 69 (1960).

by 1960, the requirement for privity in contract had been abolished for virtually all suits based on negligence and warranty. This development provided the impetus for the next landmark opinion which established products liability as a truly viable means for recovery against those in commerce.

§10.03 Strict Liability

The concept of strict liability was first adopted in January, 1962 by the California Supreme Court in *Greenman v Yuba Power Products, Inc.*[8] Mr. Greenman was using a power tool when a block of wood struck him on the head. Although the case was tried on theories of breach of implied warranties and negligence, the California Supreme Court stated that the plaintiff had made out a new and separate cause of action in strict liability against the manufacturer. The court stated:

> A manufacturer is strictly liable in tort when an article he places on the market, knowing that it is to be used without inspection for defects, proves to have a defect that causes injury to a human being . . . . To establish the manufacturer's liability it was sufficient that plaintiff proved that he was injured while using the [product] in a way it was intended to be used as a result of a defect in design and manufacture of which plaintiff was not aware that made the [product] unsafe for its intended use.[9]

The rationale for adopting this new theory of law was "to insure that the costs of injuries resulting from defective products are borne by the manufacturers that put such products on the market rather than by the injured persons who are powerless to protect themselves."[10] The California Supreme Court felt that the burden of proving negligence was too great in cases involving unreasonably dangerous products. Thus, the court, in effect, made the concept of fault or negligence unnecessary where the injured party could prove that the product was defective and dangerous.

§10.04 Section 402A

Prior to 1962, the American Law Institute had proposed a section for the *Restatement (Second) of the Law of Torts* which would establish a special liability against sellers of food where the injured party would not have to prove that the seller had been at fault or negligent. The reporter for the *Restatement*

---

[8] 59 Cal 2d 57, 377 P2d 897, 27 Cal Rptr 697 (1962). However, the concept of absolute liability for manufacturers was first enunciated in a concurring opinion by Justice Roger Traynor of the California Supreme Court in Escola v Coca-Cola Bottling Co of Fresno 24 Cal 2d 453, 150 P2d 436 (1944). Justice Traynor later wrote the majority opinion in the *Greenman* case.

[9] *Id* at 62, 63, 377 P2d at 900, 901, 27 Cal Rptr at 700, 701.

[10] *Id* at 63, 377 P2d at 901, 27 Cal Rptr at 701.

recognized the impact of the *Greenman* decision and in 1964 proposed a new version which did not limit this special liability to sellers of food. Section 402A was finally approved in 1965 and provides as follows:

(1) One who sells any product in a defective condition unreasonably dangerous to the user or consumer or to his property is subject to liability for physical harm thereby caused to the ultimate user or consumer, or to his property, if
   (a) the seller is engaged in the business of selling such a product, and
   (b) it is expected to and does reach the user or consumer without substantial change in the condition in which it is sold.

(2) The rule stated in Subsection (1) applies although
   (a) the seller has exercised all possible care in the preparation and sale of his product, and
   (b) the user or consumer has not bought the product from or entered into any contractual relation with the seller.[11]

Since the adoption of §402A in 1965, some version of strict liability has been adopted by courts and legislatures in 44 states.

Strict liability was intended to be a concept where negligence or fault was irrelevant. Therefore, the trier of fact was to look to the resultant product to determine if it was defective and unreasonably dangerous and not to be concerned with any inadequacies in the design or manufacturing process. Some courts modified §402A and eliminated the concept of unreasonable danger because this injected the concept of negligence or fault.[12] More recently, courts have recognized that the concept of fault is particularly relevant where the defective condition relates to design or warning.[13] Therefore, some courts and some of the proposed federal legislation have injected a balancing test for determining whether a resultant design was unreasonably dangerous.[14]

Today, plaintiffs bringing products liability lawsuits utilize the three major concepts of negligence, breach of warranty and strict liability. It is beyond the

---

[11] Restatement (Second) of the Law of Torts §402A (1965).

[12] Cronin v JBE Olson Corp, 8 Cal 3d 121, 501 P2d 1153, 104 Cal Rptr 433 (1972).

[13] Commentators have argued for years that judging design and warning defect cases must involve some consideration of fault. See Birnbaum, *Unmasking the Test for Design Defect: From Negligence [To Warranty] to Strict Liability to Negligence,* 33 Vand L Rev 593 (1980).

[14] This balancing approach utilizes a risk–utility analysis which weighs the design alternatives and is referred to as the *reasonably prudent manufacturer* test. The test utilizes different factors formulated by Dean John Wade and Dean Page Keeton which must be considered in the weighing process. See Wade, *On the Nature of Strict Tort Liability for Products,* 44 Miss LJ 825 (1973); Keeton, *Manufacturer's Liability: The Meaning of "Defect" in the Manufacture and Design of Products,* 20 Syracuse L Rev 559 (1969). *See also* Model Uniform Product Liability Act, *reprinted in* 44 Fed Reg 62714 (Oct 31, 1979), set forth in **app 10-1.**

scope of this chapter to discuss in depth the requirements for establishing liability under these three concepts of law or discussing the many defenses available to manufacturers.[15]

Bases of Liability against a Manufacturer

§10.05 Introduction

It is clear that there are four major ways in which a manufacturer can create liability for itself in products liability. These are: manufacturing defects, design defects, marketing defects (failure to warn or instruct) and breach of warranty. There are different tests by which each of these standards is judged and it is necessary to discuss each in order to gain an appreciation of how to reduce the possibility of defective products being manufactured.

§10.06 Manufacturing Defects

Application of strict liability, as envisioned by the drafters of §402A of the *Restatement (Second) of the Law of Torts*[16] is particularly appropriate when the defective product resulted from some inadequacy in the manufacturing process.[17] The manufacturing defect or defect in construction results in the product being different in some way from what was intended by the manufacturer.

Proving a manufacturing defect is relatively easy for the plaintiff. There is an objective standard against which the product can be measured. If the resultant product which caused plaintiff's injury is different from all other products that were manufactured at or around the same time, and this difference caused plaintiff's injuries, a manufacturing defect will most likely be presumed. In addition, if the resultant product or line of products do not meet the manufacturer's specifications and this difference caused plaintiff's injuries, a manufacturing defect will also be presumed.

Although the plaintiff's burden is reduced in that he or she need not prove the manufacturer's fault or negligence, in most instances, the existence of a manufacturing defect resulted from some fault of that manufacturer. However, the plaintiff would have a very difficult burden of proving that the manufactur-

---

[15] For more information, *see generally* Sherman, Products Liability for the General Practitioner (Shepard's/McGraw-Hill 1981); Frumer & Friedman, Products Liability (3d Supp 1981); Hursh & Bailey, American Law of Products Liability (2d ed 1974); Beasley, Products Liability and the Unreasonably Dangerous Requirement (1981); and Baldwin, Hare & McGovern, The Preparation of a Product Liability Case (1981).

[16] Restatement (Second) of Torts (1965).

[17] Wade, *On the Nature of Strict Tort Liability for Products,* 44 Miss LJ 825, 831 (1973); Model Uniform Product Liability Act analysis to §104, *reprinted in* 44 Fed Reg 62714, 62721-26 (Oct 31, 1979). See **app 10-1.**

er knew or should have known about this defect in one of its products. Therefore, most commentators feel that imposing strict liability for manufacturing defects is fair for both the injured party and the manufacturer.[18]

§10.07 Design Defects

Comments to §402A establish a consumer expectations test for determining if a particular design is defective.[19] These comments state that a product can be deemed unreasonably dangerous if it is dangerous to an extent beyond that which would be contemplated by the ordinary consumer who purchases it or if it is in a condition not contemplated by the ultimate consumer. Unlike the objective standard for manufacturing defects, negligence concepts of foreseeability and reasonableness have been injected into the test for design defect.

The consumer expectations test has been severely criticized for not providing adequate guidance to the court or jury upon which to make a decision about whether a particular design is defective. Each person on that jury is likely to have a different understanding of abstract consumer expectations. Dean John Wade stated that "[i]n many situations, particularly involving design matters, the consumer would not know what to expect, because he would have no idea how safe the product could be made."[20]

As a result of this criticism, a growing number of courts and those proposing federal legislation have recognized that a balancing test with guidelines is necessary for the trier of fact to determine whether the design is reasonable under all the circumstances.[21] The Uniform Product Liability Act sets forth the test as follows:

> In order to determine that the product was unreasonably unsafe in design, the trier of fact must find that, at the time of manufacture, the likelihood that the product would cause the claimant's harm or similar harms, and the seriousness of those harms outweighed the burden on the manufacturer to design a product that would have prevented those harms, and the adverse effect that alternative design would have on the usefulness of the product.[22]

In addition, the drafters of the Uniform Product Liability Act include examples of evidence that are especially probative in making this determination. These include a consideration of the warnings and instructions provided with the product, the technological and practical feasibility of a safer designed

---

[18] Phillips, *The Standard for Determining Defectiveness in Products Liability,* 46 U Cin L Rev 101 (1977).

[19] Restatement, *supra* note 16, at §402A, comments g and i.

[20] Wade, *On the Nature of Strict Tort Liability for Products,* 44 Miss LJ 825, 829 (1973).

[21] See discussion at **§10.04.**

[22] Model Uniform Product Liability Act, §104(B)(1), *reprinted in* 44 Fed Reg 62714, 62721 (Oct 31, 1979).

product, comparative costs involved with the alternative design and the additional harms that might result from the alternative design.[23]

Several courts have gone so far as to shift the burden to the manufacturer to prove that on balance the benefits of the challenged design outweigh the risk of danger inherent in such design.[24] This shifting of the burden arguably makes it necessary for evidence of a manufacturer's reasonable care to be admitted on the issue of design defect. The plaintiff will argue that such evidence is irrelevant in a strict liability action and is therefore inadmissible. The use of the risk-utility analysis instead of the consumer expectations test and the shifting of the burden of proof to the defendant make it much more likely that the court will accept the evidence of due care in that these concepts make relevant the concept of fault with strict liability for design defects.

§10.08 Marketing Defects

Failure to adequately warn or instruct persons who use a manufacturer's products can be the basis of products liability in both negligence and strict liability. The scope of the duty is set forth in the *Restatement* as follows:

> one who supplies a chattel for another to use for any purpose is subject to liability for physical harm caused by his failure to exercise reasonable care to give to those whom he may expect to use the chattel any information as to the character and condition of the chattel which he possesses, and which he should recognize as necessary to enable them to realize the danger of using it.[25]

As with design defect, there is a great deal of similarity between strict liability and negligence when used for alleging marketing defects. Concepts of reasonableness and foreseeablility have been injected into the criteria for determining whether there are inadequate warnings and instructions.

A manufacturer is not required to warn of the danger of a misuse which is not reasonably foreseeable. The court in *Thibault v Sears, Roebuck & Co*[26] stated in part:

> manufacturers cannot foresee and warn of all absurd and dangerous uses of their product. . . . The duty to warn is concomitant with the general duty of the manufacturer, which "is limited to foreseeing the probable results of the normal use of the product or a use that can reasonably be

---

[23] *Id,* §104(B)(2)(a)-(e), 44 Fed Reg at 62721.

[24] Barker v Lull Engg Co, 20 Cal 3d 413, 573 P2d 443, 143 Cal Rptr 225 (1978); Caterpillar Tractor Co v Beck, 624 P2d 790 (Alaska 1981).

[25] Restatement (Second) of Torts, §388, comment b (1965).

[26] 118 NH 802, 395 A2d 843 (1978).

anticipated."[27]

The manufacturer is also not under a duty to warn of an open or obvious danger. Determination of whether a danger is open or obvious includes evidence concerning the status, intelligence and training of the particular user subjected to the hazard.[28] However, some courts have held that the obviousness of the danger is only one factor to take into account in deciding whether the manufacturer had a duty to warn.[29]

Generally, the duty to warn and instruct is owed to all whom the supplier or manufacturer should expect to use the chattel or be endangered by its use including purchasers, users, consumers and handlers of the product.[30] In addition, some courts have stated that a manufacturer must insure that the warning goes to the original or ultimate purchaser of the product, employees of the purchaser and third parties exposed to foreseeable and unreasonable risk of harm by the failure to warn.[31]

Some courts and proposed federal legislation recognize an exception in situations where it is impossible or very difficult to give a warning or instruction to the person exposed to the hazard.[32] This exception revolves around the foreseeability and reasonableness of expecting that the warning will get to the ultimate person subjected to the hazard. The more reasonable or foreseeable, the more likely that a manufacturer need only warn the initial purchaser. In *Reed v Pennwalt Corp,*[33] the court held that a warning to the injured employee by the manufacturer was not necessary where the employer was a large industrial concern with its own safety programs, the manufacturer had no effective means of communicating the warnings to the ultimate user, the product was not in the original container and it was reasonable to expect that the purchaser would communicate whatever was necessary to the ultimate users.[34]

Once the duty to warn arises, the warning must be adequate under the circumstances. One court has held that an adequate warning is one that could reasonably be expected to catch the attention of a reasonably prudent person in the circumstances of its use and the content must be understandable and

---

[27] *Id* 847 (footnote omitted).

[28] Collins v Ridge Tool Co, 520 F2d 591 (7th Cir 1975).

[29] Micallef v Miehle Co, 39 NY2d 376, 348 NE2d 571, 384 NYS2d 115 (1980). The patent danger rule had been rejected in 19 states as of January 1982. See Franchetti v Intercole Automation Inc, 523 F Supp 454 (D Del 1982).

[30] Restatement, *supra* note 25, §388, comment a.

[31] McLaughlin v Mine Safety Appliances Co, 11 NY2d 62, 181 NE2d 430 (1962).

[32] Restatement, *supra* note 25, §388, comment n; Model Uniform Product Liability Act, §104(C)(5) reprinted in 44 Fed Reg 62714, 62721 (Oct 31, 1979); Reed v Pennwalt Corp, 22 Wash App 718, 591 P2d 478 (1979).

[33] 22 Wash App 718, 591 P2d 478 (1979).

[34] 22 Wash App at 724, 591 P2d at 481.

convey a fair indication of the nature and extent of the danger to that person.[35]

A warning should generally identify the seriousness of the risk involved and the probability of the risk arising, identify the nature of the risk and provide the user with information on how to avoid the risk.[36] Another factor in determining whether a particular warning label is adequate is the location of that label in relation to the level of required conspicuousness. Thus, a warning in a catalog has been held inadequate where reasonable care dictated that the warning be put directly on the product.[37]

A related concept to a defect in warnings and instructions is violation of the *post-sale* duty to warn. Various decisions over the past 20 years have recognized that a manufacturer has a duty to warn of the existence of hazards discovered after the product was sold.[38] These decisions are based on negligence and not strict liability. Therefore, the liability is for failure to exercise reasonable care in light of the discovery of the hazard after the sale of the product. The drafters of the Uniform Product Liability Act recognize such a duty and state in the act that there is an obligation on the manufacturer to act with reasonable care to learn about serious risks connected with its product after it is manufactured and when it learns of such a risk, it is to act as a reasonably prudent manufacturer to inform product users or appropriate persons about the risk.[39]

§10.09 Breach of Warranty

A cause of action in warranty can be based on an express or implied warranty by the product's seller. The Uniform Commercial Code, which was first adopted in 1957 and has since been adopted by all the states except Louisiana, provides the basis for a breach of warranty action. The Uniform Commercial Code provides that an affirmation of fact, promise, description of the goods, sample or model which becomes part of the basis of the bargain creates an express warranty of conformity.[40] The Uniform Commercial Code also creates an implied warranty of merchantability which is deemed to exist in all sales contracts[41] and an implied warranty of fitness for a particular purpose.[42]

It is beyond the scope of this chapter to further discuss actions for breach of warranty. While most products liability lawsuits allege a breach of warranty, plaintiffs usually proceed most seriously with the cause of action for strict

---

[35] Bituminous Casualty Corp v Black & Decker Mfg Co, 518 SW2d 868 (Tex 1974).

[36] *See* Borel v Fibreboard Paper Prods Corp, 493 F2d 1076 (5th Cir 1973), *cert denied*, 419 US 869 (1974); Brownlee v Louisville Varnish Co, 641 F2d 397 (5th Cir 1981); Ross, *Legal and Practical Considerations for the Creation of Warning Labels and Instruction Books*, 4 J Prod Liab 29 (1981).

[37] Griggs v Firestone Tire & Rubber Co., 513 F2d 851 (8th Cir 1975), *cert denied*, 423 US 865 (1975).

[38] La Belle v McCauley Indus Corp, 649 F2d 46 (1981) and cases cited at 49.

[39] Model Uniform Product Liability Act, §104(c)(6), 44 Fed Reg at 62721.

[40] UCC §2-313.

[41] *Id* §2-314.

[42] *Id* §2-315.

liability. In addition, there are certain restrictions on the types of people who may bring suit for breach of warranties created under the Uniform Commercial Code.[43]

§10.10 Misrepresentation

Courts have overwhelmingly held product sellers liable when their statements about their products prove to be untrue, and when the injured party has relied on these statements and suffered damage or injury.[44] Different types of misrepresentation have developed over the years, such as innocent misrepresentation, negligent misrepresentation and fraudulent misrepresentation.

The drafters of the *Restatement* established strict liability for innocent misrepresentation by providing:

> One engaged in the business of selling chattels who, by advertising, labels, or otherwise, makes to the public a misrepresentation of a material fact concerning the character or quality of a chattel sold by him is subject to liability for physical harm to a consumer of the chattel caused by justifiable reliance upon the misrepresentation, even though
>
> (a) it is not made fraudulently or negligently, and
>
> (b) the consumer has not bought the chattel from or entered into any contractual relationship with the seller.[45]

Drafters of federal legislation have adopted this concept of strict liability in the Uniform Product Liability Act.[46] Although the act utilizes the term *express warranty,* analysis within the act makes clear that this term is essentially the same as the term *misrepresentation* utilized in §402B.

Under the law of misrepresentation, a manufacturer has been held liable when a stone broke a supposedly shatterproof windshield, causing the plaintiff to receive an eye injury.[47] A drug company was held strictly liable when the plaintiff became physically dependent on a drug that was advertised as being free and safe from all dangers of addiction.[48] As a result of the application of strict liability to misrepresentation, it is important for manufacturers to verify the truthfulness of representations made by them or anyone on their behalf.

---

[43] *See Id* §2-318, Alternatives A, B and C. A majority of states have adopted Alternative A which requires privity of contract in order to sue for breach of warranty unless the injured party is a member of the buyer's family or household or a guest in the buyer's home. For more information, see Whaley, Warranties and the General Practitioner (1981).

[44] Baxter v Ford Motor Co, 179 Wash 123, 35 P2d 1090 (1934).

[45] Restatement (Second) of Torts §402B (1965).

[46] Model Uniform Products Liability Act, §104(D), *reprinted in* 44 Fed Reg 62714, 62721 (Oct 31, 1979).

[47] Baxter v Ford Motor Co, 179 Wash 123, 35 P2d 1090 (1934).

[48] Crocker v Winthrop Lab, 514 SW2d 429 (Tex 1974).

§10.11 Effect of Products Liability on Innovation

Clearly, products liability has, to a certain extent, stifled industrial innovation. Surveys have shown that some manufacturers have stopped manufacturing certain products as a result of costs of products liability, have gone out of business entirely because a profit was not possible after paying for products liability and have been reluctant to develop new products where the benefits did not clearly outweigh the risks.[49]

These responses to the growing products liability problem have been caused by uncertainty as to the potential future risks to be encountered by manufacturers. This uncertainty has in turn been caused partly by the inconsistency among court decisions around the United States and by some courts which continually expand the scope of potential liability for a manufacturer.[50] For these reasons, manufacturers in many industries view products liability as a crisis which threatens their very existence. Such manufacturers are going to be very cautious in the creation of new products.

Uncertainty has also been caused by the lack of clear definitions of the words, *defect* or *unreasonable danger.* This has made it difficult for manufacturers to make decisions concerning the required level of safety in their products. The increased use of the risk–utility analysis and the itemization of the types of evidence which should be considered by the trier of fact in balancing the risk against the utility is somewhat helpful. However, the final balancing is still generally going to be performed by a jury of laypeople. While the majority of products liability lawsuits which go through trial result in a verdict for the manufacturer, the unpredictability of a jury balancing these factors is enough to force many settlements in situations where there should be little liability.

While industrial innovation has been somewhat stifled by the advent of products liability, products liability has also caused an increase in the safety and reliability of products on the market. Innovation should not reduce the ultimate safety of a product. Therefore, this might be a healthy result. While many manufacturers are unfairly paying the cost of products liability for products that are clearly safe enough, many of those products which are not now being manufactured due to the cost of products liability were probably too unsafe anyway.

Manufacturers should be able to be innovative in the creation of new products without incurring substantial potential liability if they utilize generally recognized techniques used to avoid or reduce legal liability. These tech-

---

[49] See 1 Product Liability: Industry Study; published in conjunction with the Final Report of Interagency Task Force on Product Liability (US Dept of Commerce 1977). This volume contains results of a telephone survey of 337 manufacturers and an analysis of the results of products liability surveys conducted by 20 national trade associations.

[50] For example, *see* Sindell v Abbott Labs, 26 Cal 3d 588, 607 P2d 924, 163 Cal Rptr 132, *cert denied,* 449 US 912 (1980), for adoption of the *market share liability doctrine* which allows a plaintiff to recover without proving which defendant-manufacturer manufactured the defective product.

niques are well-established and well-known, but not widely practiced. It is incumbent upon all manufacturers and others in the chain of distribution to be conscious of practices which increase the risk of products liability exposure and to engage in activities which minimize this risk.

Pre-Accident Prevention in the Production Process

§10.12 Introduction

Although products liability did not become a highly visible, clearly identifiable legal problem until the late 1960s, most manufacturers have for many years been concerned with preventing legal liability and manufacturing safe products. While it is doubtful that there were many safety professionals employed by manufacturers in the early 1900s, there were professionals in jobs such as quality control, quality assurance and design assurance. These professionals were mainly concerned with insuring that the product met its design and manufacturing specifications. Thus, manufacturers were initially concerned with minimizing the possibility of manufacturing defects.

Quite recently, however, the engineering profession has recognized that creating a safe design is not something that comes naturally. Safety is a separate concept subject to new and different techniques. Before 1970, there were no courses taught by engineering schools in the United States on safety. Even today, only a very small percentage of engineering students receive instruction on safe design methods.[51]

Many of the concepts currently used by engineers in products liability prevention come from safety engineering methodology developed during the 1950s and 1960s by engineers working for the armed services who needed to develop ways to prevent accidents involving our nation's missile program.[52]

Starting during the 1970s, many manufacturers appointed separate product safety managers whose jobs were to analyze all functions within that manufacturing concern and coordinate activities aimed at reducing accidents and improving product safety and reliability.[53] These safety professionals chaired product safety committees, if they existed, which comprised personnel from different functions such as purchasing, marketing, legal, financial, quality assurance, engineering and manufacturing. Even manufacturers who had no substantial products liability problems formed such departments or committees so that they could insure that such problems would be unlikely to arise in the future.

The involvement of attorneys in this product safety process most likely arose out of their handling of the litigation which has plagued many manufacturers

---

[51] Hammer, Product Safety Management and Engineering 53 (1980).

[52] *Id* 55.

[53] *See generally,* McGuire, The Product-Safety Function: Organization and Operations (1979).

since 1960. Many attorneys today feel that product safety is a job for engineers and not attorneys. This is not correct. Product safety and products liability prevention efforts should have substantial attorney involvement. A descripgion of some of these methods and the ways in which attorneys can make a contribution will be described in the following materials.

The 1970s and 1980s have shown and will show a great increase in the number of in-house attorneys employed by corporations to deal, in part or totally, with defending products liability lawsuits and assisting manufacturing personnel in avoiding or reducing risks in this area. Attorneys in law firms employed by corporations have also been involved, possibly less frequently, in assisting their clients in minimizing the risks of products liability. Inside and outside attorneys should try to increase their involvement in this process. It is a mistake to assume that engineers and other professionals in a manufacturing concern have adequate resources to make the design and manufacturing decisions that can have such a great impact on their employer.

The prevention techniques described below are generally employed before an accident occurs. However, they can also be used to evaluate a product for repair, retrofit or redesign. Using these techniques will go a long way toward insuring that a significant products liability problem is unlikely to arise in the first place.

§10.13 Prevention Program Goals

A prevention program is rarely going to be successful unless it has the complete support of top management within that manufacturing concern. Practicing many of these techniques can increase the amount of time and money spent on producing products. As a result, from a purely economic standpoint, there might be resistance to engaging in such programs. Therefore, it is necessary for top management to issue a policy statement setting forth the goals of such programs and the responsibilities of the different manufacturing functions in accomplishing these goals. One corporation has stated that: "It is the intent of this Policy to assure that the products of ITT System units are safe and in compliance with all applicable laws, regulations and standards established for the protection of users of such products."[54] Another company has stated its goals as follows:

> The principal objects of this policy are to ensure that all products: 1. Are safe for their intended use and reasonably foreseeable misuse. 2. Perform their functions with minimum performance degradation over their service life. 3. Have a minimum out-of-service time during their service life. 4. Throughout their life cycles, have as few adverse environmental effects as practical under present technology.[55]

---

[54] McGuire, The Product-Safety Function: Organization and Operations 34 (1979).

[55] Product Liability of Manufacturers: Prevention and Defense 1981, at 417 (Ross & Foley eds).

These programs have been called product safety programs, product integrity programs, quality assurance programs or product reliability programs. It makes no difference what you call the program as long as it is clear that the manufacturer vigorously supports the concepts of safety and reliability.

§10.14 Design Specifications

Certainly the most obvious point at which to plan for product safety is when the product is first designed. Most manufacturers begin their design process by preparing written design specifications from information submitted by engineering, manufacturing, marketing, purchasing and legal personnel. These specifications define and document the requirements the product must meet.

Initially, the marketing function must identify the ultimate customers, their needs and reasonable expectations and the uses and possible misuses of this product. This includes identification of the actual environment in which the product will operate. The existing and pending federal, state and local legislation which might affect product design must be identified along with any standards or regulations. All pending or completed litigation involving that particular manufacturer or others in the same or similar industries should also be considered for its effect on product design.

It must be decided whether component parts will be manufactured inside or by a supplier. At this point, critical parts should be identified so that the design engineer can properly consider how these parts are to be produced.

The packaging, handling and transportation environment to be encountered by the product must be identified along with any test and inspection requirements during develpment and manufacture. Lastly, documentation required for use by the customer, for marketing and sale of the product and for internal use by the company should be identified.

Actual design of the product starts after the specifications are finished. Formal design reviews should take place at predetermined points during the design phase to insure that the design fully complies with the requirements identified during the specification process.

§10.15 Considerations for a Safe Design

One of the most important things to identify and comply with are applicable government legislation and standards and industry standards and customs. Meeting industry and government standards is often not adequate because government agencies merely establish minimum standards.[56] Failure to comply with these standards will result in liability if causation is proved. However, compliance with these minimum standards will not always result in a legally sufficient product. Therefore, a decision must be made at this point whether to exceed these standards. Once the decision has been made, someone should

---

[56] The TJ Hooper, 60 F2d 737 (2d Cir 1932).

monitor any changes in standards or advances in technology or engineering knowledge. Whatever decision is made, there should be complete documentation concerning the manufacturer's awareness of and compliance with the applicable industry and government standards and customs.

The product should be designed to incorporate simple, inexpensive safety devices to protect parts of the body from moving equipment or to incorporate interlock switches which will deenergize or stop the equipment in hazardous situations. It is necessary to determine which safety devices will be sold as standard equipment with the product and which safety devices may be offered as optional equipment.[57]

A recent decision by a New York court held that a manufacturer will not be liable for a design defect when a safety device is optional and the purchaser decides not to purchase it.[58] The court felt that the purchaser was in the best position to determine whether the optional safety equipment was necessary under the circumstances. Other jurisdictions, however, have held that the manufacturer has a nondelegable duty to provide safety guards as standard equipment.[59]

The product should be designed to minimize or prevent injuries which might be caused by reasonably foreseeable uses and misuses of that product. Foreseeable misuse does not include all possible misuses of that product, nor does the product have to be idiot-proof. However, the design engineer, with the help of counsel, should survey past misuses to decide when a particular misuse becomes foreseeable. As in other areas, there should be documentation created and retained concerning this particular analysis and decision.

§10.16 Engineering Evaluations during Design

Many different types of engineering evaluations have evolved from the United States missile program. These are identified as failure mode and effects analysis, failure mode and criticality analysis, fault hazard analysis, fault tree analysis, operating hazard analysis and circuit logic analysis.[60]

As soon as the new product is conceived, a preliminary hazard analysis should be performed. Included are analyses of the hazards which would be present in this product, the cause of the hazards, the effect of encountering these hazards and any corrective or preventive measures. Hazards should be minimized to a level that is acceptable and the design process should proceed. As more and more is learned about the design, more detailed analyses can be performed.

The failure mode and effects analysis is a technique for systematically identi-

---

[57] See Birnbaum, *Design Defects,* in Product Liability of Manufacturers: Prevention and Defense 1981, at 20 (Ross & Foley eds).

[58] Biss v Tenneco, Inc, 64 AD2d 204, 409 NYS 2d 874 (1978).

[59] Bexiga v Havir Mfg Corp, 60 NJ 402, 290 A2d 281 (1972).

[60] *See generally,* Hammer, Product Safety Management and Engineering (1980); and Product Liability of Manufacturers: Prevention and Defense 1981 (Ross & Foley eds).

fying, analyzing and documenting possible failures of the product and the effects of such failures on system performance and safety. This test examines each component part for possible failure and the effect of the failure on the total design. The fault tree analysis is a top-to-bottom analysis in that it begins with a presumed undesirable event with the product and identifies events at lower levels that can cause this system failure. An operating hazard analysis focuses on the interrelationships between the product and the user and identifies hazards to the user during the operation of the product. This analysis is one necessary step in developing warning labels and proper instructional literature to be included with the product.

Attorney involvement during this process is relatively minor in that engineers are best trained to prepare and perform these evaluations and analyses. However, attorneys should be involved to the extent that they understand what is being done, what documentation is being created and retained and what conclusions are being drawn from these tests.

§10.17 Design Decision Making

The criteria for selecting the final design for actual manufacture are based on a great number of factors such as cost, performance, safety, reliability and appearance. In order to justify why a particular design was chosen and alternative designs rejected, the design decision process should be formalized. This formalized process helps identify all the characteristics of various design alternatives and select the alternative with the best combination of characteristics.

Alternative designs should be analyzed on the basis of characteristics important to the potential customers. Such factors as initial cost, reliability, serviceability, performance, operating costs, safety, noise, vibration and convenience should be given values. After the values are totaled, the best alternative should emerge. The final decision should then be made and documented.

This process is similar to the performance of a risk–utility or cost–benefit analysis. This decision process is critical because such evidence is admissible in many courts in justifying a manufacturer's particular design. An attorney's involvement in this process is essential in making sure that the design engineer considers the same factors that a jury might consider in a lawsuit involving design defect.[61] An attorney should also assist in creating and retaining proper documents concerning this process. As was seen in the Ford Pinto gas tank litigation, the creation of documents which appear to callously devalue human life can jeopardize a manufacturer's position during litigation.

§10.18 Manufacturing

The manufacturing process includes assembly, inspection, testing and packaging. Inspection and testing throughout the production process should be

---

[61] See discussion at **§§10.04** and **10.06.**

done to insure that the product has been manufactured and assembled according to the design and manufacturing specifications. It is only necessary to perform external, nondestructive testing for component parts which are fragile or sealed.[62] However, the manufacturer is usually responsible for component parts that prove to be defective or inadequate.[63]

Prior to packaging the product, the manufacturer should inspect it to insure that all safety-related devices are in place, all warning labels are attached to the product and appropriate instructions are provided. Appropriate documentation of this inspection should be created and retained in the event a subsequent problem arises. A manufacturer could photograph the product prior to insertion into its package to show the condition of the product and that the safety guards were included. If a lawsuit revolves around the lack of safety guards, the manufacturer can use such photographs as evidence that the product was altered or modified subsequent to the sale.

The manufacturer should create and retain appropriate identification records so that component parts, either manufactured in-house or purchased from the outside, can be traced and recalled. The manufacturer should be able to identify and trace a particular component part to the exact owner. This system is necessary to minimize the number of products which may have to be inspected or retrofitted if problems do arise and also to attempt to pass on any liability to the component part manufacturer.

The manufacturer should also give due consideration to the type of packaging and the transportation environment. Packaging must be safe and must adequately protect the users and bystanders against concealed and hazardous conditions. In addition, packaging must protect the product from damage during handling and storage.

§10.19 Warnings and Instructions

Consideration must be given to hazards remaining in the product when the final design is approved. The design engineer and other appropriate personnel should be satisfied that a combination of the design features and the warnings or instructions will keep the risks and hazards to an acceptable level. A manufacturer may not just design its product with "excessive preventable risk"[64] and attach warning labels concerning such hazards. Several courts have held that even though the warning was adequate, the fact that the manufacturer could have designed out the hazards allowed the court to impose liability against that manufacturer.[65] As a result, a manufacturer should design the product to be

---

[62] O'Rourke v Day & Night Water Heating Co, 31 Cal App 2d 364, 88 P2d 191 (1939).

[63] Vandermark v Ford Motor Co, 61 Cal 2d 256, 391 P2d 168, 37 Cal Rptr 896 (1964).

[64] Barker v Lull Eng Co, 20 Cal. 3d 413, 573 P2d 443, 143 Cal Rptr 225 (1978).

[65] Uloth v City Tank Corp, 376 Mass 874, 384 NE2d 1188 (1978); Thibault v Sears, Roebuck & Co, 118 NH 802, 395 A2d 843 (1978).

as safe as possible and only then should it attach warnings or give instructions concerning the remaining hazards.

For each product line, one of the first things that should be done is to review any applicable government statutes or regulations.[66] Failure to comply with these can result in absolute liability once causation is proved. If there are no statutes or regulations pertaining to the particular hazard to be warned about, the manufacturer should choose a consistent system to apply to all of the same products of a particular division or, it is hoped, to all products that it manufactures. Consistency among warning systems between manufacturers can result in more people comprehending what is meant by the use of various words of warning and the use of symbols or pictograms on labels. In addition, it is easier to argue when defending the adequacy of warning labels that these are so commonly used that the injured person should have understood the meaning of that particular label.

There are various organizations which are promulgating guidelines for warning label systems. Associations such as the American National Standards Institute, National Electrical Manufacturers Association and Society of Automotive Engineers have existing and proposed guidelines for warnings.[67] In addition, several corporations have devised guidelines for warning systems which are being considered by some of these associations.[68]

A suggested format for the label proposed by Westinghouse Electric Corporation is shown in Appendix 10-2. The top panel of this label combines a signal word and a hazard alert symbol with an appropriate color, all intended to indicate the level of hazard intensity (seriousness of the harm and the probability of the harm occurring if the warning is ignored). Next is a pictogram or symbol which should give some indication of the nature of the hazard (e.g., electrical) and possibly indicate the hazard's severity. The first message on the proposed label is intended to reinforce the message in the symbol or pictogram by indicating the nature of the hazard. The second message is intended to reinforce the level of hazard intensity as given in the top panel. The last message on the label is intended to give the user information on how to avoid the hazard.

In addition to using a consistent format, FMC Corporation and Westinghouse Electric Corporaiton have proposed a similar use of colors and words with definitions for a label's top panel. These words are *DANGER, WARNING and CAUTION* and are defined as follows:

---

[66] For example, *see* the Federal Hazardous Substances Act, 15 USC §1261; the Flammable Fabrics Act, 15 USC §1191; and the Poison Prevention Packaging Act, Pub L No 91-601, 84 Stat 1670-74 (1970).

[67] ANSI Product Alerting Accident Prevention Signs, Z535.4 (Feb 1981), NEMA Guidelines (1977) and SAE Recommended Practice J115a (1979).

[68] FMC Corporation's proposed system is in Product Liability of Manufacturers: Prevention and Defense 1981, at 111 (Ross & Foley eds). Westinghouse Electric Corporation's proposed system is in Product Safety Label Handbook (Westinghouse Printing Div, Trafford, Pa 1981).

DANGER - Immediate hazard which will result in severe personal injury or death.

WARNING - Hazard or unsafe practice which could result in severe personal injury or death.

CAUTION - Hazard or unsafe practice which could result in minor personal injury or product or property damage.[69]

The color *red* is used with the word *danger*, the color *orange* is used with the word *warning* and the color *yellow* is used with the word *caution*. Arguably, consistent utilization of the label format, signal words and the associated colors will result in better communication of hazards on products.

In establishing guidelines for the creation of a warning system, the manufacturer, with the assistance of counsel, should indicate how to select the correct signal word and how to determine where labels are to be placed, including reading distance, viewing angle and lighting, how to write a warning label and how to select a symbol or pictogram which would be appropriate for a particular hazard.

Instructional literature is also very important in minimizing or preventing the possibility of accidents and liability. Instructions are affirmative statements of how to operate, maintain, install or repair the product safely. Writing instructions during the initial design phase will help identify hazards in the design and in operation of the product. While instructions can be much more detailed than any warning label, they should not be so long and involved that it becomes unlikely that the user will read or understand them. In addition, they must be written for the comprehension level of those persons who will be using the product.

Warnings within the instruction book should be consistent with those used on the product. Therefore, if a particular hazard is at the *danger* level, there should be a warning in the book utilizing the same signal word. The instruction book should also contain various contractual disclaimers and a summary at the front of the book of the more important hazards which might be encountered while using the product.[70]

Since the trier of fact is in a position to use hindsight in determining whether a particular warning label was necessary or, if present, was adequate, it is often very difficult to prevail in a failure to warn lawsuit. As with most other prevention techniques, the best way to prevent liability is to prevent accidents. If in doubt, the manufacturer should overwarn and overinstruct rather than underwarn and underinstruct. While it is possible that a court would hold that a particular warning was inadequate because it was too voluminous and it is therefore unlikely that anyone would read or understand it, there have been no court decisions where this has been so held.

---

[69] Product Liability of Manufacturers, *supra* note 68, at 128. The definitions proposed by Westinghouse and some trade associations are slightly modified.

[70] Ross, *Legal and Practical Considerations for the Creation of Warning Labels and Instruction Books*, 4 J Prod Liab 29, 42 (1981).

§10.20 Advertising, Marketing and Distribution

Liability is allowed against the manufacturer for misrepresentations about the product which are relied upon and result in injury or damage to the product's user.[71]
Therefore, it is important for the manufacturer to ensure that all product advertisements and sales literature do not increase its potential liability.

Exaggerations about the product should be avoided, especially absolutes concerning safety, durability or dependability. Advertisements and sales brochures should accurately present product safety features and portray safe operating practices; otherwise the manufacturer could be liable for harm to a user who used the product on the basis of such a portrayal.[72]

Marketing personnel can also make statements which amount to express or implied warranties or might allow the user to believe that the product can be used in an unsafe manner. Such personnel must be educated about what can be said to sell the product to minimize such risks.

Distribution practices for a particular product should be analyzed to ensure that the product is received by the ultimate user in a safe condition. If the product has been substantially altered, most courts hold that the manufacturer is not liable.[73] However, the courts in some states will hold a manufacturer liable where the product has been substantially altered if such modification was reasonably foreseeable.[74] Therefore, a manufacturer should try to be kept apprised of modifications, alterations or other practices which result in delivery of an unsafe product to the final user.

§10.21 Recalls, Retrofits and Repairs

The manufacturer must establish a feedback system through its customers distributors, service personnel or sales personnel to learn of problems involving products in the field. Failure to discover post-sale problems and, if discovered, to remedy them, can result in liability for negligence.[75]

Once a product problem, complaint or hazard is identified among products in the field, the manufacturer must decide what response is appropriate. There are a wide range of acceptable responses depending on the circumstances. Some of them are doing nothing, sending a warning to the product's user or owner, sending a replacement part or a replacement product to the owner or

---

[71] See **§10.10.**

[72] Leichtamer v American Motors Corp, 67 Ohio St 2d 456, 1981 Prod Liab Rptr (CCH) ¶9036 (Jeep CJ-7 overturned).

[73] Gasdiel v Federal Press Co, 78 Ill App 3d 222, 396 NE2d 1241 (1979); Restatement (Second) of Torts §402A(1)(b) (1965).

[74] Merriweather v EW Bliss Co, 636 F2d 42 (3d Cir 1980). *Contra,* Robinson v Reed-Prentice, 49 NY2d 471, 403 NE2d 440, 426 NYS2d 717 (1980).

[75] See **§10.08.**

user for free, sending the manufacturer's field personnel to retrofit or repair the hazardous product or recall all products and supply replacements.

Initially, the manufacturer must determine whether there is any state or federal legislation or regulation which might affect what response is selected.[76] Obviously, the manufacturer must follow the recall or notice procedures set forth in such legislation or regulations or mandated by the appropriate government agency.

If no legislation or regulations apply, the manufacturer should evaluate the circumstances of the particular problem, identify and analyze the available remedies, and decide the most appropriate course of conduct. Under the common law, recall of a dangerous product is not absolutely required. As a practical matter, a recall might be appropriate where the potential risk (seriousness of a potential injury and the probability of that injury) far exceeds the costs and adverse consequences involved. On the other hand, the only response that might be necessary is a notice in a trade publication. On this point, the court in *Braniff Airways, Inc. v Curtiss-Wright Corp.*,[77] stated:

> It is clear that after such a product has been sold and dangerous defects in design have come to the manufacturer's attention, the manufacturer has a duty either to remedy these or, if complete remedy is not feasible, at least to give users adequate warning and instructions concerning methods for minimizing the danger.[78]

Another court held a manufacturer had met its duty by advising an amusement park owner by letter of a product problem and forwarding a replacement part for free. The park operator failed to install the new part on the machinery and was held liable to the injured persons.[79] The drafters of the Uniform Product Liability Act stated in their analysis of the act that in some situations a general warning through advertising may be all that a manufacturer can provide and that the standard is one of reasonableness and not strict liability.[80]

The Texas Supreme Court let stand a decision that would require a manufacturer to do a great deal in warning about or remedying a post-sale safety problem.[81] Bell Helicopter discovered that its helicopter tail rotor blade had a history of inflight fatigue fracture failures. It recommended to owners of this type of blade that they have certain inspections performed after a certain number of flying hours. In addition, Bell began to design an improved blade. Prior to the date of the accident, Bell distributed to owners and operators

---

[76] For example, *see* Consumer Product Safety Act, 15 USC §§2051-2082 (1976).

[77] 411 F2d 451 (2d Cir 1969).

[78] *Id* 453.

[79] Rekab v Frank Hrubetz & Co, 261 Md 141, 274 A2d 107 (1971).

[80] Model Uniform Product Liability Act, §104(C)(6), *reprinted in* 44 Fed Reg 62714, 62721 (Oct 31, 1979), and analysis at 44 Fed Reg 62721 and 62725.

[81] Bell Helicopter Co v Phil Bradshaw, 594 SW2d 519 (Tex Civ App Corpus Christi 1979, *writ refd nre*).

information concerning the availability of a new tail rotor blade. The plaintiffs did not replace their original blade with the improved one prior to the accident. The court in *Bell* held that it did not adopt the rule in *Braniff Airways;* however, it held Bell Helicopter responsible and stated in part:

> Here, Bell assumed the duty to improve upon the safety of its helicopter by replacing the 102 system with the 117 system. Once the duty was assumed, Bell had an obligation to complete the remedy by using reasonable means available to it to cause replacement of 102 systems with 117 systems.[82]

Whatever decision is made, the manufacturer should have available guidelines for recall, retrofit or warning campaigns that can be conducted as soon as possible in order to minimize potential injuries or accidents. A good example of the benefits of quick action before a recall was ordered by the government was the Rely Tampon recall campaign of 1981.[83] Although costs were high, the manufacturer minimized the number of injuries and received intangible benefits such as good publicity for being a good corporate citizen.

Prevention through Management Techniques

§10.22 Product Safety Committee

Every manufacturing organization should have on staff an individual who is responsible for the whole range of safety considerations inherent in any design, manufacturing and marketing process.[84] This person should be clearly identified as a safety professional and should have some general understanding of safety techniques.

An important function for this safety manager is to chair any product safety committee. The significance of a particular organization's products liability problems should dictate the size and involvement of such a committee.

Where a fully operational product safety committee is organized, it should contain personnel from engineering, financial, manufacturing, marketing, law, purchasing, field sales, quality control and insurance. Possible activities for such a committee include:

1. Develop a priority list for taking action on significant product problems and identify potential costs to remedy
2. Identify nonpriority product problems, determine their causes and assign responsibility for corrective action to appropriate personnel

---

[82] *Id* 532.

[83] For a discussion of this campaign, see O'Reilly, Food and Drug Administration, §§6.08 and 18.13 (Shepard's/McGraw-Hill Supp Nov 1981).

[84] See McGuire, The Product-Safety Function: Organization and Operations (1979).

3. Review and approve product safety and quality assurance programs
4. Compile and analyze accident and safety data to assist in pinpointing product problems
5. Provide safety consultation on product design, manufacture and application
6. Aid in accident investigation, accident analysis and defense of lawsuits
7. Develop communication and feedback systems to aid product safety and loss control functions
8. Represent the company on industry committees dealing with safety matters
9. Keep company personnel informed of legal guidelines, standards and regulations for design, and other matters concerned with product safety
10. Review new product designs from a safety standpoint
11. Develop and review warning labels to be affixed to the product
12. Review warranty claims for indications of potential for personal injury or property damage
13. Educate the design engineers of ways they can avoid designing a product that creates products liability exposure
14. Develop and review appropriate warnings and instructions for instruction and operators' manuals
15. Initiate and supervise warning, recall and retrofit programs
16. Review advertising and sales promotion material for completeness, accuracy and sufficiency
17. Develop and periodically review the record retention policy to insure the preservation of evidence necessary to defend products liability claims and lawsuits
18. Establish product safety requirements and perform internal audits to insure compliance with these requirements.

§10.23 Product Safety Audit

Product safety audits should be periodically performed to review the functions of the manufacturing organization that effect product safety and identify areas for improvement. Preferably, the audit should be conducted by personnel outside the organization to be audited. If the corporation is sufficiently large, there might be a corporate-wide staff whose responsibility is to help manufacturing divisions establish product safety and products liability programs and to perform audits after these programs are established.

The first task in this audit could be sending a questionnaire to general management, marketing, engineering, quality assurance, manufacturing, purchasing and field service. After receiving answers to the questionnaire, specific interviews can be held with appropriate managers in these areas to discuss the answers to the questionnaires.[85]

After the interviews have taken place, a list of problem areas, who is responsible to correct them and a timetable for correction should be prepared and reviewed with the general management and relevant managers within that operating division. Obviously, this corporate-wide staff should follow up to make sure that the problem areas have been dealt with.

Product safety audits, as outlined above, are very time-consuming and rely on an independent corporate-wide safety staff for performing. Where this is not possible, internal managers in that division or corporation can themselves answer the questionnaire and identify problem areas.

§10.24 Educational Programs

Implementing a products liability prevention and product safety program is impossible without appropriate personnel being given some type of education in this area. The corporate-wide product safety staff should be involved in organizing and presenting programs aimed at identifying techniques that can and should be employed by each operating entity. All of these programs should include segments given by lawyers who can identify the legal reasons for complying with these established procedures.

Engineers can be instructed in products liability law; how to design and manufacture a safe product; document policy; and prevention techniques. Marketing personnel would be less concerned with how to design a safe product and more with contractual liability, advertising and promotional literature. Field service personnel have a particular need to know about oral or written statements made while performing service and investigating accidents or product failures. Selected personnel might attend a program on how to develop warning labels and instruction books.

§10.25 Document Policy

Documents are very important in products liability. They can be used to establish liability against the manufacturer, to establish a case justifying punitive damages against the manufacturer or to help a manufacturer successfully defend against the plaintiff or codefendant. The resolution of many lawsuits and claims depends on the type and quality of documentation that exists. Thus, an important aspect of products liability prevention is to establish a document program which gives guidelines on how to create documents, how to protect those documents once they are created and how long to retain such documents before destruction.

Every manufacturing organization should adopt a policy that encourages the creation of documents concerning the process of designing, manufacturing and marketing its products. Certainly, if few documents are created in this area,

---

[85]A typical product safety audit check list is contained in **app 10-3.**

it is very unlikely that any will be used against that manufacturer in a particular lawsuit. However, there are many record-keeping requirements for various local, state and federal government agencies.[86] Rather than discourage the creation of documents, a manufacturer should establish procedures and educate its employees on how to create documents which are helpful in the event of products liability problems.

Every manufacturer should insure that documents are created to indicate how and why a product design was approved for manufacture and to reflect that the product was manufactured in accordance with the manufacturer's design specifications and any applicable standards. Sales and marketing, product service and performance records should also be maintained. The manufacturer should be in a position, at any time while a particular product is still operating in the field, to describe the methods it employed in designing, manufacturing and marketing that particular product.

After the manufacturer determines what types of documents should be created, it should establish procedures to insure that harmful documents are not created.[87] All employees should be instructed to assume that every document created is going to be discoverable and possibly used against that manufacturer sometime later in litigation. Employees should keep this thought in the back of their minds when committing anything to writing. However, it is important not to alarm employees and cause them to withhold harmful information from their supervisors.

For example, if an employee feels that a particular design is in some way unsafe, the employee should try to speak with a supervisor. If the matter cannot be resolved at that point, the employee should commit the concern to writing. However, the supervisor (i.e., engineering manager) should, in turn, respond to the criticism and indicate disagreement or agreement and what remedial action is being taken. These documents should be filed together for easy retrieval.

Manufacturers should try to establish a complete product history file in the custody of one person. In this way, all documents evidencing the reasons for a particular design will be easy to retrieve and use during the defense of litigation. In addition, the manufacturer's attorney can easily review the existing documents, determine if there is something harmful, and minimize the risk of having those documents discovered in litigation by settling. Surprise during litigation caused by "smoking guns" will thus be minimized.

Once documents have been created, it is possible to protect those documents from discovery by use of the attorney–client provilege and work product

---

[86] Office of the Federal Register, General Services Administration, Guide to Record Retention Requirements (rev ed 1982).

[87] *See* Gillham v Admiral Corp, 523 F2d 102 (6th Cir 1975), *cert denied,* 424 US 913 (1976); and Hoffman v Sterling Drug, Inc, 485 F2d 132 (3d Cir 1973), for cases where the manufacturers' documents were used against them.

doctrine.[88] The manufacturer's relationship with the outside attorneys or the inside law department should be carefully established in order to create and preserve the appropriate privilege. Inside corporate counsel must be careful not to become business advisors and therefore lose any privilege attached to being an attorney. Inside attorneys should report to a general counsel, have titles indicating that they are attorneys and have these titles appear on all written material.

Legal files should be kept separate from business files and appropriate documents directed to and received by the law department should be stamped to indicate if they are confidential or privileged. Legal opinions should indicate that they are being given in response to a request for legal advice. Such opinions should not include business advice and distribution should be limited.[89]

Ideally, a manufacturer's documents should be retained and easily retrievable for as long as those particular documents might be needed in the defense of litigation. Generally, products litigation in most states has no effective ending period.[90] Therefore, a manufacturer can be sued for injury or damage caused by its product no matter how long it has been out of the control of that manufacturer. Lawsuits involving products that are 30 or 40 years old are not uncommon.

The complete product history file should include information on prior products which were used to design the currently produced product. Many products evolve over the years and manufacturers should be careful not to discard documents which justify design decisions made many years ago on prior models.

With regard to newer produced products, sales records evidencing the terms of the contract between buyer and seller should be retained in order to defend a lawsuit based on breach of warranty, breach of contract or one alleging damages which have been disclaimed in the contract.

Retaining documents for long periods of time involves storage and handling costs. The manufacturer must evaluate the types of documents that have been created and the types that might be subsequently needed, and establish a retention period for each category of document. While it is not certain that lawsuits will be lost due to the lack of documents, an employee's memory and credibility are reduced by the lack of good evidence of the manufacturer's

---

[88] *See generally,* Williams & Spoto, *Corporate Policies for the Protection of Documents: Attorney-Client Privilege and Work Product Doctrine,* in Product Liability of Manufacturers: Prevention and Defense 1981, at 747 (Ross & Foley eds).

[89] *Id* 812.

[90] Twenty states have enacted statutes of repose which attempt to effectively stop the right to sue after a certain number of years from the date of sale or manufacture. These range from 5 to 12 years. These statutes have been challenged constitutionally and have certain exceptions which have been used to try to continue actions in such states beyond the repose period. *But see* Holland v Metropolitan Life Ins Co, 1981 Prod Liab Rep (CCH) ¶9142.

reasonable care in manufacturing that product.[91]

A manufacturer's policy and program for creation, protection and retention of documents should be clearly set forth and transmitted to the appropriate employees. Without such a policy and program, the manufacturer is at the mercy of plaintiff's attorneys who will try to use the manufacturer's documents to their best advantage. At a minimum, it will be difficult for the manufacturer to explain away the existence of any harmful documents. Instead, the manufacturer should take the opportunity to minimize risks and establish a program which might possibly enhance the quality of its defense.

§10.26 Attorney Involvement

Attorneys, either corporate counsel or outside counsel retained by a manufacturer for such purposes, have a unique opportunity to practice preventive law for the client in the products liability area. Attorneys should grasp this opportunity and become involved to the fullest extent possible. Obviously, many of the preventive techniques are technically oriented and do not require much input from an attorney. However, the attorney should learn about how these techniques are being performed and documented to be sure that they are sufficient in the event of a lawsuit.

In the area of new product design or product redesign, the attorney should be involved in the preparation of design specifications by providing information on legislation, standards, regulations and previous or present litigation, and by providing guidelines for determining what is reasonably foreseeable use and misuse.

The attorney should assist in the creation of prevention programs and participate as a member of any product safety committee. The attorney should help to create the document policy and program, and should review instruction books, maintenance and operation manuals, warning labels and advertising for products liability problems and should instruct marketing personnel about ways in which these problems can be prevented or limited. The attorney should also be involved in the process of creating disclaimers, warranties, exculpatory clauses and other contractual limitations for sales contracts and for contracts for the purchase of raw materials and component parts.

Accident Prevention Programs

§10.27 Introduction

Once an accident has occurred, many new factors must be considered in minimizing products liability exposure to the manufacturer. This is particularly

---

[91] See Gover, *The Development of a Decision Model for the Retention of Documents Pertinent to Product Liability Defense,* Recs Mgmt Q (July, Oct 1975).

true for the creation of documents such as accident reports and determining the appropriateness of recalls, retrofits or repairs of similar products. In addition, the occurrence of an accident brings into question insurance coverage, claims handling procedures and litigation management techniques.

§10.28 Insurance Coverage and the Role of the Insurance Company

Most insurance for products liability comes in the form of a standard comprehensive general liability policy. This policy provides for certain coverage, specific exemptions and a particular level of coverage.[92]

Generally, the manufacturer can have pure insurance with no deductible, insurance that is retrospectively rated or no insurance. The premium for retrospectively rated insurance is based on the manufacturer's loss experience. The losses upon which the premium is based are limited and there is pure insurance for losses which exceed that limit. For self-insurance, obviously, there is no premium to pay to an insurance company.

With pure insurance, the insurance company usually has the contractual right to select outside counsel, make decisions on settlement and manage the litigation. For coverage with a retrospective rating, the manufacturer is usually much more involved due to the passthrough nature of costs for attorneys' fees, settlements or verdicts. In these situations, the manufacturer can help select the attorney, be involved in the decision to settle or try the case and act as manager of that particular litigation. Where the manufacturer is self-insured, it will usually hire a claims handling organization to investigate and assist in the defense of the claim or lawsuit. This organization is usually paid an administrative fee which is not tied to any loss experience. However, the manufacturer retains the right to control the litigation.

In all situations, the manufacturer will usually have some type of excess insurance for losses exceeding a certain amount. With such coverage, the excess insurance carrier may also become involved in coverage questions, selecting its own outside counsel and making decisions on settlement or trial.

Recently, in cases involving long-term exposure to hazardous products (asbestos), there has been a controversy over insurance coverage of particular manufacturers. The major question is whether the insurance in force at the time of the first exposure to the harmful product or that in effect at the time the injury is manifested covers the risk. There have been conflicting decisions in this area[93] and the United States Supreme Court will have to decide the issue.

---

[92] *See generally* Wallace, *Products Liability Insurance: Scope of the Protection and Coverage Issues,* in Product Liability of Manufacturers: Prevention and Defense 1981, at 423 (Ross & Foley eds).

[93] Keene Corp v Insurance Co of N Am, 667 F2d 1034 (DC Cir 1981), *revg,* 513 F Supp 47 (DDC 1981), *cert denied,* 50 USLW 3716 (Mar 9, 1982), *reh denied,* __USLW__, (Apr 27, 1982).

§10.29 Litigation Management

Every manufacturer should have a well-thought out procedure for handling products liability litigation. Once the manufacturer learns of an accident, there should be procedures for investigating the accident, taking witness statements, inspecting the product involved and gaining an understanding as to how and why the product failed and caused injury or damage. This requires coordination between the insurance company, possibly outside counsel, inside counsel and the manufacturer's engineering, marketing and field service functions to insure that the necessary investigation is performed.[94]

Once the manufacturer learns about the reasons for the accident, it must consider possible recalls, retrofits, notices or repairs to similar products.[95] Decisions made in these areas can have a great impact on the pending litigation.

Once the investigation is completed, the manufacturer must retain outside counsel, educate that outside counsel and determine a philosophy and strategy for defending the lawsuit. Inside corporate counsel should take the lead in these activities. The inside attorney can greatly enhance the defense of a lawsuit while reducing the amount of time spent by the outside attorney to learn about the company, the product and the quality of the manufacturer's defense. The inside attorney should also be a party to all strategic decisions made during the pendency of litigation and should be intimately involved in the decision to settle or try a particular lawsuit.[96]

§10.30 Recalls, Retrofits and Repairs

Evidence of any remedial actions, such as recalls, retrofits and repairs taken after an accident has occurred, may be admissible in litigation involving that particular accident.[97] The subsequent remedial measure rule as generally expressed in §407 of the Federal Rules of Evidence provides for the inadmissibility of such evidence to prove a defect in a product unless the evidence of subsequent measures is offered for another purpose such as proving ownership, control, feasibility of precautionary measures or for impeachment purposes. Much litigation has surrounded the general rule and its rather broad exceptions and it is imperative for the manufacturer's attorneys to try to determine the effect of this rule on any subsequent remedial measures engaged in by that manufacturer.

The manufacturer should always consider engaging in some remedial activi-

---

[94] *See generally* Ferry, Modern Accident Investigation and Analysis (1980).

[95] See **§10.30.**

[96] See Sheridan, *The Role of House Counsel in Defending a Products Liability Lawsuit,* and Fredman, *Post-Accident Remedial Measures and Use of Corporate Counsel,* in Product Liability of Manufacturers: Prevention and Defense 1981, at 835 and 851 (Ross & Foley eds).

[97] See Wheeler, *Selected Evidentiary Issues of Current Importance in Product Liability Litigation,* in Product Liability of Manufacturers: Prevention and Defense 1981, at 1023 (Ross & Foley eds).

ty after an accident if this will reduce the likelihood of subsequent similar accidents. This should take place irrespective of its effect on pending litigation. Failure to take such remedial actions could result in future litigation and a possibility of punitive damages.

Once a manufacturer decides to engage in post-accident remedial measures, the analysis for deciding what measure to take is very similar to that for pre-accident decisions. [98]

§10.31 Spreading the Risk

The term *spreading the risk* means trying to reallocate the risk of loss to others in the chain of distribution of a product such as the distributor, retailer, product owner, installer, independent maintenance company or employer. This reallocation can occur generally through indemnity, contribution or comparative fault.[99]

Indemnity totally shifts the loss from one party to another. The easiest way for indemnity to arise is through an express contractual indemnity. Through such contractual provisions, the manufacturer may obtain an express indemnification for damage or injury caused by the other party's negligence. However, it is unlikely that the manufacturer will be able to totally shift its liability to someone else for damage or injury caused by its negligence.[100]

The manufacturer may also argue that an implied indemnity has arisen in favor of the manufacturer and against someone else in the chain of distribution. This can happen where the manufacturer has been passively negligent and the other party in the chain of distribution has been more active in causing the injury or damage.

Contribution provides for a partial sharing of loss by the manufacturer and others in the chain of distribution. Contribution among joint tortfeasors is provided for by statute in some jurisdictions and by judicial decision in others. There are limitations on seeking contribution, especially when the other party is a governmental agency or the plaintiff's employer.

The doctrines of indemnity and contribution have been radically affected by the adoption of comparative fault. Many states have adopted that doctrine which diminishes a plaintiff's recovery to the extent that the plaintiff's fault contributed to the injuries.[101] Thus, a plaintiff's negligence is no longer a complete defense to a products liability lawsuit. The use of indemnity and contribution in connection with comparative fault often can result in an equitable apportionment of liability among the parties to the action.

---

[98] See **§10.21**.

[99] *See generally,* Crowe & Fraser, *Spreading the Risk,* and Birnbaum, *Comparative Fault, Contribution and Idemnification,* in Product Liability of Manufacturers: Prevention and Defense 1981, at 645 and 669 (Ross & Foley eds).

[100] EL White, Inc v City of Huntington Beach, 21 Cal 3d 497, 597 P2d 505, 146 Cal Rptr 614 (1978).

[101] Schwartz, Comparative Negligence (1974).

Conclusion

§10.32 General Remarks

All manufacturers and others engaged in the distribution of products in the United States have great potential for incurring some legal liability for damage or injury caused by such products. This exposure is certain to increase and expand as courts find additional ways in which to compensate persons for such injury or damage. Manufacturers must take precautions to minimize or prevent such exposure. Even though engaging in such precautionary measures can increase the cost of products, in the long run, the result should be safer products, fewer accidents and less exposure for products liability. Attorneys should be fully integrated in the whole process of designing, manufacturing and marketing of products. Their legal expertise can greatly assist the manufacturer in meeting the goals of a prevention program.

Appendix 10-1 Model Uniform Product Liability Act*

DEPARTMENT OF COMMERCE

Office of the Secretary

Model Uniform Product Liability Act

Introduction

The Department of Commerce publishes herein its "Model Uniform Product Liability Act." It is offered for voluntary use by the states.

This Model Law will help to assure that persons injured by unreasonably unsafe products receive reasonable compensation for their injuries. It should also help to stabilize product liability insurance rates.

The Model Law, if enacted by the states, would introduce uniformity and stability into the law of product liability. This, in turn, would help stabilize product liability insurance rates. Uniformity and stability in this area are needed because product liability insurance rates are set on a countrywide basis. Thus, product liability law differs from medical malpractice, automobile, and other standard lines of liability.

The current system of having individual state courts develop product liability law on a case-by-case basis is not consistent with commercial necessity. Product sellers and insurers need uniformity in product liability law so they will know the rules by which they are to be judged. At the same time, product users are entitled to the assurance that their rights will be protected and will not be restricted by "reform" legislation formulated in a crisis atmosphere. Thus, the Model Law meets the needs of product users, sellers, and insurers.

Background of the Act

The Department of Commerce chaired an 18-month interagency study on the topic of product liability and published its final report [1] on November 1, 1977.

The genesis of the "Uniform Product Liability Act" can be traced directly to the Task Force study. The Task Force found that uncertainties in the tort-litigation system were a principal cause of the product liability problem. The "Task Force Report" noted that a few courts had come

> very close to holding that the tort-litigation system should provide a recovery for persons who merely proved that they were injured by a product [and that] while these cases appear to be relatively few in number, insurers have regarded them as quite important in their pricing practices.[2]

The "Task Force Report" called for improvement in insurer ratemaking practices but indicated that even if this were done, "the specter of these cases could still serve as an arguable justification for increasing premiums." [3] Decisions supporting this view continue to appear.

On the basis of the "Task Force Report," representatives from the Office of Management and Budget and the Domestic Policy Staff of the White House asked Commerce to prepare an options paper regarding what action, if any, the Federal Government should take to address the product liability problem. That paper was published in the **Federal Register** on April 6, 1978. 43 FR 14612 (1978).[4]

One of Commerce's recommendations was that a uniform product liability law

[1] The "Final Report (cited as the "Task Force Report") of the Federal Interagency Task Force on Product Liability" (cited as the "Task Force").

[2] "Task Force Report" at p. I–27.

[3] *Id.*

[4] Public comment on the "Options Paper" was later published in the **Federal Register** on September 11, 1978. 43 FR 40438 (1978).

* 44 Fed Reg 62,714 (1979).

be prepared. The overwhelming majority of public comment received in response to the "Options Paper" supported this recommendation.[5]

On July 20, 1978, the Administration announced its program to address the product liability problem. Included in that program as the principal long-range measure was the drafting of a model uniform product liability law.

On January 12, 1979, Commerce's "Draft Uniform Product Liability Law" was published in the **Federal Register** for public comment.[6] The written commentary received by Commerce regarding the Draft Law totals approximately 1500 pages, representing 240 separate communications.[7] The Department also made a special effort to bring the Draft Law to the attention of consumers. Working with its Director of Consumer Affairs and the Office of the Special Assistant to the President for Consumer Affairs, Commerce conducted consumer forums in Washington, D.C., Detroit, Los Angeles, and Atlanta.

In addition to meeting with consumer groups, the drafters of this Act met with representatives of product seller and insurer groups which expressed an interest in the proposal. The Draft Law was also reviewed at hearings before the Subcommittee on Oversight and Minority Enterprises of the House Committee on Small Business, and before the Subcommittee on Consumer Protection and Finance of the House Committee on Interstate and Foreign Commerce.

The final version of the "Uniform Product Liability Act" has benefited substantially from its review by, and input from, the various groups affected by the product liability problem.

Other Sources of the Act

Apart from the public comment, this Act is based on the work products of the Interagency Task Force on Product Liability, including its "Final Report," its "Legal Study," its "Industry Study," and its "Insurance Study."[8] Also, a thorough review was conducted of all major case law and law review literature that had been published since the time of the Task Force's "Legal Study."

As will be apparent from the Act's section-by-section analysis, attention was given to:

(1) The findings of the extensive "Product Liability Closed Claims Survey" conducted by the Insurance Services Office in 1976–77;

(2) All product liability legislation enacted at the state level, plus major proposals introduced in state legislatures in the past two years;

(3) Congressional hearings on product liability and the Report of the House Subcommittee on Capital, Investment and Business Opportunities of the Committee on Small Business. *See* H.R. Report No. 95–997, 95th Cong., 2d Sess. (1978) (Honorable John J. LaFalce, Chairman); and

(4) Privately drafted model product liability legislation.

A bibliography of some of the major resources considered by the Department is set forth in Appendix A.

Criteria for the Act

The criteria [9] utilized in evaluating the provisions of the Model Law were:

(1) *To ensure that persons injured by unreasonably unsafe products receive reasonable compensation for their injuries.*

---

[5] 43 FR 40443.

[6] 44 FR 2996.

[7] On file, Law Library, U.S. Department of Commerce.

[8] The Task Force's reports are available from the National Technical Information Service, Springfield, Virginia 22161 (Attention: Sales Desk). Reference should be made to the appropriate accession number, and a check made payable to NTIS in the proper amount should be enclosed:

Final Report—PB 273–220, $20.00 (1 vol.) (1977).

Selected Papers—PB 278–625, $17.50 (1 vol.) (1978).

Legal Study—PB 263–601, $31.25 (7 vol.) (1977) (The Research Group, Inc.)

Legal Study:

Executive Summary—PB 265–450, $6.00 (first of 7 vol.) (1977).

Industry Study—PB 265–542, $21.25 (2 vol.) (1977) (Gordon Associates, Inc.).

Insurance Study—PB 263–600, $9.00 (1 vol.) (1977) (McKinsey, Inc.).

[9] A more extensive discussion of these criteria appears in the "Task Force Report" at VII–29. The criteria are also set forth in Subsection 103(C) of the Act and are discussed in the corresponding section-by-section analysis.

Many proposed alternatives in product liability law are primarily justified by the fact that they contain "cost-saving devices" for product liability insurers or their insureds. However, these projected "cost savings" must be balanced against the responsibility of a product seller to provide reasonable compensation to persons harmed by unreasonably unsafe products. The cost of an accident should be shifted from a claimant to a product seller when there is a logical and articulated rationale for deeming it (as compared with the injured individual or society at large) "responsible" for the claimant's injuries.

(2) *To ensure the availability of affordable product liability insurance with adequate coverage to product sellers that engage in reasonably safe manufacturing practices.*

Product liability law should attempt to create a situation in which affordable product liability insurance is available to manufacturers that follow reasonably safe manufacturing practices. The law should not, however, be modified in order to provide such insurance to manufacturers who are unwilling or unable to follow reasonably safe manufacturing practices.

(3) *To place the incentive for loss prevention on the party or parties who are best able to accomplish that goal.*

Part of the product liability problem has been caused by unsafe manufacturing practices. Obviously, it is in the interest of all groups affected by the product liability problem to reduce the number of accidents caused by products. The Task Force study showed that product liability law can help bring about this goal. The threat of tort liability and product liability judgments has prompted manufacturers to make a greater effort to produce safe products. Nevertheless, existing state product liability law does not always place the incentive for loss prevention on the party or parties who can best achieve that goal.

The decision of where to place the incentive for loss prevention was not an easy one, and at least two factors helped determine the drafters' decision. One was based on pure economics—which party can prevent the loss at lowest cost? Economic analysis of preliminary drafts of this Act were helpful in this consideration.

A second factor focused on the question of who is in the best practical position to prevent a product-related harm. This consideration may be at counterpoint with pure economic analysis. Sometimes a product seller may be in a better practical position to implement a loss prevention technique although a product user could theoretically do so at a lesser cost.

(4) *To expedite the reparations process from the time of injury to the time the claim is paid.*

Delays in the reparations process do not serve any social interest. Seriously injured claimants can ill afford to endure long delays between the time of their injury and the time they are paid. Therefore, the Act has placed emphasis on arbitration and other means that will help expedite the reparations process.

(5) *To minimize the sum of accident costs, prevention costs, and transaction costs.*

The goal of minimizing accident, prevention, and transaction costs, while worthwhile, is not easy to fulfill within the tort-litigation system. For example, one can minimize "transaction costs" by abolishing trial by jury. However, this would be at the expense of other societal values which are particularly important in product liability cases, such as the need for the individualized judgment of cases and the experience of ordinary persons in making those judgments. Nevertheless, this consideration was significant enough to weigh in formulating the Act.

(6) *To use language that is comparatively clear and concise.*

Many product liability proposals that appear sound when stated in a broad and general manner break down when one focuses on the practicality of their implementation. In drafting the Act, practicality, together with conciseness and clarity of language, were important goals. The Act was drafted as a guideline for courts, not as a detailed legal contract between product seller and user.

Other considerations were utilized in the process of formulating each of the provisions. They are highlighted in the section-by-section analysis that accompanies the Act. Again, permeating the discussion of each provision is the concern that it is fair to all groups having an interest in the product liability problem.

It is important to understand the basic philosophy that underlies the Act. Product liability law is a branch of the law of torts. Its function is to shift the cost of an accident from a claimant to a defendant when the latter is deemed "responsible" for the claimant's injuries. This "responsibility" should be defined in terms that everyone can understand. Product liability law should indicate why a particular individual product seller was sufficiently blameworthy that it should bear the cost of that injury.

Tort law is not a compensation system similar to Social Security or Worker Compensation. A product seller should not, through the medium of tort law, be asked to pay merely because its product caused an injury. If a social judgment is made that product sellers are to bear the costs of all injuries caused by their products, it would be far more efficient to make purchasers of products third-party beneficiaries of product sellers' insurance policies as is the case with other compensation systems. Such systems also utilize cost-saving devices such as limiting recovery for lost earnings, eliminating recovery for pain and suffering, and abolishing the collateral source rule. In contrast, product liability law, with its full tort law recovery, reflects the social judgment that liability should be imposed only when it is fair to hold the individual product seller responsible for an injury.

This proposal is offered in the hope that it will stabilize product liability law and benefit product sellers and users alike.

C. L. Haslam,
General Counsel, Department of Commerce..
Victor E. Schwartz,
Chairman, Task Force on Product Liability and Accident Compensation.

Outline: Uniform Product Liability Act

Uniform Product Liability Act

Code

Preamble

This Act sets forth uniform standards for state product liability tort law. It does not cover all issues that may be litigated in product liability cases; rather, it focuses on those where the need for uniform rules is the greatest. The principal purposes of the Act are to provide a fair balance of the interests of both product users and sellers and to eliminate existing confusion and uncertainty about their respective legal rights and obligations. The fulfillment of these goals should help, first, to assure that persons injured by unreasonably unsafe products will be adequately compensated for their injuries and, second, to make product liability insurance more widely available and affordable, with greater stability in rates and premiums.

* * * * *

Analysis

Preamble

The importance this Act places on increasing the degree of certainty in the product liability litigation process is tempered by the recognition that, even with nationwide adoption of a uniform code, its application may vary from state to state on some issues. One of the Act's goals, the development of a fair balance of interests, has been achieved by applying the general criteria set forth in Subsection 103(C) of the Act and by revising the first draft of the Act (*see* 44 FR 2996 (1979)) in light of the public comment it generated. A second goal is to promote a greater degree of certainty than exists under the present system. This can be achieved if the Act is adopted by the states in which a substantial majority of product liability claims are brought.

* * * * *

Code

Sec. 100. Short Title

This Act shall be known and may be cited as the "Uniform Product Liability Act."

* * * * *

Analysis

Sec. 100. Short Title

This is the customary "short title" provision. It may be placed wherever state legislative practice dictates. If a state legislature introduces parts of the "Uniform Product Liability Act" as separate measures, the short title should be adjusted accordingly.

* * * * *

Code

Sec. 101. Findings

(A) Sharply rising product liability insurance premiums have created serious problems in commerce resulting in:

(1) Increased prices of consumer and industrial products;

(2) Disincentives for innovation and for the development of high-risk but potentially beneficial products;

(3) An increase in the number of product sellers attempting to do business without product liability insurance coverage, thus jeopardizing both their continued existence and the availability of compensation to injured persons; and

(4) Legislative initiatives enacted in a crisis atmosphere that may, as a result, unreasonably curtail the rights of product liability claimants.

(B) One cause of these problems is that product liability law is fraught with uncertainty and sometimes reflects an imbalanced consideration of the interests it affects. The rules vary from jurisdiction to jurisdiction and are subject to rapid and substantial change. These facts militate against

predictability of litigation outcome.

(C) Insurers have cited this uncertainty and imbalance as justifications for setting rates and premiums that, in fact, may not reflect actual product risk or liability losses.

(D) Product liability insurance rates are set on the basis of countrywide, rather than individual state, experience. Insurers utilize countrywide experience because a product manufactured in one state can readily cause injury in any one of the other states, the District of Columbia, or the Commonwealth of Puerto Rico. One ramification of this practice is that there is little an individual state can do to solve the problems caused by product liability.

(E) Uncertainty in product liability law and litigation outcome has added to litigation costs and may put an additional strain on the judicial system.

(F) Recently enacted state product liability legislation has widened existing disparities in the law.

* * * * *

Analysis

Sec. 101. Findings

Chapters VI and VII of the "Final Report of the Interagency Task Force on Product Liability" (hereinafter cited as "Task Force Report") provide support for most of the findings made here. Additional support comes from the Report of the Subcommittee on Capital. Investment, and Business Opportunities, "Product Liability Insurance," H.R. Rep. No. 95–997, 95th Cong., 2d Sess. (1978) (Honorable John J. LaFalce, Chairman) (hereinafter cited as "LaFalce Subcommittee Report"). Among other things, the "LaFalce Subcommittee Report" called for clarification and simplification of "present tort law relating to product liability by formulating Federal standards to be adopted by the States. . . ." *Id.* at 76

Individual state studies on product liability conducted in Missouri (Report of the Senate Select Committee on Product Liability, 1977), Illinois (Judiciary I Subcommittee on Product Liability: Report and Recommendations—Part 1, undated), Georgia (Report of the Senate Products Liability Study Committee, 1978), Maine (Governor's Task Force, 1978), Michigan (Department of Commerce Task Force on Product Liability Insurance, 1978), and Wisconsin (Product Liability: An Overview, Wisconsin Legislative Council Staff, 1978) provide additional support for individual findings.

The Maine and Georgia reports emphasize that individual state tort reforms can do little to affect the product liability problem. This conclusion was reaffirmed in Governor Grasso's message vetoing a product liability tort bill passed by the Connecticut legislature in 1978, and Governor Brown's message vetoing a product liability tort bill passed by the California legislature in 1979. Both messages stressed that uncoordinated individual state tort action will not stabilize product liability insurance rates.

More specific references to the findings of this Section appear in the following citations keyed to the various findings in each Subsection:

101 (A) (1) "Task Force Report" at V–19, VI–27–28.

(2) "Task Force Report" at VI–28–32.

(3) "Task Force Report" at VI–2–26.

(4) "Options Paper on Product Liability and Accident Compensation Issues," 43 FR 14612–14 (1978); "Georgia Report", Appendix B; Johnson, "Products Liability 'Reform': A Hazard to Consumers," 56 "N.C. L. Rev." 677 (1978); Phillips, "A Synopsis of the Developing Law of Products Liability," 28 "Drake L. Rev." 317, 388 (1979); Comment, "State Legislative Restrictions on Product Liability Actions," 29 "Mercer L. Rev." 619 (1978). *See also* "Federal-State Product Liability Legislation for Client and Counsel," Federal-State Reports, Inc. (1977–79).

101(B) "Task Force Report" at I–26–28, VII–15–17; "LaFalce Subcommittee Report" at 72; "Michigan Department of Commerce Task Force on Product Liability Insurance Report" at 6 (1978).

(C) Task Force "Insurance Study" at IV–88 (citing uncertainty); "Task Force Report" at V–48–49 (relationship of premium to risk).

(D) "Task Force Report" at I–28; "Maine Report" at 23.

(E) "Task Force Report" at VII–214–16; *see also* Insurance Services Office, "Product Liability Closed Claims Survey" (hereinafter cited as "ISO Closed Claims Survey") at 118–30 (1977).

(F) *See* "Federal-State Product Liability Legislation for Client and Counsel," Federal-State Reports, Inc. (1977–79); "Product Liability Trends," at 97–98, 104–05, 157–58 (The Research Group, Inc., 1978); "Business Insurance," July 23, 1979 at 31; *see also* "Wisconsin Report" at 29–37 (describing 25 separate bills on product liability introduced in one legislative session).

Code

Sec. 102. Definitions

(A) *Product Seller.* "Product seller" means any person or entity that is engaged in the business of selling products, whether the sale is for resale, or for use or consumption. The term includes a manufacturer, wholesaler, distributor, or retailer of the relevant product. The term also includes a party who is in the business of leasing or bailing such products.

The term "product seller" does not include:

(1) A seller of real property, unless that person is engaged in the mass production and sale of standardized dwellings or is otherwise a product seller;

(2) A provider of professional services who utilizes or sells products within the legally authorized scope of its professional practice. A non-professional provider of services is not included unless the sale or use of a product is the principal part of the transaction, and the essence of the relationship between the seller and purchaser is not the furnishing of judgment, skill, or services;

(3) A commercial seller of used products who resells a product after use by a consumer or other product user, provided the used product is in essentially the same condition as when it was acquired for resale; and

(4) A finance lessor who is not otherwise a product seller. A "finance lessor" is one who acts in a financial capacity, who is not a manufacturer, wholesaler, distributor, or retailer, and who leases a product without having a reasonable opportunity to inspect and discover defects in the product, under a lease arrangement in which the selection, possession, maintenance, and operation of the product are controlled by a person other than the lessor.

(B) *Manufacturer.* "Manufacturer" includes a product seller who designs, produces, makes, fabricates, constructs, or remanufactures the relevant product or component part of a product before its sale to a user or consumer. It includes a product seller or entity not otherwise a manufacturer that holds itself out as a manufacturer.

A product seller acting primarily as a wholesaler, distributor, or retailer of a product may be a "manufacturer" but only to the extent that it designs, produces, makes, fabricates, constructs, or remanufactures the product before its sale.

(C) *Product.* "Product" means any object possessing intrinsic value, capable of delivery either as an assembled whole or as a component part or parts, and produced for introduction into trade or commerce. Human tissue and organs, including human blood and its components, are excluded from this term.

The "relevant product" under this Act is that product, or its component part or parts, which gave rise to the product liability claim.

(D) *Product Liability Claim.* "Product liability claim" includes any claim or action brought for harm caused by the manufacture, production, making, construction, fabrication, design, formula, preparation, assembly, installation, testing, warnings, instructions, marketing, packaging, storage, or labeling of the relevant product. It includes, but is not limited to, any action previously based on: strict liability in tort; negligence; breach of express or implied warranty; breach of, or failure to, discharge a duty to warn or instruct, whether negligent or innocent; misrepresentation, concealment, or nondisclosure, whether negligent or innocent; or under any other substantive legal theory.

(E) *Claimant.* "Claimant" means a person or entity asserting a product liability claim, including a wrongful death action, and, if the claim is asserted through or on behalf of an estate, the term includes claimant's decedent. "Claimant" includes any person or entity that suffers harm.

(F) *Harm.* "Harm" includes: (1) damage to property; (2) personal physical injuries, illness and death; (3) mental anguish or emotional harm attendant to such personal physical injuries, illness or death; and (4) mental anguish or emotional harm caused by the claimant's being placed in direct personal physical danger and manifested by a substantial objective symptom. The term "harm" does not include direct or consequential economic loss.

(G) *Reasonably Anticipated Conduct.* "Reasonably anticipated conduct" means the conduct which would be expected of an ordinary reasonably prudent person who is likely to use the product in the same or similar circumstances.

(H) *Preponderance of the Evidence.* "A preponderance of the evidence" is that measure or degree of proof which, by the weight, credit, and value of the aggregate evidence on either side, establishes that it is more probable than not that a fact occurred or did not occur.

(I) *Clear and Convincing Evidence.* "Clear and convincing evidence" is that measure or degree of proof that will produce in the mind of the trier of fact a firm belief or conviction as to the allegations sought to be established. This level of proof is greater than mere "preponderance of the evidence," but less than proof beyond a reasonable doubt.

(J) *Reckless Disregard.* "Reckless disregard" means a conscious indifference to the safety of persons or entities that might be harmed by a product.

(K) ***Express Warranty.*** **"Express warranty" means any positive statement, affirmation of fact, promise, description, sample, or model relating to the product.**

Analysis

Sec. 102. Definitions

(A) *Product Seller.* "Product seller" includes any party in the regular commercial distribution chain. It does not include the occasional private seller. This is in accord with the "Restatement (Second) of Torts" Section 402A, Comment f (1965). The term also includes lessors (except finance lessors) and bailors of products, in accord with the majority of decisions that have addressed the issue. *See* Annot., 52 "A.L.R.3d" 121 (1973); "Francioni v. Gibsonia Truck Corp.," 472 Pa. 362, 372 A.2d 736 (1977) (finance lessor); Fraser, "Application of Strict Tort Liability to the Leasing Industry," 34 "Bus. Law." 605 (1979).

The Act excludes the seller of real property from its coverage, except for a builder-vendor engaged in the mass production and sale of standardized dwellings, including modular homes. *See* "Schipper v. Levitt & Sons," 44 N.J. 70, 207 A.2d 314 (1965); "Berman v. Watergate West, Inc.," 391 A.2d 1351 (D.C. 1978) (extending "Schipper" to cooperative apartments); "Fuqua Homes, Inc. v. Evanston Bldg. & Loan Co.," 52 Ohio App. 2d 399, 370 N.E.2d 780 (1977) (modular homes). *But see* "Wright v. Creative Corp.," 30 Colo. App. 575, 498 P.2d 1179 (1972) (rejecting "Schipper"). The potential liability of sellers of real property not engaged in such large-scale operations is left to each state's laws governing real property transactions. However, all sellers of building materials, furnishings, appliances, and other products made part of improvements to real property are not exempted from the coverage of the Act. *See* Maldonado, "Builder Beware: Strict Tort Liability for Mass Produced Housing, 7 Real Estate L. J." 283 (1979).

The Act also excludes from its coverage the provider of professional services when a product is utilized or sold as part of the rendition of such services. *Compare* "Barbee v. Rogers," 425 S.W.2d 342 (Tex. 1968), *with* "Newmark v. Gimbel's, Inc.," 54 N.J. 585, 258 A.2d 697 (1969) (differentiating "sales-service hybrid transaction" of a beautician from "professional ministration" of a physician or dentist). The majority of current decisions look to the factual circumstances of each case and generally exclude persons exercising professional judgment within applied product liability doctrines where the provider is called upon to exercise such professional judgment. However, appropriate remedies may be available to injured parties under other theories of law, including malpractice. *See* "Batiste v. American Home Prods. Corp.," 32 N.C. App. 1, 231 S.E.2d 269 (1977).

Consequently, it is the intent of the drafters of this Act that professionals, such as pharmacists, physicians, optometrists, and opticians, should *not* be considered product sellers in those circumstances where they are selling a product while acting within their legally authorized scope of professional practice. *See* "Bichler v. Willing," 5 App. Div. 2d 331, 397 N.Y.S.2d 57 (1977); "Batiste v. American Home Prods. Corp.," *supra*. In addition, pharmacists or other professionals, employed by and working within the scope of their employment for a hospital or other health-related facility, would *not* be considered product sellers within this context since they would be rendering a part of the overall services of such facilities.

On the other hand, a pharamcist or other professional who is engaged in a commercial, non-professional sales transaction, such as the sale of perfume or photographic film, *would* be considered a product seller. Thus, the Act states that a party shall be considered a product seller when the sale of a product is the principal part of the transaction and when the essence of the relationship between the buyer and seller is *not* the furnishing of a professional skill or service. *See* Annot., 2 "A.L.R. 3d" 1425 (1970).

With respect to non-professional services, when the sale of a product is the principal part of the transaction, and the essence of the relationship is not the furnishing of judgment, skill, or services, the non-manufacturer product seller is responsible under Section 105 of this Act. Therefore, when a non-professional service, such as application, installation, or dealer preparation, is an incidental part of the overall transaction of sale of a product, the provisions of Section 105 govern the entire transaction. *See* "Newmark v. Gimbel's, Inc.," 54 N.J. 585, 258 A.2d 697 (1969); "Winters v. Sears, their legally authorized scope of practice. Thus, in the absence of any product preparation or modification, of any representation by service providers that the products are their own, or of warranty, the courts have generally not Roebuck & Co.," 89 "A.L.R.3d" 196, 554 S.W.2d 656 (Mo. 1977). However, when the non-professional service is the predominant element of the transaction, or when the service and sale elements are clearly distinguishable or contracted for separately, the Act does not apply to such non-professional service. Instead, it is governed by other applicable law of the state. *See* "Nickel v. Hyster Co.," 97 Misc. 2d 770, 412 N.Y.S.2d 273 (1978). *See also* Brook, "Sales-Service Hybrid Transactions: A Policy Approach," 28 Sw. L.J. 575 (1975); and Greenfield, "Consumer Protection in Service Transactions—Implied Warranties and Strict Liability in Tort," 1974 "Utah L. Rev." 661.

It is also the intent of the drafters that this Act include manufacturers and other product sellers of new or remanufactured products, but not ordinary commercial sellers of used products. In this context, the commercial seller of used products is one who resells a product after it has been used by a consumer or other product user and is in essentially the same condition as when it was acquired for resale. Thus, the commercial seller of used products may be differentiated from sellers of fully rebuilt or remanufactured products that are included within the definition of "manufacturer" under Subsection (B). The slight majority of decisions indicate that product liability law does apply to sellers of used products; however, such sellers are not held to the same standards as are sellers of new products, due to the different nature and condition of the used product at the time of sale. Therefore, the issue of the potential liability of sellers of used products is left for resolution not under this Act, but rather under the other law of the state. *See* "Tillman v. Vance Equip. Co.," 596 P. 2d 1299, 286 Or. 747 (1979); "Mickle v. Blackmon," 252 S.C. 202, 166 S.E.2d 173 (1969) (manufacturer); "Peterson v. Lou Bachrodt Chevrolet Co.," 61 Ill. 2d 17,

329 N.E. 2d 785 (1975) ("as is" sale by dealer).

(B) *Manufacturer.* The term encompasses those product sellers who initiate and carry out the process of production. It also includes manufacturers of component parts, "private labelers" who hold themselves out to the public as manufacturers, and those product sellers who rebuild or remanufacture products for resale in "like new" condition. It does not include independent product designers whose services are contracted for by product sellers, if such designers are not otherwise engaged in the business of selling products. *See* Subsection (A) (definition of "product seller"). Unless they actually engage in some element of manufacture, traditional wholesalers, distributors, and retailers are excluded from the definition. However, if they do participate in the actual manufacturing process, they are deemed manufacturers, but only to the extent that they engage in such activity. They do not become manufacturers of the product as a whole.

This definition is drawn in part from Arizona's product liability law. *See* "Ariz. Rev. Stat. Ann." Section 12–681 (1) (Supp. 1978). Its greatest import within this Act is in regard to the responsibility of a manufacturer for defective products, as contrasted with that of other product sellers. *See* Section 105, "The Basic Standards of Responsibility for Product Sellers Other Than Manufacturers."

(C) *Product.* "Product" means property which, as a component part or an assembled whole, is movable, and possesses intrinsic value. Therefore, included are all goods, wares, merchandise, and their components, as well as articles and commodities capable of delivery for introduction into trade or commerce.

The definition follows existing case law by including movable dwellings, such as mobile homes, campers and similar vehicles. *See* "Morrow v. New Moon Homes, Inc.," 548 P.2d 279 (Alaska 1976) (mobile home). Also included are water, natural gas, and electrical energy provided by public utilities or other product sellers. *See* "Moody v. City of Galveston," 524 S.W.2d 583 (Tex. Civ. App. 1975) (water); "Harris v. Northwest Natural Gas Co.," 284 Or. 571, 588 P.2d 18 (1978) (warnings regarding gas); "Ransome v. Wisconsin Elec. Power Co.," 87 Wis. 2d 605, 275 N.W.2d 641 (1979) (electricity). However, human biologicals, such as blood and other bodily tissues or organs, are expressly excluded from the definition. In addition, representations of value, such as money or choses in action, are excluded, although the physical document and the ink or any other material used in printing the document do fall within the definition. *See* 72 *C.J.S.* "Product" (1951), and Sections 85–92 (Supp. 1975); 63 "Am. Jur. 2d" "Products Liability" Section 5 (1972).

In a specific case, the particular product with which this Act is concerned is the "relevant product" which actually gave rise to the product liability claim. The "relevent product" may be the product as a whole or the particular component part or parts which gave rise to the product liability claim.

(D) *Product Liability Claim.* An important purpose of this Act is to consolidate product liability actions that have, at times, been separated under theories of negligence, warranty, and strict liability. This approach was suggested by the Task Force's "Legal Study" as well as by the "LaFalce Subcommittee Report." While an argument may be made that negligence theory is qualitatively different from strict liability and, therefore, should be preserved, product liability theory and practice have merged into a single entity and can only be stabilized if there is one, and not a multiplicity of, causes of action.

"Product liability claim" embraces express as well as implied warranty actions.

(E) *Claimant.* Living persons and those claiming through or on behalf of an estate in wrongful death or survival actions are both included within the meaning of the word "claimant."

Although the "Restatement (Second) of Torts" leaves open the question of whether persons other than product users should be included within the compass of product liability claims, subsequent case law has been almost uniform in ruling that such persons should be included. *See* "Giberson v. Ford Motor Co.," 504 S.W.2d 8 (Mo. 1974) (collecting cases). *See also* Annot., 33 "A.L.R. 3d" 415 (1970). The definition follows this line of decisions. *See* "Guarino v. Mine Safety Appliance Co.," 25 N.Y.2d 460, 255 N.E.2d 173, 306 N.Y.S.2d 942 (1969) (rescuer).

(F) *Harm.* Section 402A of the "Restatement" includes physical harm to persons and property. This Act provides that harm may manifest itself as damage to property, disability, including personal physical injuries and illness, death, and mental anguish or emotional harm attendant to such disability or death. It also includes mental anguish or emotional harm in circumstances in which the claimant is actually placed in personal danger, and the mental anguish or emotional harm is accompanied by substantial objective symptomatology. *See* "Wallace v. Coca-Cola Bottling Plants, Inc.," 269 A.2d 117 (Me. 1970).

The term also includes damage to the product itself. There is strong case law support for leaving direct economic loss cases to the field of commercial law. "Hawkins Constr. Co. v. Matthews Co., Inc." 190 Neb. 546, 209 N.W.2d 643 (1973); "Dennis v. Willys-Overland Motors, Inc.," 111 F. Supp. 875 (W.D. Mo. 1958); "Price v. Gatlin," 241 Or. 315, 405 P.2d 502 (1965). Courts that have allowed recovery in tort where the purchased property is "destroyed," but not where there is "mere loss of the bargain," have had difficulty applying this distinction where the product is wholly ineffective. *See* "Anthony v. Kelsey-Hayes Co.," 25 Cal. App. 3d 442, 102 Cal. Rptr. 113 (1972); "States Steamship Co. v. Stone Manganese Marine Ltd.," 371 F. Supp. 500, 504–505 (D.N.J. 1973) (collecting cases). For this reason and the fact that loss of the bargain damages are in essence a part of commercial law, claims for direct economic harm have been left to the "Uniform Commercial Code" ("U.C.C.").

The Act also does not include damages for consequential economic losses. Almost all courts have been in accord with "Seely v. White Motor Co.," 63 Cal. 2d 9, 403 P.2d 145, 45 Cal. Rptr. 17 (1965), on this issue and have left claimants with whatever rights they have under the "U.C.C." Pursuant to Section 103, the Act follows the weight of authority in this regard, and does not preempt such recovery under the "U.C.C." *See* "Brown v. Western Farmers Ass'n.," 268 Or. 470, 521 P.2d 537 (1974); "Eli Lilly & Co. v. Casey," 472 S.W.2d 598 (Tex. Civ. App. 1971); "Paul O'Leary Lumber Corp. V. Mill Equip., Inc.," 448 F.2d 536 (5th Cir. 1971).

The insurance costs of extending consequential economic losses beyond parties to a contract would be enormous. It is much less expensive and more efficient for the product purchaser to obtain insurance against consequential economic losses caused by business stoppage. Also, most courts believe that a commercial purchaser should be charged with the risk that the purchased product will not match its economic expectations unless the manufacturer agrees that it will. *See* Note, "Economic Loss in Products Liability Jurisprudence," 66 "Colum. L. Rev." 917 (1966).

(G) *Reasonably Anticipated Conduct.* The definition is based in part on Arizona product liability law. *See* "Ariz. Rev. Stat. Ann." Section 12–681(4) (Supp. 1978). "Anticipated" conduct includes occurrences which are expected, ordinary, usual, and familiar. The definition focuses on the class of persons which the product seller knows is likely to use the product. *See* "Thibault v. Sears, Roebuck & Co.", 395 A. 2d 843 (N.H. 1978). The concept of "reasonably anticipated conduct" should be contrasted with "foreseeable conduct." Almost any kind of misconduct with regard to products can be "foreseeable"—especially if the trier of fact is permitted to use hindsight, *e.g.*, that a soda bottle will be used for a hammer, that someone will attempt to drive a land vehicle on water, that perfume will be poured on a candle in order to scent it. *See* "Moran v. Faberge, Inc.," 273 Md. 538, 332 A. 2d 11 (1975). This same conduct may not be reasonably anticipated by a potential product litigant, or by the trier of fact.

The Act's reliance on the concept of "reasonably anticipated conduct" places incentives for loss prevention on both product sellers and product users. It also helps to ensure that the price of products is not affected by the liability insurance costs that would spring from providing coverage for abnormal product use.

(H) *Preponderance of the Evidence.* Unless otherwise stated, the standard of evidence under this Act is the standard utilized in most civil litigation—a preponderance of the evidence. This standard requires a showing that it is more probable than not that a fact occurred or did not occur. *See* 30 "Am. Jur. 2d" "Evidence" Section 1164 (1967).

(I) *Clear and Convincing Evidence.* Proof that is "clear and convincing" not only carries with it the power to persuade the mind as to its probable truth or correctness of fact, but also has an additional element of clinching such truth. The term is best understood in context. It requires more proof than does the preponderance of the evidence standard (the ordinary standard under this Act), but does not require proof beyond a reasonable doubt. *See, e.g.,* "Aiello v. Knoll Golf Club," 64 N.J. Super. 156, 165 A. 2d 531 (1960); "Cross v. Ledford," 161 Ohio St. 469, 120 N.E. 2d 118 (1954); "Brown v. Warner," 78 S.D. 647, 107 N.W. 2d 1 (1961).

(J) *Reckless Disregard.* As Prosser has indicated, "reckless disregard" occurs where the actor

> has intentionally done an act of an unreasonable character in disregard of a risk known to him or [one that was] so obvious that he must be taken to have been aware of it, and so great as to make it highly probable that harm would follow.

W. Prosser, "Torts" 185 (4th ed. 1971).

The term denotes aggravated conduct which represents a major departure from ordinary negligence.

(K) *Express Warranty.* This definition is based on its counterpart in the "Uniform Commercial Code" and includes any positive statement, affirmation of fact, promise, description, sample, or model relating to the product. *See* "U.C.C." Section 2–313; 63 "Am. Jur. 2d" "Products Liability" Section 93 (1972). The product seller may give an express warranty orally or in writing, or through any other actions intended as a communication. *See* "Alan Wood Steel Co. v. Capital Equip. Enterprises, Inc.," 39 Ill. App. 3d 48, 349 N.E. 2d 627 (1976); "Larutan Corp. v. Magnolia Homes Mfg. Co.," 190 Neb. 425, 209 N.W. 2d 177 (1973). It should be noted that an action based on a violation of an express warranty must include the element of reliance, and the breach must relate to a misrepresentation of material facts. *See* Subsection 104(D).

* * * * *

Code

Sec. 103. Scope of This Act

(A) This Act is in lieu of and preempts all existing law governing matters within its coverage, including the "Uniform Commercial Code" and similar laws; however, nothing in this Act shall prevent the recovery, under the "Uniform Commercial Code" or similar laws, of direct or consequential economic losses.

(B) A claim may be asserted under this Act even though the claimant did not buy the product from, or enter into any contractual relationship with, the product seller.

(C) Whenever this Act does not provide a rule of decision, reference may be made to other sources of law, provided that such reference conforms to the intent and spirit of this Act as set forth in the following criteria used as guidelines for its development:

(1) To ensure that persons injured by unreasonably unsafe products receive reasonable compensation for their injuries;

(2) To ensure the availability of affordable product liability insurance with adequate coverage to product sellers that engage in reasonably safe manufacturing practices;

(3) To place the incentive for loss prevention on the party or parties who are best able to accomplish that goal;

(4) To expedite the reparations process from the time of injury to the time the claim is paid;

(5) To minimize the sum of accident costs, prevention costs, and transaction costs; and

(6) To use language that is

comparatively clear and concise.

* * * * *

Analysis

Sec. 103. Scope of This Act

(A) The Act consolidates all product liability recovery theories into one. The approach taken is in accord with the Task Force's "Legal Study." While some have argued that for trial tactics purposes, it is useful to retain the negligence and breach of warranty causes of action as distinct from strict tort liability, a claimant's attorney can retain the essence of this utility by showing the basic wrongfulness of the product seller's conduct under Sections 104 or 105.

The Act explicitly preempts the "Uniform Commercial Code" ("U.C.C.") and similar laws in instances where such laws have governed matters within the Act's coverage. The purpose of this provision is to prevent conflicting product liability rules and remedies. *See, e.g.*, "Swartz v. General Motors Corp.," 378 N.E.2d 61 (Mass. 1978). Although the Act eliminates breach of warranty as a separate cause of action in product liability cases, express warranties continue to play an important role in the Act. *See, e.g.*, Subsections 102(K), 104(D), 105(B), 106(B)(4), 107(E)(4), 110(A)(2), 110(B)(2)(a).

Because the Act does not provide remedies for purely economic losses, Subsection (A) ensures that the Act does not prevent recovery under commercial law for loss of the use of a product. Thus, product liability actions for personal injury, illness, death, or damage to property (other than the product itself) are governed by this Act. On the other hand, recovery under the "U.C.C." or similar law for economic losses is not precluded by this Act.

(B) The Act is in accord with the "Restatement (Second) of Torts" in that it is unnecessary for the claimant to be in contractual privity with the product seller in order to recover for harm. *See* "Restatement (Second) of Torts" Section 402A, Comment l.

(C) The Act and its accompanying commentary do not purport to be an exhaustive compilation of the entire subject of product liability law. Rather, they focus on subject matter areas that the "Task Force Report" suggested have created the most problems and are of major importance.

The interstices of the Act will be filled by statutory or common law additions of the individual states. Some of these interstitial issues will be pointed out in the section-by-section analysis. Others will be discovered in the course of litigation under the Act.

It is the intent of Subsection (C) that these additions of statutory or common law should enhance, rather than conflict with, the basic purposes of this Act. Those courts or legislatures that make such additions should keep in mind the "Criteria for the Act" set forth in this Subsection. A full explanation of the criteria is set forth in the Introduction to the Act. In regard to these criteria, it is essential to note that this Act does not intend to set up a compensation system similar to Worker Compensation. Rather, the purpose of the Act is to impose liability "only where it is fair to deem the product seller responsible for an injury." "Introduction," p. 14, *supra.*

* * * * *

Code

Sec. 104. Basic Standards of Responsibility for Manufacturers

A product manufacturer is subject to liability to a claimant who proves by a preponderance of the evidence that the claimant's harm was proximately caused because the product was defective.

A product may be proven to be defective if, and only if:

(1) It was unreasonably unsafe in construction (Subsection A);

(2) It was unreasonably unsafe in design (Subsection B);

(3) It was unreasonably unsafe because adequate warnings or instructions were not provided (Subsection C); or

(4) It was unreasonably unsafe because it did not conform to the product seller's express warranty (Subsection D).

Before submitting the case to the trier

of fact, the court shall determine that the claimant has introduced sufficient evidence to allow a reasonable person to find, by a preponderance of the evidence, that one or more of the above conditions existed and was a proximate cause of the claimant's harm.

(A) *The Product Was Unreasonably Unsafe in Construction.* In order to determine that the product was unreasonably unsafe in construction, the trier of fact must find that, when the product left the control of the manufacturer, the product deviated in some material way from the manufacturer's design specifications or performance standards, or from otherwise identical units of the same product line.

(B) *The Product Was Unreasonably Unsafe in Design.*

(1) In order to determine that the product was unreasonably unsafe in design, the trier of fact must find that, at the time of manufacture, the likelihood that the product would cause the claimant's harm or similar harms, and the seriousness of those harms outweighed the burden on the manufacturer to design a product that would have prevented those harms, and the adverse effect that alternative design would have on the usefulness of the product.

(2) Examples of evidence that is especially probative in making this evaluation include:

(a) Any warnings and instructions provided with the product;

(b) The technological and practical feasibility of a product designed and manufactured so as to have prevented claimant's harm while substantially serving the likely user's expected needs;

(c) The effect of any proposed alternative design on the usefulness of the product;

(d) The comparative costs of producing, distributing, selling, using, and maintaining the product as designed and as alternatively designed; and

(e) The new or additional harms that might have resulted if the product had been so alternatively designed.

(C) *The Product Was Unreasonably Unsafe Because Adequate Warnings or Instructions Were Not Provided.*

(1) In order to determine that the product was unreasonably unsafe because adequate warnings or instructions were not provided about a danger connected with the product or its proper use, the trier of fact must find that, at the time of manufacture, the likelihood that the product would cause the claimant's harm or similar harms and the seriousness of those harms rendered the manufacturer's instructions inadequate and that the manufacturer should and could have provided the instructions or warnings which claimant alleges would have been adequate.

(2) Examples of evidence that is especially probative in making this evaluation include:

(a) The manufacturer's ability, at the time of manufacture, to be aware of the product's danger and the nature of the potential harm;

(b) The manufacturer's ability to anticipate that the likely product user would be aware of the product's danger and the nature of the potential harm;

(c) The technological and practical feasibility of providing adequate warnings and instructions;

(d) The clarity and conspicuousness of the warnings or instructions that were provided; and

(e) The adequacy of the warnings or instructions that were provided.

(3) In any claim under this Subsection, the claimant must prove by a preponderance of the evidence that if adequate warnings or instructions had been provided, they would have been effective because a reasonably prudent product user would have either declined to use the product or would have used the product in a manner so as to have avoided the harm.

(4) A manufacturer shall not be liable for its failure to warn or instruct about dangers that are obvious; for "product misuse" as defined in Subsection 112(C)(1); or for alterations or modifications of the product which do not constitute "reasonably anticipated conduct" under Subsection 102(G).

(5) A manufacturer is under an obligation to provide adequate warnings or instructions to the actual product user unless the manufacturer provided such warnings to a person who may be

reasonably expected to assure that action is taken to avoid the harm, or that the risk of the harm is explained to the actual product user.

For products that may be legally used only by or under the supervision of a class of experts, warnings or instructions may be provided to the using or supervisory expert.

For products that are tangible goods sold or handled only in bulk or other workplace products, warnings or instructions may be provided to the employer of the employee-claimant if there is no practical and feasible means of transmitting them to the employee-claimant.

(6) *Post-Manufacture Duty to Warn.* In addition to the claim provided in Subsection (C)(1), a claim may arise under this Subsection where a reasonably prudent manufacturer should have learned about a danger connected with the product after it was manufactured. In such a case, the manufacturer is under an obligation to act with regard to the danger as a reasonably prudent manufacturer in the same or similar circumstances. This obligation is satisfied if the manufacturer makes reasonable efforts to inform product users or a person who may be reasonably expected to assure that action is taken to avoid the harm, or that the risk of harm is explained to the actual product user.

(D) *The Product Was Unreasonably Unsafe Because It Did Not Conform to an Express Warranty.* In order to determine that the product was unreasonably unsafe because it did not conform to an express warranty, the trier of fact must find that the claimant, or one acting on the claimant's behalf, relied on an express warranty made by the manufacturer or its agent about a material fact or facts concerning the product and this express warranty proved to be untrue.

A "material fact" is any specific characteristic or quality of the product. It does not include a general opinion about, or praise of, the product.

The product seller may be subject to liability under Subsection (D) although it did not engage in negligent or fraudulent conduct in making the express warranty.

* * * * *

Analysis

Sec. 104. Basic Standards of Responsibility for Manufacturers

No single product liability issue has generated more controversy than the question of defining the basic standards of responsibility to which product manufacturers are to be held. *See, e.g.*, Epstein, "Products Liability: The Search for the Middle Ground," 56 "N.C. L. Rev." 643 (1978); *see also* Vetri, "Products Liability: Developing a Framework for Analysis," 54 "Or. L. Rev." 293, 310 (1975); Henderson, "Judicial Review of Manufacturers' Conscious Design Choice: The Limits of Adjudication," 73 "Colum. L. Rev." 1531 (1973); Keeton, "Product Liability and the Meaning of Defect," 5 "St. Mary's L. J." 30 (1973); Wade, "On the Nature of Strict Tort Liability for Products," 44 "Miss. L. J." 825 (1973); Walkowiak, "Product Liability Litigation and the Concept of Defective Goods: 'Reasonableness' Revisited," 44 "J. Air Law & Commerce" 705 (1978).

Much of the controversy appears to have sprung from the fact that the authors of Section 402A of the "Restatement (Second) of Torts" were focusing on problems relating to product mismanufacture or defective construction, and not on problems relating to defective design or the duty to warn. *See* "Restatement (Second) of Torts" Section 402A, Appendix (1965) (most cases cited deal with defects in construction cases); Wade, *supra*, 44 "Miss. L. J." at 830–32. In the 15 years following the publication of the "Restatement," courts have struggled to define standards of responsibility with respect to design and the duty to warn. *See, e.g.*, "Barker v. Lull Eng'r. Co.," 20 Cal. 3d 413, 573 P.2d 443, 143 Cal. Rptr. 225 (1978); "Cepeda v. Cumberland Eng'r. Co.," 76 N.J. 152, 386 A.2d 816 (1978); "Phillips v. Kimwood Mach. Co.," 269 Or. 485, 525 P.2d 1033 (1974).

In the course of this struggle, courts appear to have drifted away from rationales for imposing liability and toward verbal formulae that attempt to distinguish between negligence and strict liability.

The drafters of this Act have focused on the basic rationale of product liability and made decisions that should make it easier for courts to decide these cases. It was determined that "strict liability" is justified in two product liability areas—defects in construction and breach of express warranty.

The "Task Force Report" concluded that strict liability for defective construction can be absorbed within the existing liability insurance system. There is a degree of predictability with regard to these defective products that is not found with respect to products that are defective in design or to failure to warn. Strict liability for defective construction has also been predicated on Section 402A of the "Restatement" and implied warranty claims under commercial law. These sources support the position that consumers have the right to expect that products are free from construction defects.

Strict liability cases involving breach of an express warranty can also be justified. If a manufacturer makes a specific representation about its product, it is fair to hold the manufacturer to that promise. Moreover, the consumer has the right to expect that a product will live up to the manufacturer's representations.

While some courts have indicated that strict liability should also be applied in design and duty-to-warn cases, it is difficult to find an adequate rationale to support that result. A few courts have sought to justify the result under a theory of "risk distribution" wherein the product seller distributes the costs of all product-related risks through liability insurance; however, this rationale breaks down in practice. The application of uncertain strict liability principles in the areas of design and duty to warn places a whole product line at risk; therefore, a firmer liability foundation is needed. In terms of creating incentives for loss prevention, the approach of applying strict liability principles to design and duty-to-warn cases represents an "overkill;" a fault system will provide the needed incentive.

If the costs of product-related injuries are always to be distributed through the price of the product, the mechanism to bring about this result should be an appropriate compensation system with limited damages, not the tort-litigation system. The former is the approach taken in Worker Compensation and automobile no-fault reparations systems. While courts that have applied strict liability in design and duty-to-warn cases have often stated that they are not imposing "absolute" insurer liability, they have not been able to articulate why they draw a line short of that particular point. The reason for this is that the risk distribution rationale provides no stopping point short of absolute liability. Thus, a number of courts have plunged into a foggy area that is neither true strict liability nor negligence. The result has been the creation of a wide variety of legal "formulae," unpredictability for consumers, and instability in the insurance market.

In light of these facts, this Act places both design and duty-to-warn cases on a fault basis.

The Act utilizes the word "defective" as a generic term covering all four types of unreasonably unsafe products. A product cannot be proven defective unless the claimant shows that it is unreasonably unsafe under one or more Subsections under Section 104.[10]

This approach should not lead to problems of characterization. The claimant's pleading should indicate the theory on which it is based within the framework of Subsections (A), (B), (C), or (D). Of course, a product may be unreasonably unsafe in more than one way.

The manufacturer of a component part or parts is subject to liability under this Section for harms caused because that component part was defective. The manufacturer of a whole product is subject to liability under this Section for harms caused because the product, or any of its components, was defective.

The following commentary discusses

[10] Liability can be imposed on non-manufacturer product sellers under Section 105 when they have been negligent.

each Subsection in turn.

(A) *The Product Was Unreasonably Unsafe in Construction.* The history of imposing strict liability for products which are unreasonably unsafe in construction goes back as far as 1913 when sellers of foods were first held liable for their failure to produce a product reasonably fit for its intended use. *See* "Mazetti v. Armour & Co.," 75 Wash. 622, 135 P. 633 (1913); Prosser, "The Assault Upon the Citadel (Strict Liability to the Consumer)," 69 "Yale L.J." 1099 (1960).

Subsection (A) imposes pure strict liability on the manufacturer in accordance with Section 402A of the "Restatement (Second) of Torts." *Cf.* "Pabon v. Hackensack Auto Sales, Inc.," 63 N.J. Super. 476, 164 A.2d 773, 783 (1960) (implied warranty for defective ball bearing). The approach is also in accord with the overwhelming trend in case law in the United States. *See* Phillips, "A Synopsis of the Developing Law of Products Liability," 28 "Drake L. Rev." 317, 344–45 (1979).

In the course of its study, the Task Force found that most product sellers can absorb the financial impact of strict liablity for products which are defective in construction. *See* "Task Force Report" at VII–17.

The approach taken here is subject to theoretical criticism because

> [A] standard which judges a manufacturer by his own rather than industry standards could penalize the manufacturer who goes out of his way to build into his product a high safety level not mandated or followed by the industry.

Twerski & Weinstein, "A Critique of the Uniform Product Liability Law—A Rush to Judgment," 28 "Drake L. Rev." 221, 225 (1979).

Nevertheless, this criticism overlooks the fact that the manufacturer's self-imposition of a higher standard will function as a shield against claims alleging liability in the more costly area of defective design. By adopting a higher standard, a manufacturer may occasionally be subject to liability under Subsection (A), while another manufacturer may not be, but the first manufacturer does not place its whole product line at risk of being found unreasonably unsafe in design.

Not every minor variation from a standard will result in liability; rather, the variation must be a material one causing the claimant's harm.

In point of fact, there is no practical way to define defective construction except by the manufacturer's own standards. It is an optimal area for strict liability "because societal expectations are fairly well established with regard to such defects, and a ready gauge of acceptability exists by reference to like products that are non-defective." Phillips, "The Standard for Determining Defectiveness in Products Liability," 46 "U. Cin. L. Rev." 101, 104–05 (1977). Moreover, the imposition of strict liability with regard to defective construction is fair to the product user. Under a negligence system, the claimant would have the very difficult burden of showing that a manufacturer knew or should have known about a latent defect in one of thousands of mass-produced products.

(B) *The Product Was Unreasonably Unsafe in Design.* No court, in spite of some loose language that has been used, has imposed true strict or absolute liability on manufacturers for products which are unreasonably unsafe in design. *See* Henderson, "Manufacturers' Liability for Defective Design: A Proposed Statutory Reform," 56 "N.C. L. Rev." 625, 634–35 (1978). This is true because courts have appreciated the liability potential inherent in such cases—it is almost always posible to design a product more safely. Nevertheless, some courts in their instructions to the trier of fact have left juries "at sea." The jury is told to decide design cases by "considering" a number of factors; it is not given a guideline or formula for making the evaluation.

The approach taken here provides such a guideline. The trier of fact is instructed to impose liability if it finds that the product was unreasonably unsafe in design. The trier of fact is also given a formula to assist it in making this determination. The approach has its roots in the law of negligence [11] and has been put into modern and appropriate product liability terminology by some courts in their attempt to resolve the defective design dilemma. *See* "Hagans v. Oliver Mach. Co.," 576 F.2d 97, 99–100 (5th Cir. 1978); "Dreisonstok v. Volkswagenwerk, A.G.," 489 F.2d 1066, 1071 (4th Cir. 1974); "Jeng v. Witters," 452 F. Supp. 1349, 1356 (M.D. Pa. 1978).

Section 104 requires the court to make an initial determination as to whether the claimant has introduced enough evidence for the case to be considered by the jury. This requirement is especially important with respect to conscious design cases under Subsection (B). The dangers of the trier of fact introducing hindsight into the risk-utility analysis make it imperative for the court to apply its screening function carefully. *See* "Owens v. Allis-Chalmers Corp.," 83 Mich. App. 74, 268 N.W.2d 291 (1978); Henderson, *supra*, 73 "Colum. L. Rev." at 1531. *Cf.* Wade, *supra*, 44 "Miss. L. J." at 837–38 (suggesting that such formulae can *only* be applied by the court).

The formula set forth in Subsection (B) requires the trier of fact to balance two pairs of factors existing at the time of manufacture: (1) the likelihood that the product would cause the claimant's harm or similar harms, and the seriousness of those harms; against (2) the manufacturer's burden of designing a product that would have prevented those harms, and the adverse effect that alternative design would have on the usefulness of the product. *See* "Kerns v. Engelke," 390 N.E.2d 859, 865 (1979).

If the case is sent to the jury, the balancing formula should be placed in a jury instruction indicating that the claimant has the burden of showing, in light of the formula, that the product was unreasonably unsafe in design.

The Subsection also sets forth examples of important evidence that may be considered with respect to applying the formula.

First, warnings and instructions are important because, in an appropriate case, a product may be found not to be defective in design if the product seller has given an adequate warning about the product risk. *See, e.g.,* "Wagner v. Larson," 257 Iowa 1202, 136 N.W.2d 312 (1965); "Penn v. Inferno Mfg. Corp.," 199 So.2d 210 (La. App.) *aff'd,* 251 La. 27, 202 So. 2d 649 (1967). There are limits to this possibility, however, since a product seller will not be shielded from liability for an unreasonably unsafe product simply by indicating that the product "may be hazardous." Only a warning or instruction which, at the time of manufacture, substantially reduces the likelihood that the product would cause harm in the manner in which the claimant's harm was caused will affect the question of whether the product was defective in design.

Second, evidence relating to the technological and practical feasibility of designing and manufacturing a product which would have substantially served the expected user's needs, and would have also prevented the claimant's harm, addresses the side of the formula dealing with the manufacturer's burden. If an alternatively designed product which would have prevented the harm while preserving its usefulness could have been produced with a slight increase in cost, it is likely that the product is unreasonably unsafe. On the other hand, the manufacturer need not incorporate safety features that render a product incapable of performing some or all of the very functions that create its public demand. *See* "Hagans v. Oliver Mach., Inc.," 576 F.2d 97 (5th Cir. 1978); "Dreisonstok v. Volkswagenwerk, A.G.," 489 F.2d 1066 (4th Cir. 1974).

Third, it is important to consider evidence relating to the effect of the alternative design on the usefulness of the product. As learned scholars have observed, "it is virtually impossible to [evaluate a design defect case] without balancing risk and utility factors." *See* Twerski & Weinstein, *supra*, 28 "Drake L. Rev." at 229. *See also* "Phillips v. Kimwood Mach. Co.," 269 Or. 485, 525 P.2d 1033, 1038 (1974). It is important to note that this evidence is not directed to the general usefulness of the product in society, *i.e.*, the overall social worth of pharmaceuticals, lawnmowers, or other

[11] *See* "United States v. Carroll Towing Co.," 159 F.2d 169, 173 (2d Cir. 1947) (the "Learned Hand formula").

products.

Fourth, it is essential to consider evidence relating to the comparative costs of producing, distributing, selling, using, and maintaining the product as designed and as alternatively designed. This places the court and the trier of fact in the "real world" of design options.

Finally, if new or additional harms might have resulted if the product had been alternatively designed, the trier of fact should consider this factor in making its evaluation. Since this evidence will probably be introduced by the defendant manufacturer, it may be placed before the court and jury *after* the claimant's case has been presented.

In sum, Subsection (B) places the burden of proof on the claimant to show that in light of a balance of practical, objective factors, the product seller should bear the full cost of the injury and have the responsibility for attempting to distribute that cost through the price of its product. In light of the fact that the judgment is the equivalent of holding that the whole product line is defective, it is important that traditional tort law principles be followed and that the claimant retain the burden of proof. *See* Kalven, "Torts: The Quest for Appropriate Standards," 53 "Cal. L. Rev." 189 (1965). *But see* "Barker v. Lull Engineering, Inc.," 20 Cal. 3d 413, 573 P.2d 443, 143 Cal. Rptr. 225 (1978).

Neither the formula nor the detailed list of the more important evidentiary items includes what has been called the "consumer expectation test." The reasons for not including it are rooted in both economics and practicality. As Professor Wade, Reporter for "The Restatement (Second) of Torts," has stated:

> In many situations, particularly involving design matters, the consumer would not know what to expect, because he would have no idea how safe the product could be made.

Wade, *supra,* "Miss. L. J.," at 829.

The consumer expectation test takes subjectivity to its most extreme end. Each trier of fact is likely to have a different understanding of abstract consumer expectations. Moreover, most consumers are not familiar with the details of the manufacturing process and cannot abstractly evaluate conscious design alternatives. This has been recently recognized by the Supreme Court of Texas. "Turner v. General Motors Corp.," ,———S.W.2d———, No. B–7747 (Tex. June 13, 1979); *see also* Green, "Strict Liability Under Sections 402A and 402B: A Decade of Litigation," 54 "Tex. L. Rev." 1185 (1976).

(C) *The Product Was Unreasonably Unsafe Because Adequate Warnings or Instructions Were Not Provided.* A manufacturer may be held liable under Subsection (C) regardless of whether the product was found to be unreasonably unsafe in construction or design. Even where the lack of scientific knowledge or cost factors preclude the use of an alternative design, the manufacturer may still be required to provide a warning about the product's hazard or to provide adequate instructions about the product's use. *See* "Jacobson v. Colorado Fuel & Iron Corp.," 409 F.2d 1263, 1271 (9th Cir. 1969).

It is important to note that Subsection (C) covers two arguably separate problems. First, was there a duty to warn or instruct about a particular matter? Second, assuming that a warning or instruction was given, was it adequate? While these two problems are theoretically separate, the underlying factors that must be taken into consideration in evaluating a manufacturer's duty with respect to them overlap; therefore, the two issues have been combined in Subsection (C). As will be indicated below, there may be a different evidentiary emphasis in the two types of cases.

The basic standard of responsibility for Subsection (C) is predicated on a fault basis. A "strict liability" base in this area creates unacceptable uncertainty within the tort-litigation system. This has recently been recognized by the Supreme Court of Michigan. "Smith v. E. R. Squibb & Sons," 405 Mich. 79, 273 N.W. 2d 476 (1979).

Under Subsection (C), the trier of fact is to place itself in the manufacturer's position at the time the product was manufactured. In order to impose liability on the manufacturer, the claimant must prove that the probability that the product would cause the claimant's harm and similar harms and the seriousness of those harms rendered the manufacturer's instructions inadequate, and that the manufacturer should and could have provided the warnings or instructions which claimant alleges would have been adequate. Obviously, where harms were likely to occur and unlikely to be recognized by the product user, the necessity of adequate warnings and instructions is correspondingly acute. On the other hand, the duty to provide adequate warnings and instructions cannot go beyond the technological and other information that was reasonably available at the time of manufacture. This concept is in accord with the overwhelming majority of court decisions. *See* "Robbins v. Farmers Union Grain Terminal Ass'n.," 552 F.2d 788 (8th Cir. 1977) (collecting cases).

Unlike the basic formula for design cases, the utility of the product is not a consideration in deciding whether a warning is necessary. The focus is on the likelihood that the product would cause the claimant's harm, and the seriousness of that harm, the adequacy of the warnings that were provided, and the practical need for providing the warnings or instructions which claimant alleges would have been adequate.

Subsection (C)(2) lists examples of some of the more probative evidence that can assist the court and the jury in applying this formula.

Factor (a) focuses on the manufacturer's ability, at the time of manufacture, to be aware of the product's danger and the nature of the potential harm. This factor should not operate through hindsight.

Factor (b) involves evidence relating to the manufacturer's ability to anticipate that the likely product user would be aware of the product risk and the nature of the potential harm. The more serious the anticipated harm, the greater the duty to warn. *See* "Braniff Airways, Inc. v. Curtiss-Wright Corp.," 411 F.2d 451 (2d Cir. 1969), *on rehearing*, 424 F.2d 427 (2d Cir. 1970); "Davis v. Wyeth Laboratories, Inc.," 399 F.2d 121 (9th Cir. 1968).

Factor (c), the technological and practical feasibility of providing effective warnings and instructions, may not be significant in many cases because warnings are often relatively inexpensive to provide. However, in some situations, it may not be feasible as a practical technological matter to provide a warning or the type of warning that the claimant suggests should have been provided.

Factor (d), the clarity and conspicuousness of the warnings or instructions, is material in cases where claimant alleges that the warnings given were not adequate. As Professor Phillips has noted:

> To be adequate, warnings must be reasonably conspicuous, strong and clear. They must describe the danger and, where pertinent, the means of avoiding it.

Phillips, *supra*, 28 "Drake L. Rev." at 351.

Also highly relevant to this issue is Factor (e), evidence relating to the adequacy of the warnings or instructions that were provided. Again, as Professor Phillips has noted:

> The effectiveness of a warning may be diluted by other supplier representations of safety that lull the user into a false sense of security. While little or no warning may have to be given to the expert user, a clear and strong warning is required for dangerous products which the supplier can reasonably expect to be used by inexperienced persons. *Id.* at 351–52.

Subsection (C)(3) sets forth a basic causation link for the Subsection. The claimant must show that if adequate warnings or instructions had been provided, the harm would have been avoided either because a reasonably prudent product user would have declined to use the product (a situation which may be most pertinent with respect to a pharmaceutical with a known risk), or because the product would have been used in a manner so as to have avoided the harm. *See* "Technical Chem. Co. v. Jacobs," 480 S.W. 2d 602 (Tex. 1972); *cf.* "Potthoff v. Alms," —— Colo. App. ——, 583 P.2d 309 (Colo. Ct. App. 1978). While it has been suggested that no causation requirement should be required with respect to unavoidable pharmaceutical risks (*see* Twerski & Weinstein, *supra,* 28 "Drake L. Rev." at 236–37), the Act addresses itself to tort law, not the general public regulation of products. It is a basic requirement of tort law for the claimant to show that if the defendant had acted in a manner in which the claimant alleges was appropriate, then he or she would not have suffered harm. *See* W. Prosser, "Torts" at 236 (4th ed. 1971).

Subsection (C)(4) recognizes that a manufacturer should be able to assume that the ordinary product user is familiar with obvious hazards—that knives cut, that alcohol burns, that it is dangerous to drive automobiles at high speeds. Thus, the Act states that the manufacturer does not have to warn about dangers that are obvious. *See* "Kimble v. Waste Systems International, Inc.," —— Wash. App. ——, 595 P.2d 569 (1979). While it has been suggested that this approach may encourage the manufacture of "obviously dangerous" products, this is unlikely to occur under this Act, since the manufacturer of an unreasonably unsafe product may be subject to liability under Subsection (B).

Requiring a manufacturer to warn about matters that are obvious would tend to reduce the effectiveness of warnings. As experts in the field have observed, "Warnings, in order to be effective, must be selective." Twerski, Weinstein, Donaher, & Piehler, "The Use and Abuse of Warnings in Products Liability—Design Defect Litigation Comes of Age," 61 "Cornell L. Rev." 495, 514 (1976). It is recognized that, in some jurisdictions, there is an open-ended duty to warn about obvious dangers when they are highly likely to cause very serious injuries. *See* Marschall, "An Obvious Wrong Does Not Make a Right: Manufacturers' Liability for Patently Dangerous Products," 48 "N.Y.U. L. Rev." 1065 (1973) (collecting cases). Again, the Act approaches this problem by potentially subjecting a manufacturer of a patently dangerous product to liability for an unreasonably unsafe design.

Similarly, the product manufacturer is not under an obligation to warn about misuses or alterations of products that would not be expected by an ordinary reasonably prudent person who is likely to use the product in the same or similar circumstances. *See also* Section 112(C).

Subsection (C)(5) indicates that in the normal course of events, a manufacturer should place its warning or instruction in a place or in a form where it can be communicated to the actual product user. On the other hand, there are some situations in which this is impossible or impracticable. *See* "Bryant v. Hercules, Inc.," 325 F. Supp. 241 (W.D. Ky. 1970). Therefore, the Subsection indicates that a warning or instruction may be given to a person who may be reasonably expected to assure that action is taken to avoid the harm, or that the risk of the harm is explained to the actual product user. By way of example, the Act sets forth situations where such a process is appropriate. Thus, communication to a using or supervising expert is explicitly stated to be adequate when the product—such as a prescription drug or a radioactive material—is one which may be legally used only by, or under, the supervision of such an expert. *See* "Carmichael v. Reitz," 17 Cal. App. 3d 958, 95 Cal. Rptr. 381 (Dist. Ct. App. 1971); "Terhune v. A. H. Robbins Co.," 90 Wash. 2d 9, 577 P.2d 975 (1978).

The manufacturer of tangible goods sold or handled only in bulk or of other workplace products may communicate warnings or instructions to the employer of the claimant when that is the only practical and feasible avenue for making a warning. *See* "Reed v. Pennwalt Corp.," 22 Wash. App. 718, 591 P.2d 478 (1979).

Other situations not specifically pointed out in the Act may arise where it is permissible for the manufacturer to convey warnings to persons other than a product user, *e.g.,* warnings to the parent of a young child with respect to a food product. *Cf.* "Spruill v. Boyle-Midway Inc.," 308 F.2d 79 (4th Cir. 1962).

Post-Manufacture Duty to Warn. Subsection (C)(6) recognizes a manufacturer's duty to warn after its product has been produced. The Subsection places an obligation on a manufacturer to act with reasonable prudence to learn about serious risks connected with products after they are manufactured. When it learns of such a risk, it is to act as a reasonably prudent manufacturer in the same or similar situation. This obligation is satisfied if the manufacturer makes reasonable efforts to inform product users or appropriate persons about the risk. The Subsection is in accord with basic negligence case law. *See* "Comstock v. General Motors Corp." 358 Mich. 163, 99 N.W.2d 627 (1959); "Schenebeck v. Sterling Drug, Inc.," 423 F.2d 919 (8th Cir. 1970). The Subsection recognizes that in some situations, a general warning through an advertising medium may be all a manufacturer can provide. The standard is reasonableness, not absolute or strict liability.

(D) *The Product Was Unreasonably Unsafe Because It Did Not Conform to an Express Warranty.* When there has been an express warranty concerning the product, strict liability can readily be justified. The product seller has used specific words to induce the purchase of the product, and the Act indicates that it must be accountable for its representations. The term "express warranty" is separately defined in the Act in Subsection 102(K).

Since the case of "Baxter v. Ford Motor Co.," 179 Wash. 123, 35 P.2d 1090 (1934), courts have been virtually unanimous in imposing strict liability on product sellers when their statements about their products prove to be untrue. *See* W. Prosser, "Torts" at 652 (4th ed. 1971) ("decisions to the contrary have been amazingly few").

Thus, where a product seller advertised that its pharmaceutical was "free and safe from all dangers of addiction" and claimant, because of a rare and totally unforeseeable susceptibility, became physically dependent on the drug, strict liability was imposed. "Crocker v. Winthrop Lab., Div. of Sterling Drug, Inc.," 514 S.W.2d 429 (Tex. 1974); "Spiegel v. Saks 34th Street," 43 Misc. 2d 1065, 252 N.Y.S. 2d 852 (App. Term 1964), *aff'd,* 272 N.Y.S.2d 972 (App. Div. 1966) (perfume advertised as "non-allergenic").

While many courts use the term "express warranty" to describe the cause of action set forth in Subsection (D), "warranty" is a term of commercial law which is appropriate in the ordinary sale of goods from an immediate buyer to an immediate seller. Here, the focus is solely on tort claims for personal injury and damage to property. In order to convey the tort basis of this cause of action, Section 402B of the "Restatement (Second) of Torts" (1965), utilized the term "misrepresentation" in place of the term "express warranty." Nevertheless, this parlance has not been generally accepted by courts, although they do appreciate the distinction between the tort action and commercial law. *See* "Drayton v. Jiffee Chem. Corp.," 591 F.2d 352, 359 (6th Cir. 1978) ("sounds distinctly in tort"). In deference to this common usage, the term "express warranty," as defined in Subsection 102(K), has been utilized.

Subsection (D) applies only to representations or warranties made by the manufacturer or by a person for whom it is legally responsible. In general, the representation must have been made to the claimant, who must have relied on the warranty, and the harm must have resulted because of that reliance and because of the fact which was misrepresented. However, Subsection (D) does permit reliance by "one acting on behalf" of the claimant. Thus, if a wife purchases an automobile in reliance on a statement concerning its brakes and permits her husband to drive the car, this supplies the element of reliance although the husband, in fact, never learned of the statement. Otherwise, the Section is limited to consumers of products. *See* "Restatement (Second) of Torts" Section 402B, Comment j (1965).

The statement made by the product seller must be about a material fact or facts concerning the product. This does not include mere "puffing" or sales talk. Thus, the manufacturer of a grinding disk that broke was not deemed liable for a breach of an express warranty after stating that the product was "stronger, sharper, and longer lived than ever before available anywhere." "Jakubowski v. Minnesota Mining & Mfg.," 80 N.J. Super. 184, 193 A.2d 275 (1963), *rev'd on other grounds,* 42 N.J. 177, 199 A.2d 826 (1964); *cf.* "Berkebile v. Brantly Helicopter Corp.," 462 Pa. 83,

337 A.2d 893 (1975) (helicopter "easy to operate"). In contrast, a manufacturer was deemed strictly liable when it advertised that a golf traninig device could be operated and the "ball will not hit player." A player of limited golfing ability was indeed hit by the ball, and the defendant was held liable. "Hauter v. Zogarts," 14 Cal. 3d 104, 120 Cal. Rptr. 681, 534 P.2d 377 (1975).

* * * * *

Code

Sec. 105. Basic Standards of Responsibility for Product Sellers Other Than Manufacturers

(A) A product seller, other than a manufacturer, is subject to liability to a claimant who proves by a preponderance of the evidence that claimant's harm was proximately caused by such product seller's failure to use reasonable care with respect to the product.

Before submitting the case to the trier of fact, the court shall determine that the claimant has introduced sufficient evidence to allow a reasonable person to find by a preponderance of the evidence that such product seller has failed to exercise reasonable care and that this failure was a proximate cause of the claimant's harm.

In determining whether a product seller, other than a manufacturer, is subject to liability under Subsection (A), the trier of fact shall consider the effect of such product seller's own conduct with respect to the design, construction, inspection, or condition of the product, and any failure of such product seller to transmit adequate warnings or instructions about the dangers and proper use of the product.

Unless Subsection (B) or (C) is applicable, product sellers shall not be subject to liability in circumstances in which they did not have a reasonable opportunity to inspect the product in a manner which would or should, in the exercise of reasonable care, reveal the existence of the defective condition.

(B) A product seller, other than a manufacturer, who makes an express warranty about a material fact or facts concerning a product is subject to the standards of liability set forth in Subsection 104(D).

(C) A product seller, other than a manufacturer, is also subject to the liability of manufacturer under Section 104 if:

(1) The manufacturer is not subject to service of process under the laws of the claimant's domicile; or

(2) The manufacturer has been judicially declared insolvent in that the manufacturer is unable to pay its debts as they become due in the ordinary course of business; or

(3) The court determines that it is highly probable that the claimant would be unable to enforce a judgment against the product manufacturer.

(D) Except as provided in Subsections (A), (B), and (C), a product seller, other than a manufacturer, shall not otherwise be subject to liability under this Act.

Analysis

Sec. 105. Basic Standards of Responsibility for Product Sellers Other Than Manufacturers

Section 105 is derived in part from Tennessee law. "Tenn. Code Ann." Section 23–3706 (Supp. 1978). The Section addresses the problem of excessive product liability costs for parties other than manufacturers in the distribution chain in a way that does not compromise incentives for loss prevention. It also leaves the claimant with a viable defendant whenever a defective product has caused harm.

The "ISO Closed Claims Survey" shows that manufacturers account for 87 percent of the total amount of product liability payments, while distributors, wholesalers, and retailers account for 4.6 percent. "ISO Closed Claims Survey," Report 3, at 35 (1977). Case law suggests that these non-manufacturer product sellers can usually shift their costs to the manufacturer through an indemnity suit. *See, e.g.,* "Hales v. Monroe," 544 F. 2d 331, 332 (8th Cir. 1976); "Anderson v. Somberg," 158 N.J. Super. 384, 386 A. 2d 413, 419–20 (1978); "Litton Systems, Inc. v. Shaw's Sales & Serv., Ltd.," 119 Ariz. 10, 579 P. 2d 48, 50 (1978).

Despite their relatively small role vis-a-vis manufacturers as product liability defendants, wholesalers, retailers, and distributors are frequently brought into a product liability suit. *See, e.g.,* "Tucson Indus., Inc. v. Schwartz," 108 Ariz. 464, 501 P. 2d 936 (1972); "Vergott v. Deseret Pharmaceutical Co.," 463 F. 2d 12 (5th Cir. 1972); "Duckworth v. Ford Motor Co.," 320 F. 2d 130 (3d Cir. 1963). In light of ISO data showing that for every dollar of claims paid, at least 35 cents is spent in defense costs,[12] the net result is that wholesalers, retailers, and distributors are subject to substantial product liability costs in terms of both premiums and defense costs. These costs are added to the price of products and waste legal resources. *See* "Pender v. Skillcraft Indus., Inc.," 358 So. 2d 45 (Fla. Dist. Ct. App. 1978).

Under Section 105, product sellers other than manufacturers must exercise reasonable care in their handling of products. This obligation includes inspecting for hazards which a reasonably prudent product seller would have reason to discover. The focus of judicial inquiry will be on the opportunity the non-manufacturer product seller had to discover the hazard and on whether circumstances would permit a reasonable product seller to take corrective action. *See* "Edwards v. E. I. DuPont de Nemours & Co.," 183 F. 2d 165, 167 (5th Cir. 1950).

On the other hand, product sellers other than manufacturers are not liable for harms caused by products which are defective in construction or design, if those products were defective when they were received and the product seller had no reasonable opportunity to discover the defective condition. In addition, such product sellers are generally not liable for harms caused by breaches of the duty to warn or an express warranty unless the product seller's own conduct gave rise to the claim.

Subsection (A) provides that non-manufacturer product sellers are not subject to liability when they had no reasonable opportunity for product inspection which, in the exercise of reasonable care, would or should have revealed the existence of the defective condition. For example, if a retailer receives a defective product in a sealed container and there is no way for the retailer to be aware of the condition, the retailer will not be held liable. In general, Section 105 does not impose liability on non-manufacturer product sellers where there are defects in construction or defects in design that a reasonably prudent product seller would have had no opportunity to discover. The manufacturer can avoid many of these defects; the distributor or retailer cannot. However, a non-manufacturer product seller can waive the benefits of Subsection (A) through an express warranty that it makes or transmits to the product user. *Cf.* "Ky. Rev. Stat. Ann." Section 411.340 (Supp. 1978). The term "manufacturer" is defined in Subsection 102(B).

Under Subsection (A), non-manufacturer product sellers are required to exercise reasonable care in their handling and storage of products and to pass along warnings from the manufacturer. Non-manufacturer product sellers are also responsible for defects introduced into the product by virtue of their own negligent conduct (*e.g.,* faulty product preparation or storage). Failure to fulfill their responsibilities will subject non-manufacturer product sellers to liability for their contribution to the harm which results.

Under Subsection (B), the non-manufacturer product seller is subject to the same liability as a manufacturer if the non-manufacturer product seller makes an express warranty. In these situations, such product seller has induced the purchase of the product by these specific words, and courts have been virtually unanimous in holding them responsible if the words prove to be untrue. *See* W. Prosser, "Torts" at 652 (4th ed. 1971).

Subsection (C) addresses the justifiable concern of Justice Traynor in "Vandermark v. Ford Motor Co.," 61 Cal. 2d 256, 391 P. 2d 168, 171, 37 Cal. Rptr. 896, 899 (1964), that:

> In some cases the retailer may be the only

[12] *See* "ISO Closed Claims Survey," Report 14, at 118.

member of that enterprise reasonably available to the injured plaintiff. In other cases the retailer himself may play a substantial part in ensuring that the product is safe or may be in a position to exert pressure on the manufacturer to that end.

A majority of courts have followed the "Vandermark" case and have extended strict liability to retailers. *See, e.g.*, "McKisson v. Sales Affiliates, Inc.," 416 S.W. 2d 787 (Tex. 1967); "Housman v. C. A. Dawson & Co.," 106 Ill. App. 2d 225, 245 N.E. 2d 886 (1969). *See also* "U.C.C." Section 2–314.

Section 105 responds to Justice Traynor's concern in cases in which it is necessary to do so. If the manufacturer is not subject to service of process or has been judicially declared insolvent, or if a court determines that it would be highly likely that the claimant would be unable to enforce a judgment against the product manufacturer, the retailer, wholesaler, or distributor has the same strict liability obligations as a manufacturer. Thus, the limited liability shield set forth in Subsection (D) only operates when the product manufacturer is reasonably available for suit by the injured claimant.

Some economists may criticize the exception to the general rule set forth in Section 105. Another approach is that of a recently enacted Nebraska statute which flatly exempts non-manufacturer product sellers from liability unless they have been negligent. *See* "Neb. Rev. Stat." Section 25–21, 181 (Supp. 1978). *See also* "Shainberg v. Barlow," 258 So.2d 242, 244 (Miss. 1972) (same result under case law). However, the Nebraska approach can leave a person injured by a *defective* product (as defined in Section 104) without compensation. The Act makes clear that in these situations, the party who actually sold or commercially leased the defective product should bear the loss.

Procedurally, if the claimant commences an action against the retailer, wholesaler, or distributor, and the action is based upon a theory other than the retailer's, wholesaler's, or distributor's own negligence—such as Subsection 104(A)—the defendant would make a motion for dismissal. At that time, the claimant must show that the manufacturer is unavailable under the provisions of Subsection (C). The court would then consider and make a determination regarding the three conditions listed in Subsection (C) concerning the manufacturer. If the claimant commences an action against the retailer, wholesaler, or distributor, and one of the claimant's theories is the negligent conduct of the defendant, the defendant may move to dismiss on the other theory, but the negligence theory will be litigated. If the action is dismissed against the retailer, wholesaler, or distributor under Section 105, the claimant retains the option to proceed against the manufacturer.

Subsection (D) makes clear that the Section sets forth the entire scope of liability of non-manufacturer product sellers.

However, it should be noted that retailers, wholesalers, distributors, or others can become manufacturers for the purposes of this Act to the extent that they design, produce, make, fabricate, construct, or remanufacture a product or component part of a product prior to sale. Such parties can also become manufacturers for the purposes of this Act if they hold themselves out as a manufacturer. *See* Section 102(B) ("manufacturer") and accompanying analysis.

In order for Section 105 to operate fairly toward claimants as well as non-manufacturer product sellers, it is suggested that:

(1) The non-manufacturer product seller be treated as a party for the purposes of discovery under the applicable procedural code. If this step is not taken, the Act may place an undue burden on the claimant in his or her attempt to prove the case; and

(2) The statute of limitation vis-a-vis the non-manufacturer product seller be deemed to have tolled in case the claimant is unable to enforce his or her product liability judgment against the manufacturer.

* * * * *

Code

Sec. 106. Unavoidably Dangerous Aspects of Products

(A) An unavoidably dangerous aspect of a product is that aspect incapable, in

light of the state of scientific and technological knowledge at the time of manufacture, of being made safe without seriously impairing the product's usefulness.

(B) A product seller shall not be subject to liability for harm caused by an unavoidably dangerous aspect of a product unless:

(1) The product seller knew or had reason to know of the aspect and with that knowledge acted unreasonably in selling the product at all;

(2) The aspect was a defect in construction under Subsection 104(A);

(3) The product seller knew or had reason to know of the aspect and failed to meet a duty to instruct or warn under Subsection 104(C), or to transmit warnings or instructions under Subsection 105(A); or

(4) The product seller expressly warranted that the product was free of the unavoidably dangerous aspect under Subsection 104(D) or 105(B).

* * * * *

Analysis

Sec. 106. Unavoidably Dangerous Aspects of Products

Section 106 provides that, in general, a product seller will not be subject to liability for a product which is defective in design when a harm stems from an aspect of the product that is incapable of being made safe in light of the state of scientific and technical knowledge at the time of manufacture. The term "incapable" is meaningless unless it is placed in context. Therefore, Subsection (A) indicates that it applies when the danger cannot be avoided without seriously impairing the product's usefulness.

The approach taken in Section (A) recognizes that there may be circumstances in which a seriously injured person is left without compensation for an injury caused by an unavoidably dangerous aspect of a product; however, for reasons of policy **that consumers can appreciate, Section 106 proposes that a product seller not be held responsible for harms that are simply unavoidable. *See* Johnson, "Products Liability 'Reform': A Hazard to Consumers," 56 "N.C. L. Rev." 676, 690 (1978). If the costs of unavoidable harms are to be shifted from the individual, they should be borne by society at large. The policy predicate underlying Section 106 is that it should help encourage research and development without unleashing unreasonably unsafe products on the public. It also makes clear to policymakers that the tort-litigation system is not the means for addressing injuries caused by all product hazards.**

Section 106 is based on the "Restatement (Second) of Torts" Section 402A, Comment k (1965). *See also* "N.H. Rev. Stat. Ann." Section 507–D:4 (Supp. 1978).

With the exception of one Illinois decision, "Cunningham v. MacNeal Memorial Hosp.," 47 Ill. 2d 443, 266 N.E.2d 897 (1970), *subsequently overruled by* "Ill. Ann. Stat." ch. 91, Sections 181–84 (Supp. 1979), this approach has been followed by the common law courts throughout the United States. *See, e.g.*, "Moore v. Underwood Memorial Hosp.," 147 N.J. Super. 252, 371 A.2d 105 (1977) (serum hepatitis contracted from blood supplied); "Dalke v. Upjohn Co.," 555 F.2d 245 (9th Cir. 1977) (tooth discoloration from tetracycline); "Chambers v. G.D. Searle & Co.," 441 F. Sup. 377 (D. Md. 1975) (stroke allegedly from birth control pills); "Coffer v. Standard Brands, Inc.," 30 N.C. App. 134, 226 S.E.2d 534 (1976) (shell in nuts); "Hines v. St. Joseph's Hospital," 86 N.M. 763, 527 P.2d 1075 (1974) (blood transfusion).

Subsection (A) indicates that the time from which to judge the state of scientific and technological knowledge is the time of manufacture. *See* "Cochran v. Brooke," 243 Or. 89, 409 P.2d 904 (1966).

Subsection (B) indicates that the product seller shall not be subject to liability for harms caused by unavoidably dangerous aspects of products unless the claimant proves that one of four circumstances occurred.

The first circumstance involves the unusual situation in which the product seller knew or had reason to know of the unavoidably dangerous aspect and, with that knowledge, acted unreasonably in selling the product at all. An example is a product seller who markets a toy that is highly dangerous to children. This example can be contrasted with a pharmaceutical approved by the Food and Drug Administration, when both the government and the manufacturer know of the dangers connected with that product. In that situation, it is clear that this exception should not apply because it would not be unreasonable to sell the product.

The second circumstance deals with defects in construction. It is arguable that, even with the finest quality control, a product that is defective in construction can slip by inspection. Nevertheless, a judgment has been made by a majority of courts to impose strict liability on manufacturers for defects in construction. That judgment is preserved in this Act.

The third circumstance deals with the situation in which the product seller has failed to provide an adequate warning about an unavoidably dangerous aspect of a product. Section 106 is in accord with Subsections 104(C) and 105(A) and is predicated on a fault base—it is only to apply with respect to dangers which were known or could be discovered through the exercise of reasonable care. *See* "Dalke v. Upjohn," *supra;* "Chambers v. G.D. Searle & Co.," *supra;* "Toole v. Richardson-Merrell, Inc.," 251 Cal. App. 2d 689, 60 Cal. Rptr. 398 (1967). As these cases reflect, the factual question underlying the legal issue of whether warnings or instructions were adequate is whether a product seller has met its duty to promulgate warnings and instructions commensurate with its actual knowledge gained from research and adverse reaction reports, and its constructive knowledge as measured by scientific literature and other available means of communication. *See* "Dalke v. Upjohn Co.," *supra,* 555 F.2d, at 248; "McEwen v. Ortho Pharmaceutical Corp.," 270 Or. 375, 528 P.2d 522, 528–29 (1974). *Contra,* "Bruce v. Martin-Marietta Corp.," 544 F.2d 442 (10th Cir. 1976).

The Section's cross-reference to Subsections 104(C) and 105(A) makes clear that the product seller must meet its obligation to warn, instruct, or transmit warnings or instructions based on new information about an unavoidably dangerous aspect of a product that is discovered *after* the product has been manufactured. *See* "Love v. Wolf," 226 Cal. App. 2d 378, 38 Cal. Rptr. 183 (1964); "Sterling Drug, Inc. v. Yarrow," 408 F. 2d 1978 (8th Cir. 1969). Only those harmed after the duty to warn arose would be entitled to a claim.

Finally, Section 106 makes clear that a product seller can be liable for an unavoidably dangerous aspect of a product if the product seller expressly warranted that the product was free from such aspect. *See* Subsections 104(D) and 105(B).

* * * * *

Code

Sec. 107. Relevance of Industry Custom, Safety or Performance Standards, and Practical Technological Feasibility

(A) Evidence of changes in (1) a product's design, (2) warnings or instructions concerning the product, (3) technological feasibility, (4) "state of the art", or (5) the custom of the product seller's industry or business, occurring after the product was manufactured, is not admissible for the purpose of proving that the product was defective in design under Subsection 104(B) or that a warning or instruction should have accompanied the product at the time of manufacture under Subsection 104(C).

If the court finds that the probative value of such evidence substantially outweighs its prejudicial effect and that there is no other proof available, this evidence may be admitted for other relevant purposes if confined to those purposes in a specific court instruction. Examples of "other relevant purposes" include proving ownership or control, or impeachment.

(B) For the purposes of Section 107, "custom" refers to the practices followed by an ordinary product seller in the product seller's industry or business.

(C) Evidence of custom in the product seller's industry or business or of the product seller's compliance or non-

compliance with a non-governmental safety or performance standard, existing at the time of manufacture, may be considered by the trier of fact in determining whether a product was defective in design under Subsection 104(B), or whether there was a failure to warn or instruct under Subsection 104(C) or to transmit warnings or instructions under Subsection 105(A).

(D) For the purposes of Section 107, "practical technological feasibility" means the technological, mechanical, and scientific knowledge relating to product safety that was reasonably feasible for use, in light of economic practicality, at the time of manufacture.

(E) If the product seller proves, by a preponderance of the evidence, that it was not within practical technological feasibility for it to make the product safer with respect to design and warnings or instructions at the time of manufacture so as to have prevented claimant's harm, the product seller shall not be subject to liability for harm caused by the product unless the trier of fact determines that:

(1) The product seller knew or had reason to know of the danger and, with that knowledge, acted unreasonably in selling the product at all;

(2) The product was defective in construction under Subsection 104(A);

(3) The product seller failed to meet the post-manufacture duty to warn or instruct under Subsection 104(C)(6); or

(4) The product seller was subject to liability for express warranty under Subsection 104(D) or 105(B).

Analysis

Sec. 107—Relevance of Industry Custom, Safety or Performance Standards, and Practical Technological Feasibility

Subsection (A) adopts a fundamental principle of evidence law for the purposes of product liability cases. It excludes the showing of post-manufacture changes in the design of or warnings about a product, technological feasibility, "state of the art," or industry custom when evidence of those changes is offered to show that the product was defective at the time of manufacture. *See* Fed. R. Evid. 407 and Advisory Committee Commentary. The term "state of the art," as used in the listing of evidentiary items in Subsection 107(A), is a general, commonly used phrase whose meaning may include industry custom or the most scientifically advanced developments in the field. As will be indicated in the discussion of Subsections (B) through (E), the Act has generally eschewed the phrase because of its ambiguity. Nevertheless, it has been utilized in Subsection (A) in order to assure that all post-manufacture change is excluded from evidence.

The reasons underlying the rule on which Subsection 107(A) is based are twofold: first, subsequent changes are deemed irrelevant (all they show is that "as one gets older, one may get wiser"); and second, admission of such evidence may discourage the making of improvements or repairs. While the latter rationale has been challenged (*see* "Ault v. International Harvester Co.," 13 Cal. 3d 113, 528 P.2d 1148, 117 Cal. Rptr. 812 (1975); Schwartz, "The Exclusionary Rule on Evidence of Repair—A Rule in Need of Repair," 7 "The Forum" 1 (1971)), the relevance of such evidence on the issue of defectiveness is of very limited value. On the other hand, the prejudicial effect of showing the subsequent change or repair—particularly one undertaken by the product seller itself—is quite substantial. *See* "LaMonica v. Outboard Marine Corp.," 48 Ohio App. 2d 43, 355 N.E.2d 533 (1976); "Haysom v. Coleman Lantern Co.," 89 Wash. 2d 474, 573 P.2d 785 (1978).

Subsection (A) permits the introduction of evidence which would otherwise be excluded under the Subsection, if that evidence is highly probative and necessary to prove a relevant matter other than the fact that the product was defective at the time of manufacture. Thus, evidence of such changes may be admissible to show that the product seller knew of the defect at a certain point in time after manufacture. In appropriate circumstances, this evidence may suggest that the product seller had a duty to take reasonable steps to warn product users about a newly discovered hazard. It may also be admissible when the product seller claims the product hazard was impossible to avoid. In cases of this kind, the court should make a finding that the probative value of the evidence substantially outweighs its prejudicial nature and that there is no other equally probative proof available. Allowing the introduction of evidence in these cases should *not* become a vehicle for avoiding the basic purpose of the rule.

Subsections (B), (C), (D), and (E) address one of the major issues that has divided product sellers and consumer groups concerned about product liability. Product sellers have vigorously argued that when their products comply with the "state of the art," it is unfair to deem them defective. Further, they contend that industry custom is likely to incorporate all cost-justified product safety features. They have received some support for this contention from members of the academic community, at least in regard to purchasers, as contrasted with nonpurchaser product users, of the product. *See* R. Posner, "Economic Analysis of Law" 71 (1972).

Consumer groups respond that it is inappropriate to allow product sellers to fix indirectly their own standard of liability. *See* Johnson, "Products Liability 'Reform': A Hazard to Consumers," 56 "N.C. L. Rev." 677, 680–81 (1978).

When the issue is carefully analyzed, there is less dispute between product sellers and consumer groups than appears on the surface. Part of the sound and fury is a debate over the ambiguous term "state of the art." The approach taken in Subsections (B), (C), (D), and (E) of not using the term "state of the art" eliminates that ambiguity and focuses on what is fair to each party.

Subsection (B) defines "custom" as the "practices followed by the ordinary product seller in the defendant product seller's industry or business." *See* W. Prosser, "Torts" at 166 (4th ed. 1971).

Subsection (C) indicates that compliance with industry custom is merely evidence that the trier of fact may consider in determining whether a product was defective under Subsections 104 (B) and (C). There is strong support in the case law for this approach. *See. e.g.,* "Bruce v. Martin-Marietta Corp.," 544 F.2d. 442 (10th Cir. 1976); "Baker v. Chrysler Corp.," 55 Cal. App. 3d 710, 127 Cal. Rptr. 745 (1976); "Maxted v. Pacific Car & Foundry Co., 527 P.2d 832 (Wyo. 1974); "Roach v. Kononen," 269 Or. 457, 525 P.2d 125 (1974); "Olson v. Arctic Enterprises, Inc." 349 F. Supp. 761 (D. N.D. 1972).

Subsection (C) also permits introduction of evidence of non-compliance with custom—evidence likely to be introduced by a product liability claimant. While it might be argued that non-compliance with custom should indicate that the product was defective, situations may arise where a product seller followed an alternative procedure that was no less safe (perhaps even safer) than the custom in the industry. For that reason, non-compliance with custom does not create a situation where the trier of fact should consider the product defective *per se.*

Subsection (C) treats compliance or non-compliance with non-governmental product standards in the same manner as custom. Compliance or non-compliance with such standards is admissible in evidence when relevant to the issue before the court. Privately developed standards—for both design and performance—vary widely in their nature and quality. While almost all privately developed standards are labeled "minimum" (suggesting a standard below that which a product seller could reasonably achieve), some standards reach the best level that can be obtained as a practical technological matter. *See* "Task Force Report" at IV–13–17. On the other hand, some are clearly below the level of safety appropriate as a defense in a product liability case. *See* "Report of the National Commission on Product Safety" at 48 (1978) ("chronically inadequate both in scope and permissible level of risk").

Compliance with standards that are rigorous and objective (in that they were developed through careful, thorough product testing and a formal product safety evaluation, and up-to-date in light of the technological and scientific knowledge reasonably available at the time the product was manufactured) suggests that a product was not defective. Failure to comply with such standards, especially performance standards, suggests that it was defective. Nevertheless, because of the variance in the nature and quality of privately developed safety standards, the Act has eschewed *per se* rules in this area. *See* "Poches v. J. J. Newberry Co.," 549 F.2d 1166 (8th Cir. 1977).

Subsection (D), defining "practical technological feasibility," should be contrasted with Subsection (B)'s definition of "custom." The Subsection (D) definition contemplates a level of safety that was feasible, as a practical matter, at the time of manufacture. "Feasibility" includes economic considerations such as the ability of a product seller to price a product so that it is competitive, but excludes financial considerations peculiar to a particular product seller, such as its cash flow at the relevant time.

The definition's emphasis on *feasibility* addresses consumer concerns with "state-of-the-art" or "custom" defenses that could allow an industry or business to be shielded from liability when it has "negligently" lagged behind in both developing and utilizing safety technology.

Subsection (E) indicates that when a product seller proves by a preponderance of the evidence that it was not within practical technological feasibility to make the product safer, the product seller shall not be subject to liability. This is in accord with the great majority of case law. *See, e.g.,* "Olson v. Arctic Enterprises, Inc.," *supra;* "Wilson v. Piper Aircraft Corp.," 282 Or. 61, 577 P.2d 1322, 1326 (1978) (". . . plaintiff's *prima facie* case of a defect must show more than the technical possibility of a safe design"); "Bruce v. Martin-Marietta Corp.," *supra;* "Maxted v. Pacific Car & Foundry Co.," *supra;* "Roach v. Kononen," *supra. Cf.* "Ky. Rev. Stat. Ann." Section 411.310(2) (Supp. 1978) (includes custom). *But cf.* "Colo. Rev. Stat." Section 13–21–403 (Supp. 1978) ("state of the art" presumption only).

Significantly, only a few intermediate appellate court decisions, primarily from one state, impose liability if the product was in accord with the technological, mechanical, and scientific knowledge reasonably feasible for use at the time of manufacture. *See, e.g.,* "Gelsumino v. E.W. Bliss Co.," 10 Ill. App. 3d 604, 295 N.E. 2d 110 (1973), and its progeny. *But see* "McClellan v. Chicago Transit Authority," 34 Ill. App. 3d 151, 340 N.E.2d 61 (1975). *Compare* "Olson v. A. W. Chesterton Co.," 256 N.W. 2d 530 (N.D. 1977).

In order to meet appropriate consumer group concerns, Subsection (E) provides four exceptions to the defense. First, the defense will not apply if the product seller knew or had reason to know of the danger and, with that knowledge, acted unreasonably in selling the product at all. *See also* Subsection 106(B)(1) and analysis. This exception would only apply in very unusual situations. For example, a child's toy might comply with what was technologically feasible but, because of its danger, it would be unreasonable behavior to market the product for young children. Similarly, it would be unreasonable to manufacture a home heating unit with a radium core unless technology had reached a point at which the inhabitants of the home (and others) would be protected from radiation.

Second, although the product was in accord with the "practical technological feasibility," it may be defective in construction. It is a basic principle of this Act to apply strict liability against the manufacturer in that case. *See* Subsection 104(A) and analysis.

Third, although the product complied with the "practical technological feasibility" at the time of manufacture, the product seller may have failed to meet its post-manufacture duty to warn or instruct under Subsection 104(C)(6).

Finally, although its product was in accord with "practical technological feasibility," a product seller cannot escape strict liability for express warranty under Subsection 104(D) or 105(B).

* * * * *

Code

Sec. 108. Relevance of Legislative or Administrative Regulatory Standards and Mandatory Government Contract Specifications

(A) When the injury-causing aspect of

the product was, at the time of manufacture, in compliance with legislative regulatory standards or administrative regulatory safety standards relating to design or performance, the product shall be deemed not defective under Subsection 104(B), or, if the standard addressed warnings or instructions, under Subsection 104(C) or 105(A), unless the claimant proves by a preponderance of the evidence that a reasonably prudent product seller could and would have taken additional precautions.

(B) When the injury-causing aspect of the product was not, at the time of manufacture, in compliance with legislative regulatory standards or administrative regulatory safety standards relating to design or performance, the product shall be deemed defective under Subsection 104(B), or, if the standard addressed warnings or instructions, under Subsection 104(C) or 105(A), unless the product seller proves by a preponderance of the evidence that its failure to comply was a reasonably prudent course of conduct under the circumstances.

(C) When the injury-causing aspect of the product was, at the time of manufacture, in compliance with a mandatory government contract specification relating to design, this shall be an absolute defense and the product shall be deemed not defective under Subsection 104(B), or, if the specification related to warnings or instructions, under Subsection 104(C) or 105(A).

(D) When the injury-causing aspect of the product was not, at the time of manufacture, in compliance with a mandatory government contract specification relating to design, the product shall be deemed defective under Subsection 104(B), or, if the specification related to warnings or instructions, under Subsection 104(C) or 105(A).

Analysis

Sec. 108. Relevance of Legislative or Administrative Regulatory Standards and Mandatory Government Contract Specifications

Product sellers have contended that it is unfair to deem a product defective when the challenged aspect of that product conformed to an applicable administrative or legislative regulatory standard. They note that considerable time and thought are spent in the development of such standards and that the standards are frequently subject to intense public scrutiny prior to the time of their official adoption. Product sellers point to legislative and administrative standards as a resource to provide some predictability within the scope of product liability law. They contend that it is unfair to allow lay jurors to reevaluate a standard that has presumably been drafted by government experts. Furthermore, some product liability loss prevention experts suggest that a defense based on compliance with standards will create incentives for manufacturers to comply with such standards. *See* Task Force "Selected Papers" at 266 (Remarks of Professor Alvin S. Weinstein).

On the other hand, some consumer groups maintain that government standards are often the result of compromise decisions—decisions that are sometimes unduly influenced by industry. These consumer groups point out that some government regulatory bodies may have insufficient personnel or expertise to make independent judgments. The general approach of courts, as well as that embodied in the "Consumer Product Safety Act," 15 "U.S.C." Section 2074(A) (1976) and the "National Traffic and Motor Vehicle Safety Act," 15 "U.S.C." Section 1391 (2) (1976), is that government standards are only minimum standards, and that compliance should not be deemed an absolute defense in product liability actions. *See* "Roberts v. May,"——Colo. App.——, 583 P. 2d 305, 308 (1978). In spite of this, some states have enacted statutes that grant this effect to such compliance. *See* "N.D. Cent. Code" Section 28–01.1–05(3) (Supp. 1979); "Utah Code Ann." Section 78–15–6 (1977); "Colo. Rev. Stat." Section 13–21–403 (Supp. 1978).

Nevertheless, case law also suggests that government safety standards are often deemed sound and appropriate for application to tort law clains. *See* "Jones

v. Hittle Serv., Inc.," 219 Kan. 627, 549 P. 2d 1383 (1976) (universally accepted standards for odorizing LP gas outweigh expert opinion); "McDaniel v. McNeil Laboratories, Inc.," 196 Neb. 190, 241 N.W. 2d 822 (1976) (determination of the FDA prevails in absence of proof that the manufacturer furnished incomplete, misleading, or fraudulent information); "Simien v. S.S. Kresge Co.," 566 F. 2d 551 (5th Cir. 1978) (compliance with flammability standard); "Bruce v. Martin-Marietta Corp.," 544 F. 2d 442, 446 (10th Cir. 1976) (claimants did "not present any more stringent standards which might have been applicable at the time of manufacture").

The approach taken in Subsection (A) is based on these cases and on the "Restatement (Second) of Torts" Section 288C (1965). When the specific injury-causing aspect of the product conformed to or was in compliance with the legislative or administrative regulatory standard, the product is deemed not defective under Subsection 104(B) when design is relevant, or under Subsection 104(C) or 105(A) when the duty to warn or instruct is relevant, unless the claimant proves by a preponderance of the evidence that a reasonably prudent product seller could and would have taken additional precautions.

This approach has enabled claimants to prevail when legislative or administrative standards did not meet an appropriate level of safety. For example, in "Raymond v. Riegel Textile Corp.," 484 F. 2d 1025 (1st Cir. 1973), the claimant was able to show that a standard promulgated under the "Flammable Fabrics Act" was outdated. *See also* "Burch v. Amsterdam Corp.," 366 A. 2d 1079 (D.C. 1976) (when manufacturer knows of greater dangers not included in a statutorily mandated warning, it should bring those precautions to the attention of product users). On the other hand, it recognizes that government safety standards *may* provide an adequate basis for evaluating safety in tort law.

Subsection (B) makes it clear that when the injury-causing aspect of the product did not conform or was not in compliance with a legislative or administrative regulatory standard, the product shall be deemed defective under Subsection 104(B), when design is relevant, or Subsection 104(C), when duty to warn is relevant, unless the product seller shows by a preponderance of the evidence that its failure to comply amounted to reasonably prudent conduct under the circumstances. *See* "Restatement (Second) of Torts" Section 286 (1965). A product seller may be able to prevail under Subsection (B) by showing that compliance would have created greater dangers than non-compliance would have. *Cf.* "Davison v. Williams," 251 Ind. 448, 242 N.E. 2d 101 (1968) (violation of safety regulation may be justified in circumstances).

Subsection (C) addresses a highly specialized problem with respect to a product that had been manufactured strictly in accordance with mandatory specifications set forth in a government contract. When compliance with such a standard leads to an injury, the government, not the product seller, is the appropriate defendant. As the court in "Hunt v. Blasius," 55 Ill. App. 3d 14, 370 N.E. 2d 617, 621–22 (1977), indicated, "public policy dictates that bidders who comply strictly with governmental specifications should be shielded from liability in any respect in which the product complies." When enacting this provision, a legislature should ensure that its own state government bears financial responsibility (either through tort law or through a compensation system) for the harm it has caused by directing that the product conform to contract specifications.

Subsection (D) provides a counterweight to Subsection (C). If the bidder fails to comply with a mandatory government contract specification, and this failure to comply caused the claimant's injury, the product seller will deemed liable under Subsection 104(B) if the specification related to design, or under Subsection 104(C) or 105(A) of the specification related to instruction or warnings. If the manufacturer's compliance exceeded the government regulation and the product failed, liability should not be imposed under this Section—in that case, the technical "failure to comply" would not be the proximate cause of the injury that befell claimant.

* * * * *

Code

Sec. 109. Notice of Possible Claim Required

(A) An attorney who anticipates filing a claim shall notify all product sellers against whom the claim is likely to be made. The notice of claim shall:

(1) Identify the product as specifically as possible;

(2) State the time, place, circumstances, and events giving rise to the claim;

(3) Give an estimate of compensation or other relief to be sought.

(B) The attorney shall give notice of claim within six (6) months of the date of entering into an attorney-client relationship with the claimant in regard to the claim. For the purposes of Section 109, such a relationship arises when the attorney, or any member or associate of the attorney's firm, agrees to serve the claimant's interests in regard to the anticipated claim.

(C) If the claimant's attorney requests the information at the time the notice of claim is given, the product seller receiving the notice of claim shall promptly furnish the claimant's attorney with the names and addresses of each person whom the product seller knows to be in the chain of manufacture and distribution of the product, and who is likely to be subject to liability under Sections 104 or 105. Any product seller who fails to furnish such information may be subject to liability as provided in Subsection (E).

(D) A claimant who delays entering into an attorney-client relationship so as to delay unreasonably the notice of claim required by Subsection (A) may be subject to liability as provided in Subsection (E).

(E) Any party to the product liability claim or any attorney representing such a party who suffers a monetary loss associated with the litigation of the claim caused by the failure of a claimant or a claimant's attorney, or of a product seller or its attorney, to comply with the requirements of this Section may recover pecuniary damages, costs, and reasonable attorneys' fees from that party. Failure to comply with the requirements of Section 109 does not affect the validity of any claim or defense under this Act.

Analysis

Sec. 109. Notice of Possible Claim Required

The purpose of Section 109 is to inform product sellers at an early date that the products they produce may be defective. Under present law, a claimant can delay informing a product seller of a claim until the statute of limitation has nearly expired. In most jurisdictions, this period is two or three years. Although 77.8 percent of all bodily injury claims are reported within six months, the 22.2 percent that are not reported during this period are of concern because they represent about 68 percent of the claim payments. "ISO Closed Claims Survey" at 100 (1977).

A reasonable notice of claim requirement in product liability law promotes the interests of consumers and product users because it is a low-cost means of helping to assure product safety. Presumably, if informed about defective conditions at an early stage, a product seller is likely to take action to correct such conditions and thus prevent future injuries. This is why notice-of-claim provisions have been utilized in other contexts. *See, e.g.,* "U.C.C." Section 2–607 (warranty breaches); 18 E. McQuillan, "Municipal Corporations" Section 53.154 (3d ed. 1977) (suits against municipalities for injuries); 3 A. Larson, "Workmen's Compensation Law" Section 78.00 *et seq.* (1976) (notice of injury to employer). *See also* Comment, "Notice Requirement in Warranty Actions Involving Personal Injury," 51 "Calif. L. Rev." 586 (1963); Phillips, "Notice of Breach in Sales and Strict Tort Liability Law: Should There Be a Difference?," 47 "Ind. L. J." 457, 468–69 (1972) (observing that requiring notice of claim may encourage defendants to make reasonable settlements).

This Section is adapted from the

recently enacted "Minn. Stat. Ann." Section 604.04 (Supp. 1978). It differs from analogous notice-of-claim provisions in that it does not provide that a claim or defense will be barred by the failure of the injured party to meet its conditions. As the court noted in "Greenman v. Yuba Power Prods., Inc.," 59 Cal. 2d 57, 377 P.2d 897, 27 Cal. Rptr. 697, 700 (1963), such a provision may

> become a booby trap for the unwary. The injured consumer is seldom 'steeped in the business practice which justifies the rule,' [James, "Products Liability" (pt.2), 34 "Tex. L. Rev. 192, 197 (1955)] and at least until he has had legal advice it will not occur to him to give notice . . . .

Instead, Section 109 places a duty on the attorney to give the notice of claim. It imposes on the attorney the costs of investigation and other litigation expenses which stem from the failure to give notice.

Section 109 also places a burden on the product seller who is notified of an anticipated claim to provide the claimant's attorney with the names and addresses of others whom the product seller knows to be in the chain of manufacture and distribution of the product, and who are likely to be subject to liability under Sections 104 or 105. The product seller is not obligated, however, to supply long lists of every supplier of a component part of its product or every possible wholesale and retain purchaser of it. Product sellers may also be held liable under this Section for litigation costs that stem from the failure to provide the required names.

Subsection 109(E) is limited to losses associated directly with the litigation of the claim in issue and does not apply to possible losses suffered by persons other than a party to the product liability claim or any attorney representing such a party.

Claims arising under Section 109 can be consolidated with the principal product liability claim brought under this Act.

* * * * *

Code

Sec. 110. Length of Time Product Sellers are Subject to Liability

(A) *Useful Safe Life.*

(1) Except as provided in Subsection (A)(2), a product seller shall not be subject to liability to a claimant for harm under this Act if the product seller proves by a preponderance of the evidence that the harm was caused after the product's "useful safe life" had expired.

"Useful safe life" begins at the time of delivery of the product and extends for the time during which the product would normally be likely to perform or be stored in a safe manner. For the purposes of Section 110, "time of delivery" means the time of delivery of a product to its first purchaser or lessee who was not engaged in the business of either selling such products or using them as component parts of another product to be sold.

Examples of evidence that is especially probative in determining whether a product's useful safe life had expired include:

(a) The amount of wear and tear to which the product had been subject;

(b) The effect of deterioration from natural causes, and from climate and other conditions under which the product was used or stored;

(c) The normal practices of the user, similar users, and the product seller with respect to the circumstances, frequency, and purposes of the product's use, and with respect to repairs, renewals, and replacements;

(d) Any representations, instructions, or warnings made by the product seller concerning proper maintenance, storage, and use of the product or the expected useful safe life of the product; and

(e) Any modification or alteration of the product by a user or third party.

(2) A product seller may be subject to liability for harm caused by a product used beyond its useful safe life to the extent that the product seller has expressly warranted the product for a longer period.

(B) *Statute of Repose.*

(1) *Generally.* In claims that involve harm caused more than ten (10) years after time of delivery, a presumption arises that the harm was caused after the useful safe life had expired. This

presumption may only be rebutted by clear and convincing evidence.

(2) *Limitations on Statute of Repose.* (a) If a product seller expressly warrants that its product can be utilized safely for a period longer than ten (10) years, the period of repose, after which the presumption created in Subsection (B)(1) arises, shall be extended according to that warranty or promise.

(b) The ten- (10-) year period of repose established in Subsection (B)(1) does not apply if the product seller intentionally misrepresents facts about its product, or fraudulently conceals information about it, and that conduct was a substantial cause of the claimant's harm.

(c) Nothing contained in Subsection (B) shall affect the right of any person found liable under this Act to seek and obtain contribution or indemnity from any other person who is responsible for harm under this Act.

(d) The ten- (10-) year period of repose established in Subsection (B)(1) shall not apply if the harm was caused by prolonged exposure to a defective product, or if the injury-causing aspect of the product that existed at the time of delivery was not discoverable by an ordinary reasonably prudent person until more than ten (10) years after the time of delivery, or if the harm, caused within ten (10) years after the time of delivery, did not manifest itself until after that time.

(C) *Statute of Limitation.* No claim under this Act may be brought more than two (2) years from the time the claimant discovered, or in the exercise of due diligence should have discovered, the harm and the cause thereof.

* * * * *

Analysis

Sec. 110. Length of Time Product Sellers are Subject to Liability

Perhaps more significant than any other single factor alleged to be the cause of the nationwide product liability insurance problem are the rules governing the responsibility of product sellers for older products. Most product liability policies include claims based not only on products manufactured or sold during the given policy year, but also on products manufactured or sold in the past. In the case of sellers of durable goods, this creates an "open-ended" liability situation.

The Supreme Court of Oregon summarized the general common law rule for products with the statement: "[P]rolonged use of a manufactured article is but one factor, albeit an important one, in the determination of whether a defect in the product made it unsafe. . . ." "Tucker v. Unit Crane & Shovel Corp.," 256 Or. 318, 473 P.2d 862 (1970) (boom crane manufactured in 1956, collapsed in 1965). *See also* "Gates v. Ford Motor Company," 494 F. 2d 458 (10th Cir. 1974) (24-year-old tractor); "Kaczmarek v. Mesta Machinery Co.," 463 F.2d 675 (3d Cir. 1972) (30-year-old pickling machine); "Mondshour v. General Motors Corp.," 298 F. Supp. 111 (D. Md. 1969) (bus designed 17 years prior to accident).

Partly in response to this open-ended liability potential, a number of states have enacted statutes of repose that begin at the time a product is first sold and distributed. *See, e.g.,* "1979 Ala. Acts" No. 79–468 Section 3(b) ("10 years after the manufactured product is first put to use"); "Ariz. Rev. Stat. Ann." Section 12–551 (Supp. 1978) ("12 years after the product was first sold for use or consumption. . ."); "Fla. Stat. Ann." Section 95.031(2) (Supp. 1979) ("12 years after the date of delivery of the completed product"); "Ill. Ann. Stat." ch. 83, Section 22.2(b) (Supp. 1979) (12 years from date of first sale or 10 years from sale to actual user-claimant—strict liability); "Ind. Code Ann." Section 33–1–1.5–5 (Supp. 1978) ("10 years after the delivery . . . to the initial user"); "Neb. Rev. Stat." Section 25–224 (Supp. 1978) (10 years after first sale); "Utah Code Ann." Section 78–15–3(1) (1977) (6 years after date of initial purchase; 10 years after date of manufacture).

The advantages of these statutes are that they: (1) establish an actuarially certain date after which no liability can be assessed; and (2) eliminate tenuous claims involving older products for which evidence of defective conditions may be difficult to produce. *See* "Order of Railroad Telegraphers v. Railway

Express Agency," 321 U.S. 342, 348–49 (1944).

On the other hand, a fundamental problem with these statutes is that they *may* deprive a person injured by a product of the right to bring a claim based on a defective product before the injury has actually occurred. *See* Johnson, "Products Liability 'Reform': A Hazard to Consumers," 56 "N.C. L. Rev." 677, 689–90 (1978); "Victorson v. Bock Laundry Mach. Co.," 37 N.Y.2d 395, 335 N.E.2d 275, 373 N.Y.S.2d 39 (1975).

The limited available data show that insurers' apprehension about older products may be exaggerated. *See* "ISO Closed Claims Survey" at 105–08 (indicating that over 97 percent of product-related accidents occur within six years of the time the product was purchased and, in the capital goods area, 83.5 percent of all bodily injury accidents occur within ten years of manufacture). Nevertheless, as the "Task Force Report" indicated, the underwriters' concern about *potential* losses associated with older products may be an important factor in the recent increase in liability insurance premiums for manufacturers of durable goods. *See* "Task Force Report" at VII–21.

Section 110 attempts to provide insurers and product sellers with some security against stale claims, while preserving the claimant's right to obtain damages for injuries caused by defective products. It accomplishes this result through provisions on useful safe life, a statute of repose, and a statute of limitation.

(A) *Useful Safe Life.* The common law in most states is that "[t]he age of an allegedly defective product must be considered in light of its expected useful life and the stress to which it has been subjected." "Kuisis v. Baldwin-Lima-Hamilton Corp.," 457 Pa. 321, 319 A.2d 914, 923 (1974) (brake-locking mechanism on a crane failed after more than 20 years of use). The "Kuisis" court noted further that "in certain situations the prolonged use factor may loom so large as to obscure all others in a case." *Id.*

The basic problem has been the vagueness of the concept. Thus, while the "Task Force Report" noted that "if a useful life limitation were identified in statutory form, it might be expected that it would be given more serious attention by both judge and jury" ("Task Force Report" at VII–27), it also observed that "the concept would still lack specificity." *Id.* Subsection (A) is designed to define the concept with as much specificity as possible.

The Subsection was derived from "Minn. Stat. Ann." Section 604.03 (Supp. 1978). It serves to remind the court and the trier of fact that a product seller may be held liable only for harms caused during the useful *safe* life of the product. It does not attempt to apply fixed useful safe life standards for all products. Such an approach is not possible as a practical matter. *See* Phillips, "An Analysis of Proposed Reform of Products Liability Statutes of Limitations," 56 "N.C. L. Rev." 663, 673 (1978). Rather, it identifies factors that will assist the court and the trier of fact in determining how long a product can reasonably be expected to perform or be stored in a safe manner.

Section 110 uses the term "useful safe life" (not "useful life," a term which has already acquired a meaning in the law of taxation), because the period in which the product can have some "utility" may be well beyond the period in which the product is "safe." For example, a driver may continue to "use" tires that lack sufficient treads for safety.

A product seller may raise the "useful safe life" of a product as an affirmative defense in a product liability claim. The period of time begins at the time of delivery of the product and extends through the time during which the product would normally be likely to perform in a safe manner.

"Time of delivery" is defined as the time of delivery of a product to its first purchaser or lessee not engaged in the business of either selling such products or using them as component parts of another product to be sold. "Time of delivery" is an important concept in the Act because it marks the last time a product seller has physical control of the product and the first time that the product is in the hands of someone other than a product seller. The useful safe life and repose periods of Section 110 begin to run from this point in time.

The Section then sets forth a series of factors that will be of assistance to the court and jury in determining whether the period of time has expired. The product seller need not introduce evidence with respect to each of the listed factors. Rather, each of the litigants in a product liability claim may introduce evidence on those factors listed which, they believe, will support their respective contentions. In some cases, evidence on one or a few of the factors may be regarded as dispositive of the "useful safe life" issue and, when this occurs, the lack of evidence on one of the other factors normally will not affect the outcome of the case. The product seller has the burden of proof upon the issue.

Factors (a) through (c) under Subsection (A) are basically self-explanatory. Factor (d) refers to the useful safe life stated *by the product seller.* A product seller can place some reliance on this provision when it has indicated that a product should not be used beyond a certain period of time. However, Subsection (A) does not give the product seller absolute power to limit a product's useful safe life. While this was suggested in "Velez v. Craine & Clark Lumber Corp.," 33 N.Y.2d 117, 305 N.E.2d 750, 350 N.Y.S.2d 617 (1973), almost all courts would insist on retaining judicial power to determine whether the product seller's limitation was a reasonable one. *Cf.* "Henningsen v. Bloomfield Motors, Inc.," 32 N.J. 358, 161 A.2d 69 (1960). Further, even the "Velez" court indicated that a product seller's limitation on useful life could not bind the rights of a non-purchaser claimant. Nevertheless, where the product seller imposes a reasonable limitation, made in good faith to protect the user, the trier of fact shoud give very serious consideration to this fact in determining whether the product was used beyond its useful safe life.

Factor (e), dealing with modifications of the product by users or third parties, relates to conduct that might shorten or lengthen the useful life of the product. While the Act treats product modifications in Section 112, they are also factors in determining whether a product has been used beyond its useful safe life.

(B) *Statute of Repose.*

(1) *Generally.* Statutes of repose differ from statutes of limitation in that they set a fixed limit after the time of the product's manufacture, sale, or delivery beyond which the product seller will not be held liable. The rationale of such statutes is threefold. First, the fact that a product has been used safely for a substantial period of time is some indication that it was not defective at the time of delivery. Second, if a product seller is not aware of a claim, the passing of time may make it extremely difficult to construct a good defense because of the obstacle of securing evidence. Although the burden of proof on the issue of defectiveness remains on the claimant under the Act, a jury, as a practical matter, may demand an explanation from a product seller when the claimant has suffered a severe injury. The third rationale is that persons ought to be allowed, as a matter of policy, to plan their affairs with a reasonable degree of certainty. This goes to the heart of the product liability insurance rate-setting problem. Even though past data show that 83.5 percent of bodily injury claims arise within a ten-year period,[13] there is no safeguard in the existing law that the past will portend the future. There is always the possibility that the number of claims for older products will increase. *See* "ISO Closed Claims Survey" at 107.

On the other hand, consumers are justifiably concerned about overly broad absolute cut-offs of their right to sue. This provision recognizes consumer concerns in three basic ways:

(1) The term of the statute is ten years—beyond the term enacted or proposed in a number of states;

(2) The statute begins to run at the time of delivery, not the time of manufacture; and

(3) The statute does not contain an absolute cut-off, but rather a presumption that the product has been used beyond its useful safe life. Colorado law utilizes this approach. "Colo. Rev. Stat." Section 13–21–403(3) (Supp. 1978). Most other state product liability statutes do not.

Consumer concerns are also

[13] Figure for capital goods.

addressed by three of the four additional restrictions contained in Subsection (B)(2).

(2) *Limitations on Statute of Repose.* This Subsection contains four key limitations on its scope of operation.

First, Subsection (B) does not apply when a product seller has expressly warranted or promised that a product can be used safety for a period longer than ten years. *See* Subsection 102(K) (definition of "express warranty").

Second, the statute of repose provisions do not apply when a product seller intentionally misrepresented facts about its product and this misrepresentation was a substantial cause of claimant's harm.

Third, Subsection (B) does not affect contribution and indemnity claims. Thus, an intermediate product seller will not have to absorb a liability loss that was the true responsibility of the original manufacturer. *See* Defense Research Institute, "Products Liability Position Paper" at 22 (monograph 1976); *see also* Phillips, *supra*, 56 "N.C. L. Rev." at 670–71 (1978).

Fourth, there is an exception for products that cause perceptible harm only through prolonged exposure (*see, e.g.,* "Michie v. Great Lakes Steel Div.. National Steel Corp.," 495 F.2d 213 (6th Cir. 1974)), or that cause harms that take many years to manifest themselves. *See* "Sindell v. Abbott Laboratories," 85 Cal. App. 3d 1, 149 Cal. Rptr. 138 (1978). An exception is also made for the unusual situation in which a product contains, at the time of delivery, a hidden defect that is not discoverable by a reasonably prudent product user and does not manifest itself until after a ten-year period has expired. *See* "Mickel v. Blackmon," 252 S.C. 202, 166 S.E.2d 173 (1969) (plastic used on gearshift lost its resiliency when exposed to sunlight).

If the ten-year presumption does not apply, a product seller can still prove that the product has been utilized beyond its useful safe life under Subsection (A).

(C) *Statute of Limitation.* Tort statutes of limitation traditionally begin at the time a person is injured. This Subsection follows that approach. Nevertheless, in accord with justified consumer concerns, Subsection (C) extends the limitation period beyond the time of harm in situations where the claimant would have no reason to know about the harm or the causal connection to a defective product (*e.g.,* the case of long-term pharmaceutical harms). This reflects a general trend in both statutory and case law. *See* Birnbaum, " 'First Breath's' Last Gasp: The Discovery Rule in Products Liability Cases," 13 "Forum" 279 (1977); and Annot., 91 "A.L.R.3d" 991 (1979). The two-year period represents the length of the traditional state statute of limitation based on claims for negligence. In light of the Act's adoption of the discovery rule, this is the maximum period that seemed to be appropriate.

The underlying philosophy of this Section is congruent with Sections 105 and 106, which shield product sellers from liability for risks that they would have no reason to discover at the time of manufacture.

* * * * *

Code

Sec. 111. Comparative Responsibility and Apportionment of Damages

(A) *Comparative Responsibility.* All claims under this Act shall be governed by the principles of comparative responsibility. In any claim under this Act, the comparative responsibility of, or attributed to, the claimant shall not bar recovery but shall diminish the award of compensatory damages proportionately, according to the measure of responsibility attributed to the claimant.

(B) *Apportionment of Damages.*

(1) In all claims involving comparative responsibility, the court, unless otherwise agreed by all parties, shall instruct the jury to answer special interrogatories or, if there is no jury, the court shall make findings, indicating:

(a) The amount of damages each claimant would be entitled to recover if the comparative responsibility of each party were disregarded; and

(b) The percentage of the total responsibility of all parties to each claim that is to be allocated to each claimant:

defendant; third-party defendant; person or entity who misused, modified, or altered a product under Subsection 112 (C) or (D), [or who voluntarily and unreasonably used or stored a product with a known defective condition under Subsection 112(B)]; and person released from liability under Subsection 113(E). Under this Subsection, the court may determine that two or more persons are to be treated as a single party.

(2) When the claimant's employer's or co-employee's fault is considered, damages shall be reduced in accordance with Subsection 114(A), if applicable, or by the percentage of responsibility apportioned to such employer or co-employee, if that amount is greater. When a person released from liability under Subsection 113(E) would otherwise be liable under this Act, damages shall be reduced by the percentage of responsibility apportioned to such person.

(3) In determining the percentages of responsibility, the trier of fact shall consider, on a comparative basis, both the nature of the conduct of each person or entity responsible and the extent of the proximate causal relation between the conduct and the damages claimed.

(4) The court shall determine the award of damages to each claimant in accordance with the findings and enter judgment against each party liable. For purposes of contribution under Section 113, the court shall also determine and state in the judgment each party's equitable share of the obligation to each claimant in accordance with the respective percentages of responsibility.

(5) Damages are to be apportioned severally, and not jointly, when a party is responsible for a distinct harm, or when there is some other reasonable basis for apportioning that party's responsibility for the harm. Otherwise, judgment shall be entered against each party liable on the basis of the rules of joint and several liability.

(6) When one or more parties made a substantial contribution to an indivisible harm, or for other reasons under the common law of the state is a joint tortfeasor, upon motion made not later than one (1) year after judgment is entered, the court shall determine whether all or part of a joint tortfeasor's share of the obligation is uncollectible from that joint tortfeasor.

If the court's finding is in the affirmative, the court shall reallocate any uncollectible amount among a claimant found to be responsible and other parties who are joint tortfeasors with the party whose share is uncollectible. The reallocation shall be made according to the respective percentages of responsibility of each party.

* * * * *

Analysis

Sec. 111. Comparative Responsibility and Apportionment of Damages

(A) *Comparative Responsibility.* Subsection (A) attempts to resolve existing legal uncertainty about the relevance of a claimant's conduct and the comparative responsibility of others who contributed to the claimant's harm. It applies principles of comparative responsibility to situations where more than one person has some responsibility for the product-related incident. *Cf.* "Uniform Comparative Fault Act" ("UCFA") Section 1. The consumer-oriented fairness of pure comparative responsibility is adopted, as compared with the "non-discriminating rough justice of the modified type. . . ." "Prefatory Note," "UCFA".

Although there is no assurance that the use of comparative responsibility principles will lower the cost of product liability claims, the inherent fairness of such principles has led to their inclusion in the "UCFA" by the National Conference of Commissioners of Uniform State Laws. Section 111 borrows extensively from the "UCFA" and its accompanying commentary.

Some courts and commentators have voiced concern about the semantic and theoretical difficulties of mixing the "apples" of negligence with the "oranges" of strict liability. *See, e.g.,* "Kirkland v. General Motors Corp.," 521 P.2d 1353 (Okla. 1974); Robinson, "Square Pegs (Products Liability) in Round Holes (Comparative Negligence)," 52 "Cal. St. B.J." 16 (1977). Nevertheless, these concerns appear to be more theoretical than real.

The utility of comparative responsibility for product liability cases has been appreciated both by state legislatures[14] and courts.[15] It has also been recommended by a congressional subcommittee. "LaFalce Subcommittee Report," *supra* at 76.

Section 111 places a strong incentive for loss prevention on the party who is best able to accomplish that goal. It also avoids burdening the careful product user with liability insurance costs assessed to persons who misuse or are otherwise at fault in their handling of products. While some economic analyses indicate that a comparative responsibility system creates a risk of economic inefficiency because of an overinvestment in safety, the drafters of the Act have made a value judgment that such an "overinvestment" is worth making.

(B) *Apportionment of Damages.* In order to apply comparative responsibility principles under this Act, it is necessary for the trier of fact to supply certain information in special interrogatories. Subsection (B)(1)(a), which is based on "UCFA" Subsection 2(A)(1), indicates that the trier of fact should set forth the amount of damages a claimant would receive if each party's comparative responsibility were disregarded. This helps assure that the trier of fact does not inflate or deflate the amount of damages claimant would receive if the claimant were free from responsibility.

Subsection (B)(1)(b), which is based on "UCFA" Subsection 2(a)(2), requires the trier of fact to indicate the percentage of responsibility allocated to each claimant; defendant; third-party defendant; and person or entity who has misused, altered, or modified products under Subsections 112 (C) or (D). It also includes persons or entities who voluntarily and unreasonably used or stored a product with a known defective condition under Subsection 112(B)(2). As is noted in the analysis to that Subsection, this is an optional provision.

Although it is difficult to apportion the responsibility of an absent employer or co-employee who is immune from tort liability due to Worker Compensation laws, such an approach is necessary to ensure fairness to product sellers. It is also fair to employees because they have given up their right to sue in tort for harms caused by their employers' or co-employees' fault in exchange for their Worker Compensation benefits.

Subsection 111(B)(2) indicates that in cases of product misuse or alteration where an employer's or co-employee's fault is considered, damages should be reduced by the amount of Worker Compensation benefits the worker received or will receive in accordance with Subsection 114(A), or by the percentage of responsibility apportioned to such employer or co-employee, whichever is greater.

Thus, for example, in a case in which a manufacturer is found to have been 40 percent responsible, the injured employee's employer was 60 percent responsible, and Worker Compensation benefits constituted 15 percent of the damages, then the claimant's damages would be reduced by 60 percent, not 85 percent. On the other hand, if the manufacturer had been 80 percent responsible, the employer was 20 percent responsible, and Worker Compensation benefits constituted 30 percent of the damages, then the claimant's damages would be reduced by 30 percent, which was the amount of the Worker Compensation benefits and was greater than the employer's percentage (20 percent) of responsibility.

When a released party's fault is considered, the process is simpler. Damages are reduced according to the percentage of fault which the trier of fact attributes to that person or entity.

Subsection (B)(1)(b) also indicates that persons who have been released under Subsection 113(E) shall be included in the apportionment of responsibility. This approach is in accord with the "UCFA" and the

[14] *E.g.*, "Ark. Stat. Ann." Section 27–1763–1765 (Supp. 1977); "Me. Rev. Stat." tit. 14 Section 156 (Supp. 1978); "Mich. Comp. Laws Ann." Section 600.2949 (Supp. 1978), "Mich. Stat. Ann." Section 27A.2949; Conn. (1979 Conn. Pub. Acts 79–483, Section 4).

[15] *E.g.*, "Thibault v. Sears, Roebuck & Co.," 395 A.2d 843 (N.H. 1978); "Daly v. General Motors Corp.," 20 Cal. 3d 725, 575 P.2d 1162, 144 Cal. Rptr. 380 (1978); "Busch v. Busch Constr., Inc.," 262 N.W.2d 377 (Minn. 1977); "Butaud v. Suburban Marine & Sport. Goods, Inc.," 555 P.2d 42 (Alaska 1976).

majority of cases that have addressed this issue. Again, while it is difficult to apportion an absent person's fault, the approach helps to ensure that all releases are executed in good faith. *See* "Frey v. Snelgrove," ——— Minn. ——— 269 N.W.2d 918 (1978); "Bartels v. City of Williston," 276 N.W.2d 113 (N.D. 1979); "Pierringer v. Hoger," 21 Wis. 2d 182, 124 N.W.2d 106 (1963).

Subsection (B)(3), which is based on "UCFA" Subsection 2(b), provides a general guideline to assist the trier of fact in comparing responsibility among the parties. The "UCFA" comments (Section 2) indicate that in appropriate cases, the trier of fact may also consider:

(1) Whether the conduct was mere inadvertence or engaged in with an awareness of the danger involved;

(2) The magnitude of the risk created by the conduct, including the number of persons endangered and the potential seriousness of the injury;

(3) The significance of what the actor was seeking to attain by his conduct;

(4) The actor's superior or inferior capacities; and

(5) The particular circumstances, such as the existence of an emergency requiring a hasty decision.

"UCFA Comment" Section 2 ("percentages of fault").

"The extent of the proximate causal relation between the conduct and the damages claimed" refers to proximate cause as opposed to cause-in-fact. While, at times, the distinction may be a difficult one to draw, this Act is premised on apportioning responsibility only—pure causation in terms of cause-in-fact is irrelevant to that concept. *See* Malone, "Ruminations on Cause-in-Fact," 9 "Stan. L. Rev." 60 (1956). Proximate cause, on the other hand, is an important concept in this Section. In order for the product user's conduct to bring about a reduction or apportionment of damages, it must be a proximate cause of the harm.

The importance of the distinction between cause-in-fact and proximate cause as applied to this Section may be illustrated by the so-called "second collision case" in which the claimant sues the manufacturer for enhanced injuries due to an alleged defect in the automobile, although the initial impact of the accident was caused by the claimant's or another person's negligent driving. When the enhanced injuries caused by the "second collision" within the automobile can be reasonably separated from those injuries caused by the initial collision, damages should be approximately divided. Nevertheless, the negligent driving should not be compared with the manufacturer's conduct because it is not the proximate cause of the enhanced injuries, even though it is clearly a cause-in-fact. *See* "Austin v. Ford Motor Co.," 86 Wis. 2d 628, 273 N.W. 2d 233, 239 (1979).

On the other hand, there may be "second collision cases" in which the enhanced injuries and the initial impact injuries cannot be reasonably separated. Such a case might arise when an allegedly defective automobile was hit at high speed. *See* "Fietzer v. Ford Motor Co.," 590 F. 2d 215 (7th Cir. 1978). In such a case, the driver's conduct is considered a proximate cause of the harm for purposes of comparative responsibility.

Subsection (B)(3), which is based on "UCFA" Subsection 2(c), helps to assure that the mathematics of comparative responsibility will be correctly determined. The court must determine the award for each claimant according to the findings made under this Subsection.

Subsections (B) (4) and (5) indicate that the common law rules of joint and several liability continue to apply under this Act. In this connection, it is important for the court to determine whether the defendant *is* a joint tortfeasor, *i.e.*, if its tortious conduct was a substantial cause of an indivisible injury, or it is otherwise deemed to have that status (*e.g.*, persons acting in concert, express or implied; persons vicariously liable for the torts of another). *See* W. Prosser, "Torts" at 297–99 (4th ed. 1971). A defendant who caused only a divisible part of the claimant's harm is only severally liable for that portion. *See* "UCFA Comment" Section 4, and "Restatement (Second) of Torts" Section 433A (1965).

As with "UCFA" Subsection 2(c), the judgment for each claimant will also show the share of each party's total obligation to the claimant. This should save litigation costs and avoid the need for a special motion or a separate action on the issue.

Subsection (B)(5) follows "UCFA" Subsection 2(d) in providing for the reallocation of damages among the parties responsible when one of the parties' share is uncollectible. The reallocation procedure applies to claimants who are contributorily at fault and joint tortfeasors who have made a substantial contribution to claimant's harm.

*   *   *   *   *

Code

Sec. 112. Conduct Affecting Comparative Responsibility

(A) *Failure to Discover a Defective Condition.*

(1) *Claimant's Failure to Inspect.* A claimant is not required to have inspected the product for a defective condition. Failure to have done so does not render the claimant responsible for the harm caused or reduce the claimant's damages.

(2) *Claimant's Failure to Observe an Apparent Defective Condition.* When the product seller proves by a preponderance of the evidence that the claimant, while using the product, was injured by a defective condition that would have been apparent, without inspection, to an ordinary reasonably prudent person, the claimant's damages shall be subject to reduction. The procedural principles governing reduction of damages are set forth in Section 111.

(3) *A Non-Claimant's Failure to Inspect for Defects or to Observe an Apparent Defective Condition.* A non-claimant's failure to inspect for a defective condition or to observe an apparent defective condition that would have been obvious, without inspection, to an ordinary reasonably prudent person, shall not reduce claimant's damages.

(B) *Use of a Product With a Known Defective Condition.*

(1) *By a Claimant.* When the product seller proves, by a preponderance of the evidence, that the claimant knew about the product's defective condition, and voluntarily used the product or voluntarily assumed the risk of harm from the product, the claimant's damages shall be subject to reduction to the extent that the claimant did not act as an ordinary reasonably prudent person under the circumstances. Under this Subsection, the trier of fact may determine that the claimant should bear sole responsibility for harm caused by a defective product. The procedural principles governing reduction of damages are set forth in Section 111.

***Optional Section**

[(2) *By a Non-Claimant Product User.* If the product seller proves by a preponderance of the evidence that a product user, other than the claimant, knew about a product's defective condition, but voluntarily and unreasonably used or stored the product and thereby caused claimant's harm, the claimant's damages shall be subject to apportionment. The procedural principles governing apportionment of damages are set forth in Section 111.]

(C) *Misuse of a Product.*

(1) "Misuse" occurs when the product user does not act in a manner that would be expected of an ordinary reasonably prudent person who is likely to use the product in the same or similar circumstances.

(2) When the product seller proves, by a preponderance of the evidence, that product misuse by a claimant, or by a party other than the claimant or the product seller, has caused the claimant's harm, the claimant's damages shall be subject to reduction or apportionment to the extent that the misuse was a cause of the harm. Under this Subsection, the trier of fact may determine that the harm arose solely because of product misuse. The procedural principles governing reduction or apportionment of damages are set forth in Section 111.

(3) Under this Subsection, subject to state and federal law regarding immunity in tort, the trier of fact may determine that a party or parties who misused the product and thereby caused claimant's harm should bear partial or sole responsibility for harm caused by the product and are subject to liability to the claimant.

(D) *Alteration or Modification of a Product.*

(1) "Alteration or modification" occurs when a person or entity other than the product seller changes the design, construction, or formula of the product, or changes or removes warnings or instructions that accompanied or were displayed on the product. "Alteration or modification" of a product includes the failure to observe routine care and maintenance, but does not include ordinary wear and tear.

(2) When the product seller proves, by a preponderance of the evidence, that an alteration or modification of the product by the claimant, or by a party other than the claimant or the product seller, has caused the claimant's harm, the claimant's damages shall be subject to reduction or apportionment to the extent that the alteration or modification was a cause of the harm. Under this Subsection, the trier of fact may determine that the harm arose solely because of the product alteration or modification.

This Subsection shall not be applicable if:

(a) The alteration or modification was in accord with the product seller's instructions or specifications;

(b) The alteration or modification was made with the express or implied consent of the product seller; or

(c) The alteration or modification was reasonably anticipated conduct under Subsection 102(G), and the product was defective under Subsection 104(C) because of the product seller's failure to provide adequate warnings or instructions with respect to the alteration or modification.

The procedural principles governing reduction or apportionment of damages are set forth in Section 111.

(3) Under this Subsection, subject to state and federal law regarding immunity in tort, the trier of fact may determine that a party or parties who altered or modified the product and thereby caused claimant's harm should bear partial or sole responsibility for harm caused by the product and are subject to liability to the claimant.

* * * * *

Analysis

Sec. 112. Conduct Affecting Comparative Responsibility

(A) *Failure to Discover a Defective Condition.*

(1) *Claimant's Failure to Inspect.* Under common law, the product user had an obligation to inspect for defects; failure to do so could bar a claim. *See* "Palmer v. Massey-Ferguson, Inc." 3 Wash. App. 508, 476 P. 2d 713 (1970). However, under modern tort law, the product user is entitled to assume that the product is reasonably safe for its ordinary use. *See* "Restatement (Second) of Torts" Section 402A (1965); "Cepeda v. Cumberland Eng'r. Co.," 76 N.J. 152, 386 A. 2d 816 (1978). Subsection (A) follows these cases and does not require the product user or consumer to inspect a product for a defect. *See* "Kassouf v. Lee Bros.," 209 Cal. App. 2d 568, 26 Cal. Rptr. 276 (1962) (plaintiff, without inspection, ate a chocolate bar containing worms and maggots).

(2) *Claimant's Failure to Observe an Apparent Defective Condition.* Cases can arise where a defect would be apparent, without inspection, to an ordinary reasonably prudent person. Subsection (A)(2) incorporates the Task Force's views in this area, permitting the trier of fact to consider this conduct and reduce claimant's damages. Under Comment n to the "Restatement (Second) of Torts" Section 402A, an individual who failed to discover an apparent defective condition would, theoretically, still be allowed a full claim. On the other hand, if that person knew about the defect and proceeded anyway, the claim would be totally barred. This approach has led to considerable litigation and expense over the issue of whether a claimant knew or did not know about a particular defect. *See* "Task Force Report" at VII–51–53; *see also* "Karabatsos v. Spivey Co.," 49 Ill. App. 3d 317, 364 N.E. 2d 319 (1977); "Teagle v. Fischer & Porter Co.," 89 Wash. 2d 149, 570 P. 2d 438 (1977); "Poches v. J.J. Newberry Co.," 549 F. 2d 1166 (8th Cir. 1977). The Act eliminates this distinction and focuses on the true responsibility of the product user.

Thus, if a claimant with good eyesight ate a candy bar that had bright green worms crawling over it, Subsection (A)(2) permits the trier of fact to find that the claimant should bear some responsibility for any ill effects suffered. This example involves a defective condition that can be discovered without inspection. *Cf.* "Auburn Mach. Works Co. v. Jones," 366 So. 2d 1167 (Fla. 1979).

Subsection (A)(2) will not promote misconduct by product sellers. If they were aware of the defect in the goods at the time of sale, the punitive damages section of the Act (Section 120) would provide a strong incentive not to sell such a product.

(3) *A Non-Claimant's Failure to Inspect for Defects or to Observe an Apparent Defective Condition.* When a product seller has sold a defective product and an intervening product user negligently fails to discover the defect or to take precautions against the possible harm, case law is uniform that the product user's conduct does not relieve the product seller of liability. *See* "Ford Motor Co. v. Matthews," 291 So. 3d 169 (Miss. 1974); "Boeing Airplane Co. v. Brown," 291 F.2d 310 (9th Cir. 1961); "Comstock v. General Motors Corp.," 358 Mich. 163, 99 N.W.2d 627 (1959). In this instance, the product seller is required to bear responsibility for the defective product it placed on the market.

(B) *Use of a Product With a Known Defective Condition.*

(1) *By a Claimant.* When it is clear that a claimant voluntarily and unreasonably used a product with a known defective condition, the claimant's damages are subject to reduction. Care must be taken not to allow recovery of a claim in a situation where individuals, in effect, have created their own product liability claim. In that regard, it should be noted that consent is a defense even to intentional wrongs. *See* W. Prossser, "Torts" at 101 (4th ed. 1971).

However, there may be cases where an individual voluntarily uses a product with a known defective condition, but the reasonableness of this conduct becomes a matter of dispute. For example, if a person discovers a welt in a tire, should that person be required to stop immediately and çall for assistance, or is it reasonable to proceed to a nearby gasoline station to have the tire repaired? The answer depends on the particular fact situation. Many cases arise in this shadowy zone. *See* "Henderson v. Ford Motor Co.," 519 S.W.2d 87 (Tex. 1974); "Ford Motor Co. v. Lee," 237 Ga. 554, 229 S.E.2d 379 (1976). Subsection (B)(1) allows the trier of fact to consider the claimant's conduct in the particular fact situation and to reduce damages to the extent that it is appropriate to do so. From both the claimant's and the product seller's perspective, this mitigates the "all-or-nothing" approach that has arisen as a result of some court interpretations of Comment to Section 402A of the "Restatement (Second) of Torts."

(2) *By a Non-Claimant Product User.* Subsection (B)(2) is placed in brackets as an optional section. It would apportion responsibility for injuries caused by a defective product between the product seller and a product user who voluntarily and unreasonably exposed the claimant to the product risk. Some case law and, perhaps, proper placement of incentives for loss prevention support this result. *Cf.* "Aetna Ins. Co. v. Loveland Gas & Elec. Co.," 369 F.2d 648 (6th Cir. 1966); "Drazen v. Otis Elevator Co.," 96 R.I. 114, 189 A.2d 693 (1963). On the other hand, case law will not shift responsibility where the product was entirely unfit for its intended use,[16] and Subsection (B)(2) may create a situation in which a claimant injured by a defective product cannot bring a successful suit against anyone. This could occur when the Subsection shields the seller of the defective product, and an immunity (*e.g.,* Worker Compensation) shields the negligent third-party actor. Moreover, as Professor Phillips has observed,

> The third party rarely intends to cause the plaintiff injury. Where there is no such intent, the third party's failure to prevent the injury

---

[16] *See, e.g.,* "Clement v. Crosby & Co.," 148 Mich. 293, 111 N.W. 745 (1907); "Farley v. Edward E. Tower & Co.," 271 Mass. 230, 171 N.E. 639 (1930).

> is attributable to his inadvertence, regardless of whether he actually knew or merely should have known of the danger.

Phillips, supra, 28 "Drake L. Rev." at 372.

In light of the very close balance of equities with regard to this Section, it has been bracketed and should be regarded as optional.

(C) *Misuse of a Product.* Subsection (C) provides for a reduction or apportionment of the liability of the product seller when an injury occurs, in whole or in part, because the product user misused the product in some way that the product seller could not reasonably anticipate. *See* "Netzel v. State Sand & Gravel Co.," 51 Wis. 2d 1, 186 N.W.2d 258 (1971); "General Motors Corp. v. Hopkins," 548 S.W.2d 344 (Tex. 1977). The definition of "misuse" is based on the Act's concept of "reasonably anticipated conduct." *See* Subsection 102(G) and analysis.

Damages are reduced or apportioned "to the extent" that the misuse caused the harm and, under Subsection (C)(2), the trier of fact may determine that the harm arose *solely* because of product misuse. "Helene Curtis Indus. v. Pruitt," 385 F.2d 841 (5th Cir. 1967).

Subsection (C)(3) indicates that a third party who is not immune under state or federal law and who has misused a product may be subject to liability to the claimant.

(D) *Alteration or Modification of a Product.* Subsection (D) deals with the situation in which a person or entity other than the product seller has altered or modified the product and this has led to the claimant's harm. Alteration or modification (as contrasted with misuse) occurs when a claimant or third-party product user changes the product's design, construction, or formula, or modifies or removes instructions that accompanied or were displayed on the product.

Some courts have imposed liability on the product seller in this situation if the alteration or modification was in some manner "foreseeable." *See, e.g.,* "Blim v. Newbury Indus., Inc.," 443 F.2d 1126 (10th Cir. 1971) (machine safety guard removed by co-worker). Courts that have held the original product seller responsible in these instances have bordered on imposing absolute liability. Thus, insurers have a just concern about the broad-scale imposition of liability where intervention by another party was the principal cause of the accident. As the American Insurance Association has noted:

> It is difficult enough to calculate the risk associated with a given product even where there is access to knowledge about its basic inherent characteristics. . . . The task becomes impossible if the premium calculations must take into account not only the inherent properties of the machine, but also its transformation in the hands of others, and their neglect of repair and maintenance.

AIA, "Product Liability Legislative Package" at 16 (monograph 1977).

Moreover, if the law ignores alterations and modifications or products, it fails to place an incentive for loss prevention on those who might engage in such conduct.

The authors of the "Restatement (Second) of Torts" Section 402A, Subsection 1(b), recognized this fact and subjected the product seller to liability only when the seller's product reached "the user or consumer without substantial change in the condition in which it was sold." Comment g to that Section stated the matter more firmly:

> The seller is not liable when he delivers the product in a safe condition, and subsequent mishandling or other causes make it harmful by the time it is consumed. The burden of proof that the product was in a defective condition at the time that it left the hands of the particular seller is upon the injured plaintiff; and unless evidence can be produced which will support the conclusion that it was then defective, the burden is not sustained.

Recently, a number of state legislatures have enacted the essence of this comment into law. "Ariz. Rev. Stat. Ann." Section 12–683(2) (Supp. 1978); "Ind. Code Ann." 33–1–1.5 Section 4(b)(3) (Supp. 1978); "Ky. Rev. Stat. Ann." Section 411–320 (1978); "N.H. Rev. Stat. Ann." Section 507–D:3 (Supp. 1978); "R.I. Gen. Laws" Section 9–1–32 (1978); "Tenn. Code Ann." Section 23–3708 (1978); "Utah Code Ann." Section 78–15–5 (1977).

According to the statistics of the

Insurance Services Office, product modification only occurs in approximately 13 percent of product liability cases. Of these cases, the largest number of the product modifications (39 percent) result from the conduct of employers. *See* "ISO Closed Claims Survey" at 140–41. This raises the main problem with rules that limit a product seller's responsibility for subsequent product alterations or modifications—often the injured worker cannot sue the one who is really at fault because of the "exclusive remedy" provisions of Worker Compensation statutes. However, it is fair to state that the destruction of a tort remedy against the employer

> should not of itself create a third-party remedy against the manufacturer or distributor of the product in question. If Worker Compensation is regarded as a proper remedy in other cases of an exclusive employer's wrong, then so too should it be where that wrong involves the product acquired from third-party defendants.

AIA, "Product Liability Legislative Package," *supra* at 15–16.

Nevertheless, Subsection (D) takes account of the hardship that can result from an overly broad liability limitation in cases of product modification or alteration. Thus, the provision is very narrowly drawn.

Subsection (D)(2) ensures that the principles of comparative responsibility are correctly applied by reference to Section 111. If an alteration or modification was made by the claimant, damages are subject to reduction in accordance with the claimant's percentage of responsibility. If a party other than the claimant is responsible for the modification or alteration, damages are to be apportioned in accordance with the parties' percentages of responsibility.

In the case of an employer or co-employee immune from tort liability or a released party under Subsection 113(E), the rules set forth in Subsection 112(B)(2) would apply with regard to reduction of damages.

Under Subsection (D)(2), a product seller may avoid liability to the extent it proves that claimant's harm was proximately caused by the alteration or modification. If the harm arose solely because of the product alteration or modification, the product seller will not be liable at all. On the other hand, the rule adopted ensures that the product seller will remain liable to the extent that it is responsible for the harm. *See* "Fincher v. Surrette," 365 So. 2d 860, 863 (La. Ct. App. 1979). Subsection (D)(2) applies the comparative responsibility procedural principles set forth in Section 111.

As Subsections (D)(2) (a) and (b) indicate, the product seller cannot avoid responsibility for product alterations or modifications which the seller suggested (per instructions) or to which the seller expressly consented.

Subsection (D)(2)(c) indicates that the product seller may have a duty to warn against modifications or alterations of its product when it may reasonably anticipate that such conduct will occur on the part of persons who are likely to use the product. As Subsection 102(G) (definition of "reasonably anticipated conduct") indicates, this refers to conduct that would be engaged in by an ordinary reasonably prudent person.

Subsection (D)(2)(c) is not intended to encompass every type of act foreseeable by virtue of hindsight or otherwise. If alterations or modifications involve conduct which should have been reasonably anticipated by the product seller, such product seller will be responsible for harms that result if the failure to provide warnings against the type of alteration or modification at issue renders the product defective under Subsection 104(C). A general warning against any alterations or modifications will not suffice.

Subsection (D)(3) indicates that a third party who is not immune under state or federal law and who has negligently altered or modified a product may be subject to liability to the claimant.

* * * * *

Code

Sec. 113. Multiple Defendants: Contribution and Implied Indemnity

(A) A right of contribution exists under this Act between or among two or more persons who are jointly and severally liable, whether or not judgment has been recovered against all or any of them. It may be enforced either in the original action or by a separate action brought for that purpose. The basis for contribution is each person's equitable share of the obligation, including the equitable share of a claimant, as determined in accordance with the provisions of Section 111. For the purposes of this Act, contribution and implied indemnity are merged.

(B) If the proportionate responsibility of the parties to a claim for contribution has been established previously by the court, as provided in Section 111, a party paying more than its equitable share of the obligation may, upon motion, recover judgment for contribution.

(C) If the proportionate responsibility of the parties to the claim for contribution has not been established by the court, contribution may be enforced in a separate action, whether or not a judgment has been rendered against either the person seeking contribution or the person from whom contribution is being sought.

(D) Contribution is available to a person who enters into a settlement with a claimant only (1) if the liability of the person against whom contribution is sought has been extinguished by the settlement, and (2) to the extent that the amount paid in settlement was reasonable.

(E) A release, covenant not to sue, or similar agreement entered into by a claimant and a person liable discharges that person from all liability for contribution, but it does not discharge any other persons liable upon the same claim unless it so provides. However, the claim of the releasing claimant against the other parties is reduced by the amount of the released party's equitable share of the obligation, determined in accordance with the provisions of Section 111.

(F) If a judgment has been rendered, the action for contribution must be commenced within one year after the judgment becomes final. If no judgment has been rendered, the person bringing the action for contribution must have (1) discharged by payment the common liability within the period of the statute of limitation or repose applicable to the claimant's right of action against him and commenced the action for contribution within one year after payment, or (2) agreed while action was pending to discharge the common liability and, within one year after the agreement, have paid the liability and brought an action for contribution.

* * * * *

Analysis

Sec. 113. Multiple Defendants: Contribution and Implied Indemnity

Section 113 is based on Sections 4, 5, and 6 of the "Uniform Comparative Fault Act" ("UCFA"). Here, however, contribution and implied indemnity are merged in one section. Express indemnity—where one party has agreed to hold the other harmless for damages arising out of product liability actions—is left to commercial and common law. *See* "Task Force Report" at VII–99–103.

There is clear precedent for the merger of contribution and implied indemnity. *See* "Safeway Stores, Inc. v. Nest-Kart," 21 Cal. 3d 322, 579 P.2d 441, 146 Cal. Rptr. 550 (1978); "Skinner v. Reed-Prentice Division, Etc.," 70 Ill. 2d 1. 374 N.E.2d 437 (1977), *cert. denied,* 436 U.S. 946 (1978); "Busch v. Busch Constr., Inc.," ——— Minn. ———, 262 N.W.2d 377 (1977); "Dole v. Dow Chemical Co.," 30 N.Y.2d 143, 282 N.E.2d 288, 331 N.Y.S.2d 382 (1972); *see also* "N.Y. Civ. Prac. Law" Section 1402 (1976). This approach avoids the "all-or-nothing" aspect of implied indemnity law. In most situations, fault will be apportioned among product seller defendants. However, a situation could arise where the trier of fact could find that one product seller in the distribution chain was entirely responsible for a product harm. In that regard, this Section should be read in conjunction with Sections 111 and 112.

Subsection (A), which is based on "UCFA" Subsection 4(a), establishes a right of contribution which may be enforced in the original action or in a separate action and provides that contribution will be determined by the proportionate responsibility of the defendants. The Subsection outlines the procedure for the trier of fact to make the appropriate determination.

Subsection (B), which is based on "UCFA" Subsection 5(a), outlines a simplified procedure whereby a party adjudged liable who has paid more than its proportionate share can recover from one who has paid less.

Subsection (C), which is based on "UCFA" Subsection 5(b), provides a mechanism for apportioning responsibility when all potential defendants were not parties to the original action. It indicates that if the court has not determined the proportionate responsibility of all parties in the original action, contribution may be obtained in a separate action.

Subsection (D), which is based on "UCFA" Subsection 4(b), makes it clear that a person who has settled a claim may seek contribution from another person. However, this Subsection limits the right of contribution in such a situation by requiring that the amount paid in settlement be reasonable and that the claimant's right to recover from the other person be extinguished by the settlement.

Subsection (E), which is based on "UCFA" Section 6, deals with the effect of the release of one or more, but not all, tortfeasors. Under this Subsection, a released person is free from liability for contribution, and the liability of other persons to the claimant is reduced by the released person's equitable share of the responsibility as determined in accordance with the provisions of Section 111. Although this provision "may have some tendency to discourage a claimant from entering into a settlement, this solution is fairly based on the proportionate-fault principle." "UCFA Comment" Section 6. Furthermore, the provision that the liability of the non-settling persons is reduced by the settling person's percentage of the responsibility should discourage claimants from entering into collusive settlements with one defendant at the expense of the other parties.

The Act does not directly address the validity of "Mary Carter agreements." Such agreements are typically entered into between a claimant and one or more, but not all, of the product sellers. They are usually made in secret, and the agreeing product sellers remain as parties to the action. The agreeing product sellers' liability is decreased in direct proportion to the nonagreeing product sellers' increase in liability, and, in return, the agreeing product sellers guarantee the claimant a certain amount of money if the claimant does not obtain a judgment, or if it is less than a specified amount. While such an agreement appears collusive, some state courts currently regard "Mary Carter agreements" as valid. *See, e.g.,* Frier's, Inc. v. Seaboard Coastline Ry.," 355 So. 2d 208 (Fla. Dist. Ct. App. 1978). However, to the extent that a court chooses to regard a "Mary Carter agreement" as a *settlement,* Subsection (E) is intended to discourage such agreements.

Subsection (F) is based on "UCFA" Section 5 and sets forth the period of time in which an action for contribution may be brought.

* * * * *

Code

Sec. 114. Relationship Between Product Liability and Worker Compensation

(A) In the case of any product liability claim brought by or on behalf of an injured person entitled to compensation under a state Worker Compensation statute, damages shall be reduced by the amount paid as Worker Compensation benefits for the same injury plus the present value of all future Worker Compensation benefits payable for the same injury under the Worker Compensation statute.

(B) Unless the product seller has expressly agreed to indemnify or hold an employer harmless for harm caused to the employer's employee by a product, the employer shall have no right of subrogation, contribution, or indemnity against the product seller when the harm to the employee constitutes a product liability claim under this Act. Also, the employer's Worker Compensation insurance carrier shall have no right of subrogation against the product seller.

(C) When final judgment in an action brought under this Act has been entered prior to the determination of Worker Compensation benefits, the product seller may bring a subsequent action for reduction of the judgment by the amount of the Worker Compensation benefits, or for recoupment from the employee if the product seller has paid a judgement which includes the amount of such benefits.

* * * * *

Analysis

Sec. 114. Relationship Between Product Liability and Worker Compensation

The relationship between product liability and Worker Compensation is a major topic covered in depth in the "Task Force Report" at VII–85–113. Under current law in a number of states, the interaction of product liability and Worker Compensation law results in the manufacturer of a workplace product paying the entire out-of-pocket cost of a product-related workplace injury, plus damages for pain and suffering. This result occurs because the product manufacturer is unable to place a portion of the cost of that injury on an employer whose negligence may have helped bring about the claimant's injury. *See* "Seaboard Coast Line R.R. v. Smith," 359 So. 2d 427 (Fla. 1978); "Task Force Report" at VII–89–99.

After weighing many considerations, the Task Force and the United States Department of Commerce concluded that the development of Worker Compensation as a sole source of recovery in product-related workplace accidents would be the best solution to the problem, but only if the worker received additional benefits in the course of overall Worker Compensation reform. A model product liability law, however, is an inappropriate vehicle for making alterations of that dimension in Worker Compensation law.

The search for the next best solution is not an easy one. One approach is to permit a full contribution claim in all cases in which the employer is at fault. *See* "Dow v. Dole Chem. Co.," 30 N.Y. 2d 1, 282 N.E. 2d 288, 331 N.Y.S. 2d 382 (1972); "Skinner v. Reed-Prentice Division, Inc.," 70 Ill. 2d 1, 374 N.E.2d 437 (1977). This approach is in accordance with the principle of comparative responsibility, and it places strong incentives for loss prevention on the employer.

However, if full contribution or indemnity by the product manufacturer against the employer is permitted, the employer may be forced to pay an employee—through the conduit of the third-party tortfeasor—an amount in excess of the employer's statutory Worker Compensation liability. This thwarts a central concept behind Worker Compensation, *i.e.*, that the employer and employee receive the benefits of a guaranteed, fixed-schedule, no-fault recovery system, which constitutes the *exclusive* liability of the employer. The approach also increases transaction costs.

On the other hand, if contribution or indemnity is not allowed, the product seller might have to bear the burden of a full common law judgment, despite the possibly greater responsibility of the employer. As the Supreme Court of Minnesota recently noted, "[t]his obvious inequity is further exacerbated by the right of the employer to recover directly or indirectly from the third party the amount he has paid in compensation regardless of the employer's own negligence." "Lambertson v. Cincinnati Corp.," ——— Minn. ———, 257 N.W. 2d 679, 684 (1977).

Equally troublesome is the fact that the present system dulls employer incentives to keep workplace products safe. The "ISO Closed Claims Survey" suggests that employer negligence is involved in 56 percent of product liability workplace cases. "ISO Closed Claims Survey," Report 10 at 81 (1978).

The purpose of the solution adopted in Section 114 is to sharpen employer incentives to keep workplace products safe without undermining the limited-liability concept that is central to the Worker Compensation system. The approach is based, in substantial part, on a proposal developed by the American Insurance Association (AIA). *See* AIA, "Product Liability Legislative Package" at 75–76 (1977). This approach has also been incorporated in the

"National Workers' Compensation Standards Act of 1979," S. 420, 96th Cong., 1st Sess. (1979) and "Standards for State Product Liability Tort Litigation Act," H.R. 1675, 96th Cong., 1st Sess. (1979), and recommended by the "LaFalce Subcommittee Report" at 74.

Subsection (A) reduces the liability of a product seller by the total amount that has been or will be awarded in a state Worker Compensation proceeding. This Subsection operates regardless of whether the employer was, in fact, at fault. *See also* Subsection 112(B)(4) and accompanying analysis.

Subsection (B) abolishes the subrogation lien of the Worker Compensation carrier, or the right of a self-insuring employer to contribution or indemnity, in all workplace harm cases which are covered by this Act, except where the product seller has expressly agreed to indemnify or hold the employer harmless for harm caused by the product. As with Subsection (A), Subsection (B) operates regardless of the employer's fault or lack thereof.

Subsection (C) provides a procedural mechanism for reducing the product seller's liability when final judgment has been entered before the amount of Worker Compensation benefits has been determined.

The principal benefit of the approach adopted in Subsections (A) and (B) is a reduction in litigation transaction costs. Subrogation actions are not allowed. Furthermore, proceedings under this Act will be streamlined because in cases of employer negligence, there will be no three-party litigation as to the relative percentages of fault of employers and manufacturers. *See* AIA, "Legislative Package" at 67–68.

An additional benefit of the approach in Subsections (A) and (B) is that an injured employee will recover the same benefits received under the present system. At the same time, Section 114 cuts off the ability of the Worker Compensation carrier or self-insuring employer to shift its liability to the manufacturer. The employer, therefore, will have a greater incentive to provide a safe workplace than it has under the present system.

A third benefit is that the approach will also, to some degree, eliminate the harshness of the present system as it applies to manufacturers, by allowing product sellers to reduce their liability for workplace injuries. It will do this in all cases, even where the employer was not at fault.

While cases may arise where a product seller may pay more than its fair share under a comparative fault system, they are likely to be few in number under this Act. This is because an employer's misuse, or improper alteration or modification of the product, will in turn reduce the employee's award under Subsection 112(C). As the analysis to that Subsection indicates, this is also fair to employees because they agreed, within the context of their Worker Compensation system, to forego their right to sue when it is based on their employers' or co-employees' fault.

A closely related alternative approach to Section 114 was advanced by the Supreme Court of Minnesota in "Lambertson v. Cincinnati Corp.," *supra*. The court in that case held that the product manufacturer would be allowed limited contribution up to the amount of the Worker Compensation lien. This reduces the inequity against the product manufacturer, but preserves the employer's interest in not paying more than its Worker Compensation liability. The principal disadvantage of the "Lambertson" approach, as compared with Section 114, is that "Lambertson" does not reduce transaction costs.

Finally, it should be noted that any modification of the present tort and Worker Compensation systems, short of a sole source remedy, retains the multiple transaction costs of having two separate proceedings—the Worker Compensation proceeding and then the product liability claim—address a single injury. However, considering all the equities involved, Section 114 appears to offer the soundest solution apart from modifying Worker Compensation law to create a sole source remedy.

* * * * *

Code

Sec. 115. Sanctions Against the Bringing of Frivolous Claims and Defenses

(A) After final judgment has been entered under this Act, any party may,

by motion, seek reimbursement for reasonable attorneys' fees and other costs that would not have been expended but for the fact that the opposing party pursued a claim or defense that was frivolous. A claim or defense is considered frivolous if the court determines that it was without any reasonable legal or factual basis.

(B) If the court decides in favor of a party seeking redress under this Section, it shall do so on the basis of clear and convincing evidence. In all motions under this Section, the court shall make written findings of fact.

(C) The motion provided for in Subsection (A) may be filed and the claim assessed against a party or a party's attorney or both, depending on which person or persons were responsible for the assertion of the frivolous claim or defense.

(D) Claims for damages under this Section shall be limited to expenses incurred by parties to the action or persons under a legal or contractual duty to bear the expenses of the action.

* * * * *

Analysis

Sec. 115. Sanctions Against the Bringing of Frivolous Claims and Defenses

The ISO data indicate that substantial product liability costs are incurred in the defense of product liability claims and lawsuits. "ISO Closed Claims Survey," Report No. 14 (1977) (defense costs equal about 35 percent of claim payments). Some have placed the blame for unnecessary defense costs and needless litigation on the contingent fee system. Nevertheless, as the plaintiffs' bar properly observes, the contingent fee brings no return to a claimant's attorney when the claimant is unsuccessful. On the other hand, some have argued that the contingent fee system has a negative impact on certain product liability cases to the extent that it causes insurers to settle non-meritorious claims because the cost of defending such cases may be greater than the amount of settlement.

Analysis of the countervailing arguments suggests that the best solution to reducing unnecessary litigation costs is to address the heart of the problem by discouraging frivolous claims and defenses.

Section 115 is based, in part, on "Ill. Ann. Stat." ch. 110, Section 41 (Supp. 1979). It is also predicated on a proposal of the California Citizens Commission on Tort Reform advocating sanctions against "frivolous" claims or defenses. Report of the California Citizens' Commission on Tort Reform, "Righting the Liability Balance" at 146–47, 153–54 (1977).

The underlying purpose of Section 115 has broad support in existing statutes and court rules. For example, Rule 11 of the Federal Rules of Civil Procedure subjects an attorney to disciplinary action if the attorney knowingly files a pleading or defense where no grounds support it. *See* "Barnett v. Laborers' Int'l Union of N. Am.," 75 F.R.D. 544 (W.D. Pa. 1977). Similarly, Federal Rule of Appellate Procedure 38 permits a court to award "just damages and single or double costs" to a party who has been subject to a "frivolous" appeal.

Additionally, Federal Rule of Civil Procedure 37(c) provides sanctions for an unreasonable failure to admit averments of fact or the genuineness of documents. In the federal courts, the above rules are supplemented by 28 "U.S.C." Section 1927 (1976) (imposing costs on an attorney who "multiplies the proceedings . . . to increase costs . . ."). *See* Annot., 68 "A.L.R.3d" 209 (1976).

The Section is not intended to affect the law relating to malicious prosecution.

Under Subsection (A), the statute may be invoked by either a product liability claimant or a product seller. Recovery is limited to reasonable attorneys' fees and other costs that would not have been expended but for the fact that the opposing party pursued a claim or defense that was frivolous.

In order to make a finding that the claim was frivolous, the court must conclude that the claim was without *any* reasonable legal or factual basis. Thus, this standard allows full room for bringing claims under novel legal theories.

Subsection (B) provides additional assurances that only those who bring frivolous claims will be penalized. First, the court may only impose damages under Section 115 on the basis of clear and convincing evidence, not merely a preponderance of the evidence. *See* "State of West Virginia v. Charles Pfizer & Co.," 440 F.2d 1079, 1092 (2d Cir. 1971) ("clear showing of bad faith"). Second, the court must set forth its findings of fact in writing.

Subsection (C) gives the court latitude to impose costs on either attorney or client. As the "Task Force Report" noted, it is unlikely that many claimants will be financially able to respond to such a claim. "Task Force Report" at VII–62. It must be remembered that the attorney is in the best position to make a judgment about the reasonableness of bringing a claim or raising a defense. *See* "ABA Code of Professional Responsibility," DR 7–102(A)(1)(2); *cf.* "Acevedo v. Immigration & Naturalization Serv.," 538 F.2d 918, 921 (2d Cir. 1976).

Subsection (D) makes clear that recovery under this Section is limited to expenses incurred by claimant or defendant or persons under a legal or contractual duty to bear the expenses of the action.

*   *   *   *   *

Code

Sec. 116. Arbitration

(A) *Applicability.*

(1) Any party may by a motion institute a pre-trial arbitration proceeding in any claim brought under this Act, if the court determines that:

(a) It is reasonably probable that the amount in dispute is less than $50,000, exclusive of interest and costs; and

(b) Any non-monetary claims are insubstantial.

(2) Arbitration may not be used if both the claimant and one or more defendants state that they do not want an arbitration proceeding.

(B) *Rules Governing.*

(1) *Substantive Rules.* The substantive rules of an arbitration proceeding under this Section are those contained in this Act as well as those in applicable state law.

(2) *Procedural Rules.* These are the procedural rules of an arbitration proceeding under this Section. If this Section does not provide a rule of procedure, reference may be made to the "Uniform Arbitration Act" or other sources of law. Any reference to other sources of law must conform to the intent and spirit of this Section.

***Optional Subsection**

[(3) *Additional Rules and Administration.*

(i) The ———— (legislature to specify appropriate state agency or administrative body) is empowered to promulgate additional procedural rules for this Section.

(ii) The ———— (legislature to specify American Arbitration Association or similar organization) shall carry out the day-to-day administration of arbitration under this Section.]

(C) *Arbitrators.*

(1) Unless the parties agree otherwise, the arbitration shall be conducted by three persons: an active member of the state bar or a retired judge of a court of record in the state; an individual who possesses expertise in the subject matter area that is in dispute; and a layperson.

(2) Arbitrators shall be selected in accordance with applicable state law in a manner which will assure fairness and lack of bias.

(D) *Arbitrators' Powers.*

(1) Each arbitrator to whom a claim is referred has the power, within the territorial jurisdiction of the court, to conduct arbitration hearings and make awards consistent with the provisions of this Act.

(2) State laws applicable to subpoenas for attendance of witnesses and the production of documentary evidence apply in proceedings conducted under this Section. Arbitrators shall have the power to administer oaths and affirmations.

(E) *Commencement.* Arbitration hearings shall commence not later than thirty (30) days after the claim is referred to arbitration unless, for good cause shown, the court shall extend the period. Hearings shall be concluded promptly. The court may order the time and place of the arbitration.

(F) *Evidence.*

(1) The Federal Rules of Evidence [or designated state evidence code] may be used as a guide to the admissibility of evidence in an arbitration hearing.

(2) Strict adherence to the rules of evidence, apart from relevant state rules of privilege, is not required.

(G) *Transcript of Proceeding.* A party may have a transcript or recording made of the arbitration hearing at its own expense. A party who has had a transcript or recording made shall furnish a copy of the transcript or recording at cost to any other party upon request.

(H) *Arbitration Decision and Judgment.* The arbitration decision and award, if any, shall be filed with the court promptly after the hearing is concluded. Unless a party demands a trial pursuant to Subsection (I), the decision and award shall be entered as the judgment of the court. The judgment entered shall be subject to the same provisions of law, and shall have the same force and effect as a judgment of the court in a civil action, except that it shall not be subject to appeal.

(I) *Trial Following Arbitration.*

(1) Within twenty (20) days after the filing of an arbitration decision with the court, any party may demand a trial of fact or a hearing on an issue of law in that court.

(2) Upon such a demand, the action shall be placed on the calendar of the court. Except for the provisions of Subsection (3), any right of trial by jury that a party would otherwise have shall be preserved inviolate.

(3) At trial, the court shall admit evidence that there has been an arbitration proceeding, the decision of the arbitration panel, and the nature and amount of the award, if any. The trier of fact shall give such evidence whatever weight it deems appropriate.

(4) A party who has demanded a trial but fails to obtain a judgment in the trial court which is more favorable than the arbitration award, exclusive of interest and costs, shall be assessed the cost of the arbitration proceeding, including the amount of the arbitration fees, and—

(i) If this party is a claimant and the arbitration award is in its favor, the party shall pay the court an amount equivalent to interest on the arbitration award from the time it was filed; or

(ii) If this party is a product seller, it shall pay interest to the claimant on the arbitration award from the time it was filed.

* * * * *

Analysis

Sec. 116. Arbitration

The "Task Force Report" suggested that mandatory non-binding arbitration may result in more accurate decisions, reduce overall litigation costs, and expedite the decision process in product liability cases. *See* "Task Force Report" at VII–229–39.

A synopsis of the basis for these conclusions is that: (1) cases will be decided more accurately because a small group, with a member who is an expert in the field, should be able to comprehend the esoteric details of product liability cases; (2) over time, the process will develop a resource bank of relatively neutral experts less easily misled in technical areas than a jury of laypersons; (3) arbitrators will be less affected by the emotional aspects of the case or by the artistry of counsel; and (4) the privacy of arbitration proceedings (as compared to judicial proceedings) will prompt more complete revelation of special manufacturing designs or processes. This, in turn, will permit more accurate judgments. *See* "Task Force Report" at VII–235.

The "ISO Closed Claims Survey" suggests further that arbitration will reduce accident reparation transaction costs. Even allowing for the fact that more substantial product liability claims are litigated to a verdict than are handled by arbitration, ISO data indicate that the average expense for attorneys and other allocated loss adjustment costs are considerably lower when the case is handled by arbitration as compared with a court verdict. *See* "ISO Closed Claims Survey," Report 14 at 120.

On the other hand, costs may increase under arbitration if there are numerous demands for trial following arbitration. This potential problem may not be as serious, however, as was once thought. Data collected by the Department of Justice show that appeal rates at the state level for a trial following arbitration have ranged from 5 to no more than 15 percent of all cases arbitrated. Hearings before the Subcommittee on Improvements in Judicial Machinery of the Senate Committee on the Judiciary, 95th Cong., 2d Sess. 22 (1978) (Statement of Former Attorney General Griffin B. Bell). *See also* report of the California Citizens' Commission on Tort Reform, "Righting the Liability Balance" at 143 (1977).

Broader use of arbitration should expedite the reparations process. The "Task Force Report" showed that in the medical malpractice area, for example, the arbitration process had achieved a more expeditious resolution of claims than the jury system. *See* "Task Force Report" at VII–238.

Indeed, the benefits of arbitration have prompted the Department of Justice to recommend that mandatory non-binding arbitration be used in federal courts in all tort and contract cases. The Department of Justice reached this conclusion after its Office of Judicial Improvements made a thorough analysis of the matter in a study conducted wholly independently of the "Task Force Report."

Section 116 draws on portions of the Department of Justice's proposed bill on mandatory non-binding arbitration. *See* S. 373, 96th Cong., 1st Sess. (1979); H.R. 2699, 96th Cong., 1st Sess. (1979); the Statement of Former Attorney General Griffin B. Bell, *supra;* as well as state legislation on the topic of arbitration.

(A) *Applicability.* The Act provides for arbitration when the amount in dispute is less than $50,000. This figure is to be determined by the court under a standard of reasonable probability. In this context, note that the sanctions imposed by Section 115 will apply to any party who delays the proceedings by making a frivolous assertion that the amount in dispute is above or below $50,000. Also, the proceeding is not mandatory if both the claimant and at least one defendant state that they do not want it.

The $50,000 figure is the same as that in the Department of Justice bill, and it should cover the bulk of product liability claims. In that regard, the ISO closed claims data, trended for severity, show that the average paid claim in bodily injury cases is $26,004. While some have suggested limiting arbitration to smaller claims, it is the larger claims that have been the greater transaction cost items in product liability cases. *See* "ISO Closed Claims Survey" at 113.

While there has been no state experience with cases at the $50,000 level, Former Attorney General Bell has noted that when Pennsylvania increased the jurisdictional amount for the state's arbitration program from $3,000 to $10,000, there was no increase in the appeal rate. Statement of Former Attorney General Griffin B. Bell, *supra* at 20.

It seems relatively certain that an arbitration procedure will help expedite and reduce costs connected with smaller claims. ISO closed claims data show that the large majority of product liability payments are relatively small (more than two-thirds are under $1,000—even when trended for severity). *See* "ISO Closed Claims Survey" at 113.

(B) *Rules Governing.* Subsection (B)(1) indicates that arbitrators shall apply the product liability substantive law rules of this Act. Where the Act does not provide a rule of decision, relevant state law should be applied.

Subsection (B)(2) indicates that where a procedure is not provided in Section 116—*e.g.,* when a court can vacate a judgment—the "Uniform Arbitration Act" (enacted in a number of states), or another appropriate source of law, is to be used as a resource.

Subsection (B)(3), an optional section, permits the state to designate a supplemental source of procedural rules and to empower the American Arbitration Association or similar organizations to carry out the day-to-day administration of arbitration.

This Subsection leaves a great deal of room for the states to develop their own arbitration systems. The drafters of the Act have chosen to include only the most important procedural points concerning arbitration. The goal of substantive uniformity in product liability law will not be compromised by allowing individual states the freedom to choose the arbitration procedures that they find are the most efficient and workable.

(C) *Arbitrators.* The rules under Subsection (C) provide latitude for the parties to select a single arbitrator. Otherwise, the arbitration is to be conducted by three persons, one who is an active member of the state bar or a retired judge, one who has expertise in the subject matter area that is in dispute, and one who is a layperson. This provision differs slightly from the Department of Justice proposal in light of the needs of product liability. Having an individual who is familiar with the scientific nature of the subject matter involved will help expedite the case and serve as a deterrent to the presentation of biased expert testimony.

In addition, Subsection (C) provides for a layperson to be included to help assure that the consumer perspective regarding product safety is represented. The process of selecting a layperson should not be complicated. It is suggested that either normal jury rolls be utilized or that a list of laypersons be compiled for this purpose.

Aside from general guidelines regarding fairness and lack of bias, the Act does not outline the method of choosing arbitrators, but leaves that matter to the individual states. A state can help implement the general guidelines by requiring each arbitration panel candidate to disclose any personal acquaintance with the parties or their counsel and allow a *voir dire* examination. *See* "Mich. Comp. Laws Ann." Section 600.5045(1)(2) (Supp. 1978). Some of the better procedures include:

(1) Having the American Arbitration Association select a pool of candidates according to its established selection procedures. Each party is allowed to reject certain candidates and rate the remainder in order of preference. Additional provisions take effect if this procedure fails to produce a panel. *See* "Mich. Comp. Laws Ann." Section 600.5044(4)(5) (Supp. 1978);

(2) Having the court appoint arbitrators. "Mass. Ann. Laws" ch. 231, Section 60(B) (Supp. 1978);

(3) Having an arbitration administrator appoint arbitrators. "Wis. Stat. Ann." Section 655.02 (Supp. 1979); and

(4) Having the parties and court combine to appoint arbitrators. "Neb. Rev. Stat." Sections 44–2840, 2841 (Supp. 1978); "Ohio Rev. Code Ann." Section 2711.21(A) (Supp. 1979).

(D) *Arbitrators' Powers.* These provisions are taken from the Department of Justice proposal on arbitration. They grant the arbitrators jurisdiction and also give them powers of subpoena.

(E) *Commencement.* This provision is also derived from the Department of Justice proposal. Its purpose is to help expedite the proceeding. The Act contains a slight modification of the Justice proposal in order to allow an extension for "good cause shown." This seems appropriate in light of the fact that some product liability cases are very complex. *Cf.* "Ariz. Rev. Stat. Ann." Section 12–567(C) (Supp. 1978) (medical malpractice).

(F) *Evidence.* One method of expediting the process is to use informal means of proof. Nevertheless, some guidelines are needed. The Act follows the Department of Justice proposal in referring to the Federal Rules of Evidence as general guidelines. Strict adherence to rules of evidence is not required. *See* "Ariz. Rev. Stat. Ann." Section 12–567(D) (Supp. 1978).

(G) *Transcript of Proceeding.* With respect to the provision of a transcript of proceeding, the Act generally follows the Department of Justice draft.

(H) *Arbitration Decision and Judgment.* The Act follows the Department of Justice proposal's provisions on decisions and judgment. The parties may request a trial on issues of law or fact. If they do not so request in a timely manner, the action is at an end—there is no appeal.

(I) *Trial Following Arbitration.* The Act follows the approach taken by a number of state medical malpractice arbitration statutes. It admits the results of the arbitration proceeding into evidence before the jury. This should act as a deterrent against seeking unnecessary trials. *See, e.g.,* "Ariz. Rev. Stat. Ann." Section 12–567(M) (Supp.

1978); "Mass. Ann. Laws" ch. 231, Section 60(B) (Supp. 1978). *Cf.* "Wis. Stat. Ann." Section 655.19 (Supp. 1979) (excluding findings and order of arbitration panel).

The approach of Section 116 appears to be in accord with the Federal Constitition. *Cf.* "*Ex parte* Peterson," 253 U.S. 300, 309 (1920). Moreover, with the exception of two Ohio lower court decisions, state courts have upheld the constitutionality of provisions that do admit panel findings before the jury. *See* "Eastin v. Broomfield," 116 Ariz. 576, 570 P.2d 744, 750 (1977); "Attorney General v. Johnson," 282 Md. 168, 385 A.2d 57, 67–68 (1978)" "Paro v. Longwood Hosp.," Mass. Adv. Sh. (1977) 2353, 369 N.E.2d 985 (1977); "Prendergast v. Nelson," 199 Neb. 97, 256 N.W.2d 657 (1977); "Strykowski v. Wilkie," 81 Wis. 2d 491, 261 N.W.2d 434 (1978). *contra* "Simon v. St. Elizabeth Medical Center," 3 Ohio Op. 3d 164, 355 N.E.2d 903, 907–09 (C.P. 1976); "Graley v. Satayatham," 74 Ohio Op. 2d 316, 343 N.E.2d 832 (C.P. 1976). *See generally* Reddish, "Legislative Response to the Medical Malpractice Insurance Crisis: Constitutional Implications," 55 "Tex. L. Rev." 759, 793 (1977); Lenore, "Mandatory Medical Malpractice Mediation Panels—A Constitutional Examination," 44 "Ins. Counsel J." 416, 422 (1977).

A possible drawback of this Section's approach is that a jury may have difficulty evaluating the conclusions of the panel where the jury is not privy to the prior fact-finder's qualifications and method of operation. Also, the jury may get sidetracked from the actual evidence in the case. *See* the observations of Judge Hinton in a classic comment, 27 "Ill. L. Rev." 195 (1932); Annot., 18 "A.L.R.2d" 1287 (1951). *But see* Fed. R. Evid. 803 (22) (admitting felony convictions in a cognate civil case). Nevertheless, the benefits to be gained by a reduction in transaction costs outweigh these concerns and support the admission of the results of the arbitration proceeding into evidence at trial.

Subsection (I)(4) provides an additional deterrent against ill-considered appeals for trials following arbitration. If a party fails to obtain a judgment more favorable than the arbitration award, the court will assess the cost of the arbitration proceeding, including the amount of arbitration fees, plus interest, against that party.

In light of the fact that the present product liability system has created serious problems and the fact that mandatory non-binding arbitration has the potential for dealing with some of those problems, this slight incentive for retaining a sound arbitration award should not run afoul of constitutions in most states. *See* "Task Force Report" at VII–233. The Act does not enumerate grounds upon which a court may vacate an arbitration award. Guidance on this issue may be obtained from Section 12 of the "Uniform Arbitration Act."

* * * * *

Code

Sec. 117. Expert Testimony

(A) *Appointment of Experts.* **The court may, on its own motion or on the motion of any party, enter an order to show cause why expert witnesses should not be appointed, and may request the parties to submit nominations. The court may appoint any expert witness agreed upon by the parties, and may appoint witnesses of its own selection. The court may consult with knowledgeable individuals or with professional, academic, consumer, or business organizations and institutions to assist with the selection process. An expert witness shall not be appointed by the court unless the expert consents to serve. An expert witness appointed by the court shall be informed of his or her duties in writing, a copy of which shall be filed with the clerk, or at a conference in which the parties shall have an opportunity to participate. An expert witness so appointed shall advise the parties of any findings; shall be available for deposition by any party; and may be called to testify by the court or any party. The court-appointed expert witness shall be subject to cross-examination by each party, including the party calling that expert as a witness.**

(B) *Compensation.*

(1) Expert witnesses appointed by the court are entitled to reasonable

compensation for their services in an amount to be determined by the court. The court, in its discretion, may tax the costs of such expert on one party or apportion them among parties in the same manner as other costs.

(2) In exercising this discretion, the court may consider:

(a) Which party, if any, requested the court appointment of the expert;

(b) Which party had judgment entered in its favor; and

(c) Whether the amount of damages recovered in the action bore a reasonable relationship to the amount sought by the claimant or conceded to be appropriate by the product seller or other defendant.

(C) *Disclosure of Appointment.* In the exercise of its discretion, the court may authorize disclosure to the jury of the fact that the court has appointed the expert witness.

(D) *Parties' Selection of Own Experts.* Nothing in this Section shall limit the parties in calling expert witnesses of their own selection.

(E) *Pre-Trial Evaluation of Experts.* The court in its discretion may conduct a hearing to determine the qualifications of all proposed expert witnesses. The court may order a hearing on its own motion or on the motion of any party.

(1) *Need for Pre-Trial Evaluation.* In determining whether to grant such a motion, the court shall consider:

(a) The complexity of the issues in the case; and

(b) Whether the hearing would deter the presentation of witnesses who are not qualified as experts on the specific issues.

(2) *Factors in Evaluation.* If the court decides to hold such a hearing, it shall consider;

(a) The background and skills of the proposed witness;

(b) The formal and self-education the proposed witness has undertaken relevant to the case or to similar cases; and

(c) The potential bias of the proposed witness.

(3) *Findings of Fact.* In making a determination as to whether a proposed expert witness is qualified, the court shall state its findings of fact to the parties.

(4) *Determination.* Based upon its findings of fact regarding the qualifications of any proposed expert witness, the court, in its discretion, may limit the scope of the witness' testimony, or may refuse to permit such witness to testify as an expert.

* * * * *

Analysis

Sec. 117. Expert Testimony

In General. The Task Force's "Legal Study" demonstrated that product liability cases are often compromised because of the lack of standards with regard to selecting and presenting expert testimony. *See* Volume IV, "Legal Study" at 153–155. One part of the problem is the biased expert; another is the unqualified expert.

Even if experts are properly qualified and objective, a jury of laypersons is often in a poor position to determine which expert is correct. For this reason, this Act gives the court power to make greater use of pre-trial arbitration where an unbiased, qualified expert will serve on the panel. *See* Section 116. Where arbitration is not used, however, this Section should promote the goal of presenting objective and sound expert testimony to the jury.

(A) *Appointment of Experts.* Subsection (A) is based on Rule 706 of the Federal Rules of Evidence and similar state rules. It indicates that courts have the power to appoint experts on their own authority. A number of courts have utilized this power even without the benefit of Fed. R. Evid. 706 or a similar state rule. *See* Annot., 95 "A.L.R.2d" 390 (1964). As the "Task Force Report" noted, the presence of a court-appointed expert "has a cautionary impact on the expert for hire whose theories at trial are subject to dispute not only by an adversary expert, but also by a neutral court-appointed one." "Task Force Report" at VII–43, *citing* Mitchell, "The Proposed Federal Rules of Evidence: How They Affect Product Liability Practice," 12 "Duquesne L. Rev." 551, 557–58 (1974). *See also* 2 J. Wigmore, "Evidence" Section 563, at 648 (3d ed. 1940) (". . . this expedient would remove most . . . abuses").

Subsection (A) also encourages the court to seek the assistance of various individuals, organizations, or institutions in making the selection of a court-appointed expert. The court may wish to utilize this opportunity to establish a panel of independent experts, recognized as such by their peers, from which selections for individual cases may be made. The court may also seek the assistance of these organizations to evaluate the credentials of expert witnesses nominated by the parties.

One problem with court-appointed experts is that the trier of fact may give them an aura of infallibility which they do not deserve. Under Subsection (A), this possibility is diminished because the experts are subject to cross-examination by *each* party. Also, Subsection (C) allows the court in its discretion to decline to disclose to the jury that the expert witness is, in fact, court-appointed.

(B) *Compensation.* Under Fed. R. Evid. 706 and similar state rules, compensation of experts is left to the judge's discretion. Subsection (B) goes a step farther and provides three guidelines for compensating experts. These guidelines should serve as an added inducement for attorneys to present objective expert testimony. The guidelines suggest that the court may impose the cost of the court-appointed expert on losing parties as well as on parties whom the court finds were substantially inaccurate in their estimation of damages.

(C) *Disclosure of Appointment.* Subsection (C) follows Fed. R. Evid. 706. In most instances, it is important for the trier of fact to appreciate that the witness is court-appointed. However, circumstances may arise in which the court believes disclosure of that fact will give the witness too much credence with the jury. Therefore, the court has the discretion to withhold the information when it is appropriate to do so.

(D) *Parties' Selection of Own Experts.* Subsection (D) also follows Fed. R. Evid. 706. Precluding the parties from introducing their own experts would vest too much power in court-appointed experts.

(E) *Pre-Trial Evaluation of Experts.* A rule authorizing a court-appointed expert does not, in and of itself, provide guidance about who is properly qualified to testify in product liability cases. There are many approaches to that issue. One approach, used in some medical malpractice statutes, would require that an expert witness spend a substantial portion of his or her professional time in the actual practice of his or her area of expertise. While this approach may be appropriate in the area of medical malpractice, it was not followed here because a person may be well-versed in technical product liability matters even if he or she does devote substantial time to research or other endeavors other than actual practice. *See* "Task Force Report" at VII–44. Unfortunately, it is impractical to utilize a "standard test" for all experts in product liability cases. *See* Donaher, Piehler, Twerski & Weinstein, "The Technological Expert in Products Liability Litigation," 52 "Tex. L. Rev." 1303, 1325 (1974).

(1) *Need for Pre-Trial Evaluation.* This rule gives some guidance to the trial court in deciding whether to conduct a pre-trial hearing on the qualifications of expert witnesses. It is not necessary or cost-efficient to utilize the procedure in *all* cases. It is appropriate to do so in more complex cases and also where the pre-trial hearings would serve as a deterrent to the presentation of witnesses who were not qualified. Such a deterrent should discourage parties from prolonging the litigation needlessly and, thus, encourage the expeditious resolution of claims under this Act. Either party may bring this matter before the court by motion.

(2) *Factors in Evaluation.* The factors in evaluation are drawn from Donaher *et al., supra,* 52 "Tex. L. Rev." 1303.

The court should examine the expert witness' background and skills and determine whether they are appropriate for the purposes of the case. The court should not only review the witness' formal education, but also whether the witness had undertaken specific preparation for the litigation before the court. Finally, the court should examine a witness for bias. A witness with marginal expert skills and a strong bias should be considered unqualified.

(3) *Findings of Fact.* If it seems clear to the court that the expert's background and experience do not qualify the expert to testify, it should state this conclusion in its findings of fact to the parties.

(4) *Determination.* This provision empowers the court to limit the scope of an expert's testimony to the witness' specific area of expertise. It also allows the court to refuse to permit the witness to testify as an expert when that individual is not qualified to do so.

* * * * *

Code

Sec. 118. Non-Pecuniary Damages

(A) For the purposes of this Section, "non-pecuniary damages" are those which have no market value and do not represent a monetary loss to claimant.

(B) When sufficient evidence has been introduced, the amount of non-pecuniary damages shall be determined by the trier of fact. However, the court shall have and shall exercise the power to review such damage awards for excessiveness.

***Optional Subsection**

[(c) Non-pecuniary damages under this Act shall not exceed $25,000, or twice the amount of the pecuniary damages, whichever is less, unless the claimant proves by a preponderance of the evidence that the product caused claimant to suffer serious and permanent or prolonged (1) disfigurement, (2) impaired of bodily function, (3) pain and discomfort, or (4) mental illness.]

***Optional Subsection**

[(D) Every third year following the effective date of this Act as stated in Section 122, the ——————— Committee(s) of [each House of] the Legislature of this State shall review the monetary limitations contained in Subsection (C) to determine whether such limitations should be changed in view of the economic conditions existing at that time. Upon a finding that such change is warranted, said Committee(s) shall introduce legislation to amend the monetary limitations contained in Subsection (C).]

* * * * *

Analysis

Sec. 118. Non-Pecuniary Damages

Claimant's "non-pecuniary damages" include awards for pain and mental suffering. They have no market value and, thus, are to be contrasted with pecuniary damages which compensate victims for lost wages, medical and rehabilitation costs, and other actual expenditures brought about by an unreasonably unsafe product.

According to the "ISO Closed Claims Survey," 70 percent of claims closed with payment include amounts in addition to a claimant's pecuniary loss. *See* "ISO Closed Claims Survey" at 54. Moreover, the average amount of payment above pecuniary loss increases significantly in the higher payment range. *Id.* at 54–55. A most important reason for the difficulty in setting product liability rates is the "open-endedness" of damages for pain and suffering. *See* "Task Force Report" at VII–64–65. These escalating premiums have an inflationary impact on consumer prices as the product seller's insurance costs are passed on to the public. Damages that go beyond the amount necessary to ensure that the claimant receives reasonable compensation for the harm suffered represent an unwarranted cost.

The "Task Force Report" suggested that limits on awards for pain and suffering "would reduce uncertainty and thereby mitigate the 'apprehension factor' that has contributed to the rise in product liability insurance rates." *Id.* at VII–65. Nevertheless, such awards have deep historical roots and should not be limited in a manner that unreasonably curtails the rights of injured parties. Section 118 takes the position that a fixed limit on the amount of non-pecuniary damages is appropriate where claimant's harm is only temporary.

Section 118 is predicated on an examination of the weaknesses in the major rationales offered in support of awards for non-pecuniary damages. The award for non-pecuniary damages arose in early common law cases as a substitute for an injured claimant seeking personal "vengeful retaliation." *See* "Task Force Report," *id.* In those cases, the defendant usually committed an intentional wrong. This rationale has little application to cases arising under product liability. Under this Act, a product seller may be held liable for harm caused by products found to be defective in construction regardless of fault. The same result occurs when an express warranty is not true. In cases of harm caused by products found to be defective in design, or defective because of the absence of adequate warnings, the trier of fact must consider more sophisticated matters than would be applied under a general negligence standard.

A second rationale to support the award of damages for pain and suffering is that they have an important deterrent function. The "Task Force Report" found evidence that the general product liability problem caused manufacturers to devote more attention to product liability loss prevention techniques. *See* "Task Force Report" at VI–50. The approach taken in Section 118 retains this deterrent function while placing some reasonable limits on awards for pain and suffering.

A third rationale, supported by members of the plaintiffs' bar and some economic legal scholars, is that awards for pain and suffering are a reasonable attempt to reduce the serious discomfort endured by a claimant. *See* R. Posner, "Economic Analysis of Law" 82 (1972). On the other hand, studies have questioned whether monetary awards for pain and suffering do anything to alleviate the symptoms they are alleged to address. *See* J. O'Connell & R. Simon, "Payment for Pain & Suffering: Who Wants What, When & Why" (1972); Peck, "Compensation for Pain: A Reappraisal in Light of New Medical Evidence," 72 "Mich. L. Rev." 1355 (1974). This Section adheres to the former assumption to the extent that when a claimant suffers serious and permanent or prolonged (1) disfigurement, (2) impairment of bodily function, (3) pain and discomfort, or (4) serious mental illness, the amount of non-pecuniary damages is left to the sound discretion of the trier of fact with appropriate review by the court in cases of abuse of that discretion.

However, when the claimant has not suffered such serious and permanent or prolonged harms,[17] non-pecuniary damages are limited to $25,000 or twice the amount of pecuniary damages, whichever is less.

An ample body of case law in the area of Worker Compensation and, more recently, automobile injury reparation statutes, provides guidance for the court to determine whether a harm is permanent or prolonged. *See, e.g.*, "Falcone v. Branker," 135 N.J. Super. 137, 342 A.2d 875 (1975) (collecting cases); "Vitale v. Danylak," 74 Mich. App. 615, 254 N.W.2d 593 (1977) (summary judgment under automobile no-fault act); "*In re* Requests of Governor and Senate, Etc.," 389 Mich. 411, 208 N.W.2d 469, 480 (1973) (term "permanent" supplies basis for legal interpretation).

Courts also have defined and explained the term "mental illness" in a number of contexts. "Carroll v. Cobb," 139 N.J. Super. 439, 354 A.2d 355 (1976) (voter registration requirements); "Sachs v. Commercial Ins. Co.," 119 N.J. Super. 226, 290 A.2d 760 (1972) (insurance policy); "*In re* Humphrey," 236 N.C. 141, 71 S.E.2d 915 (1952) (incompetency proceedings); "Commonwealth v. Moon," 383 Pa. 18, 117 A.2d 96 (1955) (committal proceedings); "Interstate Life & Accident Ins. Co. v. Houston," 50 Tenn. App. 172, 360 S.W.2d 71 (1962) (insanity exclusion provision).

Objections to limits on awards for non-pecuniary damages take several forms. One is that such limits may violate due process or equal protection clauses of some state constitutions. *Cf.* "Wright v. Central Du Page Hosp. Ass'n.," 63 Ill. 2d 313, 347 N.E.2d 736, 743 (1976) (restriction on amount of general damages in medical malpractice); "Graley v. Satayatham," 74 Ohio Op.2d 316, 343 N.E.2d 832, 836 (C. P. 1976) (requiring list of collateral source benefits in medical malpractice).

[17] The Section does not address the question of whether non-pecuniary damages should be allowed or limited in wrongful death or survival actions.

Another objection involves state constitutional prohibitions on damage limitations. *E.g.*, "Ariz. Rev. Stat. Ann.," Const. Art. 18, Section 6 (1956); "Ky. Rev. Stat. Ann.," Const. Section 54 (1891); "Penn. Stat. Ann.," Const. Art. 3, Section 18 (1969). An argument that Section 118 does not violate such prohibitions is that a strict product liability cause of action did not exist at the time the state constitution was adopted and is therefore exempt from its interdictions. *See* "Rail N Ranch Corp. v. State," 7 Ariz. App. 558, 441 P.2d 786, 788 (1968).

These objections notwithstanding, Section 118 can be supported. First, the common law rule will continue to operate where injuries are serious. *Cf.* "Rybeck v. Rybeck," 141 N.J. Super. 481, 358 A.2d 828, 836 (1976), *appeal dismissed*, 150 N.J. Super. 151, 375 A.2d 269 (1977) (limited court access for pain and suffering in no-fault—"the law is permitted to treat large problems differently from small problems if there is a rational basis for the difference"). Second, some ceiling or limit on damages for pain and suffering will reduce uncertainty in one of the greatest liability insurance ratemaking problem areas. Third, the limit will help assure that there is adequate insurance capacity to provide compensation for pecuniary losses caused by unsafe products.

Subsection (D) provides a mechanism for review and evaluation of the $25,000 limitation contained in Subsection (C). It is anticipated that the legislature may wish periodically to modify this limitation in light of significant changes in economic conditions.

Because of the facts that (1) the topic of pain and suffering goes beyond product liability and is applicable to all of tort law, (2) the limitation as drafted may raise transaction costs in regard to the issue of whether a harm is "serious, permanent, or prolonged," and (3) constitutional problems that may arise in some states, Subsections (C) and (D) are placed in brackets as optional sections.

* * * * *

Code

Sec. 119. The Collateral Source Rule

In any claim brought under this Act, the claimant's recovery, or that of any party who may be subrogated to the claimant's rights under this Act, shall be reduced by any compensation from a public source which the claimant has received or will receive for the same damages. For the purposes of this Section, "public source" means a fund more than half of which is derived from general tax revenues.

* * * * *

Analysis

Sec. 119. The Collateral Source Rule

The collateral source rule is a principle of tort law under which the defendant is not permitted to take "credit" for any money that an injured claimant received from another (collateral) source. The rule embraces both payments for loss of wages and medical expenditures.

The rule may permit double recovery by the claimant and may increase transaction costs. This Section recognizes these possibilities and provides for a limited modification of the collateral source rule when the claimant has received compensation from a public source for the same damages. Its approach is similar to that followed in medical malpractice by the states of Tennessee and Pennsylvania. *See* Volume V, "Legal Study" at 146.

There are two significant arguments against proposals to modify the existing rule. The first is that the "wrongdoer" should not have the benefit of a windfall. Proponents contend that it is better that the claimant have the benefit of a windfall than the defendant.

This argument can be rebutted in the context of product liability. Under Section 104 of this Act, a manufacturer may be held liable for defects in construction or breach of an express warranty on a strict liability basis. In other cases, a product seller's liability is based on fault, but not on gross negligence or intentionally wrongful conduct. Therefore, a selective modification of the collateral source rule in the context of product liability may be justified. Indeed, some states have upheld such selective abolition or modification where liability was predicated solely on a fault basis. *See* "Eastin v. Broomfield," 116 Ariz. 576, 570 P.2d 744, 751–52 (1977); *But see* "Graley v. Satayatham," 74 Ohio Op. 2d 316, 343 N.E. 2d 832 (C.P. 1976) (holding unconstitutional selective abolition in medical malpractice context).

The second argument against modifying the collateral source rule is that a product seller should not be permitted to "externalize" the cost of an injury caused by its products. This argument is very strong when the injured claimant has purchased health and accident coverage. In that instance, the defendant product seller should not be able to benefit from the claimant's prior prudence. Nevertheless, some proposals have modified the rule in that situation. *See* "Neb. Rev. Stat." Section 44–2819 (Supp. 1978); "Prendergast v. Nelson," 199 Neb. 97, 256 N.W. 2d 657, 669 (1977); National Product Liability Council, "Proposed Uniform State Product Liability Act" Section 207 (undated). *See also* Comment, "An Analysis of State Legislative Responses to the Medical Malpractice Crisis," 1975 "Duke L.J." 1417, 1447–50. The argument is less persuasive when the claimant has not directly contributed to the insurance fund as is the case with Worker Compensation.

When the claimant has received damages from a public source, the argument loses all of its force. The benefits received were not through the claimant's pre-accident financial planning or made a part of the claimant's remuneration as a condition of employment. Rather, they were derived from public tax funds that accumulated in part by contributions from the product seller. Since the product seller would be able to distribute the cost of a judgment that included this amount among consumers through product pricing, the public may be subjected to excessive, duplicative costs.

Section 119 defines a "public source" as a fund more than half of which is derived from general tax revenues. This would include welfare benefits and other forms of medical, hospitalization, and disability benefits which are funded by state or federal tax funds. It does not include Worker Compensation payments (which are specifically dealt with in Section 114), employee group insurance plans, or other similar sources. It also does not presently include payments received pursuant to the "United States Social Security Act." Although Social Security contributions are withheld from an employee's pay, they are not "general tax revenues." However, if the method of funding the Social Security system is changed, such payments might then come within the operation of this Section. *Compare* the approach taken in Section 119 *with* the various approaches taken in state medical malpractice statutes. *See, e.g.,* "Tenn. Code Ann." Section 23–3418 (Supp. 1978) (complete abrogation of collateral source rule expressly includes Social Security benefits); "Neb. Rev. Stat." Section 44–2819 (Supp. 1978) (partial abrogation of collateral source rule does not include Social Security benefits).

A probable effect of Section 119 will be to reduce double expenditures for medical costs. The "ISO Closed Claims Survey" suggests that medical costs represent approximately 19.7 percent of product liability claims. "ISO Closed Claims Survey" at 57. Nevertheless, the cost savings generated by this Section will probably be modest. The ISO closed claims data, which were quite limited on this point, show that approximately 6.4 percent of claimants have been reimbursed by public collateral sources. *See* "ISO Closed Claims Survey" at 181. Collateral sources paid for 19.8 percent of the claims in those cases (this closely parallels the general percentage of medical benefits). Nonetheless, this Section should help reduce overall insurance costs. Liability insurers should take this matter into account when they formulate rates and premiums.

Section 119 also takes account of existing legislation that may authorize subrogation by public collateral sources. In order to reduce transaction costs and duplicative distribution costs, this Section prohibits such subrogation.

Finally, Section 119 does not alter existing law that prohibits the defendant from introducing into evidence the fact that the claimant has been indemnified by a collateral source. That alternative approach was rejected because it would leave the trier of fact in the role of balancing the delicate policy elements that surround proposals calling for abolition of the collateral source rule. This should be an issue of law for state legislatures or courts. Also, that approach would reduce the potential benefit of collateral source rule modifications in that it would increase transaction costs and lower predicatability and consistency in the allocation of collateral benefits. *See* "Task Force Report" at VII–74–75. *Cf.* Defense Research Institute, "Products Liability Position Paper" at 44–45 (1976) (advocating modification of evidentiary rules to allow trier of fact to consider all collateral benefits).

* * * * *

Code

Sec. 120. Punitive Damages

(A) Punitive damages may be awarded to the claimant if the claimant proves by clear and convincing evidence that the harm suffered was the result of the product seller's reckless disregard for the safety of product users, consumers, or others who might be harmed by the product.

(B) If the trier of fact determines that punitive damages should be awarded, the court shall determine the amount of those damages. In making this determination, the court shall consider:

(1) The likelihood at the relevant time that serious harm would arise from the product seller's misconduct;

(2) The degree of the product seller's awareness of that likelihood;

(3) The profitability of the misconduct to the product seller;

(4) The duration of the misconduct and any concealment of it by the product seller;

(5) The attitude and conduct of the product seller upon discovery of the misconduct and whether the conduct has been terminated;

(6) The financial condition of the product seller;

(7) The total effect of other punishment imposed or likely to be imposed upon the product seller as a result of the misconduct, including punitive damage awards to persons similarly situated to the claimant and the severity of criminal penalties to which the product seller has been or may be subjected; and

(8) Whether the harm suffered by the claimant was also the result of the claimant's own reckless disregard for personal safety.

* * * * *

Analysis

Sec. 120. Punitive Damages

Some product sellers and others have called for the abolition of punitive damages on the grounds that they serve no proper "tort law" purpose,[18] and at least one court has accepted these arguments in the area of product liability. *See* "Walbrun v. Berkel, Inc.," 433 F. Supp. 384–85 (E.D. Wis. 1976); "Roginsky v. Richardson-Merrell, Inc.," 378 F.2d 832 (2d Cir. 1967) (dictum).

Nevertheless, as Section 120 acknowledges, punitive damages serve an important function in deterring product sellers from reckless disregard for safety in the production, distribution, or sale of dangerous products. *See* "Toole v. Richardson-Merrell, Inc.," 251 Cal. App. 2d 689, 60 Cal. Rptr. 398 (1967); "Gillham v. Admiral Corp.," 523 F.2d 102 (6th Cir. 1975). At the same time, Section 120 recognizes and addresses punitive damages problems in the specific context of product liability.

While many product sellers have expressed great concern about the economic impact of punitive damages, the "ISO Closed Claims Survey" suggests that the number of cases in which such damages are imposed is insubstantial. "ISO Closed Claims Survey" at 183. Nevertheless, concern about punitive damages has caused some insurers to decline to provide insurance coverage for these damages. Also, a number of states and some insurers have declined to permit such coverage. They contend that a product seller should not be allowed to pass this

---

[18] *See* "Proposed Uniform State Product Liability Act" Section 206 (National Product Liability Council) (undated); *see generally* Defense Research Institute, "The Case Against Punitive Damages" (monograph 1969) (marshalling arguments).

cost on to an insurer. Transcending all of these concerns is the total lack of legal structure surrounding punitive damages. The approach taken in Section 120 is to provide such a structure, so as to reduce product sellers' reasonable concerns about punitive damages, while at the same time retaining the important deterrent function of punitive damages.

Subsection (A) addresses a basic argument against punitive damages—specifically that they apply a criminal law sanction to a civil law case, even though the defendant does not have the benefit of the constitutional protections that would be available under criminal law. Subsection (A) moves away from the ordinary "preponderance of evidence" test of civil cases and toward the criminal standard, but does not turn completely to a pure criminal standard of proof "beyond a reasonable doubt." Because the defendant is not subject to incarceration, and the "punishment" is more in the nature of a civil fine than a criminal sanction, the Act requires the claimant to prove by "clear and convincing" evidence that punitive damages are justified. *See* Subsections 102 (H) and (I) (definitions of "preponderance of evidence" and "clear and convincing evidence").

Subsection (A) also requires that the claimant prove that the product seller's conduct demonstrated reckless disregard for the safety of others. The phrase "reckless disregard"—the traditional barrier that the plaintiff must cross in order to obtain punitive damages—means a conscious indifference to the safety of persons who might be injured by the product. *See* Subsection 102(J) (definition of "reckless disregard"); *cf.* W. Prosser, "Torts" at 9–10 (4th ed. 1971). The "reckless disregard" standard is identified in statutory form to avoid any possible misinterpretation of this basic area of law. Thus, it should be clear that a product seller does *not* have to pay punitive damages under ordinary strict liability or negligence standards which fall short of reckless disregard.

Subsection (B) follows the current common law system in allowing the trier of fact to determine at its discretion whether punitive damages should be awarded. *See* Prosser, *supra* at 9. On the other hand, this Subsection draws upon a newly enacted Minnesota statute in having the court, rather than the jury, determine the amount of those damages. "Minn. Stat. Ann." Section 549.21 (Supp. 1978). This approach is in accord with the general pattern of criminal law where the jury determines "guilt or innocence" and the court imposes the sentence. This is particularly appropriate in product liability cases where, under current law, product sellers are potentially subject to repeated imposition of punitive damages for harm caused by a particular product.

Subsection (B) provides guidelines for the court in determining the amount of punitive damages. The eight factors are derived from "Minn. Stat. Ann." Section 549.20(3) (Supp. 1978). The drafters of that statute relied on a very thorough analysis of product liability punitive damages. *See* Owen, "Punitive Damages in Products Liability Litigation," 74 "Mich. L. Rev." 1257, 1299–319 (1976).

Factors (1) and (2) are self-evident. If the facts show that the product seller was actually aware of the specific hazard *and* its seriousness, *and* marketed it anyway, a higher award is in order.

Factor (3), profitability, recognizes that punitive damages may be used to attack directly the profit incentive that generated the misconduct.

Factor (4) is important regardless of the basic requirement that the product seller must have reckless disregard for the safety of others. If the product seller consciously concealed its activities, this fact argues for a higher award.

Factor (5) acknowledges that the product seller who was reckless in producing the product, but who acted quickly to remove the product from the market upon discovery of the hazard, should not be subject to as harsh a sanction as one who failed to act. Some have suggested that punitive damages should be awarded only where corporate management has either authorized, participated in, or ratified conduct that shows a conscious or reckless disregard for public safety. *See* "Task Force Report" at VII–79. Section 120 rejects that approach because it could foster legal disputes as to whether an individual stood "high enough" in the corporate structure to bear responsibility for punitive damages. Nevertheless, in circumstances where a non-management employee caused the harm and management acted quickly to mitigate that harm once it was discovered, a lower award is appropriate.

Factor (6) permits the court to consider the impact of the award on the product seller in light of its financial condition. This consideration has deep roots in common law. It is one that has been subject to criticism from product sellers and economists. Nevertheless, in light of the fact that the deterrence of wrongful conduct is the principal rationale for punitive damages, it is appropriate to consider the impact an award will have on a particular product seller.

Factor (7) is more important in product liability cases than in other liability cases because it addresses the problem of multiple exposure to punitive damages. This factor directs the court to consider both criminal and civil liability to which the product seller has been or may be subjected.

Factor (8) recognizes that the injury may also be attributable in part to the claimant's reckless disregard for personal safety. In such instances where the defendant can show such reckless disregard on the part of the claimant, punitive damages may be diminished proportionately according to the comparative responsibility according to the comparative responsibility of that claimant. The comparison, therefore, is between the reckless disregard on the part of the product seller and that of the claimant, and *not* between the reckless disregard of the product seller and mere negligence of the claimant. *See also* Section 111.

The Act indicates that the award of punitive damages goes to the claimant and not to the state. While the argument that "since the damages are non-compensatory, they should go to the state" has some merit, the approach was rejected because of constitutional problems and the fact that it might place a claimant's attorney in a potential conflict of interest situation by forcing the attorney to represent both the claimant and the state. *See* "Task Force Report" at VII–79.

When the trier of fact determines that punitive damages should be awarded, one option available to the court is to limit the award to a multiple of the compensatory damages as is done in antitrust actions. While this approach has appeal, it is not appropriate in the case of product liability because this area of the law addresses a multiplicity of different kinds of wrongful acts. Antitrust law, on the other hand, addresses one basic kind of wrongful act: an antitrust violation. Under this Section, the defendant's wrongful conduct could range from that which merely could cause damage to property to conduct which could result in the deaths of many people. Antitrust law violations involve economic injury only, and rarely, if ever, result in serious physical injury or death. Thus, the punitive damages in product liability actions do not lend themselves to quantification in the same manner as do such damages in antitrust cases.

* * * * *

Code

Sec. 121. Severance Clause

If any part of this Act shall be adjudged by any court of competent jurisdiction to be invalid, such judgment shall not affect, impair, or invalidate the remainder thereof, but shall be confined in its effect to that part of this Act declared to be invalid.

* * * * *

Analysis

Sec. 121. Severance Clause

This Section makes clear that in the event any court of competent jurisdiction declares any part of this Act to be invalid, such action shall be confined to that part alone, and shall not affect the validity of the remainder of the Act. The source of this Section is "1976 N.Y. Laws," ch. 955 (12) (McKinney), which concerns medical malpractice.

* * * * *

Code

Sec. 122. Effective Date

This Act shall be effective with regard to all product liability claims filed on or after ————, 19——.

*** Alternative Section**

[This Act shall be effective with regard to all claims accruing on or after ————, 19——. It shall be prospective in operation, and shall only apply to a product-related harm occurring on or after this date. When the facts giving rise to a claim are discovered or should have been discovered after this date, this Act shall govern the claimant's action. When the facts giving rise to a product-related harm are discovered or should have been discovered prior to the effective date of this Act, the law of this State which was applicable at the time of such discovery shall govern the claimant's action.]

* * * * *

Analysis

Sec. 122. Effective Date

The issue of the effective date of a tort-related statute has not been given intense analysis in legal periodicals—perhaps because there have been relatively few legislative initiatives in the tort area. In an attempt to balance fairness to all parties against procedural convenience, the drafters of this Act have chosen to apply the statute to all claims *filed* after the effective date.

The shortcoming of this approach is that it may create a rush to the courthouse—or, conversely, a delay—by claimants' attorneys. Also, the new law will apply to some injuries that occurred *before* the effective date.

On the other hand, since the Act does not create new causes of action, but condenses, clarifies, and balances common law development, the approach taken is a fair one. Also, it allows claimants, product sellers, and insurers to know, with precision, what law will be applied to a claim. *See* "1979 Minn. Laws" ch. 81, Section 3; "Conn. Gen. Stat." Section 52–572 1 (rev. 1979); *cf.* "Peterson v. City of Minneapolis," 285 Minn. 282, 173 N.W.2d 353 (1969); "Godfrey v. State," 84 Wash. 2d 959, 530 P.2d 630 (1975); *see also* Annot., 37 "A.L.R.3d" 1438 (1971).

Because state legislatures are likely to differ about the issue of effective date, an alternative approach has been drafted. The alternative provision would have the Act apply only to a harm actually sustained or causes of action discovered, or which should have been discovered (*e.g.*, latent disease) on or after the effective date. This approach makes the Act fully prospective in nature and recognizes that certain illnesses, such as malignancies, only become discoverable some time after the actual contact with the product. In the case of disease or other illness with delayed effects, the Act would apply if the harm or the cause thereof were discovered on or after the effective date. *See* Section 109(C); *see also* 73 "Am. Jur. 2d" "Statutes" Sections 385 *et seq.* (1974).

A significant shortcoming of this alternative approach is that for many years it will leave claimants, product sellers, and insurance ratemakers with two potential sources of product liability law—existing common law and this Act. It should be noted that the phrase "the law of this State," as used in the alternative, includes a state's conflict of law rules as well as its substantive product liability law.

* * * * *

Appendix A—A Bibliography of Major Compendium Sources Reviewed in Connection With the Model Code

Final Report, Interagency Task Force on Product Liability (NTIS, 1977).

Product Liability Legal Study, Interagency Task Force on Product Liability (NTIS, 1977).

Product Liability Industry Study, Interagency Task Force on Product Liability (NTIS, 1977).

Product Liability Insurance Study, Interagency Task Force on Product Liability (NTIS, 1977).

Selected Papers, Interagency Task Force on Product Liability (NTIS, 1978).

Insurance Services Office, Product Liability Closed Claims Survey: A Technical Analysis of Survey Results (ISO, 1977).

Product Liability Insurance, A Report of the Subcommittee on Capital, Investment and Business Opportunities of the Committee on Small Business, H.R. Rep. No. 95–997, 95th Cong., 2d Sess. (1978).

"Impact on Product Liability": Hearings before the Senate Select Committee on Small Business, 94th Cong., 2d Sess., 95th Cong., 1st Sess. (1976–77).

"Product Liability Insurance": Hearings on S. 403 before the Consumer Subcommittee of the Senate Committee on Commerce, Science and Transportation, 95th Cong., 1st Sess. (1977).

Department of Commerce Task Force Report on Product Liability Insurance (State of Michigan, 1978).

Final Report of the Governor's Task Force on Product Liability (Maine, 1978).

Product Liability: An Overview, Wisconsin Legislative Council, Research Bulletin 78 (1978).

Illinois House of Representatives, Judiciary I Subcommittee on Product Liability, Report and Recommendations (1978).

Report of the Senate Product Liability Study Committee, State Capitol, Georgia (1978).

Report of the Senate Select Committee on Product Liability, Missouri Senate (1977).

American Law Institute, Restatement (Second) of Torts, Section 402A and Appendices (1965).

Defense Research Institute, Inc., Products Liability Position Paper (1976).

The Alliance of American Insurers, Product Liability Tort Reform Proposals (1976).

Proposed Uniform State Product Liability Act (National Product Liability Council).

American Insurance Association, Product Liability Legislative Package (1977).

The California Citizens' Commission on Tort Reform, Righting the Liability Balance (1977).

A review was also conducted of all enacted state product liability laws, all proposed federal product liability laws, and major proposed state product liability laws, as well as all law review and other related literature and major case law reported or published since the completion of the Interagency Task Force's seven-volume Legal Study in December 1976. That study reviewed case law and literature published prior to that date. Consideration was also given to pre-1976 sources that were not reviewed by the Legal Study.

[FR Doc. 79-33253 Filed 10-30-79; 8:45 am]

BILLING CODE 3510-17-M

Appendix 10-2 Westinghouse Warning Label Format*

This is the preferred label format and should be used unless vertical space is severely limited.

All label formats are based on a grid which is 12 units wide. The depth of the label varies depending on the message length and whether a symbol or pictograph is used. However, the depth is always measured in whole grid units.

Labels are assembled by placing elements on the grid according to detailed instructions provided in Chapter 13. The example below explains specific guidelines and spacing.

Signal Word
Identifies the seriousness of the potential hazard. Signal words are preprinted and sized to fit the label grid.

Standard Border
A ½-unit border surrounds the label on all four sides. This simplifies printing and screening during production.

Alternate Border
A 1½-unit border can be used if space is required for mechanical fasteners.

Color
Color is used to reinforce the message and seriousness of the hazard. Each signal word has a corresponding color. Colors are specified in label layout kits.

Alert Symbol
Appears with signal words Danger, Warning, and Caution. It is pre-printed on the artwork included in the layout kits.

Symbols and Pictographs
A symbol or pictograph can be used to reinforce the verbal message. Preprinted symbols and pictographs are provided with layout kits. They are 10 units wide and 6 units in depth—sized to fit the allotted area.

Hazard Identification
The first word message following the signal word. It identifies the type of hazard and is set in bold type. (Helvetica 18 on 20)

Result of Ignoring Warning
This message tells what will happen if the warning is ignored. It is part of the first message block and is also set in bold type (Helvetica 18 on 20). In most cases, it begins on the line immediately after the hazard identification.

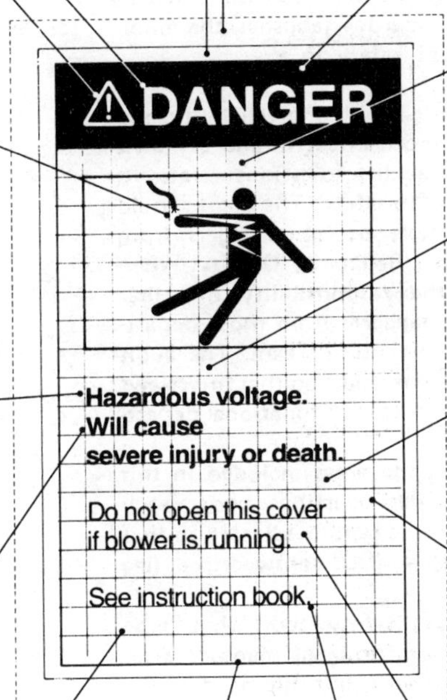

Space
A one-unit space should always separate the signal word from the symbol/pictograph or text area.

Space
A one-unit space should always be maintained between the symbol/pictograph and the following text.

Space
Always allow a one-unit space between identifying the hazard and the damage or injury it can cause and the following text.

Right Margin
The right hand margin of type is not to be aligned (justified). It will end in varying positions according to where lines are broken. This is good. It is more readable. Do not justify type. Do not center type.

Space
Two unit spaces always separate the last line of type from the border line at the bottom of the label.

Side and Bottom Rule
Extends from edge of signal word panel downward and closes off the label 2 units below the last line of type.

Punctuation
Periods should be used after each sentence or sentence fragment. Do not use exclamation points, colons, dashes or other punctuation unless it's absolutely necessary.

Avoiding the Hazard
This message tells the reader what precautions to take to avoid the hazard. It is set in regular weight type. It is separated from the preceding message by one unit. This is the last part of the label.

* Westinghouse Electric Corporation, Product Safety Label Handbook 32 (Westinghouse Printing Div, Trafford, Pa 1981).

Appendix 10-3 Product Safety Audit Check List*

PRODUCT SAFETY AUDIT CHECKLIST

Introduction

The purpose of a product safety audit is to review all operations of the division or operating unit that effect product safety and identify areas for improvement. All methods, techniques, procedures, paperwork, and personnel should be open to review.

The standard against which this audit should be performed is the "Corporate Product Safety Requirements" document, This standard establishes the minimum requirements for product safety in a division.

The audit may be combined with the reliability and quality program audit for a more effective and efficient audit since many of the checklist questions relate to reliability and quality as well as to safety. The audit should be carried out by a team made up of managers from Engineering, Manufacturing, and Reliability/Quality. When operations that are under the responsibility of other departments are audited, the managers from those departments should be included in the Audit Team. The audit team should plan and schedule the audit so that interviews may be held simultaneously with several functional departments.

The audit should cover all the areas included in this checklist and should answer the questions under each heading. The questions have been carefully selected so that a "YES" answer is desirable; a "NO" answer is a flag suggesting deeper inquiry.

The results of the product safety audit should be reported to the division manager or plant manager after each problem area has been reviewed with the head of the department responsible for correcting the problem and with his responsible members of supervision. The audit report should identify the items needing correction and the persons assigned to do the work. It should also describe the progress made in correcting the problems identified in the previous product safety audit.

---

* Product Liability of Manufacturers: Prevention and Defense, 493-510, (Ross & Foley eds 1981). The above excerpt appears in PLI's 1981 course handbook *Product Liability of Manufacturers: Prevention and Defense* and is reprinted here with their permission. Copyright 1981 by the Practising Law Institute, New York City.

QUESTION	YES	NO
A. General Management		
1. Has a formal division product safety policy statement been issued? (Ref. Management Guide Product Integrity.)		
2. Does the product safety policy statement encompass all the functional groups in the division and designate the scope of responsibility of each group?		
3. Are product safety audits conducted periodically?		
Recommended actions implemented?		
4. Has an individual or functional group been assigned the responsibility of keeping abreast of enacted and pending legislation related to the safety of the division's products and services and recommending appropriate actions?		
5. Has an individual or functional group been assigned the responsibility of periodically reviewing all situations and claims involving personal injury arising from the user of the division's products and services and recommending appropriate actions?		
6. Are division product safety objectives established as a part of the division's profit plan and strategic program plans?		
7. Is a product safety review required on each new product design and major redesign and periodically on products in production?		
If so, who has the responsibility?		
B. Marketing		
1. Are marketing personnel made aware of the product safety policy and their responsibility in connection with the policy?		
If so, how?		

Other Comments:

QUESTION	YES	NO
2. Are marketing personnel kept abreast of marketing product safety-related practices and procedures through training programs or other communication means?		
Is current government policy and laws pertaining to product safety included?		
3. Are marketing personnel made aware of their product liability exposure due to advertising, written warranties, implied warranties, and through verbal claims about the product?		
If so, by what means?		
4. Are marketing personnel kept informed in the proper use of the product, potential hazards in normal use and misuse of the product, and product limitations and the consequences of exceeding those limitations?		
If so, by what means?		
5. Are all product advertisements and sales literature reviewed by other departments for safety and legal (warranty) implications before release?		
If so, what other departments?		
6. Are product advertisements and sales literature reviewed to see that safe practices are encouraged, product limitations are outlined, inflated product claims are avoided?		
If so, by whom?		
7. Are product promotion displays reviewed to see that product safety features are presented and safe operating practices portrayed?		
Also, that adequate warning and caution labels are attached to the product?		
If so, by whom?		
8. Are written warranties reviewed by the law department for adequacy and clarity of language?		

Other Comments:

QUESTION	YES	NO
9. Are marketing personnel instructed in the procedures to follow in the event they are informed of personnel injury or property damage involving products and services?		
If so, how?		
10. Are all instruction books reviewed for adequacy, warnings against product misuse, and information on potential product hazards and how best to avoid them?		
If so, by whom?		
11. Is there a policy on use of disclaimers and hold-harmless agreements in purchase orders and contacts?		
Is it being followed?		
12. Are product design specifications reviewed for appropriate safety criteria?		
If so, by whom?		
13. Are terms and conditions in contracts and purchase orders reviewed and approved?		
If so, by whom?		
14. Are there documented marketing procedures for questions nos. 1-13?		
Are the procedures controlled and kept current?		
Procedures readily available to personnel?		
15. Are new product safety features offered to owners of earlier generation products which do not contain the safety feature?		
Is the offer documented and made a part of the job record file?		

Other Comments:

QUESTION	YES	NO
C. Engineering		
1. Are engineering personnel made aware of the product safety policy and their responsibility in connection with the policy?		
If so, how?		
2. Are engineering personnel kept abreast of engineering product safety-related practices and procedures through training programs or other communication means?		
Is current government policy and laws pertaining to product safety included?		
3. Are engineering product safety-related practices and procedures documented, made a part of the design assurance program, controlled and maintained up to date?		
4. Do personnel have access to copies of pertinent legislation, codes and standards (federal, state, local and industry) which influence product safety design?		
5. Does each product design specification include applicable safety criteria?		
Are applicable federal, state, local and industry standards included?		
Is expected product service life documented?		
6. Are the customer's intended uses and reasonably foreseeable misuses identified and considered in the product design?		
Is the product tested under those conditions of use and misuse?		
Where the product cannot be designed to be safe under misuse conditions, are those misuses documented in the user's instruction books?		
7. Are product designs analyzed for potential hazards in extreme conditions of temperature, humidity, voltage, pressure, shock, vibration, corrosive atmosphere, dust, sand radio frequency interference, nuclear radiation?		

Other Comments:

QUESTION	YES	NO
Are potential hazards identified and features incorporated in the design to protect the user against the potential hazards?		
Are those hazards which cannot be eliminated in the design documented in the user's instructions and labeled on the product?		
Are those hazards reported to general management?		
8. Is a Failure Mode & Effects Analysis (FMEA), Fault Tree Analysis (FTA) or other safety analysis regularly conducted on new product designs and major redesigns?		
Are stress analyses conducted on critical electrical and mechanical components?		
9. If a product design does not comply with applicable safety standards, are the reasons for non-compliance documented?		
Made a part of the engineering job record file?		
Reported to general management?		
10. If a decision is made not to include a specific safety feature in a product design, are the reasons documented?		
Reported to general management?		
Made a part of the engineering job record file?		
11. Are formal design reviews conducted on all new product designs and major redesigns?		
Is a product safety review a regular part of the formal design review?		
Does the design review team include representatives from other functional departments in the division?		
Does the product safety review take into account past experiences involving personal injury, property damage and environmental damage with the product or similar products?		

Other Comments:

QUESTION	YES	NO
12. Is a safety design checklist provided to your design engineers?		
13. Is a safety design review checklist provided to the design review team?		
Does the checklist include the following questions:		
Safety interlocks used?		
Sharp edges, burrs eliminated?		
Moving, rotating assemblies covered?		
Hazard warnings precise, conspicuous, draw attention? . . .		
Are controls easy to under; accessible?		
Is an emergency STOP button necessary?		
Is the emergency STOP button located "FRONT & CENTER"?		
At other accessible locations?		
Is it recessed, guarded to prevent accidental shutdown? . . .		
Will product be safe to operate in temperature extremes, high humidity?		
Will product be safe to use after mechanical shock, vibration from handling and transportation?		
Has safety of servicing personnel been considered?		
Are performance monitoring devices considered?		
Are unsafe thresholds clearly marked?		
Can they be easily read and understood?		
Will the operator, user know what to do immediately if the unsafe threshold is reached?		
Is it difficult to by-pass or deactivate safety features?		
Are rating and/or instruction plates clear, concise, complete?		

Other Comments:

QUESTION	YES	NO
Does the product comply with applicable safety laws and regulations?		
Does the product comply with safety standards and guidelines of such agencies as UL, ANSI, AHAM, ASTM, NEMA, AEC, OSHA?		
Are required safety features easy to see?		
Obvious how they operate?		
Are certification labels obvious? Easy to find?		
Brought to buyer's attention?		
Are warnings and cautions in large, contrasting print?		
Is the need for training of personnel to perform maintenance minimized?		
Can servicing be performed without the need for special tools?		
Does the design eliminate the possibility of the product being reassembled incorrectly which could result in a potential safety hazard?		
Are special instructions for safety discarding or destroying old unit considered?		
14. Are manufacturing, quality assurance and other functional groups advised of critical product parameters?		
How?		
15. Are the requirements of the product design spec and design analysis verified by test programs?		
Is there a test specification itemizing all parameters to be tested?		
Is certification of your products by an outside agency required?		
Are critical components tested under extreme environmental conditions over their entire operating range?		

Other Comments:

QUESTION	YES	NO
Are safety tests run on prototype models before production go-ahead?		
Are test results documented, analyzed and made a part of the engineering job record file?		
16. Are design changes reviewed for their effect on product safety?		
17. Is the product packaging for transportation designed to meet applicable standards?		
Is the packaged product tested under realistic transportation environmental conditions?		
18. Does engineering review quality control procedures to ensure relevant product safety considerations are included? . .		
Vendor source control procedures?		
Incoming material and parts inspection and test procedures?		
Parts in process?		
Final assembly inspection and test procedures?		
19. Does engineering review test and inspection records for product safety considerations?		
Test data and results?		
Material review board action?		
Manufacturing discrepancies and defect rates?		
Functional failures?		
20. Does engineering review field failure reports and field complaints for product safety and liability considerations? . .		
21. Does engineering review sales promotional literature for technical accuracy and to detect exaggerated performance claims?		
Is there a checklist of items to look for in the reviews? . . .		

Other Comments:

QUESTION	YES	NO
22. Does engineering review all instruction books for product safety considerations?		
Is there a checklist of items to look for in the reviews? . . .		
23. Are key design engineers designated and trained in proper handling of litigation and in how to function as an expert witness?		
24. Is there a procedure for interfacing with other functional departments in the review and analysis of product liability potential and actual claims?		
D. Quality Assurance		
1. Do the vendor and in-house quality programs give emphasis to product safety?		
2. Are quality assurance personnel made aware of the product safety policy and their responsibility in connection with the policy?		
If so, how?		
3. Are quality assurance personnel kept abreast of product safety related quality practices and procedures through training programs or other means of communication?		
Is current government policy and law pertaining to product safety included?		
4. Do quality assurance personnel have access to current copies of applicable governmental regulations, codes and standards (federal, state, local and industry) pertaining to product safety-related quality assurance activities?		
5. Do quality assurance personnel utilize documented engineering analyses (such as FMEA's, FTA's) in planning production tests and inspection of critical components and functions?		
6. Are results of field reports, accidents, customer problems, litigation on prior products or similar products considered in planning the production test program?		
Do test conditions simulate customer/user conditions as close as possible?		

Other Comments:

QUESTION	YES	NO
7. Are test and inspection procedures reviewed with engineering for relevant product safety considerations?		
8. Does final assembly inspection include checks to ensure all safety-related features (such as guards) are in place, that required cautions and warnings are attached to the product, and that the designated instructions are provided? . .		
9. Is product packaging checked for conformance to specifications?		
10. Are test problems and failures documented, analyzed?		
Are test failures analyzed from a customer safety viewpoint?		
Analysis results communicated to appropriate functional departments for corrective action?.		
11. Is there a system which documents safety problems in the field and relays them back to the division?		
Are all field reports logged in, investigated?.		
Is there a procedure which triggers problem investigation after a predetermined number of failures or incidents occur?		
Are failed parts, materials or assemblies recovered and returned from the field for failure analysis?.		
Is the knowledge gained from failure analysis and accident investigation communicated to the appropriate functional departments for corrective action?		
12. Is a product identification system in effect which provides traceability of the product to the ultimate user for possible product recall?		
Are periodic practice "dry runs" conducted in locating a particular batch of products after they have been released to the distribution system or are in the hands of the customer?		
13. Does quality assurance review sales promotional literature for quality-related technical accuracy and to detect exaggerated performance claims?.		

Other Comments:

QUESTION	YES	NO
14. Does quality assurance review all instruction books for quality-related product safety considerations?		
15. Are quality assurance product safety-related practices and procedures documented, controlled, maintained up to date?		
Readily available to personnel?		
E. Manufacturing & Purchasing		
1. Are manufacturing and purchasing personnel made aware of the product safety policy and their responsibility in connection with it?		
If so, how?		
2. Are manufacturing and purchasing personnel kept abreast of product safety-related practices and procedures through training programs or other means of communication?		
Is current government policy and laws pertaining to product safety included?		
3. Does manufacturing work from approved and up-to-date drawings?		
4. Is there an effective system for bringing engineering's attention to inadequacies on drawings and specifications? . . .		
For requesting engineering action to correct drawings or specifications?		
5. Does manufacturing have procedures which assure that equipment and facilities are maintained properly?		
6. Can the equipment and facilities meet engineering drawings and specifications necessary to assure product safety?		
7. Are work instructions used to the extent necessary to assure controlled conditions?		
8. Is manufacturing planning done considering quality and product safety requirements?		

Other Comments:

QUESTION	YES	NO
9. Are there standards for workmanship?		
10. Are tolerances defined for parts and processes?		
11. Is inspection and test coverage adequate?		
12. Are people adequately trained for their jobs?		
13. Is there a program to assure adequate housekeeping?		
14. Does manufacturing management show an interest in product safety and insist on quality work from the work force?		
15. Is there a continuing corrective action program with the participation of manufacturing management?		
16. Are measures of manufacturing quality performance regularly reported to management?		
17. Are detailed work instructions to feeder, process, assembly, inspection, test and packaging reviewed for product safety-related considerations?		
18. Are manufacturing's and purchasing's files of drawings and specifications kept current?		
Are critical parts identified for proper attention during the purchasing and manufacturing process?		
19. Is there a formal system for supplier selection?		
Are supplier quality system audits performed?		
Is past or potential supplier quality performance considered in selection?		
Is supplier qualification documented?		
Are receiving inspection and test pre-planned?		
Are records of supplier quality performance maintained? . .		
Is there a periodic analysis of supplier quality performance?		

Other Comments:

QUESTION	YES	NO
Are troublesome suppliers identified and is corrective action taken?		
Are there procedures for assuring that design changes are promptly sent to suppliers?		
Are there systems for assuring that suppliers do not change designs or processes without informing		
Are Corporate standards used so far as practical in the purchasing of materials and parts?		
Are Purchasing Department Specifications (PDS) and Material cards (M cards) used?		
Are there systems for assuring that needed certifications and test reports are requested and received?		
20. Do drawing Distribution and Recall Systems assure the proper revision at workplace?		
Are process and material specifications included in control systems?		
Are procedures and work instructions included?		
Is the change procedure documented?		
Are the changes approved by Manufacturing, Quality Assurance and others as necessary?		
Is there controlled distribution of changes?		
Is disposition of material affected specified on change notices?		
Do changes affecting suppliers get controlled distribution?		
Are there procedures for handling customer directed changes?		
Is effectivity specified?		

Other Comments:

QUESTION	YES	NO
21. Are material handling methods adequate and well defined? . .		
Are storerooms controlled and periodically inspected? . . .		
Is material stored to prevent damage and deterioration? . .		
Is inspection and test status identifiable?		
Are critical operations, inspections, tests and treatments records maintained?		
Are there systems for shelf life control?		
22. Are there written procedures covering handling of non-conforming materials?		
Is non-conforming material identified and segregated? . . .		
Is non-conforming material promptly dispositioned?.		
Are deviations handled by formal procedures, documented, and approved by Quality Assurance and Engineering?		
Is corrective action required on recurring or serious deviations?		
Are repair procedures required and used when finished item deviates from requirements?		
Are periodic analyses of rejections performed?		
Are scrap, rework and costs identified, recorded and periodically reported?		
23. Do process specifications define process conditions and quality criteria?		
Are process specifications distributed in controlled way? . .		
Are process specifications available and used in factory? . .		
Are processes performed by vendors controlled?		
Are process conditions monitored for control purposes? . .		

Other Comments:

QUESTION	YES	NO
Are process and operator qualification requirements documented and controlled?		
Are process conditions and part or material records maintained?		
Are process procedures prepared and used where necessary?.		
Is checking by chemical or physical testing laboratories defined and regularly scheduled?		
24. Is regular measuring and test equipment calibration performed?		
Are records kept and periodic analysis made of calibration frequencies?.		
Are calibration procedures and tolerances documented? . .		
Are there handling, storage and use procedures for measuring and test equipment?		
Is there identification of calibration status?		
Are calibration standards identified, properly cared for, and regularly checked to National Standards?		
Is there control of measuring and test instruments and gages at suppliers and in field?		
Is special processing instrumentation included?		
Are fixtures and tools used as media of inspection calibrated?		
25. Do packaging specifications exist and are they used?.		
Are shipping and environmental tests performed?.		
26. Do purchase orders contain hold harmless agreements or indemnification clauses under which vendors and sub-contractors assume liability for their products and/or performances?		

Other Comments:

QUESTION	YES	NO
F. Field Service		
1. Are field service personnel made aware of the product safety policy and their responsibility in connection with the policy?		
2. Are field service personnel kept abreast of product safety-related field service practices and procedures through training programs or other means of communication?		
Is current government policy and laws pertaining to product safety included?		
3. Are field service personnel made aware of their product liability exposure due to written warranties, implied warranties and through verbal claims made about the product?		
Are field service personnel adequately informed in the proper use of the product, potential hazards in normal use and misuse of the product, and product limitations and the consequences of exceeding those limitations?		
4. Are field service personnel instructed in the procedures to follow in the event they are informed of personal injury or property damage involving products and services?		
5. Are field service personnel instructed to record any defective or unsafe conditions observed during servicing of the product, which is not included in the authorized work, in their service report?		
Are they instructed to obtain the customer's signature on the service report as evidence that the conditions have been brought to his attention?		
6. Are field failures and customer complaints documented in field service reports?		
Are these reports provided to the appropriate functional departments for corrective action?		

Other Comments:

QUESTION	YES	NO
7. Are all instruction books reviewed for:		
Simple and easy to follow instructions?		
Properly stressed safe methods and practices?		
Potential hazards identified and information included on avoiding or lessening the hazard?		
Warnings against misuses included?		
Possible dangers of using unauthorized repair parts stressed?		
Dangers and consequences of not following maintenance instructions required for safe operations stressed?.		
8. Are field service personnel instructed that *personal opinions* which are not fully supported by facts are not to be included in correspondence or other documentation?		

Other Comments:

The Inside Counsel's Role in Innovation: A Conceptual Framework

Edward S. Binkowski

11

Overview: The Law and Innovation

§11.01 Introduction

Innovation may be thought of as any change possessing both novelty and value. In a narrow sense, there is a body of law of innovation, which is concerned with the ways in which the benefits of innovation may be secured for the innovator through some form of legal sanction, such as a patent.[1]

More broadly speaking, however, the law and innovation are bound together in many other ways. Assumptions and prejudices about the progress and utility of technological change inform the legal treatment of areas where innovation is involved;[2] various enforcement and regulatory procedures find special application in the face of innovation;[3] and innovation itself may prompt developments of the law, just as certain legal consequences may guide as well as shelter innovation.[4]

The legal structure of the United States embraces the notion that innovation is a good in itself. This may be seen in the strength of the protection given innovation through the patent system, a level of protection which runs to some degree counter to other predispositions toward competition and free markets.[5] The encouragement of innovation is one of the cornerstones of antitrust law, and, further, a finding of innovative behavior on the part of a firm may serve as an effective defense to charges of anticompetitive behavior.[6] On some

---

[1] For a full discussion of these issues, see **ch 4.**

[2] Kaysen & Turner, Antitrust Policy (1959).

[3] *E.g.*, Antitrust Division, US Dept of Justice, Antitrust Guide Concerning Research Joint Ventures (1980); *reprinted in* 1980 Antitrust & Trade Reg Rep (BNA) No 992 (special supp Dec 4, 1980).

[4] *E.g.*, California Computer Prods v IBM, 613 F2d 727 (9th Cir 1979); Berkey Photo, Inc v. Eastman Kodak Co, 603 F2d 263 (2d Cir 1979), *cert denied,* 444 US 1093 (1980).

[5] Bowman, Patent and Antitrust Law (1973).

[6] California Computer Prods v. IBM, 613 F2d 727 (9th Cir. 1979); see **ch 3,** especially **§3.05.**

occasions novelty alone without any demonstrated value may be regarded as sufficient. Direct governmental support in terms of funding,[7] tax benefits[8] and limitations of liability[9] also provides obvious encouragement of innovation. Indeed, there often seems to be such a strong identification between technological progress and national pride that absence of innovation may become evidence of illegality.[10]

In the past decade, however, these pro-innovation attitudes have been combined with growing environmental concerns and active resistance to applications of specific technologies.[11] It must be noted that many of these critics maintain that their stance is at base protechnology[12] not anti-, and in fact demand accelerated innovation in some areas. Shifts in public perception of an economy of scarcity and an era of lowered expectations in fact work to *increase* support and protection of selected innovation.[13]

In other circumstances, innovation may be simply accepted as a fact which renders existing precedent inapplicable since the historical behavior supporting the precedent is made irrelevant by the innovation.[14] In some cases, the innovation itself may be claimed to be of sufficient value to justify breaking new ground despite the special market advantages it may receive;[15] in others, perceived dangers associated with innovation may serve as a legal cloak for some other social concern which is being advanced.[16] In the most extreme situations,

---

[7] Terleckyj, Effects of R&D on the Productivity Growth of Industries (National Planning Assn 1974).

[8] See **ch 5** and **§11.16.**

[9] Gordon v New York Stock Exchange, 422 US 659 (1975); Flood v Kuhn, 407 US 258 (1972); for an example limiting liability for nuclear power plants, *see* the Price-Anderson Act, Pub L No 88-703, 71 Stat 576 (1957) (codified at 42 USC §2012).

[10] United States v United Shoe Mach Corp, 110 F Supp 295, 330 (D Mass 1953), *affd,* 347 US 521 (1954).

[11] *See, e.g.,* Thompson, At the Edge of History (1970), and Passages About Earth (1973); Mumford, The Myth of the Machine II: The Pentagon of Power (1970); Lovins & Price, Non-Nuclear Futures (1975).

[12] *E.g.,* Schumacher, Small is Beautiful (1973); Lovins, Soft Energy Paths (1977).

[13] Meadows, Donella, Meadows, Randers & Behrens, The Limits to Growth (1972); Mesarovic & Pestel, Mankind at the Turning Point (1974); Brandt, *et al,* North-South: A Program for Survival (1980); Barney, *et al,* The Global 2000 Report to the President (1980).

[14] Berkey Photo, Inc v. Eastman Kodak Co, 603 F2d 263, 281-84 (2nd Cir 1979), *cert denied,* 444 US 1093 (1980); California Computer Prods v IBM, 613 F2d at 742-44 (9th Cir 1979).

[15] Berkey Photo, Inc v. Eastman Kodak Co, 603 F2d 263, 281-84 (2d Cir 1979), *cert denied,* 444 US 1093 (1980); California Computer Prods v IBM, 613 F2d at 742-44 (9th Cir 1979).

[16] MacPherson v Buick Motor Co, 217 NY 382, 391 (1916), rejecting the claim that an automobile manufacturer is immune from liability for a driver's injuries: "The law does not lead us to so inconsequential a conclusion. Precedents drawn from the days of travel by stage coach do not fit the conditions of travel today. . . . They are whatever the needs of life in a developing civilization require them to be."

radical innovation is used as an argument to claim that the controversy is outside the legal arena altogether.[17]

Finally, the law catches up with innovation through extended application to new forms of conduct, creation of novel remedies, modification of regulatory and enforcement mechanisms and occasionally even formation of new areas of law.[18] On a day-to-day basis, technological innovation has been a motivation for extremely creative legal innovation in presentation of certain activities as not so much beyond as between more familiar concepts, as in the use of technological advances to create loopholes in regulatory oversight.[19]

§11.02 Triggering Events for Inside Counsel

Legal issues involving innovation which confront inside counsel will typically be of two kinds. In the first, the innovation itself will appear to be one of the dominant characteristics of the problem, such as a new product development requiring substantial horizontal expansion. However, on closer inspection, it will become clear that innovation in its own right is not really the problem; other concepts will be more effective for both description and justification of a state of events, such as an economies of scale defense to an antitrust charge in the situation mentioned above.[20] In the second, familiar corporate operations with their own well-established areas of subject matter law may not seem to involve questions of innovation directly but will be seen on closer inspection to profit from a consideration of innovation, or to be ineluctably drawn into any discussion of corporate innovation iniatives, as in such straightforward applications as financing options and tax planning.[21]

The first set of concerns discussed in this chapter, where innovation seems to be of more importance than deserved, includes:

1. Does evidence of innovation support or subvert a claim of economies of scale[22] or scope[23] as a justification of anticompetitive behavior?
2. Does innovation play a role in the organization of the specialized production and distribution of products that can affect market definition for legal purposes?[24]
3. To what extent can evidence of radical innovation in its own right out-

---

[17] California Computer Prods v IBM, 613 F2d 727, 744 (9th Cir 1979).

[18] See Industrial Concentration: The New Learning (Goldschmid, Mann & Weston eds 1974).

[19] See Yuckman, *The National Public Radio Satellite System: FCC Jurisdiction Over a New Communications Technology,* 8 Rut J Computer Tech & Law 135 (1980).

[20] See **§§11.05-.10.**

[21] See **§§11.11-.16.**

[22] See **§11.05.**

[23] See **§11.06.**

[24] See **§11.07.**

weigh associated anticompetitive effects?[25]

4. Does innovative behavior affect the planning and tax consequences of expense–capital tradeoffs?[26]
5. Does the changing contemporary character of innovation undermine all the above interactions?[27]

The second set of concerns discussed here, where innovation deserves more attention than it may routinely obtain, includes:

1. Does innovation suggest new forms of financing and alternative modes of measuring return?[28]
2. Does innovation provide a justifying spark for mergers and acquisitions?[29]
3. What is the appropriate relation between innovation and advertising?[30]
4. Does the innovative activity create special liabilities?[31]
5. Should (or can) innovation be promoted or managed by particular internal decision-making structures?[32]
6. Are there special tax planning options for innovative activity?[33]

§11.03 Long-Run Impact of Innovation

While many innovations are claimed, usually in retrospect, to be overnight or miraculous, significant advances are more frequently the product of years of research followed by an often even longer period of development, refinement and implementation.[34] The true value of the discovery may take yet more time to become realized in the presence of future related innovations or specific needs.[35] Therefore, determination of the impact of innovation and its effect on counsel's responsibility, or even the degree to which innovation in a given industry may exist, is necessarily a long-run analysis.

Understanding of the ways in which the process of innovation unfolds and interacts with the development of the firm itself serves two useful purposes.

---

[25] See **§11.08.**

[26] See **§11.09.**

[27] See **§11.10.**

[28] See **§11.11.**

[29] See **§11.12.**

[30] See **§11.13.**

[31] See **§11.14.**

[32] See **§11.15.**

[33] See **ch 5** and **§11.16.**

[34] Holton, *Scientific Research and Scholarship: Notes Toward the Design of Proper Scales,* 91 Daedalus 362 (Spring 1962).

[35] See Clarke, Profiles of the Future (1963).

First, various patterns are routinely associated with the progress of innovation and its impact. Familiarity with these patterns on the part of inside counsel can act as an early warning system for the identification of legal issues before they become emergencies. Second, these patterns of innovation and effect can also serve as guides for structuring evidence and issues of proof in support of a contested course of action.

Problems associated with the identification and measurement of innovation include:

1. What criteria should be used to determine the appropriate time interval for demonstrating proof of innovation?[36]
2. What measurable surrogates for innovation constitute evidence of its existence?[37]
3. Can new arrangements of old ideas constitute evidence of innovation?[38]
4. Can critical mass arguments be employed to justify the size of an enterprise as a prerequisite for innovation?[39]
5. How can differential rates of innovation be demonstrated?[40]
6. To what extent are spin-offs successful as evidence of innovation?[41]

Questions addressing the impact of innovation on an industry and the place of a firm within an industry, discussed within this chapter, include:

1. Can the existence of competition within an industry be shown to have a positive effect on the level of innovation?[42]
2. Does innovation itself have procompetitive effects?[43]
3. Can the level of innovation within an industry be used to defend or explain industry concentration?[44]
4. Can growth within an industry be used as evidence of innovation?[45]
5. Does innovation promote industry growth and in that way neutralize otherwise anticompetitive advantages given an innovator?[46]

---

[36] See §11.17.
[37] See §11.18.
[38] See §11.19.
[39] See §11.20.
[40] See §11.21.
[41] See §11.22.
[42] See §11.23.
[43] See §11.23.
[44] See §11.24.
[45] See §11.25.
[46] See §11.25.

6. What quantitative models can be used as evidence of innovation?[47]

§11.04 Extracorporate and Extralegal Background

The twentieth century is readily characterizable by extremely rapid technological innovation, and the pace of this innovation seems to continue to accelerate, though unevenly.[48] While many take it for granted that this pace and the advantages it brings will remain unabated, there are some contrary forces at work.

For example, in many ways the "easy" innovations are over. The lone inventor in a home garage has less and less to contribute as innovation becomes more and more complex. Enormous capital investment, the need for specialized talents and equipment and often lengthening lead times seem to be inescapable attributes of innovation today and in the foreseeable future.[49] Further, there is increasing skepticism about the notion of technological innovation as automatically and inherently good in itself. Whether this is based on a reaction to past and anticipated technological disasters, to the tunnel vision of adoption of counterproductive innovation without regard for adverse consequences or to the wide gap between scientific achievement and political wisdom, this skepticism may be expected to grow in the future.[50] Finally, in an age of economic scarcity and indeed at times lowered expectations, the risk taking associated with innovation may become more limited if resources contract.[51]

One response to such problems is to try to turn back the clock, to reverse the rate and impacts of innovation.[52] However noble such a goal might appear to some in attempting to mitigate further problems, this response offers little hope for dealing with the structures of past and current innovations, which tend to have a life of their own. A more practicable version of such an approach is the effort to scale down and decentralize some forms of innovation, in the belief that more human-sized technology will alleviate concern in some cases.[53]

---

[47] See **§11.26.**

[48] Toffler, Future Shock (1970).

[49] This was claimed to be true as early as Veblen, The Instinct of Workmanship (1914). More recent references include Dirlam & Thorkleson, *Implications of the Individualist Theory of Invention,* 6 Antitrust Bull 173 (1961); and Bell, The Coming of Post-Industrial Society, ch 3 (1973).

[50] *See, e.g.,* Thompson, At the Edge of History (1970), and Passages About Earth (1973); Mumford, The Myth of the Machine II: The Pentagon of Power (1970); Lovins & Price, Non-Nuclear Futures (1975).

[51] Meadows, Donella, Meadows, Randers & Behrens, The Limits of Growth (1972); Mesarovic & Pestel, Mankind at the Turning Point (1974); Brandt, *et al,* North-South: A Program for Survival (1980); Barney, *et al,* The Global 2000 Report to the President (1980).

[52] Ellul, The Technological Society (1964); Reich, The Greening of America (1970); Roszak, Where The Wasteland Ends (1972).

The dominant view can be summarized as: "if innovation got us into this, it will have to get us out." A greater, though more careful and creative, reliance on technology is claimed to be the most appropriate way to deal with difficulties caused by technology.[54]

Regardless of which view is espoused, it seems clear that there is a duty to deal with these extracorporate and extralegal concerns (with their concomitant dangers). Ways to begin within industry and law are offered in this chapter through discussion of the following topics:

1. Technological forecasting[55]
2. The relation of innovation to pure research[56]
3. Industry and academy interactions[57]
4. The problems of lead time for innovation[58] and limits to growth arguments[59]

Alternative Business Contexts of Innovation Having Legal Implications

§11.05 Economies of Scale

The business and technical context for innovation can have vital consequences (e.g., under the antitrust laws)[60] and must be taken into account by counsel. One such instance is the presence of scale economies. A process is said to exhibit economies of scale where the average cost of production decreases while the quantity produced increases.[61] More loosely, the term is used to refer to efficiencies associated with large-scale operation, however far from the underlying mass production manufacturing model. These economies may be of four types: financial, due to the larger firm's decreased risk and increased ability to raise capital; administrative, due to the centralization of some staff functions, permitting increased expertise; marketing, due to use of bargaining power or combined promotion; and economies associated with the factors of production themselves, that is, efficiencies directly realized in the production

---

[53] *E.g.*, Schumacher, Small is Beautiful (1973); and Lovins, Soft Energy Paths (1977).

[54] Ramo, Century of Mismatch (1970); O'Neill, 2081 (1981).

[55] See **§11.27.**

[56] See **§11.28.**

[57] See **§11.29.**

[58] See **§11.30.**

[59] See **§11.30.**

[60] See **ch 3.**

[61] Bain, Industrial Organization (1959).

process or in storage and transport.[62]

Economies of scale are often used as a justification of horizontal integration, mergers, expansion to monopoly-sized market shares and other behaviors with antitrust implications and thus their existence assumes legal significance.[63] In this broader context, the relationship between innovation and economies of scale becomes more important, especially if it can be shown that innovative (rather than anticompetitive) activity is aided by the size of the firm.

This relationship flows in two directions. First, economies of scale may be seen to have a strong impact upon innovation. Unless decentralization from within occurs, the very size of an organization may prevent the communication needed for innovation; bureaucratic entanglements may strangle novelty. The larger the organization, the more entrenched institutional conservatism may be; the larger-scale the process, the greater the financial conservation associated with writing off an outmoded investment.[64] Such processes may in fact even tend to be resistant to change of all kinds, ranging from correction of errors to environmental incursions and market shifts, and may make it difficult to argue for the social benefits of promoting innovation.[65]

On the other hand, it can be argued that scale of a process promotes innovation by the high returns given for small improvements. The size of the organization itself may provide a low-risk environment favoring innovation; lack of uniformity within an organization or organizational slack may be an opportunity for random, unintended, but valuable innovation.[66]Finally, size itself may provide the critical mass of personnel and problems which is sometimes necessary for innovation.[67]

Innovation may also have a direct effect on economies of scale. While innovation is more than mere efficiency, many familiar aspects of innovation are characterized by an increase in capacity in return for fixed cost. Such an increase provides the motivation for and the justification of economies of scale simultaneously.[68] Early stages of innovation, however, are typically associated with the expensive design of one-shot prototypes, and some claim any large-scale process is by definition not innovative but merely the result of past innovation.[69]

---

[62] *Id.*

[63] See **ch 3.**

[64] Williamson, *The Economics of Antitrust: Transaction Cost Considerations,* 63 Am Econ Rev 316 (1973).

[65] *Id.*

[66] Williamson, *Economies as an Antitrust Defense Revisited,* 125 U Pa L Rev 699 (1977).

[67] See **§11.26.**

[68] Bain, *supra* note 61. *See also* United States v Jerrold Elecs Corp, 187 F Supp 545 (ED Pa 1960), *affd per curiam,* 365 US 567 (1961); Dehydrating Process Co v AO Smith Corp, 292 Fd2 653 (1st Cir), *cert denied,* 368 US 931 (1961); Flicker, *Newcomer Defenses,* 18 Stan L Rev 457 (1966).

[69] Williamson, *supra* note 64.

§11.06 Economies of Scope

Economies of scope exist where a fixed quantity of two *different* products is produced for a given cost.[70] The traditional example is that of mutton and wool; one of more current interest would be that of petroleum and other heavier oil products.[71] Strictly termed *joint production costs,* economies of scope are usually more loosely used to refer to any by-product situation.[72]

Unlike economies of scale, study of these economies has been largely limited to factors of production;[73] like economies of scale, the direct relevance of innovation has an applications viewpoint, not a legal one. However, economies of scope are often used as a justification of market expansion, product-tying, predatory pricing and other anticompetitive behavior.[74] In such cases, the ties between innovation and economies of scope assume added importance in showing the appropriate intention and impact of a firm's actions. As with economies of scale, influences in both directions are significant.

A search for economies of scope may often provide a healthy incentive for innovation. The need for effective use of all elements of the production inputs will value both efficiency and novelty simultaneously.[75] Resource conservation concerns and goals of self-sufficiency also argue well for the rate of innovation under an economies of scope perspective.[76] A way in which an inordinate emphasis on economies of scope and an accompanying standardization of the process might retard innovation is where the focus on different outputs from fixed inputs blinds the would-be innovator to benefits obtainable from slight, though innovative, variation of those inputs.[77] (It is, of course, also true that a more general shortsightedness may also prevent full exploitation of inputs under any specific circumstance, no matter how keen the search.)

Innovation also serves more often than not as a basis for economies of scope. The very terminology in this area attempts to capitalize on the unanticipated economies which innovation may bring: *synergy,*[78] suggesting interactions and "whole is greater than the sum of its parts" arguments; *serendipity,*[79] emphasiz-

---

[70] Panzar & Willig, *Economies of Scope,* 71 Am Econ Rev 268 (1981).

[71] Hanson, A Dictionary of Economics and Commerce 247 (1965).

[72] Panzar & Willig, *supra* note 70.

[73] *Id.*

[74] United States v Corn Prods Ref Co, 234 F 964, 974-77 (SDNY 1916); Otter Tail Power Co v United States, 410 US 366, 368-70, 372-74 (1973).

[75] Baumol, *On the Proper Cost Tests for Natural Monopoly in a Multiproduct Industry,* 67 Am Econ Rev 809 (Dec 1977).

[76] Meadows, Donella, Meadows, Randers & Behrens, The Limits to Growth (1972); Mesarovic & Pestel, Mankind at the Turning Point (1974); Brandt, *et al,* North-South: A Program for Survival (1980); Barney, *et al,* The Global 2000 Report to the President (1980).

[77] Box & Draper, Evolutionary Operation (1969).

[78] Bode, Synergy (1971).

[79] Burke, Connections (1978).

ing the accidental aspects of discovery; and *spinoffs,*[80] underscoring tangential resulting applications. The unintended miracle has become almost a cliché of technological advance. The scenario where innovation would not entail economies of scope would occur only in that situation where the consequences of the innovation were so novel and recent that the unforeseen applications had not yet come into view.[81]

§11.07 Levels of Specialization

Economies of scale and scope may provide efficiencies which promote innovation and thereby have antitrust impacts; as discussed above, however, they may in some cases lead to actual disefficiencies through overcentralization of control and standardization of outputs. In such cases, increasing specialization, rather than standardization, and decentralization of control may provide an answer, not just to the fostering of innovation but also to legal questions of market definition and determination of corporate policy.

A process or an organization may grow more specialized in two ways: internally, or with respect to its inputs, by becoming more decentralized operationally and by associating every quantity of input with, ideally, exactly one item of output;[82] and externally, or with respect to its outputs, by becoming more decentralized distributionally and by associating every item of output with, ideally, exactly one consumer or user.[83]

This matching of work to product or product to user runs counter to the economies of scale and scope discussed above, both of which attempt to take advantage of some global qualities of production. Increasing levels and degree of specialization give up the elements of homogeneity and control on which the economies of uniformity rest. In return, increasing specialization may provide greater adaptability and fine tuning of resources to needs; however, it must still be assumed that such decentralization and local optimization will counter the efficiencies bought at a high cost of monitoring and standardizing of information by a central authority not making the best use of it.[84]

To the extent that levels of specialization may play a part in market definition, labor substitutability in investment decisions and incremental pricing strategies, innovation's impact on specialization deserves attention.[85]

Specialization may promote innovation in its own right, as in the case of

---

[80] See *Spin-offs and High Technology Industry,* NY Times, Jan 19, 1981, §IV, at 6.

[81] See Van Duijn, *Fluctuations in Innovation Over Time,* 13 Futures 264 (Aug 1981).

[82] Williamson, Markets and Hierarchies: Analysis and Antitrust Implications (1975).

[83] *Id.*

[84] Williamson, *The Economics of Antitrust: Transaction Cost Considerations,* 63 Am Econ Rev 316 (1973).

[85] United States v Corn Prods Ref Co, 234 F 964 (SDNY 1916); Tampa Elec Co v Nashville Coal Co, 365 US 320 (1961).

Henry Ford's assembly line.[86] Such internal specialization, however, will always carry with it a price with respect to external specialization unless options are built into the planning process, i.e., "any color you want so long as it's black."[87] External specialization in its logical extreme of total differentiation for each user may encourage radical innovation, but only at the cost of risking loss of large submarkets altogether.[88] (Innovation may serve to redefine and differentiate either the product or the market itself; in the latter case, the new developments may not fit the old configuration of the firm and may be readily transferable to another market or competitor too independently of the product to allow a firm to retain direction over them.)[89]

Innovation tends to prompt specialization, both internal and external, since the novel aspects of the advance will often require special treatment.[90] Further, since innovation typically feeds on itself, i.e., rapid developments follow rapid developments, innovation will also typically work to separate and thus specialize initially indistinguishable functions and applications made available by unsuspected or previously uneconomical uses.[91] Innovation can in fact become so identified with specialization as to undercut the viability of any parallel developments or technology transfers.[92]

§11.08 Radical Innovation

Innovation which alters the means of production altogether is often termed radical innovation, i.e., a complete replacement of the process and/or inputs.[93] Radical innovation is usually the hoped-for outcome of governmental support as well as the public stereotype of invention.[94]

Radical innovation may also serve as the basis for a finding of "superior skill, industry, and foresight"[95] in defense to a charge of monopoly.[96] There is, of

---

[86] See Boorstin, The Americans: The Democratic Experience 548-52 (1973).

[87] *Id* 551.

[88] Kotler, Marketing Decision Making: A Model Building Approach (1971).

[89] *Id* 536-38.

[90] Williamson, *supra* note 84.

[91] Resnick, Turney & Mason, *Marketeers Turn to "Countersegmentation,"* 79 Harv Bus Rev 100, (Sept-Oct 1979).

[92] For a current example, see *Videodiscs: A Three-Way Race for a Billion Dollar Jackpot,* Bus Week, July 7, 1980, at 72, and Zuckerman, *Who'll Survive the War of the Videodiscs,* 77 Mechanics Illustrated 56 (Jan 1981).

[93] Schumpeter, Business Cycles (1939).

[94] Compare the national reaction to the Manhattan Project in developments of the atomic bomb during World War II with the critical response to the perceived failure of the "War on Cancer" in not producing an immediate panacea. See Sobel, *Seeking Not 7 But 100 Cancer Cures,* NY Times, Dec 30, 1979, §XII, at 9.

[95] See ch 3; *see also,* United States v Aluminum Co of America, 148 F2d 416, 430 (2d Cir 1945), *on remand,* 91 F Supp 333 (SDNY 1950) (Alcoa).

[96] *Alcoa,* 148 F2d at 430.

course, an essential tension between the goals and the application of antitrust and patent law in that radical innovation coupled with strong patent protection may result in a monopoly or near-monopoly position for the innovative firm.[97] The current judicial environment seems to accept that sufficiently innovative behavior may overcome its own anticompetitive impact.[98] An allied argument is that incentives and capacity for innovation sheltered by a patent or regulatory monopoly position may provide compensatory benefits to the industry as a whole and to the nation.[99]

Radical innovation may also play a part in market definition concerns, for it will often create an entirely new market and almost as a matter of course provide control of that market.[100] Thus, sufficiently creative developments may bring about their own downfall. Excessive zeal in protecting innovation through a circle of patents may also have a counterproductive impact through retarding innovation by other parties or promoting complacency on the part of the original innovator.[101]

To a lesser extent, radical innovation may be an important issue of proof in predatory pricing cases where sharp cost changes can be associated with the innovation.[102] The identification of successive generations of a product or of specific stages of a product life cycle through key innovative initiatives may be sufficient for such a purpose.[103] (Conversely, the rapidity of generations or short length of product life cycles may also serve as proof of innovative behavior, to be carefully distinguished from charges of planned obsolescence and arbitrary product differentiation.)[104]

Finally, whether particular industry structures are conducive to radical innovation is a matter of some controversy.[105] It is sometimes argued that large capital-intensive enterprises with many homogeneous customers are particularly resistant to such innovation because of built-in inertia.[106] Such circumstances, however, may themselves provide the relative freedom from risk necessary to promote radical innovation. In any event, main-frame computing equipment and telecommunications, two especially large and capital-intensive

---

[97] *E.g.*, SCM Corp v Xerox Corp, 645 F2d 1195 (2d Cir 1981), *cert denied*, — US — (1982).

[98] *See* Berkey Photo, Inc v Eastman Kodak Co, 603 F2d 263, 281-84 (2d Cir 1979), *cert denied*, 444 US 1093 (1980); California Computer Prods v IBM, 613 F2d 727, 742-44 (9th Cir 1979).

[99] See *Weinberger Feared Breakup of AT&T Would Hurt Units*, Wall St J, Apr 10, 1981, at 2, for a national security defense.

[100] Telex Corp v IBM, 367 F Supp 258 (ND Okla 1973), *revd*, 510 F2d 894 (10th Cir 1975).

[101] *See* SCM Corp v Xerox Corp, 645 F2d 1195 (2d Cir 1981).

[102] California Computer Prods v IBM, 613 F2d 727 (9th Cir 1979).

[103] Van Duijn, *Fluctuations in Innovation Over Time*, 13 Futures 264 (Aug 1981).

[104] See Mueller, *Sources of Monopoly Power: A Phenomenon Called Production Differentiation*, 18 Am U L Rev 1 (1968).

[105] See **§§11.21** and **11.23-.26.**

[106] Bain, Industrial Organization 31 (1959).

industries with enormous markets, are often considered to have excellent records of innovation.[107] (Not coincidentally, these two industries are also characterized by a market-dominant industry leader and frequent antitrust activity.)

§11.09 Expense-Capital Tradeoffs

The operations of a firm (or of the economy as a whole) may be characterized as an input–output table in which every output or product of the firm is associated with expense and capital components of inputs.[108] Innovation may change these fixed components with some overall benefit to production, often in one of the ways described above.[109]

In some cases, however, merely the relation of the expense and the labor components may change. Some believe innovation should not be said to exist under such circumstances; rather, there is a simple tradeoff between expense and capital, converting one into the other.[110]

Nonetheless, such a tradeoff is not a trivial exercise. For example, a conversion of existing capital to expense, either directly through depreciation or indirectly by externalizing particular activities (i.e., purchasing products or services outside the firm), may serve to clear the way for true innovation by disposing of outmoded investment.[111] A similar tactic may also be effective in ensuring work force stability during labor negotiations where innovative processes are used to replace invested plant with new employees for new functions; an expense–capital tradeoff without immediate economic benefit to the firm may in such a case still prove to be quite profitable.[112]

Conversely, a (partial) conversion of existing expense to capital might represent the internalization of certain functions to strengthen control, ensure standardization, smooth variation in supply and generally improve information flows within the firm.[113]

Finally, any expense–capital tradeoff may be rationally made where differential growth rates in either quantity or the rate of innovation itself are foreseen, that is, conversion of a function from a capital item to an expensed one, or vice versa, is really an investment in the future if it minimizes the discounted costs of total expenditures over time. No matter what the motivation, however, any tradeoff will create ambiguities as to what the true cost of any output may be.[114]

---

[107] Bode, Synergy (1971).

[108] Leontief, Input-Output Economics (1966).

[109] *Id.*

[110] Mansfield, Microeconomics 132-37 (1970).

[111] Mensch, Coutinho & Kaasch, *Changing Capital Values and the Propensity to Innovate,* 13 Futures 276 (Aug 1981).

[112] See *Labor and Multifactor Productivity by Industry,* pt I, in New Developments in Productivity Measurement and Analysis (Kendrick & Vaccara eds 1975).

[113] Williamson, *The Economics of Antitrust: Transaction Cost Considerations,* 63 Am Econ Rev 316 (1973).

[114] See Mensch, Coutinho & Kaasch, *supra* note 111.

There is a close tie between innovation and such tradeoffs in the treatment of innovation itself. While research and development costs are often treated as expenses and related to the level of sales, much financial analysis suggests that they may be better thought of as an investment in the firm itself and should be scaled with respect to the firm's earnings for nontax accounting purposes, even if not necessarily for tax purposes, to encourage innovation and to estimate its actual worth.[115] Such an investment might then be depreciated over the life of the innovative product.[116] Desire for an accurate impression of the true value of a firm as well as recent tax legislation support such a view.[117] This may also be applicable to public sector research and development expenditures, which should in some cases be treated as investments rather than current expenditures.[118]

§11.10 Innovation as Information

One traditional view of innovation is that it occurs where some feature of an object or step in a process is changed. This change may then be protected by a patent, which describes the change, or by trade secret status, while the secret is *about* the innovation.[119]

This traditional view becomes less and less applicable to the cutting edge of innovation as time goes on.[120] Increasingly often, an advance is not represented by a mere better mousetrap, but by information and abstract ideas. Design, configuration, organization and structure are all words that are often better associated with innovation today than the images conjured up by the cotton gin, the telegraph or even the laser, and counsel must learn to deal with these abstractions.[121]

Innovation as a novel structuring of ideas is also a far broader notion that that receiving traditional protection. Some analogies with copyright law are possible[122] and some recent developments continue to expand the concept of

---

[115] Fruhan, Financial Strategy 35-39, 58-60 (1979); *see also* **ch 5** for discussion of the tax aspects.

[116] Fruhan, *supra* note 115.

[117] See **ch 5** and **§11.16.**

[118] See **ch 12.**

[119] See **ch 4.**

[120] See **§11.19.**

[121] Reznick, *Synercom Technology, Inc v University Computing Co: Copyright Protection for Computer Formats and the Idea/Expression Dichotomy,* 8 Rut J Computer Tech & Law 65 (1980).

[122] International News Serv v Associated Press, 248 US 215 (1918); Note, *Copyright for Uncopyrightables,* 108 U Pa L Rev 699 (1960); Bodenhausen, *Protection of "Neighboring Rights,"* 19 Law & Contemp Prob 156 (1954).

patentability.[123] The treatment of computer software is a particularly pressing problem.[124] Some businesses may seek to confront such problems through license or franchise agreements, effectively rendering those who deal directly with the novel information part of the firm, with their own stake in keeping the information secure.[125]

Another issue may be found in approaches to interpretation and enforcement of postemployment restraints governing the behavior of key personnel.[126] Highly innovative industries tend to be highly mobile ones, and fairly stringent geographical and temporal restraints have been upheld if tied closely enough to specific subject matter.[127] One countervailing aspect of this trend is that as innovation in the form of reconfigured information becomes more and more complex, transferability of such information becomes more and more difficult.[128]

There are other assaults on innovation as information over and above issues of availability of traditional protections. Governmental regulation of the last two decades has made unprecedented amounts of information available to the public (including competitors) on the internal workings of corporations and on the government's own deliberations as well.[129] Information concerning experimental results and developmental dead ends is a particularly significant instances.[130]

Finally, it is possible to argue that the development of innovation as information is relatively risk-free, since the value of the investment that produces it,

---

[123] Diamond v Chakrabarty, 447 US 303 (1980); *In re* Sarkar, 588 F2d 1330, 200 USPQ 132 (BNA) (CCPA 1978); *In re* Bergy, Coats & Malik, 596 F2d 952, 201 USPQ (BNA) 352 (CCPA 1979); *In re* Phillips, Eiben, & Morgan, 608 F2d 879, 203 USPQ (BNA) 971 (CCPA 1979); *In re* Sherwood, 613 F2d 809, 204 USPQ (BNA) 537 (CCPA 1980).

[124] Gottschalk v Benson, 409 US 63 (1972); Dann v. Johnston, 425 US 219 (1976); Soltysinski, *Computer Programs and Patent Law: A Comparative Study,* 3 Rut J Computer Tech & Law 1 (1973); Gemignani, *Legal Protection for Computer Software: The View from '79,* 7 Rut J Computer Tech & Law 269 (1980).

[125] Thompson, Franchise Operations and Antitrust (1971); Painton & Co v. Bourns, Inc, 442 F2d 216 (2d Cir 1971).

[126] Kniffin, *Employee Non-Competition Covenants: The Perils of Performing Unique Services,* 10 Rut-Cam LJ 25 (Fall 1978). Two excellent review articles which also treat antitrust applications are Kolowrat, Stack & Lynch, *Restraints on Technology Access: Protection of Trade Secrets and Confidential Information,* 49 Antitrust LJ 735 (1981); and Blake, *Employee Agreements Not to Compete,* 73 Harv L Rev 625 (1960).

[127] Reed, Roberts Assocs v Strauman, 40 NY2d 303 353 NE2d 590 (1976); Continental Group, Inc v Kinsley, 422 F Supp 838 (D Conn 1976).

[128] Kniffin, *supra* note 126.

[129] Steinberg, The Information Establishment (1980). *See, e.g.,* Amchem Prods Inc v Costle, 481 F Supp 195 (SDNY 1979); and Chevron Chem Co v Costle, 499 F Supp 732 (D Del 1980).

[130] See **§11.14.**

experience, may not be greatly diminished by failure to innovate.[131] Rather, it may be necessary to follow some paths to their conclusion, even if that conclusion is unsuccessful.

Corporate Planning and Innovation

§11.11 Financing Options and Pay-Out Parameters

The effectiveness of most traditional debt and equity raising methods may be enhanced for a firm with a record or image of innovation.[132] Innovation may also be a factor in creating or favoring specific financing options; estimation of the return on such options can be especially complex.

With respect to external financing options, governmental support of innovation at both the federal and local levels provides a number of opportunities. Direct funding,[133] copartnership,[134] low-cost loans,[135] tax benefits,[136] access to other sponsored research and development results,[137] loan guarantees[138] and favorably interpreted enforcement guidelines[139] are among the special opportunities afforded innovative activities on the part of the firm. These opportunities either provide debt and equity capital in their own right (funding, copartnership and loans), reduce the need for such capital (tax benefits and research and development access), or ease its acquisition elsewhere (loan guarantees and selective enforcement).[140]

Joint ventures, while a wholly general mechanism for raising capital and distributing risk, find many of their applications in areas of innovation.[141] Research and development joint ventures (though not joint production and

---

[131] Morton, Organizing for Innovation (1971).

[132] Brooks, The Go-Go Years (1973).

[133] See Terleckyj, Effects of R&D on the Productivity Growth of Industries (National Planning Assn 1974); Subcommittee on Priorities and Economy in Government, Joint Economic Committee, 94th Cong, 2d Sess, Priorities and Efficiency in Federal Research and Development (1976).

[134] *E.g.*, the Communications Satellite Act of 1962, Pub L No 87-624, 76 Stat 419 (codified at 47 USC §700).

[135] *E.g.*, the Reconstruction Finance Corporation Act 47 Stat 5 (1932) (codified at 15 USC §601); see also **ch 12.**

[136] See **§11.16.**

[137] Public Technology (Mercer & Philips eds 1981).

[138] Schnitzer, Contemporary Government and Business Relations 338-47 (1978).

[139] Antitrust Division, US Dept of Justice, Antitrust Guide Concerning Joint Ventures (1980); *reprinted in* 1980 Antitrust & Trade Reg Rep (BNA) No 992 (special supp Dec 4, 1980).

[140] Additional discussion of these and related topics may be found in **chs 3, 9.**

[141] Young & Bradford, Joint Ventures: Planning and Action (1979).

marketing) are especially favored by antitrust law, both in the United States[142] and abroad.[143] In a related option, venture capital funding is almost by definition associated with novel enterprises.[144] Finally, licensing and royalty agreements with respect to past or pending innovation may be available as an external financing option in this context.[145]

Internal financing options are typically a combination of two forms. In the first, there is a direct link to sales on a product line-by-product line basis where every area is expected to justify its research on an ongoing basis; this provides an immediate incentive for innovation and rewards past performance.[146] In the second, research and development costs are treated as a investment made by the corporation in itself out of retained earnings which are then "plowed back" without regard for the identity of the contributing area but on the basis of overall corporate goals.[147] In the first type, inefficiencies may result from duplication of effort; in the second, from absence of direction of effort. Both suffer from the fact that poor sales or low earnings may indicate the need for more and not less innovation.[148] Still, the myopic nature of most business school training usually results in such expenditures being cut first, not last.[149] A more reasonable approach in theory might be that research and development costs represent a fixed proportion of *investment*.[150]

No commonly accepted procedures exist for estimation of rewards from financing of innovative options, but innovation does throw four significant obstacles in the path of more general methods. First, innovation adds a strongly probabilistic all-or-nothing element into the already great uncertainty surrounding such calculations, that is, across a range of projects there will be swings between many total losses and a few overwhelming successes.[151] Next, innovation costs are not simply additive, at least within the same application, that is, if some amount of funding is optimal for some unit of innovation, 10 times that amount will not necessarily produce 10 units or produce one unit 10 times faster, as might be the case in a completely standardized and centralized process.[152] Third, the innovation itself will often change the relevant time

---

142 Pitofsky, *Joint Ventures Under the Antitrust Laws,* 82 Harv L Rev 1007 (1969); **§3.10.**

143 Hawk, United States, Common Market, and International Antitrust (1979).

144 Young & Bradford, *supra* note 141.

145 *See, e.g.,* Bischel, *Intercompany Transfers of Technology,* 5 Intl Tax J 181 (1979).

146 *E.g.,* The Economics of Drug Innovation (Cooper ed 1969).

147 Fruhan, Financial Strategy 35-39, 58-60 (1979).

148 Francis, Principles of R&D Management (1977).

149 Hayes & Abernathy, *Managing Our Way to Economic Decline,* 58 Harv Bus Rev 67 (Sept-Oct 1980).

150 Fruhan, *supra* note 147.

151 Makridakis, *If We Cannot Forecast, How Can We Plan?,* 14 Long Range Planning 10 (June 1981).

152 New Developments in Productivity Measurement and Analysis (Kendrick & Vacarra eds 1975).

frame over which return should be measured.[153] Finally, the uncertainty innovation brings to intended results is small with respect to uncertainty of unintended results, both positive and negative.[154]

§11.12 Mergers and Acquisitions

Many mergers and acquisitions are claimed to be based on some economies of scope and/or scale, that is, the resulting combination is said to be more efficient and effective than its components.[155] In other cases, the rationale advanced is more specific and plausible: expansion in an existing market, entry into a new one, assurance of supply or demand, diversification of risk or even just convenient fianancial attributes.[156] Very rarely is even a naked powerplay publicly announced as such; conversely, many mergers and acquisitions attempt to associate themselves with the proposition that the combination will promote innovation.

In what ways could the union of two firms be more innovative than their simple sum? As discussed earlier, it is quite problematic when a mere increase in total resources above a threshold level of sufficiency will have a positive effect on rate of innovation, especially if the total level of research and development expenditures remains unchanged.[157] It is possible that the acquisition of specific nonfungible assets—e.g., key personnel, innovative reputation, geographic location—might improve the innovative capability of the merged firm,[158] with two reservations: this might at times be done only at the cost of diluting the applied value of these assets, so that their total worth remains constant, and the factors which make these assets nonfungible may impede their ready transferability.[159]

On the affirmative side, the needs and resources of the two firms might be complementary, so that weaknesses in one firm are shored up by strengths in another. This will be most likely to occur where whole states of production are absent, however, and the resulting forward or backward integration again will have little direct bearing on innovation[160] (except, perhaps, where the ability to implement anticipated innovation is advanced). Synergy, the notion that the whole is greater than the sum of its parts, as applied at the corporate level to a merger or acquisition, may have an effect on the innovation process only

---

[153] Brock, The US Computer Industry: A Study of Market Power 100 (1975).

[154] Makridakis, *supra* note 151.

[155] See **§§11.05-11.06.**

[156] Steiner, Mergers (1975).

[157] See **§11.05.**

[158] Williamson, *Economies as an Antitrust Defense Revisited,* 125 U Pa L Rev 699 (1977).

[159] Steiner, *supra* note 156.

[160] Williamson, *The Vertical Integration of Production: Market Failure Considerations,* 61 Am Econ Rev 112 (1971).

where research programs are centered in related but distinct fields.[161] Finally, it can be argued that the prospect of being acquired has a negative impact on innovation due to the fear of the entrepreneur of losing control of what has been created, a fear not wholly allayable by patent protection.[162]

These points suggest that the defense of a prospective merger or acquisition (to be distinguished from a research and development joint venture) on the basis of promotion of innovation *alone* may be very difficult. (Further, except for presumed freedom from antitrust sanctions, the same caveats apply to any *intra*-corporate mergers or acquisitions for such a purpose.[163]

§11.13 Advertising and Marketing

Although advertising is sometimes castigated as a pure societal waste, the sole purpose of which is to *create* demand for products, it has judicially recognized informational value.[164] In some industries advertising costs are commensurate with production costs, and heavy reliance upon advertising is claimed to be a weapon for oligopolistic competition.[165] Paradoxically, however, the product most in need of effective advertising, at least at the primitive level of announcing its presence, would be a truly novel one.

How can advertising and other marketing options be used to advance such a product? The immediate aim of some industrial changes is mere product differentiation, entirely apart from the grounds for differentiation.[166] If innovation becomes more and more complex, the case for explaining it also becomes so, with even greater difficulties associated with showing the superiority of this complexity;[167] innovation may, however, simplify the product or its use as well.[168] Stringent limits on claims which can be made in advertising are an additional constraint.[169]

An approach to dealing with these problems is to change the emphasis on a new product to an emphasis on a new market or a new use.[170] A change in distribution can achieve the latter more easily than an advertising campaign the former;[171] indeed, a legitimately innovative product defines a new market of

---

[161] *Id.*

[162] See **§11.20.**

[163] Williamson, *supra* notes 158 and 160.

[164] Thurow, The Zero-Sum Economy 174-49 (1980).

[165] *Id.*

[166] Mueller, *Sources of Monopoly Power: A Phenomenon Called Product Differentiation,* 18 Am UL Rev 1 (1968).

[167] Kotler, Marketing Decision Makings: A Model Building Approach (1971).

[168] *Id.*

[169] See Posner, Regulation of Advertising by the FTC (1974).

[170] Kotler, *supra* note 167; Resnick, Turney & Mason, *Marketeers Turn to "Countersegmentation,"* 79 Harv Bus Rev 100 (Sept-Oct 1979).

[171] Kotler, *supra* note 167; Resnick, *et al, supra* note 170.

its own.[172]

Innovation presents special problems for cost passthroughs in pricing structures. Placing development costs on the front end of an innovative product may delay or prevent any significant market penetration;[173] failure to do so may give rise to charges of predatory pricing,[174] as may selectively associating these costs with outmoded products (effectively forcing them to fund their successors and/or justify their existence) as part of a migration strategy.[175]

Legitimate innovative intentions may also be waylaid by accepted advertising and marketing practices in the promotion of innovation for its own sake without regard for social utility of the new need which is created,[176] capitalization upon planned obsolescence[177] and excessive trading on a past history of innovation.[178] These practices may undercut the viability of any argument that innovation provides benefits which outweigh its possible anticompetitive effects.

§11.14 Testing and Experimentation Liability

Testing and experimentation are a necesary part of all product development, whether or not the product might be considered innovative. There is a duty on the part of a manufacturer to test even routine results,[179] and in many specific applications the protocols for such testing are set out in great detail by government regulation.[180] The consequences of liability incurred during the process of or on the basis of such testing or experimentation are discussed elsewhere.[181]

The whole notion of liability for initial stages of development raises some general issues concerning the promotion of innovation. It is not suggested that testing and experimentation in their own right impede advance; rather, they are integral information-providing steps in the innovative process.[182] Further, despite efforts to demonstrate the cost of governmental regulation in this field,

---

[172] Berkey Photo, Inc v Eastman Kodak, 603 F2d 263 (2d Cir 1979), *cert denied,* 444 US 1093 (1980); California Computer Prods v IBM, 613 F2d 727 (9th Cir 1979).

[173] The Economics of Drug Innovation (Cooper ed 1969).

[174] *In re* IBM Peripheral EDP Devices Antitrust Litigation, 481 F Supp 965 (ND Cal 1979).

[175] *AT&T's Price Tactics Prod Firms to Replace Old Phone Equipment,* Wall St J, Apr 24, 1981, at 1.

[176] Galbraith, Economics and the Public Purpose, ch XIV (1973).

[177] Francis, Principles of R&D Management, ch 4 (1977).

[178] California Computer Prods v IBM, 613 F2d 727, 742 (9th Cir 1979).

[179] Nicklaus v Hughes Tool Co, 417 F2d 983 (8th Cir 1969).

[180] Mitucek, The Political Economy of Regulation (1980).

[181] See **ch 10.**

[182] Francis, *supra* note 177.

no convincing studies exist to show a decrease in level or rate of discovery.[183] The fear of liability despite compliance with known regulations is another matter, however. In some areas, absolute liability is claimed to act as a practicable bar to related research topics.[184] There have even been suggestions advanced that such research may itself constitute a prohibited act.[185] Research in certain areas must be carried on under direct or indirect government supervision.[186] These interventions clearly delay innovation to some unknown degree (though perhaps with good purpose).

A truly innovative product also raises legal and ethical problems with respect to informed consent by a user in the face of large uncertainty.[187] At the same time, there may be a duty on the part of a manufacturer to provide the latest product, despite very unclear guidelines for its release.[188]

Advertising of products in the experimental stage is often frowned upon,[189] yet an obligation may exist to provide even negative information to appropriate parties.[190] However, legitimate concern over liability for innovative products may give a protective antitrust defense in some resale restraint or product tie-in cases.[191]

Finally, even where the innovative product itself may not be dangerous or cause special liability, the fact of *being* innovative may have indirect and hidden costs of liability for a firm, as in the cases of dislocation and safety factors affecting the work force and of disruptions to the manufacturing process.[192]

§11.15 Internal Decision Structures

Innovation has become an increasingly complex process over the past two decades and is sometimes achievable only through very large and complex enterprises.[193] The need for this magnitude of resources is not so remarkable

---

[183] *E.g.*, Magat, *The Effect of Environmental Regulation on Innovation,* 43 Law & Contemp Prob 4 (1979); Braeutigan, *The Effect of Uncertainty in Regulatory Delay on the Rate of Innovation,* 43 Law & Contemp Prob 98 (1979).

[184] *New Hope for Going to Market With DNA,* Bus Week, Sept 24, 1979, at 64.

[185] *US Eases Safety Standards for Recombinant DNA Work,* NY Times, Jan 30, 1980, at 16.

[186] For such restriction on encryption research, see *National Science Foundation Ties With National Security Agency,* NY Times, Aug 27, 1980, at 1, and *Code Makers and Breakers, id,* Aug 17, 1981, §A, at 14.

[187] Fiorentino v Wenger, 19 NY2d 407, 227 NE2d 296 (1967).

[188] Pontifex v Sears, Roebuck & Co, 226 F2d 909 (4th Cir 1955).

[189] Gibson v California Spray Chem Corp, 29 Wash 2d 611, 188 P2d 316 (1948).

[190] See Schnitzer, Contemporary Government and Business Relations 328-47 (1978).

[191] Tripoli Co v Wella Corp, 425 F2d 932 (3rd Cir 1970), *cert denied,* 400 US 831 (1970); *see generally* **ch 3.**

[192] See 1 The Quality of Working Life (Davis & Cherns eds 1975).

[193] Johnson, Kast & Rosensweig, The Theory and Management of Systems Pt 1 1973).

as the fact that they can be utilized effectively to promote creativity at all.[194] In simple processes, many activities can be carried on in parallel using standardized decision-making procedures so that additional resources can provide an immediate gain in efficiency. In moderately complicated situations, some activities must be carried on in series with very careful organization required for their ordering, so that additional resources may or may not be able to increase efficiency along the critical path. In very complex systems, however, there is a need for each activity to communicate with every other one so that added resources may either slow down the process by their need for coordination or else be wasted.[195]

The solution to this seemingly fundamental limit to productivity and innovation has come in the form of systems analysis, which has been responsible among many other applications for the success of space exploration and the growth of the data processing industry over the past generation.[196] It is, in fact, appropriate to think of systems analysis itself as *innovation* and some other achievements as mere technological embodiments of its consequences.[197]

The systems approach is the model for internal decision making in almost all major research centers.[198] It is not an infallible guide to organizing and managing research, however, nor is it a completely effective one. In its emphasis on reliability and overall control, it depends critically on a high level of redundancy in design.[199] It cannot deal completely effectively with the fact that advances will be associated with effects that are not merely unknown but unknowable in nature, not to mention unintended.[200] Excessive overall control may itself dampen initiatives that can lead to successful discoveries.

The innovation process almost by definition cannot be entirely organized. In any institution, bureaucratic influences will work to combat or delay creativity.[201] The desire to control and predict the process will limit the potential utility of its outputs; the need to protect confidentiality (and not incidentally forestall liability) will war with the need for free and complete information flows; and a dependence upon smooth operation of the business will necessarily run counter to the hopes for radical breakthroughs.[202]

The most effective means of managing innovation has been described[203] as systematized "happy anarchy." Imposition of stricter structures will tend to produce clearer structures without clear gains.

---

[194] Brooks, The Mythical Man-Month (1975).

[195] *Id* chs 1-2.

[196] Brock, The US Computer Industry: A Study of Market Power (1975); Webb, *NASA as an Adaptive Organization,* in Technological Change and Management 109-44 (Ewing ed 1970).

[197] Churchman, The Design of Inquiring Systems (1971).

[198] Francis, Principles of R&D Managment ch 1 (1977).

[199] Bode, Synergy ch 2 (1971).

[200] Johnson, Kast & Rosensweig, *supra* note 193, ch 1.

[201] Morton, Organizing for Innovation (1971), especially chs 5-7.

[202] *Id. See also* Francis, *supra* note 198, ch 3.

[203] Morton, *supra* note 201, ch 7.

§11.16 Tax Planning and Innovation

Tax planning in the context of innovation has some very specific applications.[204] Among these are:

1. Timing issues in the writing off of losses associated with failed innovation[205]
2. Income spreading devices for successful innovation[206]
3. Treatment of royalties and licensing arrangements[207]
4. Analogies with intellectual property generally[208]

These important problems, however, pale before the challenges and opportunities presented by the changes in the tax law embodied in the Economic Recovery Tax Act of 1981.[209] The act has very far-reaching implications for promotion of innovation, both directly and indirectly.[210]

The major direct consequence of the act is the establishment of an analogue to the investment tax credit for research and development expenses.[211] The provisions of the act allow a tax credit of 25 per cent of the increase over the base period[212] in any year of wages and other expenses[213] associated with qualifying research and development. The massive spur to innovation that this credit may produce can only be guessed at. It is also a major step forward in the recognition of innovative research as a capital investment.

The indirect consequences of the act may be even more significant. Other provisions substantially lower the effective corporate tax rate[214] and improve the favorable treatment of capital gains.[215] These cuts, taking place at the same

---

[204] See **ch 5.**

[205] *See* IRCode §§248, 269, 381 and 382. *See also* Weithorn & Suwalsky, Research and Development Expenses (BNA Tax Management Portfolio 42-2d 1975).

[206] *See* IRC §§174, 312, 316 and 532.

[207] *See* IRC §§1235, 1249, 1253 and 7001.

[208] See Bischel, *Intercompany Transfers of Technology,* 5 Intl Tax J 181 (1979).

[209] Pub L No 97-34, 95 Stat 176 (codified at 26 USC §1 *et seq).*

[210] The impact of the act is discussed also in **ch 5.**

[211] Pub L No 97-34, §221, 95 Stat at 241 (codified at 26 USC §44F). Other relevant provisions include a stepped-up basis for charitable contributions of research equipment, *id* §222, 95 Stat at 196 (codified at 26 USC §170), and a two-year suspension of IRC §861 allocation rules for research and development, *Id* §223, 95 Stat at 249 (codified at 26 USC §174).

[212] The base period for 1981 is 1980; for 1982, the average of 1980 and 1981 expenditures; for 1983, 1984, and 1985, the average of the preceding three years. The provision expires January 1, 1986. Pub L No 97-34, §221(c), 95 Stat at 241 (codified at 26 USC §44F).

[213] All rental and materials expenses, and 65% of all contracted expenses. *Id* §221(b), 95 Stat at 241 (codified at 26 USC §44F).

[214] *See generally id* §§201-267, 95 Stat at 204-67 (codified at 26 USC *passim).*

[215] The maximum long-term capital gains rate is effectively reduced to 20%. *Id* §102, 95 Stat at 186 (codified at 26 USC §1201).

time as the research tax credit, may work for a substantial increase in retained earnings for most affected corporations. How best to deal with these earnings? Reinvestment in further research and development, for a further credit, and so on, may be attractive.

The same factors may simultaneously cause a large drop in corporate requirements for external financing just as such financing may become more available along with the capital gains rate cut; a likely result is that more and more research projects will be able to receive funding, projects which because of initial cost or speculative return would not have been pursued earlier.[216]

Measuring Innovation

§11.17 Time Frame

Many firms tend to regard any activity specifically associated with research and development as innovation and to downplay the innovative aspects of actual implementation and of changes in operations not initiated by corporate research and development. Of course, claiming that one organization has responsibility for innovation (which cannot occur elsewhere) will not make it so. It is necessary to be able to define and measure innovation in a much more specific way in situations ranging from determination of a tax credit to demonstration of the procompetitive effects of innovation.

Many problems exist, however, in construction of evidence which will support the existence of innovation. There is a peculiar difficulty in the selection of the proper time frame in which to estimate whether (or claim) innovation has taken place. When the time frame chosen is very small, i.e., a snapshot of the industry at a given moment, it is difficult to appreciate any change at all. If the time frame chosen is very large, it is likely that the specific instance of innovation hunted for may be supplanted, perhaps repeatedly, by still later innovations, if in fact conditions persist to make any comparison valid at all.[217]

An intermediate range, one tied to the life of the innovation itself and its implementation, seems a commonsensical response,[218] but there are several pitfalls with such a choice as well. An advance will by definition have novel effects and possible unintended spinoffs which may be difficult to observe.[219] The discovery also will itself change the relevant time frame for inspection, and past experience may be a poor guide in its selection.[220]

In justifying a course of action, three different types of time frame frequently come into play. In looking at the degree of competition in an industry as a

---

[216] In certain troubled areas, however, such as the automotive and steel industries, these funds are much more likely to be used for the replacement of outmoded plants.

[217] Toffler, Future Shock (1970).

[218] Jones & Twiss, Forecasting Technology for Planning Decisions 94-95 (1978).

[219] *Id* 19-26.

[220] See Brock, The US Computer Industry: A Study of Market Power (1975).

whole, the courts often examine the fairly long time frame in which the germinal needs for innovation arise, that is, the length of time required for research to be initiated and reach some tentative conclusion.[221] A shorter time frame, associated with the first implementation of the innovation, may be examined in studying a single firm's aggressiveness in pursuing or meeting competition.[222] A typically still shorter interval, the period in which the innovation does or does not receive public acceptance, is sometimes used in measuring market power.[223]

It is not clear whether appropriate time frames for viewing current innovation are growing longer (though it does seem that implementation lead times, at least, are increasing on average)[224]. Every generation tends to claim an unprecedented rate of growth of innovation,[225] perhaps with reason, but the coinciding phenomena of the information explosion[226] and prospective space industrialization[227] seem to add special credence to present optimistic estimates. What does seem unfortunately certain is that the *need* to forecast further and further ahead seems to become more important as the ability to do so decreases.[228]

§11.18 Choice of Surrogates for Innovation

Innovation is an abstract term incapable of direct experience or measurement. Evidence to support claims of novelty may of course be available through the use of expert witnesses[229] but there are readily accessible surrogates for innovation with which such testimony may be guided and supplemented.

Cost is the most frequently used surrogate for the efficiency and value aspects of innovation.[230] On closer examination, however, any advance is too easily confounded with other cost-eliminating devices[231] and claims of extraor-

---

[221] United States v Aluminum Co of America, 148 F2d 416 (2nd Cir 1945), *on remand,* 91 F Supp 333 (SDNY 1950) (Alcoa).

[222] United States v United Shoe Mach Corp, 110 F Supp 295 (D Mass 1953), *affd,* 347 US 521 (1954).

[223] California Computer Prod v IBM, 613 F2d 727 (9th Cir 1979).

[224] See Galbraith, Economics and the Public Purpose (1973).

[225] Bell, The Coming of Post-Industrial Society 318-22 (1973).

[226] Evans, The Micro Millenium (1980); Barron, The Future With Microelectronics (1979).

[227] O'Neill, 2081 (1981).

[228] See **§11.29.**

[229] Schuck, *Techniques for Proof of Complicated Scientific and Economic Facts,* 40 FRD 33 (1967).

[230] Mansfield, Industrial Research and Technological Innovation (1968).

[231] Williamson, *Economies as an Antitrust Defense Revisited,* 125 U Pa L Rev 699 1977; Williamson, *The Vertical Integration of Production: Market Failure Considerations,* 61 Am Econ Rev 112 (1971).

dinary savings due to innovation, if not believed, may form the basis for charges of unwarranted price-cutting.[232] Since innovation typically changes or enhances the function of a product, finding a common basis for cost comparisons may be difficult as well.[233] On the other hand, finding a perfect match for function and value of the product would be defining the problem away.[234]

Labor equivalents are rarely used as surrogates for innovation, due to their difficulty of estimation and the loaded connotations of automation.[235] Simple counting of patents and innovative implementations may be appropriate for tracking the record of a single company over time, but such a method is largely useless for comparison across companies, let alone across industries, due to varying investment resources and opportunities.[236]

The soundest approach, although somewhat complex, is the use of cost-performance curves, where an attempt is made to match the cost of a particular product with its capacity for performance.[237] In such a comparison, absolute costs are unimportant; what is significant is the tradeoff with performance. Where the relationship is highly leveraged—a small increase in cost for a large gain in performance *or* a large decrease in cost for a small loss in performance—true innovation is strongly suggested.[238]

The concept of measuring innovation by any form of surrogates may be attacked on several grounds. The act of discovery may be by nature as unmeasurable as it is unpredictable.[239] Creativity introduces strong discontinuities into the production process; attempts to model its progress by quasi-biological metaphors of organic growth may be particularly misleading.[240] All such surrogates are conditioned on some notion of substitutability while innovative products are precisely nonsubstitutable.[241] The only available response to these counters is the inevitable continuity, and associated inefficiency, of implementation of the advance.

§11.19 Technology versus Organization

The national economies of all industrialized countries have moved resolutely from a primary sector (mining, agriculture) to a tertiary sector (services)

---

[232] California Computer Products v IBM, 613 F2d 727 (9th Cir 1979).

[233] *Id.*

[234] Linstone & Sahal, Technological Substitution: Forecasting Techniques and Applications (1976).

[235] Automation and Technological Change (Dunlop ed 1962).

[236] Tilton, International Diffusion of Technology 56-63 (1971).

[237] Ayres, *Envelope Curve Forecasting,* in Technological Forecasting for Industry and Government (Bright ed 1968).

[238] *Id.*

[239] Makridakis, *If We Cannot Forecast, How Can We Plan?* 14 Long Range Planning 10 (June 1981).

[240] Jones & Twiss, Forecasting Technology for R&D Decisions 212-14 (1978).

[241] Linstone & Sahal, *supra* note 234.

orientation, and such progression appears likely for developing countries as well.[242] There is even speculation about the genesis of a dominating quarternary meta-service sector, particularly in dense urban centers.[243] As the economy has evolved in this manner, so has relevant innovation, from an emphasis on manufacturing to one on communications.[244] The structures governing the strength and form of protection given such innovation may be outmoded, but it is not clear that any easy modifications are available.

Innovation has become more and more concerned with organizing and distributing information rather than with the creation of mechanical devices.[245] In a period dubbed[246] the "Information Age" or the "Third Wave," this shift is well exemplified by the dependence of almost all economic activity on the electronic data processing industry, a dependence based not, as was naively first expected, on computer hardware but on the software representing the organization and manipulation of information within such machines.[247] (Indeed, more powerful computers are not those with greater blunt-instrument calculating resources but those allowing greater sophistication of instruction.)[248]

How can innovation be measured given such a trend away from any physcial representations? Two indirect approaches are avilable. The first lies in the realization that the capacity for such innovation rests in human minds and experience rather than in concrete investment.[249] The level of technical training of employees,[250] mobility of such employees[251] and the expense–capital ratio[252] in research may therefore be measures of at least the potential for innovation, no matter how confounded with other factors they may be.[253]

Another approach is associated with the notion that technology-centered innovation is achievement-oriented,[254] i.e., how many widgets an hour are produced. Organization-centered innovation is more process-oriented,[255] i.e.,

---

[242] Bell, The Coming of Post-Industrial Society (1973).

[243] *Id* ch 2; Smith, *Info City,* New York Magazine 21 (Feb 9, 1981).

[244] Evans, The Micro Millenium (1980); Barron, The Future With Microelectronics (1979).

[245] Toffler, The Third Wave (1980).

[246] *Id.*

[247] Brooks, The Mythical Man-Month (1975).

[248] *Id.*

[249] Thurow, The Zero-Sum Economy (1980).

[250] Machlup, The Production and Distribution of Knowledge in the United States (1962).

[251] Kidder, The Soul of a New Machine (1981).

[252] New Developments in Productivity Measurement and Analysis (Kendrick & Vaccara, eds 1980).

[253] *Id.*

[254] Florman, The Existential Pleasures of Engineering (1976).

[255] London, The People Side of Systems (1976).

in how many ways a process can be adapted or fine-tuned. The more complex and variable the process, the greater the level of innovation represented.

§11.20 Critical Mass Arguments

One of the problems in trying to describe the level of innovation in a given firm or industry is the all-or-nothing quality of the outputs of the innovation process; either something new was achieved or it was not.[256] If examination drops back a step and tries to simply study the level of effort going into research and development, another binary argument may manifest itself relating to the size of such resources.

The critical mass argument[257] claims that a certain threshold level of resources (personnel or investment) is required for any true innovation to take place, and that support below such a level is ineffective at best and may represent a complacent, noncompetitive posture at worst.

In practice, the application of this argument usually takes one of three forms. The first approach uses a radiation analogy[258] and is the only one conforming to the scientific use of the term.[259] Just as a piece of radioactive material will not explode until it reaches a critical mass and achieves fission, too small a group cannot master and organize the quantity of research required for an innovative explosion.[260] While successful lone wolf activity is admittedly rare, this form of the argument probably does not recognize the fact that most bursts of inspiration are associated with an individual or a small group.[261]

A second, weaker form of the argument recognizes the existence of the isolated spark but, relying on a cross-fertilization analogy, contends that the probability of such a spark is increased by the presence of like-minded individuals with complementary interests in constant communication.[262] While reasonable for supporting the notion of a lower bound for research resources, this approach also implies the existence of a relatively low upper bound as well, beyond which the costs of intra-group communication exceed its benefits.[263]

The weakest form of the argument is merely a blunderbuss approach relying on the stereotype of the absent-minded professor and the presumption that a large amount of completely undirected research effort provides the greatest

---

[256] Francis, Principles of R&D Management (1977).

[257] See Scherer, Industrial Market Structure and Economic Performance ch 4 (1970).

[258] Morton, Organizing for Innovation (1971).

[259] Other approaches are typically disguised economies of scale; see Mansfield, Monopoly Power and Economic Performance (1964).

[260] McPhee, The Curve of Binding Energy (1973).

[261] Jewkes, Sawers, & Stillerman, The Sources of Innovation 82-89 (1958).

[262] Bode, Synergy (1971).

[263] Williamson, *The Economics of Antitrust: Transaction Cost Considerations,* 63 Am Econ Rev 316 (1973); Brooks, The Mythical Man-Month (1975).

opportunity for the inherently unlikely breakthrough.[264] In actuality, many applications of such an approach seem to be rewards for past innovation rather than guarantees of future success.[265]

§11.21 Rate of Innovation

All of the problems discussed above associated with any attempt to measure the existence of innovation are compounded when differential rates across firms or industries are the area of interest. The biggest conceptual obstacle to examining change in rates of innovation is that the surrogates typically used depend themselves on an assumption of a common backdrop or standard against which changes can be measured and thereby identified as innovation.[266] Estimation of changes in the changes then will require a standard model of innovation, a model that is sadly lacking.[267] Changes in the rate of innovation might be indications themselves of especially significant advances.[268]

The difficulty can be ameliorated by tracking those factors which might be reasonably associated with changes in the rate of innovation, even if a clear tie cannot be demonstrated. It is plausible to assume that there are macroeconomic effects relating to overall availability of capital and expansionary markets which provide an opportunity-push promotion of innovation.[269] While these effects typically have no differential industry or firm impacts in their own right, governmental intervention and special legislation and spending policies may focus them sharply.[270] The defense and aerospace industries are cases in point.[271]

The case for different innovation rates for different industries has not been convincingly proven.[272] The direct effect of firm-specific resources upon rate of innovation is also unknown.[273] Demonstrations of such variation can usually be traced to market- or field-specific causes. Large markets for fungible

---

[264] Mansfield, *supra* note 259.

[265] Marschak, *Strategy and Organization in a System Development Project,* in The Rate and Direction of Inventive Activity (National Bureau of Economic Research 1962).

[266] Jones & Twiss, Forecasting Technology for Planning Decisions 34-52 (1978).

[267] *Id.*

[268] See Bell, The Coming of Post-Industrial Society (1973); Toffler, Future Shock (1970).

[269] Jones & Twiss, *supra* note 266, at 22-24.

[270] *Id.*

[271] See Mansfield, Industrial Research and Technological Innovation 9-11 (1968); Thurow, *The Relationship Between Defense-Related and Civilian-Oriented Research and Development Priorities,* in Subcommittee on Priorities and Economy in Government, Joint Economic Committee, 94th Cong, 2d Sess, Priorities and Efficiency in Federal Research and Development (1976).

[272] Mansfield, *supra* note 271, p II; Technological Change (Gold ed 1975).

[273] Mansfield, *supra* note 271.

products, for example, do not support rapid change.[274] Certain areas of science may be hot due to intellectual ferment and controversy at a particular time and generate their own momentum for discovery.[275] Both these aspects of variation are examples of demand-pull promotion of innovation, which may also arise from altered characteristics of consumers resulting in altered needs.[276]

The soundest approach to considering such changes on a time-scale is comparative examination of the intervals between announcement of an innovation and its marketable implementation, two dates which can be at least roughly determined.[277] Variation in such intervals, though certainly confounded with other discretionary corporate policies and expectations of market reaction, may serve as a guide to *relative* rates of innovation.[278]

§11.22 Spinoffs and Synergy

The unintended consequences of innovation may be turned into an asset, at least in hindsight, in confronting the problems of measuring it. If the historical record is long and broad enough to allow recognition of such consequences, a strong case may be made for the presence of innovative behavior.[279] Two of the favorable ways in which these consequences may manifest themselves are spinoffs and synergy.

A spinoff may be narrowly defined as the offering of an internally developed and supported product or service outside the developing organization, often through a spunoff entity.[280] The decision for the spinoff may take place for several reasons. If the product appears to be inherently unprofitable the organization may choose to externalize its production and expend its own efforts more efficiently.[281] Contrarily, an organization may also decide to spinoff a line of business that seems especially profitable in order to maximize its growth and make efficient use of the parent's resources.[282] Where the product, however, is spun off largely because it seems alien to the producing organization, it is very likely due to the highly innovative nature of the source of the process

---

[274] Galbraith, Economics and the Public Purpose (1973).

[275] See **§11.28.**

[276] *E.g.,* The Effects of the Demographics of Individual Households on Their Telephone Usage (Brardon ed 1981).

[277] Van Duijn, *Fluctuations in Innovation Over Time,* 13 Futures 264 (Aug 1981).

[278] United States v United Shoe Mach Corp, 110 F Supp 295 (D Mass 1953), *affd,* 347 US 521 (1954).

[279] Bode, Synergy (1971).

[280] McGuire, Spin-Off Products and Services: The Commercialization of Internally Supported Resources (1976).

[281] Williamson, *The Vertical Integration of Production: Market Failure Considerations,* 61 Am Econ Rev 112 (1971).

[282] McGuire, *supra* note 280.

which produces both the familiar and the novel.[283] This is particularly the case where it would be inappropriate or illegal for the initiating organization to enter the new field.[284] The paradigm for the innovation-based spinoff is the application of government-sponsored space research to other fields,[285] although antitrust motivations are common as well in suggesting toehold market entries.[286]

Synergy is the phenomenon of two inputs producing more output in combination than their simple sum would achieve, usually unexpectedly.[287] These inputs may be at the level of firms themselves.[288] The explanation of synergy usually rests on arguments associated with economies of scale,[289] economies of scope[290] and critical mass.[291] While it has been argued previously[292] that such factors will not necessarily produce an innovative result, and so will not give sufficient justification for such a combination in the first place, the converse is not true; in the rare event that synergy is achieved it can stand as a strong marker for innovative behavior.[293]

Innovation and Industry Structure

§11.23 Innovation and Competition

There are many competing models of the economic activity and power of a firm used in antitrust law to determine the existence and estimate the effects of anticompetitive behavior. Innovation often plays an important role in the theories supporting the models, but no single model or theory can account for all the ways in which innovation interacts with industry structure and behavior. Arguments relating the two must always be tailored to the history and configuration of the specific industry in question. The types of available arguments are sketched below.

Innovation has been claimed to influence and be influenced by competition within an industry in both a positive and a negative fashion. While popular folklore favors the positive impacts, every possible conclusion has been ad-

---

[283] Mansfield, The Economics of Technological Change 224-28 (1968); *Spin-Offs and High Technology Industry,* NY Times, Jan 19, 1981, §IV, at 6.

[284] United States v Eversharp, Inc, 1967 Trade Cas (CCH) ¶72,221.

[285] Webb, *NASA as an Adaptive Organization,* in Technological Change and Management 109-144 (Ewing ed 1970).

[286] United States v Falstaff Brewing Corp, 410 US 526 (1973).

[287] Bode, *supra* note 279.

[288] Steiner, Mergers (1975).

[289] See **§11.05.**

[290] See **§11.06.**

[291] See **§11.20.**

[292] See **§11.05, 11.06 and 11.22.**

[293] Mansfield, Microeconomics 132-37 (1970).

vanced at some time. In the right context, innovation can be a defense to or explanation of charges of anticompetitive behavior.

Competition has traditionally been considered a spur to innovation in two ways. In the most simplistic form of free-market analysis, every firm will be earnestly striving to increase its competitive advantage.[294] Since creative behavior may enhance its productivity and market attractiveness, each firm will attempt to be highly innovative, and in fact the competition among firms will be stepped up a level to determine which can be most imaginative.[295] In a less abstract formulation, innovation might also be a reaction to competition, born out of fear of a competitor's research and development potential.[296]

If the competition is perceived as exceptionally fierce, however, it may have a negative effect upon innovation. The inevitable risks and lead time associated with any speculative venture may preclude active innovation; no single firm will dare to be the first to make a change.[297] A state of perfect competition could also be conceived such that all potential advances have been exhausted and every firm is locked in a zero-sum struggle for its share of a market of fungible products.[298]

Innovation can be a creator of competition in its own right, even when not produced by competition. It serves to expand old markets, to open up new ones and to give any firm, at least potentially, the opportunity to be a significant factor.[299] Successful research may also function as a wild card, upsetting entrenched industry positions.[300]

However, just as very active competition may in practice stifle innovation, so may very active innovation eliminate competition. A completely successful firm may assume a dominant position in the market; to the extent that its creative behavior formed the market in the first place, it may even obtain a monopoly.[301] In a less extreme form, active innovation may also create a noncompetitive niche in which each former competitor has been just novel enough to place itself into a specialized location in the market and become relatively independent of actions in the remainder of the market.[302]

---

[294] United States v United Shoe Mach Corp, 110 F Supp 295 (D Mass 1953), *affd* 347 US 521 (1954).

[295] United States v Aluminum Co of America, 148 F2d 416 (2d Cir 1945), *on remand,* 91 F Supp 333 (SDNY 1950).

[296] Mansfield, Industrial Research and Technological Innovation 97-100 (1968).

[297] Schumpeter, Capitalism, Socialism, and Democracy, chs vii-viii (3d ed 1950).

[298] MacLaurin, Invention and Innovation in the Radio Industry 251 (1947).

[299] Mansfield, *supra,* note 296, at 100-07.

[300] Picard v United Aircraft Corp, 128 F2d 632, 642-43 (2d Cir 1942); Hale & Hale, *Monopoly in Motion,* 41 Va L Rev 431 (1955).

[301] *See* United States v Aluminum Co of America, 148 F2d 416 (2d Cir 1945); United States v EI du Pont de Nemours & Co (cellophane), 118 F Supp 41 (D Del 1953), *affd,* 351 US 377 (1956).

[302] Turner, *The Definition of Agreement Under the Sherman Act: Conscious Parallelism and Refusals to Deal,* 75 Harv L Rev 655 (1962); Shenefield, Memorandum on Identification and Challenge of Pricing Practices in Concentrated Industries (Antitrust Division, Dept of Justice, May 26, 1978).

§11.24 Innovation and Industry Concentration

As with the relationship between innovation and competition, many conflicting claims have been made with respect to the presumptive association of level of innovation and degree of concentration in an industry. No single theory can prevail for all factual situations; each, with its own special weaknesses, can again be used to defend against or explain away seemingly monopolist behavior.

It has been argued that a highly concentrated industry will tend to be an innovative one on the rationale that a few active competitors will seize any means of differentiating themselves from their competitors.[303] With the prize of market dominance clearly in sight, each will strive to be as creative as possible. A less agressive model for concentration prompting innovation would suggest that in a shared oligopoly situation, each firm will not feel threatened by its fellows and will have the relative luxury of indulging in novel behavior, though perhaps not so actively as when prodded by competition.[304]

Of course, if the competition is relatively high, i.e., the oligopoly is a cutthroat one, then each firm will again be effectively coerced into inaction.[305] Alternatively, absent competition, members of a concentrated industry may feel no need even to experiment with innovation.[306] Further, if the industry has become concentrated through a process of elimination over time, then the few survivors are likely to be institutionally conservative and restrained by their embedded investment from indulging in radically new alternatives.[307]

Such a process of elimination itself might be based on a record of innovation, i.e., only those who change, live. Innovation then would be a mechanism for creating concentrated industries, with the most imaginative firm driving out (almost) all competitors.[308] Even if the survival process is not specifically predicated on innovation, other qualities of innovative firms may give them an advantage.[309] Innovation might also lead to industry concentration where each of a small number of firms develop an isolatable, specialized product without direct competition.[310]

Finally, the theories that claim innovation promotes competition can also be used to demonstrate that innovation will instead be an antidote to industry concentration. Technical advances can stimulate demand and market growth

---

[303] Scherer, Industrial Market Structure and Economic Performance (1970).

[304] *Id* ch 15.

[305] Shenefield, Memorandum on Identification and Challenge of Pricing Practices in Concentrated Industries, (Antitrust Division, Dept of Justice, May 26, 1978).

[306] United States v United Shoe Mach Corp, 110 F Supp 295, 347 (D Mass 1953), *affd,* 347 US 521 (1954).

[307] United States v EI du Pont de Nemours & Co (cellophane), 118 F Supp 41 (D Del 1953), *affd,* 351 US 377 (1956).

[308] United States v Von's Grocery Co, 384 US 270, 287-89 (1966).

[309] Thurow, The Zero-Sum Economy (1980).

[310] Weiss, *The Concentration-Profits Relationship and Antitrust,* in Industrial Concentration: The New Learning 189-93 (Goldschmid, Mann & Weston eds 1974).

to accommodate a new entrant.[311] New product developments can also dramatically alter standing within a concentrated industry, inciting intragroup competition and welcoming newcomers.[312]

Empirical results examining any of these contentions are quite mixed.[313]

§11.25 Innovation and Growth

It will now come as no surprise to learn that the correlation between innovation and the growth of a firm or an industry is the subject of a plethora of conflicting theories, even with the additional focus upon economic return.

Growth itself may encourage innovation. Growth in national economic activity provides growth in available investment capital as well as an expansion in demand.[314] Market growth will stimulate innovation as competitors scramble for a share.[315] Overall growth of the firm for whatever reason provides additional earnings; this higher return on investment encourages reinvestment of those earnings in some equally high return modality. The most obvious candidate is the firm itself, and where direct expansion of production is limited, reinvestment in the form of research will be encouraged.[316] Growth in terms of the opening up of new markets will also promote creativity in the search for a position in such markets.[317]

Periods of rapid growth may also be associated with such high returns that advances are not perceived as necessary.[318] To the extent that growth is the result of an intent to maximize short-run profits (perhaps without attention given to longer-run impacts), the need or the opportunity to innovate may be ignored.[319] A period of sustained growth or the same shortsighted mentality may also lead to a discounting of the value of creative interventions due to a fear that the risk and lead time aspects of change will interrupt such growth.[320]

Innovation may be seen to stimulate growth in its creation of wholly new

---

[311] Williamson, *Dominant Firms and the Monopoly Problem: Market Failure Considerations,* 85 Harv L Rev 1512 (1972).

[312] Picard v United Aircraft Corp, 128 F2d 632 (2d Cir 1942); Hale & Hale, *Monopoly in Motion,* 41 Va L Rev 431 (1955).

[313] Wilder, *The Electronic Data Processing Industry: Market Structure and Policy Issues,* 20 Antitrust Bull 25 (1975); Scherer, *supra* note 303, at 374-76; Weiss, *supra* note 310; Blair, Economic Concentration: Structure, Behavior, and Public Policy (1972).

[314] Forrester, *Innovation and Economic Change,* 13 Futures 323 (1981).

[315] Bain, Pricing, Distribution, and Employment (1953).

[316] Fruhan, Financial Strategy (1979).

[317] Bain, *supra* note 315.

[318] Forrester, *supra* note 314.

[319] Hayes & Abernathy, *Managing Our Way to Economic Decline,* 58 Harv Bus Rev 67 (Sept-Oct 1980).

[320] Schumpeter, Capitalism, Socialism, and Democracy (3d ed 1950).

markets and in its stimulation of demand through creation of new needs.[321] It may also be claimed that innovation at some level is the basis of all growth, i.e., something new must have entered the process at some point.[322]

If the lead time for implementation of a discovery is long, then innovative behavior may actually retard growth (at least relative to industries with shorter lead time).[323] Where a firm is a minor competitor trying to catch up with more developed organizatins, or where the degree of competition throughout the industry is accelerating, distinctive behavior under such circumstances constitutes an expenditure of effort and resources which might take the place of immediate growth.[324]

§11.26 Production Functions and Learning Curves

There are two concepts helpful in evaluating the effects of innovation on the structure of an organization from inside the organization: the production function and the learning curve.

The production function[325] is a model for relating levels of required capital and labor inputs to levels of output. Investment decisions are then made on the basis of the most effective mix of these components. Innovation simply changes the mix, and a standardization of output over time by the associated inputs over time will reveal any instances of innovative behavior.[326] The simplicity of this notion is appealing, but it has several serious drawbacks. First, it depends crucially on the assumption that capital and labor are perfectly substitutable, and that such substitution will always efficiently take place when required.[327] Many factors other than a true advance can change the mix of the function: different materials, lowered interest rates, better principles of organization, and so forth.[328] Some theorists claim that the production function approach is perfectly neutral to innovation, that is, innovation is just another point in a group of possible mixes and represents no real change in the function itself.[329] The production function also assumes perfect comparability of outputs before and after innovation and does not allow for consideration of different stages of product life cycles.[330] Finally, the method runs into grave

---

[321] Mansfield, Industrial Research and Technological Innovation 97-100 (1968).

[322] Schumpeter, *supra* note 320.

[323] Mensch, Coutinho, & Kaasch, *Changing Capital Values and the Propensity to Innovate,* 13 Futures 276 (Aug 1981).

[324] Villard, *Competition, Oligopoly, and Research,* 66 J Pol Econ 483 (1958).

[325] Mansfield, Industrial Research and Technological Innovation 97-100 (1968).

[326] *Id.*

[327] Mansfield, Microeconomics (1970).

[328] New Developments in Productivity Measurement and Analysis (Kerdrick & Vaccara eds 1975).

[329] Mansfield, *supra* note 327.

[330] Hayes & Wheelwright, *The Dynamics of Process-Product Life Cycles,* 57 Harv Bus Rev 127 (Mar-Apr 1979).

difficulties in dealing with a firm with multiple outputs and in associating particular inputs with their products.[331]

The learning curve[332] is a method of examining gains in the efficiency of a process over time by examining changes in the average cumulative cost of an output. This approach is useful in that it typically does not attempt to associate decreases in cost on the curve directly with innovation. Instead it recognizes that a creative incursion will usually sharply *increase* average costs at first until the organization learns how to implement it effectively.[333] Innovative periods are therefore those associated with any steep periods of change in such a curve.[334]

While a much sounder approach generally than the production function, the learning curve shares with it (as with most theoretical models) an inability to capture more than a small, isolated part of a complex organization's activities.[335]

Innovation and Decision Making

§11.27 Technological Forecasting

Is technological forecasting possible at all? Is significant innovation unpredictable by its very nature? The performance record of most technological forecasting methods and their practitioners has generally been quite poor.[336] Not only has much innovation been missed altogether,[337] but the applications of known and foreseeable technology have also been consistently missed;[338] there have been both "failures of imagination" and "failures of nerve."[339]

Should such an effort to predict then be indulged in? Over and above continuing improvements in forecasting methodology, especially for the near term,[340] such a process is still useful for several reasons. First, past errors in forecasting have great instructive value, as much in promoting general skepticism as in providing specific suggestions for avoiding their repetition.[341] Second, forecasting methods do reasonably well in at least indicating a range of

---

[331] Mansfield, *supra* note 325.

[332] Porter, Competitive Strategy (1980).

[333] Thurow, The Zero-Sum Economy (1980).

[334] Jones & Twiss, Forecasting Technology for Planning Decisions 163-64 (1978).

[335] Spence, *The Learning Curve and Competition,* 12 Bell J Econ 49 (1981).

[336] Clarke, Profiles of the Future (1963).

[337] *E.g.*, Martin, Future Development in Telecommunications 10-21 (1977).

[338] Toffler, Future Shock (1970).

[339] Clarke, *supra* note 336, chs 1-2.

[340] Bell, The Coming of Post-Industrial Society 196-221 (1973).

[341] Martin, *supra* note 337.

possible worlds and an estimate of their variability.[342] Third, such forecasts, no matter how inaccurate, serve as effective exercises in contingency planning, the results of which are often relevant for unanticipated innovation as well.[343] Finally, forecasting need not be a wholly passive activity but can itself initiate prospective innovation in a constructive way.[344] Accordingly, counsel should view the context of industry and firm forecasts very warily. The general types of forecasts and their usual weaknesses are described below.

Most forecasting methods can be classified as one of three types: subjective, structural and historical.

Subjective forecasting methods,[345] such as brainstorming,[346] Delphi,[347] and scenarios,[348] are all essentially just organized means of getting allegedly informed opinion on a given topic. The best applications of such methods rely heavily on cross-fertilization and controversy among the experts.[349] To the extent a compromise conclusion is forced on the group, however, the one radical yet accurate opinion may be ignored;[350] if the compromise is not forced, such an opinion lies buried in the long list.[351] (In either event, these methods are still useful for the planning purposes mentioned above.) The overall success of this approach is probably strongly limited by individual conservatism and may be summarized by Clarke's law: "When a distinguished but elderly scientist states that something is possible, he is almost certainly right. When he states that something is impossible, he is very probably wrong."[352]

Structural methods,[353] such as morphology,[354] cross-impact analysis[355] and

---

[342] Jones & Twiss, Forecasting Technology for Planning Decisions 14-15 (1978).

[343] Kahn & Wiener, The Year 2000 1-5 (1967).

[344] Francis, Principles of R&D Management 166-70 (1977).

[345] Jones & Twiss, *supra* note 342, at 219-25.

[346] Gordon, Synectics: The Development of Creative Capacity (1961); de Bono, Lateral Thinking for Managers (1971); Twiss, Managing Technological Innovation (1974).

[347] Linstone & Taroff, The Delphi Method: Techniques and Applications (1975); Bright, A Brief Introduction to Technology Forecasting: Concepts and Exercises ch 5 (1972).

[348] Kahn & Wiener, *supra* note 343; Meadows, Donella, Meadows, Randers & Behrens, The Limits to Growth (1972); Bright, *supra* note 347, ch 10.

[349] Jones & Twiss, *supra* note 342, at 226-27.

[350] Martino, Technological Forecasting for Decision Making 16-18 (1972).

[351] Hill & Fowles, *The Methodological Worth of the Delphi Forecasting Technique,* 7 Tech Forecasting & Soc Change 179 (1975).

[352] Clark, *supra* note 336, at 14.

[353] Jones & Twiss, *supra* note 342, at 106-36, 243-357.

[354] Jantsch, Technological Forecasting in Perspective, 174-80 (1967).

[355] Martino, *supra* note 350, 15 271-86; Enzer & Alter, *Cross-Impact Analysis and Classical Probability,* 10 Futures 227 (1978).

contextual mapping,[356] all presume that if a problem is examined very carefully, if all the relevant factors are spelled out, then innovation will be much easier to foresee. In practice, the true worth of these methods is their aid in avoiding blind spots when used in conjunction with other forecasting methods rather than stimulating actual creativity.[357]

Finally, historical methods,[358] such as envelope curves,[359] time series analysis[360] and dynamic modeling,[361] all attempt to project the future from the past. Since the past itself always reflects some process of ongoing change, such an exercise is not as conceptually pointless as it may at first seem, but an inability to predict disruption is very damaging.[362]

Only the relatively smooth implementation of innovation, as opposed to its discontinuous discovery, gives all three types of methods any credibility.[363]

§11.28 Innovation and Pure Research

It is usually taken as an article of faith that applied, i.e., mission-directed, research of the sort that produces recognized innovations is critically dependent upon the level of basic, nondirected, research.[364] Many corporate investment decisions in research and development as well as in market expansion are justified in this way; counsel should be prepared to treat this justification critically.

It is not immediately clear why this should be so. Arguments in support of such a proposition usually center around two observations: the difficulty of guiding basic research[365] and the serendipity of applications of basic research.[366] Both attributes, however, concentrate on the *accidental* value of basic research, leaving the door open to arguments that were the nature of such research understood more precisely, then much more effective and selective

---

[356] Martino, *supra* note 350, at 159; Jantsch, *supra* note 354, at 170.

[357] Jones & Twiss, *supra* note 342, at 122-27, 254-56.

[358] *Id* 155-56.

[359] Ayres, *Envelope Curve Forecasting,* in Technological Forecasting for Industry and Government (Bright ed 1968).

[360] Box & Jenkins, Time Series Analysis (1970).

[361] Forrester, Industrial Dynamics (1961); Coyle, Management System Dynamics (1977).

[362] *But see* Thom, Structural Stability and Morphogenesis: An Outline of a General Theory of Models (1975), for a treatment of the new branch of mathematics called *catastrophe theory.*

[363] Jones & Twiss, *supra* note 342, at 22-26; Kahn & Wiener, *supra* note 343, at 5-8, 34-39.

[364] Nelson, *The Simple Economics of Basic Scientific Research,* 67 J Pol Econ 297 (1959).

[365] Morton, Organizing for Innovation (1971); Francis, Principles of R&D Management (1977).

[366] Bode, Synergy (1971); Burke, Connections (1978).

choice of research areas could be made and most basic research ignored.[367]

A more successful description of the manner in which basic research supports innovation rests on the timing of innovation.[368] The perceived coincidental or accidental nature of much innovation may be seen as almost mechanistic in situations where many components of the innovation have been in existence independently for some time; then the final factor becomes available and initiates the complete innovation.[369] (This is often the basis of the common simultaneous discovery phenomenon.)[370] Basic research is that which develops and maintains those elements of innovation outside of any time constraints or requirements for immediate application.[371] If only needed research were carried on, it would quickly run out of working material.

Much of such development probably depends on the character of science itself.[372] Competing theories evolve over time aiming towards axiomatization, that is, reduction of the understanding of the field to a few basic concepts. To the degree that this simplification process is successful, it will guide the rate of innovation: faster in times of competing theories, slower when complete axiomatization seems to have been reached.[373] With the effort to become more concrete and precise, more and more anomalies and exceptions appear which cannot be explained away; the structure ultimately *collapses* and the process begins all over again.[374] If applied research provides the answers, then basic research supplies the questions. Apparent cycles in innovation and success of research are probably associated with the swings in such a process;[375] in a quiet field it will be difficult to justify expansion of the firm on the grounds of the need for additional basic research.

§11.29 The Corporation and the University

The business and the academic worlds have many relationships which support the innovative process despite very real conflict, on occasion, between the values of the two communities.[376]

The most obvious way in which universities support innovation in industry

---

[367] Nelson, *supra* note 364.

[368] Jewkes, Sawers, & Stillerman, The Sources of Innovation (1958).

[369] *Id.*

[370] *E.g.*, Ogburn, Social Changes 99-102 (1922); Kuhn, The Essential Tension: Selected Studies in Scientific Tradition and Change ch 4 (1977).

[371] Nelson, *supra* note 364.

[372] Kuhn, The Structure of Scientific Revolutions (1962); Feyerabend, Against Method (1975).

[373] Kuhn, *supra* note 372, chs III-V.

[374] *Id* chs VI-IX; Feyerabend, *supra* note 372, at chs 2-5, and 18.

[375] Paradigms and Revolution pt IV (Gutting ed 1980); Judson, The Eighth Day of Creation (1979); for a parallel discussion with respect to the arts, see Kubler, The Shapes of Time (1962).

[376] The Embattled University (Graubard & Ballotti eds 1970).

is through ongoing academic involvement in basic research.[377] Areas of inquiry that never would (and perhaps never should) be a part of an ongoing business are routinely examined under the strict competition of the marketplace of ideas.[378] In a more direct manner, universities also train the research personnel who will be employed in industry;[379] they hold themselves open as well to retraining of corporate personnel[380] and provide valuable, although often restricted, access to research facilities and the researchers themselves.[381]

American business recognizes its debt to the academic establishment in many ways. There is considerable direct funding of research by industry,[382] as well as the substantial contribution made through corporate taxes to governmental support;[383] large charitable contributions are common;[384] and corporations certainly provide a reciprocal service to educational institutions in offering employment to graduates[385] (and the far from occasional professional consulting contract).[386] Corporate applications of academic innovation also provide both psychic and monetary income.[387]

There have been frequent federal tax proposals for stimulating corporate support of universities, including a stepped-up basis for equipment donated for research purposes and a tax credit for all corporate-sponsored university research.[388] A partial credit for incremental support made its way into the Economic Recovery Tax Act of 1981.[389] A dramatic increase in corporate funding, no matter what the stimulating mechanism or attitude, may be required in the face of recent federal spending cuts.[390]

Finally, a new modality for an industry–academy partnership has been recently pioneered in the form of corporate funding of university research in exchange for favored licensing treatment in the event of any resulting innovation.[391] This symbiotic relationship, initiated aptly enough in the field of

---

[377] Bowen, Investment in Learning 296-308 (1977).

[378] Breneman & Finn, *Agenda for the Eighties,* in Public Policy and Private Higher Education 419-20 (Breneman & Finn eds 1978).

[379] Bell, The Coming of Post-Industrial Society, 221-32 (1973).

[380] *Back to School,* Wall St J, Mar 3, 1981, at 1.

[381] *Joining Hands Against Japan,* Bus Week, Nov 10, 1980, at 108.

[382] US Dept of Education, Digest of Educational Statistics, tables 199-200 (1980).

[383] *Id* tables 196-200; US Dept of Commerce, Statistical Abstract of the United States, table 483 (1980).

[384] Watson, Industry Aid to Education (1965).

[385] US Dept of Education, *supra* note 382, tables 173-77.

[386] Bowen, *supra* note 377, at 320-32.

[387] Spaght, The Bright Key (1965).

[388] *E.g.,* HR 1864, 97th Cong, 1st Sess, 127 Cong Rec H469 (Feb 16, 1981); S 692, 97th Cong, 1st Sess, 127 Cong Rec S2125 (Mar 12, 1981).

[389] Pub L No 97-34, §221(b), 95 Stat 176, 241 (codified at 26 USC §44F).

[390] *Private Giving,* Cong Q, May 30, 1981, at 935.

[391] *Genetech Reports Interferon-Yeast Tie,* NY Times, Feb 27, 1981, §IV, at 3; *Waltham Seeks Ordinance to Control Gene Research,* NY Times, Dec 24, 1980, at A10.

recombinant DNA analysis,[392] may prove to be a model for the future of similarly loose working arrangements.[393]

§11.30 Lead Time and the Limits to Growth

Effective legal counsel may be an asset in decision making with respect to innovation which is typically exacerbated by two factors. The first is the long lead time before the results of the decision can be effectively evaluated (and hence the extended fragility of the assumptions upon which the decision is made).[394] Planning under such circumstances is perceived as gambling under great uncertainty as to even the identity of the possible outcomes.[395]

Curiously enough, the rapid pace of recent innovation requires planners to work with longer and longer lead times; a process of frequent change needs to be followed down more alterative paths than a reasonably stable one.[396] As other factors[397] increase the lead time of implementation of innovation, so the lead time of planning for innovation must increase as well. A highly inflationary economy also places a premium upon the correct allocation of resources over the long run.[398] The increasing complexity of the innovative process works for longer planning lead times as well.[399]

The other crippling burden placed on planning for innovation in these last decades of the twentieth century is the increasing volatility of the impacts of technological decisions.[400] Loudly proclaimed limits to growth and era of scarcity arguments[401] contend that resources to support innovation (not to mention civilization itself) are nearing exhaustion and that therefore the right planning decisions must be made for the sake of efficiency or survival. At the same time, evidence of failures of past technological planning abounds,[402]

---

[392] *Genetech Reports Interferon-Yeast Tie, supra* note 391; *Waltham Seeks Ordinance to Control Gene Research, supra* note 391.

[393] Cooper, *The Patent System and the "New Biology,"* 8 Rut J Computer Tech & Law 1 (1980); the September, 1981 issue of Scientific American is also dedicated to present and potential applications of industrial microbiology.

[394] Makridakis, *If We Cannot Forecast, How Can We Plan?,* 14 Long Range Planning 10 (June 1981).

[395] Twiss, Managing Technological Innovation (1974).

[396] *Id.*

[397] See Galbraith, Economics and the Public Purpose pt 3 (1973).

[398] See Thurow, The Zero-Sum Economy (1980).

[399] Twiss, *supra* note 395; Francis, Principles of R&D Management (1977).

[400] Toffler, Future Shock (1970).

[401] Meadows, Donella, Meadows, Randers & Behrens, The Limits to Growth (1972); Mesarovic & Pestel, Mankind at the Turning Point (1974); Brandt, *et al,* North-South: A Program for Survival (1980); Barney, *et al,* The Global 2000 Report to the President (1980).

[402] *E.g.,* Nader, Unsafer at Any Speed (1965); Carson, Silent Spring (1962); Fuller, We Almost Lost Detroit (1975); Commoner, The Closing Circle (1971).

unanticipated environmental incursions proliferate[403] and some technological infrastructures appear to be ready to collapse of their own weight.[404] The hand of the decision-maker must surely hesitate a moment, and perhaps even tremble, in considering that the consequences of his or her decision may encompass destruction of the ozonosphere.[405]

Some small comfort may be had from the historical familiarity of these problems, even if they seem to be occurring on a much more grander scale today.[406] Technological crises in the past have always generated their own solutions, only to engender new crises.[407] Although it may be claimed that the crises confronted today represent such an exaggerated swing in the process that a return to equilibrium is not possible, the more successful response[408] is probably the recognition that such solutions typically take the form of moving out of a closed system, of generalizing the process out of the box.

Such possibilities exist today as well. Space exploration and potential industrialization explode the basis of limits to growth arguments.[409] The microrevolution and the information economy transform the notion of critical resources altogether.[410] Potential solar energy development has the capacity to decentralize whole industry structures and transform towards stability political and economic processes as well, world-wide.[411]

In the face of such possibilities, a decision-maker should be comforted and not threatened by unpredictability:[412]

> An optimist is someone who believes the future is uncertain.

---

[403] Ward & Dubos, Only One Earth (1972).

[404] Heilbroner, An Inquiry Into the Human Prospect (1974).

[405] Chlorofluorocarbons in the Environment: The Aerosol Controversy (Sugder & West eds 1980).

[406] Dubos, The Wooing of Earth (1980).

[407] Toffler, *supra* note 400.

[408] Fuller, Synergetics (1975).

[409] O'Neill, The High Frontier (1977), and 2081 (1981).

[410] Evans, The Micro Millenium (1980); Toffler, *supra* note 400.

[411] Commoner, The Politics of Energy (1979).

[412] Attributed to Edward Teller, in Dubos, *supra* note 406 at 41.

Major Policy Options

Richard A. Givens

12

§12.01 Options for the Innovator

Innovation necessarily threatens people because it disturbs the status quo. In particular, those with a special interest at stake, such as competitors and any group threatened with displacement or deprivation of a bottleneck position, are likely to seek to block it. Inertia is an even greater opponent. Inertia is manifested in the myriad requirements of the legal system, and in the internal bureaucratic obstacles within organizations. It is also manifested in delays and difficulties in accumulating external financial or other support where risks are high, where requirements are "lumpy" so that small increments cannot be used alone and where benefits are likely to be deferred.

In an effort to overcome these obstacles, the innovator has several options. Each of them is likely to be of use at different times. Consultation, for example, as a strategy may enable the innovator to win over potential opponents at an early stage by co-opting them into the process. Obviously, insofar as patentable technical information is concerned, others can be brought in only after a patent application has been filed.[1] Advantage may often be obtained, however, by approaching regulatory bodies, labor groups, suppliers, customers, financing sources and the like at a fairly early stage and seeking input and support.

---

[1] See **ch 4**.

In some agencies, formal procedures for this will exist. This will not normally apply, however, to competitors or any groups where implacable opposition exists. Indeed, contacts with competitors may produce antitrust problems under many circumstances.[2]

If opposition through creation of regulatory, legislative or other legal obstacles at the instance of competitors, or others affected is expected, a counter-strategy is imperative for success. Potential allies must be identified and brought into the fight if possible.[3] If legal conflict is anticipated, positions should be explored and researched *before,* not merely *after* it erupts. Such questions as these may be considered in advance:

1. What kinds of arguments are available to each side in each expected area of dispute?
2. What kinds of conduct can be developed or avoided that will increase or decrease the chances of success?
3. What facts should be developed in advance?
4. What is the most cost-effective way to conduct any litigation that occurs?
5. What kinds of fact patterns will make tribunals *want* to decide in the innovator's favor?
6. What kinds of outside amicus or other support many be potentially available?
7. What will be the strategy options of the opponents?

If major conflict occurs, the attitude of the public is likely to be decisive in the end. In that event if may be important to ask questions such as these at an early stage:

1. Will the innovation benefit the public or segments of it and if so, how?
2. How can the public become informed of these benefits?
3. Can the public be informed of the barriers to these benefits and of the effects of inertia?
4. Should an effort be made to inform the public of the opposition's strategies and tactics? Will this exacerbate the conflict? Does this matter? Will lawsuits be generated?
5. What other interest groups might assist in informing the public? What credibility will these groups be likely to have?
6. What harmful side-effects, if any, might the innovation cause?
7. Can these be avoided or reduced? Can they be offset by companion innovations, e.g., to use in a valuable manner otherwise harmful by-products?

---

[2] See **ch 3.**

[3] See Givens, Advocacy ch 16 (1980).

8. Can the public be made aware of efforts to reduce or avoid any harmful side-effects, or perhaps that they may not occur at all as in the instance of the window-shattering noise from the Concorde supersonic transport which never materialized?
9. If side-effects are inevitable can they be dealt with, and if so, can the public be informed of this?
10. Are basic values such as freedom of expression, privacy, economic survival, safety or national security involved with the innovation and if so, how?

One of the risks commonly ignored in publicizing innovation is that of permitting representatives of an interest group, industry or agency to make statements which are later unsupportable if attacked and which must be retracted or so modified as to destroy credibility.[4] Divorce of substantive strategy from public information strategy is a serious danger to any innovation which becomes subject to public debate.

Assuming the public or governmental agencies are to become involved, one option is to seek to minimize visibility likely to engender controversy. High visibility public or government involvement may, however, be inherent in the innovation, for example in the case of space industrialization or perhaps controlled nuclear fusion.

If visibility is desirable or inevitable, one option is to make a strong case for the innovation early, seeking affirmative support, and perhaps inviting participation in developing the shape and implementation of the innovation. Affirmative support may include as one element the acceleration, consolidation or even preemption[5] of otherwise serious legal obstacles or required bureaucratic approvals[6] should this be found warranted.

§12.02 Need for Innovation from the Public Viewpoint

Society's need to facilitate innovation springs from many sources, including among others:

---

[4] See Givens, Advocacy ch 4 (1980), for courtroom analogies.

[5] *See, e.g.,* Flood v Kuhn, 407 US 258 (1972); San Diego Bldg Trades Council v Garmon, 359 US 236 (1959); Garner v Teamsters Union, 346 US 485 (1953); Local Div 732 v Metropolitan Atlanta Rapid Transit Auth, 667 F2d 1327 (11th Cir 1982). Express antipreemption clauses are increasingly included in legislation in an effort to clarify this subject. *See* Prudential Ins Co v Benjamin, 328 US 408 (1946), discussed in Dowling, *Interstate Commerce and State Power: Revised Version,* 47 Colum L Rev 547 (1947). This does not always meet with complete success. See Note, 46 Va L Rev 297 (1960).

[6] Absent legislation, federal interstate commerce clause limitations may also apply. *See, e.g.,* Raymond Motor Transp v Rice, 434 US 429 (1978); Hunt v Washington Apple Advertising Commn, 432 US 333 (1977); Southern Pac Co v Arizona, 325 US 761 (1945); see Givens, *The Commerce Clause and the Interdependent Economy,* 53 ABAJ 719 (1967).

1. The need for an expanding economic pie to support increasing demands on resources, such as:
   a. Expectations of productivity-based increased in living standards and in society's safety net
   b. Efforts to overcome historic lags in economic advancement of disadvantaged groups in society
   c. Requirements necessitated by the international situation, including relations with developing areas and national security-related costs
   d. The cost of overcoming environmental pollution
   e. The need for enhancing workplace, consumer and environmental safety or the consequences of injuries and the needs of larger numbers of older citizens.
2. The need to develop substitutes for:
   a. Scarce exhaustible resources
   b. Environmentally or otherwise harmful existing technologies
   c. Technologies made obsolete by international competition
3. The need for research to deal with emerging hazards such as:
   a. Dependence for essential services on interruptible central sources (oil, for example) on an increasing scale
   b. Increasing resistance of disease agents to antagonists
   c. The possibility of increasing temperatures on earth leading to a rise in ocean levels
4. Such opportunities as:
   a. The possibility that relatively open societies might develop a sufficient lead in technological, economic and other forms of power as to make it in the interests of emerging leaders of totalitarian adversaries to end the unstable mutual balance of terror and enter a new competitive or cooperative relationship before an error or malfunction brings about catastrophe[7]
   b. Discovery of dramatic new energy sources and other now unknown technologies that may overcome problems not yet resolved by what will later be termed our present primitive methods in many fields[8]
   c. The possibility boldly proclaimed by John Quincy Adams on be-

---

[7] See **§1.06.**

[8] *E.g., Power Made From Ocean Heat,* NY Times, July 5, 1980, at 28; Bernstein, *Recreating the Power of the Sun,* NY Times Magazine, Jan 3, 1982, at 18; Spencer, *et al, Coal Gasification for Electric Power Generation,* 215 Sci 1571 (Jan 26, 1982); *see also* Berry, The Next Ten Thousand Years (1974); Clarke, The Promise of Space (1968); Hubbard, The Search Is On (1969); compare 42 USC §§9101-9311.

half of the new nation in the early nineteenth century[9] and by numerous others since, of seeking to satisfy the underlying excitement of our limitless curiosity about the universe through exploration of its mysteries, from the very small to the very large and the infinitely complex, and of the paradoxes in its basic laws insofar as now discerned.[10]

§12.03 Primary Means of Promotion of Innovation

Society's means of seeking to promote innovation are varied and frequently inconsistently applied. The primary means are provision of the requisite economic environment and, alternatively, affirmative promotion.

Furnishing the Basic Infrastructure and Allowing the Marketplace to Function

Where consumer demand will make an innovation profitable, the most effective social policy is generally to allow the marketplace to encourage it. This requires, however, a basic infrastructure of political stability and security of the person, as well as basic education, transportation and communication systems and similar services. Erosion of any of these can tend to defeat innovation.

Part of a market-oriented strategy for promoting innovation may be removal of excessive regulatory or tax barriers which may stifle innovation.[11] This may be sought either by overall economy-wide policies or by more specific efforts aimed at innovation generally or at particular types of innovation.

More specific efforts may take the form of tax incentives or simplification of layers of regulatory approval required for an innovation to proceed.[12] Such simplification is, however, counterproductive if relaxation is poorly designed and thus leads to errors with serious consequences that result in a climate of greater hostility. Careful but not overly vexatious control dependent upon the character of the risks to be controlled is, of course, the optimum. In many areas this may mean no specific interference.

---

[9] First Annual Message, Dec 6, 1825, 1 State of the Union Messages of the Presidents 249 (1966).

[10] *See, e.g.*, Weinberg, *The Forces of Nature*, 29 Bull Am Acad Arts & Sci No 4, at 13, 29 (Jan 1979); Teilard de Chardin, The Phenomenon of Man pt III (1959); Feinberg, The Prometheus Project (1969); Heisenberg, Physics and Beyond (1971); Givens, *Technology and Mankind's Future*, 123 America 254 (Oct 10, 1970); Einstein & Infeld, The Evolution of Physics (1942); Waldrop, *Bubbles Upon Tides of Time*, 215 Sci 1082 (Feb 26, 1982).

[11] See **chs 5-11.**

[12] On layers of approval, see Givens, Advocacy ch 14 (1980) and references cited; *compare, e.g.*, HR Rep No 575, 95th Cong, 1st Sess (1977).

Affirmative Promotion

Affirmative promotion may be necessary where there is no consumer-oriented market for the direct or intermediate product or service, or where investment is deterred. The latter may occur because the risks involved are high, and thus, tend to chill investment for the same reason individuals, enterprises and institutions buy insurance, or because payout is postponed, contrary to what most investors wish.

Investment may also be deterred because much of the benefit from the innovation escapes the innovator to society in general (which must occur to some extent in all instances or no one would have an interest in buying the resulting product or service at the price charged).[13] This is especially applicable to infrastructure innovations which are often subsidized, and to the creation of entirely new industries. Furthermore, the investment required may be large and "lumpy" in the sense that large amounts of capital must be secured before any of the investment can be employed profitably at all. This is often an expensive and even fatal barrier.

If decision-makers, with or without active interest on the part of the general public, can be convinced that affirmative promotion of a type of innovation is urgent, this can be accomplished in a wide variety of ways. Governmental contracting can be used for products, including intermediate research, as is normal in the case of defense and other governmental procurement.

Govermentally funded research, in the form of pure research contracts, grants or use of governmental employees can also be used, with the results at times used for governmental procurement and at times made available through extension services (generic technology). In the case of applied research and development, momentum is often lost because the people and institutions doing the research are usually not the ones to use it. Further, the goal often becomes the use of X dollars rather than the achievement of Y result. In the case of basic research, these options often suffer from excessively detailed restrictions[14] and short-term commitments.

Credit or credit guarantees may be provided for private sector research designed to achieve goals defined as vital. This should properly be defined as an investment, not a current expense, and could be handled through a congressionally created corporation such as the Reconstruction Finance Corporation,[15] the Export-Import Bank[16] or Alexander Hamilton's First Bank of the

---

[13] Terleckyj, Effects of R&D on the Productivity Growth of Industries (National Planning Assn 1974).

[14] *See, e.g.*, Freedman, *Yale Bars a US Study Grant, Refusing to Account for Time,* NY Times, Mar 23, 1982, at 1; *compare also* Merck & Co v Staats, 665 F2d 1236 (DC Cir 1981); Norman, *Small Business [Set-Aside] Bills Upset Universities,* 214 Sci 1003 (Nov 27, 1981); Gutowsky, *Federal Funding of Basic Research: The Red Tape Mill,* 212 Sci 636 (May 8, 1981).

[15] 47 Stat 5 (1932); see Jones & Angly, Fifty Billion Dollars (1951); *see also* 61 Stat 105, 934 (1957); 88 Cong Rec 1009, 1632-38 (1942). *Compare* Pozen, *Synthetic Fuels Corporation-Investment Bank or Government Agency,* 36 Bus Law 953 (1981).

[16] For the original statute, *see* Export-Import Bank Act, 59 Stat 526 (1945).

United States.[17] It could also be done through the central banking system itself.[18] This has been done before during the Great Depression and defense emergencies.[19]

Such an approach differs radically from govermental credit assistance for ongoing economic activity. The following drawbacks apply in the latter case:

1. Competitors who receive such assistance have an unjustified advantage over others in the continuing conduct of business[20]
2. The tendency will be to attract people and investment into a continuous activity which is not justified by the marketplace and thus requires further subsidy[21]
3. The temptation to use such programs involving ongoing activity for political patronage is likely to be irresistible[22]
4. Any industries picked as winners at a particular time will be outflanked by further technological advance, thus forcing the continued subsidization of what will then become losers[23]
5. The likelihood that massive losses in bailouts will tarnish the reputation of even constructive uses of credit for purposes for which it is actually needed[24]

Perhaps for these reasons, there has been a somewhat wider spectrum of support for credit for research funding than for other forms of activity.[25]

---

[17] First Bank of the United States, 1 Stat 191 (1791); see Hurst, *Alexander Hamilton: Law Maker,* 78 Colum L Rev 483 (1978); Hacker, Alexander Hamilton ch 8 (1957). Compare Note, *The Communications Satellite Act of 1962,* 76 Harv L Rev 388 (1962).

[18] *See* 88 Cong Rec 1632-38 (1942); 38 Stat 264 (1913), *as amended,* (codified at 12 USC §§263, 289, 290, 347(c), 355, 413, 462, 462(a)).

[19] Jones & Angly, *supra* note 15. The medium and short-run economic effects depend on external circumstances and on the amounts involved compared with the magnitude of the national product. See Arenson, *Economic Scene,* NY Times, Mar 26, 1982, at D2; *compare* Clinger, *Capital-Budgeting,* NY Times, Feb 2, 1982, at A25; Cowan, *Balancing the Budget By Law,* NY Times, Mar 7, 1982, at 6F.

[20] *Compare* Caro, The Power Broker (1975); Clark, *Making the Fed A Branch of the Treasury,* Wall St J, Mar 16, 1982, ed page.

[21] This problem is also inherent in price guarantees and related subsidies for ongoing activity. See Givens, Advocacy ch 15 (1980); *The Search for Alternatives to Protection,* 30 Fordham L Rev 17 (1961); Herzog & Jenkins, *Price Guarantees and Purchase Agreements Under the Synthetic Fuels Corporation Act,* 36 Bus Law 995 (Apr 1981).

[22] *Compare* Williams, Huey Long (1969).

[23] See Griorson, *Reindustrialization's Poor Record,* Wall St J, Sept 30, 1980, at 39; *compare* Cowan, *Trying to Resolve the Productivity Issue,* NY Times, Jan 18, 1982, at D3; Wiedenbaum, *Subsidies Business Does Not Need,* NY Times, Mar 31, 1981, at A19.

[24] The Reconstruction Finance Corporation, however, ended in the black. See Jones & Angly, *supra* note 15.

[25] Much of the discussion occurred at the time of the first gas-line crisis of the mid-1970s. *See, e.g.,* former Secretary of the Treasury John Connally, quoted in Safire, *And I Will Be Heard!,* NY Times, May 22, 1975, at 39; editorial, 133 America 84-85 (Aug

A critical advantage of use of an investment approach is the ability to secure long-term commitments for vital research.[26] In the case of government contracting, this may also be sought by multiyear appropriations[27] or multiyear contractual obligations.[28]

§12.04 Specific Policy Options

Specific measures to improve the rate of industrial innovation include those listed below among others.

Consolidation of Approval Requirements

The number of agencies, layers within agencies and levels of government which must become involved for a given course of action to go forward is often an almost impenetrable barrier aside from any substantive objections. One approach to this problem is for all reviews to be conducted by a single respon-

---

30, 1975); Special Committee on Consumer Affairs, *A National Energy or Technology Development Bank,* 29 Rec AB City NY 718 (1974); Letters from New York Academy of Sciences to Sens. Javits & Buckley (May 19, 1976). *Compare also* Luncheon Address by Richard L Abbott, Vice President of Montgomery Ward, before the American Apparel Mfrs Assn (Feb 21, 1974); US Congressional Budget Office, The Productivity Problem: Alternatives for Action 83 (1981); Worsinger, *New Technologies and Antitrust,* 47 NY St BJ 651 (1975), *reprinted in* 81 Case & Com 33 (1976); HR 10259, 94th Cong, 1st Sess (1975); S 3111, 94th Cong, 1st Sess (1976), Cong Rec at S2981 (daily ed Mar 9, 1976).

[26] *Federal Incentives for Innovation, Hearings before the Special Subcommittee on Science, Technology & Commerce, Senate Commerce Committee,* 93d Cong, 1st Sess, Serial No 57 at 47 (1973); *National Science Policy, Hearings before the Subcommittee on Science, Research & Development of the House Committee on Science & Astronautics,* 91st Cong, 2d Sess 20 (1970); Marshall, *NOAA Buffeted By Budget Storms,* 215 Sci 1482 (Mar 19, 1982); Hiroka, *Letter to the Editor,* NY Times, Mar 12, 1982, at A28. *Compare also* Kalberer, Grant Length and Budget Stability at NIH, 211 Sci 675 (Feb 13, 1981).

[27] See Coolidge, State of the Union Message, Dec 7, 1926, 3 State of the Union Messages of the Presidents 2690 (1966); *e.g.,* 80 Stat 1259-1264 (1966); 72 Stat 438 (1958); 42 USC §2459(a). *Compare* Penner, *The Budget Process Is Tied in Knots,* NY Times, Feb 28, 1982, at F3. The sole constitutional limitation on length of appropriations relates to armies. US Const art I, §8, cl 12.

[28] *See, e.g.,* HR Rep No 410, 97th Cong, 1st Sess 30, 39 (1981); compare 61 Stat 105, 934 (1947) (Greek-Turkish aid through use of the Reconstruction Finance Corporation); for up-to-date information, see Federal Contracts Reporter (BNA). *Compare also* Mohr, *Fears About Deficit Impair Effort to Improve Acquisitions,* NY Times, Mar 14, 1982, at 1, 32; Hershey, *Congress Cuts too Close to Bone,* NY Times, Feb 21, 1982, at 14; Weinberger, *Creativity and Collaboration in Government—The Budget Process,* in the Frontiers of Knowledge 235, 246 (1975); Farnsworth, *US Agencies Rush to Spend Before Cutoff,* NY Times, Aug 16, 1979, at 1. For Defense Department views, see Department of Defense Statement on Multiyear Procurement by Hon Richard D DeLauer, Undersecretary of Defense for Research and Engineering, before the Subcommittee on Defense, House Appropriations Committee, 97th Cong, 1st Sess (June 16, 1981); Statement on Multiyear Contracting and Other Procurement Provisions of HR 3519 by Hon. Caspar W. Weinberger, Secretary of Defense before the House Armed Services Committee, 97th Cong., 1st Sess (June 23, 1981).

sible agency on behalf of—or in lieu of—all others.[29]

For example, federal review may constitutionally preempt state and local barriers if interstate commerce is involved and if Congress so intended.[30] Judicial intervention through preliminary injunctive relief prior to rulings on the merits may also be limited.[31] Such an approach might, of course, be more controversial where the agency logically designated becomes engaged in overregulation or is regarded as captured by those regulated, or where local interests are intimately affected.[32]

Encouragement of Research by Contractors

A reasonable amount of independent research and development could be allowed as part of overhead on government contracts if relevant and actually performed. Appropriation bills could be worded so as to permit this.

Similarly, contracts themselves on major complex projects could be made covering several years so as to permit effective research.[33] This may otherwise be stymied by the uncertainty of dependence upon annual budgets.[34] Such contracts could be authorized either through permitting multiyear contracts aside from specific appropriations by merely creating governmental liability,

---

29 *See, e.g.,* HR 10259, 94th Cong, 1st Sess (1975); HR 1225, 96th Cong, 1st Sess (1979). This may be analogous to supersession of jurisdiction of one agency by another, but not merely prior resort to one agency before having to deal with another, a distinction discussed in Note, 58 Colum L Rev 673 (1958). *Compare generally* Ramo, *The Regulation of Technological Activities: A New Approach,* 67 ABAJ 1456 (1981). Jurisdiction limited to one agency should be distinguished from science court proposals which contemplate an additional entity, utilizing experts in variegated fields and attempting to make authoritative pronouncements on controversial matters. *See, e.g.,* HR 638, 97th Cong, 1st Sess (1981); 19 Jurimetrics J No 4 (Summer 1979).

30 Concerning the power of Congress to preempt state or local law where federal activities are involved, *see* First Agricultural Bank v State Tax Commn, 392 US 339 (1968); Kern-Limerick, Inc v Scurlock, 347 US 110 (1954); Mayo v United States, 319 US 441 (1943); Johnson v Maryland, 254 US 51 (1920); McCulloch v Maryland, 4 Wheat 316 (1819). Similarly, Congress may preempt state or local regulation where interstate commerce is affected and it chooses to exercise such power. *See* Perez v United States, 402 US 146 (1971); Northern Natural Gas Co v State Commn, 372 US 84 (1963); Cloverleaf Butter Co v Patterson, 315 US 148 (1942); Garner v Teamsters Union, 346 US 485 (1953); Dowling, *Interstate Commerce and State Power: Revised Version,* 47 Colum L Rev 547 (1974). Some local regulation is, of course, vulnerable under the commerce clause even absent congressional action. *See, e.g.,* Raymond Motor Transp v Rice, 434 US 429 (1978); Hunt v Washington Apple Advertising Commn, 432 US 333 (1977); Bibb v Navajo Freight Lines, 359 US 520 (1959); HP Hood & Sons v DuMond, 336 US 525 (1949); Southern Pac Co v Arizona, 325 US 761 (1945).

31 *See* Lauf v EG Shinner & Co, 303 US 323, 330 (1938), upholding the Norris-LaGuardia Act, 47 Stat 70 (1932) (codified at 29 USC §101 *et seq*).

32 See Jaffe, *The Effective Limits of the Administrative Process: A Reevaluation,* 67 Harv L Rev 1105 (1954).

33 See Miller, *Long-Term Contracts Proposed as Cost Measure,* NY Times, Mar 4, 1981, at A20; See **§12.03.**

34 See **§12.03.**

or by multiyear appropriations to remain available until expended.[35]

Patent rights for contractors and grantees is another area where greater incentives could be developed.[36]

A National Technology Development Corporation

The use of credit or credit guarantees through a specialized financial institution might be appropriate where affirmative promotion of vital technologies is critical. Areas within which credit or credit guarantees could be considered might be determined by a broadly based scientific advisory panel, assisted by recommendations from citizens, industry and government agencies, utilizing the criteria of societal urgency and of conditions inhibiting private investment, such as high risk.[37] Congress might be given the opportunity to disapprove such categories, even by one-house action, inasmuch as internal governmental matters[38] and not rules having the force of law as to private parties would be involved.

Loans or loan guarantees through such a corporation might be conditioned on the applicant:

1. Showing that maximum feasible investment from other sources has been obtained but by itself is insufficient
2. Agreeing that a percentage of any profits resulting from successful research will be paid to the corporation to reimburse it for the high risk involved and generate a revolving fund for other research
3. Agreeing that while patents on assisted research would belong to the innovator, they would remain enforceable only as long as the results of the assisted research were actually exploited. In other words, assisted research could not be used solely to obtain blocking patents.

The funds of such a corporation could be generated in the same way as by prior specialized financial institutions such as the Reconstruction Finance Cor-

---

[35] *Id.*

[36] See US Dept of Commerce, Draft Report on Patent Policy of the Advisory Committee on Industrial Innovation Established as Part of Domestic Policy Review (PB No 290403, 1978).

[37] See **§12.03.**

[38] On prior legislative veto provisions involving internal governmental organization, see 5 USC §906(a); 91 Stat 32 (1977); 85 Stat 574 (1971); 80 Stat 394 (1966); 71 Stat 611 (1957); 63 Stat 205 (1949); 47 Stat 1518 (1933); United States v Paramount Publix Corp, 73 F2d 103 (CCPA 1934); *A Study of the Constitutionality of the Congressional Veto,* Cong Rec H2587 (daily ed Apr 6, 1978); Committee on Federal Legislation, *The Legislative Veto,* 34 Rec AB City NY 208 (1979); Javits & Klein, *Congressional Oversight and the Legislative Veto,* 52 NYU L Rev 455 (1977). Pending litigation on the issue of legislative vetoes and their constitutionality relates to rules directly affecting citizens. See 81 Colum L Rev 1721 (1981).

poration,[39] but for more limited and focused purposes. To provide congressional control over amounts involved without subjecting a research-promoting institution to the uncertainties of annual appropriations, obligations could be limited by law to a fixed percentage of the gross national product.[40] As an alternative, such a corporation could notify Congress 90 days in advance of the total of all additional obligations to be issued, giving one or both houses of Congress the opportunity to disapprove if it deemed the amount excessive.[41]

Such an approach could provide the advantages of long-term commitments and affirmative promotion of socially vital new technologies utilizing the initiative of the private sector. This combination, together with the robust[42] debate of an open society so vital to correct mistakes,[43] might well lead to an immense accretion of technological and economic power.

§12.05 Research Priorities

Priorities which may lend themselves to successful efforts to obtain public-sector support that have national implications may embrace some of the following by way of example:

First, research in basic physics should be continued. There are numerous indications that we do not yet understand the basic building blocks of the universe, and that vast energies below our current reach remain to be understood and even tapped.[44] In this regard, space science and astronomy beyond the atmosphere are also vital, since the very small and very large are inextricably connected by what we know of the basic laws.

Second, space industrialization should be an important priority. Solar energy, collected in space 24 hours a day, 365 days a year without interruption, could be a limitless resource by terrestrial standards, and is waiting to be

---

[39] Such means in the past have included the guarantee of obligations of such institutions by the government, or purchase of them by the Treasury or the central banking authorities. See authorities cited, Worsinger, *New Technologies and Antitrust,* 47 NY St BJ 651, *reprinted in* 81 Case & Com 33 (1976); First Bank of the United States, 1 Stat 191 (1791).

[40] This concept was taken from, *e.g.,* numerous proposed budgetary control amendments although the context is entirely different. *Compare generally* Committee on Federal Legislation, *Budget Balancing by Constitutional Amendment,* 34 Rec AB City NY 428 (1979).

[41] See **§12.03.**

[42] New York Times v Sullivan, 376 US 254, 270-72 (1964).

[43] *See* Abrams v United States, 250 US 616, 630 (1919) (Holmes, J, dissenting); Dowling, *The Methods of Mr. Justice Stone in Constitutional Cases,* 41 Colum L Rev 1160 (1941).

[44] *See, e.q.,* Feinberg, What is the World Made Of? (1977); Goldhaber, *Is the Proton Stable?,* 210 Sci 851 (Oct 21, 1980); Kaufmann, The Cosmic Frontiers of General Relativity (1977); Heisenberg, Physics and Beyond (1971); Bergmann, The Riddle of Gravitation (1968); Vaucouleurs, *The Case for a Hierarchical Cosmology,* 167 Sci 1203 (Feb 7, 1970); Gal-Or, *The Crisis About the Origin of Irreversibility and Time Anistrophy,* 176 Sci 11 (Apr 7, 1972).

tapped.[45]

Third, development of other new power sources and environmentally sound uses of power should be a high priority. Nuclear fusion, the source of the sun's energy, seems close to becoming a nearly limitless, virtually non-polluting power source within decades, seemingly deserving high priority.[46] A patent has already been granted on the first device to obtain energy from temperature gradients at differing ocean depths.[47] Once fusion or collection of solar energy in space utilizing space-manufactured equipment or other new power sources is available, it should be possible to utilize electricity for large-scale desalinization of water. Hydrogen might be developed for use as a vehicle fuel (producing water as the end-product), thus making petroleum far more valuable for chemical uses.[48]

Fourth, substitutes must be developed to replace scarce and strategically vulnerable materials. With physical and human energy, the basic building blocks of the universe can be transformed into almost any desired form. Research to develop substitutes for items which may run out, are already scarce or are subject to politically inspired or cartel-generated interruption should thus be a high priority.

Fifth, environmentally favorable technologies and substitutes for environmentally problematic processes must be developed. Rather than cutting back economic activity due to environmental dangers, we can positively improve the environment while removing some of the risks by development of alternatives. Some may involve simpler but previously undiscovered means, such as improved biological control of pests.

---

[45] See O'Neill, The High Frontier (1977); Clarke, The Promise of Space (1968); Michener, *The Advantages of a Space Program,* 182 Sci 358 (Oct 26, 1973); O'Neill, *Space Colonies and Energy Supply to the Earth,* 190 Sci 943 (Dec 5, 1975); *see also* O'Neill, *Colonies in Orbit,* NY Times Magazine, Jan 18, 1976, at 10; *Proposal for Colonies in Space Hailed by Scientists as Feasible,* NY Times, May 13, 1974, at 1; Sullivan, *Princeton Gathering Makes Detailed Assessment of Problems of Establishing a Colony in Space,* NY Times, May 12, 1975, at 54; Blakeslee, *Experts Press for Space Colony,* NY Times, Aug 23, 1975, at 47; McElheny, *Electricity Transmitted by Radio Beam on Coast,* NY Times, Oct 10, 1975, at 40; O'Neill, *Space: A Place in the Sun,* NY Times, June 12, 1975, at 37; *Symposium on the Future of Space Science and Space Applications, Hearing before the Subcommittee on Science, Technology and Space, Senate Commerce Committee,* Serial No 58, 95th Cong, 2d Sess (1978); Sullivan, *Engineer Proposes Giant Space Mirrors,* NY Times, Feb 6, 1977, at 24; *compare also* S 3599, 95th Cong, 2d Sess (1978); S12497 and S3541, 95th Cong, 2d Sess (1978) (Space Industrialization Act).

[46] *E.g.,* Bernstein, *Recreating the Power of the Sun,* NY Times Magazine, Jan 3, 1982, at 18; Spencer, *et al,* Coal Gasification for Electric Power Generation, 215 Sci 1571 (Jan 26, 1982); *see also* Berry, The Next Ten Thousand Years (1974); Clarke, The Promise of Space (1968); Hubbard, The Search Is On (1969); *compare* 42 USC §§9101-9311.

[47] See **§12.01;** *Power Made From Ocean Heat,* NY Times, Aug 5, 1980, at 28.

[48] On another potential transportation breakthrough, see *Propelled Lighter-Than-Air Vehicles, Hearings before the Subcommittee on Science, Technology and Space, Senate Committee on Commerce, Science and Transportation,* Serial No 6, 96th Cong. 1st Sess (1979). Oil, however, will undoubtedly become even more critical for chemical and materials uses.

Each of these efforts may well generate spinoffs which could be even more valuable than the results hoped for originally.

§12.06 Long-Term Implications

Those goals listed in the preceding section are only short-term goals. Further away are other "nested" challenges. In this regard, the most reliable assumption would seem to be that future technological capability will bear the same relationship to today's capability as today's bears to that of a thousand years ago.[49]

Physicist Gerald Femberg has contended that with long-term commitments there is virtually *nothing* that cannot be done.[50] In any event, without accepting any such proposition or its full implications,[51] it is clear that expanding power

---

[49] See Berry, The Next Ten Thousand Years (1974); Clarke, The Promise of Space (1968); Hubbard, The Search Is On (1969); Teilhard de Chardin, The Phenomenon of Man (1959); Givens, *Technology and Mankind's Future,* 123 America 254 (Oct 10, 1970).

[50] Feinberg, The Prometheus Project (1969).

Such motivation can be harnessed by recognition of the depth and severity of the challenges "nested" ahead of us. As explained by Winston Churchill: "It is a curious fact about the British Islanders . . . that as danger comes nearer and grows they become progressively less nervous; when it is imminent they are fierce, when it is mortal they are fearless. These habits have led them into some very narrow escapes." *Quoted in* Never Give In! 10 (1965).

Positive, affirmative motivations can perform as well once the true extent of the possibilities to be realized are recognized, and the possibility of participation in whatever degree in events of such significance is glimpsed, according to e.g., Holmes in the discussion of incentives for the highest levels of performance in the legal field in Collected Legal Papers (1921); May, The Courage to Create (1975); Teilhard de Chardin, The Phenomenon of Man (1959).

It is possible that concentration on an external goal of sufficient importance will draw energy from normal internal self-concerns that ordinarily take up most of our energy, thus freeing large amounts of vitality and attention for the task itself, as suggested by practitioners of oriental martial arts and by the "inner sports" analyses of factors influencing success in certain kinds of contests requiring precise movements. See Westbrook & Ratti, Aikido and the Dynamic Sphere (1970); Herrigel, Zen and the Art of Archery (1953); Durkheim, The Japanese Cult of Tranquillity (1974); William, Martial Arts of the Orient (1975); Gallwey, Inner Tennis (1976); Zabriski, The Handbook of Inner Sports (1973).

[51] Presumably, these would include uncovering previously unknown laws (see Einstein & Infeld, The Evolution of Physics (1942)) just as previously unknown particles were predicted and then created where they had not existed at all (Pagels, The Cosmic Code: Quantum Physics as the Language of Nature 322 (1982); even unfolding new hitherto dormant dimensionalities as a result of Einsteinian curvature of space-time (Kaufmann, The Cosmic Frontiers of General Relativity (1977)); and, indeed, creating of new universes has been posited in Sato, *et al, Multi-Production of Universes,* 108B Physics Letters 103 (Jan 14, 1982). *Compare* Kaufmann, Black Holes and Warped Spacetime (1979); Jastrow, Until the Sun Dies (1977).

entails expanding stewardship and responsibility in its use,[52] without which destruction becomes the result in one guise or another. Thus, ever more energizing vistas and objectives must guide advancing technological power if it is to survive.[53]

Many, such as Robert Jastrow, believe that diffused hydrogen gas condensed to form stars, triggering nuclear reactions yielding heavier elements, which in turn made possible complex molecules, life, intelligence and, ultimately, the complexity of a global human society.[54] At each stage, more intense interactions are generated. If so, this dilemma of wise use of power may have been inherent in the basic laws to the extent they permit the evolution of awareness sufficiently acute to alter the environment, thus creating power that grows ever more rapidly as knowledge begets new questions and new answers. This power is, of course, subject to wise or unwise use.

The power to modify circumstances made possible at the level of human society has never been systematically developed with foresight, energy, perseverance and attention to both objectives and side-effects. Our nascent technology assessment movement has begun to focus on review of potentially harmful effects of innovation.[55] In some instances this has caused excessive layers of multiple approvals and delay. In other instances, it has not yet controlled serious consequences that can flow from some kinds of activity; this in turn causes unjustified hostility to technology generally.[56]

Technology assessment has to date been almost solely negative. The development of *affirmative* technology assessment involving promotion of socially vital innovation can have far-reaching positive consequences. Options other than continuing to use some vulnerable technologies or to forgo needed economic activity can be more often made available.

The marketplace will benefit through the infusion of new industries and new alternatives. The political system will give the people an opportunity to participate in choosing what alternatives should be made available where market

---

[52] See Thomas, *How Should Humans Pay Their Way,* NY Times, Aug 24, 1981, at A15; *compare* Gale, *The Anthropic Principle,* 245 Sci Am 154 (1981). In the legal field, *see* Steele v Lousiville & R, 323 US 192 (1944); Stone, *Public Influence of the Bar,* 48 Harv L Rev 1 (1934); Mason, Harlan Fiske Stone: Pillar of the Law 344-84 (1956).

[53] See Teilhard de Chardin, The Phenomenon of Man (1959). On increases in effectiveness based upon enhanced commitment, see wartime experiences reported in, *e.g.,* Cookridge, They Came From the Sky (1965).

[54] Jastrow, Until the Sun Dies (1977).

[55] *See* Technology Assessment Act, 2 USC §471 *et seq. See also* National Environmental Policy Act, 83 Stat 852 (1970) (codified at 42 USC §4321 *et seq*); Toxic Substances Control Act, 90 Stat 2003 (1976) (codified at 15 USC §2601 *et seq*); Consumer Product Safety Act, 86 Stat 1207 (1972), *as amended* (codified at 15 USC §1021 *et seq*). For background, see Toffler, Future Shock (1970).

[56] A typical imbroglio involving failure to use research as a tool instead of "crash" regulatory demands, followed by policy reversals is the children's sleepwear fiasco recounted in Riegel Textile Corp v Celanese Corp, 649 F2d 849 (2d Cir 1981), warned of by Richard L. Abbott in Luncheon Address before the American Apparel Mfrs Assn Feb 21, 1974.

incentives are chilled by such factors as "lumpiness" and high risk. The market can then determine the best allocation of resourses once vital new technologies are developed, both by market incentives themselves and by affirmative promotion where needed.

§12.07 Historical and Contemporary Context of Innovation

There is a sense in the United States that a new, giant step forward in technology could alter the world power balance and change the course of history dramatically, but that we lack the means to bring it about in peacetime. Highly traditional tools have already been mentioned, however, which might suffice for the purpose. These must be examined in their historical and contemporary context.

During the period from the founding of the American republic until the mid-1970s, the United States seemed to possess a limitless cornucopia of resources to be had for the taking—including, of course, old-fashioned hard work.[57] In fact, the frontier, followed by the mass production of which the automobile became the symbol in the 1920s and then the vast technological advance spurred by World War II, created an expectation of virtually automatic annual productivity increases.

The challenge which became the center of attention was not lagging productivity but that of distributing what could be produced.[58] When not enough money was in circulation to buy what was put on the market, the channels of trade became clogged and large numbers went jobless. To solve this, jobs were created by stimulating demand in the economy, first slowly through civilian programs, then massively by World War II. In that conflict energies were galvanized because of a perception that the survival of the nation and the American idea of freedom were at stake. Our announced objectives were correspondingly high, to extend liberty ultimately "everywhere in the world,"[59] jointly with our allies. Winston Churchill proclaimed that "blood, sweat and tears" if undergone bravely could lead the world to "broad, sunlit uplands."[60]

Seemingly impossible goals of industrial production were set and surpass-

---

[57] See Stone, Men to Match My Mountains: The Opening of the Far West 1840-1900 (1956); 2 Morrison & Commager, Growth of the American Republic 189-213 (4th ed 1961).

[58] See Loeb, Full Production Without War (1946).

[59] See 3 State of the Union Messages of the Presidents 2860 (1966); 2 Hofstadter, Great Issues in American History 398-99 (Vintage ed 1958), *quoted in* 52 ABAJ 1046 (1966); *see also* Harnsberger, The Lincoln Treasury 67-68 (1950) (quoting 6 Nicolay & Hay, Complete Works of Abraham Lincoln 156 (1894)).

[60] Address before the House of Commons, June 18, 1940, *quoted in* Never Give In! (Hallmark ed 1965); see Churchill, The Unrelenting Struggle: War Speeches of The Right Hon Sir Winston Churchill (1942); Humes, Churchill, Speaker of the Century (1980).

ed.[61] Because management needed every resource to lift output, employees were given greater latitude to suggest job improvements; because jobs were assured for everyone, workers were willing to cooperate in seeking more efficient methods without fear of "working themselves out of a job."[62] Federal law preempted local and even other federal approvals otherwise required before work could be done.[63] Red tape was circumvented.

The construction of defense plants was paid for in large part through the Reconstruction Finance Corporation (RFC),[64] initially created at the instance of President Herbert Hoover,[65] and used to rescue failing banks in 1933. The RFC obtained its funds the same way the central banking system does, by transactions in which a participant can issue paper guaranteed by the United States, the same process which enables the Federal Reserve Board routinely to target a specified percentage increase in the money supply for a given period.

When the physical product of the economy and money in circulation balance each other and go up or down together, the money in circulation enables the goods and services to change hands. If the money supply rises more than does the amount of goods and services on the market, obviously the money will lose some of its value through inflation. To prevent this in World War II, temporary price controls and rationing were used. Huge bond issues also soaked up dollars that might otherwise have chased scarce goods that had gone to war and were not on retail shelves. At the end of the war, many of these bonds represented assets in citizens' hands that helped to make it possible for cars, homes and much else to be purchased on a larger scale than ever before.

With the war came many new technological innovations with civilian applications that helped to make the United States the undisputed technological leader in the first postwar decades. Some, like DDT, were later banned as harmful. Most of the others are still in use: jet aviation radar, rocketry used for space exploration, antibiotics, nuclear energy and numerous others.

The Apollo mission to the moon similarly stimulated technological innovation and brought earthbound spinoffs such as teflon, as well as promoting development of communications and resource-finding satellites.

With the automatic annual productivity dividend taken for granted also came waste. We did things in the most expensive manner. Many additional social objectives were added to each basic function. Multiple agencies and legislative bodies at multiple levels of government laid down requirements, overlapping

---

[61] See Thomsen, Vote of Censure (1968); 3 State of the Union Messages of the Presidents 2861-65 (1966); *but see* Speer, Inside the Third Reich (1970).

[62] See Drucker, Concept of the Corporation (1953). Such participation promotes incremental productivity gains as well as major innovation. See Abernathy, The Productivity Dilemma (1979), (reviewed, NY Times Jan 31, 1979, at D4).

[63] See **§12.04.**

[64] Jones & Angly, Fifty Billion Dollars (1951).

[65] 47 Stat 5 (1932) (Codified at 15 USC §601).

and crisscrossing each other with little concern for the cost passed on to the consumer.

Technology showed its dangerous side, too, as thalidomide caused deformities and DDT threatened a silent spring. The technology assessment movement thus stressed a second look at technologies that might be harmful before they were implemented.

Coincident with the advent of the OPEC oil cartel came a new cost–push inflation that created a novel and most uncomfortable situation: the combination of unemployment and inflation, dubbed stagflation. Unlike the war-created demand–push inflation in which more money chased scarce goods, this kind of inflation stemmed from a real or cartel-generated scarcity of basic resources, including but far from limited to the most fundamental one, energy.

As businesses had to pay more for what they needed, they had to raise prices to be able to meet costs. When customers had difficulty paying the higher tabs for cars, refrigerators and numerous other items, sales fell, and with them employment. The combination of joblessness and rising costs was unprecedented. It brought profound consternation and an assumption by many that economic decline and with it strategic decline might be inevitable.

Manipulating the demand for goods and services obviously would not cure this sickness. Thus, the concept of supply side economics was born: if you could enhance the supply of goods and services, there would once again be an expanding economic pie, and both inflation and unemployment could be overcome. To do this, we sought to cut burdensome taxes and excessively burdensome government regulation. This alone, however, would not be enough unless it also stimulated the basic engine of the economic advance: technological innovation. Hence, the 1981 tax revisions provide incentives for private research and development.[66]

However, research does not attract investment to the full extent that the profits which might be earned would justify. Furthermore, underinvestment in research is hard to remedy through any single solution. It is difficult for tax incentives or similar devices to overcome the major causes of underinvestment in research, especially where very big, lumpy but critical projects are concerned, such as space industrialization or development of new energy sources.[67]

The consequences of underinvestment are far-reaching. A lag in technology and productivity contributes to cost–push inflation. With limited resources, pressures for maintenance of social benefits clash with demands for more incentives for the private sector, leading to unacceptable alternatives. Few expect these tax reductions and reduction of over-regulation alone to overcome the technological lag without other measures. For example, among the alternatives discussed in this chapter, affirmative promotion of research, defined as vital to national purposes, could offer hope of assisting to overcome

---

[66] See **ch 5.**

[67] But see Surrey, *Tax Incentives as a Device for Implementing Government Policy,* 83 Harv L Rev 705 (1970).

the economy of scarcity and its ill effects.[68]

A major advance in technology, however, could not only lay the foundation for a new convergence of interests of nontotalitarian nations, but allow technological jobs not requiring advanced degrees to be opened to disadvantaged citizens. The Reverend Leon Sullivan of Philadelphia proved that this could be done when many uneducated, unemployed individuals—including many who could hardly read initially—were trained and put to work on skilled tasks for the Apollo project. The key, Sullivan explained in 1969 in *Build, Brother, Build,* was enthusiastic and sympathetic leaders, and the assurance of jobs, for which the training was expressly designed. Jobs are the most powerful means of raising the bottom tier of society, as World War II demonstrated. The initiative of youth such as high school physics students who helped NASA plan orbits around Mercury,[69] the latent abilities of the untrained tapped by Sullivan for Apollo and the knowledge of the retired as well as the knowledge of the trained would all be needed in such a context, with important social as well as technical benefits.[70]

---

[68] See Childe, What Happened in History (1946); Mansfield, *Basic Research and Productivity Increases in Manufacturing,* 70 Am Econ Rev 863 (Dec 1980); Starr & Rudman, *Parameters of Technological Growth,* 182 Sci 358 (Oct 26, 1973); Mansfield, *Research and Development, Productivity and Inflation,* 209 Sci 1091 (Sept 5, 1980); David, *Science Futures: The Industrial Connection,* 203 Sci 837 (Mar 2, 1979); Hutten, *On the Importance of Productivity Change,* 69 Am Econ Rev 126 (1979); Slichter, *The Industry of Discovery,* 128 Sci 1610 (Dec 22, 1958); Terleckyj, *Effects of R&D on the Productivity Growth of Industries* (National Planning Assn 1974); Jones, *Progress, Remember?,* NY Times, Aug 24, 1979, at A25; Lewis, *Technology, Enterprise and American Economic Growth,* 215 Sci 1204 (Mar 5, 1982). *See also Destinies for American Research, Hearing before the Subcommittee on Energy Research and Production, House Committee on Science & Technology,* 96th Cong, 1st Sess (1979); House Committee on Science & Technology, 96th Cong, 2d Sess, Technology and Innovation for Manufacturing: Proceedings of Conference (Serial FF, May 4, 1979). For congressional recognition, *see* 15 USC §3701. *See also* Letters, NY Times, April 26, 1981, at 22E.

[69] Wilford, *Craft Speeds Toward Mercury on Wings of Student's Idea,* NY Times, Nov 4, 1979, §1, at 64; *compare* in former periods, Stone, Immortal Wife (1944); *compare also* Givens, *Toward Fuller Use of Human Resources,* 17 Labor LJ 621 (1966); Wirtz, The Boundless Resource (1975).

[70] See McCarroll, *Rehiring Retirees,* NY Times, Nov 22, 1981, at 8F.

Cases

A

D

F

G

H

Heart of Atlanta Motel, Inc v United States, 379 US 241 (1964) **§3.15**

Heatransfer Corp v Volkswagenwerk, AG, 553 F2d 964 (5th Cir 1977), *cert denied,* 434 US 1087 (1978) **§3.06**

Heheman v EW Scripps Co, 661 F2d 1115 (6th Cir 1981) **§2.02**

Heisse v State, 519 F Supp 36 (D Vt 1980) **§8.27**

Helvering v Lazarus Co, 308 US 252 (1939) **§5.04**

Henderson v Hayden, Stone, Inc, 461 F2d 1069 (5th Cir 1972) **§§6.25, 6.47**

Henningsen v Bloomfield Motors, Inc, 32 NJ 358, 161 A2d 69 (1960) **§10.02**

Heublein, Inc, FTC Complaints & Orders (CCH) ¶21,783, Antitrust & Trade Reg Rep (BNA) F-1 (Oct 23, 1980) (FTC Oct 7, 1980) **§3.02**

Hildreth v Mastoras, 257 US 27 (1921) **§4.15**

Hill v A-T-O, Inc, 535 F2d 1349 (2d Cir 1976) **§3.05**

Hill v Der, 521 F Supp 1370, [1981-82 Transfer Binder] Fed Sec L Rep (CCH) ¶98,311 (D Del 1981) **§6.42**

Hines v Davidowitz, 312 US 52 (1941) **§8.28**

Hirsh v Dupont, 396 F Supp 1214 (SDNY 1975) **§6.06**

Hitchman Coal & Coke Co v Mitchell, 245 US 229 (1917) **§2.03**

HK Porter Co v Black & Decker Mfg Co, 182 USPQ 401 (ND Ill 1974), *affd,* 518 F2d 1177, 187 USPQ 476 (7th Cir 1975) **§4.01**

Hobbs v United States Atomic Energy Commn, 451 F2d 849, 171 USPQ 713 (5th Cir 1971) **§4.06**

Hoffman v Sterling Drug, Inc, 485 F2d 132 (3d Cir 1973) **§10.25**

Holland v Metropolitan Life Ins Co, 1981 Prod Liab Rep (CCH) ¶9142 **§10.25**

Home Box Office v Directors Guild of America, 531 F Supp 578 (EDNY 1982) **§2.03**

Hospital Bldg Co v Trustees of Rex Hosp, 425 US 738 (1976) **§§3.02, 3.15**

Hotchkiss v Greenwood, 11 How (52 US) 248 (1851) **§§4.01, 4.16**

Hotel Sec Checking Co v Lorraine Co, 160 F 467 (2d Cir 1908) **§4.02**

HP Hood & Sons v DuMond, 336 US 525 (1949) **§§3.02, 12.04**

Hunt v Crumboch, 325 US 821 (1945) **§2.03**

Hunt v Mobil Oil Co, 550 F2d 68 (2d Cir), *cert denied,* 434 US 984 (1977) **§3.12**

Hunt v Washington Apple Advertising Commn, 432 US 333 (1977) **§§1.01, 3.02, 3.15, 12.01, 12.04**

Hunter v City of Pittsburgh, 207 US 161 (1907) **§8.02**

Hunt-Wesson Foods, Inc v Ragu Foods, Inc, 627 F2d 919 (9th Cir 1980) **§3.05**

Huron Portland Cement Co v Detroit, 362 US 440 (1960) **§3.15**

Huset v JI Case Threshing Mach Co, 120 F 865 (8th Cir 1903) **§10.01**

Hutchinson, City of, v Otto, 306 Minn 136, 235 NW2d 604 (1975) **§8.17**

Hvidsten v United States, 185 F Supp 856 (DND 1960) **§5.02**

I

IAM v OPEC, 649 F2d 1354 (9th Cir 1981) **§§1.01, 1.02, 3.12**

IBM Corp v United States, 298 US 131 (1936) **§3.06**

IBM Peripheral EDP Devices Antitrust Litig, *In re,* 481 F Supp 965 (1979 ND Cal) **§§3.05, 3.07, 11.13**

ILC Peripherals Leasing Corp v IBM (Memorex), 458 F Supp 423 (ND Cal 1978), *affd,* 636 F2d 1188 (9th Cir 1981) **§§3.04, 3.05**

Industrial Assn v United States, 268 US 64 (1925) **§2.03**

Ingersoll-Rand Co v Brunner Lay, Inc, 474 F2d 491, 177 USPQ 112 (5th Cir 1973) **§4.18**

Interamerican Ref Corp v Texaco Maracaibo, Inc, 307 F Supp 1291 (D Del 1970) **§3.12**

Interlego AG v FAO Schwarz, Inc, 191 USPQ 129 (ND Ga 1976) **§4.08**

Interlego AG v FAO Schwarz, Inc, 187 USPQ 580 (ND Ga 1975) **§4.08**

International Air Indus, Inc v American Excelsior Co, 517 F2d 714 (5th Cir 1975), *cert denied,* 424 US 943 (1976) **§3.07**

International Mfg Co v Landon, Inc, 336 F2d 723 (9th Cir), *cert denied,* 379 US 998 (1964) **§3.10**

International News Serv v Associated Press, 248 US 215 (1918) **§11.10**

International Salt Co v United States, 332 US 392 (1947) **§3.03, 3.06**

Interstate Cigar Co v Sterling Drug, Inc, 655 F2d 29 (2d Cir 1981) **§3.07**

Interstate Circuit, Inc v United States, 306 US 108 (1939) **§§3.03, 3.13, 3.14**

Island Tobacco Co, Ltd v RJ Reynolds Indus, Inc, 513 F Supp 726 (D Hawaii 1981) **§3.07**

IT&T v General Tel & Elecs Corp, 449 F Supp 1158 (D Hawaii 1978) **§3.07**

J

Jack Kahn Music v Baldwin Piano & Organ Co, 604 F2d 755 (2d Cir 1979) **§3.09**

Jack Winter, Inc v Koratron Co, 375 F Supp 1 (ND Cal 1974) **§3.08**

James B Lansing Sound, Inc, 46 Fed Reg 29928 (FTC June 4, 1981) **§3.16**

Janich Bros v American Distilling Co, 570 F2d 848 (9th Cir 1977), *cert denied,* 439 US 829 (1979) **§3.07**

Japan Line Ltd v County of Los Angeles, 441 US 434 (1979) **§8.29**

Jarvis Surgical Co v Davis, 15 Misc 2d 1035, 186 NYS2d 357 (Sup Ct 1958) **§2.05**

Jefferson v State, 527 P2d 37 (Alaska 1974) **§8.02**

JI Case v Borak, 377 US 426 (1964) **§6.67**

JI Case Co v NLRB, 321 US 332 (1944) **§2.05**

John Nuveen & Co v Sanders, 619 F2d 1222 (7th Cir 1980), *cert denied,* 450 US 1005 (1981) **§§6.42, 6.47**

Johnson v Maryland, 254 US 51 (1920) **§12.04**

Johnson Constr Co v Township of White Lake, 351 Mich 374, 88 NW2d 426 (1958) **§8.17**

Jones, *In re,* 373 F2d 1007 (CCPA 1967) **§4.02**

Joseph E Seagram & Sons v Hawaiian Oke & Liquors, Ltd, 416 F2d 71 (9th Cir 1969), *cert denied,* 396 US 1062 (1970) **§§3.03, 3.07**

Jou-Jou Designs, Inc v ILGWU, 643 F2d 905 (2d Cir 1981) **§§2.01, 2.02, 2.03, 2.05**

J Radley Metzger Co v Fay, 4 AD2d 436, 166 NYS2d 87 (1st Dept 1957) **§2.05**

K

Kahn, *Ex parte,* 132 USPQ 359 (POBA 1961) **§4.18**

Kearney & Trecker Corp v Guiddings & Lewis, Inc, 452 F2d 579 (7th Cir 1971), *cert denied,* 405 US 1066 (1972) **§3.08**

Keene Corp v Insurance Co of N Am, 667 F2d 1034 (DC Cir 1981), *revg* 513 F Supp 47 (DDC 1981), *cert denied,* 50 USLW 3716 (Mar 9, 1982), *reh denied,* __ USLW __ (Apr 27, 1982) **§10.28**

Keller v Commissioner, 77 TC No 70 (1981), *appeal filed* (10th Cir, Apr 5, 1982) **§5.02**

Kennedy v Hazelton, 128 US 667 (1888) **§4.04**

Kern-Limerick, Inc v Scurlock, 347 US 110 (1954) **§12.04**

Kewanee Oil Co v Bicron Corp, 416 US 470 (1974) **§3.08**

Keynard v Nego Corp, 33 F2d 1026 (2d Cir 1980) **§2.05**

Kiefer-Stewart Co v Joseph E Seagram & Sons, 340 US 211 (1951) **§3.07**

Kingsberry Homes v Gwinnett County, 248 F Supp 765 (ND Ga 1965) **§8.17**

Kingsley Intl Pictures Corp v Regents, 360 US 684 (1959) **§3.15**

Kinnett Dairies, Inc v Dairymen, Inc, 512 F Supp 608 (MD Ga 1981) **§3.07**

Klor's, Inc v Broadway-Hall Stores, Inc, 359 US 207 (1959) **§§3.03, 3.13**

Kodshey Elecs, Inc v Mechanex Corp, 486 F2d 449 (10th Cir 1973) **§2.06**

Koehring Co v National Automatic Tool Co, 362 F2d 100, 149 USPQ 887 (7th Cir 1966) **§§4.06, 4.08**

Kotrich v County of Du Page, 19 Ill 2d 181, 166 NE2d 601 (1960) **§8.20**

L

La Belle v McCauley Indus Corp, 649 F2d 46 (1981) **§10.08**

Lafayette, City of, v Louisiana Power & Light Co, 435 US 389 (1978) **§3.11**

Laitram Corp v King Crab, Inc, 244 F Supp 9 (D Alaska 1965), *modified,* 245 F Supp 1019 (1966) **§3.08**

Lambert v California, 355 US 225 (1957) **§3.15**

Lancaster Colony Corp v Alden Accessories, Ltd, 506 F2d 1197, 184 USPQ 193 (2d Cir 1974) **§4.02**

Landmark Communications, Inc v Virginia, 435 US 829 (1978) **§3.15**

Lane, *In re,* 58 Cal 2d 99, 372 P2d 897, 22 Cal Rptr 857 (1962) **§8.02**

M

N

O

P

R

S

T

Talley v California, 362 US 60 (1960) **§3.15**

Tampa Elec Co v Nashville Coal Co, 365 US 320 (1961) **§§3.05, 3.07, 11.07**

The Telephone Cases, 126 US 1 (1887) **§4.18**

Telex Corp v IBM, 367 F Supp 258 (ND Okla 1973), *revd,* 510 F2d 894 (10th Cir 1975) **§11.08**

Telex Corp v IBM Peripheral EDP Devices Litig, 481 F Supp 965 (ND Cal 1979) **§3.04**

Texas v EPA, 499 F2d 289 (5th Cir 1974) **§7.56**

Texas Gulf Sulphur Co v JR Simplot Co, 418 F2d 793 (9th Cir 1969) **§3.07**

Textile Workers v Lincoln Mills, 353 US 448 (1957) **§2.04**

Theatre Enters, Inc v Paramount Film Distrib Corp, 346 US 537 (1954) **§§3.03, 3.13, 3.14**

Theis, *In re,* 610 F2d 768, 204 USPQ 188 (CCPA 1979) **§4.08**

Thibault v Sears, Roebuck, & Co, 118 NH 802, 395 A2d 843 (1978) **§§10.08, 10.19**

Thomas v Winchester, 6 NY 397 (1852) **§10.01**

Thuau, *In re,* 135 F2d 344, 57 USPQ 324 (CCPA 1943) **§4.02**

Timberlane Lumber Co v Bank of America, 549 F2d 597 (9th Cir 1977) **§3.12**

Timely Prods Inc v Costanzo, 465 F Supp 91, 201 USPQ 567 (D Conn 1979) **§4.19**

Times Picayune Pub Co v United States, 345 US 594 (1953) **§2.03**

Timken Roller Bearing Co v United States, 341 US 593 (1951) **§§3.03, 3.07, 3.12**

Tinnerman Prods, Inc v George K Garrett Co, 185 F Supp 151 (ED Pa 1960), *affd,* 292 F2d 137 (3d Cir), *cert denied,* 368 US 833 (1961) **§3.08**

The TJ Hooper, 60 F2d 737 (2d Cir 1932) **§10.15**

Tlapek v Chevron Oil Co, 407 F2d 1129, 161 USPQ 193 (8th Cir 1969) **§4.07**

Tose v First Pa Bank, 648 F2d 879 (3d Cir 1981) **§3.07**

Touche Ross v SEC, 609 F2d 470 (2d Cir 1979) **§6.63**

Touche Ross & Co v Redington, 442 US 560 (1979) **§§6.50, 6.56**

Train v Natural Resources Defense Council, Inc, 421 US 60 (1975) **§§7.23, 7.24**

Transamerica Computer Co v IBM, 481 F Supp 965 (ND Cal 1979) **§3.04**

TransAmerica Mortgage Investors, Inc v Lewis, 444 US 11 (1979) **§§6.50, 6.56**

Transparent-Wrap Mach Corp v Stokes & Smith Co, 329 US 637 (1947) **§§3.08, 4.19**

Trenton, City of, v New Jersey, 262 US 182 (1923) **§8.02**

Tripoli Co v Wella Corp, 425 F2d 932 (3d Cir 1970), *cert denied,* 400 US 831 (1970) **§§3.09, 11.14**

Truax v Corrigan, 257 US 312 (1921) **§2.03**

TSC Indus, Inc v Northway, Inc, 426 US 438 (1976) **§§6.50, 6.56**

U

Uloth v City Tank Corp, 376 Mass 874, 384 NE2d 1188 (1978) **§10.19**

United States v Cotton Baking Co, 513 F Supp 223 (ED Pa 1981) **§3.07**

United States v Crescent Amusement Co, 323 US 173 (1944) **§§3.06, 3.08, 3.14**

United States v Dairymen, Inc, 1978-1 Trade Cas (CCH) ¶62,053 (WD Ky 1978), *affd in part, revd in part on other grounds,* 660 F2d 192 (6th Cir 1981) **§§3.07, 3.16**

United States v Darby, 312 US 100 (1941) **§3.15**

United States v Dubilier Condenser Corp, 289 US 178 (1933) **§4.03**

United States v EI duPont de Nemours & Co (cellophane), 118 F Supp 41 (D Del 1953), *affd,* 351 US 377 (1956) **§§1.02, 3.03, 3.05, 3.10, 4.13, 4.19, 11.23, 11.24**

United States v EI duPont de Nemours & Co (GM), 353 US 586 (1957) **§3.16**

United States v Eli Lilly & Co, 1959 Trade Cas (CCH) ¶69,536 (DNJ 1959) **§3.11**

United States v Ethyl Gasoline Co, 309 US 435 (1940) **§3.08**

United States v Eversharp, Inc, 1967 Trade Cas (CCH) ¶72,221 (Pa 1967) **§11.22**

United States v Falstaff Brewing Corp, 410 US 526 (1973) **§11.22**

United States v Gard, 344 F2d 120 (2d Cir 1965) **§2.05**

United States v General Dynamics Corp, 415 US 486 (1974) **§§3.05, 3.16, 11.05**

United States v General Dynamics Corp, 258 F Supp 36 (SDNY 1966) **§3.06**

United States v General Elec Co, 115 F Supp 835 (DNJ 1953) **§3.08**

United States v General Elec Co, 80 F Supp 989 (SDNY 1952) **§3.08**

United States v General Elec Co, 272 US 476 (1926) **§3.08**

United States v General Motors Corp, 384 US 127 (1966) **§§3.09, 3.16**

United States v Griffith, 334 US 100 (1948) **§§3.03, 3.05, 3.08, 3.14**

United States v Grinnell Corp, 384 US 563 (1966) **§3.03**

United States v Hooker Chem & Plastics Corp, 16 Envt Rep Cas (BNA) 1268 (SDNY 1981) **§7.51**

United States v Hutcheson, 312 US 219 (1941) **§2.03**

United States v Imperial Chem Indus, 105 F Supp 215 (SDNY 1954) **§3.08**

United States v IT&T, 324 F Supp 19 (D Conn 1970), *appeal dismissed,* 404 US 801 (1971) **§3.06**

United States v Jelenko, 23 F2d 511 (4th Cir 1927) **§5.04**

United States v Jerrold Elecs Corp, 187 F Supp 545 (ED Pa 1960), *affd per curiam,* 365 US 567 (1961) **§§3.05, 3.06, 4.19**

United States v Johns-Mansville Corp, 259 F Supp 440 (ED Pa 1966) **§3.10**

United States v Koppers Co, 652 F2d 290 (2d Cir 1981) **§3.03**

United States v Krasnov, 143 F Supp 184, 110 USPQ 411 (ED Pa 1956), *affd per curiam,* 355 US 5 (1957) **§§3.08, 4.19**

United States v Lehigh Valley RR, 254 US 225 (1920) **§§2.03, 3.14**

United States v Lever Bros, 216 F Supp 886 (SDNY 1963) **§3.08**

United States v Line Material Co, 333 US 287 (1948) **§§3.08, 3.10**

V

W

Y

Z

Treaties

International Convention for the Protection of Industrial Property (1883) (The Paris Convention, Conference of Stockholm, July 14, 1967), 3 Manual for the Handling of Applications for Patents, Design and Trade Marks Throughout the World (Octrooibureau Los en Steiger, Amsterdam, Holland) **§4.11**

Patent Cooperation Treaty, 3 Manual for the Handling of Applications for Patents , Design and Trade Marks Throughout the World (Octrooibureau Los en Steiger, Amsterdam, Holland) **§4.14**

Statutes

United States Constitution

United States Code

See Statutes at Large table for references to the following statutes:

Statutes at Large

See United States Code table for references to the following statutes:

Internal Revenue Code

Model Codes and Acts

Uniform Commercial Code

State Constitutions

State Statutes

Administrative Materials

Code of Federal Regulations

Federal Register

Internal Revenue Service

Treasury Regulations

Revenue Rulings

Revenue Procedures

Letter Rulings

Securities and Exchange Commission

Releases—Securities Act of 1933

Releases—Securities Exchange Act of 1934

Securities Exchange Act of 1934 Release No 15664 (Mar 21, 1979) **§6.66**

Securities Exchange Act of 1934 Release No 15719 (SEC Staff Report on Transactions in the Marine Protein Corporation Industrial Development Revenue Bonds) (Apr 11, 1979) **§§6.20, 6.22, 6.68**

Securities Exchange Act of 1934 Release No 18114 (Sept 23, 1981) **§6.53**

Securities Exchange Act of 1934 Release No 18189 (Proposed System for Classification of Smaller Issuers for Purposes of Exempting Them from Certain Reporting and Other Requirements) (Oct 20, 1981) **§§6.01, 6.16**

Securities Exchange Act of 1934 Release No 18417 (Jan 13, 1982) **§6.47**

Securities Exchange Act of 1934 Release No 18418 (Jan 13, 1982) **§6.47**

Securities Exchange Act of 1934 Release No 18419 (Jan 13, 1982) **§6.47**

Release—Investment Company Act of 1940

Investment Company Act of 1940 Release No 11552 (Jan 14, 1981) **§6.41**

Administrative Proceedings

Public Investigation in the Matter of the Hot Issues Securities Markets, SEC Administrative Proceeding File No 4-148 **§6.13**

Public Fact-Finding Investigation and Rule-making Proceeding in the Matter of "Going Private" Transactions by Public Companies or Their Affiliates, SEC Administrative Proceeding File No 4-178 **§6.13**

Summary of Comments Relating to: Small Business Hearings and Proposed Form S-18, SEC File No S7-734 **§6.16**

Reports

Form S-18, A Monitoring Report on the First 18 Months of its Use (SEC Mar 1981), summarized in 595 Sec Reg L Rep (BNA) A-17 (Mar 18, 1981) **§6.37**

Report of the Securities and Exchange Commission on Questionable and Illegal Corporate Payments (submitted to the Senate Banking, Housing and Urban Affairs Committee, May 12, 1976) **§6.57**

SEC Staff Report on Corporate Accountability (printed for the Senate Committee on Banking, Housing and Urban Affairs, 96th Cong, 2d Sess, Sept 4, 1980) **§6.58**

News Digests and Releases

SEC Monthly Statistical Review, Release PR-32 (July 1981) **§6.13**

SEC News Digest, Nov 17, 1980, at 1 **§6.66**

SEC News Digest, Issue 81-46, July 30, 1981, at 1 **§6.37**

Legislative Materials (Documents and Legislative History)

S Rep No 316, 96th Cong, 1st Sess (1979) **§2.06**

Advisory Committee on Corporate Disclosure, 95th Cong, 1st Sess, Report to the Securities and Exchange Commission, (Comm Print No 29, House Comm on Interstate and Foreign Commerce 1977) **§§6.01, 6.59**

House Committee on Science and Technology, 96th Cong, 2d Sess, Technology and Innovation for Manufacturing: Proceedings of Conference (Serial FF, May 4, 1979) **§12.07**

3 Institutional Investor Study Report of the Securities and Exchange Commission, HR Doc No 64, 92d Cong, 1st Sess (Mar 10, 1971) **§6.46**

Subcommittee on Priorities and Economy in Government, Joint Economic Committee, 94th Cong, 2d Sess, Priorities and Efficiency in Federal Research and Development (1976) **§§11.11, 11.21**

The Whistleblowers: Report Prepared for the Senate Committee on Governmental Affairs, 95th Cong, 2d Sess (Comm Print 1978) **§2.07**

Bank Holding Company Act Amendments, Hearings before the House Committee on Banking and Currency, 91st Cong, 1st Sess (1969) **§§3.08, 3.13**

Destinies for American Research, Hearing before the Subcommittee on Energy Research and Production, House Committee on Science and Technology, 96th Cong, 1st Sess (1979) **§12.07**

Federal Incentives for Innovation, Hearings before the Special Subcommittee on Science, Technology and Commerce, Senate Commerce Committee, Serial No 57, 93d Cong, 1st Sess (1973) **§12.03**

Federal Trade Commission, Regional Office Operations, Hearings before a Subcommittee of the House Committee on Government Operations, 95th Cong, 1st Sess (1977) **§2.07**

House Judiciary Committee, Hearings on Bills to Amend the Clayton Act, 74th Cong, 2d Sess (1975) **§3.07**

National Science Policy, Hearings before the Subcommittee on Science, Research & Development of the House Committee on Science and Astronautics, 91st Cong, 2d Sess (1970) **§12.03**

Propelled Lighter-Than-Air Vehicles, Hearings before the Subcommittee on Science, Technology and Space, Senate Committee on Commerce, Science and Transportation, Serial No 6, 96th Cong, 1st Sess (1979) **§12.05**

Symposium on the Future of Space Science and Space Applications, Hearings before the Subcommittee on Science, Technology and Space, Senate Commerce Committee, Serial No 58, 95th Cong, 2d Sess (1978) **§12.05**

Joint Hearing of the Senate Finance Committee and the Assembly Ways and Means Committee on the State Operations and Capital Projects Budgets, 125th Annual Legislative Sess (Albany, NY, Feb 4, 1982) (testimony of George M Low, President, Rensselaer Polytechnic Institute) **§9.16**

Accounting Materials

Authorities

A

Aaron, *Reflections on the Legal Nature and Enforceability of Seniority Rights,* 75 Harv L Rev 1532 (1962) **§2.02**

1 ABA Antitrust Sec, The FTC as an Antitrust Enforcement Agency: The Role of Section 5 of the FTC Act in Antitrust Law (Monograph No 5, 1981) **§§2.07, 3.03**

ABA Antitrust Sec, The Robinson-Patman Act: Policy and Law (Monograph No 4, 1980) **§3.16**

ABA Antitrust Sec, Vertical Restraints Limiting Intrabrand Competition (Monograph No 2, 1977) **§3.09**

Adams, *The Automobile—A Luxury Becomes a Necessity,* in Hamilton, Price and Price Policies 27-82 (1938) **§§2.01, 3.07**

Adams, 1 State of the Union Messages of the Presidents 246 (1966) **§§1.06, 12.02**

Administration of Building Regulations (Claiborne ed 1973) **§8.16**

Advisory Commission on Intergovernmental Relations, Building Codes: A Program for Intergovernmental Reform (1966) **§8.12**

Aggressive Innovation and Antitrust Liability, 53 S Cal L Rev 1469 (1980) **§3.02**

Agricultural Services Division, Food and Agricultural Organization, United Nations, Worldwide List of Textile Research Institutes (1974) **§9.03**

Allport, Becoming: Basic Considerations for a Psychology of Personality (1955) **§2.08**

Aloi, *Legal Problems in Planned Unit Development: Uniformity, Comprehensive Planning Conditions, and the Floating Zone,* 1 Real Est LJ 5 (1972) **§8.20**

American Assembly, Automation and Technological Change (1962) **§2.02**

American Bar Foundation, Overcoming Legal Uncertainties About Use of Solar Energy Systems (1978) **§8.06**

American National Standards Institute, Product Alerting Accident Prevention Signs (Feb 1981) **§10.19**

Andersen, *The Antitrust Consequences of Manufacturer-Suggested Retail Prices,* 54 Wash L Rev 763 (1979) **§3.09**

Antitrust Division, US Dept of Justice, Antitrust Guide Concerning Research Joint Ventures, *reprinted in* 1980 Antitrust & Trade Reg Rep (BNA) No 992 (special supp Dec 4, 1980) **§§3.10, 3.11, 11.01, 11.11**

Areeda, *Predatory Pricing,* 49 ABA Antitrust LJ 897 (1980) **§3.07**

3 Areeda & Turner, Antitrust Law (1978) **§3.07**

5 Areeda & Turner, Antitrust Law (1980) **§3.06**

Areeda & Turner, *Predatory Pricing and Related Practices Under Section 2 of the Sherman Act,* 88 Harv L Rev 697 (1975) **§3.07**

Aronson, *Economic Scene,* NY Times, Mar 26, 1982, at D2 **§12.03**

Argyris, Organization and Innovation (1965) **§2.07**

Arthur D Little, Inc, Residential Solar Heating and Cooling Constraints and Incentives, A Review of the Literature (1976) **§§8.06, 8.07**

Asch, *Effects of Group Pressure Upon the Modification of Judgments,* in Readings in Social Psychology 2 (rev ed 1952) **§2.07**

Ashford & Heaton, *Regulation and Technical Innovation,* 5 EPAJ 32 (1979) **§7.57**

Ashworth, Green, Pollack, Odland, Saltonstall & Perelman, The Implementation of State Solar Incentives: A Preliminary Assessment (1979) **§8.26**

AT&T's Price Tactics Prod Firms to Replace Old Phone Equipment, Wall St J, Apr 24, 1981, at 1 **§11.13**

Atwood & Brewster, Antitrust and American Business Abroad (Shepard's/McGraw-Hill, 2d ed 1981) **§3.02**

Austin, Price Discrimination (2d rev ed 1959) **§3.07**

Automation and Technological Change (Dunlop ed 1962) **§11.18**

Ayres, *Envelope Curve Forecasting,* in Technological Forecasting for Industry and Government (Bright ed 1968) **§§11.18, 11.27**

Axelrod, *The Evolution of Cooperation,* 211 Sci 1309 (Mar 27, 1981) **§2.01**

Axinn, *Predatory Pricing: Chaos in California,* NYLJ 1 (Nov 20, 1979) **§3.07**

B

Bachman, *Cushioning the Impact of Technological Change,* 13 Labor LJ 731 (1962) **§2.02**

Back to School, Wall St J, Mar 3, 1981, at 1 **§11.29**

Bain, Industrial Organization (1959) **§§11.05, 11.08**

Bain, Pricing, Distribution, and Employment (1953) **§11.25**

Baker, *Agenda for the Reagan Administration: It's Time to Rethink Assumptions,* National LJ, Feb 23, 1981, at 32 **§3.11**

Baker, *Cheese & Macaroni,* NY Times, June 19, 1979, at A 21 **§§3.03, 3.08**

Baker, *Combinations and Conspiracies—Is There A Difference?,* 14 Antitrust Bull 71 (1969) **§3.05**

Baker, *Terminal Education,* NY Times Magazine, Nov 9, 1980, at 29 **§1.01**

Bakke, *Joint Efforts in Developing Standards,* 44 ABA Antitrust LJ 337 (1975) **§3.10**

Baldwin, Hare & McGovern, The Preparation of a Product Liability Case (1981) **§10.04**

Banschich, *High Tech Fever Grabs States, Cities,* High Tech 18 (Mar/Apr 1982) **§9.09**

Barmat, *Two Trees or One: The Problem of Intra-Enterprise Conspiracy,* 23 Mont L Rev 158 (1962) **§3.07**

Barnett, Machinery and Labor (1926) **§§2.02, 2.03**

Barney, *et al,* The Global 2000 Report to the President (1980) **§§11.01, 11.04, 11.06, 11.30**

Barron, The Future With Microelectronics (1979) **§§11.17, 11.19**

Baumol, *On the Proper Cost Tests for Natural Monopoly in a Multiproduct Industry,* 67 Am Econ Rev 809 (Dec 1977) **§11.06**

Baxter, *Legal Restrictions on Exploitation of the Patent Monopoly: An Economic Analysis,* 76 Yale LJ 267 (1966) **§3.08**

Baxter, Remarks before the Senate Judiciary Committee, *reported in* Antitrust & Trade Reg Rep (BNA) at A-10 (Mar 2, 1981) **§§3.03, 3.09, 3.10**

Beasley, Products Liability and the Unreasonably Dangerous Requirement (1981) **§10.04**

Bell, The Coming of Post-Industrial Society (1973) **§§11.04, 11.17, 11.19, 11.21, 11.27, 11.29**

Bergmann, The Riddle of Gravitation (1968) **§12.05**

Berle, Power Without Property (1959) **§§2.03, 3.05, 3.16**

Bernstein, *Recreating the Power of the Sun,* NY Times Magazine, Jan 3, 1982, at 18 **§§12.02, 12.05**

Berry, The Next Ten Thousand Years (1974) **§§1.06, 12.05, 12.06**

Bezdek, Margolin, Sparrow, Sponsler, Ezra, Spongberg, Miller, Meeker, Roseman & Misa, Analysis of Policy Options for Accelerating Commercialization of Solar Heating and Cooling Systems (1977) **§8.07**

Bickel, The Least Dangerous Branch (1962) **§§3.15, 3.16**

Bickner, Science at the Service of Government: California Tries to Exploit an Unnatural Resource (Public Policy Research Organization) University of California at Irvine, 1971) **§9.09**

Birnbaum, *Comparative Fault, Contribution and Indemnification,* in Product Liability of Manufacturers: Prevention and Defense 1981 (Ross & Foley eds) **§10.31**

Birnbaum, *Design Defects,* in Product Liability of Manufacturers: Prevention and Defense 1981 (Ross & Foley eds) **§10.15**

Birnbaum, *Unmasking the Test for Design Defect: From Negligence [To Warranty] to Strict Liability to Negligence,* 33 Vand L Rev 593 (1980) **§10.04**

Bischel, *Intercompany Transfers of Technology,* 5 Intl Tax J 181 (1979) **§§11.11, 11.16**

Black, *The Lawfulness of the Segregation Decisions,* 69 Yale LJ 421 (1960) **§3.15**

Black, The People and the Court (1960) **§§3.15, 3.16**

Blair, Economic Concentration: Structure, Behavior, and Public Policy (1972) **§11.24**

Blake, *Employee Agreements Not to Compete,* 73 Harv L Rev 625 (1960) **§11.10**

Blake & Jones, *Toward A Three-Dimensional Antitrust Policy,* 65 Colum L Rev 422 (1965) **§3.02**

Blake & Rahl, Business Regulation in the Common Market Nations (1969) **§3.02**

Blakeslee, *Experts Press for Space Colony,* NY Times, Aug 23, 1975, at 47 **§12.05**

Blechman, *Conscious Parallelism, Signalling and Facilitating Devices: The Problem of Tacit Collusion Under the Antitrust Laws,* 24 NYL Sch L Rev 881 (1979) **§§3.03, 3.13**

Blumrosen, *Seniority Rights and Industrial Change,* 47 Minn L Rev 505 (1963) **§2.02**

Bock, *Antitrust Blueprint for the 1980s,* 55 NYU L Rev 574 (1980) **§3.15**

Bock, The Innovator as an Antitrust Target (Conference Board Bull No 74, 1980) **§3.04**

Bode, Synergy (1971) **§§11.06, 11.08, 11.15, 11.20, 11.22, 11.28**

Bodenhausen, *Protection of "Neighboring Rights",* 19 Law & Contemp Prob 156 (1954) **§11.10**

Boorstin, The Americans: The Democratic Experience (1973) **§11.07**

Bork, The Antitrust Paradox (1978) **§§3.03, 3.13, 11.01**

Bowen, Investment in Learning (1977) **§11.29**

Bowman, *The Prerequisites and Effects of Resale Price Maintenance,* 22 U Chi L Rev 825 (1955) **§3.16**

Box & Draper, Evolutionary Operation (1969) **§11.06**

Box & Jenkins, Time Series Analysis (1970) **§11.27**

Braeutigan, *The Effect of Uncertainty in Regulatory Delay on the Rate of Innovation,* 43 Law & Contemp Prob 98 (1979) **§11.14**

Bramson, Coping With Difficult People (1981) **§2.01**

Brandt, *et al,* North-South: A Program for Survival (1980) **§§11.01, 11.04, 11.30**

Brebbia, *The Role of Advisory Opinions and the Business Review Procedure,* 18 Antitrust Bull 191 (1973) **§3.11**

Breneman & Finn, *Agenda for the Eighties,* in Public Policy and Private Higher Education (Breneman & Finn eds 1978) **§11.29**

Brewster, Antitrust & American Business Abroad (1958) **§3.12**

Bright, A Brief Introduction to Technology Forecasting: Concepts and Exercises (1972) **§11.27**

Briloff & Engler, Accountancy and the Merger Movement: A symbiotic Relationship, 5 J Corp L 81 (Fall 1979) **§6.63**

Brock, The US Computer Industry: A Study of Market Power (1975) **§§11.11, 11.15, 11.17**

Brodsky, Kaufman & Tooker, University/Industry Cooperation: A Preliminary Analysis of Existing Mechanisms and Their Relationship to the Innovation Process (Center for Science and Technology Policy, Graduate School of Public Administration, New York University 1980) **§9.07, 9,10, 9.16**

Bromberg, Securities Law: Fraud, SEC Rule 10b-5 (Shepard's/McGraw-Hill 1967) **§6.67**

Brooks, The Go-Go Years (1973) **§11.11**

Brooks, The Mythical Man-Month (1975) **§§11.15, 11.19, 11.20**

Brown, Franchising: Realities and Remedies (1978) **§3.09**

Brown, Laying Waste: The Poisoning of America by Toxic Chemicals (1981) **§7.51**

Browne, *Is Problem Solving America's Lost Art?,* NY Times, July 29, 1980, at C1 **§1.01**

Bruff, *Judicial Review in Local Government Law: A Reappraisal,* 60 Minn L Rev 669 (1976) **§8.17**

Bunnell & Kahn, Removing Obstacles to Building Reuse and Community Conservation at the Local Level (1979) **§8.15**

Burka, *The Truth About John Connally,* 7 Texas Monthly 156 (Nov 1979) **§9.04**

Burke, Connections (1978) **§§11.06, 11.28**

Burger, An Analysis of the National Science Foundation's Innovation Centers Experiment (National Science Foundation, July 1977) **§9.15**

Burstein, *A Theory of Full-Line Forcing,* 55 Nw UL Rev 62 (1960) **§3.06**

Buxbaum, *Antitrust Regulation Within the European Economic Community,* 61 Colum L Rev 402 (1961) **§3.02**

Byrnes, Lowry & Bordinant, *Product Shortages, Allocation and the Antitrust Laws,* 20 Antitrust Bull 713 (1975) **§3.08**

C

Code Makers and Breakers, NY Times, Aug 17, 1981, §A, at 14 **§11.14**

2 Collected Works of Abraham Lincoln **§12.07**

College, Company Sign Research Pact on Artificial Cells, Washington Post, Sept 13, 1981, at 21 **§6.06**

Colosimo, Wilkof & Duga, The State Development Foundation: Meeting Societal Needs Through Centers of Innovation (Ohio Dept of Economic and Community Development 1974) **§§9.05, 9.06**

Comment, *Contract and Conditioning Zoning: A Tool for Zoning Flexibility,* 23 Hastings LJ 825 (1972) **§8.20**

Comment, *The Home Buyers' Protection Act: An Alternative to Building Codes for Single Family Homes,* 54 S Cal L Rev 529 (1981) **§§8.12, 8.13, 8.14, 8.15**

Comment, *Software Taxation: A Critical Reevaluation of the Notion of Intangibility,* 1980 BYU L Rev 859 **§8,09**

Commerce Clearing House, State Tax Guide (1979) **§§8.05, 8.10**

Commerce Clearing House, State Tax Laws (1979) **§8.09**

Committee on Corporation Law, *Special Interest Directors,* 35 Rec AB City NY 26 (1980) **§§2.01**

Committee on Developments in Business Financing, *Resale of Debt Securities Acquired in Private Placements,* 34 Bus Law 1927 (July 1979) **§§6.48, 6.49**

Committee on Federal Legislation, *Budget Balancing by Constitutional Amendment,* 34 Rec AB City NY 248 (1979) **§12.04**

Committee on Federal Legislation, *The Legislative Veto,* 34 Rec AB City NY 208 (1979) **§12.04**

Committee on Labor and Social Security Legislation, *Report Concerning (1) The Role of Judge and Arbitrator in Labor Arbitration and (2) Injunctions Against Strikes in Breach of Contract,* 20 Rec AB City NY 37 (1965) **§2.04**

Committee on Labor Law, Federal Bar Council, *Sweetheart Contracts,* 23 Indus & Lab Rel Rev 105 (1959), *reprinted in* 115 Cong Rec S6318 (daily ed June 12, 1969) **§2.05**

Commoner, The Closing Circle (1971) **§11.30**

Commoner, The Politics of Energy (1979) **§11.30**

Commons, *American Shoemakers, 1648-1895,* in Labor and Administration 219 (Commons ed 1913) **§3.15**

The Communications Satellite Act of 1962, 76 Harv L Rev 388 (1962) **§12.03**

The Conference Board, Report No 791, Leasing: Experiences and Expectations (1980) **§6.08**

Cookridge, They Came From the Sky (1965) **§§2.07, 12.07**

Coolidge, State of the Union Message, Dec 7, 1926, 3 State of the Union Messages of the Presidents 2690 (1966) **§12.03**

Cooper, *The Patent System and the "New Biology",* 8 Rut J Computer Tech & Law 1 (1980) **§11.29**

D

Durkheim, The Japanese Cult of Tranquillity (1974) **§12.06**

E

The Economics of Drug Innovation (Cooper, ed 1969) **§§11.11, 11.13**

Eddy, *Effects of the Magnuson-Moss Act Upon Consumer Product Warranties,* 55 NCL Rev 835 (1977) **§3.09**

Editorial, 133 America 84-85 Aug 30, 1975 **§12.03**

Editorial, New Deal for Construction, 119 America 93 (Aug 17, 1968) **§2.02**

Education for Invention, Indus Research/Dev 121 (Apr 1978) **§9.16**

Edwards, The Price Discrimination Law (1959) **§3.07**

The Effects of the Demographics of Individual Households on Their Telephone Usage (Brandon ed 1981) **§11.21**

Egan, *To Err is Human Factors,* 85 Tech Rev 23 (Feb 1982) **§2.04**

Einstein & Infeld, The Evolution of Physics (1942) **§§12.02, 12.06**

Eisenberg, *Private Ordering Through Negotiation,* 89 Harv L Rev 673 (1976) **§2.01**

Ellul, The Technological Society (1964) **§11.04**

Elman, *The New Constitutional Right to Advertise,* 64 ABAJ 206 (1978) **§3.15**

Elman, *Petrified Opinions and Competitive Royalties,* 66 Colum L Rev 625 (1966) **§3.09**

The Embattled University (Graubard & Ballotti eds 1970) **§11.29**

Environmental Law Institute, Legal Barriers to Solar Heating and Cooling of Buildings (1978) **§8.09**

Enzer & Alter, *Cross-Import Analysis and Classical Probability,* 10 Futures 227 (1978) **§11.27**

Evans, The Micro Millenium (1980) **§§11.17, 11.19, 11.30**

Ewing, The Human Side of Planning (1969) **§§2.01, 2.07, 3.16**

F

Falvey, *The Benefits of Working With the Best Workers,* Wall St J, May 11, 1981 **§2.07**

Fanning, *The Challenge of Automation in the Light of the Natural Law,* Labor LJ 875 (1960) **§2.02**

Fantini & Weinstein, Toward a Contact Curriculum (B'nai B'rith 1969) **§1.01**

Farnsworth, *House Panel Approves Foreign Takeover Curb,* NY Times, July 17, 1981, at D1 **§3.16**

Farnsworth, *US Agencies Rush to Spend Before Cutoff,* NY Times, Aug 16, 1979, at 1 **§12.03**

Feder, *Gene Splicing Licensed,* NY Times, Aug 19, 1981, at D-1 col 4 **§6.06**

Federal Energy Administration, II Solar Heating and Cooling of Buildings, Commercialization Report, Opinions and Strategies (1977) **§§8.06, 8.07**

Federal Spending: *The North's Loss is the Sunbelt's Gain,* Natl J 878 (June 6, 1976), **§9.10**

Federal Trade Commission New York Regional Office, Report on Hearings on Obstacles to Expansion of the Building and Construction Industry (1973) **§2.02**

Federal Trade Commission, Trade Regulation Rule on Negative Option Sales in Commerce, 16 CFR pt 425 **§2.08**

The Federalist No 10 (Madison) **§12.07**

Feiker, The Training of Men for the Textile Industry (1934) **§9.03**

Feinberg, The Prometheus Project (1969) **§§1.06, 12.02, 12.07**

Feinberg, What is the World Made Of? (1977) **§12.05**

Feller, Long & Nelson, Economic Aspects of the Relationship Between Research and Development Activities and State Economic Growth (Institute for Research on Human Resources, Pennsylvania State University 1972) **§9.12**

Ferry, Modern Accident Investigation and Analysis (1980) **§10.29**

Feyerabend, Against Method (1975) **§11.28**

Field, Building Regulatory Practices and the Courts (1980) **§8.13**

Field, Judicial Solutions to the Tensions Between Construction Codes and Innovation, 4 Urb L & Policy 107 (1981) **§§8.02, 8.12, 8.13, 8.14, 8.15, 8.17**

Field & Rivkin, The Building Code Burden (1975) **§§8.02, 8.12, 8.15**

Field & Ventre, Local Regulation of Building Agencies, Codes and Politics, in Municipal Yearbook (International City Management Assn 1971) **§§8.12, 8.14, 8.16**

The Fifth Legislative Climates, 14th Annual Report, Indus Dev 5 (Jan/Feb 1980) **§9.07**

The Fifth Legislative Climates: What a Difference a Decade Makes, Indus Dev 11 (Nov/Dec 1975) **§9.07**

Finkin, *Limits of Majority Rule in Collective Bargaining,* 64 Minn L Rev 183 (1980) **§2.05**

Fiske, *Reading Analysis is Called Lacking,* NY Times, Dec 14, 1980, at F1 **§1.01**

Flicker, *Newcomer Defenses,* 18 Stan L Rev 457 (1966) **§§3.05, 3.06, 11.05**

A Flood of Red Ink Swamps the Airlines, Intl Business Week, 89 (Nov 16, 1981) **§6.08**

Florida Solar Energy Center, Operation of the Collector Certification Program (1978) **§8.26**

Florman, The Existential Pleasures of Engineering (1976) **§11.19**

Forkosch, *Take It or Leave It as a Bargaining Technique,* 18 Labor LJ 676 (1967) **§2.01**

Forrester, Industrial Dynamics (1961) **§11.27**

Forrester, *Innovation and Economic Change,* 13 Futures 323 (1981) **§11.25**

Fox, *Monopoly and Competition; Tilting the Law Towards a More Competitive Economy,* 37 Wash & Lee L Rev 49 (1980) **§3.03**

Francis, Principles of R&D Management (1977) **§§11.11, 11.13, 11.15, 11.20, 11.27, 11.28, 11.30**

Frank, Courts on Trial (1963) **§3.09**

Frank & Frank, Not Guilty (1957) **§3.09**

Frankfurter & Greene, The Labor Injunction (1930) **§2.03**

Fredman, *Post-Accident Remedial Measures and Use of Corporate Counsel,* in Product Liability of Manufacturers: Prevention and Defense 1981 (Ross & Foley, eds) **§10.28**

Freedman, *Yale Bars a US Study Grant, Refusing to Account for Time,* NY Times, Mar 23, 1982, at 1 **§12.03**

Freidel, FDR: Launching the New Deal (1973) **§3.17**

Freund, Police Power (1904) **§8.02**

Freund & Greene, *Substance Over Form S-14: A Proposal to Reform SEC Regulation in Negotiated Acquisitions,* 37 Bus Law 1483 (July 1981) **§6.39**

Friedman, *Environmental Checklist and Outline of Impacts of CAA Amendments of 1977,* 14 Real Prop Prob & Tr J 873 (1979) **§7.57**

Friedman, A History of American Law (1979) **§9.01**

Friendly, *In Praise of Erie—and of the New Federal Common Law,* 39 NYU L Rev 383 (1964), *reprinted in* 19 Rec AB City NY 64 (1964) **§3.15**

Frost, *Legal Incidents of Non-Use of Patented Inventions,* 14 Geo Wash L Rev 273 (1946) **§3.08**

Fruhan, Financial Strategy (1979) **§§11.09, 11.11, 11.25**

Frumer & Friedman, Products Liability (3d Supp 1981) **§10.04**

Fugate, Foreign Commerce and the Antitrust Laws (2d ed 1973) **§3.12**

Fulda & Schwartz, Regulation of International Trade and Investment (1970) **§3.02**

Fuller, Synergetics (1975) **§11.30**

Fuller, We Almost Lost Detroit (1975) **§11.30**

G

Galbraith, Economics and the Public Purpose (1973) **§§11.13, 11.17, 11.21, 11.30**

Givens, *Litigation as a Means of Economic Competition,* 27 Rec AB City NY (1972) **§1.04**

Givens, *Manhart in the Light of Bakke,* 4 Employee Rel LJ 330 (1978) **§2.05**

Givens, *National Emergency Labor Disputes,* 34 Temple LQ 17 (1960) **§2.01**

Givens, *Parallel Business Conduct Under the Sherman Act,* 5 Antitrust Bull 273 (1960) **§§3.03, 3.11, 3.13**

Givens, *Power to Define the Constitutional Rights of Defendants: Congress and the Federal Courts,* 46 Fordham L Rev 383 (1977) **§3.15**

Givens, *The Search for an Alternative to Protection,* 30 Fordham L Rev 17 (1961) **§§2.02, 3.10**

Givens, *Section 301, Arbitration and the No-Strike Clause,* 11 Labor LJ 1005 (1960) **§2.04**

Givens, *Technology and Mankind's Future,* 123 America 254 (Oct 10, 1970) **§§12.02, 12.06, 12.07**

Givens, *Toward Fuller Use of Human Resources,* 17 Labor LJ 621 (1966) **§§2.08, 12.07**

Givens, *The Validity of Delayed Awards Under Section 301, Taft Hartley Act,* 9 Arb J 161 (1961) **§2.04**

Givens & Worsinger, *Vertical Restraints After Repeal of Fair Trade,* 45 Fordham L Rev 1093 (1977) **§§3.03, 3.07, 3.09, 3.16**

Glaser, *Participatory Management,* 210 Sci 962 (Nov 28, 1980) **§§2.07, 3.16**

Goebel, Cases and Materials on the Development of Legal Institutions (1946) **§3.12**

Goldhaber, *Is the Proton Stable?,* 210 Sci 851 (Oct 21, 1980) **§12.05**

Goldschmid, *Antitrust's Neglected Stepchild: A Proposal for Dealing With Restrictive Covenants Under Federal Law,*73 Colum L Rev 1193 (1973) **§2.06**

Goldstein, *Electronic Journalism and Union Rivalry,* 29 Labor LJ 137 (1978) **§§2.02, 2.04, 2.05**

Gomberg, *The Work Rules and Work Practices Problem,* 12 Labor LJ 643 (1961) **§2.02**

Gomes, *et al,* At Work in Copper (INFORM, Inc 1979) **§2.02**

Gordon, Synectics: The Development of Creative Capacity (1961) **§11.27**

Gover, *The Development of a Decision Model for the Retention of Documents Pertinent to Product Liability Defense,*Recs Mgmt Q (July, Oct 1975) **§10.25**

Grabowski & Mueller, *Life Cycle Effects on Corporate Returns on Retentions,* 57 Rev Econ Statistics 400 (1975) **§3.16**

Greer, *A Critique of Areeda & Turner's Standard for Predatory Practices,* 24 Antitrust Bull 233 (1979) **§3.07**

Gribbon, *Does the Law of Monopolization Inhibit Innovative Behavior?,* 49 ABA Antitrust LJ 925 (1980) **§3.02**

Griorson, *Reindustrialization's Poor Record,* Wall St J, Sept 30, 1980, at 39 **§12.03**

Gutowsky, *Federal Funding of Basic Research: The Red Tape Mill,* 212 Sci 636 (May 8, 1981) **§12.03**

H

Haasman, *Method of Doing Business,* 50 J Pat Off Socy 503 (July 1968) **§4.02**

Hacker, Alexander Hamilton (1957) **§12.03**

Hagman, Urban Planning and Land Development Control Law (1971) **§§8.19, 8.20**

Hale & Hale, *Monopoly in Motion,* 41 Va L Rev 431 (1955) **§§11.23, 11.24**

Hale & Hale, *Reciprocity Under the Antitrust Laws,* 113 U Pa L Rev 69 (1914) **§3.06**

Hall, The Competence Process (1980) **§2.07**

Hall, *The History of Section 6 of the Federal Trade Commission Act,* 27 Rec AB City NY 221 (1972) **§3.03**

Halloran, Going Public (3d ed Sorg Printing Co Aug 1979) **§6.13**

Hammer, Product Safety Management and Engineering (1980) **§§10.11, 10.16**

Handler, Antitrust in Perspective (1957) **§§3.09, 3.16**

Handler, *Antitrust—1978,* 78 Colum L Rev 1363 (1978) **§3.09**

Handler, Cases and Materials on Labor Law (1st ed 1944) **§2.03**

Handler, *Labor and Antitrust: A Bit of History,* 40 ABA Antitrust LJ 233 (1971) **§2.03**

Handler, Blake, Pitofsky & Goldschmid, Trade Regulation: Cases & Materials (1975) **§§3.07, 3.08**

Handler & Steuer, *Attempts to Monopolize and No-Fault Monopolization,* 129 U Pa L Rev 125 (1980) **§§3.03, 3.06**

Handler & Zifchak, *Collective Bargaining and the Antitrust Laws* 81 Colum L Rev 495 (1981) **§§2.02, 2.03**

Hanson, A Dictionary of Economics and Commerce (1965) **§11.06**

Hartz, Economic Policy and Democratic Thought (1948) **§9.01**

Hawk, *Act of State Doctrine, Noerr-Pennington Abroad and Foreign Government Compulsion Defense,* 47 ABA Antitrust LJ 987 (1978) **§3.12**

Hawk, United States, Common Market and International Antitrust (1979) **§§3.02, 11.11**

Hayes & Abernathy, *Managing Our Way to Economic Decline,* 58 Harv Bus Rev 67 (Sept/Oct 1980) **§§2.07, 3.16, 11.11, 11.25**

Hayes & Wheelwright, *The Dynamics of Process-Product Life Cycles,* 57 Harv Bus Rev 127 (Mar/Apr 1979) **§§3.16, 11.26**

Hayes & Wheelwright, *Linked Manufacturing Process and Product Life Cycles,* Harv Bus Rev 133 (Jan/Feb 1979) **§3.16**

Hechinger, *Psychologist Sees a Key to Learning in Managing "Unsolvable" Problems,* NY Times, Aug 18, 1981, at C5 **§1.01**

Heilbroner, *An Inquiry Into the Human Prospect* (1974) **§11.30**

Heinz, *The Partial Closing Conundrum,* 1981 Duke LJ 710 **§2.02**

Heisenberg, Physics and Beyond (1971) **§§12.02, 12.05**

Hellerstein, *State Income Taxation of Multijurisdictional Corporations: Reflections on Mobile, Exxon & HR 5076,* 79 Mich L Rev 113 (1980) **§8.29**

Herrigel, Zen and the Art of Archery (1953) **§12.06**

Hersey, Zest for Work (1955) **§2.08**

Hershey, *Congress Cuts Too Close to the Bone,* NY Times, Feb 21, 1982, at 14 **§12.03**

Herzberg, Mausner & Snyderman, The Motivation to Work (1959) **§2.08**

Herzog & Jenkins, *Price Guarantees and Purchase Agreements Under the Synthetic Fuels Corporation Act,* 36 Bus Law 995 (Apr 1981) **§12.03**

Hill & Fowles, *The Methodological Worth of the Delphi Forecasting Technique,* 7 Tech Forecasting & Soc Change 179 (1975) **§11.27**

Hoffman, *Joint Industry Research and Other Cooperative Programs: Antitrust Hazards and Lawful Opportunities,* 32 Food Drug Cosm LJ 444 (1977) **§3.10**

Holden, *Scientist with Unpopular Data Loses Job,* 210 Sci 749 (Nov 14, 1980) **§2.07**

Holton, *Scientific Research and Scholarship: Notes Toward the Design of Proper Scales,* 91 Daedalus 362 (Spring 1962) **§11.03**

Holusha, *Ford Workers Accept New Contract By Wide Margin,* NY Times Mar 1, 1982, at 1 **§2.02**

Hommers, *Protection of Employees Affected by Railroad Consolidations,* 15 Labor LJ 736 (1964) **§2.02**

Houseman, Run-Through (1972) **§2.08**

How to Prove You're Clean (Sun Information Services Co 1980) **§7.56**

Howell, *The Judicial Conservatives of Three Decades Ago: Aristocratic Guardians of the Preogatives of Property and the Judiciary,* 49 Va L Rev 1447 (1963) **§3.15**

Hubbard, The Search Is On (1969) **§§12.02, 12.05, 12.06**

Humes, Churchill, Speaker of the Century (1980) **§12.07**

Hursh & Bailey, American Law of Products Liability (2d ed 1974) **§10.04**

Hurst, *Alexander Hamilton: Law Maker,* 78 Colum L Rev 483 (1978) **§12.03**

Hutten, *On the Importance of Productivity Change,* 69 Am Econ Rev 126 (1979) **§§3.03, 3.11, 12.07**

Hutter, *Drafting Enforceable Employee Non-Competition Agreements,* 45 Alb L Rev 311 (1981) **§2.06**

Hyson, *The Problem of Relief in Developer—Initiated Zoning Litigation,* 12 Urb L Ann 21 (1976) **§8.23**

I

Industrial Concentration: The New Learning (Goldschmid, Mann, & Weston eds 1974) **§§11.1, 11.24**

Industrial Relations Research Assn, Collective Bargaining and Productivity (1965) **§2.02**

Initial Public Offerings of Common Stock: The Role of the Regional Broker-Dealer in the Capital Formation Process (US Small Business Administration and Securities and Exchange Commission, Phase I Report, Mar 1980) **§6.47**

*Injunctive Enforcement of Arbitration Awards Prohibiting Strikes,*17 Labor LJ 292 (1966) **§2.04**

In technology, simplicity is the ultimate sophistication, NY Times, Oct 10, 1977, at 29, *quoted in* Practising Law Institute, Drafting Documents in Plain Language 87 (1979) **§2.07**

Interface of National Labor and Antitrust Policies, 23 Labor LJ 115 (1982) **§2.03**

Isaacson, *The Grand Equation: Labor Arbitration and the No-Strike Clause,* 48 ABAJ 914 (1962) **§2.04**

J

Jacobs, The Economy of Cities (1969) **§§2.07, 3.10, 3.16**

Jaffe, *The Effective Limits of the Administrative Process: A Reevaluation,* 67 Harv L Rev 1105 (1954) **§12.04**

Jantsch, Technological Forecasting in Perspective (1967) **§11.27**

Jarrett & Jones, Creating Jobs: Connecticut's Product Development Corp (Council of State Governments 1978) **§9.13**

Jastrow, Until the Sun Dies (1977) **§12.06**

Javits & Klein, *Congressional Oversight and the Legislative Veto,* 52 NYU L Rev 455 (1977) **§12.04**

Jay, Management and Machiavelli (1968) **§2.07, 3.09, 3.16**

Jentes, *Assessing Efforts to Challenge Aggressive Competition as an Attempt to Monopolize,* 49 ABA Antitrust LJ 937 (1980) **§3.05**

Jewkes, Sawers & Stillerman, The Sources of Innovation (1958) **§§11.20, 11.28**

John XIII, Mater et Magistra (1961) **§§2.08, 3.16**

Johnson, *Antitrust in France,* 22 Rec AB City NY 455 (1967) **§3.02**

Johnson, Kast & Rosensweig, The Theory and Management of Systems (1973) **§11.15**

Joining Hands Against Japan, Bus Week, Nov 10, 1980, at 108 **§11.29**

Jones, *Guaranteed Pay Proposed by United States,* NY Times, July 27, 1966, at 35, col 1 **§2.02**

Jones, *Progress, Remember?,* NY Times, Aug 24, 1979, at A25 **§12.07**

Jones & Angly, Fifty Billion Dollars (1951) **§§12.03, 12.07**

Jones & Twiss, Forecasting Technology for Planning Decisions (1978) **§§11.17, 11.18, 11.21, 11.26, 11.27**

Josko & Klevorick, *A Framework for Analyzing Predatory Pricing Policy,* 89 Yale LJ 218 (1979) **§3.07**

Judson, The Eighth Day of Creation (1979) **§11.28**

K

Kahn & Wiener, The Year 2000 (1967) **§11.27**

Kalberer, Grant Length and Budget Stability at NIH, 211 Sci 675 (Feb 13, 1981) **§12.03**

Kamien & Schwartz, *Market Structure and Innovation,* 13 J Econ Lit 9 (1975) **§3.16**

Kanwit, Federal Trade Commission (Shepard's/McGraw-Hill 1979) **§§2.07, 3.07, 3.11**

Kaplan, *Israeli Antitrust Law and Practice,* 26 Rec AB City NY 389 (1971) **§3.02**

Katz, *Harvey and Medical Research,* 160 JAMA 1137 (1956) **§3.10**

Kaufmann, Black Holes and Warped Spacetime (1979) **§12.06**

Kaufmann, The Cosmic Frontiers of General Relativity (1977) **§§12.05, 12.06**

Kaye, Plaia & Hertzberg, International Trade Practice (Shepard's/McGraw-Hill 1981) **§3.02**

Kaysen, *Collusion Under The Sherman Act,* 65 QJ Econ 263 (1951) **§3.07**

Kaysen & Turner, Antitrust Policy (1959) **§11.01**

Kayton, *The Low and High of It,* NY Times, Nov 8, 1978, at A27 **§12.04**

Kayton, Lynch & Stern, *Fraud in Patent Procurement,* 43 Geo Wash L Rev 1 (1974) **§3.08**

Keeton, *Manufacturer's Liability: The Meaning of "Defect" in the Manufacture and Design of Products,* 20 Syracuse L Rev 559 (1969) **§§10.04, 10.07, 10.16**

Kennedy, The Burden and the Glory (Nevins ed 1964) **§§2.03, 3.16**

Kessler & Stern, *Competition, Contract and Vertical Integration,* 69 Yale LJ 1, (1959) **§3.05**

Kidder, The Soul of a New Machine (1981) **§11.19**

Kintner, A Robinson-Patman Primer (1975) **§3.07**

Kintner & Goldstein, *Section 2 of the Sherman Act: The Attempt to Monopolize Offense,* 24 NY L Sch L Rev 573 (1979) **§3.03**

Kintner & Griffin, *Jurisdiction Over Foreign Commerce Under the Sherman Antitrust Act,* 18 BC Ind & Com L Rev 199 (1977) **§3.12**

Kirshy, Automation and Collective Bargaining (1964) **§2.02**

Klein, *Some Firms Fight Ills of Bigness By Keeping Employee Units Small,* Wall St J, Feb 5, 1982, at 1, col 1 **§2.07**

Kniffin, *Employee Non-Competition Covenants: The Perils of Performing Unique Services,* 10 Rut-Cam LJ 25 (Fall 1978) **§11.10**

Koestler, Insight and Outlook (1949) **§2.07**

Koller, *When is Pricing Predatory?,* 24 Antitrust Bull 283 (1979) **§3.07**

Kolowrat, Stack & Lynch, *Restraints on Technology Access: Protection of Trade Secrets and Confidential Information,* 49 Antitrust LJ 735 (1981) **§11.10**

Kotler, Marketing Decision Making: A Model Building Approach (1971) **§11.13**

Kramer, *Air Quality Modeling: Judicial, Legislative and Administrative Reactions,* 5 Colum J Envtl L 236 (1979) **§7.56**

Kraemer, Solar Law (Shepard's/McGraw-Hill 1978) **§8.05**

Kripke, *Required Periodic Disclosure Under the Securities Acts and the Proposed Federal Securities Code,* 33 U Miami L Rev 1425-648 (Summer 1979) **§6.63**

Kripke, *Where Are We On Securities Disclosure After the Advisory Committee Report?,* 6 Sec Reg LJ 99 (Summer 1978) **§6.63**

Kross, American Economic Development (1900) **§9.04**

Kubler, The Shapes of Time (1962) **§11.28**

Kuenzel & Schiffres, *Making Sense of Robinson-Patman: The Need to Revitalize its Affirmative Defenses,* 62 Va L Rev 1211 (1976) **§§3.07, 3.16**

Kuhn, The Essential Tension: Selected Studies in Scientific Tradition and Change (1977) **§11.28**

Kuhn, The Structure of Scientific Revolutions (1962) **§11.28**

L

Labor and Multifactor Productivity by Industry, in New Developments in Productivity Measurement and Analysis (Kendrick & Vaccara eds 1975) **§11.09**

Lamm, *Photovoltaic Commercialization: An Analysis of Legal Issues Affecting a Government-Accelerated Solar Industry* (Solar Energy Research Institute 1979) **§8.25**

Landau, *Alabama Power Co v Costle: An End to a Decade of Controversy Over the Prevention of Significant Deterioration of Air Quality?,* 10 Envtl L 585 (1980) **§7.31**

LaPierre, *Technology-forcing and Federal Environmental Protection Statutes,* 62 Iowa L Rev 771 (1977) **§7.57**

Lawrence & Minan, *The Role of Warranties and Product Standards in Solar Energy Development,* 34 Vand L Rev 537 (1981) **§§8.03, 8.25, 8.26**

Lederer, *Wright Line or Spur Track?,* 34 Labor LJ 67 (1982) **§2.06**

Lenmark-Ellis, *The Professional Troublemakers,* NY Times, Aug 22, 1981, at 23 **§2.07**

Leontief, Input-Output Economics (1966) **§11.09**

Lesnick, *Gravamen of the Secondary Boycott,* 62 Colum L Rev 1363 (1962) **§2.03**

Lesnick, *Job Security and Secondary Boycotts: The Reach of NLRA §§8(b)(4) and 8(e),* 113 U Pa L Rev 1000 (1965) **§2.03**

Lesnick, *State Court Injunctions and the Federal Common Law of Labor Contracts,* 79 Harv L Rev 757 (1966) **§2.04**

Letwin, Law and Economic Policy (1965) **§3.03**

Levenstein, Why People Work (1964) **§§2.08, 3.16**

Levi, *The Parke, Davis-Colgate Doctrine: The Ban on Resale Price Maintenance,* 1960 Sup Ct Rev 258 **§§3.03, 3.09**

Levy, *Judicial Review of Zoning Cases—New Rules,* 6 Tulsa LJ 1 (1969) **§8.23**

Levy, Origins of the Fifth Amendment (1968) **§3.12**

Lewin, *Group Decision and Social Change,* in Readings in Social Psychology 2 (rev ed 1952) 459 **§2.07**

Lewis, *Technology, Enterprise and American Economic Growth,* 215 Sci 1204 (Mar 5, 1982) **§12.07**

Lewis, *The Profits in Common Sense,* NY Times, Sept 26, 1980, at 35 **§§2.07, 3.16**

Linstone & Sahal, Technological Substitution: Forecasting Techniques and Applications (1976) **§11.18**

Linstone & Taroff, The Delphi Method: Techniques and Applications (1975) **§11.27**

Litwak & Daniels, *Innovations in Development Finance* (Council of State Planning Agencies 1979) **§9.13**

Lockhart, *A Revolution in State Taxation of Commerce,* 65 Minn L Rev 1025 (1981) **§8.29**

Loeb, Full Production Without War (1946) **§12.07**

London, The People Side of Systems (1976) **§11.19**

Long, *et al,* Appropriate Technology and Social Values (1978), reviewed, 212 Sci 38 (Apr 3, 1981) **§1.04**

Lopata, *Research Triangle—A Far Out Concept That Worked,* Iron Age 105 (Nov 23, 1981) **§9.10**

Louis, *Vertical Distributional Restraints After Sylvania: A Postscript and Comment,* 76 Mich L Rev 226 (1978) **§3.09**

Lovins, Soft Energy Paths (1977) **§§11.01, 11.04**

Lovins & Price, Non-Nuclear Futures (1975) **§§11.01, 11.04**

Lublin, *Watchdog Has Hard Time Hearing Whistles,* Wall St J, Oct 17, 1980, at 30 **§2.07**

Luce, *The Rule of Law: Its World Implementation is Man's Hope,* 49 ABAJ 727 (1963) **§12.02**

Lusky, *Minority Rights and the Public Interest,* 52 Yale LJ 1 (1942) **§3.15**

Lynch, *Polaroid Plans "Entirely New" System for Another Market; Details Are Secret,* Wall St J, Apr 22, 1981, at 3, col 1 **§6.59**

M

Machlup, The Production and Distribution of Knowledge in the United States (1962) **§11.19**

Mack, Controlling Retailers (1936) **§3.16**

MacKenzie, The Time Trap (1972) **§§2.07, 3.10, 3.16**

Magat, *The Effects of Environment Regulation on Innovation,* 43 Law & Contemp Prob 4 (1979) **§§7.06, 11.14**

Makridakis, *If We Cannot Forecast, How Can We Plan?,* 14 Long Range Planning 10 (June 1981) **§§11.11, 11.18, 11.30**

Mandelker, *Deregulation of Power and Function in Zoning Administration,* 1963 Wash ULQ 60 **§8.22**

Mansfield, *Basic Research and Productivity Increase in Manufacturing,* 70 Am Econ Rev 863 (Dec 1980) **§§3.03, 3.11, 12.07**

Mansfield, The Economics of Technological Change (1968) **§11.22**

Mansfield, Industrial Research and Technological Innovation (1968) **§§11.18, 11.21, 11.23, 11.25, 11.26**

Mansfield, Microeconomics (1970) **§§11.09, 11.22, 11.26**

Mansfield, Monopoly Power and Economic Performance (1964) **§11.20**

Mansfield, *Research and Development, Productivity, and Inflation,* 209 Sci 1091 (Mar 2, 1979) **§§3.03, 12.07**

Masnfield, *Tax Policy and Innovation,* 215 Sci 1365 (Mar 12, 1982) **§1.02**

Mansfield, Technological Change (1971) **§9.01**

Manual for the Handling of Applications for Patents, Design and Trade Marks Throughout the World (Octrooibureau Los En Steiger, Amsterdam, Holland) **§4.05**

Manual of Patent Examining Procedures (Patent and Trademark Office, US Dept of Commerce) **§4.14**

Mara & Engel, *Institutional Barriers to Solar Energy: Early HUD Demonstration Experiences,* 1 Solar L Rep 1095 (1980) **§8.13**

March & Simon, Organizations (1967) **§2.07**

Markey, *Some Patent Problems,* 80 FRD 203 (1979) **§§4.15, 4.17**

Marrow, Management by Participation (1967) **§2.07**

Marschak, *Strategy and Organization in a System Development Project,*in The Rate and Direction of Inventive Activity (National Bureau of Economic Research 1962) **§11.20**

Marshall, *New Perspectives on National Emergency Disputes,* 18 Labor LJ 451 (1967) **§2.01**

Marshall, *NOAA Buffeted By Budget Storms,* 215 Sci 1482 (Mar 19, 1982) **§12.03**

Martin, Future Development in Telecommunications (1977) **§11.27**

Martin *Transshipping,* Dealerscope 8 (Mar 1979) **§3.09**

Martino, Technological Forecasting for Decision Making (1972) **§11.27**

Maslow, *Colgate Updated: Resale Price Maintenance and Antitrust Liability,* 48 NY St BJ 626 (1976) **§§3.03, 3.09**

Maslow, Toward a Psychology of Being (1962) **§2.08**

Mason, Harlan Fiske Stone: Pillar of the Law (1956) **§12.06**

Massachusetts Technology Development Corporation, Annual Report **§9.14**

May, The Courage to Create (1975) **§12.06**

McCarroll, *Rehiring Retirees,* NY Times, Nov 22, 1981, at 8F **§12.07**

Mc Donald, *California Pushing Research in Microelectronics,* Chron Higher Educ (Apr 7, 1982) **§9.09**

McElheny, *Electricity Transmitted by Radio Beam on Coast,* NY Times, Oct 10, 1975, at 40 **§12.05**

Mc Gregor, The Human Side of Enterprise (1960) **§§2.08, 3.16**

McGuire, The Product-Safety Function: Organization and Operations (1979) **§§10.11, 10.12, 10.22**

McGuire, Spin-Off Products and Services: The Commercialization of Internally Supported Resources (1976) **§11.22**

McLaurin, Invention and Innovation in the Radio Industry (1947) **§11.23**

McPhee, The Curve of Binding Energy (1973) **§11.20**

McQuade, *Conspiracy, Multicorporate Enterprises and Section 1 of the Sherman Act,* 41 Va L Rev 183 (1955) **§3.07**

2 McQuillin, The Law of Municipal Corporations (1979) **§8.02**

Meadows, Donella, Meadows, Randers & Behrens, The Limits to Growth (1972) **§§1.06, 11.01, 11.04, 11.06, 11.24, 11.27, 11.30**

Mendelssohn, The Search for Absolute Zero (1966) **§§1.06, 3.10**

Mensch, Coutinho & Kaasch, *Changing Capital Values and the Propensity to Innovate,* 13 Futures 276 (Aug 1981) **§§11.09, 11.25**

Mesarovic & Pestel, Mankind at the Turning Point (1974) **§§11.01, 11.04, 11.06, 11.30**

Michener, *The Advantages of a Space Program,* 182 Sci 358 (Oct 26, 1973) **§12.05**

Milgrim, Trade Secrets **§4.20**

Miller, *Better Ideas Need More Watts,* NY Times, Nov 2, 1980, at E9 **§2.07**

Miller, *Long Term Contracts Proposed as Cost Measure,* NY Times, Mar 4, 1981, at A20 **§12.04**

Miller, *Small Business Investment Companies; Licensing, Tax and Securities Considerations,* 36 Bus Law 1979 (July 1981) **§§6.41, 6.44**

Miller, *Solar Energy and Land Use in Colorado,* 6 Envtl L Rep (Envtl L Inst) 50039 (1977) **§8.19**

Minan & Lawrence, *State Tax Incentives to Promote the Use of Solar Energy,*56 Tex L Rev 835 (1978) **§§8.05, 8.06, 8.07, 8.08**

Mitucek, The Political Economy of Regulation (1980) **§11.14**

Moch, *The Role of a State Science Commission,* in Science for Society, Proceedings of the National Conference on Goals, Policies and Programs of Federal, State and Local Science Agencies Atlanta 1970 (Moch, ed, Georgia Science and Technology Commission 1971) **§9.08**

Moesteler, *Innovation and Evaluation,* 211 Sci 881 (Feb 27, 1981) **§1.04**

Mohr, *Fears About Deficit Impair Effort to Improve Acquisitions,* NY Times, Mar 14, 1982, at 1 **§12.03**

More, *US Anti-trust Laws and Territorial Provisions in Licensing Foreign Patents and Know-How for Foreign Use,* 24 Rec AB City NY 512 (1969) **§3.12**

2 Morrison & Commager, Growth of the American Republic (4th ed 1961) **§12.07**

Morton, Organizing for Innovation (1971) **§§11.10, 11.15, 11.20, 11.28**

Mueller, *Sources of Monopoly Power: A Phenomenon Called Product Differentiation,*18 Am UL Rev (1968) **§§11.08, 11.13**

Mumford, The Myth of the Machine II: The Pentagon of Power (1970) **§§11.01, 11.04**

Murphy, *Agreement on the Railroads: The Joint Railway Conference of 1926,* 11 Labor LJ 823 (1960) **§2.01**

N

Nader, Unsafe at Any Speed (1965) **§11.30**

National Bureau of Economic Research, The Rate and Direction of Inventive Activity (1962) **§11.20**

National Electrical Manufacturers Assn, Guidelines (1977) **§10.19**

National Institute of Building Sciences, States' Energy Conservation Standards for New Construction: State Legislative Analysis (1978) **§8.12**

National Science Foundation, An Analysis of the National Science Foundation's Innovation Centers Experiment (1977) **§9.13**

National Science Foundation, NSF Publication No 78-313, National Patterns of R&D Resources, Funds and Personnel in the United States, 1953-1978-79 **§9.04**

National Science Foundation Ties With National Security Agency, NY Times, Aug 27, 1980, at 1 **§11.14**

Naumoff, *An Analysis of the Contract Bar Doctrine,* 7 Labor LJ 197 (1956) **§2.05**

Naylor, *Subcontractors Penalty Clauses and Plant Removals,* 30 Drake L Rev 203 (1980) **§2.02**

Nelson, Railroad Transportation and the Public Interest (1959) **§3.14**

Nelson, *The Simple Economics of Basic Scientific Research,* 67 J Pol Econ 297 (1959) **§11.28**

New Developments in Productivity Measurement and Analysis (Kendrick & Vaccara eds 1975) **§§11.11, 11.19, 11.26**

New Hope for Going to Market With DNA, Bus Week, Sept 24, 1979, at 64 **§11.14**

New York State Agricultural Experiment Station, Ninety-Seventh Annual Report (1979) **§9.02**

New York State Atomic and Space Development Authority, Ten Years of Atomic Progress in New York State (1969) **§9.09**

New York State Science and Technology Foundation, CID Information Guide, (Oct 1981) **§9.14**

New York Stock Exchange Company Manual **§§6.39, 6.53, 6.65**

Nierenberg, The Art of Negotiating (1968) **§2.01**

Nirenberg, Getting Through to People (1963) **§2.01**

Nordhaus, Patent-Antitrust Law **§4.19**

Nordhaus, Patent License Agreements **§4.19**

Norstrom, Law of Sales (1971) **§8.25**

Norman, *Small Business [Set-Aside] Bills Upset Universities,* 214 Sci 1003 (Nov 27, 1981) **§12.03**

North Carolina's Research Triangle, The Chapel Hill Newspaper, Feb 1980 **§9.10**

Note, *Aggressive Innovation and Antitrust Liability,* 53 S Cal L Rev 1469 (1980) **§3.04**

Note, *The Communications Satellite Act of 1962,* 76 Harv L Rev 388 (1962) **§12.03**

Note, *Copyright for Uncopyrightables,* 108 U Pa L Rev 699 (1960) **§11.10**

Note, *Labor Law Problems in Plant Relocation,* 77 Harv L Rev 1100 (1964) **§2.02**

Note, *Joint Research Under the Antitrust Laws,* 39 Geo Wash L Rev 1112 (1971) **§3.10**

Note, *Work Assignment Disputes Under the National Labor Relations Act,* 73 Harv L Rev 1150 (1960) **§2.02**

Noun, *Product Liability and Wind System Manufacturers: Proposals for Protection,* 2 Solar L Rep 73 (1980) **§8.25**

Nowak, Rotunda & Young, Constitutional Law (1977) **§8.29**

Nye, *Can Conduct Oriented Enforcement Inhibit Conscious Parallelism?,* 44 ABA Antitrust LJ 206 (1975) **§§3.03, 3.13**

NY Times, Feb 16, 1959, at 1 **§2.05**

NY Times, Dec 19, 1959, at 50, col 1 **§2.01**

NY Times, Apr 11, 1960, at 18, col 1 **§2.02**

NY Times, Feb 4, 1966, at 14, col 1 **§2.02**

NY Times, Apr 7, 1981, at D-1, col 3 **§6.15**

O

Office of the Federal Register, General Services Administration, Guide to Record Retention Requirements (rev ed 1981) **§10.25**

Ogburn, Social Changes (1922) **§11.28**

Oldman & Schoettle, State and Local Taxes and Finance (1974) **§8.10**

O'Neal, *In Search of High-Brow Industry,* The South Magazine 33 (Oct 1979) **§9.10**

O'Neill, *Colonies in Orbit,* NY Times Magazine, Jan 18, 1976, at 10 **§12.05**

O'Neill, The High Frontier (1977) **§§1.06, 11.30, 12.05**

O'Neill, *Space: A Place in the Sun,* NY Times, June 12, 1975, at 37 **§12.05**

O'Neill, *Space Colonies and Energy Supply to the Earth,* 190 Sci 943 (Dec 5, 1975) **§12.05**

O'Neill, 2081 (1981) **§§11.04, 11.17, 11.30**

Ongman, *"Be No Longer a Chaos": Constructing a Normative Theory of the Sherman Act's Extra-Territorial Jurisdictional Scope,* 71 Nw UL Rev 733 (1977) **§3.12**

O'Reilly, Food and Drug Administration (Shepard's/McGraw-Hill supp 1981) **§10.21**

Orrick, *The Clayton Act: Then & Now,* 24 ABA Antitrust LJ 44 (1964) **§3.16**

Oster & Quigley, *Regulatory Barriers to the Diffusion of Innovation: Some Evidence from Building Codes,*8 Bell J of Econ 361 (1977) **§§8.02, 8.12, 8.13, 8.14**

Ouchi, Theory Z: How American Business Can Meet the Japanese Challenge (1981) **§§2.07, 3.16**

P

Pagels, The Cosmic Code: Quantum Physics as the Language of Nature (1982) **§12.06**

Palamountain, The Politics of Distribution (1955) **§3.09**

Panzar & Willig, *Economics of Scope,* 71 Am Econ Rev 268 (May 1981) **§11.06**

Paradigms and Revolution (Gutting ed 1980) **§11.28**

Parkinson, Parkinson's Law (1957) **§§2.07, 3.16**

Pascale & Athos, The Art of Japanese Management (1981) **§§2.07, 3.16**

Patents at Colleges and Universities, Ad Service Supplement 2:4:1:1 (Mar 1978) **§9.15**

Penner, *The Budget Process Is Tied in Knots,* NY Times, Feb 28, 1982, at F3 **§12.03**

Pennsylvania Science and Engineering Foundation, PSEF—Funding Strategies **§9.06**

Perkins, The Roosevelt I Knew (1946) **§2.01**

Perlman, A Theory of the Labor Movement (1928) **§§2.01, 2.02, 2.03**

Phillips, *Observations on the Hales' Comment,* 113 U Pa L Rev 77 (1914) **§3.06**

Phillips, *The Standard for Determining Defectiveness in Products Liability,* 46 U Cin L Rev 101 (1977) **§10.06**

Pitofsky, *Beyond Nader: Consumer Protection and the Regulation of Advertising,* 90 Harv L Rev 661 (1977) **§3.07**

Pitofsky, *Joint Ventures Under the Antitrust Laws,* 82 Harv L Rev 1007 (1969) **§§3.05, 3.10, 11.11**

Pitofsky, *The Sylvania Case: Antitrust Analysis of Non-Price Vertical Restrictions,* 78 Colum L Rev 1 (1978) **§§3.09, 3.16**

Pollack, *Funds Outgrow Investment Opportunities,* NY Times, June 17, 1981, at D-1 **§6.05**

Porter, Competitive Strategy (1980) **§11.26**

Porter, Job Property Rights (1954) **§2.01**

Posner, Regulation of Advertising by the FTC (1974) **§11.13**

Power Made From Ocean Heat, NY Times, July 5, 1980, at 28 **§§12.02, 12.05**

Pozen, *Synthetic Fuels Corporation—Investment Bank or Government Agency?,* 36 Bus Law 953 (1981) **§§3.14, 12.03**

Practising Law Institute, Domestic and International Licensing of Technology (1980) **§1.02**

Practising Law Institute, Drafting Documents in Plain Language (1979) **§§2.08, 3.15**

Practising Law Institute, Marketing and Distribution: Regulations and Restraints (1978) **§§3.03, 3.09, 3.16**

Practising Law Institute, 19th Annual Advanced Antitrust Seminar: Price Discrimination (1979) **§3.07**

Practising Law Institute, The Repeal of the Fair Trade Laws: Imapct on Product Distribution (1976) **§§3.09, 3.16**

President's Commission on Urban Problems, Building the American City (1968) **§8.12**

Previant, *Boycotts Under the 1959 Amendments: The Full Circle From Duplex v Deering?,* Proc 13th Ann NYU Conf on Labor 141 (1960) **§2.03**

Pribram, Cartel Problems (1935) **§§3.07, 3.09**

Private Ordering Through Negotiation, 89 Harv L Rev 673 (1976) **§2.01**

1 Product Liability: Industry Study, published in conjunction with the Final Report of Interagency Task Force on Product Liability (US Dept of Commerce 1977) **§10.11**

Product Liability of Manufacturers: Prevention and Defense 1981 (Ross & Foley eds) **§§10.13, 10.16, 10.19, app 10-3**

Product Safety Label Handbook (Westinghouse Printing Div 1981) **§10.19, app 10-2**

Proposal for Colonies in Space Hailed by Scientists as Feasible, NY Times, May 13, 1974, at 1 **§12.05**

Public Technology (Mercer & Phillips eds) **§11.11**

Q

1 The Quality of Working Life (Davis & Cherns eds 1975) **§11.14**

Q&A: Small Business and the SEC (SEC Mar 1980) **§6.01**

R

Rahl, *Conspiracy and the Antitrust Laws,* 44 Ill L Rev 743 (1950) **§3.13**

Ramo, Century of Mismatch (1970) **§11.04**

Ramo, *The Regulation of Technological Activities: A New Approach,* 67 ABAJ 1456 (1981) **§12.04**

Raskin, *The Cooperative Economy,* NY Times, Feb 14, 1982, at 1 **§2.02**

Raskin, *Pay by the Hour? The Week? The Year? For Life?,* NY Times Magazine, Sept 4, 1966, at 6 **§2.02**

Rasmussen, *The Impact of Technological Change on American Agriculture, 1862-1962,* J Econ History (1972) **§9.01**

Rayhill, *The Robinson-Patman Act: A Revival of Defenses,* in Marketing and Distribution: Regulations and Restraints 210 (Practising Law Institute 1978) **§3.07**

"R&D" and Gene Amdahl form the foundation for a $55 million tax shelter, Wall St J, July 28, 1981, at 1 col 5 **§6.06**

Reich, *Consumer Protection and the First Amendment,* 61 Minn L Rev 705 (1977) **§§3.03, 3.15**

Reich, The Greening of America (1970) **§11.04**

Reichenbach, Philosophic Foundations of Quantum Mechanics (1946) **§4.24**

Report of Legislative Committee, Antitrust Section, NY State Bar Assn, Good Faith Cost Justification, Trade Reg Rep (CCH) 484 (Mar 23, 1981), *reprinted in* 22 Antitrust Law Inst 647 (Practising Law Institute 1981) *and* 53 NY St BJ 376 (Aug 1981) **§3.07**

Report of the ABA Commission to Study the Federal Trade Commission (1969) **§3.07**

Report of the Attorney General's National Committee to Study the Antitrust Laws (1955) **§§3.07, 3.08**

Report of the National Commission for Review of Antitrust Laws and Procedures (1979) **§§3.02, 3.04, 3.07, 3.10**

Report of the White House Task Force on Antitrust Policy (1968), *reprinted in* 2 Antitrust L & Econ Rev 11 (Winter 1968-69) **§3.16**

Resnick, Turney & Mason, *Marketeers Turn to "Countersegmentation",* 79 Harv Bus Rev 100 (Sept-Oct 1979) **§§11.07, 11.13**

Restatement (Second) of the Law of Torts (1965) **§§10.04, 10.08, 10.10**

Reznick, *Synercom Synercom Technology, Inc v University Computing Co: Copyright Protection for Computer Formats and the Idea/Expression Dichotomy,* 8 Rut J Computer Tech & Law 65 (1980) **§11.10**

Rich, *Monopolies, "Invention" and Section 103,* in The Encyclopedia of Patent Practice and Invention Management (Calvert ed 1964) **§§4.03, 4.16**

Roark, *Academic Scientists Eye the "Industry Connection",* Chron Higher Educ 1 (Apr 30, 1979) **§9.15**

Robinson, *Reconciling Antitrust and the First Amendment,* 48 ABA Antitrust LJ 1335 (1979) **§3.15**

Robinson & Eppler, Going Public (2d ed rev 1978) **§§6.13, 6.35**

Roberts, The Competition Law of Canada and Antitrust Law of the United States (1980) **§3.02**

Roosevelt, First Inaugural Address, Mar 4, 1933, NY Times, Mar 5, 1933, *reprinted in* 2 Hofstadter, Great Issues in American History 353 (1958) **§2.08**

Ross, *Legal and Practical Considerations for the Creation of Warning Labels and Instruction Books,* 4 J Prod Liab 29 (1981) **§§10.08, 10.19**

Rostow, *The New Sherman Act,* 14 U Chi L Rev 567 (1974) **§3.16**

Roszak, Where the Wasteland Ends (1972) **§11.04**

Rottenberg, *Property in Work,* 15 Indus & Lab Rel Rev 402 (1962) **§2.01**

Rowe, *Administration and Enforcement of the Periodic Reporting Provisions of the Securities Exchange Act of 1934,* 25 Okla L Rev 157 (May 1972) **§6.63**

Rowe, *Cost Justification of Price Differentials Under the Robinson-Patman Act,* 59 Colum L Rev 584 (1959) **§3.07**

Rowe, Price Discrimination Under the Robinson-Patman Act (1962) **§§3.07, 3.16**

Rowe, *SEC Clarifies "Safe Harbor" Projection Rules,* Legal Times of Washington, July 16, 1979, at 9 **§6.56**

Rowe, *SEC Enforcing Through Administrative Proceedings,* Legal Times of Washington, June 2, 1980, at 15, col 1 **§§6.63, 6.66**

Rowe, *SEC Shifts to Administrative Enforcement, Remedies,* Legal Times of Washington, Sept 14, 1981, at 23, col 1 **§§6.63, 6.66**

Rowe, *"US Steel" Reflects Disclosure Governance Trends,* Legal Times of Washington, Nov 15, 1979, at 10, col 1 **§6.57**

Rupert, *The Relationship of Patent Law to Antitrust Law,*49 Antitrust LJ 755 (1981) **§3.08**

Russell, *The Mass Extinctions of the Late Mesozoic,* 240 Sci Am 58 (1982) **§12.02**

S

Safire, *And I Will Be Heard!,* NY Times, May 22, 1975, at 39 **§12.03**

Sakharov, Progress, Coexistence and Intellectual Freedom (1968) **§1.06**

Salpukas, *Coal Strategy: Long Contracts,* NY Times, May 13, 1981, at D1 **§3.07**

Sanderson, Codes and Code Administration **§8.13**

Sato, *Municipal Affairs in California,* 60 Cal L Rev 1055 (1972) **§8.02**

Sato, *et al, Multi-Production of Universes,* 108B Physics Letters 103 (Jan 14, 1982) **§12.06**

Schapiro, *The German Law Against Restraints of Competition—Comparative and International Aspects,* 62 Colum L Rev 201 (1962) **§3.02**

Scheiber, Vatter & Faulkner, American Economic History (9th ed 1976), **§§9.01, 9.02**

Schelling, The Strategy of Conflict (1960) **§§1.05, 2.01, 3.07, 3.16**

Scherer, Industrial Market Structure and Economic Performance (1970) **§§3.05, 3.16, 11.24, 11.20**

Scherer, *Predatory Pricing and the Sherman Act,* 89 Harv L Rev 869 (1976) **§3.07**

Schneider, *The Elusive Definition of a Security,* 14 Rev Sec Reg 981 (Jan 23, 1981) **§6.17**

Schneider, *The Statutory Law of Private Placements,* 14 Rev Sec Reg 869 (Aug 26, 1981) **§6.23**

Schneider, Manko & Kant, Going Public, Practice, Procedure and Consequences (Packard Press reprint 1981) **§§6.13, 6.35**

Schneider & Shargel, *"Now That You Are Publicly Owned . . .",* 36 Bus Law 1631 (July 1981) **§6.52**

Schneider & Zall, *Section 12(1) and the Imperfect Exempt Transaction: The Proposed I & II Defense,* 28 Bus Law 1011 (1973) **§6.49**

Schnitzer, Contemporary Government and Business Relations (1978) **§11.11**

Schuck, *Techniques for Proof of Complicated Scientific and Economic Facts,* 40 FRD 33 (1967) **§11.18**

Schumacher, Small is Beautiful (1973) **§§4.18, 11.01, 11.04**

Schumpeter, Business Cycles (1939) **§11.08**

Schumpeter, Capitalism, Socialism and Democracy (3d ed 1959) **§§1.06, 11.23, 11.25**

Schwartz, Comparative Negligence (1974) **§10.31**

Schwartz & Schwartz, Engineering Evidence (Shepard's/McGraw-Hill 1981) **§1.04**

Scientific American 55-57 (Aug 1967) **§4.18**

Scientific American (Sept 1981) **§11.29**

The Search for Alternatives to Protection, 30 Fordham L Rev 17 (1961) **§12.03**

Seidel, Housing Costs and Government Regulations (1978) **§§8.02, 8.13, 8.14**

Selander, *Is Annual Style Change in the Automobile Industry an Unfair Method of Competition? A Rebuttal,* 82 Yale LJ 691 (1973) **§3.16**

Seldim, *When Stock Is Not a Security: The Sale of Business Doctrine Under the Federal Securities Laws,* 37 Bus Law 637 (Jan 1982) **§6.19**

Severo, *Should Firms Screen The Workplace or the Worker?,* NY Times, Sept 28, 1980 **§2.06**

Shapp & Jurkat, *New Growth . . . New Jobs for Pennsylvania (The Schapp Foundation 1962)* **§9.04**

Sharav, *Cost Justification Under the Robinson-Patman Act,* Mgmt Acct 15 (July 1978) **§3.07**

Shenefield, Memorandum of Identification and Challenge of Pricing Practices in Concentrated Industries (Antitrust Div, Dept of Justice, May 26, 1978) **§11.23**

Sheridan, *The Role of House Counsel in Defending a Products Liability Lawsuit* in Products Liability of Manufacturers: Prevention and Defense 1981 (Ross & Foley eds) **§10.29**

Sherman, Products Liability for the General Practitioner (Shepard's/McGraw-Hill 1981) **§1.02**

Shils, *Industrial Unrest in the Nation's Rail Industry,* 15 Labor LJ 81 (1964) **§2.01**

Should Firms Screen the Workplace or the Worker?, NY Times, Sept 28, 1980 **§2.06**

Siegan, Economic Liberties and the Constitution (1980) **§3.15**

Silverman, Trends in Research and Development: New York State and the Nation, Chronica 1 (Sept/Oct 1978) **§9.12**

Skillern, Environmental Protection: The Legal Framework (Shepard's McGraw-Hill 1981) **§§7.12, 7.15, 7.23, 7.36, 7.43, 7.52**

Slichter, *The Industry of Discovery,* 128 Sci 1610 (Dec 22, 1968) **§§1.06, 12.07**

Smit, The Law of the European Economic Community (looseleaf, updated) **§3.02**

Smith, *Info City,* New York Magazine (Feb 9, 1981) **§11.19**

Smith, Wealth of Nations (1812 ed) **§3.07**

Snell, Annual Style Change in the Automobile Industry as an Unfair Method of Competition, 80 Yale LJ 567 (1971) **§3.16**

Sobel, *Seeking Not 7 But 100 Cancer Cures,* NY Times, Dec 30, 1979, §VII, at 9 **§11.08**

Sobel, *Testing the Psyche Can Strain Belief,* NY Times, Sept 28, 1980 **§2.06**

Society of Automotive Engineers, Recommended Practice (1979) **§10.19**

Solar Energy Research Institute, Land-Use Barriers and Incentives to the Use of Solar Energy (1979) **§8.04**

Solar Energy Research Institute, Legal and Institutional Implications of Providing Financial Incentives to Encourage the Development of Solar Technologies (1979) **§§8.07, 8.09, 8.10**

Solar Energy Research Institute, National Conference of State Legislatures, Solar Warranties Workshop: A Summary (1978) **§8.25**

Solar Energy Research Institute, Problems in the Administration of State Solar Legislation (1979) **§8.04**

Solar Energy Research Institute, A Survey of State Approaches to Solar Energy Incentives (1979) **§§8.04, 8.05, 8.06, 8.07, 8.08, 8.09, 8.10**

Solar Energy Research Institute, Wind Energy: Legal Issues and Institutional Barriers (1979) **§8.04**

Soltysinski, *Computer Programs and Patent Law: A Comparative Study, 3 Rut J Computer Tech & Law 1 (1973)* **§11.10**

Somers, *et al,* Adjusting to Technological Change (1963) **§2.02**

Sonnenschein, *Federal Securities Law Coverage of Note Transactions: The Antifraus Provisions,* 35 Bus Law 1567 (July 1980) **§6.18**

Spaght, The Bright Key (1965) **§11.29**

Special Committee on Consumer Affairs, Assn of the Bar of the City of New York, *Consumer Products Warranty Fulfillment,* NYLJ, Nov 28, 1980, at 1 **§3.09**

Special Committee on Consumer Affairs, *Creation of Consumer Fraud Units in United States Attorney's Offices in Major Metropolitan Areas,* 32 Rec AB City NY 371 (1977) **§2.07**

Special Committee on Consumer Affairs, *A National Energy or Technology Development Bank,* 29 Rec AB City NY 718 (1974) **§12.03**

Speer, Inside the Third Reich (1970) **§§1.06, 2.08, 3.10, 12.07**

Spence, *The Learning Curve and Competition,* 12 Bell J Econ 49 (1981) **§11.26**

Spencer, *et al, Coal Gasification for Electric Power Generation, 215 Sci 1571 (Jan 26, 1982)* **§§12.02, 12.05**

Sullivan, *Engineer Proposes Giant Space Mirrors,* NY Times, Feb 6, 1977, at 24 **§12.05**

Sullivan, *Princeton Gathering Makes Detailed Assessment of Problems of Establishing a Colony in Space,* NY Times, May 12, 1975, at 54 **§12.05**

Sullivan & Shanoff, *Air Quality Assessments: Dispersion Modeling,* 17 Trial 50 (Dec 1981) **§7.56**

Summary of State Legislation, 3 Solar L Rep 294 (1981) **§8.05**

Survey of Institutional Patent Policies and Patent Administration, Ad Service Supplement 2:4:2:1 (Mar 1978) **§9.15**

Surrey, *Tax Incentives as a Device for Implementing Government Policy,* 83 Harv L Rev 705 (1970) **§12.07**

Sweetheart Contracts, 23 Ind Lab Rel Rev 105 (1969) **§2.05**

Symposium, 45 Antitrust LJ 72 (1976) **§3.09**

Symposium, 49 ABA Antitrust LJ 1185 (1980) **§3.12**

Symposium, Regulation and Innovation, 43 Law & Contemp Prob 1 (1979) **§7.06**

T

Taggart, Cost Justification (1959) **§3.07**

Taylor, *Antitrust Nominee Says He Plans to Limit Agency's Role, Prosecute Price Fixers,* Wall St J, Mar 20, 1981, at 4 **§3.07**

Taylor, Rethink: A Paraprimitive Solution (1973) **§1.06**

Technological Change (Gold ed 1975) **§11.21**

Teilhard de Chardin, The Future of Man (1964) **§2.08**

Teilhard de Chardin, The Phenomenon of Man pt III (1959) **§§12.02, 12.06**

Telser, *Why Should Manufacturers Want Fair Trade?,* 3 JL & Econ 86 (1960) **§§3.09, 3.16**

Terleckyj, Effects of R&D on the Productivity Growth of Industries (National Planning Assn 1974) **§§1.03, 3.03, 3.11, 3.14, 3.16, 11.01, 11.11, 12.03, 12.07**

Testing the Psyche Can Strain Belief, NY Times, Sept 28, 1980 **§2.06**

Thom, Structural Stability and Morphogenesis: An Outline of a General Theory of Models (1975) **§11.27**

Thomas, *How Should Humans Pay Their Way,*NY Times, Aug 24, 1981, at 15 **§12.06**

Thomas, Miller & Robbins, Overcoming Legal Uncertainties About Use of Solar Energy Systems (1978) **§8.09**

Thompson, At the Edge of History (1970) **§§11.01, 11.04**

Thompson, Franchise Operations in Antitrust (1971) **§11.10**

Thompson, Passages About Earth (1973) **§§11.01, 11.04**

Thomsen, Vote of Censure (1968) **§§1.06, 12.07**

Thurow, The Zero-Sum Economy (1980) **§§11.13, 11.19, 11.24, 11.26, 11.30**

Tilton, International Diffusion of Technology (1971) **§11.18**

Toffler, The Third Wave (1980) **§11.19**

Toffler, Future Shock (1970) **§§11.04, 11.17, 11.21, 11.27, 11.30, 12.06**

Townsend, Up the Organization (1970) **§3.16**

Tribe, American Constitutional Law (1978) **§8.29**

Triburs, Applying Science to Industry, 70 US News & World Report 35 (Jan 8, 1971) **§§2.07, 3.10**

Tuchman, Stillwell and The American Experience in China (1970) **§2.07**

Turner, *Conglomerate Mergers and Section 7 of the Clayton Act,* 78 Harv L Rev 1313 (1965) **§3.07**

Turner, *The Definition of Agreement Under the Sherman Act: Conscious Parallelism and Refusals to Deal,* 75 Harv L Rev 655 (1962) **§§3.13, 11.23**

Turner, *Patents, Antitrust and Innovation,* 12 Antitrust Bull 227 (1967) **§§3.08, 3.10**

Twiss, Managing Technological Innovation (1974) **§11.30**

U

UK Commission Will Investigate Legality of Manufacturer's Refusal to Supply Discounters, Antitrust & Trade Reg Rep (BNA) A-19 (May 21, 1981) **§3.02**

United Automobile Workers, Ford and General Motors Agreements of 1982, described in UAW—Ford Report (spec ed Feb, 1982) **§2.02**

US Bureau of the Census, 1977 Census of Governments **§8.01**

US Bureau of the Census, 1978 Statistical Abstract of the United States **§8.02**

US Congressional Budget Office, The Productivity Problem: Alternatives for Action (1981) **§12.03**

US Dept of Commerce, Statistical Abstract of the United States (1980) **§11.29**

US Dept of Education, Digest of Educational Statistics (1980) **§11.20**

US Dept of Energy, Psycho-Economic Factors Affecting the Decision Making of Consumers and the Technology Delivery System (1978) **§8.26**

US Dept of Housing & Urban Development, HUD Solar Status (1978) **§8.05**

US Dept of Justice, Antitrust Guide for International Operations, *reprinted in* 1977 Trade Reg Rep (CCH) No 226 (special supp Feb 1, 1977) **§§3.09, 3.12**

US Eases Safety Standards for Recombinant DNA Work, NY Times, Jan 30, 1980, at 16 **§11.14**

Wasserstein, *British Merger Policy from an American Perspective,* 82 Yale LJ 656 (1973) **§3.02**

Watson, Industry Aid to Education (1965) **§11.29**

Wayne, *Joys of Fleeing the Corporate Stable,* NY Times, Nov 15, 1981 3, at 1 **§2.02**

Webb, *NASA as an Adaptive Organization,* in Technological Change and Management (Ewing ed 1970) **§§11.15, 11.22**

Weinberger, *Creativity and Collaboration in Government—The Budget Process,* in The Frontiers of Knowledge (1975) **§12.03**

Weinberger Feared Breakup of AT&T Would Hurt Units, Wall St J, Apr 10, 1981, at 2 **§11.08**

Weinberg, *The Forces of Nature,* 29 Bull Am Acad Arts & Sci No 4, at 13 (Jan 1979) **§12.02**

Weiss, *The Concentration-Profits Relationship and Antitrust,* in Industrial Concentration: The New Learning (Goldschmid, Mann & Weston eds 1974) **§11.24**

Weithorn & Suwalsky, Research and Development Expenses (BNA Tax Management Portfolio 42-2d 1975) **§11.16**

Wellington, *The Nature of Judicial Review,* 36 Rec AB City NY 360 (Oct 1981) **§3.15**

Wertheimer, Productive Thinking (1959) **§§1.01, 2.07**

Wertheimer, *Small Offerings: Recent SEC and Legislative Initiatives,* in Twelfth Annual Institute on Securities Regulation 507 (PLI Corporate Law & Practice, Course Handbook Series No 348 Fleischer, Lipton & Mundheim eds 1980) **§6.01**

Westbrook & Ratti, Aikido and the Dynamic Sphere (1970) **§12.06**

Westin, Whistle Blowing: Loyalty and Dissent in the Corporation (1981) reviewed, Bus Week 8 (Nov 24, 1980) **§2.07**

Weyand, *Majority Rule in Collective Bargaining,* 45 Colum L Rev 556 (1945) **§2.05**

Whaley, Warranties and the General Practitioner (1981) **§10.09**

Wheeler, *Selected Evidentiary Issues of Current Importance in Product Liability Litigation,* in Product Liability of Manufacturers: Prevention and Defense 1981 (Ross & Foley eds) **§10.30**

When Has Substantial Agreement Been Reached for Application of the Contract Bar Rule, 33 Labor LJ 121 (1982) **§2.05**

White, *Participant Governmental Action Immunity from the Antitrust Laws: Fact or Fiction?* 50 Tex L Rev 474 (1972) **§3.11**

Whittlesen, *Protection Against Discharge Without Cause,* NY St BJ 101 (Feb 1981) **§2.06**

Wiedenbaum, *Subsidies Business Does Not Need,* NY Times, (Mar 31, 1981, at A19) **§12.03**

Wilcox & Shepard, Public Policies Toward Business (5th ed 1975) **§3.16**

Wilder, *The Electronic Data Processing Industry: Market Structure and Policy Issues,* 20 Antitrust Bull 25 (1975) **§11.24**

Wilford, *Craft Speeds Toward Mercury on Wings of Student's Idea,* NY Times, Nov 4, 1979, at 64 **§12.07**

William, Martial Arts of the Orient (1975) **§12.06**

3 Williams, American Land Planning Law (1975) **§8.20**

Williams, Huey Long (1969) **§12.03**

Williams & Spoto, *Corporate Policies for the Protection of Documents: Attorney-Client Privilege and Work Product Doctrine,* in Product Liability of Manufacturers: Prevention and Defense 1981 (Ross & Foley eds) **§10.25**

Williamson, *Dominant Firms and the Monopoly Problem: Market Failure Considerations,* 85 Harv L Rev 1512 (1972) **§11.24**

Williamson, *Economics as an Antitrust Defense Revisited,* 125 U Pa L Rev 699 (1977) **§§11.12, 11.18**

Williamson, *The Economics of Antitrust: Transaction Cost Considerations,* 63 Am Econ Rev 316 (1973) **§§11.05, 11.07, 11.09, 11.20**

Williamson, Markets and Hierarchies: Analysis and Antitrust Implications (1975) **§§3.16, 11.07**

Williamson, *Predatory Pricing: A Strategic and Welfare Analysis,* 87 Yale LJ 284 (1977) **§3.07**

Williamson, *The Vertical Integration of Production: Market Failure Considerations,* 61 Am Econ Rev 112 (1971) **§§11.12, 11.18, 11.22**

Willis & Pitofsky, *Antitrust Consequences of Using Corporate Subsidiaries,* 43 NYU L Rev 20 (1968) **§3.07**

Winfield, *Joint Ventures in the Implementation of New Technology,* 2 Legal Notes & Viewpoints 59 (Feb 1982) **§3.10**

Winston, A Builder of the New South (1920) **§9.03**

Wirtz, The Boundless Resource (1975) **§12.07**

Wolfson, Phillips & Russo, Regulation of Brokers and Dealers and Securities Markets (1977) **§6.47**

Work in America: Report of a Special Task Force to the Secretary of Health, Education and Welfare (1972) **§§2.08, 3.16**

Works & Morgan, *The Land Grant Colleges* (US Dept of Agriculture, Staff Study No 10, 1939) **§9.02**

Worsinger, *New Technologies and Antitrust,* 47 NY St BJ 651 (1975) *reprinted in* 81 Case & Com 33 (1976) **§§1.02, 3.03, 3.11, 3.14, 12.03, 12.04**

Y

Young & Bradford, Joint Ventures: Planning and Action (1979) **§11.11**

Yuckman, *The National Public Radio Satellite System: FCC Jurisdiction Over a New Communications Technology,* 8 Rut J Computer Tech & Law 135 (1980) **§11.01**

Z

Zabriski, Handbook of Inner Sports (1973) **§12.06**

Zausner, Unvarnished (1941) **§§2.01, 2.02, 2.03**

Zuckerman, *Who'll Survive the War of the Videodiscs,* 77 Mechanics Illustrated 56 (Jan 1981) **§11.07**

Index

C

D

E

F

U

W

Z

LEGAL STRATEGIES FOR INDUSTRIAL INNOVATION

1985 Cumulative Supplement
Current through July, 1985

Richard A. Givens, Editor
Botein, Hays & Sklar
New York, New York

Edward S. Binkowski
Bruce M. Kramer
Frank J. Mauro
Lloyd McAulay
Kenneth Ross
Richard H. Rowe
Frank F. Skillern
Robert S. Taft

Insert in the pocket at the back of the bound volume. Discard supplement dated 1984.

Shepard's/McGraw-Hill
P.O. Box 1235
Colorado Springs, Colorado 80901

Supplement ISBN 0-07-023332-2

Preface to 1985 Supplement

Amid debate on methods of enhancing United States competitiveness, governmental, public, and industry recognition of the importance of scientific and technological advance and innovation to the economy has sharpened even further since the publication of the original volume.

This supplement provides updated information relevant to how innovators can best deal with the existing legal climate and to major policy options.

New recognition of the importance of encouragement of private innovation is reflected in the National Cooperative Research Act of 1984, discussed in **§3.10** of this supplement, and in other changes in the national climate.

Another significant development was enactment of the Local Government Antitrust Act ("LOGATA"), which sets a precedent in barring antitrust damage suits against those who act in reliance on directives of local governments. Injunctions, however, are still permitted to stop illegal behavior. As discussed in **§3.11** of this supplement, LOGATA could lead to similar results involving directives of other levels of government.

Section 12.02 of this supplement also discusses a new Bar Committee proposal based on the example of the New York State Office of Business Permits, which seeks to assist private enterprises engaged in development important to defined national goals in navigating through the maze of legal barriers, and **§12.07** contains revised recommendations for a national agenda for promotion of industrial innovation.

Richard A. Givens
June, 1985

New sections appearing in this supplement

Legal Issues Affecting Innovation

Richard A. Givens

1

§1.01 Introduction

A broadening consensus has emerged on the importance to the nation and the economy of technological development; it is becoming equally clear that innovation in "high" technology must be accompanied by scientific and technical advances on a broad base. *See generally State of the Union Message of the President,* NY Times, Jan 26, 1983, at A14, col 5; B8, col 3-4; Business—Higher Education Forum, America's Competitive Challenge: The Need for a National Response (Apr 1983); *Cover Story,* Bus Week, July 4, 1983, at 62; Altman & Garten, *Stimulate Commercial Technology—Or Else,* Bus Week, Mar 5, 1984, at 18; Kilborn, *Competitiveness Panel Reports on US Lag,* NY Times, Feb 14, 1985, at D1; RGK Foundation, Initiatives for Accelerating the Technology Diffusion Process (1983); Konecci et al, Initiatives for Technology Venturing (IC2 Inst, Univ of Texas at Austin 1984); HR Rep 98-693, 98th Cong, 2d Sess (1984); National Academy of Sciences, *International Competition in Advanced Technology: Decisions for America* (1983), discussed in Boffey, *U.S. Priority on Technology is Urged,* NY Times, Apr 15, 1983, at A3; *A Technology Lag That May Stifle Growth,* Bus Week, Oct 1, 1982, at 126; Dickson, *Technology and Cycles of Boom and Bust,* 219 Sci 933 (Feb 25, 1983); HR 1243, 98th Cong, 1st Sess (1983) (Economically Strategic Industrial Research and Development Act); *compare also* Tsongas, *Op-Ed Article,* NY Times, Mar 1, 1983, at A23; and Pollack, *The Birth of Silicon Statesmanship,* NY Times, Feb 27, 1983, §3, at 1, *with Rise of New Products: But How New Are They?,* NY Times Dec 13, 1984, at D28; *High-Technology Jobs Going Overseas as U.S. Costs Rise,* NY Times, Mar 19, 1983, at 1, 33; *America Rushes to High Tech for Growth: Many Call It A Panacea, But Hopes for New Jobs Are Overly Optimistic,* Bus Week, Mar 28, 1983, at 84; Fiske, *Leaders of Companies and Educators Stress Need to Compete,* NY Times, May 16, 1983, at Al, A13.

The importance of innovation is becoming increasingly clear. *See, e.g.,* Technology and Trade: Some Indicators of the State of U.S.

Industrial Innovation, Subcommittee on Trade of the Committee on Ways and Means, U.S. House of Representatives, U.S. Government Printing Office, Washington, D.C., 1980, at 2; Science Indicators—1980; Report of the National Science Board, National Science Foundation, Washington, D.C., 1981; Two Decades of Research on Innovation: Selected Studies of Current Relevance Prepared For the Use of the Special Study on Economic Change of the Joint Economic Committee of the Congress of the United States, U.S. Government Printing Office, Washington, D.C., 1980, at 138, 148; *Industrial Policy,* Congressional Quarterly Weekly Report, Washington, D.C., Aug 20, 1983. Secretary of Commerce Malcolm Baldridge has stated that "the Federal Government must fund R&D necessary for our national defense and basic, long-term, research in the nondefense sector. . . ." Testimony before the Subcommittee on Savings, Pensions & Investment Policy, Senate Finance Committee, *Government Policies to Promote High Growth Industries Based on New Technologies and to Increase U.S. Competitiveness,* 98th Cong, 1st Sess 10 (1983).

Similarly, statistics suggest that capital investment alone, without innovation, does not necessarily result in enhanced productivity. See Abernathy, The Productivity Dilemma (1978), *reviewed in* Andrews, *Technology: Productivity or Innovation?* NY Times, Jan 31, 1979, at D4; Robinson, The Accumulation of Capital (1966); *New Data Explain Lag in Productivity,* NY Times, Apr 7, 1983, at D6 (new Bureau of Labor Statistics "Multi-Factor Productivity Index").

Trade competitors of the United States support high-risk research on a long-term basis. See Assistant Secretary of Commerce D. Bruce Merrifield, *quoted in* 45 Bureau of National Affairs, Antitrust & Trade Reg Rep 296 (Feb 23, 1984) (a "whole nation targeting its industry"); Bus Week, Jan 10, 1983, at 45, 47 (France); *Where Japan Has Research Edge,* Bus Week, Mar 14, 1983, at 116; Dickson, *Britain Rises to Japan's Computer Challenge: In Departure from its Free-market Principles, the Government will Spend $300 Million to Subsidize Research on Information Science,* 220 Sci 799 (May 20, 1983). *Compare generally* Gibney, Miracle by Design: The Real Reasons Behind Japan's Economic Success (1982); Magaziner & Reich, Minding America's Business (1982); Johnson, MITI and the Japanese Miracle (1982); Organization for Economic Cooperation, The Industrial Policy of Japan (Paris 1972).

United States trade balances are generally better in R&D intensive goods and services broadly defined than elsewhere. Guzzardi, *How to Foil Protectionism,* 107 Fortune 6, Mar 21, 1983, at 76, 79; *compare* Reich, *Beyond Free Trade,* 61 Foreign Affairs 4, at 773 (Spring 1983); Weil, *Anti-Targeting as Bad for Trade as Smoot-Hawley Tariffs in the 1930s,* NY Times, Feb 18, 1984, at A23.

In note 1, page 1, fifth line, delete " __ US __ " and substitute:

456 US 985

§1.02 Legal Promotion of Innovation

For legislative recognition of the importance of innovation, *see* Stevenson-Wydler Technology Innovation Act of 1980, 15 USC §3701 *et seq.* This follows from a long-accumulating literature on its increasing importance. *See, e.g.*, Mansfield, *Basic Research and Productivity Increase in Manufacturing,* 70 Am Econ Rev 863 (Dec 1980); Hutten, *On the Importance of Productivity Change,* 69 Am Econ Rev 126 (1979); Shanklin & Ryans, *Organizing for High-Tech Marketing,* 84 Harv Bus Rev 164 (Nov Dec 1984); Flint, *Lagging Productivity,* NY Times, Jan 27, 1979, at 27; Kelman, *Whatever Happened to U.S. Innovation,* NY Times, Feb 14, 1979, at 14F; Mullaney, *The Spreading Malaise,* NY Times, July 16, 1978, at F15; Sullivan, *Adviser Warns U.S. Lags in Scientific Growth Rate,* NY Times, Oct 6, 1973, at 10; Javits, *The Need for a National Productivity Improvement Policy,* Cong Rec S1100 (daily ed Feb 6, 1979); Starr & Rudman, *The Parameters of Technological Growth,* 182 Sci 358 (Oct 26, 1973); David, *Science Futures: The Industrial Connection,* 203 Sci 837 (Mar 2, 1979); Freeman, Unemployment and Technological Innovation (1982); Terleckyj, Effects of R&D on the Productivity Growth of Industries (1974).

§1.03 —Individual Enhancement of Pro-Innovation Policies

According to Reich, *The Next American Frontier* (1983) and his summary in *The Next American Frontier: What's Really Wrong With the American Economy,* 251 Atlantic 43 (Mar 1983), insufficient attention is often given to the advantages of flexible production offering custom-made options to users, in addition to mass production of interchangeable items. The needs of the marketplace would presumably dictate the best approach if close interaction between the enterprise and actual or potential customers occurs. Reich says this opportunity is often missed because "product entrepreneurship," involving primarily technical considerations, has temporarily lost the spotlight that attracts the best minds in many industries to what he refers to as "paper entrepreneurship." *See also* Pione & Sabel, The Second Industrial Divide (1984); Heilbroner, *Beyond Mass Production,* NY Times Book Rev, Jan 6, 1985, at 7.

Effective technological management includes both motivation of managers and employees for "product enterpreneurship," as discussed in **§§2.07-2.08,** and study of available sources to learn what others are doing, which may suggest technical clues, sources of supply, or markets. *See, e.g.*, Japan Engineering News, *Patents and Licensing* (ISSN 0388-7081); Walsh, *Translating Japanese Technology into US Terms,* 224 Sci 265 (Apr 20, 1984); McGraw-Hill, Biotechnology Newswatch (periodical); Breakthrough: The Newsletter of Innovation by Boardroom Reports. Patent records must also be searched (see **ch 4**).

Technological management also includes adequate internal feedback, as discussed in **§§2.07-2.08;** *compare* Goldscheider, Technology Management Handbook (1984); *Special Report: A New Era for Management,* Bus Week, Apr 25, 1983, at 50; Givens, Advocacy, ch 14 (Shepard's/McGraw-Hill 1980); *see generally* Bacon & Butler, Planned *Innovation* (2d ed 1981); 1982 Technology Management Handbook (Clark Boardman). On licensing procedures and problems, see Practising Law Institute, Technology Licensing (1982); Patent Law Association of Chicago, Tax Guide for Patents, Trademarks and Copyrights (1983).

On challenges in particular industries, see O'Reilly, Gaynor, Carroll, & Cronin, Jr, Federal Regulation of the Chemical Industry (Shepard's/McGraw-Hill 1980); Krueger & Yarema, *New Institutions for New Energy Technology: The Case of Ocean Thermal Energy Conversion,* 54 S Cal L Rev 767 (1981); Dula, *Regulation of Private Commercial Space Activities,* 23 Jurimetrics J 156 (1983); Zammit, *Computers, Software and the Law,* 68 ABAJ 970 (1982); Marshall, *Law of the Sea and the New U.S. Oceans Policy,* 55 NY St BJ 8 (Apr 1983).

On capital sources, see Petillon & Hill, R&D Partnerships: Structuring the Transaction (1984); Tomczak, Corporate & Commercial Finance Agreements (Shepard's/McGraw-Hill 1984); Practising Law Institute, Selected Problems in the Initial Public Offering for High-Technology Companies (1984); Halloran, Venture Capital and Public Offering Negotiation (1982); Term Loan Handbook (McCann ed 1983); *compare* Worcester & Boggs, *Capital for Geothermal Energy Projects,* 14 Nat Resources Law 713 (1982). On state and local involvement, see **chs 8-9;** Practising Law Institute, State Regulation of Capital Formation and Securities Transactions (1983); Carey, A Strategy for Developing Technology-Based Industry in New York (Governor's Office 1982); see generally **chs 5-6.**

§1.04 Legal Obstacles to Innovation

See generally Burr, *Disincentive to be Inventive,* 70 ABAJ 93 (May 1984).

Add to note 18, page 4:

See **ch 10;** Perrow, Normal Accidents: Living With High-Risk Technologies (1984), *reviewed in* Pfeiffer, *Courting Disaster,* NY Times Book Rev, Apr 8, 1984, at 19; Practising Law Institute, Product Design Litigation (1982); Wheeler, *Product Liability: Civil or Criminal?,* 17 F of ABA Sec on Tort & Ins Prac 250 (1981); Schaden & Heldman, Product Design Liability (1982); Practising Law Institute, Product Liability of Manufacturers (1981); Baldwin & McGovern, Preparation of a Product Liability Case (1981); Johnson v International Harvester Co, 702 F2d 492 (4th Cir 1983); Lieb v American Motors Corp, 538 F Supp 127 (SDNY 1982); Givens, *Product Improvement: Evidence of Liability?,* NYLJ,

July 18, 1983, at 1; Sun, *Lawyers Flush Out Toxic Shock Data,* 224 Sci 132 (Apr 13, 1984).

Add to note 20, page 4:

See also BNA, International Environment Reporter (looseleaf); Practising Law Institute, Environmental Law and Practice (1982); Note, *Technology Based Emission and Effluent Standards and the Achievement of Ambient Environmental Objectives,* 91 Yale LJ 792 (1982).

§1.05 —Strategies for Dealing with Obstacles

Add to note 27, page 5:

An essential ingredient in overcoming any kind of obstacle is to understand in advance that difficulties will arise, and hence not derail the effort merely for that reason, as pointed out in Davis, *New Projects: Beware of False Economies,* 85 Harv Bus Rev 95 (Mar/Apr 1985).

On the risk that coalitions will form against innovations, see Olson, The Rise and Decline of Nations (1982), *reviewed in* 219 Sci 163 (Jan 14, 1983); Olson, The Logic of Collective Action: Public Goods and the Theory of Groups (1971). This tendency may be counteracted by taking advantage of external demands and opportunities to build alternative coalitions. If an expanding economic pie can be generated, the proceeds can be divided in a "non-zero sum game" to give all an incentive to cooperate. See Schelling, The Strategy of Conflict (1960).

Exceptions from otherwise applicable requirements can often be obtained for specific high-priority efforts. See Givens, Advocacy, chs 16-17 (Shepard's/McGraw-Hill 1980). Such a process may be seen most clearly in wartime or where new industries create entirely new interest group constellations. The "creative confusion" caused by change can frequently provide an opening for innovative approaches.

In some instances action may be accelerated most simply by discretionary cutting of "red tape" in the interest of national goals through favorable interpretation of existing rules, as most commonly occurs in wartime. *See, e.g.,* Jones & Angly, Fifty Billion Dollars (1951)(role of RFC in World War II).

In other instances, legislation may authorize certain conduct "notwithstanding any other provision of law, including subsequent enactments not specifically referring to this section." These approaches can be combined with use of a "lead agency" which can act for others or coordinate their efforts, because of cooperation in the national interest or because of statutory authority to do so. *See* Richardson, *Cutting Regulatory Tape,* NY Times, Dec 13, 1984, at A31.

Innovation and Labor Relations

Richard A. Givens

2

§2.01 Impact of Innovation on Labor Relations

The need for innovation, as well as the problems it creates, is receiving increasing labor-management attention even as it is recognized that implementation of changes often requires more commitment than initially believed. See Lawler & Mohrman, *Quality Circles After the Fad,* 85 Harv Bus Rev 64 (Jan/Feb 1985); *The Trouble With Managing Japanese-Style,* Fortune, Apr 2, 1984, at 50.

While it is clear that employers must bargain concerning the impact of technological change, the extent of statutory obligations to bargain on technological change itself is unclear. *See NLRB v Island Typographers, Inc,* 705 F2d 44 (2d Cir 1983) and cases cited. Of course, not to discuss contemplated changes and seek acquiescence, cooperation, or suggestions on avoiding problems can be a major error entirely aside from statutory constraints.

Add to note 5, page 9:

See also Lundine, *The Jamestown Experience: A Case Study in Labor-Management Cooperation,* in Lunstedt & Colglazier, Managing, ch 12 (1982); Javits, *For Labor, Employers,* NY Times, Jan 21, 1983, at A27; Caplan, *Social Support, Person-Environment Fit, and Coping,* in Mental Health and the Economy 89, 127-28 (Ferman & Gorman eds 1979).

§2.02 Options Concerning Employees Threatened with Displacement

On application of statutory requirements for employee protection in transportation mergers and other regulated actions, *see, e.g., Jackson Transit Authority v Local Division 1285, Amalgamated Transit Union,* 102 S Ct 2902, 72 L Ed 2d 639 (1982); *Railway Labor Executives Association v United States,* 675 F2d 1248 (DC Cir 1982); *Braniff Master Executive*

Council v CAB, 693 F2d 220 (DC Cir 1982); *United Transportation Union v Consolidated Rail Corp*, 535 F Supp 697 (Spec Ct), *cert denied*, 457 US 1133 (1982).

On remedies where federal funding is accompanied by employee protection requirements, see Note, *Injunctive Relief for State Violations of Federal Funding Conditions*, 82 Colum L Rev 1236 (1982).

On treatment of collective bargaining agreements where the firm is under the supervision of a bankruptcy court, *see NLRB v Bildisco & Bildisco*, 104 S Ct 1188 (1984), *modified by* 98 Stat 353 (1985), enacting Bankruptcy Code §1113, *discussed in* 4 Bankr Law Letter # 12 (Dec 1984); *See also Teamsters Local 20 v Brada Miller Freight System*, 702 F2d 89 (11th Cir 1983).

On long-term employment assurances as a means of promoting employee effort, see Bolt, *Job Security: Its Time Has Come*, Harv Bus Rev 199 (Nov-Dec 1983); discussion in *id*, (Mar-Apr 1984) 199-200; *compare* Franklin D. Roosevelt, 5th Ann Message Jan 3, 1938, 3 State of the Union Messages of the Presidents 2840-41 (Chelsea House 1966):

> Definite strides in collective bargaining have been made and the right of labor to organize has been nationally recognized. Nevertheless, in the evolution of the process difficult situations have arisen in localities and among groups. Unfortunate divisions relating to jurisdiction among the workers themselves have retarded production within given industries and have, therefore, affected related industries. The construction of homes and other buildings has been hindered in some localities not only by unnecessarily high prices for materials but also by certain hourly wage scales. For economic and social reasons our principal interest for the near future lies among two lines: First, the immediate desirability of increasing the wages of the lowest-paid groups in all industry, and, second, in thinking the terms of regularizing the work of the individual worker more greatly through the year—in other words, in thinking more in terms of the worker's total pay for a period of a whole year rather than in terms of his remuneration by the hour or by the day.

Page 10, text, following third paragraph add:

Fourth, an offer of early retirement bonuses or incentives for voluntary separation may trim the work force. If the offer is tinged with compulsion, however, illegal age discrimination may be found under 29 USC §621 *et seq* or state laws. *Compare* McCarroll, *Rehiring Retirees*, NY Times, Nov 22, 1981, at 8F.

Page 11, text, following first full paragraph add:

Contributions to trust funds may be made by a segment of an industry which gains from innovation in order to protect jobs in another which loses, as in the case of the recording industry trust fund to benefit performing musicians.

In note 19, page 11, first line, delete "__ US __" and insert:

452 US 666

Add to note 19, page 11:

See also Milwaukee Spring Div of Ill Coil Spring Co (Milwaukee Spring II), 268 NLRB No 87 (Jan 28, 1984); United Technologies, Otis Elevator Co, 269 NLRB (No 162), Lab L Rep (CCH) ¶632 (Apr 13, 1984); Baderian & Goodman, *Plant Relocations and Union Contracts,* National LJ, Feb 27, 1984, at 24; Irving, *Closings and Sales of Businesses: A Settled Area?,* 33 Lab LJ 218 (1982); Krotseng, *Plant Relocations and Consolidations: Employers' Obligations Regarding Employee Transfer and Union Recognition,* 33 Lab LJ 609 (1982). On severance pay on sale of a business, *see* Livernois v Warner-Lambert Co, 723 F2d 1148 (4th Cir 1983).

§2.03 Conflicts over Job Security Arrangements

The Taft-Hartley Act, in 29 USC §158(b)(6), makes it an unfair labor practice for a union "to cause or attempt to cause an employer to pay or deliver or agree to pay or deliver any money or other thing of value, in the nature of an exaction, for services which are not performed or not to be performed." This section, however, is rarely successfully invoked. *See Cotton's, Inc v Teamsters Local No 5,* 547 F Supp 1336 (MD La 1982) and cases cited; Note, 52 Colum L Rev 1020 (1952). This might be altered in a context where jobs were guaranteed or employment generally was sufficiently high to have a similar effect, *e.g.,* during the Second World War. *Compare* Shaiken, Work Transformed: Automation & Labor in the Computer Age (1985), *reviewed in* NY Times, Apr 14, 1985, at F15; Givens, *Dealing With National Emergency Labor Disputes,* 34 Temple LQ 17 (1960); Gibney, Miracle by Design (1982) (Japanese experience); Aaron, *Restraints on Featherbedding,* 5 Stan L Rev 680 (1953); Daykin, *Featherbedding,* 7 Lab LJ 699 (1956); Van der Water, *Industrial Productivity and the Law: A Study of Work Restrictions,* 43 Va L Rev 155 (1957); Kossoris, *Working Rules in West Coast Longshoring,* 84 Monthly Lab Rev 1 (1961); NY Times, Apr 11, 1960, at 18, col 1; NY Times, June 30, 1960, at 62, col 1.

Add to note 33, page 13:

See also Jacksonville Bulk Terminals, Inc v ILA, 457 US 702 (1982).

Add to note 51, page 15:

See also JF Hoff Electric Co v NLRB, 642 F2d 1266 (DC Cir 1980), *cert denied,* 451 US 918 (1981); Lewis v New Orleans Clerks & Checkers, 724 F2d 1109 (5th Cir 1984) (work acquisition as distinct from protection); *compare* Freedman, *US and Britain in Actor Trade War,* NY Times, Feb 23, 1984, at C17.

Add to note 53, page 15:

See also Jim McNeff, Inc v Todd, 75 L Ed 2d 830 (1983) (upholding enforceability of pre-hire agreements in the construction industry); Woelke & Romero Framing, Inc v NLRB, 456 US 645 (1982) (construction).

§2.04 Jurisdictional Disputes Involving Union Representation

Bargaining unit descriptions do not necessarily restrict work transfers to nonunit employees; an employer can bargain to impasse on such transfers and then proceed with them, according to *Newspaper Printing Corp v NLRB,* 692 F2d 615 (6th Cir 1982).

In note 58, page 16, second line, delete "__ US __" and insert:

452 US 615

§2.05 The Free Election Model Contrasted With Specialized Groups of Employees: The Construction and Garment Industries

Add to note 71, page 17:

Compare Woelke & Romero Framing, Inc v NLRB, 456 US 645 (1982); Carpenters Local Union No 1846 v Pratt-Farnsworth, Inc, 690 F2d 489 (5th Cir 1982), *cert denied,* 104 S Ct 335 (1983).

Add to note 74, page 18:

See also NLRB v Mariott Corp, Bob's Big Boy Family Restaurants, 693 F2d 904 (9th Cir 1982); Atleson, Values and Assumptions in American Labor Law (1984).

§2.06 Initial Considerations When Innovation Creates Jobs

See generally Fasman, *Legal Obstacles to Alternative Workforce Designs,* 8 Employee Rel LJ 256 (1982); Perkins, Managing Creation: The Challenge of Building A New Organization (1983).

Add to note 77, page 19:

See also EEOC v Board of Trustees of Wayne County College, 723 F2d 509 (6th Cir 1983); Equal Employment Advisory Council, Employee Selection (Monograph No 2, Potter ed 1983); Silbergeld, *New Affirmative Action Regulations for Government Contractors,* 33 Lab LJ 230 (1982); Norris, Contract Compliance Under the Reagan Administration (1982); Bartholet, *Application of Title VII to Jobs in High Places,* 95 Harv L Rev 945 (1982); *compare* Note, *Title VII on Campus: Judicial Review of University Decisions,* 82 Colum L Rev 1206 (1982); Personal Liability of Managers and Supervisors for Corporate EEO Policies and Decisions (Levinson ed 1982); *compare also* Technology and Handicapped People, Joint Hearing, Subcommittee on Science, Research & Technology, House Committee on Science & Technology and Senate Committee on Labor & Human Services, 97th Cong, 2d Sess 163 (1982). There is no general common law duty with regard to employment of the handicapped. Bruffett v Warner Communications, Inc, 692 F2d 910 (3d Cir 1982).

Add to note 78, page 19:

When new jobs are being created with new skill demands and new employees are recruited, existing employees should be offered the opportunity to qualify. Otherwise, charges of discrimination based on age or other illegal factors may result. *Compare* Age Discrimination in Employment Act (Lake ed 1982); Coleman, *The Aging Mind Proves Capable of Lifelong Growth,* NY Times, Feb 2, 1984, at Cl, C5.

This may also promote productivity. See McCarrell, *Rehiring Retirees,* NY Times, Nov 11, 1981, at 8F.

On the other hand, structuring pension and other incentives in such a way as to encourage employees to remain, apart from their regular job benefits and satisfactions alone, like discouraging lateral hiring at senior levels in order to provide a like incentive or morale benefit for those below on the ladder, may be counterproductive in the end, according to McDowell, *Job Loyalty: Not the Virtue It Seems: Veteran employees provide continuity, but all too often they are insecure, dull and dependent,* NY Times, Mar 3, 1985, §3, at F1, F26.

Add to note 80, page 19:

Business necessity for a job criterion, a broader concept than "bona fide occupational qualification" (BFOQ) is available as a defense where there is no intent to discriminate on invidious grounds. *See* Griggs v Duke Power Co, 401 US 424 (1971); Nashville Gas Co v Satty, 434 US 136 (1977); Wright v Olin Corp, 697 F2d 1172, 1184-85 & n 21 (4th Cir 1982); *compare* Connecticut v Teal, 102 S Ct 2525, 73 L Ed 2d 130 (1982), discussed in Symposium, *Employee Selection Procedures,* 8 Employee Rel LJ 572 (1983). On criteria for tests, *see* Baker v City of Detroit, 483 F Supp 930, 971 (ED Mich 1979), discussing the Uniform Guidelines for Employee Selection Procedures under various federal laws, 29 CFR §1607.

Add to note 81, page 19:

For an example of employee objections to work being done beyond what was in a job description, *see* facts in NRLB v Magnetics Intl, Inc, 699 F2d 806, 807 (6th Cir 1983); *compare Work Revolution in U.S. Industry: More Flexible Rules on the Job Are Boosting Productivity,* Bus Week, May 16, 1983, at 100.

Add to note 83, page 19:

See Boise Cascade Corp v Secretary of Labor, 694 F2d 584 (9th Cir 1982) (issue of technological feasibility of noise level reduction); Nothstein, The Law of Occupational Safety and Health (1982); Sweltz, *Genetic Testing in the Workplace: An Analysis of the Legal Implications,* 19 Forum of ABA Tort & Ins Prac Sect 323 (1984); Brommer, *Occupational Disability Litigation Under the FELA,* 19 Forum of ABC Tort & Ins Prac Sect 205 (1984); Logan, *Adapting Fetal Vulnerability Programs to Title VIII,* 9 Employee Rel LJ 605 (1984); Auchter, *OSHA: A Year Later,* 33 Lab LJ 195 (1982); Smith, *Arbitrating Safety Grievances,* 33 Lab LJ 238 (1982); Note, *Regulation of Toxic Substances Under the Occupational Safety and Health Act,* 33 Syracuse L Rev 887 (1982).

Add to note 86, page 20:

See also Practising Law Institute, Protecting Trade Secrets (1982); 1982 Trade Secrets Handbook (Clark Boardman).

Add to note 87, page 20:

See also Wegmann v London, 648 F2d 1072 (5th Cir 1981).

§2.07 The Importance of Employee Involvement in Innovation

Lack of internal two-way communication and feedback from operating levels can cause serious technological errors. See Walton, *From Control to Commitment in the Workplace,* 85 Harv Bus Rev 77 (Mar/Apr 1985); Stevenson & Gumpert, *The Heart of Entrepreneurship,* 85 Harv Bus Rev 85 (Mar/Apr 1985); Lewis, *For Want of a Nail,* NY Times, Aug 31, 1981, at A17; Biddle, *Crippling of 2d Space Shuttle Tied to Design Flaw,* NY Times, Mar 2, 1983, at 1, 24; Egan, *To Err is Human Factors,* 85 Tech Rev 22 (Feb/Mar 1982); *compare* Janowitz, *Disarticulation,* NY Times, Apr 26, 1981, at E23.

Add to note 89, page 20:

See also Gibney, Miracle By Design: The Real Reasons Behind Japan's Economic Success (1982); Kanter, The Change Masters: Innovation and Productivity in the American Corporation (1983); Work in America: Report of a Special Task Force to the Secretary of Health, Education and Welfare (1972); Pope John XXIII, Mater et Magistra, pt II §82-83 (1961); Gardner, Excellence, ch XIV (1961); Hersey, Zest for Work (1955); McGregor, The Human Side of Enterprise (1960); Levenstein, Why People Work (1964); Holmes, Collected Legal Papers 30-33, 270-71 (1921); Peters & Waterman, In Search of Excellence: Lessons from America's Best-Run Companies (1982); *Why Cap Cities is Sitting Pretty,* Bus Week, Mar 21, 1983, at 121; The Economist (London), Feb 12, 1983, at 67.

Add to note 92, page 21:

See also Pinchot, Intrapreneuring (1985), *reviewed in* NY Times Apr 14, 1985, at F15; Lewis, *Technology, Enterprise and American Economic Growth,* 215 Sci 1204 (Mar 5, 1982); *Five Ways to Go Bust,* The Economist (London), Jan 8, 1983, at 11; *Intrapreneurial Now,* The Economist (London) Apr 17, 1982, at 67; The Human Factor in Innovation and Productivity, Report of Subcommittee, House Committee on Science & Technology, 97th Cong, 2d Sess (1982), Serial FF; Lohr, *Japan's New Nonconformists: Technology Spurs Change,* NY Times, Mar 8, 1982, at D1, D27.

Add to note 94, page 22:

See also Givens, Advocacy, reference notes to ch 14 (Shepard's/McGraw-Hill Supp 1984).

Add to note 96, page 22:

On product coverage of remedial orders, *see* Sears, Roebuck & Co v FTC, 676 F2d 385 (9th Cir 1982); American Bar Assn, Antitrust

Section (Monograph No 5, Vol 1), The FTC as an Antitrust Enforcement Agency 60-62 (1981); Kanwit, Federal Trade Commission §11.05 (Shepard's/McGraw-Hill 1979).

Antitrust and Innovation
Richard A. Givens

3

Antitrust Risks for Innovators

§3.03 Judicial Interpretation of Relevant Antitrust Provisions

Concerning current approaches to the antimerger statute, §7 of the Clayton Act, 15 USC §18, as indicated by the 1982 Justice Department Guidelines and Federal Trade Commission pronouncements, contained in Trade Reg Rep (CCH) No 546 (extra ed, June 16, 1982), and Antitrust & Trade Reg Rep (BNA) No 1969 (special supp June 17, 1982), see Practising Law Institute, New Merger Guidelines (1982); Fox, *The New Merger Guidelines: A Blueprint for Microeconomic Analysis,* 27 Antitrust Bull 519 (1982).

Add to note 25, page 32:

In Monsanto Co v Spray-Rite Serv Corp, 104 S Ct 1464, 79 L Ed 2d 775 (1984), the Court declined to consider overturning this rule, as suggested by the Department of Justice, on the grounds that the issue was not presented by the parties. The *per se* rule against resale price maintenance combinations appears to have the support of a majority of the business community according to Business Week-Harriss Poll, *Resounding Support for Price Competition,* Bus Wk, Nov 14, 1984, at 24.

In 1985, the Department of Justice issued guidelines setting forth its interpretation of when nonprice "vertical" restraints between suppliers and customers, or others at differing levels in a chain of distribution, will call forth its enforcement efforts or be regarded by it as violating the antitrust laws under Continental TV, Inc v GTE Sylvania, Inc, 433 US 36 (1977). US Department of Justice, Vertical Restraints Guidelines, Jan 23, 1985, 4 Antitrust & Trade Reg Rep (BNA) No 1199 (Spec Supp Jan 24, 1985). These guidelines are based on the general approach that nonprice vertical restraints by firms other than market leaders are frequently, although not always, benign. *See also* authorities cited in Givens, Antitrust: An Economic Approach, ch 8 (1982 & supplements).

Add to note 27, pages 32-33:

In Jefferson Parish Hosp Dist No 2 v Hyde, 52 USLW 4385 (US Mar 27, 1984), the Court applied a requirement of practical effect of foreclosing competition on the merits in the market for the tied item. Four Justices would require a significant overall anticompetitive effect in that market.

In note 29, page 33, fourth line, delete "__ US __" and substitute:

456 US 985

§3.05 Innovation as Monopolization

In note 49, page 36, second line, delete "__ US __" and substitute:

456 US 985

Add to note 49, page 36:

See also Gribbon, *Does the Law of Monopolization Inhibit Innovative Behavior?*, 49 ABA Antitrust LJ 925 (1980); Jente, *Assessing Recent Efforts to Challenge Aggressive Competition as an Attempt to Monopolize*, 49 ABA Antitrust LJ 961 (1980).

In note 54, page 37, second line, delete "__ US __" and substitute:

456 US 985

§3.06 Extension of Power from One Market to Another

In *Jefferson Parish Hospital District No 2 v Hyde,* 52 USLW 4385 (US Mar 27, 1984), the majority of the Supreme Court held that a hospital's exclusive anaesthesia contract did not significantly prevent anaesthesia users from any benefits of competition on the merits among anaesthesiologists. Four Justices stated that a significant anticompetitive impact in the tied (anaesthesia) market should be required in addition to market power in the tying market for a violation to be found under the rule of reason which they would have applied.

Add to note 86, page 41:

For a finding of no liability, *see* Shop & Save Food Markets, Inc v Pneumo Corp, 683 F2d 27 (2d Cir), *cert denied,* 459 US 1038 (1982).

§3.07 Predatory Pricing

For a defense of below-cost pricing to expand markets at an early stage of technological development, see Reich, *Beyond Free Trade,* 61 Foreign Affairs 4, at 773, 800 (Spring 1983).

For a rare decision sanctioning a local predatory pricing claim, albeit only by denial of summary judgment, *see Sunshine Books, Ltd v Temple University,* 697 F2d 90 (3d Cir 1982). *See generally* Areeda, *Predatory Pricing,* 49 ABA Antitrust LJ 897 (1980).

In note 104, page 44, second full paragraph, seventh line, delete "__ F2d __" and insert:

660 F2d 192

In note 107, page 45, fourth line, delete "__ F Supp __" and insert:

532 F Supp 983

§3.08 Patent-Antitrust Strategies

The Department of Justice believes that a rule of reason should apply to patent license restrictions. Patent Licensing: A Fresh Look at Antitrust Principles in a Changing Economic Environment, Remarks by J. Paul McGrath, Assistant Attorney General, before the Seminar Services International Conference on U.S. Patent Practice, Crystal City, Va, Apr 5, 1984. *But see Pitney Bowes, Inc v Mestre,* 701 F2d 1365 (11th Cir 1983) (hybrid patent and know-how license). *See generally* Symposium, *Special Considerations Concerning International Patent and Know-How Licensing, and Joint Research and Development Activities,* 50 ABA Antitrust LJ 497 (1981); Practicing Law Institute, Technology Licensing (1984). *Compare Transparent Wrap Machine Corp v Stokes & Smith Co,* 329 US 637 (1947); *Automatic Radio Manufacturing Co v Hazeltine Research, Inc,* 339 US 827 (1950); *United States v Westinghouse Electric Corp,* 648 F2d 642 (9th Cir 1981).

According to *Brunswick Corp v Riegel Textile Corp,* 752 F2d 261 (7th Cir 1984)(Posner, J), fraud on the patent office is only an antitrust violation if no patent could be issued to anyone, since, if patentable, an invention could be monopolized by someone, and hence there is no legitimate competition to be protected. This result is based on the increasingly clear line of authority that the antitrust laws seek to protect competition rather than individual competitors, *id* 266.

Accordingly, the outcome might be altered if evidence were introduced that an improperly obtained patent had a specifically anticompetitive effect because of the position of the patent holder in the industry, what other related patents that party had, and what related patents were or were not being exploited. The mere fact that the plaintiff would license the patent and the defendant does not, however, is insufficient according to the decision, *id* 267.

Add to note 111, page 45:

See also Practising Law Institute, Patent-Antitrust (1982).

In note 112, page 45, second line, delete "__ US __" and insert:

456 US 985

Page 46, line 4, delete "illegal" and insert:

antitrust and other

Add to note 114, page 46:

See also, for Justice Department views, note 122 *infra,* text p 48.

In note 126, page 49, second line, delete "__ US __" and insert:

456 US 985

Add to note 127, page 49:

In most countries a patent must be exploited within a specified period, or justification shown for failure to do so, in order for the patent to be enforced or licenses to others denied. See Manual for the Handling of Applications for Patents, Design and Trade Marks Throughout the World (Octrooibureau Los En Steiger, Amsterdam, Holland).

In note 129, page 49, first line, delete "__ US __" and insert:

456 US 985

In note 130, page 49, last line of second paragraph, delete "__ US __" and insert:

456 US 985

Add to note 131, page 50:

A "hybrid" agreement licensing patent rights and trade secrets was held unenforceable beyond expiration of the patent in Pitney Bowes, Inc v Mestre, 701 F2d 1365 (11th Cir 1983).

§3.09 Marketing Restrictions

Add to note 140, page 51:

Concerning territorial restrictions in the soft drink industry, *see* Soft Drink Interbrand Competition Act, 15 USC §3501 *et seq*, discussed in note 156 *infra*, text p 53.

§3.10 Joint Research

The National Cooperative Research Act of 1984, Pub L No 98-462, 98 Stat 1815, 15 USC §§4301-4305, makes it clear that joint research and development activity is subject to the rule of reason under the antitrust laws and is never illegal per se. The primary purposes of this was not to change the law, but to make clear what had always been the case. *See* Kobak, *Application of Antitrust Laws to Joint Research, Development,* (pt 1), NYLJ, Dec 10, 1984, at 1.

In a new departure in this area, the Act further provides that should a defendant prevail against an antitrust claim based on conduct of a joint research and development venture, reasonable attorney's fees may be recovered from the plaintiff "if the claim, or the claimant's conduct during the litigation of the claim, was frivolous, unreasonable, without foundation, or in bad faith." 15 USC §4304(a)(2). According to the Conference Report on the Act, HR Rep No 1044, 98th Cong, 2d Sess 15 (1984), 47 Antitrust & Trade Reg Rep (BNA) 588, 592 (Sept 27, 1984), the test is intended to adopt *Christiansburg Garment Co v EEOC,* 434 US 412, 417 (1978). The concept includes a duty to discontinue a suit which turns out to be baseless in the course of further discovery, *Nemeroff v Abelson,* 704 F2d 652 (2d Cir 1982).

In addition, where a specified notice is filed with the Attorney General and the Federal Trade Commission and a publication is made in the *Federal Register,* activities of a research and development joint venture are protected from trebling of damages under §4 of the Clayton Act (15 USC §15) in subsequent lawsuits. The protection against trebling does not apply to violations of subsequent decrees or orders, by virtue of 15 USC §4303(e), but protection from trebling does apply under similar state laws, as well as under the federal antitrust laws. 15 USC §4303(c). Prejudgment interest is authorized. 15 USC §4303(d).

Section 1 of the Act contains a definition of *research and development joint venture* designed to prevent the Act from protecting extraneous or separable subsequent activities. Section 1(b) excludes collusion relating to subsequent marketing of the product itself as distinct from licensing of proprietary information such as patents or trade secrets. It also excludes exchanges of information not reasonably required to conduct the research and development. The objective is to prevent a "spillover" from the protection of the Act to other nonresearch and development combinations. Conf Report at 11, 47 Antitrust & Trade Reg Rep (BNA) at 590.

A novel feature of the Act defines the rule of reason applicable to joint research and development ventures as "taking into account all relevant factors affecting competition, including, but not limited to, effects on competition in properly defined, relevant research and development markets." 15 USC §4302. According to the Conference Report, the concern here is with "over-inclusive" ventures which may unnecessarily chill research efforts that could compete with those sponsored by the venture itself; the possibility of over-inclusiveness is said to be present where the potential does not exist for research on a comparable scale to be done by at least three or four entities or groups. Conf Report at 10, 47 Antitrust & Trade Reg Rep (BNA) at 590.

A parallel risk, however, is that barring competitors from a joint research venture of substantial size and scope could exclude them from the opportunity to utilize the discoveries made, not by virtue of the patent laws, but by virtue of intervention under the antitrust laws limiting the scope of the joint venture. Thus, antitrust intervention could lead to monopolization of an industry, not merely "properly defined, relevant research and development markets." In such an instance, antitrust objectives would truly be turned on their heads! Taking into account "all relevant factors" would appear to mean weighing this risk.

The extent to which "over-inclusiveness" is in fact anticompetitive may depend in part on whether members of the venture are required to contribute all knowledge they acquire on the subject matter during the life of the venture to the central pool, thus precluding a secret race to get there first in the same or a related area even while the venture is proceeding. Such a restriction covering some industry members might be acceptable to get the venture going, but harmful if industry-wide.

As pointed out by James Kobak, Jr. in *Application of Antitrust Laws to Joint Research, Development,* pt I, NYLJ, Dec 10, 1984, at 1, defining a research and development market at all may be difficult in any event. Since the Department of Justice was one of the chief advocates of the concept of such market definition in connection with its concern with "over-inclusiveness," some guidance may be found in its prior positions on the subject. *See Senate Hearings on S 1841,* 98th Cong, 2d Sess 22-23, 197-08 (1984); Remarks of J Paul McGrath at the 18th Annual New England Antitrust Conference 9-11 (Nov 2, 1984).

According to legislative history, if "perhaps four comparable efforts are under way or successfully completed or there is substantial potential for such efforts, anticompetitive effects are unlikely." Conf Report at 10, 47 Antitrust & Trade Reg Rep (BNA) at 590; S Rep No 427, 98th Cong, 2d Sess (1984); see Kobak, *Application of Antitrust Laws to Joint Research, Development,* NYLJ, Dec 10, 1984, at 1.

Assistant Attorney General Baxter has indicated that a research joint venture involving no more than half an industry and leaving room for five or more other competing ventures should be per se legal. Hershey, *Antitrust Chief Offers Guidelines for Research,* NY Times, May 11, 1983, at D1; 44 Antitrust & Trade Reg Rep (BNA) 1114, at 963 (May 12, 1983); see Baxter interview, 52 ABA Antitrust LJ 1, at 32 (1983); presentation before National Association of Manufacturers, Washington, DC, 5 Trade Reg Rep (CCH) ¶50, 447 (May 10, 1983).

A problem in defining research and development markets at all is that large numbers of small innovations occurring continuously in an industry may be predictable enough to permit a market for them to be framed, but rarely "lumpy" enough to make joint ventures worthwhile —unless the objective were to *suppress* innovation, in which case the combination would obviously be anticompetitive under the rule of reason. *See United States v Automobile Manufacturers Association,* 307 F Supp 617 (CD Cal 1969), *affd sub nom City of New York v United States,* 397 US 248 (1970); *Berkey Photo, Inc v Eastman Kodak Co,* 603 F2d 263, 301-03 (2d Cir 1979), *cert denied,* 444 US 1093 (1980).

A further difficulty is that, almost by definition, major breakthroughs can hardly be predicted at all. Indeed, they may come from outside an industry altogether, for example, from academic laboratories or spinoffs from unrelated research. Thus, the more significant the research, the less clear the delineation of *any* market, much less one that can be characterized meaningfully as "properly" defined as required for its consideration under the Act.

Indeed, a surprise entrant may sometimes be more likely to make a major breakthrough than a major factor in the industry, which may focus on previously prominent concepts that lead to blind alleys. This is especially so if internal bureaucratic machinery in one or more industry leaders tends to inhibit feisty innovative efforts. Perhaps it is for this reason that innovators which became industry leaders did not always start out as such. "If too many approvals are required before an idea can be born, the probabilities are overwhelming against it." Tribus, *Applying Science to Industry,* 70 US News & World Rep 35 (Jan 8, 1971); *see* Jacobs, The Economy of Cities (1969); Givens, Advocacy, ch 14 (Shepard's/McGraw-Hill 1980).

This factor may indicate that large joint research and development ventures are not always the most likely to succeed, and that an outsider may win the race more often than is suspected. Forcing an outsider to remain outside by an antitrust ban on "over-inclusiveness" may, however, be counterproductive in the very situations where the "lumpiness" of research is a barrier because of the need for expensive equipment.

The Conference Report (at 10) recognizes that joint ventures are most important when size is a requirement for efficiency. Since this is the chief situation where inclusiveness makes economic sense, the concepts of efficiency and of avoiding "over-inclusiveness" necessarily conflict.

It may thus turn out that in the long run the statutory concept of "properly defined, relevant research and development markets" and the Conference Report's expression of concern about "over-inclusiveness" will be chiefly pertinent where a large venture requires a total pooling of relevant information on the part of its participants. Absent this or some purpose or effect to inhibit rather than advance innovation, there may be no adverse impact on competition under the rule of reason.

This belief finds support in the fact that joint research ventures were never held violative of the antitrust laws even prior to the Act, although patent pooling covering the bulk of an industry may be subject to antitrust attack, as in the case of *Hartford-Empire Co v United States,* 323 US 386 (1945) and as discussed in **§3.08.** The Act does not as such make any change in the legality of patent pools other than to assure that if part of a covered joint venture, they will not be illegal per se. A bare agreement to exchange patents and information without affirmative efforts to engage in research or development might fall outside the definitions of the Act, and would certainly fall outside its purpose.

The deeper implications of the National Cooperative Research Act of 1984 are much more far-reaching than its specific provisions. The enactment of the bill (as HR 5041, approved in HR Rep No 656, 98th Cong, 2d Sess (1984)) by a vote of 417 to 0 in the House (130 Cong Rec H3216-17 (daily ed May 1, 1984)) signals an unprecedented national commitment to a change in overall attitudes toward industrial innovation, parallel to the virtually unanimous repeal of the antitrust exemption for fair trade laws by the Consumer Goods Pricing Act of 1975, 89 Stat 801, *discussed in California Retail Liquor Dealers Association v Midcal Aluminum, Inc,* 455 US 97 (1980).

Joint research itself, as pointed out in the bound volume, had generally held minimal risks even apart from the Act. The use within corporate board rooms of the hobgoblin of antitrust liability to chill proposals for joint research efforts was the specific target of the Act. How much new joint research will occur remains, of course, to be seen. Frequently, of course, joint research will not be the most profitable course for a feisty innovator who hopes to beat out the rest of the pack.

Major national efforts of a highly concentrated kind, whether purely private or involving governmental contracts, may require joint endeavors, and if the underlying message of the Act is needed, these should now receive encouragement rather than being subjected to hostile treatment.

One risk in this regard is that filings to preclude treble damage liability will lead to governmental demands that various detailed recordkeeping, reporting, and other requirements be accepted by the venturers as the price for avoiding possible antitrust challenge on the part of the enforcement agencies. Minutes of all meetings, records of all contacts among the venturers, further "voluntary" filings and the like would tend to defeat the original objectives of the legislation. A combination of such a trend now or in the future with overemphasis on the risk of over-inclusiveness (which has no pre-Act case law support) could result in the Act as a technical piece of legislation inhibiting rather than advancing joint research. It is up to the bar, the enforcement authorities, and, if necessary, the courts to preclude this result, if the underlying objective of the Act is to be fulfilled.

Beyond its specific provisions, the Act indicates that a national objective for the legal system is now to assist rather than retard innovation by removing unnecessary, counterproductive legal barriers that are not truly necessary to serve other vital interests. The Act may thus be cited in different contexts as well, where unnecessary bureaucratic obstacles may tend to stifle technological advances.

The Act may also stimulate efforts to devise new means of opening pathways for needed innovation through Byzantine mazes of legal barriers at all levels of government (see **§12.03** of this supplement).

The extent to which the mood expressed by the Act extends to marketing and financial as well as technological innovation may ultimately depend on whether nontechnological innovations are perceived by the public as sufficiently important to justify encouragement rather than merely neutrality. *Cf* Williams, *Glory Days End for Pharmaceuticals,* NY Times, Feb 24, 1985, §3, at F1, F11 (characterizing "plans for . . . tinkering with products to change the frequency of a dosage or packaging them differently" as a "denial of reality"); *Rise of New Products: But How New Are They?,* NY Times, Dec 13, 1984, at D28; Hanan, *How to Grow by Accelerating Profits on Existing Products,* McGraw-Hill Conference on Corporate Innovation, Dec 6, 1979; Hanan, Accelerated Growth Planning: Profit Improvement Strategies for Consumer, Industrial and Service Business Game Plans (1978).

Financial or marketing innovations which can be shown to be necessary for technological innovation and productivity gains should fare well in the climate accentuated by the National Cooperative Research Act as well as other trends since 1974.

Informal approval has been given to cooperative research by members of the semiconductor chip industry, Bus Week, May 3, 1983, at 84.

See generally "National Cooperative Research Act of 1984," Conf Report 1044, 98th Cong, 2d Sess (Sept 21, 1984).

On governmental approval generally, see **§3.11.**

Add to note 191, page 56:

See also Axinn & Stoll, *Research Joint Ventures—Smoother Sailing Under the Antitrust Laws,* NYLJ June 1, 1983, at 1; Ginsberg, *Antitrust, Uncertainty and Technological Innovation,* 24 Antitrust Bull 635 (1979); Starr, *The Electric Power Research Institute,* 219 Sci 1190 (Mar 11, 1983); Symposium, *Permissible Cooperation Among Competitors,* 44 ABA Antitrust LJ 329 (1975).

Add to note 192, page 56:

See also Brodley, *Joint Ventures and Antitrust Policy,* 95 Harv L Rev 1521 (1982).

§3.11 —Governmental Approval or Participation

The Small Business Act, 15 USC §638(d) permits approval of joint research and development among small business firms.

The Local Government Antitrust Act of 1984 (LOGATA), Pub L No 98-544, 98 Stat 2750, 15 USC §34 note, protects from antitrust damages both local governments, or their employees acting in an official capacity, and others "in any claim . . . based on any official action directed by a local government or official or employee thereof acting in an official capacity." Injunctions remain available under 15 USC §26, however, where violations are found.

The Conference Report, HR Rep No 98-1158, 98th Cong, 2d Sess (1984), *reprinted in* 1984 US Code Cong & Admin News pamphlet 10A, indicates that the intent is to immunize private parties from damages to the same extent as under the "state action" doctrine stemming from the leading case of *Parker v Brown,* 317 US 341 (1943). This does not exempt private action which is merely permitted but not supervised or required by state law, as is illustrated by such cases as *California Retail Liquor Dealers Association v Midcal Aluminum, Inc,* 445 US 97 (1980).

LOGATA may have the long-run effect of encouraging the courts to avoid imposing damages for conduct reasonably believed to have been directed by state or federal, as well as local authorities. It would hardly make sense to give greater immunity to those following municipal directives than to those following directives from higher levels of government, especially since state action immunity historically exceeds that available to municipalities. *See Community Communications Co v City of Boulder,* 455 US 40 (1982), discussed in **§8.29** of this supplement; Shenefield, *The Parker v. Brown State Action Doctrine and the New Federalism of Antitrust,* 51 ABA Antitrust LJ 337 (1982); Slater, *Local Government and State Action Immunity after City of Lafayette and City of Boulder,* 51 ABA Antitrust LJ 349 (1982).

Nor would it make sense to afford more protection from damages for action in reliance on state or local than on federal action pursuant to a valid congressional enactment. Where the courts believe damages to be unwarranted, they may look very carefully at proof of actual economic loss, or examine the more difficult question of whether other facts can lead to denial of damages. *See* Givens, Antitrust: An Economic Approach §27.01 n 29 (1983); Areeda, *Antitrust Violations Without Damage Recoveries,* 89 Harv L Rev 1127 (1976); *cf J Truett Payne Co v Chrysler Motors Corp,* 451 US 557 (1981) (denying "automatic damages" for Robinson-Patman price discrimination violations). Reasonable reliance on federal approval would be absent, however, where Congress specifically denied any approval authority, as under the National Cooperative Research Act discussed in **§3.10** of this supplement in its final or, for that matter, its originally proposed form.

In *Southern Motor Carriers Rate Conference, Inc v United States,* 53 USLW 4422 (Mar 27, 1985), the Supreme Court held that clearly state-authorized private activity was exempt from antitrust attack as state action under *Parker v Brown,* 317 US 341 (1943) if adequately supervised, even though not compulsory. The extent of supervision required thus becomes the critical issue.

The supervision in *Southern Motor Carriers* involved ineffectiveness of agreed-upon rates unless approved, since the function of the rate conference was to submit proposed joint rates to governmental authorities. If approved and merely generally supervised conduct could become effective without specific approval of each step, the claim of exemption might fail, as in *California Retail Liquor Dealers Association v Midcal Aluminum, Inc,* 445 US 97 (1980), relied upon in *Southern Motor Carriers* as stating the applicable test.

On the other hand, tight supervision leading to a burden on interstate commerce and unsupported by adequate proof of local need might be vulnerable to attack under the interstate commerce clause, under *Raymond Transport, Inc v Rice,* 434 US 429 (1978); *see also Hunt v Washington Apple Advertising Commission,* 432 US 333 (1977); *Bibb v Navajo Freight Lines, Inc,* 359 US 520 (1959); *Southern Pacific Co v Arizona,* 325 US 761 (1945); *cf Flood v Kuhn,* 407 US 258 (1972); see also **§3.15.**

Legislation discussed in **§3.10** of this supplement provides an exemption from treble damage liability (but not from single damages or injunctive relief) for joint research and development programs to the extent that the challenged conduct has been disclosed to the Attorney General and the Federal Trade Commission. Joint research and development programs also would not be deemed illegal per se under the antitrust laws (which reflect existing law as described in **§3.10** of the text).

The scope of "joint research and development program" under the Act is narrow and does not encompass actual production or marketing. If the contrary had been the case, the proposal could have had far-reaching consequences in exempting conduct not normally considered as research or development activities from any per se provisions of the antitrust laws or treble damages should disclosure be made.

Several other proposals would have provided for governmental certification of the legality of proposed research joint ventures; others would have conferred exemption upon notification to antitrust enforcement authorities under difficult-to-satisfy conditions. See S 568, 737, 1383, and 1561, 98th Cong, 1st Sess (1983); S 3116, 97th Cong, 2d Sess (1982); S 2717, 97th Cong, 2d Sess (1982); HR 6262, 97th Cong, 2d Sess (1982); Cong Rec S 15485 (daily ed Dec 18, 1982).

These proposals might have tended to lead cautious counsel and management to seek such certification or qualification routinely as a "safe harbor," and could lead to legislatively unintended but hard-to-preclude claims that failure to do so created liability for lack of good judgment. The discretionary features of certification may also lead to governmental imposition of legally unnecessary conditions. Delay would, of course, also often be inevitable.

The exemption by mere notification offered by, for example, S 3116, was conditioned on, *e.g.*, allowing all United States and some foreign firms to participate and receive all results as of right, inclusion of nonparticipant representatives on managing boards, and numerous other extremely onerous conditions. These conditions might, regardless of disclaimers, be regarded as indicative of legislative intent as to how joint research should be carried on so as to avoid possible antitrust liability; the impact of statutory provisions is never limited to their precise text. *Compare generally Silver v New York Stock Exchange,* 373 US 341 (1963); *Commissioner v LoBue,* 351 US 243, 249 (1956); *Southern Pacific Co v Arizona,* 325 US 761, 773 (1945); *Parker v Brown,* 317 US 341, 367 (1943); *United States v Hutcheson,* 312 US 219 (1941); *Electrolux v Val-Worth, Inc,* 6 NY2d 556, 569, 161 NE2d 197, 204, 190 NYS2d 977 (1959); *Schuster v City of New York,* 5 NY2d 75, 85-86, 154 NE2d 534, 540, 180 NYS2d 265 (1958); *Zendman v Harry Winston, Inc,* 305 NY 180, 189 n 3, 111 NE2d 871, 876 n 3 (1953); Stone, *The Common Law in the United States,* 50 Harv L Rev 4, 12-18 (1936); Farnsworth, *Implied Warranties of Quality in Non-Sales Cases,* 57 Colum L Rev 653, 654 (1957); Note, 82 Yale LJ 258 (1972).

Affirmative encouragement of major, expensive, and highly risky or long-term technological ventures vital to the national interest are most likely to require joint efforts that might need additional insulation from antitrust risks. If a federal agency requests or requires joint efforts in a way needed to carry out its mandate, it may already be argued that no violation exists under current law, as discussed at pp 59-60 of the text & notes 206-208. Specific statutory immunity could also be provided to secure cooperation in such major ventures vital to the national interest, when requested or required in connection with affirmative promotion of technological development (see **§§12.03-12.04**).

§3.12 International Aspects

The Export Trading Company Act of 1982, Pub L No 97-290, 96 Stat 1333 (1982), adding 15 USC §6a, imposed the following new limitation on the Sherman Act:

> This title shall not apply to conduct involving trade or commerce (other than import trade or import commerce) with foreign nations unless—
> (1) such conduct has direct, substantial, and reasonably foreseeable effect—
> (A) on trade or commerce which is not trade or commerce with foreign nations, or on import trade or import commerce with foreign nations; or
> (B) on export trade or export commerce with foreign nations, of a person engaged in such trade or commerce in the United States; and
> (2) such effect gives rise to a claim under the provisions of this Act, other than this section. If this Act applies to such conduct only because of the operation of paragraph (1) (B), then this Act shall apply to such conduct only for injury to export business in the United States.

Under §303(a) of the Export Trading Company Act, 15 USC §4013, certificates of review may be obtained from the Secretary of Commerce with the concurrence of the Attorney General for export activities which are found to:

> (1) result in neither a substantial lessening of competition or restraint of trade within the United States nor a substantial restraint of the export trade of any competitor of the applicant,
> (2) not unreasonably enhance, stabilize, or depress prices within the United States of the goods, wares, merchandise, or services of the class exported by the applicant,
> (3) not constitute unfair methods of competition against competitors engaged in the export of goods, wares, merchandise, or services of the class exported by the applicant, and
> (4) not include any act that may reasonably be expected to result in the sale for consumption or resale within the United States of the goods, wares, merchandise, or services exported by the applicant.

Section 305 of the Act, 15 USC §4015, authorizes judicial review of grants or denials of certification. If a certificate is in effect, trebling of damages is barred, actions must be brought within two years, and any liability must be based on violation of standards (1) through (4). Under §306 of the Act, 15 USC §4016, conduct described in a certificate is presumed not to violate the quoted standards; attorneys' fees can be awarded to successful defendants.

Section 306(b)(5), 15 USC §4016(b)(5), permits the Attorney General to seek injunctive relief against "conduct threatening clear and irreparable harm to the national interest."

See generally Symposium, *Current Antitrust Law and United States International Trade Practices,* 15 Vand J of Transnational L 4 (Fall 1982). For legislative history, see 1982 US Code Cong & Ad News 2431.

Add to note 217, page 61:

See Fox, *Updating the Antitrust Guide on International Operations—A Greener Light for Export and Investment Abroad,* 15 Vand J of Transnational L 709 (1982). On treatment of claims of foreign governments, *see* 96 Stat 1964 (1982), amending 15 USC §15.

In note 221, page 62, second line, delete "__ F2d __" and insert:

666 F2d 6

Antitrust Policy Seeking To Promote Innovation

§3.13 Direct Enforcement

Add to note 238, page 64:

See also Blair, Economic Concentration, ch 10 (1972).

Page 64, text, following item 4, add new paragraph:

5. Breach of contract, or of other legally enforceable obligations, which has the effect of consolidating market power by inhibiting innovation. *See Associated Radio Service Co v Page Airways, Inc,* 624 F2d 1342, 1354 (5th Cir 1980), *cert denied,* 450 US 1030 (1981); Givens, Antitrust: An Economic Approach §3.03 (1983)

§3.15 Limitations on Economic Regulation

Major sources of constitutional restarint on economic regulation include:

1. The Commerce Clause, most clearly set forth as dealing with more than mere discrimination against interstate commerce in *Raymond Motor Transport, Inc v Rice,* 434 US 429 (1978) and which seeks to protect a "national common market" as set forth in *Hunt v Washington Apple Advertising Commission,* 432 US 333, 350 (1977)

2. The privileges and immunities clause as protective of the right to practice a calling, *Supreme Court v Piper,* 53 USLW 4338 (Mar 4, 1985); the First Amendment as protecting (subject to restrictions) both content and access with regard to commercial as well as other speech, *see Preferred Communications, Inc v City of Los Angeles,* 754 F2d 1396 (9th Cir 1985) and cases cited
3. Procedural due process, *see, e.g., Gibson v Berryhill,* 411 US 564 (1973)
4. Protection against confiscation, *see Loretto v Teleprompter Manhattan CATV Corp,* 458 US 419 (1982)

Add to note 265, page 68:

National League of Cities v Usery, 426 US 833 (1976), imposing judicial limitations on federal control of state activities under the commerce power, was narrowly overruled by a 5-to-4 decision in Garcia v Donovan, 53 US 4135 (Feb 19, 1985). The *Garcia* decision held that the balance of federalism was to be held true at least primarily and in the labor standards area through the political influence of the states in the federal system, rather than by judicial intervention. *See also* before *Garcia,* Transportation Union v LIRR, 455 US 678 (1982) (interference with functions not essential to state as a state, *i.e.,* nontraditional governmental functions, here running a railroad, permissible under commerce clause); City of Rome v United States, 446 US 156 (1980) and cases cited (Civil War amendments permit congressional implementation to deal even with traditional governmental functions of states and localities).

Add to note 267, page 68:

On unconstitutional confiscation, *see* Loretto v Teleprompter Manhattan CATV Corp, 458 US 419 (1982); cases cited in Pruneyard Shopping Center v Robins, 447 US 74 (1980). On impairment of the obligation of contracts, *see* Allied Structural Steel Co v Spannaus, 438 US 234 (1978); United States Trust Co v New Jersey, 431 US 1 (1977); cases cited in Casey v Depetrillo, 697 F2d 22 (1st Cir 1983).

Add to note 290, page 70:

On balancing tests in freedom of expression cases, *see* Bolger v Youngs Drug Prod Corp, 103 S Ct 2875, 77 L Ed 2d 469 (June 24, 1983). *See generally* Easterbrook, *The Court and the Economic System,* 98 Harv L Rev 4 (1984).

Innovations, Patents and Trade Secrets

Lloyd McAulay

4

§4.01 Basic Patent Requirements

Add to discussion of Novelty:

Secret prior use by another, even though it be a commercial use, will not undercut novelty. The Federal Circuit Court upheld a patent on the process for making a waterproof fabric sold under the trademark Goretex against the claim by an infringer that the process was not novel. There was a prior use of the process by a third party wholly unrelated to the inventor or the patent owner. This third-party use was a secret commercial use of the process to make the waterproof fabric. It occurred more than one year prior to the filing of the process patent application. In *WL Gore & Associates v Garlock, Inc,* 721 F2d 1540, 220 USPQ 303 (CAFC 1983), the court held that a prior secret use by a party who had no connection to the patentee could be ignored and the patent would be valid. Since the patent is valid, this raises the interesting question as to whether or not the third-party user is an infringer of the patent. That issue is unsettled.

Effective November 8, 1984, Congress amended 35 USC §103 to provide that where a development by a company's employees or consultants entailed a number of intermediate stages to a final invention, those intermediate stages would not count against the novelty of the final invention. That is, the work of the inventors' colleagues and co-workers is not prior art to the final invention, even if those co-workers are not co-inventors of the final invention. Congress recognized that it inhibits development within a company to hold that the final invention is unpatentable over intermediate stage inventions which were known to the final inventors. As long as those intermediate stage inventions were not put into public use, there is no public policy reason for them to be considered prior art.

Add to discussion of Nonobviousness:

A strong and clear policy statement in support of sticking with the nonobviousness criteria as the standard for evaluating an invention over prior art was made by the CAFC in *Richdel, Inc v Sunspool Corp,* 714 F2d 1573, 219 USPQ 8 (Fed Cir 1983). The CAFC stated that all relevant evidence of nonobviousness has to be taken into account. The case strongly implies a rejection of all subtests of nonobviousness. The case shows an appreciation that all the evidence has to be put together, its relevance determined and on the basis of the totality of that evidence, a decision made as to whether or not the invention was obvious at the time the invention was made over the art known at that time.

The CAFC recognizes that there is no formula for these criteria of nonobviousness. These are the criteria which the parties and their lawyers will argue. What is or is not obvious at the time that an invention is made cannot be set forth as a series of objective criteria. The objective facts include a range of considerations and there usually is a great variety of relevant evidence, but the assessment of that evidence by the decision maker is ultimately a subjective one. At least genius is not required, *Vandenberg v Dairy Equipment Co,* 740 F2d 1561, 1565, 224 USPQ 195, 197 (CAFC 1984).

Determinations of utility and novelty are to a great extent objective determinations, as is the determination of whether or not a patent application has been timely filed. In contrast, the determination that an invention is or is not obvious over the prior art known at the time is relatively subjective.

§4.02 Patentable Subject Matter

Add to note 25, page 82:

The newly created Court of Appeals for the Federal Circuit (CAFC) has cited *In re* Miller, 418 F2d 1392 (CCPA 1969) favorably in finding patentable subject matter in an educational mathematical device which includes printed matter (digits) on one or more concentric bands. The CAFC declared that printed matter can constitute the novel feature which provides the nonobvious relationship supporting patentability. The court found that the particular sequence of numbers on the band was not suggested by the prior art and thus held that a patent may issue. The court said, "Differences between an invention and the prior art cited against it cannot be ignored merely because those differences reside in the content of the printed matter." *In re* Gulack, 703 F2d 1381, 1385, 217 USPQ 401, 403 (Fed Cir 1983).

The CAFC warns, however, that the printed matter must have some structural or functional relation to the rest of the structure of the claim (703 F2d at 1385). What this means is that printed matter on a substrate is not patentable where the substrate does no more than carry the printed matter. Otherwise, any page of printed matter would be patentable. *Gulack* appears to settle many of the issues involved in the relation of printed matter to patentable subject matter.

§4.02A —Computer Programs and Computer-Assisted Inventions (New)

The electronic revolution, beginning with the advent of the telephone, in large part involves the processing of information. The manner in which information is processed is becoming an increasingly important component of modern technology. Manufacturing process controls, computer data processing, and robotics all involve such processing. The information may be used to entertain, to inform, to control industrial processes, to provide security, or to provide a basis for decision making.

It is not always clearly recognized that all this information processing is related to technology. As a consequence, there have been attempts to distinguish between patent protection in these areas. Thus, there was an attempt to extend the bar against the patenting of a mathematical formula to include a bar against the patenting of computer programs and even of computer program-assisted inventions. Much of the uncertainty in that area was settled by the Supreme Court in *Diamond v Diehr & Lutton,* 450 US 175 (1981); see **§4.02.**

A range of options are open to the innovator, including patent protection, copyright protection, and trade secret protection as well as various combinations thereof.

Because the computer program is subject to being set down as a writing or the equivalent, there is the availability of copyright protection instead of, and sometimes supplementing, the usual choice between patents and trade secrets. The availability of copyright protection is recognized in a number of cases, including *Tandy Corp v Personal Microspace Computers Co, Inc,* 524 F Supp 171, 214 USPQ 178, (ND Cal 1981); *GCA Corp v Chance,* 217 USPQ 718 (ND Cal 1982); and *Apple Computer, Inc v Formula International, Inc,* 562 F Supp 775, 218 USPQ 47 (CD Cal 1983).

In addition, where the computer programs and computer-assisted inventions are not mass marketed, there is often the option of obtaining protection through contractual restraints on customer use and on the right to reproduce the software (see **ch 3**). Such contractual restraints are effective means of maintaining the trade secret nature of the software.

There are several present legal uncertainties with respect to software protection that make it difficult to determine an optimum strategy. The greatest of these uncertainties is whether or not one can obtain copyright protection (or in some cases patent protection) while at the same time maintaining the software as a trade secret.

The patent laws require full disclosure of the preferred mode for practicing the invention (35 USC §112) in order to obtain a patent. But most software is not patentable in that it fails to meet the nonobvious criteria of 35 USC §103. If there is a patentable aspect of a software arrangement, it may well be possible to separate it from the bulk of the software arrangement, so that full disclosure of the inventive feature can be made without disclosing all of the details that constitute the results of routine but expensive programming. Accordingly, it might be possible to obtain a patent on the novel and inventive feature of the software while maintaining the bulk of the program as a trade secret.

Most software is probably not patentable, in that most software is an obvious application of the skill of the art, even though it be an expensive application and a high level quality of that skill. Accordingly, copyright protection has recently come into favor as a means for protecting software. Since 1980, the rate at which software programs have been registered in the United States Copyright Office has increased dramatically, so that by mid-1983 about 75 registration applications were being filed each week. Comments by Jodi Rush, Assistant Chief of the Examining Division of the Copyright Office, at the Practising Law Institute Seminar on Software Protection and Marketing, June 9, 1983.

A major advantage of copyright protection of a program over patent protection is that a preliminary injunction is generally available for copyright infringement, but is generally not available for patent infringement. For example, in *Apple Computer, Inc v Formula International, Inc,* 562 F Supp 775, 218 USPQ 47 (CD Cal 1983), a preliminary injunction was granted on the claim for copyright infringement, whereas the claim for infringement of two of Apple's patents was not even considered in connection with the request for preliminary injunction.

Furthermore, *Apple v Formula* holds that a program held in "read only memory" (ROM) is copyrightable. The case also stands for the proposition (apparently conceded by both sides in the case) that the registration of the "source code" also affords copyright protection for the "object code." The source code is a program language which is structured to be read and understood by a programmer. The object code is the series of zeros and ones into which the computer translates the source code so that it can be operated on by the computer circuitry. Thus the object code is derived from the source code in the sense that the computer translates the source code into an object code. Copyright protection on the source code will thus afford protection for the translation into the object code.

An important recent case has specifically stated that the affixing of a copyright notice to work and limited publication of that work will not undercut a claim that the subject matter constitutes a trade secret. Only if there has been a general publication (that is, one without restriction on access) will the subject matter cease to be a trade secret. *Technicon*

Medical Information Systems Corp v Green Bay Packaging, Inc, 687 F2d 1032, 215 USPQ 1001 (7th Cir 1982). In other words, the nature of the publication determines whether or not the trade secret is still a trade secret. The fact that the publication is or is not copyrighted is not crucial to the existence of a trade secret.

The reason for the substantial increase in computer program copyright registrations since 1980 is the combined effect of:

(1) the cost and even unavailability of patent protection

(2) the recognition of the legitimacy of copyright protection for computer programs through the Computer Software Copyright Act of 1980, which amended 17 USC §§101 and 117

(3) the availability of special relief concerning the deposit of information incorporating trade secrets in the Copyright Office, as well as the existence of certain partial deposit requirements which often permit registration of a copyright where trade secrets are involved by deleting that portion of the material which would reveal the trade secrets. 37 CFR §§202.19(c)(5) and 202.20(c)(2)(vii), Copyright Office Circular R61.

(4) the recognition by some courts that obtaining copyright protection does not necessarily undercut trade secret protection. *Technicon Medical Information Systems Corp v Green Bay Packaging, Inc,* 687 F2d 1032, 215 USPQ 1001 (7th Cir 1982). The matter has not been fully settled; there remains some risk of being deemed to have waived a claim to trade secrets on the basis that the claim of copyright protection constitutes an admission of publication. However, recent practice indicates that such a risk is on the decline

It must be kept in mind, though, that copyright protection for a computer program does not operate to bar or limit another from developing a program that performs the exact same function or purpose as does the copyrighted program.

The distinction between expression and idea is crucial to this limitation on the scope and value of copyright protection. Ideas can be covered by patent protection. The expression of an idea, and only the expression, can be covered by copyright protection. For example, the maze game idea embodied as Pac-Man cannot be protected by copyright even though the symbols and graphics may be so protected. *Atari, Inc v Williams,* 217 USPQ 746 (ED Cal 1981) (Pac-Man game not infringed by Jawbreaker game); *Atari, Inc v North American Philips Consumer Electronics Corp,* 217 USPQ 1265 (ND Ill 1981) (Pac-Man game not infringed by K.C. Munchkin game).

Thus, a copyright does not prevent another from independently creating a program to provide the same function as a copyrighted program. Indeed, if a copyrighted expression of a programming idea were, in a particular case, the only way to express that idea, then copyright protection would probably not be available, because it would give the author what is in effect a patent on the idea, and such protection must be obtained through the patent laws. However, one of the values of a copyright on software is that a substantial investment is normally required for another party to develop a program independently.

If an analysis of a flow chart from which the computer program is developed shows a nonobvious idea or relationship, then patent protection should at least be considered. In addition, copyright protection might be sought for the computer program expression of that idea.

An important development is the rejection of a legal distinction between programs which interact with the human user and programs which interact solely with the computer system. In two cases, *Apple Computer, Inc v Franklin Computer Corp,* 714 F2d 1240, 219 USPQ 113 (3d Cir 1983) and *Apple Computer, Inc v Formula International Inc,* 725 F2d 521 (9th Cir 1984), it was held that copyright protection can extend to computer programs in all their aspects. Copyright protection extends not only to application programs which directly interact with the computer user but also to operating systems programs which are designed to manage the computer system. Both cases interpret the relevant statute, 17 USC §§101 and 117, broadly to reject the doctrine that copyright protection is limited only to those computer programs which produce works that themselves qualify for copyright protection. These cases hold that copyright can exist on a program expressed in machine-readable language even if embedded in a microchip. This may make it impossible for others to build fully compatible equipment. Until that issue is further litigated, it is important to keep in mind that there are some unsettled issues on the scope and significance of copyright protection for computer programs.

§4.04 Identification of the Inventor

Effective November 8, 1984, Congress amended 35 USC §§116 & 120 to make the inventorship requirement less of a trap. Thus, a joint application for a patent may be filed even though (1) the inventors did not physically work together or at the same time, or (2) each inventor did not make the same type or amount of contribution, or (3) each inventor did not make a contribution to the subject matter of every claim of the patent. It is still important to include as co-inventors anyone who did make a significant contribution to the invention, but good faith inclusion is not likely to invalidate a patent. Thus, the better practice is to include rather than to exclude co-inventors.

§4.06 Effect of an Offer for Sale

In a most significant case, the Court of Appeals for the Federal Circuit held that an offer to sell a product made by an inventive method started the one-year grace period running on the filing of a patent application to cover that method, *DL Auld Co v Chroma Graphics Corp,* 714 F2d 1144, 219 USPQ 13 (CAFC 1983). This *Auld* case probes the limits of the on-sale bar. The samples (decorative foil emblems) were produced by hand using the inventive method. The samples were not to be sold. They were presented to prospective customers to elicit sales. If orders were obtained, Auld would then tool up for a mass production method that had yet to be completely developed. No orders were in fact obtained prior to the critical date. Further experimentation was required and undertaken to adapt the hand method to mass production. However, the handmade samples were the result of successful operation of the method and thus an offer to sell the product based on those samples established the critical date for starting the one-year grace period.

However, as was well put in one court's instructions to a jury on this on-sale issue, an offer for sale must be of "a completed invention that has been shown to be commercially useful for the purpose intended." The Federal Circuit approved this language in *Shatterproof Glass Corp v Libbey-Owens Ford Co,* 728 F2d 613, 225 USPQ 633, 639 (CAFC 1985).

It is of some comfort to the small inventor that an offer to sell the invention (as distinct from a product embodying the invention) does not constitute an on-sale bar, *Scott Paper Co v Moore Business Forms,* 594 F Supp 1051, 224 USPQ 11, 29 (D Del 1984). In *Scott,* the inventor Davis entered into an agreement with a company (Royal McBee) to test out the invention in a carbonless recording paper. The tests were made prior to the on-sale bar date at the McBee plant by McBee employees in secrecy. No product was made for commercial sale.

§4.07 —Confidentiality Exception

Add to note 65, page 88:

See also DL Auld Co v Chroma Graphics Corp, 714 F2d 1144 (CAFC 1984), discussed under in **§4.06** of this supplement. As recently stated by the Federal Circuit, "secrecy alone does not necessarily negate public use," *Kinzenbaw v Deere & Co,* 741 F2d 383, 390, 222 USPQ 929, 934 (CAFC 1984).

§4.08 —Experimental Use Exception

Because of a half-dozen cases decided by the Court of Appeals for the Federal Circuit (CAFC), the guidelines as to what constitutes experimental use have become somewhat clearer.

In re Corcoran, 640 F2d 1331, 208 USPQ 867 (CCPA 1981) provides a useful discussion of policy and also sets forth a detailed standard for judging whether or not a customer installation is a sale or an experimental use. Furthermore, this case is significant precedent for all cases throughout the country because the new Court of Appeals for the Federal Circuit (CAFC) will be handling all appeals on patent cases from the district courts as well as from the Patent Office, and the CAFC has adopted the Court of Customs and Patent Appeals (CCPA) cases as its precedents. The *Corcoran* court adopted the following statement from an earlier Second Circuit case:

> We accordingly conclude that [35 USC] Section 102(b) bars an application for patent filed more than one year after the solicitation of an order for a specific article to be produced later, where the following requisites are present:
>
> (1) The complete invention claimed must have been embodied in or obvious in view of the thing offered for sale. . . . Complete readability of the claim on the thing offered is not required because whatever is published (or on sale) more than one year prior to the filing of a patent application becomes part of the prior art over which the claim must be patentable. . . .
>
> (2) The invention must have been tested sufficiently to verify that it is operable and commercially marketable. This is simply another way of expressing the principle that an invention cannot be offered for sale until it is completed, which requires not merely its conception but its reduction to practice. . . .
>
> (3) Finally, the sale must be primarily for profit rather than for experimental purposes. . . . 640 F2d at 1333-34.

The court applied these principles to this case involving an invention in a roller blind. The patented blind was the same as a blind manufactured prior to the critical date except for the fact that the earlier product did not have a horizontal roller and a tab. Both the roller and the tab were set forth in the patent claims. But those two features, the roller and the tab, were found to be sufficiently standard and obvious in window shades so that the claims of the patent application were obvious in view of the product placed into commerce prior to the critical date.

This case highlights the fact that it is important to file a patent application on the first commercial embodiment of an invention even though it is quite certain that the design is not final and that changes will be made after field experience.

Barmag Barmer Maschinenfabrik AG v Murata Machinery, Ltd, 559 F Supp 491, 217 USPQ 1027 (WDNC 1983), *affd,* 731 F2d 831, 221 USPQ 561 (CAFC 1984) provides another recent example in which the experimental use exception failed to save a patent because the experimental purpose for the installation of two machines sold to a customer was not adequately proved. The customer was not required to submit performance reports on the machine. The customer was not required to keep its knowledge of the machine confidential. No restrictions were placed on the customer's use or control of the machines. The testing that was undertaken was deemed to be for the purpose of making commercial refinements rather than to establish that the invention was operable.

The CAFC addressed the distinction between the three following situations:

1. An offer to sell where there is product available to meet orders
2. An offer to sell where there is no product actually on hand, but where there exists a physical embodiment of what is offered
3. An offer to sell without the existence of a physical embodiment of what is being offered

Clearly, the first situation starts the one-year statutory bar running. The CAFC made clear in its opinion that the second situation also starts the statutory bar running, but indicated that the third situation would probably not. However, there might be circumstances, particularly with a simple product, where the third situation starts the statutory bar running, because of the bar's purpose. The CAFC emphasized that the purpose of the statutory bar is to preclude commercial exploitation of an invention which has the effect of expanding the period of exclusive rights granted by the patent statute, and it is that purpose which will determine the rule to be applied in a particular situation.

Although many people would label the testing of a device in order to refine it and make it commercially acceptable as part of experimental testing, the *Barmag* case makes it clear that such is not the kind of experimental use which avoids the start of the one-year grace period.

In a somewhat related situation, *In re Smith,* 714 F2d 1127, 218 USPQ 976 (CAFC 1983), makes clear that where the tests are not solely for experimental purposes, then the tests are of a commercial nature and do start the one-year grace period running. In the *Smith* case, a composition for antistatic and antisoil treating of carpets and for deodorizing carpets was tested by 76 consumers in their own homes for two weeks. After those tests, experimentation was continued on different formulations. The court considered that the testing was in large part market testing to determine the consumer demand for the invention. Probably crucial to the court's decision was that data on the invention's operation and usefulness could have been easily obtained in the labs of Airwick, the owner of the invention. The assistance of the 76 typical housewives was not needed for that purpose. Yet a two-judge dissent, citing Murphy's law, pointed out that it is unlikely that lab tests can duplicate a field test. They doubted that lab technicians could anticipate all the varieties of carpets, varieties of vacuum cleaners, and unforeseen manners in which untrained users could produce malfunctions. Nevertheless, the majority believed that the main purpose of the test was to determine if the typical housewife would buy the product and at what price. Market testing to determine consumer acceptability is a commercial use and not experimentation, even though additional experimentation on other formulations followed from the consumer testing.

Thus, it is clear from the *Smith* and *Barmag* cases that, to be deemed experimental testing, the testing had better be directed solely to determining the operability of the invention. Still, there are circumstances where public testing, even by customers who will eventually use the product, may be an experimental use where the conditions of the test are likely to be difficult to duplicate in the laboratory. Thus Caterpillar Tractor tested a device to seal oil into the joints of a track of a tractor by having tracks placed on tractors owned and run by independent customer parties. These field tests for durability, life, and reliability were not capable of being made in the laboratory. Citing the *City of Elizabeth* case (discussed at page 90 of the main text), the court held these tests to be experimental and held the patent valid, *Caterpillar Tractor Co v Berco, SPA,* 215 USPQ 948 (D Wyo 1982), *affd without discussion of this issue,* 714 F2d 1110, 219 USPQ 185 (CAFC 1983).

In *Pennwalt Corp v Akzona, Inc,* 740 F2d 1573, 222 USPQ 833 (CAFC 1984), the Federal Circuit held that the testing of an insecticide in order to meet the regulatory requirements of the EPA to obtain a temporary permit was not experimental use but was for commercial purposes, and thus started the one-year grace period running. However, in *Pennwalt,* the court made a useful statement that "a use or sale is experimental . . . if it represents a bona fide effort to perfect the invention or to ascertain whether it will answer its intended purpose." One would think that the testing in *Smith* (where continued experimentation occurred) or the testing in *Pennwalt* (the purpose of which was to determine whether the insecticide could be legally marketed by showing it was safe or otherwise met EPA requirements) were bona fide efforts to either perfect the invention or to ascertain whether it would answer its intended purpose. Unlike the *Corcoran* and *Barmag* situations, where a completed invention was put into commercial use, the *Smith* and *Pennwalt* situations would seem to exemplify the purposes of the experimental use exception. However, the rulings of the Federal Circuit in these cases makes it clear that the prudent practice is to proceed with the patent application before it gets out of one's shop, even if a continuation-in-part application has to be filed later to cover the final product.

There is one recent case which indicates that the Federal Circuit recognizes the continued vitality to the experimental use exception where testing is done outside of the lab. The case of *TP Laboratories, Inc v Professional Positioners, Inc,* 724 F2d 965, 220 USPQ 577 (CAFC 1984) shows a somewhat more sympathetic understanding of the plight of the innovator faced with the difficult problem of juggling experimentation and product refinement at the same time that it is attempting to establish a business or at least to determine consumer interest. In this case, an orthodontic appliance device was used by the dentist inventors over a period of years on patients who were charged a professional fee. This use was deemed consistent with an experimental purpose and thus did not start the one-year time bar running. The court found that the inherent control by the inventor dentist, coupled with the magnitude of the commercial use (on a relatively few patients), indicated the experimental context. This case may indicate an understanding of the experimental nature inherent in the initial use of most complex technological inventions.

§4.16 —Synergism

Add to note 114, page 101:

The synergism doctrine appears headed for the last roundup. Apart from the criticisms, if not repudiation, of the doctrine made by the Second, Third, Seventh, Ninth, and Tenth Circuit cases cited in note 114 of the text, there appears to be a willingness to simply ignore the concept. For example, in Shackelton v J Kaufman Iron Works, Inc, 689 F2d 334, 217 USPQ 98 (2d Cir 1982), a patent on a security window gate was held valid because it met a real commercial need even though the novel structure was in the combination of known gate elements. Synergism was not discussed. The Second Circuit in *Shackelton* provided a more useful insight in its statement, "A change of function for a well known element of a combination patent is a benchmark of nonobviousness." 689 F2d at 339, 217 USPQ at 102.

In Compro-Frink Corp v Valk Mfg Co, 216 USPQ 531 (ED Pa 1982) and in Preemption Devices, Inc v Minnesota Mining & Mfg Co, 559 F Supp 1250 (ED Pa 1983), the courts explicitly construed *Rengo v Molins* as being a rejection of the synergism standard with respect to the validity of combination patents. The *Preemption Devices* court made the perceptive statement that "it is comparatively rare in this modern technological world that anything is comprised of totally new elements. . . ." 559 F Supp at 1250.

More important is a major Ninth Circuit en banc decision in which synergism was disapproved as being determinative on the issue of nonobviousness. Sarkisian v Winn-Proof Corp, 688 F2d 647 (9th Cir 1982). This en banc determination was a rehearing of three separate cases. The court analyzed the concept of synergism and found it inappropriate in view of the statutory standard of nonobviousness. Unfortunately, the court attempted to find another magic helper to aid in the application of the statutory nonobviousness standard. The Ninth Circuit indicated that an "unusual or surprising result" is a prerequisite to a finding of nonobviousness in a combination patent. It is not clear how such a standard explicates nonobviousness. It seems to provide no guidance different from or better than provided by the statutory nonobviousness standard itself. The "unusual result" standard may be preferable to the synergism standard because the dictionary definition of synergism imposes constraints that are more limiting than are imposed by the statutory standard of nonobviousness. However, the basic issue of patentability is the obviousness of bringing together the parts that make up the combination.

Perhaps most important of all, all appeals in patent matters since October 1, 1982 must be taken to a single court of appeals, the Court of Appeals for the Federal Circuit (CAFC), whose judges include those from the former Court of Customs and Patent Appeals (CCPA). The CCPA has been notably more sympathetic to patentability than have many of the circuit courts. The CAFC has held CCPA precedents as persuasive before it. South Corp v United States, 690 F2d 1368, 215 USPQ 657 (Fed Cir 1982).

The CAFC has just ruled that synergism is not an appropriate standard for determining nonobviousness (patentability) under the patent statute, Chore-Time Equip, Inc v Cumberland Corp, 713 F2d 774, 781, 218 USPQ 673, 677 (CAFC 1983). This should about complete the demise of the synergism doctrine. Other opinions in the CAFC and earlier opinions of its predecessor courts suggest that the CAFC is unlikely to impose new language as a standard in a hope to somehow clarify the statutory nonobviousness standard. Thus the "unusual result" standard set forth by the en banc Ninth Circuit is likely to face a deserved oblivion before wreaking its semantic damage.

§4.17 Evaluation of Patentability and Identification of the Patentable Feature

Problem solving rather than structural uniqueness as the key to what constitutes patentable invention may be coming into its own. In *Shackelton v J Kaufman Iron Works, Inc,* 689 F2d 334, 217 USPQ 98 (2d Cir 1982), the Second Circuit reversed the lower court holding of invalidity of a safety window gate patent. Two structural features were key. One is a vertical post that locks the gate with a pin and slot combination. The second is a metal plate to cover the latch bar handle. These structural features are found in prior lockers, window shutters, and elevator gates. Shackelton used these features to solve the particular problem of permitting ready opening of the window gate from the inside in case of fire while preventing opening of the gate from the outside by a would-be burglar.

The essence of the invention lay in the solution of the problem of assuring ready opening from one side and security from the other side. This problem had existed for a long time and had defied others prior to Shackelton. To be patentable, the combination had to be novel in some structural respect; otherwise the claim would cover prior art and thus be invalid because of overbreadth. As long as there is a minimum of novelty, the nature and significance of the problem to be solved is more important to nonobviousness than is the uniqueness of the structure.

§4.17A —Recordkeeping (New)

Maintaining records (including workbooks, preferably bound) of the inventor is of critical importance to provide evidence many years later of what the situation was at the time the invention was made. It is hard many years later, after an invention has become a success, widely used and widely known, to recapture the element of surprise, unexpectedness, and nonobviousness at the time the invention was made. It is hard to recapture the feeling of confoundment and sometimes even of mystery in trying to solve the problem before the solution was invented. Among the records that are worth keeping include any notes, records, or drawings kept by the inventor or those working with the inventor. Other records can also be very useful: contemporaneous response or reaction by customers, competitors, or others in the field should be kept. Often those skilled in the art will express doubt that the problem can be solved prior to its solution. Those of ordinary skill in the art usually have ordinary imaginations, and their responses can provide the kind of background which shows that the invention was indeed nonobvious at the time that it was made, even though later familiarity has bred a feeling of obviousness.

§4.19 Licenses

A change has come over the Justice Department in its approach toward the legal constraints on patent licensing. In a series of speeches, officials of the Justice Department have taken the position that the market power derived from patents should not be treated any differently than the market power derived from other lawful economic assets or other lawful legal status. *Competition Policy and the Patent Misuse Doctrine,* remarks of Roger B. Andewelt, Chief, Intellectual Property Section, Antitrust Division, before the Bar Association for the District of Columbia Patent, Trademark and Copyright Section (Nov 3, 1982); *Current Antitrust Division Views on Patent Licensing Practices,* remarks of Abbott B. Lipsky, Jr., Deputy Assistant Attorney General, Antitrust Division, before the American Bar Association Antitrust Section (Nov 5, 1981); *Basic Principles to Apply at the Patent-Antitrust Interface,* remarks by Roger B. Andewelt, Chief, Intellectual Property Section, Antitrust Division, before the Houston Patent Law Association (Dec 3, 1981).

Accordingly, the Department considers such matters as tie-ins, tie-outs, mandatory package licensing, or royalty based on unpatented product as proper unless their effect is to lessen competition in violation of the antitrust laws.

One suspects that this opposition to per se prohibitions may have little real impact unless the courts evaluate and, where proper, appreciate the extent to which market considerations do in fact limit the extent to which a patent owner can extract concessions from a licensee. In general terms, the value of a patent is a function of the advantage the invention covered by the patent provides over the prior art. That advantage normally dictates a limit on the royalty obtainable. If other concessions are required in the license, these concessions will normally have an economic value that will reflect itself in a reduced royalty. Just as a nonexclusive license tends to carry a lower royalty rate than an exclusive license, so too basing the royalty on the entire product line, regardless of whether or not the individual items contain a patented component, will tend to provide a lower royalty rate than basing the rate on only those individual products having the patented component.

The Department's position is that where other considerations, such as overall market domination, render these "free-market" economic considerations minor or secondary, then the antitrust laws apply in the same fashion as they do in connection with any market dominance situation. The Justice Department's Assistant Attorney General in charge of the Antitrust Division, J. Paul McGrath, has stated a current Department of Justice position that patent license restrictions should be analyzed under a rule of reason, and expressed skepticism about the value of per se rules. He believes that a license which restricts the licensee's competitive conduct, even if it affects products outside the scope of the patent, is not inherently suspect nor per se unlawful, and that these restrictions are often reasonably necessary to achieve a pro-competitive benefit. Increased foreign competition is one factor in bringing about this shift in position. The address entitled *Patent Licensing: A Fresh Look at Anti-Trust Principles in a Changing Economic Environment* was delivered on April 5, 1984 and appears at 27 Pat, Trademark & Copyright J (BNA) 624. In part, he relies on the reasoning of the Supreme Court in *Continental TV, Inc v GTE Sylvania, Inc,* 433 US 36 (1977). More particularly, McGrath finds nothing inherently wrong with either tie-ins of unpatented product or field-of-use restrictions.

However, in *Pitney Bowes, Inc v Mestre,* 701 F2d 1365 (11th Cir 1983), the court held a clause in an agreement calling for payment of royalties past the expiration date of the patents involved to be unenforceable. In *Brulotte v Thys Co,* 379 US 29, 143 USPQ 264 (1964), the United States Supreme Court held a royalty agreement unenforceable where it provided for payment after the patents expired. By contract, state law can enforce payment under a trade secret license agreement on an indefinite basis. By license agreement, payments could be required even after the technology became known, and thus where there was no longer a trade secret. *See Aronson v Quick Point Pencil Co,* 440 US 257, 201 USPQ 1 (1979) and *Kewanee Oil Co v Bicron Corp,* 416 US 470, 181

USPQ 673 (1974). In *Pitney Bowes,* the license covered both patent and trade secrets. The expiration date was the later of the death of the patent owner or the last patent expiration date. The patent owner, Mestre, died one and a half years after the last licensed patent expired. The court held that as long as patents were involved, the fact that there were also trade secrets could not warrant payment of royalties after the patents expired. The court stated that the *Brulotte* rule applied. The court also noted that when patents actually issue, the potential exists for conflict between federal patent law and trade secret law. When such a conflict occurs, federal law is supreme.

In *Boggild v Kenner Products,* 576 F Supp 533 (SD Ohio 1983), the Court rejected the reasoning of the Eleventh Circuit in the *Pitney Bowes* case and held that a license agreement entered into before a patent application was deemed allowable could call for the payment of royalties to a date which turns out to be subsequent to the expiration date of the patent which eventually issues.

§4.20 Trade Secrets

Further recognition of the value and scope of the trade secret is afforded by the Third Circuit in *Rohm & Haas Co v Adco Chemical Co,* 689 F2d 424, 215 USPQ 1081 (3d Cir 1982). The case held that, under the common law of Pennsylvania and New Jersey, a trade secret may exist even though every element of the trade secret is known to the industry as long as the combination of these elements produces a product superior to that of competitors. The appropriator of a trade secret runs quite a risk under a further rule set forth in this case. The Third Circuit held that the appropriator can lose the benefits of its future independent development work because of the difficulty of determining how much of the improvement is attributed to those independent efforts and how much to the information gained by the wrongdoing.

The Seventh Circuit has also underscored the currently favorable attitude of courts toward trade secrets. *Goldberg v Medtronic, Inc,* 686 F2d 1219, 216 USPQ 89 (7th Cir 1982) held that there was misappropriation of trade secrets even though the misappropriator could have obtained the information lawfully. There was a breach of confidence and it is no defense that the defendant could have independently obtained the information if in fact he did not do so.

§4.20A —Employee Agreements and Trade Secrets (New)

To some extent, employee agreements to maintain confidentiality and to limit postemployment competition can be considered unnecessary because the law (1) implies a fiduciary duty by an employee to the employer, and (2) requires employees to keep trade secrets confidential. *See Tlapek v Chevron Oil Co,* 407 F2d 1129, 161 USPQ 193 (8th Cir 1969). One of the elements of the fiduciary duty of employees is that they cannot conspire to set up a competing business while still working for their employer. See the discussion of the New York law set forth in *Arnold's Ice Cream Co v Carlson,* 170 USPQ 324 and 171 USPQ 333 (EDNY 1971).

However, it is prudent to have signed written agreements even if they do no more than state the underlying rule of law, because it constitutes fair notice to fair-minded employees of their obligations and therefore tends to reduce the risk of litigation. Signed agreements also tend to reduce the range of uncertainty in the application of the underlying law, since the underlying law is state law, varies from state to state, and, in certain states, is not fully developed. The signed written agreement may be the only basis for relief in countries outside the United States.

In addition, in appropriate technological circumstances and with personnel of high technical skills, a signed written agreement can be an important technique in laying a basis for distinguishing between those developments which are proprietary to the company and the background technology which constitutes expertise that employees are free to carry with them and use in other jobs.

It must be kept in mind that the confidentiality agreement, without more, may be of little value. What is more critical than the executed confidentiality agreement is that there be actual trade secrets and, most important of all, that they be treated as trade secrets, in that access to them is limited to those with the need to know and that those with access understand that these are trade secrets which must be kept confidential. See the discussion in *Dynamics Research Corp v The Analytic Sciences Corp,* 400 NE2d 1274, 209 USPQ 321, 327-33 (Mass App Ct 1980).

Vigitron, Inc v Ferguson, 419 A2d 1115, 215 USPQ 354 (NH 1980) is a recent case standing for the proposition that an employee may be enjoined from using or disclosing information obtained as a result of a confidential relationship even if that information could be found in the public domain. This case also makes clear that the former employee cannot be prevented from using general expertise and that the injunction to protect trade secrets must be limited so as to avoid stifling competition. It is a long-recognized rule that an employee may use the general skills and knowledge acquired during employment. *Schulenburg v Signatrol, Inc,* 33 Ill 2d 279, 387, 212 NE2d 865, 147 USPQ 167, *cert denied,* 383 US 959, 148 USPQ 772 (1965).

Avoid constraints that make an employee unemployable elsewhere. The courts will not uphold an agreement that makes an employee a virtual hostage. *Reed Roberts Associates v Strauman,* 440 NY2d 303, 386 NYS2d 671 (1976).

The different purposes and legal consequences of employee confidentiality agreements and employee covenants not to compete suggest the desirability of having separate agreement documents. There is normally a wide range of employees who may be exposed to trade secrets and thus will be required to sign confidentiality agreements. There may even be merit in having all employees sign a confidentiality agreement, even though procedures are established, as they should be, to limit access to confidential information to those with a need to know. By contrast, a covenant not to compete is appropriate only for a limited number of employees, and an extension of such a covenant to employees where the restraint is not fully justified may invalidate the entire program or agreement.

Another reason to keep employee confidentiality agreements separate from covenants not to compete is that the former should be enforceable for as long as the trade secret exists, and thus may have an indeterminate length, while the latter normally must be limited in time. This problem is discussed in *Water Services, Inc v Tesco Chemicals, Inc,* 410 F2d 163, 162 USPQ 321 (5th Cir 1969).

In general, a covenant not to compete would seem to call for analysis similar to that used in the antitrust laws. Therefore, it is important that any such covenant be limited to those employees where there is meaningful justification for the covenant and also be limited in time and scope. Time and the nature of the postemployment activities must be defined narrowly in a fashion that is commensurate with the nature of the employee's status, the nature of the employee's access to confidential information, and the need for the employee to remain employable in his or her field. A United States Supreme Court concurring opinion in a somewhat different context, involving an exclusive distributor dealing arrangement, is appropriate here. Justice Brennan said: "The problem is not simply whether some justification can be found but whether the restraint so justified is more restrictive than necessary, or excessively anticompetitive, when viewed in light of the extenuating interest." *White Motor Co v United States,* 372 US 253, 270 (1963). He also said, "If the restraint is shown to be excessive for the manufacturer's needs, then its presence invites suspicion. . . ."

§4.21 Fostering Innovation

The use of small entities independent or within large organizations properly motivated and minimally interfered with appears to be the best current thinking on how to foster innovation. See Peters & Waterman, In Search Of Excellence (1982). The reason seems to be not so much the greater incentives at small entities as the fewer disincentives at small entities. See *Lab Extraordinaire,* 123 Sci News 9, Jan 1, 1983; *Intrapreneurial Now,* The Economist, Apr 17, 1982, at 67; *Why Small Companies Will Survive,* Chemical Week, Dec 1, 1982, at 60.

The Administration has proposed changes in the law affecting the doctrine of misuse as it relates to the licensing of patents. National Productivity and Innovation Act of 1983, HR 3878, S 1841, 26 Pat

Trademark & Copyright J 456. The changes are intended to foster innovation and in particular joint R&D ventures. If adopted, certain presently suspect licensing constraints will be safer to employ. Such adoption will probably accelerate the present judicial shift away from the harsher applications of the misuse doctrine.

§4.25 —Increased Risks of Infringing a Patent (New)

In a significant development, the Court of Appeals for the Federal Circuit has expanded the risk that a patent infringer will have to pay treble damages or attorneys' fees. In *Underwater Devices, Inc v Morrison-Knudsen Co,* 717 F2d 1380, 219 USPQ 569 (Fed Cir 1983) and *Rosemount, Inc v Beckman Instruments, Inc,* 727 F2d 1540, 221 USPQ 1 (Fed Cir 1984), the court held that there is a duty to obtain competent legal advice before initiating possible infringing action. Therefore, the court reasoned, willfullness in infringing, sufficient to warrant the award of treble damages and attorneys' fees under 35 USC §§284 and 285, might arise out of failure to obtain competent legal advice. The theory is that the failure to obtain competent legal advice often means that the infringer has no reasonable basis for believing it had a right to proceed with the infringing act, and having no reasonable basis constitutes grounds under the rule of *Stickle v Heublein, Inc,* 716 F2d 1550, 219 USPQ 377 (Fed Cir 1983) for the award of treble damages and attorneys' fees.

Although there has long been a risk that treble damages and attorneys' fees would be awarded the injured patent owner, that risk has been small absent a "smoking gun" type of documentation showing a wanton willingness to infringe on rights which the infringer believed were valid. These recent cases indicate a willingness by the court to award such additional damages on the basis of actions somewhat closer to carelessness about another's patent rights than to a conscious decision to infringe rights recognized to be valid.

Of course, if one does not know that a relevant patent exists, this rule does not apply. There is no duty to look for possible patents which might be infringed by a new product, but lack of knowledge of a patent does not excuse infringement. Lack of knowledge of a patent may simply avoid the peril of an award of treble damages and attorneys' fees for *willful* infringement.

Tax Planning for Innovation

Robert S. Taft

5

§5.02 Basic Elements of Federal Corporate Income Taxation

On August 19, 1982, Congress passed the Tax Equity and Fiscal Responsibility Act of 1982 (TEFRA), Pub L No 97-248, 96 Stat 324. While TEFRA limited a number of tax incentives useful to innovators, it did not disturb such items as research and development credit and deduction provisions (which are rarely used, although they can, at times, provide tremendous benefits). Though TEFRA made innovative tax planning more difficult in some ways, it also made it more necessary.

This new law was designed to produce $98.3 billion in additional revenue over a three-year period through a combination of federal spending cuts, tax increases, and so-called reform measures. TEFRA includes restrictions on safe-harbor leasing, cutbacks in the use of accelerated depreciation and investment credits and a tightening of the tax rules governing mergers and liquidations.

Rather than encourage innovative tax planning, TEFRA deliberately discouraged it. Many corporate tax incentives have been enacted over the years. In view of budget deficits and increased tax benefits available through accelerated depreciation (ACRS) and the investment tax credit, certain tax benefits and corporate tax preferences were trimmed by TEFRA by 15 per cent generally for tax years beginning after 1982. Also, to prevent the combination of these incentive cutbacks and the corporate minimum tax from reducing the net tax benefits from these preferences even further, adjustments were likewise made to the corporate minimum tax. The adjustments to the corporate minimum tax are accomplished by including in the minimum tax base only 71.6 per cent of the reduced tax preferences that are subject to the minimum tax on corporations. IRC §§57(b) and 291. These changes apply to all corporations other than S corporations. IRC §291(e)(2).

In the past, gain on the sale or other disposition of depreciable realty was treated as ordinary income to the extent of the excess of depreciation claimed over straight line depreciation, and any remaining gain was treated as long-term capital gain under IRC §1231. Now, for corporations other than S corporations, IRC §291(a)(1), added by TEFRA, reduces the portion of the gain which may be treated as long-term by 15 per cent in the case of sales or other dispositions of §1250 depreciable realty made after 1982 in fiscal tax years ending after 1982. Therefore, if the property disposed of after 1982 has been depreciated under an accelerated method, the amount subject to recapture as ordinary income is the sum of the additional depreciation and 15 per cent of the depreciation that would have resulted if the straight line method had been used.

Similarly, TEFRA repealed post-1985 accelerated methods of depreciation. TEFRA §206, amending IRC §168(b)(1). Under accelerated cost recovery system (ACRS), the capital cost of most tangible depreciable property is recovered using accelerated methods of cost recovery over statutory recovery periods. IRC §168, added by the Economic Recovery Tax Act of 1981 (ERTA), Pub L No 97-34, 95 Stat 176. ACRS deductions are determined by applying the statutory percentage for the respective class of property to the unadjusted basis of the recovery property. As enacted by ERTA, three separate series of ACRS statutory percentages were provided for three-year, five-year, 10-year and 15-year public utility property classes: one series was for property placed in service in 1981-1984; one was for property placed in service in 1985; and one was for property placed in service after 1985. *Id.*

The ACRS deduction for personal property placed in service in 1981-1984 approximated the benefits of using the 150 per cent declining balance method with a change to the straight line method at such time as would maximize such deduction. The percentage was to be increased to 175 per cent on the declining balance method for items placed in service in 1985 and then further increased to 200 per cent (declining balance method) with a change to the sum of the years-digits method for items placed in service after 1985.

TEFRA repealed the scheduled increases in the cost recovery rates for personal property placed in service after 1984. TEFRA §206. In addition, it limited ACRS benefits by requiring that the basis of an asset is to be reduced by 1/2 of the amount of the investment credit allowed on the asset unless the taxpayer elects to claim a reduced investment credit. TEFRA §205, adding IRC §48(g). As a result, the current cost recovery rate for personal property based on 150 per cent declining balance and the straight line method is made permanent, and the basis of the asset being depreciated is reduced by 1/2 of the amount of the investment credit allowed on the asset. It should be noted that if there is a disposition resulting in recapture of the investment credit on property as to which a downward basis adjustment was made, the basis of such property immediately before the event resulting in recapture will be increased by 50 per cent of the recapture amount. Where an investment credit for which a downward basis adjustment was made does not result in a tax benefit because it remains unused at the end of a 15-year carryover period, a deduction is allowed to the taxpayer for 50 per cent of the unused credit.

When a lessor elects to pass through the investment tax credit to a lessee under IRC §48(d), the reduction in basis otherwise required under the rules specified above does not apply. Instead, the lessee must include an amount equal to 50 per cent of the credit allowable in income ratably over the ACRS recovery period for the property. However, a lessee has the option to elect a two percentage point reduction in its regular credit, in which case it is not required to include in income 50 per cent of the credit allowable. IRC §48(d) and (q)(4). This 2 per cent reduction is TEFRA's alternative to reducing an asset's basis with respect to the investment credit. It is available to all taxpayers involved with depreciable property.

In 1982, the tax law permitted the investment tax credit to offset the first $25,000 of income tax liability plus 90 per cent of tax liability in excess thereof. For tax years beginning after 1982, TEFRA reduces the percentage to 85 per cent of the tax liability above $25,000. TEFRA §205(b), amending IRC §46(a)(3)(B).

Despite these reduced benefits, careful balancing of depreciation and investment credit will still result in innovative tax-benefit structuring.

In the area of mergers and acquisitions, TEFRA repealed the rules which allowed nonrecognition of gain or loss by a corporation upon distribution of its assets in a partial liquidation and the taxation of the distribution to shareholders as full payment for their stock. TEFRA §222, amending IRC §§302, 311(d), and 346. It retained capital gains treatment for partial liquidations of stock of noncorporate shareholders by treating a distribution in partial liquidation to such shareholders as full payment for their stock. IRC §302(b)(4) and (e), as amended by TEFRA.

TEFRA also repealed IRC §334(b)(2), which treated a purchase and liquidation of a subsidiary corporation as a purchase of that corporation's assets. However, TEFRA added a new §338, which provides that a purchasing corporation, within 75 days after it makes a "qualified stock purchase" with respect to the acquired corporation (a "target corporation"), may make an irrevocable election to treat the target corporation as if the target had purchased and sold its assets in a complete liquidation at the close of the acquisition date, for an amount generally equal to the purchasing corporation's basis in the target's stock. This gives the purchasing corporation a basis in the target's assets equal to its basis in the target's stock without the necessity of liquidating the target corporation. The target corporation itself is deemed to have sold its assets for an amount equal to the purchasing corporation's "grossed up basis" in the target's stock on the acquisition date, as adjusted under the regulations. The purchasing corporation's grossed up basis is an amount equal to its basis in the target's stock multiplied by a fraction having 100 per cent as its numerator and the percentage of value of the target's stock held by the purchasing corporation on the acquisition date as its denominator.

This amount is allocated among the target corporation's assets. The target corporation generally realizes no gain or loss on a deemed sale because this is treated as a distribution of the assets in complete liquidation of the target. IRC §336. However, the target corporation will recognize gain on certain "recapture items" (*e.g.*, §§1245 and 1250 depreciation recapture). This gain will now be recognized by the target on its separate return before it becomes a member of the acquiring corporation's affiliated group, so that this recapture income may not be included on a consolidated return of that affiliated group and used to offset its losses. In addition, if the purchasing corporation owns less than 100 per cent of the target's stock, the target's nonrecognition of gain or loss is limited to an amount determined by multiplying the gain or loss by the maximum percentage by value of stock in the target held by the purchasing corporation during a one-year period, beginning on the acquisition date. This rule does not apply if the target corporation is liquidated during the one-year period. IRC §338.

Since this chapter is not intended to be a full explanation of TEFRA, it should be sufficient to point out that TEFRA has limited many of the tax benefits in the Code extant prior to the advent of TEFRA. Obviously, this affects innovative tax planning and requires more, not less, innovation.

Corporate Tax Simplification

Much has been said during the past months about simplifying the tax law for individuals. However, not to be forgotten amid all the discussion about tax simplification is the fact that there is also a corporate income tax, which is also in need of simplification.

Corporate income tax was enacted four years before the introduction of the individual income tax. Thus, we find corporations first taxed in 1909. Congress levied the tax as an excise on the privilege of doing business as a corporation, in order to avoid a constitutional issue. The law was challenged in *Flint v Stone Tracy Co,* 220 US 106 (1910), and the Supreme Court upheld the authority of the federal government to impose such a tax.

The corporate income tax produced more revenue than the individual income tax in 17 of the 28 years before 1941 when the latter was greatly expanded as a source of wartime revenue. From 1941 through 1967, corporation income tax receipts were second only to those of the individual income tax, but they were overtaken by payroll taxes in fiscal 1968 and have since been declining in importance. The corporation income tax accounted for about 6 per cent of federal receipts in fiscal 1982, compared with 19 per cent in 1968. In spite of this large drop in tax collection, there has been relatively little movement in the statutory corporate income tax rate during the past three decades. The rate was reduced from 52 per cent to 48 per cent by the 1964 tax cut under President Johnson; raised temporarily by the 10 per cent Vietnam War surcharge and lowered again in 1978 to its current level of 46 per cent.

While the nominal tax rate on corporations has remained constant, the last two decades have seen a steady erosion of effective tax rate paid on corporate earnings because of the multitude of special tax exemptions, credits, and deductions added to the corporate tax structure. These preferential changes in the tax law have been part of an overall federal government industrial policy to help American businesses compete in the international marketplace. Recently, former Treasury Secretary Regan conceded that the United States had an extensive set of industrial policies which discriminate between industries and activities and provide a competitive advantage (or disadvantage) to some industries through these choices. Thus, for example, companies which do defense business with the federal government earn billions of dollars upon which taxes are deferred until their "contracts are completed." It is very easy to keep the books open in such arrangements and not pay taxes on earnings for years and years. Today, the federal government collects roughly $60 billion from the corporate income tax, but gives away $80 billion a year to corporations in the form of "tax expenditures." Every concession, credit, deduction, and loophole which gets enacted as an "economy shaping" device serves to reduce the revenues available to the federal government and to shift the tax burden onto other taxes. By so doing, we make simplification of the individual income tax structure far more difficult than it would normally be. Accordingly, what is really needed is an overall tax simplification of both individual and corporate tax provisions. The industrial policy features of the tax code with respect to corporations are expensive, badly targeted, inefficient, and inflexible. It is understood that some assistance from the federal government to individual industries and sectors is both a practical necessity and a competitive advantage, but this assistance should rarely be provided via the Tax Code.

Cleaning up the tax base through loophole closing and the termination of unnecessary credits and deductions would restore a greater measure of neutrality to the Tax Code and improve the overall efficiency of our economy. Such base-broadening would also permit a lowering of the overall corporate tax rate, a change which should have an overall positive impact on the economy.

It is suggested that with respect to corporations, tax simplification could be established by:

1. *Ending sectoral preferences.* A simplified and neutral Tax Code would eliminate preferences for the various different sectors of

business, such as defense contractors, etc. Income would be taxed in each sector in the same fashion

2. *Terminating activity preferences.* Thus, all the loopholes designed to favor specific activities (such as the FISC, which encourages exports) would be treated in the same tax-neutral fashion
3. *Eliminate the distinction between the taxation of assets.* Thus, the distortions caused by depreciation allowance in excess of actual economic depreciation would end. The tax life of assets would be matched to their actual economic life, thereby creating no artificial preferences on different assets, under a simplified and neutral Tax Code
4. *Modify the provisions relating to financial structure preferences.* A neutral Tax Code would eliminate such preferences by eliminating the preferential treatment of capital gains, and integrating the individual and corporate income taxes with respect to dividends to eliminate the double taxation thereof

The ultimate benefit of ending the above tax distortions or variations would be the reduction of the corporate tax rate. It may be hoped that by creating a neutral Tax Code with broadbased reforms, one could reduce the corporate rate by 50 per cent.

1984 Business Tax Changes

The Tax Reform Act of 1984, Pub L No 98-369, makes, among others, the following changes in business tax law:

1. It limits the amount of the 85 per cent dividends received deduction applicable to debt finance portfolio stock (IRC §246A)
2. Graduated rates are established for corporations with taxable income in excess of $1 million by subjecting them to an additional 5 per cent tax (up to $20,250) on taxable income in excess of $1 million (IRC §§11(b), 1561(a))
3. Extraordinary dividends paid on stock may reduce the corporation's basis in the stock (IRC §1059)
4. Corporations will now generally recognize gain on nonliquidating distributions of appreciated property (IRC §§301(e), 311(d))
5. The holding period requirement for purposes of the dividends received deduction is moved from 16 to 46 days (IRC §§246(c), 7701(f))
6. The accumulated earnings tax now covers widely held companies (IRC §532(c))
7. Corporate tax preference items, except for coal or iron ore depletion, are reduced by 20 per cent (IRC §§57(b), 291). Under TEFRA, it was a 15 per cent reduction

8. Excess payments under "golden parachute" contracts will generally not be deductible and will subject the recipient to a 20 per cent excise tax (IRC §§280G, 4999, 275(a), 34121(v))
9. Rules will be developed to conform the determination of earnings and profits of a corporation to the corporation's real income (IRC §312)
10. The net operating loss carryover provisions will not go into effect until 1986 (instead of 1984) (1984 Act, §62, amending Tax Reform Act of 1976 §806(g))
11. Stock ownership thresholds of 80 per cent apply in determining affiliated groups status (IRC §338(h)(1)(A), (3)(B))
12. Depreciation schedules have been changed, as has the investment tax credit. For all real estate except low-income housing, the depreciation period has been moved up from 15 to 18 years. More than 50 per cent business use is required for the ITC and the ACRS on "listed property." Movies are not recovery property and generally are not eligible for the ITC; a three-year recovery period is provided for sound recordings. Only $4,000 in depreciation may be taken on automobiles in the year placed in service ($6,000 thereafter) and a maximum $1,000 ITC (IRC §§48(g), 57(a), 168, 312(k), 1245(a))

§5.05 Independent Business Purpose Requirement

Corporations may benefit from an off-beat Tax Court case ruling. In *Anthony J. Ditunno,* 80 TC 12 (1983), the Tax Court ruled that individuals do not have to hold themselves out to others as engaged in the selling of goods or services in order to be engaged in a trade or business. The court found that the test it formulated in *Alfred A. Gentile,* 65 TC 1 (1975), which focused on the provisions of goods or services, was overly restrictive. It will now use a facts and circumstances test to determine whether a taxpayer is carrying on a trade or business. As a result, the court found that an individual who was a professional gambler was engaged in a trade or business.

From 1977 through 1979, Ditunno was a full-time gambler with no other profession or full-time employment. He gambled only on horse races and went to the race track six days a week, year round for that purpose. He had no other income except for a small amount of interest. Each year he reported winnings of about $60,000, but he also deducted a substantial portion of these winnings on Schedule C as gambling losses. The Internal Revenue Service argued that he was not in the trade or business of gambling, and so his gambling losses were not deductible under IRC §162 in computing adjusted gross income. In their view, the gambling loss deductions were tax preference items, so that he was subject to the minimum tax for 1977 through 1979. Mr. Ditunno argued that the losses were deductible in computing his adjusted gross income because he was in the business of gambling and, therefore, the losses were not tax preference items.

The Tax Court applied the facts and circumstances test, focusing on Mr. Ditunno's level of activity. Since he devoted full time to his gambling activities, it concluded that he was engaged in a business. Thus, the court concluded that he was in the business of gambling and his losses were deductible and were not tax preferences.

Accordingly, the interpretation of "carrying on a trade or business" under IRC §162 should logically be extended to corporate activities. This should enable corporations to participate in activities which would not normally have been called "carrying on a trade or business" in the past, and still be deemed to have an independent business purpose for their activities.

§5.10 —Affiliated Corporations

Before the Miscellaneous Revenue Act of 1980, Pub L No 96-605, 94 Stat 3521, "common control" definitions allowed corporate partners with a 50 per cent or less interest in the partnership to exclude certain partnership employees from the retirement plans of the corporate partners. IRC §414(m) closed this discrimination loophole by requiring employees of an "affiliated service group" (ASG) to be treated as employed by a single employer and lumped together for plan coverage purposes.

The Internal Revenue Service has proposed new standards to clarify what constitutes an ASG and provided limited administrative exemptions. Proposed Treasury Reg §1.414(m)-1, reprinted in 1983-13 IRB 81-89. The Service says taxpayers may rely on these proposals pending adoption of final regulations.

Under these standards, an ASG consists of any one of three possible corporate combinations. First, it can consist of a "service organization" (first service organization or FSO) and a shareholder or partner in the FSO regardless of the ownership percentage (A Organization). An A Organization must regularly associate with the FSO in performing services for others. Second, an ASG exists when an organization (B Organization) generates a "significant portion" of its business by performing services for an FSO, an A Organization, or both. B Organization's services are the same type of services historically performed by employees in the FSO or A Organization's service field. Moreover, at least 10 per cent of the B Organization must be held by FSO or A Organization officers, owners, or highly paid employees. Third, an ASG will exist when a FSO is combined with at least one A Organization and one B Organization. An A Organization with only a minimal interest in a FSO can avoid mandatory aggregation if it does not attempt to avoid plan coverage requirements. However, professional service corporations do not qualify for this exemption from the aggregate requirement.

The IRS proposals do not cover TEFRA's extension of the IRC §414(m) ASG rules to "affiliated management organizations." These rules are effective for post-1983 taxable years and will be covered by separate regulations. However, the general intent of these rules is the same as discussed above.

Therefore, in planning to structure pension benefits for key employees, one can no longer utilize an affiliated service company to avoid broader coverage. To do so, one must make the affiliate a C Organization. One that does call for ASG treatment is an A or B Organization. Such a C Organization would work independently from the FSO, though it could work on part of the same contract, as a subcontractor, perhaps, for an independent purchaser. The key seems to be that the C Organization, in some respects more like an independent contractor than an employee, should not limit its work to that involving the FSO or A Organization. This type of innovative planning might, in proper cases, avoid the imposition of the ASG label.

§5.12 —Impact of the Economic Recovery Tax Act on Research and Development

Under Pub L No 98-369, a tax credit remains available for incremental research expenses paid or incurred after June 30, 1981 and before January 1, 1986. Thus, the 25 per cent credit added as IRC §44F by the Economic Recovery Tax Act of 1981 basically goes out of existence at the end of 1985.

While removing this credit is consistent with the general goal of tax simplification, and recognizes the reality that much "research" may have been recategorized activity that went on before under other labels (*cf* Surrey, *Tax Incentives as a Device for Implementing Government Policy,* 83 Harv L Rev 705 (1970)), if tax incentives are to continue to be used to promote national objectives, R&D is a priority object of such encouragement, as was recommended by the President's Commission on Industrial Competitiveness in its Summary of Recommendations at 52 (1985), which stated: "The R&D tax credit should be strengthened and clarified to include expenses involved in developing and implementing innovative manufacturing processes, machinery, and facilities." If the §44F credit is not renewed, other measures should be devised to promote the same objectives, and to do so making maximum use of free market incentives (see **ch 12** of this supplement).

If the credit is renewed, however, then (as recommended by the President's Commission) attention might well be given to making it as effective as possible. Those who would make use of it while it exists would do well to be aware of its limitations in its present form.

As mentioned in **§5.02,** TEFRA did not amend the research and development credit enacted by ERTA as IRC §44F. However, this credit has proved to be of limited benefit for several reasons.

Most importantly, the credit only applies to research expenditures incurred in carrying on a trade or business in which a taxpaying entity is already engaged. No credit is available for expenses relating to a potential trade or business. Thus, new businesses conducting research for future production and sales as a result of technology developed through such research, and ongoing businesses that undertake research geared to the development of a new business activity, may not claim the credit. Proposed Treasury Reg §1.44F-2(a)(2). The IRS position in these regulations finds some support in the legislative history to the provision. *See* the House Conference Report, HR Conf Rep No 215, 97th Cong, 1st Sess, at 223-24 (Aug 1, 1981) and Senate Conference Report S Rep No 144 97th Cong, 1st Sess, at 77-78 (July 6, 1981). *Compare* the Supreme Court result in *Snow v Commissioner,* 414 US 1128 (1974), holding that §174 deductions are allowable where the expenditures are incurred "in connection with" a trade or business. The legislation should be amended in these respects if its full potential is to be realized.

In addition, there are other important limitations. Prepaid contract research expenses may not be taken into account when paid; they are treated as paid or incurred, for these purposes, during the period when the qualified research is conducted. IRC §44F(b)(3)(B). Here again, the purpose of the credit may be impaired.

Finally, the credit may never exceed the taxpayer's tax liability reduced by other credits and, in the case of unincorporated businesses and S corporations, may not exceed the amount of tax attributable to that portion of taxable income which is allocable to that taxpayer's interest in such trade or business or entity. IRC §44(g)(1). However, the credit may be carried back three years and carried forward 15 years.

Therefore, the innovative tax planner must guide the client within the confines of the rules. While it will be difficult for a new enterprise to qualify for the credit, taxpayers carrying on some trade or business may be able to avail themselves of this credit, to the extent that the research developed could be used in the particular business carried on by the taxpayer. The taxpayer must be careful, however, not to transfer all rights to the resulting technology before the project is completed and the taxpayer uses it in its business to a realistic extent, since research might then not be treated as having been incurred for the taxpayer's business.

Venture capital is frequently used for financing arrangements for research and development (R&D) programs. Partnerships are often most useful as financing structures for a corporation seeking to conduct R&D. They provide the corporation with outside capital and off-balance-sheet treatment of the R&D expense, while allowing the corporation to share in the benefits if the R&D effort is successful. Moreover, they allow an investor to benefit from deductions by the R&D effort, while providing the prospect of a very substantial return. The flow of tax benefits to investors and the parallel sheltering of their income reduce the amount of capital the investors have placed at risk in the R&D venture and increase their return from the venture. Thus, the use of a partnership may cause an investment in a R&D venture to be more attractive to investors than an investment in a straight venture capital deal.

Typically, a research and development partnership is established by a corporation (the sponsor) which creates a wholly owned subsidiary to serve as a general partner. Usually, 99 per cent of the profits and losses of the partnership are allocated to the limited partners and 1 per cent to the general partners, as in other venture capital limited partnerships. The sponsor contributes any existing technology relating to the intended research project to the subsidiary, which in turn contributes such technology to the partnership. The partnership then contracts with the sponsor and perhaps other corporations to conduct research.

If the research is successful, the payout to the investor can take several forms and be structured in various ways. The methods of payout include:

1. *Equity method*—general partner and limited partner incorporate the partnership by exchanging their partnership interests for stock in a newly formed corporation pursuant to IRC §351. Incorporation facilitates the acquisition of additional capital which may be necessary to market products utilizing the new technology, and will allow the venture to accumulate earnings without subjecting the investors to income tax liability. Stock of the newly incorporated company may be registered to allow easy transferability of shares
2. *Technology-purchase method*—sponsor exercises an option granted to it at the inception of the partnership either to purchase technology for a lump sum or to acquire a license to use the technology for royalties
3. *Partnership-purchase method*—sponsor exercises an option granted to it at the inception of the partnership to purchase partnership interests of the limited partnership. This may produce capital gains subject to IRC §751 ordinary income consequences

There are, of course, other alternatives, such as joint ventures, etc.

R&D expenditures must be "in connection with a trade or business" to be deductible under IRC §174. This is usually not a problem because the "in connection with" standard is more liberal than the "carrying on" standard of IRC §162. However, there may be risks if no trade or business relationship is established. It is important to have at least the feasibility of marketing products or of engaging in ongoing R&D activities (as opposed to a single project). Otherwise, there may be no deduction and it may be that the only business found will be receipt of royalty income. The research must be in the experimental or laboratory sense. It may be contracted out, but the partnership must bear the risk that the research will not be successful.

Interest expenses incurred by the partnership will be deductible by the partners, subject to three potential relevant limitations:

1. IRC §163(d) limits the amount of investment interest paid or incurred by a partnership which a partner may deduct. *Investment interest* is interest paid on indebtedness incurred or continued to purchase or carry property held for investment. As a rule, this would not be an R&D problem, since most of the partnership's assets will not be used to carry property held for investment, but it may arise if partners borrow funds to purchase their partnership interests
2. The partnership may also be limited as to the amount of interest it can deduct if it is using a cash basis method of accounting, because of the IRS position that a taxpayer has not "paid" interest for purposes of IRC §163 if it paid such amount from proceeds borrowed from the lender. (*Wilkerson v Commissioner,* 6 F2d 980 (9th Cir 1981)). The partnership should be sure it has sufficient nonborrowed funds available to pay interest in order to preserve the best argument for deductibility
3. IRC §265 may prevent a partner from deducting partnership interest expenses where a partner also possesses tax-exempt securities. Rev Proc 71-18, 1972-1 CB 740, *clarified by* Rev Proc 74-8, 1974-1 CB 419, stated that an individual possessing tax-exempt securities who borrows to purchase a limited partnership interest will be presumed to have incurred that indebtedness to carry his tax-exempt securities

The R&D partnership may be structured so that the partnership pays the R&D contract expense at the time the contract is entered into, even though the development work may extend beyond the year of payment. There are no particular cases on this point but substantial authority exists in other areas which address the deductibility of prepaid cattle expenses, such as feed and intangible drilling costs. The authority suggests that such prepaid expenses will be deductible provided that the prepayment is a nonrefundable payment rather than a deposit. *See* Rev Rul 79-220, 1979-2 CB 210. There must also be a business purpose for the prepayment other than tax advantage. Finally, the prepayment may not materially distort income. In this regard there is a one-year safe harbor rule. The Court of Appeals for the Ninth Circuit held that "material distortion of income does not occur where the taxpayer will not benefit from the expenditure for more than 12 months after the close of the taxable year." *Van Raden v Commissioner,* 650 F2d 1046, 1050 (9th Cir 1981).

It is unlikely that the IRC §44F research credit will be available to the partnership. IRC §44F requires that research expenditures be paid or incurred in carrying on a trade or business. This is a more difficult test than the "connected with" test, and usually eliminates the credit for the partnership.

Under IRC §1235, long-term capital gain treatment is available for sale of patentable property. This is only available to "holders," among other requirements. Holders may include partners who are individuals, but the partnership should acquire patentable property directly from the inventor. IRC §1221 poses a problem in that technology can constitute property "primarily held for sale to customers in the ordinary course of business," and this would produce ordinary income. It could be argued that if the R&D partnership intends to use the technology to develop products, then the technology is not held "primarily" for sale in the ordinary course of business. *See Ofria v Commissioner,* 77 TC 524 (1981), *non acqd in* 1983-27 IRB 4 and 1983-2 IRB 5. The holding period for long-term capital gains begins when technology is "reduced to practice," a point that can be difficult to ascertain.

On the sale of a partnership interest, IRC §741 provides that except for proceeds attributable to IRC §751 assets, the sale of a partnership interest will be deemed to be the sale of a capital asset. IRC §751 is an exception to capital gain treatment and provides that any amounts received from the sale of a partnership interest which are attributable to either (1) unrealized receivables of the partnership, or (2) substantially appreciated partnership inventory, will be treated as ordinary income. Substantially appreciated inventory generally includes any property other than IRC §1231 property. To qualify under IRC §1231, one must meet essentially the same requirements as for IRC §1221. Thus, if the technology does not constitute an IRC §1221 or §1231 asset, sale of a partnership interest will not provide an alternative method of obtaining capital gains treatment.

One of the potential traps in this area is the possibility that a R&D partnership could be recharacterized as merely a loan to the sponsor. *Estate of Helliwell v Commissioner,* 77 TC 964 (1981). In order to avoid an adverse characterization of the partnership based on *Helliwell,* it is highly important that the partnership bear the risk of ownership of the technology. In addition, the sponsor should not guarantee loans to the partnership. If the sponsor has any options, they must be exercised at current fair market value of the technology in the event the R&D effort is successful. If the option is not exercised, the partnership should possess an absolute right to use the developed and contributed technology in any manner it wishes.

§5.14 Deferred Compensation Issues

The investment based tax credit for employee stock ownership plans has long favored capital-intensive businesses. It allows an additional 1.5 per cent credit if plan contributions are in the form of qualifying employer securities. This credit garners employers a tax benefit for contributions to their tax credit stock ownership plans (TRASOPs). IRC §44G. In 1981, ERTA extended these benefits to labor-intensive businesses by replacing the investment based credit with a payroll based credit (frequently called a PAYSOP). ERTA §331(a), Pub L No 97-34, 95 Stat 176, amending IRC §44G. This change is phased in over a five-year period for compensation paid or accrued after December 31, 1982. The Internal Revenue Service has issued final regulations on the basic TRASOP investment based credit, which will self-destruct beginning in 1983. The regulations do not reflect changes made with respect to PAYSOPs.

A TRASOP must invest primarily in the employer's securities under §301(d) of the 1975 Tax Reduction Act. IRC §409A(a)(2). This is a continuing TRASOP requirement and therefore, any transaction that changes a corporation's status as employer, such as a reorganization, may require conversion of qualifying employer securities. Cash (or other assets) realized from such a conversion must be reinvested in employer securities within 90 days.

Under IRC §46(a)(2), a sponsor can claim an additional credit of up to 1-1/2 per cent (making the total credit 11-1/2 per cent of the year's qualified investment) with respect to periods beginning on January 1, 1977 and ending on or before December 31, 1982, if an amount equal to the additional credit is contributed to the TRASOP by employees. These contributions are deemed mandatory rather than voluntary, so they are not subject to the 10 per cent cap on voluntary employee contributions. Treasury Reg §1.46-9.

Therefore, a corporation in a cash-flow squeeze which is taxed as profitable can utilize the tax credit ESOP to insulate profits from taxation. By establishing such a plan and contributing its own securities to the ESOP trust, it will not only obtain a deduction for the value of such stock, but also a tax credit of 11.5 per cent. This will reduce taxes and, if it creates a carryback under IRC §44G(b)(2)(A)(i), could even provide a refund of taxes from a prior year.

TEFRA makes numerous other changes with respect to employee benefits. These changes primarily affect qualified pension and profit-sharing plans. The maximum annual addition for profit-sharing and other defined contribution plans has been reduced from $45,475 to $30,000. The maximum annual retirement benefit under a defined benefit pension plan has been reduced from $136,425 to $90,000. The reduction in the overall limits is accompanied by a freeze on cost-of-living adjustments for three years. TEFRA §235(a) and (g), Pub L No 97-248, 96 Stat 324, amending IRC §415(c)(1)(A) (b)(1)(A) and (b)(7).

The age below which actuarial reductions are required in the maximum benefit limit for defined benefit plans is increased from 55 to 62 years and, where an employee participates in both a defined contribution plan and a defined benefit plan of the same employer, the aggregate limit of allowable deductible contributions is lowered from 140 per cent to 125 per cent of the separate maximum dollar amount. TEFRA §235, amending IRC §415(e).

Loans from tax qualified plans have been substantially limited. A loan is to be treated as a distribution to the extent that it exceeds certain limits. However, a loan from a qualified plan which is repaid within five years is treated as a distribution only to the extent that the amount of the loan, when added to the outstanding loan balance of the employee under all other loans from that plan, whether made on, before, or after August 13, 1982, exceeds the lesser of (1) $50,000, or (2) 1/2 of the present value of the employee's nonforfeitable accrued benefit under such plan, but not less than $10,000.

TEFRA eliminates distinctions in the tax law between qualified plans of corporations and those of self-employed individuals (Keogh plans). TEFRA §§238 and 237(a). However, there is a phase-in period for this, and it will take two years before there is actual equality. Furthermore, there will continue to be some benefits available in corporate plans as opposed to the Keogh-type plans. (*E.g.*, Keogh plans will be prohibited from making loans to owner-employers).

TEFRA also imposes certain penalties on a plan when it becomes a "top heavy plan." TEFRA §240(a), adding IRC §416. If a plan is found to be "top heavy," it must provide for a more rapid vesting schedule, provide for minimum nonintegrated contributions or benefits for plan participants who are nonkey employees, limit the amount of a participant's compensation which may be taken into account for benefits, and reduce the aggregate limit on contributions and benefits for certain key employees. Any *defined benefit plan* is a top-heavy plan for a plan year if, as of the determination date: (1) the present value of the accumulated accrued benefits for participants who are key employees for the plan year exceeds 60 per cent of the present value of the accumulated accrued benefits for all employees under the plan, or (2) the plan is part of a top-heavy group.

Any *defined contribution plan* is a top-heavy plan for a plan year if, as of the determination date: (1) the sum of the account balances of participants who are key employees for the plan year exceed 60 per cent of the sum of the account balances of all employees under the plan, or (2) the plan is a part of a top-heavy group.

The determination date for any plan year is (1) the last day of the preceding plan year, or (2) in the case of the first plan year of any plan, the last day of such plan year.

TEFRA provides rules under which two or more plans of a single employer are aggregated to determine whether the plans, as a group, are top-heavy. If they are, all plans are subject to the "top heavy penalties" as set forth above. It should be noted that a key employee is deemed any participant in an employer plan who, at any time during the year or any of the four preceding plan years, is (1) an officer of the employer, (2) one of the 10 employees owning the largest interests in the employer, (3) a 5 per cent owner of the employer, or (4) a 1 per cent owner of the employer having an annual compensation of more than $150,000. TEFRA §240(a), adding IRC §426(i).

TEFRA has made other changes with respect to deferred benefits. For example, effective August 13, 1982, TEFRA requires that partial surrenders or cash withdrawals from any annuity policy prior to the annuity starting date will be treated first as income, to the extent that the cash value of the contract exceeds the investment in the contract, with the remainder treated as a return of capital. Loans against the contract, or pledging the contract, will be treated as amounts withdrawn from the contract. TEFRA §265(a) and (c), amending IRC §72(e).

An aggregate limit of $100,000 is placed on the estate tax exclusion for benefits payable to a beneficiary (other than the executor) of a deceased employee under a qualified plan, a tax-sheltered annuity, IRA or certain military retirement plans. Previously, the estate tax exclusion for such plans was unlimited. TEFRA §245(a), adding IRC §2039(g).

Clearly then, innovative tax counsel, in planning for key employees of a corporation, must seek vehicles outside of the qualified plan area. The primary vehicle which has been developed is the supplemental pension contract for the key employee. This contract is a straight deferred pension benefit which picks up the difference between the amount the employee receives on retirement from the qualified program and a certain percentage of his or her average annual compensation during the last several years prior to retirement or average of several years of highest compensation. More often than not, this percentage is 50 per cent of his or her final average compensation. Cost-of-living adjustments are a possible addition in the contract, to become effective subsequent to retirement. As the law tightens up more and more on qualified plan benefits, the supplemental type benefit becomes important to key employee benefits. Deductions are not available until the benefits are paid out and the employee will not be taxed until benefits are received.

§5.17 Industrial Development and Arbitrage Bonds

Effective for most industrial development bonds (IDBs) issued after 1982, interest paid on such bonds will not be exempt from tax unless, after notice and public hearing, the bond issue is approved by an elected legislative body or an elected public official. TEFRA §215(a), Pub L No 97-248, 96 Stat 324, adding IRC §103(k). Hearing and approval by public officials is not required if the issue is approved by a referendum of the voters. This means that every issue of IDBs must be approved, after public hearing, by voters, a legislative body, or an official of the issuing jurisdiction.

In addition, if the facilities to be financed by the issue are to be located in whole or in part outside of the geographical jurisdiction of the issuer, the issue must be approved after public hearing by the voters, a legislative body, or an official of each jurisdiction in which any part of the facility is to be located. For facilities located entirely within a single state, determination of jurisdiction is left to state law. Obviously, the use of IDBs has now been crimped by this new provision which was put into the tax law by TEFRA. TEFRA has politicized the IDB and the tax planner must develop a knowledge of and relationship with local politics and politicians in order to effectuate the use of the IDB. This is obviously a benefit to local politicians, but not necessarily for taxpayers or the purposes of IDBs.

In addition to these political limitations, TEFRA has provided that, for IDBs issued after 1982, interest paid on such bonds will not be exempt if the average length of maturity of the bonds exceeds by more than 20 per cent the average expected economic life of the facilities financed. IRC §103(b)(14). The economic life of an asset is measured from the later of (1) the date on which the bonds are issued, or (2) the date on which the facility is placed in service (or is reasonably expected to be placed in service if it has not been placed in service when the bonds are issued). Another limitation inserted by TEFRA relates to a small issue exemption. Under the old law, an issue of $1 million or less could have qualified for the small issue exemption of IRC §103(b)(6) even though it is part of a larger issue which would qualify for exemption, for the most part, by virtue of the use to which the proceeds would have been put. TEFRA terminates this and states that such an issue will no longer qualify for the $1 million small issue exemption. IRC §103(b)(6)(n). However, the alternative $10 million exemption under IRC §103(b)(6)(D) will continue to be available if elected.

Further changes by TEFRA include the following: research expenses have been excluded from capital expenditures for purposes of the elective small issue exemption; the small issue exemption will be denied for certain specified activities; and finally, the small issue exemption is scheduled to expire as to obligations issued after 1986.

Overall, the use of IDBs for innovators' tax planning has been greatly hampered by TEFRA. The innovative tax planner might look to financing from qualified pension and profit-sharing plans as an alternative to the IDB. This is so because interest received by such plans is tax-exempt to such plans. Thus, it is possible to negotiate a rate of interest more closely aligned to the IDB interest than to traditional financing interest.

§5.21 Employee Welfare Benefit Programs (New)

An educational benefit plan can be a valuable tool for an innovative corporation to attract top quality employees. If the corporation intends to take current deductions for contributions to such a plan, much care and attention must be taken in setting up the plan, or the corporation may end up having to wait until payments are made to the plan beneficiaries before it is permitted a deduction.

In *Greensboro Pathology Associates, PA v United States*, 698 F2d 1196 (Fed Cir 1982), a recent decision by the new United States Court of Appeals for the Federal Circuit held that an educational benefit plan was a welfare benefit plan under IRS Reg §1.162-10, contributions to which were then deductible under IRC §162. It was thus not a deferred compensation plan with contributions deductible under IRC §404 when payments are made to beneficiaries.

The corporation involved deducted costs of an educational benefits plan which covered its four shareholder employees and two non-shareholder employees. Under the plan, every employee was eligible for $4,000 in benefits for up to four years for each child between the ages of 18 and 29 pursuing a course of undergraduate or graduate study at an accredited college or university. All employees could participate without restriction or qualification. At inception, 11 children of employees were designated to receive benefits; nine were children of shareholders, two were not. Each child was eligible without regard to academic achievement or financial need. A child would continue to be eligible if the parent left the company's employ due to retirement or disability reasons. If a parent left for any other reason, benefits would cease and funds remaining in the child's account would be applied to reduce the corporation's future contributions. An independent trustee administered the plan; it could be amended or terminated by the corporation's directors, but no part of the trust corpus could revert or inure to the benefit of the corporation or its shareholders.

The Internal Revenue Service argued that a plan is a deferred compensation plan if it involves benefits too large to be considered fringe benefits, or if loss of benefits results if an employee leaves the employer corporation. The Court of Appeals, however, pointed to the following factors as controlling in determination of whether the plan is deferred compensation or a benefit for the employee's welfare:

1. Is it a welfare plan (*i.e.*, is it concerned with the employee's well-being)
2. Are benefits based on the employee's earnings (indicating compensation)
3. Do benefits increase with length of employment (indicating compensation)
4. Are benefits available to all employees
5. Are benefits a substitute for salary
6. Does the plan serve its stated purpose or is it a sham
7. Does the employer lose control of contributed funds
8. Can the funds revert to the employer

9. Is the plan independently administered

As noted above, factors 2 and 3 indicate compensation. Factor 4 was found essential to a welfare benefit plan. A company's retention of control over the funds indicates that in reality it has not spent the money. The court concluded that Greensboro's plan was a welfare benefit plan and that its plan contributions were deductible when made.

The *Greensboro* case is important since, as indicated in §**5.14** of this supplement, qualified benefits for key employees have been greatly limited by TEFRA. This type of program, if it continues to be upheld, as seems likely absent legislative change, may make up for some of the benefits lost in the qualified plan area.

Golden Parachutes

Corporations frequently enter into agreements with key personnel that have come to be called "golden parachutes," under which the corporation agrees to pay these individuals certain amounts, frequently in excess of their usual compensation, in the event that the control of the corporation changes. The reason for this is that these key personnel are generally nervous about a possible takeover where the acquiring company has its own personnel and will replace or diminish the job of the acquired company's personnel. Under the 1984 Tax Reform Act, Pub L No 98-369, §67, payments under such agreements are designated *parachute payments,* and a portion of such a payment—an "excess parachute payment" made to "disqualified individuals"—is not deductible by the corporation. Furthermore, the individual receiving the payments is subject to an excise tax (in addition to income taxes) of 20 per cent of the entire parachute payment in the event that there is an excess payment. The excise tax is to be withheld by the employer in the case of payments that constitute wages, and the tax is not deductible for income tax purposes by the recipient of the parachute payment. Such payments are also subject to FICA taxes when paid.

For purposes of this provision, a *disqualified individual* is any individual who (1) is an employee, independent contractor, or other person specified in regulations who performs personal services for the corporation and (2) is an officer, shareholder, or highly compensated individual. A personal service corporation (or similar entity) is treated as an individual under this provision.

An *excess parachute payment* is an amount equal to the excess of any parachute payment over the portion of the base amount allocated to the payment. A *parachute payment* is any payment in the nature of compensation to a disqualified individual if: (1) payment is contingent on the change in the ownership or effective control of the corporation, or in the ownership of a substantial portion of the corporation's assets and (2) the aggregate present value of such contingent payments equals or exceeds three times the base amount. The present value of these payments is to be determined under IRC §1274, which was added by the 1984 Act. Any payment in the nature of compensation to a disqualified individual that violates any securities law or regulation is also considered a parachute payment, whether excess or not.

For purposes of the contingency requirement, any payment made under an agreement entered into within one year of a change of control or ownership, or under an amendment made within one year of a previous agreement, is presumed to be contingent on the change of control or ownership unless the contrary is established by clear, convincing evidence. Any transfer of property is considered a payment in the amount of the property's fair market value.

The term *base amount* referred to in this provision is an individual's annualized includable compensation for a base period, consisting of the most recent five taxable years ending before the date on which the ownership or control of the corporation changed, or the portion of this period during which the individual was an employee of the corporation. *Annualized includable compensation* for the base period is the average annual compensation that was payable by the corporation and includable by the individual in gross income for the taxable years of the base period. The portion of the base amount applicable to a parachute payment is an amount that bears the same ratio to the entire base amount as the present value of the payment bears to the aggregate present value of all such payments.

The amount of any excess parachute payment is reduced by any part of a parachute payment that, by clear and convincing evidence, is shown to be a reasonable compensation for personal services actually rendered. For purposes of this rule, reasonable compensation is to be offset first against the base amount.

This provision applies to payments made under agreements entered into or renewed after June 14, 1984, in taxable years ending after that date. One of the real problems with respect to this provision is that, in looking at golden parachute payments, one must include all the income of the disqualified individual, including that attributable to stock options, deferred compensation, etc. The calculation is doubly important because if the parachute payment turns out to be an excess parachute payment, the 20 per cent excise tax is not only on the excess amount but on the entire amount of the parachute payment. This is a very onerous position and makes it very difficult to protect key personnel in the event of a takeover.

It seems to this writer that effective tax planning in this area requires that golden parachute contracts be carefully limited to avoid any excess parachute payment. Secondary contracts can be drafted with respect to a key executive who terminates employment for any reason other than change in control which could pick up such excess payments. As long as the two contracts are not in any way tied together, it may be possible to avoid the unfavorable tax treatment put in by the 1984 Act with respect to golden parachute payments. However, very careful drafting is going to be required.

§5.22 S Corporations (New)

The S corporation, which replaces in title the former "Subchapter S" corporation, constitutes innovative tax planning in and of itself. The regular corporation is now designated a C corporation. The S corporation comes into existence as a result of the Subchapter S Revision Act of 1982, Pub L No 97-354, 96 Stat 1669 (the Act).

The underlying philosophy of the Act allows a fully operating corporate structure to have the nontax benefits of that structure (such as limited liability), while having more of the tax aspects of a partnership. This creates potential innovative planning for many closely held enterprises. The S corporation bypasses corporate taxation, and its shareholders (like partners) have the full benefits of all losses, tax-exempt interest, charitable contributions, depletion and depreciation, etc. The tax characteristics of the corporate items pass through to and are retained in the hands of the shareholders.

In summary, the Act increases the number of shareholders of an S corporation from 25 to 35. It still requires that the corporation have only one class of stock, but now permits voting differences in that class. Straight debt is excluded. Under the old law, an election to be a Subchapter S corporation had to be made in the preceding tax year or on or before the first 75 days of the current tax year. Under the new law, IRC §§1362-1377, the election is to be made in the preceding tax year or on or before the 15th day of the third month of the current tax year. Under the old law, with respect to the effect on investment tax credit recapture, an election was treated as a disposition unless an agreement was signed under Treasury Reg §1.47-1(b). Under the new law, it is not a disposition but corporate liability in case of early disposition.

Before enactment of the revisions, all shareholders had to consent to revocation of an S election. Under the new provisions, owners of more than one-half of the shares must consent. Revocation formerly had to be made by the end of the first month of the tax year; under the new law, revocation by the 15th day of the third month of the tax year and revocation dated prospectively are now permitted. The S corporation's election will be terminated as of the date of revocation specified, as opposed to as of the beginning of the tax year, under former law.

Under prior law, there was an automatic termination of Subchapter S status if there was passive income (*i.e.*, from outside investment) amounting to 20 per cent of gross receipts for one tax year, effective retroactively. Under the new sections, there is automatic termination of S status if there is passive income amounting to 25 per cent of gross receipts for each of three consecutive tax years; the corporation must also have earnings and profits. Such a termination is effective on the first day after the end of the third tax year.

Under the old law, a new shareholder who affirmatively refused to consent within 60 days of becoming a shareholder would cause a termination of the election. There would also be a termination when foreign income equalled or exceeded 80 per cent of the gross receipts. There are no such provisions or restrictions under the new law. Also, under prior law there was no provision concerning inadvertent termination, such as happened when the corporation suddenly found itself with an excess of shareholders due to the death of a shareholder whose holdings devolved to many individuals, pushing the total number of shareholders to more than 25. Under the new law, the Internal Revenue Service may treat such as a nonterminating event. There were no provisions under prior law, upon termination, for the resulting short tax year. Under the new law, two short tax years are established—an S short tax year and a C short tax year, except for carryover items.

Under the old law, there was no tax on excessive passive income (although it meant loss of Subchapter S status). Currently, there is a tax on a portion of net passive income only where the corporation has C earnings and profits.

Formerly, undistributed taxable income was passed through to shareholders owning stock on the last day of the corporate tax year; long-term capital gain passed through on a per share, per day basis. Also, there was no loss carry-forward provision; loss in excess of basis in stock and debt was lost. Under the revised sections, each item that can affect liability of any shareholder is passed through separately to shareholders on a per share, per day basis. There is now also a carry-forward provision to a subsequent tax year.

Under the old law, stock basis was increased by share of undistributed taxable income and reduced by share of net operating loss and certain distributions. Under the Code as amended, it is increased by share of income and decreased by share of expenses, loss and certain distributions. S corporations under the new provisions can only elect a calendar year period and must get approval from the Internal Revenue Service for other business-related periods. Audit proceedings are conducted only on the corporate level and not on the shareholder level as under the old law.

Obviously, in situations where one wishes to avoid corporate tax and pass tax items through to the shareholders, the S corporation election is of enormous advantage. With respect to losses, it is usually advantageous to be able to pass losses down to shareholders. Under the new law, if a shareholder is unable to deduct the applicable share of the S corporation's loss for the current tax year that is in excess of his aggregate basis in the S corporation stock and in loans made to the corporation, the shareholder may now carry forward such losses until the tax year in which the shareholder's basis is restored, and at that time deduct such loss. This can be a substantial advantage. For additional input in this area, please see Grant, Subchapter S Taxation (Shepard's/McGraw-Hill, 1980 & Supp 1983).

There are certain other minor differences between the new and the old law, so that the entire law should be carefully studied by anyone contemplating using the S corporation for innovative tax planning. By 1984, TEFRA, Pub L No 97-248, 96 Stat 324, will have equalized qualified plan benefits between corporate and noncorporate entities so that such benefits for an S corporation will be substantially the same as those for a C corporation. With such possibilities available, the new S corporation should be added to the tools of the innovative tax planner.

An S corporation may be used instead of a partnership in a venture capital transaction with basically the same tax considerations as if the partnership were used. If there are no 5 per cent shareholders, IRC §341 collapsible corporation problems do not apply. However, because of step transaction problems and basis for internal debt problems, a partnership is more viable than a S corporation in the area of venture capital.

§5.23 Tax Considerations Relative to Venture Capital Transactions (New)

The tax aspects of venture capital programs shape the very form of the transaction. The first step in initiating a venture capital program is to determine the nature of the vehicle to be used, and the location of the vehicle. The choices include a general partnership, a limited partnership, a C corporation, an S corporation, or a joint venture; location means within the United States or without. Far and away the most popular form of venture capital vehicle is the domestic limited partnership.

A venture capital limited partnership is a pool of capital provided primarily by large corporations and institutions, as limited partners, managed by individual general partners who are experienced in making and managing venture capital investments. The partnership form is used because this form of business venture is not subject to federal or state income tax (but may be subject to various state, city, or county unincorporated business taxes). It also enables the venture capital fund managers to have complete autonomy from the providers of capital in the conduct of the fund's business, and limits the risk of the providers of capital to the investment they have committed to the fund. It facilitates the creation of a "carried interest" for the venture fund managers in the realized gains and capital appreciation of the fund's investments. *Carried interest* is the interest of professional employees of the partnership who are not partners, for sharing in the profits of one or more investments made by the partnership. Finally, it facilitates the use of certain exceptions or exemptions under federal and state securities law. It normally has a minimum life of from seven to ten years, during which the partners are not permitted to withdraw their capital and new partnership interests will not be issued.

Partnerships are not taxed as such unless they are characterized as "an association" taxable as a corporation. Four major characteristics typical of an association taxable as a corporation are:

1. Continuity of life
2. Centralization of management
3. Limited liability
4. Free transferability of interests

An entity organized as a valid limited partnership under state law will ordinarily be treated as a partnership for tax purposes provided it has not more than two of the above characteristics. IRC §7701; Reg §301.7701-2(b)(1). The partnership format is used primarily to obtain the flow-through of items of taxable income, gain, loss, deduction, and credit under IRC §701. For example, there is no recognition of gain on distributions in kind of appreciated securities from partnership to partners under IRC §731.

As a general rule, venture capital funds do not request private rulings as to status from the Internal Revenue Service. However, there is a procedure for doing so. Rev Proc 74-17, 1974-1 CB 438. This procedure provides that in order to obtain a favorable ruling on partnership status, the general partners of a limited partnership must at all times have at least 1 per cent interest in each material item of partnership income, gain, loss, deduction, and credit. This does not mean that the general partner must make a contribution equal to 1 per cent of the partnership's contributed capital, but it is customary for general partners to make substantial capital contributions. *See also* Rev Proc 75-16, 1975-1 CB 676. The imbalance of general versus limited partner capital contribution has raised the specter of the possibility of the treatment of income to a carried interest as ordinary income subject to tax under IRC §61, as opposed to characterization based upon the flow-through of a particular type of partnership income. Thus, any income to the carried interest would be deemed to be compensation to such interest, not partnership income. However, under present tax law, a carried interest should get the flow-through characterization benefit regardless of whether there is an allocation of such income that is disproportionate to historical capital contribution. Such special allocations of partnership income and losses are generally permitted but they will be denied if they do not have "substantial economic effect." IRC §704(b)(2).

In general, investments or contributions made by the limited partners will be paid in cash, though occasionally contributions will take the form of appreciated securities. No gain or loss is recognized to a partnership or any of its partners on a contribution of property or money to the partnership in exchange for a partnership interest. IRC §721. The partnership's basis for securities contributed by a partner is the same as the basis for such securities in the hands of the contributing partner. IRC §723. In order to give a contributing partner credit for the current value of appreciated securities at the time of contribution without penalizing the other partners for the built-in potential tax liability, the partnership agreement should provide that upon disposition of the contributed securities by the partnership, any gain recognized, to the extent of the difference between the partnership's basis and the value at the time of contribution, will be allocated to the contributing partner. IRC §704(c)(2).

It should be noted that IRC §721 contains an exception to the "nonrecognition of gain on the transfer of property to a partnership" rule if the partnership is deemed an investment company under IRC §851. This prevents the partnership from being a "swap fund," giving the partners diversification without tax recognition.

On distributions out of the partnership, there is nonrecognition of gain or loss to a partner upon receipt of securities from the partnership. IRC §731. Gain is recognized upon receipt of money in excess of the partner's adjusted basis (immediately before the distribution). IRC §731(a)(1). The partnership does not recognize gain or loss on distribution of property to a partner, including money. IRC §731(b).

A transfer of a capital interest in a partnership in exchange for services constitutes capital income. Reg §1.721-1(b)(1). However, a transfer of an interest in future profits in a partnership has historically *not* been treated as taxable income. On the other hand, a transfer of a future profits interest in respect of past services may constitute taxable ordinary income. *Sol Diamond,* 56 TC 530 (1971).

If a venture capital partnership uses borrowed funds, a portion of the income or gain from such investment may be treated as unrelated debt-financed income. Retirement plans and charitable organizations described in IRC §501(c) are taxed upon their unrelated business taxable income under IRC §512. Unrelated business taxable income includes unrelated debt-financed income under IRC §514. *Unrelated business taxable income* is that income derived from any trade or business the conduct of which is not substantially related (aside from the need of such organization for income or funds or the use it makes of the present funds derived) to the exercise or performance by such organization of its function constituting the basis for its tax exemption. *Unrelated debt-financed income* is income of an unrelated business derived from debt-financed property. *Debt-financed property* means any property which is held to produce income and with respect to which there is an acquisition indebtedness at any time during the taxable years, with some minor exceptions.

A tax-exempt partner may recognize taxable income on receipt of unrelated business taxable income which may arise from debt-financed property. IRC §512. If this potential exists, the partnership should refrain from borrowing. To avoid such a problem and still leverage the equity of the partners, a corporate subsidiary can be organized by the partnership, which is licensed as a small business investment company (SBIC) under the Small Business Investment Act of 1958. The SBIC is a separate and distinct entity from its stockholder partnership for tax purposes. Borrowings therefore would usually not be attributed to the tax-exempt limited partners of the partnership stockholder of the SBIC, and this should not give rise to unrelated business taxable income. However, the SBIC may have taxable income and only post-tax profits will flow to the partnership as dividends. The use of a SBIC not only avoids unrelated business income issues, but also allows for corporate tax benefits, pension plan, insurance, etc., within its own structure.

The Employment Retirement Security Act of 1974 (ERISA) makes qualified pension and profit-sharing plans poor limited partners in a venture capital fund because of a 1979 Labor Department regulation which defines the term *plan asset* in such a manner as to possibly include the venture capital fund investments as part thereof. Plan trustees would be personally liable as fiduciaries for the venture capital fund activities unless there is compliance with certain specifics. The regulation is not final, but must be watched.

No gain or loss is recognized either to the partnership or to any of its partners upon a contribution of property to the partnership in exchange for a partnership interest. IRC §721. However, a partner must recognize any gain realized on the transfer of appreciated property to a partnership that would be treated as an investment company if the partnership has been incorporated. IRC §721(b). The basis of an interest in a partnership acquired by a contribution of property, including money, to the partnership is the amount of such money and the adjusted basis of such property in the hands of the contributing partner at the time of the contribution, increased by the amount (if any) of gain recognized to the contributing partner at such time. IRC §722. The basis of property contributed to a partnership by a partner shall be the adjusted basis of such property to the contributing partner at the time of the contribution, increased by the amount (if any) of gain recognized to the contributing partner at such time. IRC §723.

If the contributed property is subject to debt or if liabilities of the partner are assumed by the partnership, the basis of the contributing partner's interest is reduced by the portion of the indebtedness assumed by the other partners. Such assumption of the partner's debt by others is treated as a distribution of money to him, and is treated by the others who assumed the debt as a contribution of money by them. Reg §1.722-1.

The basis of property received in a distribution, other than in liquidation of a partner's interest, will ordinarily be the same as the basis in the hands of the partnership immediately prior to distribution.

In no case may the basis of property in the hands of the distributee exceed the basis of his partnership interest reduced by the amount of money distributed to him in the same transaction. IRC §732(a). The basis of property distributed in liquidation of a partner's interest is the basis of the distributee's partnership interest less any money received in the same transaction. IRC §732(b). On liquidation, there is a gain or loss only on money, not on securities. IRC §731(a).

If the basis of distributive assets in the hands of the distributee partner is less than the basis of the assets in the hands of the partnership, there may be an *unused basis.* The basis is said to be unused because no loss is allowed on distribution of the assets to the partner. The Internal Revenue Code permits the partnership to elect to adjust the basis of its remaining assets to take up this unused basis. IRC §§734, 754, and 755.

If gain is recognized to a partner because of a distribution of money, a similar increase in the partnership's basis of its remaining assets may be made by election. If an election has been made, the partnership may have to decrease the basis of its remaining assets in some cases. The decrease would be required for the excess of the basis of distributed assets to the partner over the basis which the partnership had for those assets, in the event of a distribution in liquidation of a partner's interest. Decrease would also be required to the extent that any distribution to a partner resulted in loss to the partner. IRC §734(b)(2).

Finally, there is an election for adjustment of basis when there is a change in the membership of the partnership. IRC §743. Elections are made in a written statement filed with the partnership return for the first year to be covered by the election. Reg §1.754-1.

Syndication fees are nondeductible. Organization fees and other start-up costs are not expensed but must be capitalized. IRC §§195 and 709. Organization expenses can be amortized over at least 60 months (syndication fees may not).

The fiscal year of the partnership ordinarily must be the same as the fiscal year of the principal partners. IRC §706(b)(1). A principal partner, for this purpose, is a partner having an interest of 5 per cent or more of partnership profits *or* capital. However, a newly formed partnership may obtain Internal Revenue Service consent to adopt a fiscal year ending as much as three months prior to the end of the fiscal years of the principal partners. Rev Proc 72-51, 1982-2 CB 832.

Any fixed or guaranteed payments made to partners for services or for the use of capital are generally treated as though paid to an outsider for purposes of computation of gross income and business expense deductions. IRC §707(c). Such payments will be taxed as ordinary income to the partners.

The holding period of an interest in a partnership is measured from the date the partner acquired his interest, not from the date of acquisition by the partnership of specific partnership assets. The holding period of property received from a partnership includes the period during which it was held by the partnership. IRC §735(b).

The use of corporate form rather than limited partnership is preferred when the objective is long-term capital appreciation and *not* current income. In such event, there are tax consequences at the corporate level and gain at the shareholder level, and there is a tax problem potential with the unreasonable accumulation of income in the corporation.

§5.24 Off-Shore Investment Company (New)

The primary purpose of an off-shore investment company is the deferral of tax on trading gains and possibly the conversion of short-term capital gain into long-term capital gain. The off-shore investment company is generally organized under the laws of a tax-haven jurisdiction (i.e., one that does not have an income, gift, estate, or transfer tax), such as the Cayman Islands. The company that is so established is usually a holding company, which then establishes an off-shore investment company for the actual trading in stocks, bonds, and commodity futures. Earnings are ultimately repatriated in the form of distributions and redemption of shares, in respect of which gain is hopefully accorded capital treatment.

Usually, these companies have many diverse shareholders, thus avoiding the personal holding company penalty tax with respect to any United States source income. Foreign investment companies are subject to tax on certain income from United States sources. For example, dividends paid by a domestic corporation, other than a corporation less than 20 per cent of whose gross income is derived from United States sources, are treated as United States source income. IRC §861(a)(2)(A). Dividends paid by a foreign corporation, 50 per cent or more of whose gross income is effectively connected with a United States trade or business, are treated as United States source income, in an amount which bears the same ratio to such dividends as the effectively connected income bears to the gross income of the corporation from all sources. IRC §861(a)(2)(B).

Interest paid by United States residents is sourced in the United States. IRC §861(a). The term *United States resident* would include a foreign corporation or partnership engaged in a United States trade or business. There are four exceptions to this rule:

1. Interest paid by a domestic corporation less than 20 per cent of whose income is from United States sources
2. Interest received from a foreign corporation less than 50 per cent of whose income is effectively connected with a United States trade or business; if 50 per cent or more of such corporation's income is effectively connected, only that portion of interest paid which bears the same ratio to such interest as the effectively

connected income bears to all gross income will be treated as from United States sources

3. Bank interest, unless such interest is effectively connected with the recipient's United States trade or business
4. Original issue discount on treasury bills with a maturity of six months or less, which will be excluded from gross income of the foreign investment company

Under the general scheme of United States taxation, a foreign corporation is taxable, in the absence of treaty protection, at the rate of 30 per cent on its United States source fixed or determinable income, and at graduated rates on its income which is effectively connected with the conduct of its United States trade or business.

Unless a foreign corporation is engaged in a United States trade or business, United States source income, other than fixed or determinable income, is not taxable. Therefore, unless the activities of the investment holding company and its operating subsidiary constitute the conduct of a United States trade or business, any gain realized from its trading activities (other than incidental United States source dividend and interest income) will not be subject to United States taxation.

Trading in stocks or securities through a resident broker, commission agent, or other independent agent does not constitute the conduct of a United States trade or business provided the trader has no fixed place of business in the United States. IRC §864(b)(2)(A)(i). A similar rule is provided with respect to trading in commodities of a kind customarily dealt in on an organized exchange if the transaction is of a kind customarily consummated at such place ("traded on an established exchange"). IRC §864(b)(2)(i)(iii).

With respect to the criteria applied for purposes of determining whether a foreign corporation has a fixed place of business ("principal office"), the Code regulations provide that the corporation is considered to have only one principal office; a corporation will not be treated as having its principal office in the United States if all or a substantial portion of all of the following functions are performed at an office or offices outside the United States:

1. Communicating with its shareholders (including the furnishing of financial reports)
2. Communicating with the general public
3. Soliciting sales of its own stock
4. Accepting the subscriptions of new stockholders
5. Maintaining its principal corporate records and books of account

6. Auditing its books of account
7. Disbursing payments of dividends, legal fees, accounting fees, and officers' and directors' salaries
8. Publishing or furnishing the offering and redemption price of the shares of stock issued by it
9. Conducting meetings of its shareholders and board of directors
10. Making redemptions of its own stock

Reg §1.964-2(c)(2)(iii)

Two of the pitfalls in off-shore dealing include foreign personal holding company characterization and controlled foreign corporation characterization. A United States shareholder is subject to tax on his pro rata share of undistributed income (after taxes), as well as on distributed income of a foreign personal holding company. IRC §551. A United States shareholder of a controlled foreign corporation is taxed on certain of the corporation's undistributed earnings (subpart F income) as well as its distributed earnings. A foreign corporation will meet the stock ownership requirement for treatment as a foreign personal holding company if, at any time during the taxable year, more than 50 per cent in value of its stock is owned directly or indirectly by five or fewer individuals who are citizens or residents of the United States. IRC §551(a)(2).

A foreign corporation will meet the stock ownership test for treatment as a controlled foreign corporation when more than 50 per cent of the total combined voting power of all classes of stock entitled to vote is owned by United States shareholders (including corporations), each owning 10 per cent or more of the voting stock of the corporation. IRC §957(a).

As indicated above, the gross income of a foreign corporation is generally limited to United States source and effectively connected income. IRC §882(b). With respect to a corporation not engaged in a United States trade or business, the regulations specify that gross income consists of only United States source income (fixed or determinable income). In computing taxable income of a foreign corporation, deductions are only available to the extent allocable to effectively connected income, and are allowable only if a return is filed.

In most instances, a disposition of an investor of his shares in an off-shore investment company is tax-effective if through a share redemption, resulting either in a complete termination of the shareholder's interest or in a substantially disproportionate redemption, in that the percentage of the company's outstanding shares directly or constructively held by the shareholder, immediately after the redemption, is 80 per cent or less of such shareholder's percentage interest immediately prior to the redemption and less than 50 per cent of the total outstanding shares. Assuming that the corporation is not a controlled foreign corporation, any gain realized on the share redemption should be accorded capital treatment, provided that the corporation is not a foreign investment company within the meaning of IRC §1246. That definition includes a foreign corporation which is registered under the Investment Company Act of 1940, as amended, either as a management company or as a unit investment trust or engaged primarily in the business of investing, reinvesting, or trading securities at a time when more than 50 per cent of the total combined voting power of all classes of stock entitled to vote, or the total value of shares of all classes of stock, was held, directly or indirectly by United States persons.

To avoid classification as a foreign investment company, the corporation must structure its holdings in such a fashion that it will not be considered an investor in securities. The level of control stock owned, the degree of participation, and the frequency of portfolio trading will all influence such a determination. Any corporation engaged in commodities trading should not be considered as meeting the business activities test. Major ownership by a foreign individual or corporation will avoid the 50 per cent test.

In the event it is determined that the off-shore company is a foreign investment company within the meaning of IRC §1246(b), any gain realized upon a sale or exchange of shares, a distribution and redemption accorded sales treatment, or a distribution in liquidation will be treated as ordinary income to the extent of the shareholders' share of earnings and profits accumulated for the taxable years beginning after December 31, 1962. The ordinary income treatment applies to all United States shareholders (irrespective of the nature of the entity owning the shares) disposing of stock in a foreign corporation which was a foreign investment company at any time during the period the shareholder held the stock. For purposes of the collapsible partnership provision of IRC §751, pursuant to which gain realized on the sale of a partnership interest or the receipt of a disproportionate distribution from a partnership is accorded ordinary treatment, shares of a foreign investment company are treated as inventory items. IRC §751(d)(2)(C).

Another problem is the application of the accumulated earning tax. The tax on accumulated earnings under IRC §531 is applicable to foreign corporations which are formed or availed of for the purpose of avoiding the income tax with respect to its shareholders by allowing earnings and profits to accumulate rather than be distributed to such shareholders. An unreasonable accumulation of earnings or the existence of a mere investment company is sufficient to evidence the purpose to avoid income tax. The tax is applicable to foreign corporations, with respect to any income derived from sources within the United States, if any of the shareholders are subject to income tax on distributions by the corporations. The accumulated earnings tax will apply to all United States source income of the corporation even though less than all the stock is owned by persons subject to tax upon distribution. The minimum accumulated earnings credit is $250,000. That amount is the maximum credit for an investment company. An off-shore company should be able to avoid the accumulated earnings tax by distributing its profits to a holding company and filing a United States tax return for informational purposes showing no United States tax liability. The holding company should not be subject to the accumulated earnings tax since it is not engaged in a trade or business in the United States, and the dividend would constitute foreign source income not effectively connected with a United States trade or business, which is the reason for having the holding company parent and the operating company subsidiary, as originally outlined.

From the individual shareholder's point of view, United States citizens will be taxed on all income from whatever source derived, but the character of the income will vary. Dividends received by United States individual shareholders will be taxed regardless of the source of the dividend. In the case of United States shareholders, sale or exchange of stock of an off-shore investment company will generally be treated under the capital asset rule. With respect to redemption, the rules of IRC §302(b)(3) relating to a termination of the shareholder's interest will ordinarily control in situations where an off-shore investment corporation redeems all of the stock of any shareholder. If less than all of the shares are redeemed, or if the redemption does not qualify under IRC §302(b)(3), the transaction will be treated as a sale or exchange rather than a dividend, if the other IRC §302(b) requirements are met. Thus, one may see that if the managers of a venture capital fund intend to continue managing the fund after their shares are redeemed, the off-shore fund may not be the way to go for them because their continued management violates the termination of interest rules under IRC §302(c)(3).

§5.25 Corporate Stock Options (New)

The 1984 Tax Reform Act, Pub L No 98-369, §57, provides that a corporation does not recognize gain or loss upon the lapse or acquisition of an option to buy or sell its stock (including treasury stock). This provision is effective with respect to options required or lapses after the date of enactment in taxable years ending after that date. Previously, a corporation recognized no gain or loss on the receipt of money or other property in exchange for its stock, and did not recognize gain or loss upon redemption of its stock with cash for an amount different from that received upon the issuance of the stock.

Under IRC §61(a)(12), income from the discharge of indebtedness is realized when a taxpayer's debts are satisfied at less than their face value. However, courts have held that a corporation does not realize such income upon the satisfaction of its debts with its own stock, even though the value of the stock is less than the amount of the debt. The Tax Reform Act of 1984 generally eliminates the "stock for debt" exception to the rules relating to income from discharge of indebtedness. Under said Act, a corporation that satisfies a debt by transferring corporate stock to its creditor is treated as if it had paid the creditor with money equal to the fair market value of the stock. Thus, the corporation will have income from discharge of indebtedness to the extent that the principal of the debt exceeds the value of the stock (and any other property transferred). The provision does not apply to insolvent corporations or corporations that are debtors in Title 11 (bankruptcy) cases.

This provision is generally effective for transfers made after the date of enactment and taxable years ending after such date. There are special transitional rules and workout rules which provide an exception. There are also two exceptions with respect to transfers to certain controlling shareholders and transfers pursuant to debt restructure agreements. Absent the availability of these exceptions for a corporation using its own stock to discharge debt, the corporations will now be faced with the realization of income therefrom. Accordingly, this avenue of tax strategy has been effectively shut down by the 1984 Act.

§5.26 The Foreign Sales Corporation (New)

The system of Domestic International Sales Corporations (DISC) generally will be replaced after December 31, 1984, with a new system of Foreign Sales Corporations (FSC) as a result of the 1984 Act, Pub L No 98-369, §801. Under the FSC system, a portion of the foreign trade income of an FSC will be exempt from tax at the corporate level, provided it is derived from the foreign presence and economic activity of the FSC. In contrast, under the DISC system, there was no corporate income tax imposed on DISC income, and there was a partial deferral of taxes at the shareholders' level. Although DISCs have not been abolished by the 1984 Act, their tax benefits are limited and an interest charge for tax-deferred amounts is imposed on DISC shareholders.

Under the FSC provisions, a small business meeting certain requirements may elect to be treated as a "small FSC" or to remain a DISC.

To qualify as an FSC, a corporation must meet six requirements designated to insure that it has adequate foreign presence. If a corporation meets all six requirements and makes an election that complies with the procedural requirements of IRC §926(f)(i), it will be treated as an FSC. The first requirement for FSC status is that the corporation be a foreign corporation created or organized under the laws of a qualified foreign country. To be so qualified, at the time of the organization or creation of the FSC, the United States must have had in effect with the country of incorporation, a bilateral or multilateral agreement, or an income tax treaty with respect to which the secretary certifies that the exchange of information is adequate to prevent tax avoidance. However, the foreign country cannot be a country that has a statute (or other stated policy) which denies the

Internal Revenue Service access to the home office of the FSC for audit purposes.

A corporation will be considered to be a foreign corporation if organized under the laws of any United States possession. For the purposes of this Act, Puerto Rico is considered to be part of the United States.

The second requirement for FSC status is that the corporation not have more than 25 shareholders.

The third requirement is that the FSC not have any preferred stock. There can be more than one class of common stock, but the second or third class cannot look like or in effect be a preferred stock.

The fourth requirement is that the corporation contain an office located outside the United States (or any United States possession) at which there is a permanent set of tax records, including invoices. Additionally, the FSC is required to maintain, at a location within the United States or in any possession of the United States, permanent books of account or records which are sufficient to establish the amount of gross income and deductions, credits, and other matters required to be shown on any return.

The fifth FSC requirement is that the board of directors of the FSC always include at least one individual who is not a resident of the United States. The nonresident member may, however, be a United States citizen, according to committee reports.

Under the sixth FSC requirement, an FSC cannot be a member of any control group of corporations of which a DISC is a member. For purposes of the FSC rules, the term *control corporation* has the meaning given in IRC §1563(a), except that the percentage of control needed is 50 per cent instead of 80 per cent.

Once the above six requirements are met, a portion of the foreign trade income of the FSC is treated as foreign source income, which is not effectively connected with the conduct of a trade or business within the United States; as such, it is exempt from United States income tax. *Foreign trade income* is defined as gross income derived from foreign trading gross receipts.

The portion of an FSC's foreign trade income that is exempt depends on whether the amount of foreign trade income is computed under (1) the general corporate rules relating to transactions between unrelated parties or the IRC §482 rules for arm's-length pricing and transactions between related parties, or (2) the optional administrative pricing rules. If the income is computed under category 1, 32 per cent of the foreign trade income so derived is exempt foreign trade income. However, such income cannot be derived from intangibles which do not constitute export property. The remaining 68 per cent of foreign trade income from such transactions is not exempt income that is treated as taxable foreign source income.

If an FSC's foreign trade income is determined under administrative pricing rules, 16/23 of this foreign trade income is exempt income. Nonexempt income is treated as United States source income.

Deductions of an FSC related to foreign trade income earned from a transaction are to be allocated proportionately between the exempt and the nonexempt foreign trade income derived from such transaction. Because exempt foreign trade income is not taxed by the United States, no credits are allowed to any FSC other than credits allowable under IRC §§27(a), 33, and 34.

Generally, foreign trade income other than exempt foreign trade income is treated as income effectively connected with the conduct of a trade or business through a permanent establishment in the United States. In addition to such nonexempt income, interest, dividends, royalties, and other investment income received or accrued and carrying charges received or accrued are treated in the hands of an FSC as income effectively connected with the trade or business in the United States. Further, nonexempt foreign trade income generally is treated as United States source income. Therefore, nonexempt foreign trade income, investment income, and carrying charges will be taxed currently and nonexempt foreign trade income will be treated as United States source income for purposes of the foreign tax credit limitation. However, nonexempt foreign trade income earned in a transaction with an unrelated party, or determined under the arm's-length pricing method, is not deemed to be United States income. Rather, the source and taxation of such income will be determined in the same manner as under existing law.

Foreign trading gross receipts, from which foreign trading income (and hence, exempt foreign trading income) is derived, generally are attributable to the export of certain goods and services. In order for receipts to be classified as foreign trading gross receipts, an FSC must meet foreign management and foreign economic process requirements. Thus, foreign trading gross receipts can be derived only from the following five types of activities: (1) the sale, exchange, or other disposition of export properties; (2) the lease or rental of export property for use by the lessee outside the United States; (3) services which are related and subsidiary to any sale, exchange, or other disposition of export property by the FSC or any lease or rental of export rental for use by the lessee outside the United States; (4) engineering or architectural activities for construction projects located outside the United States; and (5) the performance of managerial services for an unrelated FSC or DISC in furtherance of the production of foreign trading gross receipts of the first three types. However, at least 50 per cent of the gross receipts of the FSC must be derived from the first three types of exporting activity before income from managerial services provided to an unrelated FSC produces foreign trading gross receipts.

There are two categories of receipts that do not constitute foreign trading gross receipts under the Act. They are certain receipts from transactions between related parties and investment income and carrying charges.

An FSC has foreign trading gross receipts only if its management takes place outside of the United States. An FSC meets this test only if it fills the following three requirements during its taxable year: (1) all meetings of its board of directors and its shareholders are outside the United States; (2) its principal bank account is maintained outside the United States at all times; and (3) all dividends, legal and accounting fees, and salaries for officers and members of the board are disbursed out of bank accounts maintained outside the United States.

In order for receipts to qualify as foreign gross trading receipts from a transaction, an FSC or its agent must have participated outside the United States in the solicitation (other than advertising), the negotiation, or the making of the contract relating to the transaction. In addition, either foreign direct costs incurred by the FSC must equal or exceed 50 per cent of the total direct cost incurred in making the transaction, or if the direct foreign cost incurred by the FSC in the transaction is equal to or exceeds 85 per cent of the costs attributable to any two of the five activities relating to the disposition of export properties (as explained above), then the foreign economic process requirement for the FSC is met. For purposes of the direct cost tests, there are five categories of economic activities relating to the disposition of export property, which include advertising and sales promotion; processing of customer orders and arranging for delivery of the export properties; transportation from the time of acquisition by the FSC to the time of delivery to the customer; determination and transmittal of a final invoice or statement of account and the receipt of payment; and the assumption of credit risk.

Export property must be manufactured, produced, grown, or extracted in the United States by a person other than an FSC. It must be held primarily for sale, lease, or rental in the ordinary course of a trade or business, by or to an FSC, for direct use, consumption, or disposition outside the United States; and it must not have more than 50 per cent of the fair market value of the property attributable to articles imported into the United States.

A corporation can choose to be treated as a small FSC if two requirements are met. First, the corporation must make an election to be so treated; second, if the corporation is a member of a control group of corporations that include an FSC, the corporation will not be treated as a small FSC unless the other FSC also elects to be a small FSC. The principal advantage of the small FSC election is that foreign management and economic process requirements do not apply to small FSCs. A disadvantage of small FSC status is that a small FSC's foreign trading gross receipts in excess of $5 million will not be taken into account in determining the amount of its exempt foreign trade income. If this $5 million limitation is exceeded, the corporation may decide which receipts will fall within the limitation.

There are specific arm's-length pricing rules as well as other rules related to the operation of FSCs. The most important rule for purposes of this paragraph is that relating to distribution to shareholders. Any distribution by an FSC or former FSC to its shareholders from earnings and profits is to be treated as made first out of profits from foreign trade income and then out of any other earnings and profits. Distributions made from profits attributable to foreign trade income are treated as income effectively connected with the conduct of a trade or business in the United States, and as United States source income in the case of any such distribution to a shareholder of an FSC or former FSC that is a foreign corporation or nonresident alien individual. This will subject such income to United States income tax.

Thus, there is now a new modus operandi for doing business abroad. Although the 1984 Tax Reform Act replaces the DISC system with the FSC system, the DISC is not abolished per se. Instead, the Act makes them all interest-deferred DISCs. Thus, a shareholder of a DISC must pay interest on the amount of the shareholder's DISC-related deferred tax liability. This interest is due at the same time as is the shareholder's regular tax, and there is a penalty for late payment. Taxable income of a DISC attributable to qualified export receipts which exceed $10 million is deemed attributed as a taxable dividend. In most instances, it will probably be wise to terminate a DISC and replace it with either an FSC or a small FSC for doing business abroad. Transitional rules on the Tax Reform Act of 1984 make it relatively easy to move from DISC status to FSC status.

§5.27 Employee Fringe Benefits (New)

The 1984 Tax Reform Act, Pub L No 98-369, §531(b), also limited a corporation's ability to compensate employees via the fringe benefit method. Effective January 1, 1985, an employee's gross income will not, for federal income tax purposes, include any fringe benefit which qualifies as a "no additional cost service," qualified employee discount, working condition fringe or de minimis fringe. Qualified tuition reductions are also excluded from an employee's gross income for federal income tax purposes with respect to education furnished after June 30, 1985.

No additional cost services, qualified employee discounts, working condition fringes, de minimis fringes, and qualified tuition reductions are not only excluded from an employee's gross income, but also from his or her wage base for purposes of income tax withholding, FICA, FUTA, and RRTA.

Employers providing fringe benefits must meet certain nondiscrimination requirements in order for such benefits to apply to favored groups. Officers, owners, and highly compensated employees cannot exclude these benefits unless such fringe benefits are available on substantially the same basis to each person in a "group" of employees which is defined under a reasonable classification set up by the employer that does not discriminate in favor of officers, owners, or highly compensated employees. The nondiscrimination rule does not apply to working condition fringe benefits or de minimis fringes (other than subsidized eating facilities).

The value of a no additional cost service provided to employees or their spouses or dependent children by employers is excluded from tax provided that it is granted to the employee as part of the line of business of the employer. This means that the service provided to the employee must be one that is offered for sale to customers in the ordinary course of the line of business of the employer for which the employee is working. However, there can be reciprocal arrangements between or among employers which would still qualify for the exclusion from taxable income. With respect to qualified employee discounts, the discount can be excluded only to the extent that such discount does not exceed gross profit percentage of the price at which the property is offered by the employer to customers. In the case of a qualified service, the excludable amount cannot exceed 20 per cent of the price at which the service is offered by the employer to nonemployee customers in the ordinary course of business.

All other fringe benefits will now be taxed to employees. The Treasury Department has been given regulatory authority to provide the time and manner for collection or payment by an employer of the employment taxes on nonexcludable, noncash fringe benefits. This will mean more recordkeeping requirements for employers and probably less fringe benefits for employees. There seems no doubt that corporations will have to start increasing wages for employees to make up for the loss of various incentive and fringe benefits previously exempt from taxation. This will work out, provided that there is tax simplification and possibly a flat tax rate in the future for individuals.

§5.28 Time Value of Money (New)

The Tax Reform Act of 1984, Pub L No 98-369, §91(a), (g), for the first time puts in writing recognition of the time value of money. Prior to the Act, under the "all events test" of Regulation §1.461-1(a)(2), an accrual basis taxpayer was generally entitled to deduct the face amount and accrued expense in the taxable year in which (1) all of the events have occurred that determine the fact of liability and (2) the amount of the liability can be determined with reasonable accuracy. This test was interpreted by courts to permit the current deduction of expenses that are related to activities to be performed or amounts to be paid in future years. The current deduction of future expenses resulted in an overstated deduction to the extent that the time value of money was not taken into account. Thus, a corporation that claimed a current deduction for the undiscounted amount of an expense that it is obligated to pay in a future tax year has overstated the deduction by the amount it exceeds the present value of the future expense.

The 1984 Act modified the all events test in a way that prevents such overstatement of advance deductions. Generally effective on the day after the date of enactment, all of the events that establish liability for an amount, for the purpose of determining whether such amount has been incurred with respect to any item, are treated as not occurring any earlier than the time that economic performance occurred with respect to that item. In the case of a liability of the taxpayer that requires a payment for property or services, economic performance is deemed to occur as the property or services are provided to the taxpayer. The liability arises out of the taxpayer's use of property; economic performance occurs as the taxpayer uses the property. If the liability of the taxpayer requires him or her to provide services or property, then economic performance occurs as the taxpayer provides the service or property. There is an exception for certain reoccurring items.

Similarly, with respect to the time value of money, the 1984 Act, in §41(b), has in effect eliminated low interest and interest-free loans. It has done so by imputing an interest rate dependent upon the nature and term of the loan. Loans of $10,000 or less are considered de minimis and are excluded from this, but all other loans are caught in the net. Thus, tax planning in the future, with respect to industrial innovation, will require an appreciation of the time value of money.

Innovation and Securities Law

Richard H. Rowe

6

Introductory Materials

§6.05 —Venture Capital

Add to note 4, page 151:

1983 Guide to Venture Capital Sources.

§6.06 —Partnerships

In note 6, page 152, second line, delete "__ US __, 50 USLW 3278 (Oct 13, 1981)" and insert:

454 US 897 (1981)

Add to note 7, page 152:

Potts, *PRC, E.F. Hutton Join to Offer New High-Tech Tax Shelter Plan,* Wash Bus, Wash Post, June 27, 1983, at 3, col 1; *Research and Development Limited Partnerships,* 38 Bus Law 1917 (Aug 1983); Levinson, *General Partnership Interests and the Securities Act of 1933: Recent Judicial Development,* 10 Ohio NUL Rev 463 (1983).

Page 153, delete last two sentences of carryover paragraph and substitute:

Accounting and financial reporting requirements applicable to public companies involved in these types of arrangements could limit their use in some circumstances. Financial Accounting Standards Board, Statement No 68, *Research and Development Arrangements* (1982); *see also* Securities Act of 1933 Release No 6476 (Aug 8, 1983) (SEC rules for accounting for internal costs of developing computer software for sale or lease to others). To the extent that accounting and financial reporting requirements, such as those discussed above and those requiring expensing of research and development expenditures, inhibit research and development, they should be reconsidered. Permitting capitalization of research and development expenditures and their amortization over an appropriate period, along with adequate disclosure, might encourage greater expenditure on research and development. This should not affect the tax treatment of such expenditures (see **ch 5**).

§6.08 —Leasing

Page 154, at end of section add:

The recent OPM debacle evidences many of the securities laws and other legal pitfalls which can be present in lease financing arrangements. *See generally Crocker National Bank v Rockwell International Corp,* 555 F Supp 47 (ND Cal), Fed Sec L Rep (CCH) ¶99,006 (ND Cal Oct 22, 1982) (alleged violations of the Racketeer Influenced and Corrupt Organizations Act, 18 USC §§1961-68 by computer leasing firm); *Crocker National Bank v Rockwell International Corp,* 555 F Supp 47 (ND Cal 1982), *later proceeding,* Fed Sec L Rep (CCH) ¶99,007 (ND Cal Nov 19, 1982) (nonrecourse notes and security agreements offered by computer leasing firm together constituted a security).

§6.09 —Profit Sharing

Page 154, at end of section add:

The SEC staff has become concerned about accounting for transactions between parents and subsidiaries involved in business combinations and corporate restructurings, including carve-outs and spin-offs, techniques sometimes used by employees to acquire technology which a large company does not desire to develop.

Where a company purchases substantially all of the common stock of another company, the staff has taken the position that a new basis of accounting is established for the purchased assets and liabilities, and the acquiring company's acquisition costs should be "pushed down" to be used to establish a new basis for accounting in the acquired company's separate financial statements. Staff Accounting Bulletin Release No 54 (Nov 3, 1983). The staff will not insist or "push down" accounting where the acquired company has publicly held debt or preferred stock or there is a substantial minority interest in that company.

The staff also takes the position that in a spin-off or carve-out transaction, the income statement of the spun-off or carved-out entity should reflect all of its costs of doing business, including expenses incurred by its parent on its behalf, such as salaries, rent, depreciation, advertising, professional services and selling, and general and administrative expenses. Staff Accounting Bulletin Release No 55 (Nov 30, 1983). Pro forma data also may be required as to terminated or revised cost-sharing agreements, such as tax-sharing agreements. *See also In re Pro-Mation, Inc*, Securities Act of 1933, Release No 6522 (Mar 30, 1933) (stop order proceeding relating to a company which spun off operations with heavy research and developmental expenses; former owner forced by SEC to reverse interest on loans to new company and write down receivables from new company when that company experienced serious financial difficulties).

§6.13 —Public Offerings

In note 10, page 155, third line, delete "(July 1981)" and substitute:

(June 1983)

Add to note 10, page 155:

Coleman & Seldin, A Businessman's Guide to Capital-Raising Under the Securities Laws (Packard Press 1982); Klein & Reese, Taking High Tech Companies Public (Packard Press Dec 1983).

§6.14 —Industrial Development Bonds

Page 156, at end of section add:

For example, the Tax Equity and Fiscal Responsibility Act of 1982, Pub L No 97-248, 96 Stat 324, 466, (1982), repealed tax exemptions for small industrial revenue bond issues issued after 1986, and the Mortgage Subsidy Bond Tax Act of 1980, Pub L No 96-499, 94 Stat 2660 (1980), limits the use of bonds to finance certain residential housing. The Deficit Reduction Act of 1984, Pub L No 98-369, Title VI, would impose further limitations and might retroactively eliminate the exemption from registration under §3(a)(2) of the 1933 Act, 15 USC §77c(a)(2), for certain bonds. The SEC has indicated that bonds issued before the enactment of the statute but after the retroactive effective date of the provisions in question, generally Oct 31, 1983, would not have been issued in violation of the 1933 Act. *See, e.g.*, New Jersey Economic Development Authority (SEC no action letter available Mar 1, 1984).

§6.15 —Offshore Financing

Page 157, at end of section add:

The increasing internationalization of the securities markets should enhance the ability of United States companies to raise capital abroad and the SEC appears to be disposed toward facilitating the process. See **§6.33.** However, these developments may have a significant impact on the SEC's ability to enforce the United States securities laws. See **§6.66.**

Applicability of the Securities Laws

§6.16 Overview

Page 158, after carryover sentence, add:

The Secondary Mortgage Market Enhancement Act of 1984, Pub L No 98-440, 98 Stat 1689 (to be codified in scattered sections of 12 USC and 15 USC) preempts certain state registration requirements for certain mortgage-backed securities.

The Federal Trade Commission has sought approval of the Office of Management and Budget, under the Paperwork Reduction Act (44 USC §3501 *et seq*), to conduct a study of blue sky securities laws and survey of state registration of securities for the purpose of evaluating the effect of different types of state securities regulations, particularly on smaller businesses. 45 Fed Reg 9580 (1983).

Add to note 22, page 158:

The North American Securities Administrators Association has adopted policy guidelines for a uniform limited offering exemption based, in part, on the SEC's Regulation D. NASAA Uniform Limited Offering Exemption, 1 Blue Sky L Rep (CCH) ¶5294A. These policy guidelines differ in several substantial respects from Regulation D and are not binding upon the states, in any event. The SEC's Government-Business Forum on Small Business Capital Formation, Sept 23-25, 1982, Arlington, Virginia, organized pursuant to §19(c)(4) of the 1933 Act, 15 USC §77s(c)(4), recommended that the states adopt uniform exemptions from registration of securities under state law which parallel exemptions under federal law. *Final Report,* SEC Government-Business Forum on Small Business Capital Formation (Nov 1982), at 60. The Forum did not recommend federal preemption of state securities laws, although it was considered. The Forum's recommendations are under consideration by the SEC for purposes of reporting to Congress. The SEC together with the North American Securities Administrators Association also is seeking public comment on means for increasing uniformity in securities regulation. See Securities Act of 1933 Release No 6474 (July 22, 1983).

Add to note 23, page 159:

SEC Government-Business Forum on Small Business Capital Formation, Sept 23-25, 1982, Arlington, Virginia; Final Report, SEC Government-Business Forum on Small Business Capital Formation (Nov 1982). The SEC held local forums on small business capital formation during 1983 in Washington, D.C., Houston, Texas, St. Louis, Missouri, San Francisco, California, and Boston. The SEC Today 4 (May 9, 1983).

The SEC also publishes a rulemaking agenda for rules which might affect smaller businesses. *See, e.g.*, Securities Act of 1933 Release No 6567 (Feb 28, 1985).

Add to note 24, page 159:

There are also texts on the subject. *See, e.g.*, Fass & Wittner, Blue Sky Practice for Public and Private Limited Offerings (1985); Long, 1985 Blue Sky Law Handbook.

§6.17 Is a Security Involved?

Add to note 26, page 159:

Schneider, *Definition of a "Security"—1983-1984 Update,* 17 Rev Sec Reg 851 (Sept 12, 1984). In addition, the Business Lawyer publishes an annual review of case law in this area. *See, e.g.*, Subcommittee on Annual Review, *Federal Securities Regulation (Case Law Section)*, 40 Bus Law 159 (Nov 1984).

§6.18 —Commercial Loans

In note 33, page 161, third line, delete "__ US __, 50 USLW 3486 (Dec 15, 1981)." and insert:

454 US 1124 (1981);

In note 33, page 161, add at end of first sentence:

Crocker Natl Bank v Rockwell Intl Corp, 555 F Supp 47 (ND Cal), *later proceeding,* Fed Sec L Rep (CCH) ¶99,007 (ND Cal Nov 19, 1982) (nonrecourse notes and security agreements offered by computer leasing firm together constitute a security).

In note 33, page 161, fifth line, delete "__ US __, 50 USLW 4285 (Mar 8, 1982)" and insert:

455 US 551 (1982)

In note 33, page 161, seventh and eighth lines, delete "__ US __, 50 USLW 3668 (Feb 23, 1982)," and insert:

455 US 939 (1982);

Add to note 34, page 161:

But see Crocker Natl Bank v Rockwell Intl Corp, 555 F Supp 47 (ND Cal), *later proceeding,* Fed Sec L Rep (CCH) ¶99,007 (ND Cal Nov 19, 1982).

Page 161, at end of section add:

A pledge of securities as collateral for a commercial loan may involve a sale of security for purposes of the federal securities laws. *Rubin v United States,* 449 US 424 (1981).

§6.19 —Sale of a Business

Delete entire section and substitute:

When an entire closely owned business is sold to another person by an individual or closely related group and the sale is accomplished by a transfer of stock, even though the transfer of such stock probably was exempt from registration under the 1933 Act, the transaction would be subject to the antifraud provisions of the federal securites laws. *Landreth Timber v Landreth,* 85 L Ed 2d 692 (1985).

Exemptions under the Securities Laws

§6.22 Exempt Securities

Page 165, add at end of section:

The SEC Government-Business Forum on Small Business Capital Formation recommended that the SEC exempt public securities offerings of $5 million or less from the registration provisions of the 1933 Act. Final Report, SEC Government-Business Forum on Small Business Capital Formation (Nov 1982), at 49.

§6.24 Definition and Significance of an Offering

Add to note 60, page 166:

For an example of the application of these factors, *see* SEC v Murphy, 626 F2d 633 (9th Cir 1980).

Page 167, add after carryover paragraph:

The integration doctrine has been particularly troublesome for public offerings by partnerships with common sponsors, particularly those engaged in real estate syndications. The Third Annual SEC Government-Business Forum on Small Business Capital Formation has recommended that the SEC adopt a safe harbor rule clarifying the conditions under which such offerings would not be integrated. Final Report, 1984 SEC Government-Business Forum on Small Business Capital Formation (Jan 1985).

The Forum endorsed a proposal which was developed by the Subcommittee on Partnerships, Trusts and Unincorporated Associations, Committee on Federal Regulation of Securities, Section of Corporation, Banking and Business Law, American Bar Association. Subcommittee on Partnerships, Trusts and Unincorporated Associations, *Integration of Partnership Offerings: A Proposal for Indentifying a Discrete Offering,* 37 Bus Law 1591 (July 1982). The proposal would provide a safe harbor if: (1) the partnerships are separate legal entitites; (2) the funds received by the partnerships are not commingled; (3) the partnerships have an opportunity to meet their economic objectives without substantial dependence on another entity having a common sponsor; and (4) where the major portion of the proceeds of an earlier offering by another partnership is not committed, the intended use of at least 50 per cent of the proceeds of a subsequent offering by the second partnership must be specifically identified to investors.

Two states have substantially adopted these guidelines. Ore Admin R, as amended, Blue Sky Regulations, Rule 815-36-015(3) (1982); Louisiana Blue Sky Reg, Reg 1, art D, Rule 1.7.B.

The SEC staff also intends to depart from the practice it adopted in 1980 of refusing to issue no action letters relating to integration issues. The staff will respond to requests for no action letters based on the five factors listed in Rule 502(a) (17 CFR §230.502(a)) described above. *Corp Fin Activities,* 17 Sec Reg & L Rep (BNA) No 10, at 403 (Mar 8, 1985).

§6.26 —Regulatory Interpretation (Rule 506)

Add to note 71, page 169:

The SEC has issued a comprehensive interpretive release relating to Rule 506 and the other rules under Regulation D. Securities Act of 1933 Release No 6455 (Mar 4, 1983).

The SEC staff is now considering an interpretive release on the subject of "general solicitation." *Corp Fin Activities,* 17 Sec Reg & L Rep (BNA) No 10, at 403 (Mar 2, 1985).

A report prepared by the SEC's Directorate of Economic and Policy Analysis indicated that $15.5 billion of securities (about 17% of the amount of securities offered publicly for cash during the period) were included in 7,222 filings under Regulation D during the first year of the operation of that Regulation; $12.5 billion of that amount under Rule 506 (17 CFR § 230.506). *An Analysis of Regulation D (SEC May, 1984),* Fed Sec L Rep (CCH) ¶83,631 (Executive Summary). The average size of a proposed Regulation D offering during that period was $2.1 million, indicating that the Regulation was being used primarily by small issuers with five or fewer employees, an operating history of two or fewer years, revues and assets of less than $500,000 and shareholders' equity of less than $25,000. *Id.* About 25% of the proposed offerings, but only 3% of the dollar amount, were to be made under Rule 504 and 13% under Rule 505 (about 15% of the dollar amount). *Id.* Expenses were estimated to average from 1% to 5% of the amount to be offered by corporations and from 5% to 7.5% of the amount to be offered by partnerships, and legal and accounting expenses were estimated to be less than for comparable Form S-18 offerings (see **§6.37**). *Id.*

§6.30 —Circumstances under which the Regulation A Exemption Should Be Used

Page 177, in numbered paragraph 3, last line, delete "made" and insert:

sold

§6.31 —Rule 504

Add to note 120, page 177:

The SEC has issued a comprehensive interpretive release relating to Rule 506 and other rules under Regulation D. Securities Act of 1933 Release No 6455 (Mar 4, 1983). The Sec also has on its rulemaking agenda possible revisions of Form D (17 CFR §230.500). Securities Act of 1933 Release No 6567 (Feb 27, 1985).

§6.33 Offshore Offerings

Add to note 125, page 179:

See generally Evans, *Offerings of Securities Solely to Foreign Investors,* 40 Bus Law 69 (Nov 1984).

Page 179, add at end of section:

Certain United States companies have used offerings by domestic finance subsidiaries to finance the operations of their foreign subsidiaries. Public offerings by these finance subsidiaries subjected many of them to regulation under the Investment Company Act of 1940, as amended, 54 Stat 789 (codified at 15 USC §§80a-1 to 80a-64) (see **§6.41**), although the SEC exempted those companies from many of the provisions of that act and has recently indicated a willingness to relieve them from certain of the conditions of those exemptive orders.

The SEC has asked for comments on two concepts to facilitate transnational securities offerings taking place in the United States, Canada, or the United Kingdom. Securities Act of 1933 Release No 6568 (Mar 1, 1985). The SEC is exploring the feasibility of developing a common prospectus for use in all three jurisdictions or reciprocal arrangements which would permit a prospectus prepared under one of the countries' rules to be used for simultaneous offerings in either of the other two countries. There is some thought of expanding these concepts to cover Japan and perhaps other countries. The SEC also issued a further concept release on other aspects of the internationalization of the securities markets, such as 24-hour trading; consolidated market information systems; international regulatory coordination; and broker-dealer regulation. Securities Exchange Act of 1934 Release No 21958 (Apr 18, 1985).

The SEC previously had rescinded exemptions from registration and reporting under the 1934 Act for foreign companies whose securities are quoted in the National Association of Securities Dealers, Inc. Automated Quotation System ("NASDAQ"). Securities Exchange Act of 1934 Release No 20784 (Mar 22, 1984) (amending Rules 12g-3, 12g-4, 12h-3, and 15d-6, (17 CFR §§240.12g-3, 240.12g-4, 240.12h-3, and 240.15d-6, respectively)). This may reduce somewhat the advantage of those foreign companies in competing for capital with United States companies.

Compliance with the Securities Laws

§6.34 Assistance of the Securities and Exchange Commission

Page 181, add at end of section:

Under certain circumstances a court may treat good faith reliance on an SEC no action or interpretive letter as tantamount to good faith reliance on an SEC rule. *See, e.g., Greene v Dietz,* 247 F2d 689, 694 (2d Cir 1957); *Colema Realty Corp v Bibow,* 555 F Supp 1030 (D Conn 1983). No liability may be imposed for actions in reliance upon such rules. Securities Act of 1933, §19(a), 15 USC §77s(a); Securities Exchange Act of 1934, §23(a), 15 USC §78x(c).

§6.35 Registration of Securities

In note 131, page 181, third line, delete "1981)." and insert:

June 1983).

Page 182, add at end of section:

The SEC has endeavored to reduce some of the burdens of the registration process by permitting a wide variety of "shelf offerings" under its Rule 415. 17 CFR §230.415. *See generally* Securities Act of 1933 Release No 6423 (Sept 2, 1982). That rule, together with the SEC's simplified registration forms which rely on incorporation by reference of 1934 Act reports, permit some companies to move rapidly in response to a "market window" and raise needed capital relatively quickly and inexpensively. This puts tremendous pressure on underwriters who may be unable to perform their traditional, in-depth due diligence investigations. As a result, underwriters are concerned about the added liability exposure involved in some Rule 415 shelf offerings. The SEC moved to alleviate some of these concerns by adopting a rule setting forth factors to be considered in determining whether a person has conducted a "reasonable investigation" for purposes of §11(c) of the 1933 Act, 15 USC §77k(c). *See* 17 CFR §230.176.

The SEC's Division of Corporation Finance was reorganized early in 1984 in an attempt to speed up this review process. The SEC also intends to establish a two-year pilot project for an electronic data gathering, analysis, and retrieval system (EDGAR) for electronic filing, processing, and dissemination and has asked for comments on, and volunteer participants in, such a system. Securities Act of 1933, Release No 6519 (Mar 22, 1984. *See also* Ingersoll, *SEC Asks Companies to Volunteer in Test of Electronic Filings,* Wall St J, Mar 23, 1984, at 6, col 3; Webster,

SEC Aims to Expedite Reviews, Legal Times of Wash, Mar 5, 1984, at 1, col 2; Sanger, *SEC's Computer Revolution,* NY Times, Apr 4, 1984, at D1, col 1.

The SEC, in June and September 1984, adopted temporary rules and forms to implement EDGAR and commenced the pilot program in September, 1984 with about 150 of the approximately 10,000 companies which file reports with the SEC participating. Securities Act of 1933 Release Nos 6539 (June 27, 1984); 6540 (June 27, 1984); and 6557 (Nov 13, 1984). All reporting companies are expected to be part of the EDGAR system by 1988 with phase-in beginning in 1986.

The system presently lacks capability for automatic data gathering or selection of corporate disclosure reports for analysis on the basis of financial profiles or financial ratios. *See, e.g.,* Ingersoll, *SEC Pilot Project On Firms' Reports Assailed at Hearing,* Wall St J, May 15, 1984, at 12, col 4; Ross, *Hill Panel Criticizes SEC for New Data System,* Wash Post, Mar 15, 1984, at D3, col 1. These capabilities may be in place by the third and last phase of the project.

The SEC has also requested proposals for permanent funding for the EDGAR system by bulk users of the information in SEC filings who could recover their costs be repackaging the information and selling it to other users. Securities Act of 1933 Release No 6548 (Sept 5, 1984). Commercial use of the EDGAR system raises significant issues about access and rates, but the possible instantaneous availability of such information would be one of the more important technological developments in the securities industry. *See, e.g.,* Office of Technology Assessment, Report to House Committee on Banking, Finance and Urban Officers, 98th Cong, 2d Sess, Effects of Information Technology on Financial Services Systems (Sept 1984).

There is also an issue as to whether most foreign issuers, which are not now eligible to participate in the EDGAR pilot project, will ultimately be permitted to participate in the final system. Securities Act of 1933 Release No 6568 (Mar 1, 1985).

§6.37 Which Registration Form?

Add to note 157, page 183:

The SEC also has integrated disclosure requirements for foreign private issuers. Securities Act of 1933 Release No 6437 (Nov 19, 1982).

Add to note 161, page 184:

Rule 415 has been adopted as a permanent rule. Securities Act of 1933 Release No 6499 (Nov 17, 1983).

Page 185, at end of fourth complete paragraph add:

The Final Report of the SEC Government-Business Forum on Small Business Capital Formation (Nov 1982), at 4, recommended that the Form S-18 ceiling be increased to $10 million. The SEC amended Form S-18 to increase the ceiling for offerings of securities registered on that form to $7.5 million. Securities Act of 1933 Release No 6489 (Sept 23, 1983).

Page 186, add after carryover paragraph:

The SEC also has on its regulatory agenda the development of a uniform registration form for registration of securities to be offered by partnerships, which should facilitate offerings by entities such as R&D partnerships. Securities Act of 1933 Release No 6567 (Feb 28, 1985).

§6.38 Offerings to Employees

Add to note 180, page 187:

See also 17 CFR §§240.16a-6, -8, and -9, and 240.16b-6. The SEC has undertaken to perform a sunset review of all its rules under §16 of the 1934 Act, 15 USC §78p. Securities Act of 1933 Release No 6567 (Feb 27, 1985).

§6.38A —Leveraged ESOPS (New)

Recent amendments to the federal income tax laws and absence of registration requirements under the 1933 Act have provided the incentive for a new means of financing the acquisition by employees of a company or a substantial portion of a company's common stock—the leveraged employee stock ownership plan ("ESOP"). An ESOP is permitted to borrow money from a bank or other lending institution to purchase securities of the employer company. The lender may exclude from its income for federal income tax purposes 50 per cent of the interest earned on the loan, thus making a lower rate of interest on the loan possible. 12 USC §133. The loan will be guaranteed by the employer company and secured by the employer company securities purchased by the ESOP. The loan will be repaid by employer contributions to the plan, which are deductible by the employer company for federal income tax purposes, or by dividends on the employer company securities held by the ESOP. Cash dividends on the employer company securities held by the ESOP which are not used to repay the bank loan are passed through to the employee participants and are deductible by the employer company for federal income tax purposes. 12 USC §404.

A leveraged ESOP is not only an important method of compensating employees or of financing, but, by enabling a company to place a large block of its stock in friendly hands, if handled properly, it may be used as an anti-takeover device. *See e.g., Norlin Corp v Rooney, Pace, Inc,* Fed Sec L Rep (CCH) ¶91,564 (2d Cir 1980); *Klaus v Hi-Shear Corp,* 528 F2d 225 (7th Cir 1975); *Donovan v Biewirth,* 680 F2d 263 (2d Cir 1982); *see generally* Vartanian, Schechter, West & Levine, *Leveraged ESOP Can Aid Thrift Takeover Defense,* Legal Times, Mar 25, 1983, at 18, col 1. Recent concern about takeovers and anti-takeover measures and, more particularly, efforts to generally reform our tax system, has generated consideration of legislative proposals to curb the tax incentives for establishing leveraged ESOPs. *See, e.g.,* Treasury Department Report to the President, *Tax Reform for Fairness, Simplicity and Economic Growth* (Nov 27, 1984); and a number of bills introduced in the first session of the 99th Congress, including HR 800 & S 409 ("Bradley-Gephardt"), HR 777 & S 325 ("Kemp-Kaston"), and HR 373 & S 411 ("Roth-Moore").

Since an ESOP is a "bonus plan," the shares of employer company stock acquired by the ESOP are not required to be registered with the SEC under the 1933 Act. *See generally* Securities Act of 1933 Release Nos 6188 (Feb 1, 1980) and 6281 (Jan 15, 1981).

§6.39 Business Combinations and Other Restructurings

Page 187, after third paragraph of section add:

Rather than take the time and incur the cost and the drain on earnings of developing technology, some companies may make a cash tender offer with borrowed funds for the securities of another company which possesses the desired technology. The making of such offers and their regulation under the so-called "Williams Act," originally enacted as Pub L No 90-439, 82 Stat 454 (1968) (amending Securities Exchange Act of 1934, 78 Stat 568), as codified at §§13(d) and (e) and 14(d), (e), and (f) of the 1934 Act, 15 USC §§78m(d) and (e) and 78n(d), (e), and (f)), have been the subject of considerable controversy and study. *See generally* Final Report of Recommendations of SEC Advisory Committee on Tender Offers (July 8, 1983). *See generally* Fishbein, Wishingrad & Garcia, *Only Congress Can Implement Tender Offer Proposals,* Legal Times of Wash, Feb 20, 1984, at 15, col 1.

These statutory provisions and the SEC's regulations thereunder, Regulation 13D-G, 17 CFR §§240.13d-1 to 240.13d-102, Regulations 14D and E, 17 CFR §§240.14d-1 to 240.14d-100, 240.14e-1 to 240.14e-3, 240.13e-1, and 240.14f-1 were intended by Congress to protect investors without tipping the balance between bidders and subject companies. *Piper v Chris Craft Industries, Inc,* 430 US 1 (1977). State regulation of tender offers which imposes burdens greater than those imposed by the SEC's regulations has been found to be an unreasonable burden on interstate commerce. *Edgar v Mite Corp,* 457 US 624, 102 S Ct 2629 (1982). Various states have revised their takeover and corporate laws as a rusult of that decision. *See generally* Warren, *Developments in State Takeover Regulation: MITE and Its Aftermath,* 40 Bus Law 671 (Feb 1985). For general texts on the subject of tender offers, see Lipton & Steinberger, Takeovers & Freezouts (Law Journal Seminars-Press 1978); Fleischer, Tender Offers: Defenses, Responses, and Planning (Law & Business, Inc.; Harcourt Brace Jovanovich 1981); Axinn, Fogg & Stoll, Acquisitions Under The Hart-Scott-Rodino Antitrust Improvements Act (Law Journal Seminars-Press 1980).

In part due to the extremely high costs of tender offers, some have pursued an alternate method of gaining control or influence over a company—the proxy contest—in an effort to control the company's board of directors. Proxy contests are regulated under the SEC's proxy rules, 17 CFR §§240.14a-1 to .14a-100, particularly Rules 14a-9, 14a-10, 14a-11, and 14a-12, 17 CFR § 240.14a-9, 240.14a-10, 240.14a-11, and 240.14a-12.

Persons who acquire securities of a company in an unsuccessful attempt to obtain control may be trapped by the short-swing profit recapture provisions of §16(b) of the 1934 Act, 15 USC §78p(b). *Compare Heublein, Inc v General Cinema Corp,* 722 F2d 29 (2d Cir 1983), *cert denied,* 104 S Ct 1416 (1984) *with Texas International Airlines v National Airlines,* 714 F2d 533 (5th Cir 1983), *cert denied,* 104 S Ct 1326 (1984). The SEC, as part of its review of its rules under §16 of the 1934 Act, is considering a rule which might provide some relief from the short-swing profit recapture provisions in the case of certain change of control situations. Securities Act of 1933 Release No 6567 (Feb 27, 1985).

There has been considerable criticism in the financial press and from Congress relating to various techniques for achieving or preventing changes in corporate control. *See generally* Rohatyn, *Junk Bonds and Other Securities Swill,* Wall St J, Apr 18, 1985, at 30, col 1.

Some critics maintain that takeovers divert resources from research and development and other innovation or divert management from long-range strategic planning to concentrating on showing short-term results. *See, e.g.,* Lipton, *Takeover Abuses Mortgage the Future,* Wall St J, Apr 5, 1985, at 16, col 15; Sansweet, *Hazardous Course, Carter Hawley's Trials Show Risk in Adopting Long-Range Strategy,* Wall St J, June 19, 1984, at 1, col 1. There is great concern that the financing of takeovers with high-yield, low-rated debt instruments ("junk bonds") will have a severe impact on our credit markets and the financial institutions which purchase junk bonds if there is an economic downturn. Others maintain that takeovers, or the mere threat of takeovers, are means of correcting corporate inefficiency and reallocating resources productively. Despite these and other criticisms, no legislation is expected in

1985. *See generally* Vise, *Experts Disagree on Takeovers,* Wash Post, Mar 13, 1985, at F3, col 4.

There is no definitive study which shows whether the ability of a company to take over easily another company would benefit or harm the United States economy, although there are recent studies which deal with discrete aspects of takeovers. One such study indicates that certain anti-takeover measures adopted by corporations (so called "shark repellants"), such as "poison pill" warrants (warrants whose exercise provisions are triggered by a change in control and which permit acquisition of the acquiring company's securities at bargain prices), tend to have an adverse affect on their shareholders. *See, e.g.,* Securities Exchange Act of 1934 Release No 15230 (Oct 13, 1978); Office of Chief Economist, SEC, Shark Repellants: the Role and Impact of Antitakeover Charter Amendments (Sept 1984); and *see generally* Heard & Pound, *Shark Repellants Fail to Benefit Shareholders,* Legal Times, Mar 25, 1985, at 12, col 1 (summarizes various studies); Ferrara & Phillips, *Opposition to "Poison Pill" Warrants is Mounting,* Legal Times, Oct 15, 1984, at 13, col 1. The SEC has become so concerned with poison pill warrants, which it believes can result in a total disenfranchisement of shareholders, that it has taken the unusual step of filing a brief *amicus curiae* in a state court case, *Moran v Household International, Inc,* No 7730 (Del Ch Jan 29, 1983), challenging the issuance of such warrants. *See SEC Urges Delaware Court to Void Poison-Pill—Preferred Takeover Defense,* 17 Sec Reg & L Rep (BNA) No 12, at 481 (Mar 22, 1985).

Another study indicates that one form of tender offer (pejoratively labeled "two-tier front-end load" tender offers) may have benefited shareholders of the target companies. Securities Exchange Act of 1934 Release No 21079 (June 22, 1984). Still another study tends to show that a company's buying out a large shareholder who is threatening a takeover (so-called "greenmail") also has an adverse affect on shareholders of the company. *SEC Staff Says Greenmail Payments Depress Price of Target's Stock,* 16 Sec Reg L Rep (BNA) No 39, at 1585 (Oct 5, 1984).

A committee of the board of directors of the New York Stock Exchange, in connection with reconsideration of that exchange's listing standards, has issued a report on another form of antitakeover device, namely creation of dual-class capitalization in which a class of common stock owned primarily by insiders has a substantially greater number of votes per share than the class of common stock owned by other shareholders. Initial Report of the Subcommittee on Shareholder Participation and Qualitative Listing Standards, Public Policy Committee of the Board of Directors of the New York Stock Exchange, Dual Class Capitalization (Jan 3, 1985). The National Association of Securities Dealers, Inc has asked for views on the same issue in connection with its consideration of standards for including securities in the NASDQ National Market System. NASD Notice to Members 85-20 (Mar 28, 1985).

Use of another anti-takeover device, the golden parachute (substantial compensation paid to a corporate executive upon a change in control or ownership of his or her corporation), may have been limited somewhat by provisions of the Tax Reform Act of 1984 (Pub L No 98-369, 98 Stat 494) that preclude deduction of certain golden parachute payments and impose a 20 per cent excise tax on the recipients of such payments. *See* 12 USC §280G.

Legislation has been introduced to curb some of the more controversial takeover and anti-takeover practices, but has not been enacted. *See, e.g.*, Tender Offer Reform Act of 1984, tit II, HR 5693, 98th Cong, 2d Sess; Dingell, Equity in Foreign and Domestic Credit and Tender Offer Reform, HR Rep No 1028, 98th Cong, 2d Sess (Sept 7, 1984).

Page 190, at end of carryover sentence add:

The SEC has adopted amendments to its rules for disclosure relating to business combination transactions and adopted a new form, Form S-4 (17 CFR §239.25), for registration of securities to be issued in connection with such transactions. Securities Act of 1933 Release No 6578 (Apr 23, 1985). A companion release adopted a similar form for use by foreign private issuers, Form F-4 (17 CFR §239.34), which also governs the disclosure to be made about a foreign company being acquired by a United States domiciled company. Securities Act of 1933 Release No 6579 (Apr 23, 1985). A recent proposal would govern the discolsure required about mergers between registered investment companies. Securities Act of 1933 Release No 6570 (Mar 8, 1985).

§6.39A —Accounting for Business Combinations and Other Restructurings (New)

The SEC's regulations governing accounting for business combinations and restructurings may inhibit such transactions or subsequent financing by the parties involved. The SEC staff has taken the position that where a company has a made a significant acquisition of another business (or in some instances where such an acquisition is "probable") and the company is unable to provide the audited financial statements of the acquired business in a current report on Form 8-K, 17 CFR §249.308, or in a pending or "shelf" 1933 Act registration statement as required by 17 CFR 210.3-05, the company is precluded from filing a 1933 Act registration statement (or an amendment to a registration statement) for most types of primary offerings, and its affiliates are precluded from reselling the company's securities under 1933 Act Rule 144, 17 CFR §230.144. *See, e.g., STV Engineers, Inc* (SEC no action letter available Apr 28, 1983). The company would still be permitted to file registration statements for secondary offerings by nonaffiliates, offerings to employees, offerings under dividend or interest reinvestment plans, or offerings of securities to be issued on the exercise or conversion of outstanding securities. The SEC recently codified these positions in instructions to Form 8-K, 17 CFR §249.308. Securities Act of 1933 Release No 6578 (Apr 23, 1985).

The SEC staff also has been concerned about other aspects of accounting for transactions involved in business combinations and corporate restructurings, including carve-outs and spin-offs. See **§6.09.** Where a company purchases substantially all of the common stock of another company, the acquiring company's acquisition costs may be required to be "pushed down" to be used to establish a new basis for accounting in the acquired company's separate financial statements. Staff Accounting Bulletin Release 54 (Nov 3, 1983). The SEC staff also takes the position that in a spin-off or carve-out transaction, the income statement of the spun-off or carved-out entity should reflect all of its costs of doing business, including expenses incurred by its parent on its behalf, such as salaries, rent, depreciation, advertising, professional services and selling, and general and administrative expenses. Staff Accounting Bulletin Release No 55 (Nov 30, 1983).

§6.40 Trust Indentures

Page 190, at end of section add:

The SEC and the Securities Bar have taken steps to simplify the process of preparation and qualification of trust indentures. *See* Securities Act of 1933 Release No 6090 (July 11, 1979) (techniques for drafting trust indentures); Trust Indenture Act of 1939 Release No 605 (Jan 8, 1981) (simplified form of trust indenture); McDaniel, *A Simplified Indenture,* 13 Rev of Sec Reg 16 (Sept 24, 1980), at 871; and *Report, Model Simplified Indenture,* 38 Bus Law 741 (Feb 1983); Trust Indenture Act of 1939, Fed Sec L Rep (CCH) ¶77,345 (SEC interpretive letter available Sept 24, 1982).

§6.42 Liability

Page 192, in second sentence, delete "unless the Supreme Court should rule otherwise," and at the end of the second sentence add:

Herman & MacLean v Huddleston, 456 US 914 (1983).

In note 220, page 192, third line, delete "— US —, 50 USLW 3668 (Feb 23, 1982)" and insert:

455 US 938 (1982)

Note 229, page 193, delete text and insert:

Id §77k. Purchasers who acquire registered securities after the issuer has made generally available an earnings statement covering a period of at least 12 months beginning after the effective date of the registration statement, must prove reliance on misleading statements in, or omission from, the registration statement. 15 USC §77k(a). The SEC has adopted a safe harbor rule to provide standards as to when such an earnings statement has been made generally available. Rule 158, 17 CFR §230.158; Securities Act of 1933 Release No 6485 (Sept 23, 1983).

Add to note 234, page 194:

The Supreme Court has reserved decision on this issue. Herman & MacLean v Huddleston, 456 US 914 (1983); International Bhd of Teamsters v Daniel, 489 US 551, 557 n 9 (1979). However, the Supreme Court has addressed §17(a) of the 1933 Act in the context of criminal prosecution. Rubin v United States, 449 US 424 (1981); United States v Naftalin, 441 US 768 (1979).

Add to note 237, page 194:

Landry v All Am Assurance Co, 688 F2d 381 (5th Cir 1982).

Page 194, add at end of section:

In recognition of the potential liability which could be imposed by the incorporation by reference of "stale" 1934 Act reports in 1933 Act registration statements, the SEC has adopted a rule which permits statements in later filed documents to supersede statements in earlier filed documents for liability purposes. 17 CFR §230.412.

Regulation of Investors and Investments

§6.43 Direct Regulation of Investors—Overview

Add to note 243, page 195:

The Board of Governors of the Federal Reserve System has recently amended its Regulations G, T, U, and X which govern such extensions of credit. 12 CFR §§207.1 to 207.7, 220.1 to 220.18, 221.1 to 221.8, and 224.1 to 224.3, respectively. Further amendments have been proposed or are being considered which would permit the self—regulatory organizations (registered national securities exchanges and the National Association of Securities Dealers, Inc) to set margin requirements subject to the oversight of the SEC, the Department of the Treasury, and the board of governors of the Federal Reserve System. *See, e.g.,*Board of Governors of the Federal Reserve System, Docket No R-0538, Fed Sec L Rep (CCH) ¶83,735 (Feb 6, 1985). If these proposals are adopted, technological advances such as electronic funds transfers may eliminate the use of margin by brokerage houses to finance purchases of securities. Office of Technology Assessment, Report to House Committee on Banking, Finance and Urban Affairs, 98th Cong, 2d Sess, Effects of Information Technology on Financial Services Systems (Sept 1984).

§6.44 —Investment Companies

Page 196, at end of first complete paragraph, add:

The SEC Government-Business Forum on Small Business Capital Formation, September 23-25, 1982, Arlington, Virginia, recommended that a federally chartered, privately owned and managed capital bank be created to provide funds for small business investment companies. The Forum also recommended an 80 per cent exclusion for capital gains on small business equity investments held for five years or more. Final Report, SEC Government-Business Forum on Small Business Capital Formation (Nov 1982), at 29, 83-84.

§6.45 —Business Development Companies

Add to note 256, page 196:

Bergman, *Business Development Companies,* 17 Rev Sec Reg 877 (June 20, 1984).

Page 197, add at end of section:

At least one venture capital company, the Heiser Corp, which might have continued to take advantage of the relief from the provisions of the 1940 Act provided to business development companies, has elected to distribute its portfolio to its shareholders and go private. In an effort to encourage smaller venture capital investments, the SEC has amended its Reg E, 17 CFR §§230.601 to 230.610a, which provides an exemption from registration for securities issued by certain small business investment companies, to raise the ceiling under that exemption from $500,000 to $5 million and to otherwise make the provisions of that exemption correspond more closely to those of Regulation A. 17 CFR §§230.251 to 230.264. See **§§6.29** and **6.30.**

§6.46 —Investment Advisers

Add to note 268, page 198:

The SEC has interpreted the provisions of the Advisers Act in a manner to make those provisions applicable to a broad spectrum of financial planners and advisers. Investment Advisers Act of 1940 Release No 770 (Aug 13, 1981).

The SEC's interpretations of the Advisers Act sometimes have an Alice-in-Wonderland quality. For example, the staff of the SEC's Division of Investment Management concluded that reports issued by Venture Capital Network, Inc ("VCN"), a nonprofit corporation established by the Business and Industry Association of New Hampshire, and processed by the New Hampshire Business Development Program, an instrumentality of the State of New Hampshire (and thus exempt from registration under the Advisers Act by §202(b) (15 USC §80b-2)) which matches investors with entrepreneurs of "start up," "infant," "young," and "established" firms on the basis of the level of investment desired by the potential investor and the type of financing required by the entrepreneur, are reports concerning securities, but are not provided as part of a regular business, and therefore VCN is not subject to the Act. Venture Capital Network, Inc (available May 5, 1984). Notwithstanding the occasional lack of clarity as to the standards governing the registration requirements of the Advisers Act, the SEC is bent on enforcing those provisions even at the risk of losing its case. *See, e.g.*, SEC v Lowe, 86 L Ed 2d 130 (1985).

The North American Securities Administrators Association is also concerned about the activities of financial planners and several of its members. *See, e.g., NASAA Financial Planners Hearing Focuses on State Uniformity*, 17 Sec Reg L Rep (BNA) No 11 at 452 (Mar 15, 1985).

Add to note 270, page 198:

The SEC has adopted a safe harbor under Rule 203(b)(3)-1 (17 CFR §275.203(b)(3)-1) which would determine the circumstances in which a general partner of a limited partnership will be considered to be the investment adviser to the partnership, not the limited partners.

Page 198, add to end of third full paragraph:

The Supreme Court has provided some guidance in this regard, at least as to publishers of investment newsletters, holding that such publishers are not subject to registration as investment advisers. *SEC v Lowe*, 86 L Ed 2d 130 (1985).

Page 198, add at end of section:

The SEC had proposed a rule which would permit investment advisers to wealthy sophisticated clients to charge performance-related fees. Investment Advisers Act of 1940 Release No 865 (June 9, 1983) (proposed Rule 205-3). In March 1984, the SEC deferred action on that proposal and withdrew it. Now the Sec has reconsidered and re-proposed Rule 205-3 for comment. Investment Advisers Act of 1940 Release No 961 (Mar 14, 1985).

§6.47 —Financial Intermediaries

Add to note 274, page 198:

There also is a unique private nonprofit organization, The National Development Council, which acts as an investment banker for the public sector by assisting in raising financing for local and regional businesses.

Page 199, add after carryover paragraph:

Banks and other financial institutions are attempting to get more involved in the business of underwriting securities, but the impact of their involvement on our capital markets cannot be assessed at this time. *See generally Securities Industry Association v Board of Governors of Federal Reserve System,* 104 S Ct 2979, 468 US __ (1984) (Bankers Trust); *Securities Industry Association v Board of Governors of Federal Reserve System,* 104 S Ct 3003, 468 US __ (1984) (Charles Schwab). The SEC has interjected itself into the controversy by adopting Rule 3b-9 (17 CFR §240.3b-9), which defines the circumstances in which a bank, although otherwise excepted from the definitions of *broker* and *dealer* by §3(a)(4) and (5) of the 1934 Act (15 USC §78c(a)(4) and (5)), would not be considered a *bank* as defined in §3(a)(6) of that Act (15 USC §78c(a)(6)) and would be required to register as a broker under the 1934 Act. The validity of the rule has been challenged by the American Bankers Association.

Page 199, after fourth paragraph add:

These legal risks are in addition to the very substantial market risks underwriters must incur. These market risks may be ameliorated by information technology developments such as EDGAR (see **§6.35**) which may decrease the time between the initiation of an underwriting and the actual sale of the securities. Office of Technology Assessment, Report to House Committee on Banking, Finance and Urban Affairs, 98th Cong, 2d Sess, Effect of Information Technology on Financial Services Systems (Sept 1984).

Add to note 282, page 200:

The SEC determined to adopt Rule 415, 17 CFR §230.415, which permits shelf offerings, despite concerns about the effects on the ability of underwriters to conduct adequate due diligence investigations because of the speed at which an offering may be brought to market under the rule. Securities Act of 1933 Release No 6499 (Nov 17, 1983).

Page 200, add at end of section:

The SEC has also adopted a rule setting forth factors to be considered in determining whether a person, including an underwriter, has conducted a reasonable investigation for purposes of §11 of the 1933 Act, 15 USC §77k(c). *See* 17 CFR §230.176.

It may not always be clear when an entrepreneur or his or her employees are participating in the financing of a new venture to such a degree that they are performing functions equivalent to those of securities brokers, and should register as such under the 1934 Act and applicable state laws. The SEC, in an attempt to clarify the situation, has adopted Rule 3a4-1 (17 CFR §240.3a4-1) which provides a safe harbor from broker-dealer registration for persons who comply with its terms and conditions.

§6.49 —Safe Harbor Rule (Rule 144)

Page 201, in second paragraph of section, seventh line, delete parenthetical clause and footnote referent.

Delete note 294, page 201.

Add to note 299, page 202:

Rule 145 has been amended so that its resale provisions correspond more closely to those of Rule 144. Securities Act of 1933 Release No 6508 (Feb 22, 1984).

Page 202, delete fourth full paragraph.

Page 202, fifth full paragraph, second and third lines, delete "there is current information about the company available to the public and".

Page 203, add at end of section:

The SEC has announced that it will study the effects of the three-year holding period under Rule 144(k), 17 CFR §230.144(k). The SEC Today, Vol No 83-185, at 3 (Sept 23, 1983).

Although the SEC has taken a number of steps to facilitate resale of restricted securities, its staff administratively has taken the position that the resale provisions of Rule 144, 17 CFR §240.144, are not available to affiliates of an issuer which has failed to provide the financial statements of an acquired or to be acquired business required by Rule 3-05 of the SEC's accounting regulations, 17 CFR §210.3-05, to be included in a current report on Form 8-K, 17 CFR §249.308 or in certain registration statements. See **§6.63**; *see, e.g., STV Engineers, Inc* (SEC no action letter available Apr 28, 1983). The SEC has proposed to codify its staff's position Securities Act of 1933 Release No 6534 (May 9, 1984).

§6.50 —Other Methods of Resale

Page 203, at end of first paragraph of section add:

The SEC had proposed to raise the ceiling for secondary offerings of securities registered for resale on Form S-18 to $5 million, but later announced it was not taking further action on that proposal, pending further study. Securities Act of 1933 Release No 6489 (Sept 23, 1983).

Page 204, at end of item 3 add:

The SEC staff is reportedly considering a proposed interpretive release on resale of restricted securities. 3 Bus L Memo No 3 (Jan/Feb 1983), at 4.

Page 205, at end of first complete paragraph after indented quotation add:

See Dupuy v Dupuy, 551 F2d 1005, 1016 (5th Cir 1977), *cert denied,* 434 US 911 (1977), for an example of a face-to-face transaction.

Add to note 319, page 205:

But see Merrill Lynch, Pierce, Fenner & Smith, Inc v Curran, 456 US 353 (1982) (relying on the existence of judicial decisions finding implied remedies under the Commodities Exchange Act, 17 USC §1 *et seq* (1976 ed & Supp III) at the time of reenactment of that Act to uphold such remedies); and Herman & MacLean v Huddleston, 456 US 914 (1982) (same reenactment doctrine applied under 1934 Act). The Supreme Court has remanded another case raising the issue of the scope of implied remedy under §10(b) of the 1934 Act. Chemetron Corp v Business Funds, Inc, 682 F2d 1149 (5th cir 1982), *vacated,* — US —, 103 S Ct 1245, *cert denied,* — US —, 103 S Ct 1254 (1983) (vacating decision below and remanding for consideration in light of *Herman & MacLean, supra*).

Add to note 321, page 205:

However, scienter need only be proved by a preponderance of the evidence. Herman & MacLean v Huddleston, 456 US 914 (1982).

Add to note 322, page 205:

See also Burks v Lasker, 441 US 471 (1979) (extent to which directors' decisions may be protected by the "business judgement" rule depends upon state law, not the federal securities laws).

§6.50A —Reports by Investors

Investors who make significant acquisitions and dispositions of securities registered under §12 of the 1934 Act, 15 USC §78(1), are required to report information about those transactions, in accordance with SEC regulations, to the SEC, the issuers, and the stock exchanges upon which the securities are listed for trading.

1. Officers, directors of an issuer, and persons beneficially owning more than 10 per cent of the outstanding shares of a class of equity securities of that issuer registered under §12 of the 1934 Act are required to report their ownership of, and transactions in, any equity securities of that issuer. 1934 Act §16(a), 15 USC §78p(a). In addition, issuers are required to report information about such owners in registration statements, proxy and information statements, and reports filed with the SEC. *See generally* 17 CFR §229.403. The SEC, after a very sporadic history of enforcement of the provisions of §16(a), recently has commenced dozens of enforcement actions to compel compliance. *See, e.g.*, Litigation Release Nos 10308 (Mar 13, 1984) and 10315 (Mar 25, 1984); *SEC, Hospital Corp Settle on Reporting of Insider Trading*, Wall St J, Mar 27, 1984, at 53, col 3.
2. Persons who acquire, or have at the end of a calendar year, beneficial ownership of more than 5 per cent of the outstanding shares of a class of equity securities registered under §12 of the 1934 Act are required to report information about their acquisitions and ownership to the SEC, the issuers, and the stock exchanges upon which such securities are listed. 1934 Act §§13(d) and (g), 15 USC §78m(d) and (g); SEC Regulation 13D-G, 17 CFR §§240.13d-1 to 240.13d-7, 240.13d-101, and 240.13d-102. In addition, issuers are required to report information about such ownership in registration statements, proxy and information statements, and reports. *See generally* 17 CFR §229.403.
3. Institutional investment managers are required to report to the SEC information about the ownership in portfolios they

manage of equity securities registered under §12 of the Exchange Act. 1934 Act, §13(f), 15 USC §78m(f); 17 CFR §240.13f-1.

Failure to file such reports can result in SEC enforcement actions or private litigation.

Disclosure Requirements

§6.52 —What Must Be Reported and When?

Add to note 325, page 207:

Greenfield v Heublein, Inc, 575 F Supp 1325 (ED Pa 1983), 742 F2d 751 (3d Cir 1984), *cert denied,* US No 84-1025, slip op (US Feb 19, 1985). *See generally* Rowe, *Disclosure of Negative News,* 17 Rev Sec Reg 804 (Nov 21, 1984); Rowe, *How to Deal with Negative News—A Legal Analysis,* 5 Legal Notes & Viewpoints (Feb 1984); Brodsky, *Timing of Corporate Disclosure,* NYLJ, Nov 7, 1984, at 1, col 1.

Add to note 326, page 207:

See also Dirks v SEC, 103 S Ct 3255 (1983).

Add to note 328, page 207:

Lipman, The Personal Duties of Directors, Officers and Principal Stockholders Under the Securities Exchange Act of 1934 and Rule 144 of the Securities Act of 1933 (Packard Press Jan 1981).

§6.53 —Insider Trading

Add to note 331, page 208:

Rule 10b-18 provides a safe harbor from certain of the antimanipulative provisions of the 1934 Act. 17 CFR §240.10b-18.

Page 209, at end of carryover paragraph add:

On the propriety of trading on rumor, *see Dirks v SEC,* 103 S Ct 3255 (1983).

Add to note 339, page 209:

Schlanger v Four-Phase Sys, Inc, 528 F Supp 128, Fed Sec L Rep (CCH) ¶99,724 (SDNY Mar 6, 1984) (while there is no obligation for an issuer to disclose material information to the public if it is not trading in its securities, issuance of a press release, after an inquiry from a stock exchange, stating that the issuer was not aware of any developments which would affect the market price for its stock when the issuer was actually engaged in merger negotiations violated Rule 10b-5, 17 CFR §240.10b-5).

Add to note 345, page 209:

After SEC v Dirks, ___ US ___, 103 S Ct 3255 (1983), corporate officials may feel more comfortable talking with security analysts, as they will not be in danger of violating Rule 10b-5, 17 CFR §240.10b-5, for inadvertently revealing material nonpublic information unless they derive a personal benefit (which need not be monetary) from such disclosure. *See generally* Gonson & Butler, *In Wake of "Dirks", Courts Debate Definition of Insider,* Legal Times of Wash, Apr 2, 1984, at 16, col 1; Levine & Ferrigno, *Corporate Image Suffers if Lesson of "Dirks" is Ignored,* Legal Times of Wash, Apr 2, 1984, at 15, col 1.

See SEC complaint in SEC v Thayer, No 84 Civ 5066, [Current] Fed Sec L Rep (CCH) ¶99,718 (SDNY complaint filed Jan 5, 1984) for a discussion of what the SEC has alleged to be personal benefits in that case.

§6.53A —Disclosure to Persons Other than Investors and Market Professionals (New)

The SEC has taken the position that the antifraud provisions of the federal securities laws, principally §17(a) of the 1933 Act, 15 USC §77q(a) and Rule 10b-5, 17 CFR §240.10b-5, under the 1934 Act, apply to all public statements by persons speaking on behalf of a company which can be reasonably expected to reach investors and the securities trading markets, despite the nature of the primary audience (e.g., creditors, labor unions). Securities Act of 1933 Release No 6504 (Jan 13, 1984). The SEC indicated in that release that, in its view, to the extent that the standards for completeness and accuracy embodied in such antifraud provisions are not met, the persons responsible may be liable under such provisions. Imposing a securities law standard of disclosure on statements made in a nonsecurities investor context appears to be an unreasonable burden: for example, labor negotiators may now have to be advised by securities lawyers.

§6.54 —Periodic Disclosure

Page 211, add at end of section:

The SEC's requirement to provide a "Management's Discussion and Analysis of Financial Position" and "Results of Operations" section in annual and quarterly reports may also require some disclosure of proposed capital expenditures and the resources available to finance those expenditures, to the extent that material commitments have been made during the period for which the report is made. 17 CFR §229.303.

The SEC has expressed dissatisfaction with the disclosure being provided by companies experiencing troubled economic circumstances, particularly that provided in response to the SEC's requirement for a management discussion and analysis of the company's financial position and results of operations. *See generally SEC Enforcement Chief Warns Companies, CPAs On Management Discussion,* Analysis, 15 Sec Reg L Rep (BNA) No 2, at 95 (Jan 13, 1984); Rowe & Eppler, *Preparation of Management's Discussion and Analysis of Financial Condition and Results of Operations Under Uncertain Economic Conditions,* Prentice-Hall Sec Reg Rev ¶1106 (1983). The SEC has proposed to amend its interim reporting rules, 17 CFR §§210.10-01 and 229.303, to require segment disclosure, comparative balance sheets, and disclosure of short-term debt under those rules, and has given notice of possible rule proposals relating to additional segment information and uniform off-balance sheet debt disclosure requirements. Securities Act of 1933 Rel No 6514 (Feb 15, 1984).

SEC periodic disclosure requirements impose a relatively heavier cost burden on smaller companies. The SEC's Government-Business Forum on Small Business Capital Formation, September 23-25, 1982, Arlington, Virginia, recommended that the SEC should substantially increase the size test for the smallest companies which must file periodic reports under the 1934 Act (currently $3 million in assets and/or 300 or 500 security holders of record). Final Report, SEC Government-Business Forum on Small Business Capital Formation (Nov 1982), at 50-51. The SEC has taken steps to facilitate the termination of reporting obligations by these companies. Securities Exchange Act of 1934 Release No 210784 (Mar 22, 1984).

§6.55 Disclosure of Material Information—Overview

In note 358, page 211, delete "17 CFR §230.485." and insert:

17 CFR §230.406. The SEC rescinded Rule 485, 17 CFR §230.485, and amended Rule 406 to make the standards for confidential treatment of information required to be filed under the 1933 Act more closely

parallel to those under the 1934 Act. Securities Act of 1933 Release No 6520 (Mar 23, 1984).

§6.56 —What is Material?

Add to note 371, page 213:

The SEC has also proposed to mandate disclosure of projections made to specified outsiders in the context of a going private transaction, but the proposals have never been adopted or withdrawn. Securities Act of 1933 Release No 6101 (Aug 2, 1979). The proposed requirements are often followed in practice.

Add to note 377, page 213:

But see Merrill Lynch, Pierce, Fenner & Smith, Inc v Curran, 456 US 353 (1982) (Commodities Exchange Act, 7 USC §1 *et seq* (1976 ed & Supp III)); Herman & MacLean v Huddleston, 456 US 914 (1983); Chemetron Corp v Business Funds, Inc, 682 F2d 1149 (5th Cir 1982), *vacated,* __ US __, 103 S Ct 1245, *cert denied,* __ US __, 103 S Ct 1254 (1983).

§6.57 —Regulatory-Related Disclosure in General

Add to note 393, page 215:

See also 17 CFR §229.303 (requirement to disclose material commitments for capital expenditures).

Page 215, add at end of section:

The SEC's enforcement posture toward disclosure about violations of law which are not material in an economic sense may have shifted, and failure to disclose such violations may no longer be the basis for an SEC enforcement action. *See, e.g.,* Feders, "Illegal Conduct Materiality — An Enforcement Overview," address by Director, SEC Division of Enforcement, to the Federal Regulation of Securities Committee, Section of Corporation, Banking and Business Law, American Bar Association, Washington, DC, Nov 19, 1982, *reprinted in* 14 Sec Reg & L Rep (BNA) No 46, at 2057.

However, failure to disclose "qualitative" or "integrity" information about management in a proxy statement to the effect that a corporate officer participated in, and covered up, a bribery scheme during an internal corporate investigation into questionable payments may be an indictable offense. *United States v Dole,* 601 F Supp 430 (EDNY 1984); Stevens, *Two Southland Officials Cleared in Bribery Case,* Wall St J, Feb 28,

1985, at 7, col 1; Jackson, *Southland Corp Probe is Endorsed by US After Five Years and Some Convictions,* Wall St J, Mar 11, 1985, at 12, col 1.

§6.59 Disclosure of "Soft" Information

Add to note 409, page 218:

See Note, *Mandatory Disclosure of Corporate Projections and the Goals of Securities Regulation,* 81 Colum L Rev 1525 (Nov 1981). The SEC has proposed to mandate disclosure of projections made to specified outsiders in the context of going private transactions, but the proposals have never been adopted or withdrawn. Securities Act of 1933 Release No 6101 (Aug 2, 1979). Many companies, however, have followed the proposals in practice.

Add to note 410, page 219:

Financial Accounting Standards Board, Statement No 68, *Research and Development Arrangements* (1982).

§6.61 —Duty to Update Information

Page 221, at end of first paragraph add:

There are also benefits from voluntary updating. Updating of information in 1934 Act reports will supersede previously filed statements incorporated by reference in 1933 Act registration statements for liability purposes under the Act. 17 CFR §230.412.

§6.62 —Other Considerations

Page 222, add at end of section:

It is also possible that the SEC's requirement for a "Management's Discussion and Analysis of Financial Position and Results of Operations" section will evolve into a mandatory requirement for providing forward looking information. Note, *Mandatory Disclosure of Corporate Projections and the Goals of Securities Regulation,* 81 Colum L Rev 1525, 1541-1542 (Nov 1981). This could be detrimental to companies engaged in innovative activities, which could be faced with a choice between forgoing financing or risking liability or competitive disadvantages as the result of forecasting.

The SEC has stepped up its enforcement of its accounting and financial reporting requirements with respect to reporting companies and their auditors. *See generally* Olson, *SEC Enforcers to Focus on Accounting Fraud in 1984,* Legal Times of Wash, Jan 23, 1984, at 12. The director of the SEC's Division of Enforcement has warned in various fora that his staff will be "prospecting" in four areas:

1. Disclosure about a company's liquidity in its financial statements and management's discussion and analysis of those financial statements
2. Disclosure about operating trends. *See, e.g., SEC v Channel Industries, Ltd* (consent judgement), Civ Action No 83-3572 (DDC 1983), Litigation Release No 10213 (Nov 30, 1983) (allegations of failure to disclose unfavorable operating trends occurring after period covered by financial statements included in a registration statement)
3. Disclosure about problem loans
4. Public statements by failing companies

This prospecting will be done by enforcement division accountants after the fact, not during the Division of Corporation Finance review of a filing (unless referred by that division to the enforcement division).

The SEC has also increased its enforcement activity with respect to inadequate financial reporting. *See, e.g., In re Ronson Corp,* Securities Exchange Act of 1934 Release No 19212 (Nov 4, 1982); *SEC v Ronson Corp,* Civ Act No 83-3030 (DNJ Aug 5, 1983); *SEC To Go After Firms That Stretch Accounting Principles, Treadway Warns,* 16 Sec R L Rep (BNA) No 9, at 446 (Mar 2, 1984); *SEC Takes Dim View of Accounting "Shenanigans,"* The SEC Today, vol 45, at 1 (Mar 6, 1984); *Accounting Shenanigans and the Commission's 1984 Response,* Remarks by James C. Treadway, Jr., Commissioner, SEC, to the Financial Executives Institute, Chicago, Ill, Feb 16, 1984 (reprinted in SEC News as press handout).

Despite stepped-up SEC enforcement activity with respect to inadequate financial reporting and auditing, the accounting profession and the SEC's oversight role have come under Congressional criticism for not adequately protecting investors. *See, e.g., SEC, AICPA Defend Current Regulations; Legislators Hint of Congressional Action,* 1 Acct Prac & Reg No 5, at 1 (Mar 11, 1985).

§6.63 Accounting

Page 222, at end of second paragraph add:

The Supreme Court has expressed the view in strong dicta that the independent auditors have a public responsibility. *United States v Arthur Young & Co,* 104 S Ct 1906 (1984).

Add to note 426, page 222:

Companies should be aware that there is no accountant-client privilege under federal law similar to the attorney-client privilege, nor is there an accountants' work product doctrine. United States v Arthur Young & Co, __ US __, 52 USLW 4355 (US Mar 21, 1984).

In note 430, page 233, delete last sentence and add:

The SEC also took a rather rigid position on accounting for research and development expenditures funded by limited partnerships until the accounting profession promulgated its own standards. Financial Accounting Standards Board, Statement No 68, *Research and Development Arrangements* (1982). *See also* FASB Staff Technical Bulletin 84-1 (Mar 16, 1984) (accounting for stock issued to acquire results of a research and development arrangement).

Note 431, page 233, delete note and substitute:

The SEC has rescinded its long-standing accounting interpretation which had prevented companies which received "going concern" qualifications from the independent accountants from registering securities, subject to adequate disclosure of a company's financial difficulties and its plans to overcome those difficulties. Securities Act of 1933 Release No 6512 (Feb 15, 1984).

Page 223, delete referent to and text of note 432.

In note 433, page 223, delete text of note and insert:

17 CFR §210.305. The SEC staff, historically, however, has shown some flexibility in administering Rule 305 and its predecessors. *See, e.g.*, Price Communications Corp (SEC interpretative letter available Feb 25, 1983). The SEC staff has taken the position that where a company does not provide the required financial statements of a business it acquires, it may be precluded from filing a registration statement covering securities to be offered directly to the public to raise capital. See **§6.49;** *see, e.g.*, STV Engrs, Inc (SEC no action letter avail Apr 28, 1983). The SEC has proposed to codify its staff's position. Securities Act of 1933 Release No 6534 (May 9, 1984).

Page 224, at end of section add:

The SEC requirement that all public companies, regardless of size, adhere to generally accepted accounting principles imposes relatively significantly higher costs on smaller, developing companies. The accounting profession has had the issue of possible different accounting principles for smaller companies on its agenda for a number of years. See "Invitation to Comment, Financial Reporting by Private and Small Public Companies," Financial Accounting Standards Board (1981); Tentative Conclusions and Recommendations of the Special Committee on Accounting Standards Overload, New York, AICPA (1981) (calls for reconsideration in areas of leasing, income taxes, interest expense, and pro forma financial data for business combinations). *See generally,* Stanger & Gunther, *"Big GAAP—Little GAAP": Should There be Different Financial Reporting for Small Businesses?,* 56 NYU L Rev 1207 (1981). The issue of different accounting principles for smaller companies, including principles applicable to accounting for research and development expenditures, should be given serious consideration. The SEC Government-Business Forum on Small Business Capital Formation, Final Report (Nov 1982), at 52, rejects the "Big GAAP - Little GAAP" concept, but recommends subjecting accounting requirements to critical evaluation.

Pending SEC rule proposals would require more disclosure and analysis, on both an annual and on a quarterly basis. of industry segment information cash requirements, and off-balance sheet financing arrangements (such as operating leases, certain joint ventures and partnerships, take-or-pay contracts, and trusts used as debt defeasance devices). "Proposals Regarding Industry Segment and Other Interim Financial Reporting Matters, Managements Discussion and Analysis, and Off Balance Sheet Financing Disclosures," Securities Act of 1933 Release No 6514 (Feb 15, 1984); Spencer & Olson, *SEC's Quarterly Reporting Release Packs a Surprise,* Legal Times of Wash, Apr 2, 1984 at 18, col 1. The SEC also expressed its views that cost-contigent stock purchase warrants issued to customers in connection with sales agreements to purchase specified amounts of a company's products should be measured when the warrants become exercisable, not when they are issued. SEC Staff Accounting Bulletin No 57 (Jul 18, 1984). See **§6.62** for a discussion of the SEC's current enforcement posture with respect to accounting and financial reporting matters.

§6.63A —Accounting for Employee Compensation and Benefits

Newer companies which desire to conserve cash for research and development or other purposes may adopt various methods to compensate key employees with deferred benefits or rights to benefit from hoped-for appreciation in the company's stock. Companies which use methods such as stock appreciation rights of junior common stock may be required to make substantial charges to their reported earnings to account for these rights.

The SEC, the Internal Revenue Service, and the FASB are concerned about executive compensation plans developed principally by high-tech companies which involve issuance of *junior common stock* (stock with considerably lower dividend and voting rights than common stock, issued at prices significantly below the market price for common stock) exchangeable for common stock if certain performance goals are realized by the company. *See generally* Schrage, *Agencies Probe High-Tech Firms' Stock Plans,* Wash Post, Dec 25, 1983, at F-1; Junior Common Stock, VIII The Corporate Counsel No 6 (Nov-Dec 1983); Moore, *Junior Stock Plans Could Force Deadly Interpretations by IRS,* Legal Times, Jul 23, 1984, at 1, col 1. The SEC staff informally has taken the position that these plans involve compensation which should be recognized as having a significant impact on the results of operations of some companies. The FASB on December 16, 1983, issued, for limited exposure, a draft staff technical bulletin on the subject and subsequently determined to address the subject by interpreting Accounting Principles Board Opinion No 15, Accounting For Stock Issued to Employees. The interpretation may apply to rights granted on or after March 14, 1984, and there is some indication that the SEC also will grandfather plans in effect prior to that date.

Even though it has taken no final action, by far the most controversial activity by the FASB in this area has been with respect to accounting for pension and other postemployment benefits. FASB Discussion Memorandum of Issues Related to Employers' Accounting For Pension and Other Post Employment Benefits (1983). *See also* Staff Accounting Bulletin Release No 52 (Mar 16, 1983) (accounting for termination of overfunded defined benefit plans). The FASB will focus on pension accounting matters and defer consideration of accounting for health and welfare plans. The FASB issued a proposed statement on this subject in 1984, *Disclosure of Post Retirement Health Care and Life Insurance Benefits Information* (FASB July 3, 1984), and issues another exposure draft on pension accounting. *Employer's Accounting for Pensions* (FASB Mar 22, 1985). Because of the potential adverse impact of adoption of new pension accounting rules on a company's financial position, consideration should be given to renegotiating loan agreements to prevent technical defaults and potential adverse affects on credit ratings. *See, e.g., Hedging Against New Accounting Pronouncements,* Bus Rev 1984-2, at 2 (Price Waterhouse 1984); *Changing Accounting Principles Unsettle Financial Covenants,* 5 Bus Law Memo No 1, at 5 (Sept/Oct 1984).

A final statement on pension accounting is expected to be issued in 1985. *FASB Should Complete Work on Pension Project by End of 1985, Board Chairman Donald Kirk Says,* 1 Acct Prac & Reg No, 1 at 4 (Feb 11, 1985).

§6.64 Foreign Corrupt Practices Act

Page 224, delete text of note 446 and add:

S 430, 99th Cong, 1st Sess 131 Cong Rec 1264 (Feb 7, 1985). The SEC remains committed to enforcing the Foreign Corrupt Practices Act as long as it remains on the books. *See, e.g.*, Feders, "Illegal Conduct-Materiality: An Enforcement Overview," address by Director, SEC Division of Enforcement, to Federal Regulation of Securities Committee, Section on Corporation, Banking and Business Law, American Bar Association, Washington, DC, Nov 9, 1982, *reprinted in* 14 Sec Reg & L Rep (BNA) No 46, at 2057; *Sampson Says SEC Will Not Reconsider Decision to Eliminate Two Releases on Non-Audit Services,* 1 Acct Prac & Reg No 5, at 2 (Mar 11, 1985).

§6.65 Stockholder Involvement

Add to note 448, page 225:

See also Burks v Lasker, 441 US 471 (1979).

Add to note 452, page 225:

Also, the SEC's "going private" rules require that information be provided to security holders of a company proposing to go private which is equivalent to the information required by the SEC's proxy rules, 17 CFR §§240.14a-1 to 14a-101, even if the company is not subject to those rules, so long as it is a reporting company. 17 CFR §240.13e-3.

Add to note 454, page 225:

The SEC has substantially revised its rules relating to shareholder proposals. See Hudson, *SEC Tightens Annual Meeting Proposal Rules,* Wall St J, Aug 17, 1983, at 4, col 1; Securities Exchange Act of 1934 Release No 20091 (Aug 16, 1983).

§6.66 Securities and Exchange Commission (SEC) Administration and Enforcement—1934 Act

Page 226, add to third paragraph:

The SEC has established a pilot project for an electronic data gathering, analysis, and retrieval system (EDGAR) for electronic filing, processing, and dissemination of registration statements and reports filed under the securities laws. See **§6.35** and Securities Act of 1933 Release No 6519 (Mar 22, 1984); Securities Act of 1933 Release No 6539 (June 27, 1984); Securities Act of 1933 Release No 6540 (June 27, 1984); Securities Act of 1933 Release No 6548 (Sept 5, 1984); Securities Act of 1933 Release No 6557 (Nov 13, 1984).

Add to note 461, page 226:

Rowe, *Disclosure Statements in SEC Proceedings,* 16 Rev Sec Reg No 3 (Feb 2, 1983), at 1.

Add to note 467, page 227:

See generally Rowe, *Settlement of an SEC Enforcement Action,* 1 Sec Reg Guide (PH) ¶1121 (1980).

Page 227, add at end of section:

The SEC is seeking additional enforcement authority. The Insider Trading Sanctions Act of 1984, No 98-376 provided the SEC with authority to proceed administratively against persons who violate its tender offer regulations, 17 CFR §§240.14d-1 to 240.14d-100, and proxy rules, 17 CFR §§240.14a-1 to 14a-101. *See also SEC to Seek Expansion of Authority to Bring Administrative Proceedings,* 16 Sec Reg & L Rep (BNA) No 7, 267 (Feb 17, 1984); Ingersoll, *SEC Is Seeking To Punish Firms on False Data,* Wall St J, Feb 16, 1984, at 4; *SEC Votes to Increase Enforcement Authority,* SEC Today, No 84-33, at 1 (Feb 16, 1984) (SEC to seek legislation to permit it to proceed administratively against officers and directors who cause violations of the Exchange Act, but not to seek cease and desist authority); *SEC Review of Enforcement Remedies, Proposed Legislation on Administrative Proceedings, FOIA, Privilege,* 16 Sec Reg & L Rep (BNA) No 9, 456 (Mar 2, 1984) (full text of SEC statement to Congress). The SEC also intends to use its electronic filing system in enforcement investigations. *See generally Foreign "Blocking Laws" Hinder U.S. Enforcement, Fedders Says,* 16 Sec Reg & L Rep (BNA) No 13, at 578 (Mar 30, 1984).

The SEC also is concerned about effects of foreign secrecy and blocking laws on its enforcement program and has requested comments on the concept of foreign persons who use United States securities markets automatically waiving their rights under foreign secrecy laws, intended to protect the rights of individuals ("waiver by conduct"), but not blocking laws which are intended to protect sovereign interests. Securities Exchange Act of 1934 Release No 21186 (July 30, 1984).

§6.67 Liability

Add to note 472, page 227:

Herman & MacLean v Huddleston, 456 US 914 (1983); Chemetron Corp v Business Funds, Inc, 682 F2d 1149 (5th Cir 1982), *vacated,* __ US __, 103 S Ct 1245, *cert denied,* __ US __, 103 S Ct 1254 (1983).

Page 228, add at end of section:

A person may be convicted of criminal violations of the federal securities laws for misappropriation of material nonpublic information, but still not be subject to civil liability for the use of that information in securities transactions. *Compare United States v Newman,* 664 F2d 12 (2d Cir 1981), *affd after remand,* No 82-1273 (2d Cir), *cert denied,* __ US __, 104 S Ct 193 (1983) *and SEC v Materia,* 745 F2d 197 (2d Cir 1984), *cert pending,* Docket 84-1036 *with Moss v Morgan Stanley,* 719 F2d 5 (2d Cir 1983), *cert denied,* __ US __, 52 USLW 3611 (US Feb 21, 1984). Although the court in *Moss v Morgan Stanley* held for the defendants, in dicta it strongly supported the notion that violations of the antifraud provisions of the federal securities laws, principally 17 CFR §240.10b-5 and 15 USC §77q(a), can be a basis for liability under RICO, Racketeer Influenced and Corrupt Organizations Act, ch 96, Pub L No 91-452, §901(a), 84 Stat 941 (1970) (codified as amended at 18 USC §§1961-1968 (1982)), as would violations of the mail and wire fraud statutes, 18 USC §§1341 and 1343 (1982), which often are alleged in criminal prosecutions along with securities law counts for the same conduct. RICO, unlike the securities laws, provides for treble damages. 18 USC §1964(c). Accordingly, plaintiffs' lawyers now invariably include a RICO count in their complaints of violations of the securities laws.

Conclusion

§6.68 Professional Responsibility

Add to note 486, page 231:

The SEC took no action as a result of the comments it received.

Page 231, add at end of section:

The uncertainty as to the extent of the SEC's authority in this area, and as to its actual position as to the steps an attorney should take if the attorney suspects a client of being involved in a violation of the federal securities laws, also raises questions about the extent of the protection afforded by the attorney-client privilege and ethical rules precluding the divulging of client confidences; and also about who speaks for the client when an attorney represents an organization, such as a corporation. *See, e.g., Upjohn Co v United States,* 449 US 383 (1981); "Midyear Meeting of American Bar Association," 51 USLW No 32 (Feb 22, 1983), at 2488 (report of action by American Bar Association's House of Delegates on proposed Model Rules of Professional Conduct, including rules relating to confidentiality of information and organizations as clients). The House of Delegates action has drawn an angry reaction from some quarters. *See, e.g.*, Lawyers Duty of Disclosure Act of 1983, S 485, 98th Cong, 1st Sess (to amend mail fraud statute, 18 USC §1341, to make it a crime to use the mails in connection with a failure to timely disclose criminal or fraudulent acts committed or intended to be committed by clients).

§6.69 Commentary—Suggested Improvements in the Securities Laws

Page 231, add to item 2:

.The Final Report of the SEC Government-Business Forum on Small Business Capital Formation (Nov 1982), at 4, made a similar recommendation

Page 231, add at end of section:

6. 1934 Act reporting obligations should be reduced or eliminated for smaller companies, although the appropriate category of company may be difficult to define. 17 CFR §§230.157 and 240.0-10 (definition of small entities for purposes of the Regulatory Flexibility Act, Pub L No 96-354, §3, 94 Stat 1164 (1980)); 17 CFR §240.12g-1 (exemption from 1934 Act registration provisions for securities of issuers with total assets not exceeding $3 million); 17 CFR §§240.12g-4(a)(2) and .15d-6(a)(2) (exemption from certain 1934 Act reporting requirements for companies with fewer than 500 security holders and total assets not exceeding $3 million); see Securities Exchange Act of 1934 Release No 20784 (Mar 22, 1984)
7. Further consideration should be given to reducing the accounting overload on smaller businesses
8. The short-swing profit recapture provisions of §16(b) of the 1934 Act, 15 USC §78p(b), should be repealed and actions under SEC Rule 10b-5, 17 CFR §240.10b-5, and the Insider Trading Sanctions Act of 1984, No 98-376 should be relied upon to police insider trading, and *insider trading* should be defined by statute
9. The securities transaction reporting provisions of §16(a) of the 1934 Act, 15 USC §78p(a), should be repealed if §16(b) of that Act, 15 USC §78p(b), is repealed, in light of the duplications between §16(a) and §§13(d) and (g) of the 1934 Act, 15 USC §§78m(d) and (g), and SEC disclosure rules such as 17 CFR §229.403
10. The holding period for resales of unlimited amounts of securities by nonaffiliates of an issuer pursuant to SEC Rule 144, 17 CFR §230.144, should be reduced from three years to two years

11. The Racketeer Influenced and Corrupt Organizations Act, 18 USC §§1961-1968 (1982), should be amended to eliminate duplicative and treble damage remedies for securities fraud
12. The SEC's rules and administrative positions governing financial statements of an acquired business, principally 17 CFR §210.305, should be revised to remove absolute prohibitions on primary financings and resales by affiliates under Rule 144, 17 CFR §230.144, where the acquiring company does not obtain certified financial statements of the acquired or to-be-acquired business
13. SEC Rule 415, 17 CFR §230.415, should be liberalized to permit greater experimentation in methods of public financing of all businesses
14. Proposed Rule 205-3 should be adopted to permit investment advisers to charge performance-based fees to wealthy, sophisticated clients
15. Securities issued in connection with employee benefit plans should be exempt from registration under the 1933 Act and the 1934 Act
16. Congress should not enact legislation either to curb or to facilitate changes in corporate control without a foundation laid by a study of the scope of the Special Study of Securities Markets (Report of Special Study of Securities Markets of the Securities and Exchange Commission, HR Doc No 95, 88th Cong, 1st Sess (Aug 8, 1963)) that considers the economic impact of hostile takeovers and anti-takeover devices

Appendix 6-1 Comparison of 1933 Act Statutory Exemptions, Exemptive Rules and Small Issuer Registration Form

Page 234, under Regulation A, item 2, Limitations on General Solicitation or Advertising, add:

If in accordance with Rule 256(a)(1) or 256(c) and Rule 258. 17 CFR §§230.256(a)(1) and (c), 230.258

Page 234, under Rule 257, item 3, Nature of Offerees or Purchasers, delete "yes" and substitute:

No standards

Page 234, under Form S-18, item 5, Amount Which Can Be Sold, delete "$5,000,000" and substitute:

$7,500,000

Appendix 6-2 Comparison of 1933 Act Registration Forms

Page 236, under S-3, item 4, Information requirements, delete "Basically the same as old Form S-16" and substitute:

Relies on incorporation by reference of 1934 Act reports

Page 236, under Form S-8, item 2, Types of offerings covered, delete "no secondaries" and substitute:

secondaries permitted using Form S-3 prospectus, 17 CFR §239.16b.

Page 236, under Form S-18, item 3, Amount which can be registered, delete "$5,000,000" and substitute:

$7,500,000

Innovation and Environmental Law 7

Frank F. Skillern

General Introduction

§7.05 —Enforcement Stage

Add to note 1, page 241:

The Supreme Court recently reaffirmed a trial court's discretion to grant injunctive relief when a party is violating a regulatory statute. *See* Weinberger v Romero-Barcelo, 456 US 305 (1982); *accord,* Wisconsin v Weinberger, 745 F2d 412 (7th Cir 1984).

The National Environmental Policy Act

§7.12 Overview

Add to note 13, page 246:

A recent symposium explored in some detail the interactions between science, technology, and the law. See Symposium, *Law-Science Cooperation Under the National Environmental Policy Act,* 15 Nat Resources Law 569-662 (1983).

Add to note 14, page 246:

What constitutes "significantly affecting" and "the human environment" has been frequently litigated. Recently the Supreme Court refused to require that the NRC consider the damage to psychological health of individuals and the community from a decision to restart one unit at Three Mile Island, which had been closed in light of the 1979 accident. The Court required that there be a "sufficiently close" connection between the harm and the effect on the physical environment for psychological factors to necessitate an EIS. The Court believed that the causal relationship between the risk of a nuclear accident and psychological fears of area residents based on that risk was too remote. *See* Metropolitan Edison Co v People Against Nuclear Energy, 103 S Ct 1556, 18 Envt Rep Cas (BNA) 1985 (1983), discussed further in **§7.15**. *See also* Council on Environmental Quality, Final Regulations on Implementing NEPA, 40 CFR §1500 *et seq* (1982); *Forty most asked questions concerning CEQ's NEPA Regulations,* 46 Fed Reg 18026 (1981).

§7.14 —Federal Actions

Add to note 22, page 247:

One area of research getting special attention recently is genetic engineering concerning recombinant DNA. In one case, the D.C. Circuit concluded that, since NEPA imposed duties on agencies to consider the environmental consequences of their funding activities, the National Institutes of Health had to consider the environmental consequences of approving university research on recombinant DNA. The environmental analysis is required because of the potential for significant consequences for the human environment. Foundation on Economic Trends v Heckler, 756 F2d 143 (DC Cir 1985). *See also Gene Therapy: Research in Public,* 227 Sci 493 (Feb 1, 1985).

§7.15 —Comprehensive and Supplemental Environmental Impact Statements

Add to note 24, page 248:

Various situations may require a supplemental EIS on long-range projects or on projects done in phases. If plans change in a manner which results in different environmental effects, a supplemental EIS may be required. National Wildlife Fedn v Marsh, 721 F2d 767, 20 ERC 1158 (11th Cir 1983) (supplemental EIS needed even when the changes are beneficial). An additional EIS may be needed if inaccuracies are subsequently discovered or information becomes dated. Sierra Club v United States Corps of Engrs, 701 F2d 1011, 17 ERC 1748 (2d Cir 1983). If original estimates or projections for a project change dramatically, the initial EIS must be supplemented. Massachusetts v Watt, 716 F2d 946, 18 ERC 1145 (1st Cir 1983) (oil estimate for OCS leases on Georges Bank reduced by approximately 97%).

Page 248, at end of section add:

People Against Nuclear Energy (PANE) v NRC, 678 F2d 222 (DC Cir 1982), *revd,* 103 S Ct 1556, 18 Envt Rep Cas (BNA) 1985 (1983) illustrates the problem for an agency in determining when a supplemental EIS is needed. In *PANE* the NRC had decided not to prepare a supplemental EIS when permitting the utility to restart one of the undamaged Three Mile Island units that had been closed after the 1979 accident. The plaintiffs argued that a supplemental EIS was necessary in light of the accident to consider the psychological impact of restart on individuals in the area and on community well-being, including governmental, business, and commercial activities. The DC Circuit concluded that psychological factors were embraced by NEPA when they are combined with physical effects resulting from fears of a recurring disaster.

In *Metropolitan Edison Co v People Against Nuclear Energy (PANE),* 103 S Ct 1556, 18 Envt Rep Cas (BNA) 1985 (1983), the Supreme Court reversed the DC Circuit's decision. The Court said that psychological damage based on a risk of a nuclear accident does not have to be considered in an EIS. The Court noted that psychological factors are aspects of the human environment within the meaning of NEPA. However, when the psychological harm is based merely on a risk—an unrealized physical injury—the Court held that its relationship is too remote from an impact on the physical environment to be included in an EIS. According to the Court, the application of NEPA does not turn on the gravity of harm but on the connection between the harm and the physical environment. In the words of the Court, that connection must be "sufficiently close" for NEPA to apply.

In *NRDC v NRC,* 685 F2d 459 (DC Cir 1982), *revd sub nom Baltimore Gas & Elec Co v NRDC,* 103 S Ct 2246 (1983), the NRC adopted a generic rule to determine the environmental effects of fuel-cycle activities. Under the NRC's "Table S-3 rule," each proposed plant would not have to consider individually the impacts of the uranium fuel-cycle, but could use the numbers assigned environmental effects on Table S-3. That table gave no value to the spent fuel processing after final deposit of high-level radioactive waste in salt domes, reflecting the NRC's belief that such storage is safe and flawless.

The NRDC challenged the rule, which was based on this zero-release assumption, on the grounds that it inadequately considered the uncertainties of permanent storage for high-level radioactive waste, as well as the health and socioeconomic impacts of waste management activities. The DC Circuit concluded that the NRC had not adequately considered relevant factors under NEPA and hence invalidated the rule. The court remanded the rules to the NRC for consideration of health, socioeconomic, and cumulative effects of the fuel-cycle activities.

The Supreme Court in *Baltimore Gas & Electric Co v Natural Resources*

Defense Council, 103 S Ct 2246 (1983), reversed the DC Circuit's decision in *NRDC v NRC.* The Court concluded that the agency had taken the requisite "hard look" at the environmental ramifications of the rule. In addition, the Court held that the zero-release assumption was a policy judgment applied for limited purposes. As such, it was within the bounds of reasoned decisionmaking under the APA. The Court did not agree with the DC Circuit that the rule precluded licensing boards from considering all of the environmental effects as required by NEPA.

The Court also warned lower courts of intruding improperly into matters appropriately within an agency's expertise. Those areas, like "new scientific frontiers" about which the agency must make projections or predictions, are ones particularly deserving of great deference from a reviewing court.

§7.20 —Incomplete or Unavailable Information

Page 251, add at end of section:

Several cases have interpreted the so-called worst case analysis required under the CEQ regulations, 40 CFR §1522 (1983). In a significant case, *Sierra Club v Sigler,* 695 F2d 957, 18 ERC 1649 (5th Cir 1983), the court concluded that the worst case analysis was authorized both under NEPA itself and by prior case law interpreting NEPA. In *Sigler,* the Sierra Club argued that the Corps of Engineers' EIS was inadequate because it did not include a worst case analysis of a catastrophic, total cargo loss by a supertanker in Galveston Bay, as hypothesized by the plaintiffs.

The court proceeded to determine whether the hypothetical catastrophe had to be covered in an EIS by considering if it was the type of incident the regulations required an agency to examine. The court concluded that the low-probability-but-catastrophic event was precisely the type of accident intended to be included in the worst case situation by the regulations. The court also distinguished between subparagraphs (b)(1) and (b)(2) of the regulations. Subparagraph (b)(1) concerns situations where essential or unknown information is missing. Subparagraph (b)(2) deals with situations requiring information about adverse impacts that may be significant and important to a decision, but whose control is beyond the state of the art. The court concluded that, under the second category, a total oil spill by a supertanker would constitute a significant impact, and means to control such an event are beyond the state of the art. Hence, the agency has to prepare a worst case analysis concerning the impact in the EIS.

The Fifth Circuit also had to consider the problem of the remoteness or likelihood of the catastrophic event ever occurring. The agency argued that the event was too unlikely to occur, and a worst case analysis scenario would be too speculative. The court concluded that the regulations require an analysis of the worst case *and* a discussion of the probability of its occurrence. Thus, remoteness or likelihood of an event occurring does not except it from inclusion in the EIS.

Similar results have been reached by other courts. *See Southern Oregon Citizens Against Toxic Sprays, Inc v Clark,* 720 F2d 1475, 20 ERC 1001 (9th Cir 1983) (worst case analysis required on Bureau of Land Management's proposed spraying of pesticides). *But see Conservation Law Foundation v Watt,* 560 F Supp 561, 18 ERC 1904 (D Mass 1983); *Village of False Pass v Watt,* 565 F Supp 1123, 18 ERC 2129 (D Alaska 1983) (worst case analysis on oil spills not required for OCS lease sales which do not authorize actual drilling).

Under the mandate in *Southern Oregon Citizens Against Toxic Sprays,* the BLM prepared a worst case analysis for one of its districts in Oregon. In *Save Our Ecosystems v Clark,* 747 F2d 1240, (9th Cir 1984), the Ninth Circuit reviewed the BLM's EIS containing that worst case analysis, and concluded that the analysis did not adequately cover a "spectrum of events" required under the CEQ regulations and other cases, including *Sigler.Id* 1245. In addition, the court concluded that the Forest Service could not rely on the EPA's registration of herbicides under FIFRA to justify not preparing an adequate EIS. The court also modified the injunctive relief to enjoin all aerial spraying in any BLM or Forest Service district in Oregon.

The cases concerning worst case analysis have presented special problems to the CEQ. The CEQ has issued an advance notice of proposed rulemaking concerning modifying §1502.22 of the Final Regulations. The CEQ's concern is whether the judicial interpretation is reflected in the questions on which comments were solicited:

1. Under what circumstances and to what extent must a federal agency engage in forecasting or speculation when confronted with scientific uncertainty or gaps in information concerning the environmental effects of a proposed action?
2. How can an analysis be structured to present reasonable forecasting in the face of scientific uncertainty or information gaps about the effects of proposed action to provide more useful and understandable information for decisionmakers and other interested parties?
3. Does the type of analysis called for in 40 CFR §1502.22 require federal agencies to go beyond the "rule of reason," as traditionally expressed in judicial decisions interpreting NEPA?
4. Should a threshold standard be established which would trigger the preparation of the type of analysis identified in response to question 1, such as a threshold of severe consequences, a threshold of probability, or a threshold of scientific credibility?

5. Is the term "worst case" appropriate for this type of analysis? If so, how should it be defined? If not, what is the most appropriate term for this type of analysis, and how should it be defined?

49 Fed Reg 50744 (Dec 31, 1984) (advance notice of proposed rulemaking).

§7.22 —Avoiding Bias

Page 253, add at end of section:

Courts continue to be wary of agencies merely rubberstamping an EIS to support a proposed project. An extreme example was *Sierra Club v United States Army Corps of Engineers,* 701 F2d 1011, 17 ERC 1748 (2d Cir 1983). In that case the district court found that the agencies had not made good faith efforts to comply with applicable laws. The agencies had been notoriously dilatory in complying with NEPA, left out relevant information in the EIS, and generally were nonresponsive to critical comments from other agencies or public interest groups. The district court had ordered the appointment of a special master to oversee the preparation by the agency of an adequate supplemental EIS in a timely fashion.

The Second Circuit agreed that the EIS was inadequate and required that the agency prepare a supplemental EIS in good faith. It disagreed, however, that the situation was so severe as to justify invoking the extraordinary equitable remedy of appointment of a special master. *See also Sierra Club v Sigler,* 695 F2d 957, 18 ERC 1649 (5th Cir 1983), in which the Fifth Circuit indicates its suspicions that the agency was merely rubberstamping a project and not impartially evaluating the environmental data.

The Clean Air Act

§7.23 Technology-Forcing Standards

Add to note 41, page 253:

The EPA recently revoked the NAAQS for hydrocarbons. 48 Fed Reg 628 (1983).

Add to note 42, page 253:

The EPA was ordered to issue hazardous emission pollutant standards for radionuclides in Sierra Club v Gorsuch, __ F Supp __, 17 ERC 1748 (ND Cal 1982). Those standards were proposed in 1983.

48 Fed Reg 15076 (Apr 6, 1983). A similar result was reached more recently with respect to inorganic arsenic for which the EPA was unwilling to issue standards. In New York v Gorsuch, 554 F Supp 1060, 18 ERC 1575 (SDNY 1983), the EPA was ordered to issue appropriate standards within 180 days. Emission standards for inorganic arsenic were promulgated on July 20, 1983. 48 Fed Reg 33112.

Add to note 46, page 253:

The acid precipitation or acid rain problem remains controversial. The EPA continues to maintain its position that additional studies must be conducted on the nature and amounts of sulfur emitted and their effects when transported interstate or internationally. To that end, the EPA has published a two volume, draft study entitled, "The Acidic Deposition Phenomenon and Its Effects: Critical Assessment Review Papers." The Critical Assessment Review Papers were issued for public review and comment in May 1983. *See also* Lucas, *Acid Rain: The Canadian Position,* 32 Kan L Rev 165 (1983).

The states of New York and Connecticut have tried to address the transboundary air pollution problem by forcing the EPA to be more critical in its review of revisions to state SIPs. Specifically, the states have challenged the EPA's policy of approving revisions to SIPs which will allow increases in emissions of sulfur dioxide by stationary sources based only on an evaluation of the impact of the increased emissions on the immediate area around the source. Other states have also contended that the EPA should consider the effects of the emissions in other states as well.

The Second, Sixth, and Seventh Circuits consistently have dismissed petitions for review of the EPA's approval of revisions to SIPs. Those courts have concluded that the EPA does not have to consider the interstate effect of pollution in determining whether to approve a revision to an individual state's SIP. The courts determined that the appropriate forum and proceeding regarding interstate pollution effects is the §7426 proceeding. *See* New York v EPA, 710 F2d 1200, 19 ERC 1367 (6th Cir 1983) (revision to Tennessee's SIP); Connecticut v EPA, 696 F2d 147, 18 ERC 1418 (2d Cir 1982) (revision to New York's SIP); New York v EPA, 716 F2d 440, 19 ERC 1662 (7th Cir 1983) (revision to Illinois's SIP). *See also* 42 USC §7426.

The problem of acid rain has also received considerable academic commentary and attention. *E.g.,* Gallogly, *Acid Precipitation: Can the Clean Air Act Handle It?,* 9 BC Envtl Aff L Rev 687 (1981);. Kramer, *Transboundary Air Pollution and the Clean Air Act: An Historical Perspective,* 32 Kan L Rev 181 (1983); Lutz, *Managing a Boundless Resource: United States Approaches to Transboundary Air Quality Control,* 11 Envtl L 321 (1980); Ostrov, *Interboundary Stationary Source Pollution—Clean Air Act §126 and Beyond,* 8 Colum J Envtl L 37 (1982).

The EPA also has published its draft study of acid rain. *See*

Environmental Protection Agency, The Acidic Deposition Phenomenon and its Effects: Critical Assessment Review Papers (vols 1 & 2, 1983). These draft papers have been circulated for public review. *See also* Crawford, *The Problems of Causation in Private Legal Remedies for Damage from Acid Rain,* 17 Nat Resources L 413 (1984); Maugh, *Acid Rain's Effect on People Assessed,* 226 Sci 1408 (Dec 21, 1984).

§7.25 —Emission Limitations

Add to note 52, page 255:

The EPA has been ordered to issue emission limitations for radionuclides as a hazardous air pollutant. Sierra Club v Gorsuch, 17 ERC 1748 (ND Cal 1982).

§7.30 —Innovative Technology Waiver

Add to note 70, page 258:

The EPA's noncompliance regulations were reviewed in Duquesne Light Co v EPA, 698 F2d 456, 18 ERC 1489 (DC Cir 1983). In an extensive opinion, the DC Circuit sustained for the most part the EPA's rule. *See also* 48 Fed Reg 26627 (June 9, 1983) (proposed interruptive rulemaking).

§7.31 Bubble Concept

In note 72, page 258, second line, delete "603.07 and 623.28" and insert:

603-07 and 623-28

Add to note 72, page 258:

The use of the bubble concept generally was expounded on by the EPA in its proposed Emissions Trading Policy Statement. 47 Fed Reg 15076 (1982). *See* NRDC v Gorsuch, 685 F2d 718 (DC Cir 1982), which disallowed the use of the bubble concept in nonattainment areas. That decision was subsequently reversed *sub nom* Chevron USA, Inc v Natural Resources Defense Council, Inc, 104 S Ct 2778 (1984). In *Chevron,* the Court reviewed extensively the history of the bubble concept and the differing interpretations of it. The Supreme Court concluded that the statutory term *stationary source* was not defined in the statute, and gave substantial deference to the EPA's interpretation of the statute. The Court stated, "We hold that EPA's definition of the term 'source' is a permissible construction of the statute which seeks to accommodate progress in reducing air pollution with economic growth." *Id* 2793-94. Hence, the DC Circuit's opinion in *Gorsuch* was overruled, and the EPA's plantwide definition for nonattainment areas sustained. See also **§7.35.**

§7.32 —Prevention of Significant Deterioration

Add to note 73, page 259:

See also Emissions Trading Policy Statement, 47 Fed Reg 15076 (1982) (EPA's proposed policy statement and accompanying technical issues document).

§7.34 —Nonattainment Areas

Add to note 81, page 260:

The EPA issued a policy statement discussing further permissible use of the bubble concept and emission trade-offs generally in nonattainment and clean air areas. Emissions Trading Policy Statement, 47 Fed Reg 15076 (1982).

Add to note 82, page 260:

The 1981 regulations authorizing use of the bubble concept in nonattainment areas were invalidated in NRDC v Gorsuch, 685 F2d 718 (DC Cir 1982). That decision also will affect the use of the bubble concept discussed in the proposed Emissions Trading Policy Statement, 47 Fed Reg 15076 (1982).

§7.35 —Current Environmental Protection Agency Regulations

Add to note 83, page 260:

See also Emissions Trading Policy Statement, 47 Fed Reg 15076 (1982) (proposed policy statement).

Add to note 84, page 260:

See also Emissions Trading Policy Statement, 47 Fed Reg 15076 (1982) (proposed policy statement).

Pages 260 and 261, delete last two sentences in section and insert:

Use of the bubble concept in nonattainment areas was proposed by the EPA after the *Alabama Power* decision. That action was challenged judicially by the NRDC. The DC Circuit in *NRDC v Gorsuch,* 685 F2d 718 (DC Cir 1982), *revd sub nom Chevron USA, Inc v Ruckelshaus,* 104 S Ct 2778 (1984), relied on its prior decisions in *Alabama Power* and *ASARCO* to invalidate use of the bubble concept in nonattainment areas. The court relied on the distinction between programs designed to maintain existing air quality and those designed to improve air quality. According to the court, the nonattainment program is designed to improve air quality, and hence the bubble concept is incompatible with it.

In *Chevron USA, Inc v Ruckelshaus,* 104 S Ct 2778 (1984), the Supreme Court reversed the *Gorsuch* holding. The Court concluded that the EPA's plantwide definition of source for nonattainment areas was a permissible construction of the Clean Air Act. The Court reviewed extensively the legislative history and the statute, and concluded that Congress had not expressed an intent concerning the bubble policy. Hence, it was a judgment appropriately made by the agency, not a court. The Court stated that the agency's interpretation was entitled to substantial deference and that its definition of the word *source* was a permissible construction of the Act.

The Clean Water Act

§7.36 Technology-Forcing Aspects

In note 93, page 261, delete "Id §1311(1)(A)-(B) and (2)(B)." and insert:

Id §1311(b)(1)(B).

§7.37 —Technological Factors

In note 95, page 262, delete text of note and insert:

33 USC §1311(b)(2)(A).

Page 262, delete fourth line in second paragraph of section and insert:

conventional pollutant control technology (BCT), but are given a longer time to achieve

In note 96, page 262, delete text of note and insert:

Id §1311(b)(2)(E). EPA's original conventional pollutant standards were reviewed in American Paper Inst v EPA, 660 F2d 954 (4th Cir 1981). The Fourth Circuit invalidated the standards because the EPA had not considered all of the statutorily required factors. Subsequently the EPA proposed new, revised effluent limitation guidelines for best conventional pollutant control technology (BCT). 47 Fed Reg 49176 (1982) (proposed rule). *See also* 33 USC §1314(a)(4).

§7.39 —Variances

Add to note 104, page 263:

Accord Appalachian Power Co v EPA, 671 F2d 801 (4th Cir 1982); Crown Simpson Pulp Co v Costle, 642 F2d 323 (9th Cir 1981).

Add to note 106, page 264:

More recently, the fundamentally different factors variance provisions for the CWA were sustained in Chemical Mfrs Assn v NRDC, 105 S Ct 1102 (1985). This was the first time the Supreme Court had looked at the variance provision as applied to toxic pollutants. The Court rejected the argument the §1321(1)(1) prohibited variances for effluent limitations for toxic pollutants, adhering to its now frequently invoked principle of substantial deference to an agency's interpretation (*see* Chevron USA, Inc v Ruckelshaus, 104 S Ct 2778 (1984), discussed in **§7.35**), and sustained the EPA's interpretation of the Act. The Court concluded that the agency's interpretation was sufficiently rational to preclude a reviewing court from substituting its judgment for that of the agency.

Regulation of Chemicals and Wastes

§7.44 Registration of Pesticides

Add to note 129, page 268:

Proprietary claims to data submitted in support of registration of pesticides either on or after the 1972 and 1978 amendments to FIFRA have been asserted by registrants. The registrants have argued that use of submitted data by the EPA to support registration of the chemical for a new use by another company if that company will compensate the original submitter constitutes a taking of property for private purposes without just compensation. These claims and related ones concerning required public disclosure of data supporting a registration under the act usually have been rejected by the courts that have considered them. *See, e.g.*, Chevron Chem Co v Costle, 641 F2d 104 (3d Cir 1981) (manufacturer lacks proprietary interest in data submitted to EPA), *cert denied*, 452 US 961 (1981); Mobay Chem Co v Costle, 517 F Supp 254 (WD Pa 1981) (equal protection and due process taking challenges rejected), *affd sub nom* Mobay Chem Co v Gorsuch, 682 F2d 419, *cert denied*, 103 S Ct 343 (1982); Petrolite Corp v EPA, 519 F Supp 966 (ED Pa 1981) (use of submitted data to support other applications not a taking). *But see* Monsanto Co v EPA, 564 F Supp 552, 18 ERC 2081 (ED Mo 1983), *probable jurisdiction noted*, 104 S Ct 231 (1983) (submitter has property rights under state law in health and safety data and mandatory disclosure would be an unconstitutional taking). The district court in *Monsanto* enjoined enforcement of the disclosure provisions of the statute. The EPA's application for stay of the injunction pending review by the Supreme Court was denied. Ruckelshaus v Monsanto Co, 104 S Ct 3 (1983) (Justice Blackmun).

In Ruckelshaus v Monsanto Co, 104 S Ct 2862 (1984), the Supreme Court addressed the problem of the company's proprietary interest in submitted data and the confidentiality requirements of FIFRA. The Court concluded that whether information was confidential as a trade secret would be determined under state law. It found that whether the information was "taken" in a constitutional sense by a required disclosure would turn on the provisions of the Act in effect when the application was filed, and concluded that for registration sought under the original FIFRA, there was no reasonable expectation of confidentiality. Moreover, the Court noted that under the Pesticides Act of 1978, notice was given to applicants for registration that the data could be disclosed and used by the EPA in conjunction with other applications. For data submitted between 1972 and 1978, however, the Court held that there could be a taking by disclosure. Whether a taking had actually occurred would turn on how the compensation provisions of the Act were applied, which had not yet been done. Hence the Court did not resolve whether in fact a taking had occurred in the instant case.

In a subsequent case, Monsanto Co v Ruckelshaus, 753 F2d 649 (8th Cir 1985), the Eighth Circuit held that the EPA did not have to disclose the name of a registration applicant or the active ingredient of a pesticide for which registration was sought to a manufacturer.

Currently under the Act the data submitted to register pesticides after September 30, 1978, is protected for 10 years from use to support another application unless consent of the original submitter is obtained. With respect to data submitted after December 31, 1969, the EPA can use the data originally submitted without the consent of the submitter during a 15-year period after submission if the other applicant is willing to compensate the original submitter. 7 USC §136a(c)(1)(D)(i) and (ii).

Registration may be required even if the product is not being marketed as a pesticide. In N Jonas & Co, v EPA, 666 F2d 839 (3d Cir 1981), the EPA assessed a penalty for failing to register a swimming pool oxidizer product as a pesticide. Even though the company did not market or sell the product as a pesticide, and its unregistered label disclaimed any pesticide characteristics of the product, the fine was upheld by the court. The court applied an objective standard to determine that the product was within the statutory meaning of *pesticide* and hence had to be registered.

§7.52 Resource Conservation and Recovery Act

Add to note 149, page 272:

The Resource Conservation and Recovery Act was amended by the Hazardous and Solid Waste Act Amendments of 1984, Pub L No 98-616, 98 Stat 3221. These amendments extensively changed the requirements and regulations for disposing and handling of hazardous and solid waste. In particular, new regulations are required for land and surface disposal facilities. Moreover, minimum technology requirements are imposed on disposal facilities under the Act. More stringent restrictions are placed on the burning or blending of hazardous wastes. The amendments also contain new provisions concerning listing and delisting of hazardous wastes which would be subject to regulation.

The focus of the 1984 Amendments is on a more stringent regulatory program. The EPA recently issued regulations to implement parts of the 1984 Amendments. *See* 50 Fed Reg 21772 (May 28, 1985) (proposed rule—underground storage tanks).

§7.53 —Hazardous Waste

Add to note 150, page 272:

See 40 CFR pt 262 (generator standards), pt 263 (transporter standards), and pt 264 (treatment, storage, and disposal standards) (1982).

The requirements for generators, transporters, and treatment, storage, and disposal facilities were changed significantly in 1984. *See* Hazardous and Solid Waste Act Amendments of 1984, Pub L No 98-616, 98 Stat 2227.

Miscellaneous Considerations

§7.56 Need for Innovation in Testing, Monitoring and Other Compliance and Standard-Setting Techniques

Add to note 161, page 274:

Interesting constitutional issues also can be presented by different techniques for enforcement of regulatory statutes. In Dow Chem Co v United States, 536 F Supp 1355 (ED Mich 1982), the company challenged the EPA's inspection by use of aerial photography to determine that a violation existed. The company argued that the aerial photography gave access to information that would not be available by entry or a visual inspection and hence was an unreasonable search and seizure. It also argued that aerial photography inspections were not authorized by the Clean Air Act.

The district court agreed with the company's arguments. It concluded that the aerial photography and overflights constituted an unreasonable search and were not authorized by the Clean Air Act. As such, they were an unreasonable interference with the company's constitutionally protected property rights.

See also Mobil Oil Corp v EPA, 716 F2d 1187, 19 ERC 2043 (7th Cir 1983) (CWA authorizes the EPA to conduct administrative searches allowing collection of samples of a discharger's untreated waste water, as well as of waste water required to be treated pursuant to a permit, if necessary to enforce the permit).

State and Local Regulation of Innovation

Bruce M. Kramer

8

§8.19 Direct Governmental Intervention: Land Use Controls—In General

One of the problems with direct governmental intervention in the marketplace is conflicting objectives or goals. In the case of solar energy, there can be a substantial conflict in the land use area between active and passive solar energy systems. For example, a passive solar energy system would encourage the planting of tall trees to provide shade in the summer. These same trees, if located near a lot line, may block the sun during the summer from solar collectors on a neighbor's roof. In California, the state has attempted to deal with the problem of how to reconcile these two innovative technologies through the passage of the California Shade Control Act, Cal Pub Res Code §§25980-86 (West Supp 1982).

The legislature could have preferred one of the innovative technologies over the other, but instead delegated to the judicial system the problem of reconciling disputes on an ad hoc basis. In most cases, the provisions of the Shade Control Act favor the one who has built an active solar system by entitling that party to have access to the sun. Cal Pub Res Code §25982 (West Supp 1982). But another party planning to use a passive solar technology may petition a court to allow an exemption from the general access requirements, if the court finds that the passive technology would provide a greater net energy savings than the active system which would be impacted. *Id* §25986.

The statutory mechanism, in attempting to coordinate the development of two potentially competing new technologies, has created a great deal of uncertainty because of the ad hoc nature of the litigation process. However, an opposite result would have been to favor active systems over passive systems, at least where they are in conflict, which, given the state-of-the-art knowledge of the two systems, was probably undesirable.

Add to note 156, page 307:

For a general discussion of the problems caused by land use control systems and solar access, see Eisenstadt, *Access to Solar Energy: The Problem and its Current Status,* 22 Nat Resources J 21 (1982).

§8.20 —Substantive Provisions

Page 309, at end of section add:

The spread of innovative alternative energy products and systems has led to some problems with the extant land use regulatory system, necessitating the use of the variance system. In *Ormond Beach v State ex rel Del Marco,* 426 So 2d 1029 (Fla 1983), the problem of conflicting state and local regulatory provisions and the proper standards to apply in the issuance of a variance were discussed. The landowner sought a variance from the height and setback requirements of the local zoning ordinance in order to erect a windmill on his oceanfront residential property. The variance was denied by the board of adjustment on the ground that there was no showing of hardship, a requirement of both the state enabling act and the local ordinance. The board of adjustment ignored another Florida statute which encouraged the use of alternative energy systems by preventing local legislative and administrative bodies from adopting measures which drive up the cost or make impossible the owning or operating of an alternative and innovative system or product of energy production. Fla Stat Ann §163.04 (West Supp 1985). The two state statutes gave conflicting commands to the decisionmakers. The board of adjustment followed the variance statute requiring it to issue variances only upon a showing of unnecessary hardship. The trial court concluded that the statute dealing with innovative energy products prevailed. The appellate court, however, tried to reconcile the two statutory directives and concluded that one should not prevail to the exclusion of the other. While cities and their boards of adjustment cannot prohibit the construction and maintenance of renewable energy sources, they do not have to grant every variance application merely because it would be inconvenient to locate the innovative energy source elsewhere. The court remanded the case for further factual exploration of where the device could be located on the property to minimize or eliminate the need for a variance if it would make the product economically or technologically infeasible to operate. 426 So 2d at 1032.

§8.21 —Modification of the Regulatory Scheme

Page 309, at end of section add:

A land use regulatory system can be modified to deal with innovative products or systems, although in the process local governments are likely to add minimum performance standards regarding the operation of the innovative product or system. In *Rose v Chaikin,* 187 NJ Super 210, 453 A2d 1378 (1982), the problems inherent in an ex post facto regulation of an innovative product appeared. The defendants wanted to erect a windmill as an alternative energy source on their property. At the time of construction neither the defendants nor apparently, the building permit officer knew about an ordinance which limited the noise emanating across property lines to 50 decibels (dBa). A permit was issued and the windmill, during operation, created noise in excess of the 50 dBa limit. Plaintiffs were the neighbors who sought to enjoin the continued operation of the windmill on both common law nuisance and statutory grounds. The municipal authorities, who had been notified of the alleged violation, had apparently done nothing to enforce the noise limits in their zoning ordinance. 453 A2d at 1381-82.

The court upheld the validity of the 50 dBa limit in the zoning ordinance, which had been enacted to deal with certain types of alternative energy systems, including windmills. The municipality's failure to enforce its own ordinance allowed private parties to seek enforcement if they received a special injury. In this case, the injury was clear and the windmill had to comply with the noise limits. The court did not make a serious inquiry into the 50 dBa limit, choosing instead to take a very deferential view of the validity of the "windmill" ordinance. 453 A2d at 1384. It is interesting to note that the plaintiff's heat pump also violated the 50 dBa limit, because of its location, but the court suggested that because the heat pump was rarely used it was not an actionable nuisance, although it could be subject to an action to enforce the municipality's noise ordinance, which also applied to heat pumps.

§8.29 —Dormant Commerce Clause

Page 319, text, at end of first full paragraph add:

It is important to note that the Supreme Court has recently limited the application of the dormant commerce clause to state *regulatory* activity. If the state is a market participant rather than a market regulator, the issue of undue burdens or discrimination against interstate commerce does not apply. *Reeves, Inc v Stake,* 447 US 429 (1980); *Hughes v Alexandria Scrap Co,* 426 US 793 (1976). See **§3.09.**

§8.29A State and Local Franchising and Contracting (New)

State and local governments can have an indirect impact on innovative technologies or products through the exercise of their ability to franchise or contract for the delivery of goods and services. A principal example of how local governments can affect an innovative technology by its policy in the franchising field is cable television (CATV). The tremendous expansion of CATV has brought with it a corresponding strain in some existing legal doctrines and has provided a fertile source of litigation. Much of that litigation deals with the problem of franchising municipal CATV operators, through exclusive or nonexclusive contracts, for the provision of CATV services to municipal residents. The result of this litigation may be a stifling of the proliferation of CATV and its more recent offshooots, to the detriment of both the industry and the public.

Two recent cases explore the pitfalls that underlie the municipal franchising agreement as it relates to an innovative technology. They illustrate the complexity of involving state and local governments in the distribution of a new technology, even where the governmental body is seeking to assist or aid in the dissemination to the public of that technology.

The landmark case is *Community Communications Co, Inc v City of Boulder,* 455 US 40 (1982). (A major law review has already published a symposium on the *Boulder* decision. *See, e.g.*, Williamson, *Commentary: The Reagan Administration's Position on Antitrust Liability of Municipalities,* 32 Cath UL Rev 371 (1983); Civiletti, *The Fallout From Community Communications Co. v. City of Boulder: Prospects for a Legislative Solution,* 32 Cath UL Rev 379 (1983); Vanderstar, *Liability of Municipalities Under the Antitrust Laws: Litigation Strategies,* 32 Cath UL Rev 395 (1983); Note, *Community Communications Co. v City of Boulder: The Emasculation of Municipal Immunity from Sherman Act Liability,* 32 Cath UL Rev 413 (1983).)

The facts in *Boulder* were rather simple. In 1966, CCC had purchased from another party a 20-year, revocable, nonexclusive franchise agreement to provide cable television services within the Boulder city limits. Prior to 1980, neither the economics nor the technology would apparently support anything more than a limited use of the franchise in one part of the city with restricted channel options. In 1979, new technology was available so that CCC wanted to expand its services and service area. The City Council, in response to the new technology and solicitations of interest from potential competitors of CCC, reviewed its policy and decided to enact a moratorium on CCC's right to expand its cable service operation. The ostensible purpose was to foster competition, because the Council had determined that once a cable distribution system is set up a natural monopoly tends to occur.

The sole issue in *Boulder* was whether or not the city was entitled to an exemption from the application of the antitrust laws under *Parker v Brown,* 317 US 341 (1943). *Parker* had created an exemption for "state action" that immunized certain types of state activity from potential antitrust liability. The *Parker* exemption was held not to apply per se to municipalities in *City of Lafayette v Louisiana Power & Light Co,* 435 US 389 (1978). See **§3.11.** The test for a state action exemption that had emerged prior to *Boulder,* which was confirmed by the majority decision therein, involves a two-step analysis. The first requires that the state's policy relating to anticompetitive behavior must be "clearly articulated and affirmatively expressed." 455 US at 52. The second prong requires active state supervision of the anticompetitive policy. This second prong was not discussed since the Court determined that Boulder had failed the first test.

Boulder argued that Colorado, by constitutionally mandated non-preemptible home rule, had given Boulder sufficient authority to regulate in a potentially anticompetitive manner where it was an otherwise reasonable exercise of the police power. The Court disagreed and concluded that the mere granting of home rule authority on local matters to substate units would not satisfy the first prong of the *Parker* state action exemption.

One must remember that the Court never got to the merits of the antitrust allegation. Its sole holding was that home rule status alone will not provide the clearly articulated and affirmatively expressed state policy necessary under prior decisions to engage in regulatory behavior which may have an anticompetitive effect.

An excellent example of how to avoid dealing with governmental bodies in an anticompetitive manner is illustrated by the facts of *Affiliated Capital Corp v City of Houston,* 700 F2d 226 (5th Cir 1983). In that case, CATV was again involved. The city wanted to license private operators to provide CATV service to the residents of the city. However, the mayor and city council implemented that goal in a way which violated the antitrust laws. After the voters of Houston, in 1972, had rejected a monopoly franchise system covering the entire city, several potential operators began informal negotiations with the mayor over how to best provide service. The result of these negotiations led to the mayor requesting the competitors to come up with a plan by themselves for dividing up service areas and present it to the council and the mayor for action. The plaintiff in this case wanted to make an independent presentation on the merits, but was referred to the private consortium by a member of the city council. Plaintiffs refused to join the consortium and were frozen out of the eventual franchise package.

This kind of division of operating areas falls within the classic horizontal territorial restraint cases which, as described in **§3.03,** are per se violations of the Sherman Antitrust Act. The fact that the horizontal restraint was given a stamp of approval by the mayor and city council did not render the restraint exempt from operation of the antitrust laws. The activities of the other CATV operators, had there been no municipal approval, would have violated §1 of the Sherman

Antitrust Act, according to the majority opinion. Thus the operators and the city were liable for treble damages, since the city did not fit within the *Parker* or *Boulder* state action exemption.

The panel decision of the 5th Circuit in *Affiliated Capital Corp v City of Houston,* 735 F2d 1555 (5th Cir 1984), was reversed by an en banc decision which reinstated part of the original jury verdict finding that there was anticompetitive behavior by the private parties involved in the CATV franchising process. The court held that the jury could have found that the defendants had acted in a conspiracy to deny the plaintiff's access to the CATV franchising decision, and upheld the jury's inference that the plaintiff would have received the franchise were it not for the anticompetitive behavior of the defendant. The city itself was dismissed from the litigation voluntarily although the mayor was a defendant at the time of the en banc decision. The court concluded that the mayor was not acting in a legislative capacity during the time in which the city was carved into separate territories, and therefore was not entitled to absolute immunity under *Hernandez v City of Lafayette,* 643 F2d 1188 (5th Cir 1981), *cert denied,* 455 US 901 (1982). Instead, the mayor would be entitled to qualified immunity under the standards set out in *Harlow v Fitzgerald,* 457 US 800 (1982).

Municipal regulation or franchising operations must contend with antitrust limitations after *Boulder.* Unless there is a clearly articulated state policy and active state supervision relating to an anticompetitive policy, municipalities which engage in anticompetitive behavior may be subject to potential liability under the various federal antitrust laws. See generally **ch 3.**

The reality of potential antitrust liability of local governments for their actions has come to pass with a multi-million dollar judgment being entered against several Illinois local governmental units for activities relating to the exercise of their land use and other police powers. *Unity Ventures v County of Lake,* __F Supp__ (ND Ill 1984) (No 81-C-2745). The $28.5 million judgment awarded is presently on appeal, but it represents some 6000 per cent of the tax receipts of the Village of Grayslake and 150 per cent of the tax receipts of the County of Lake, both of whom would be principally responsible for paying the judgment.

The problem combines the provision of sewage collection services in this suburban Chicago area and the county's efforts to regionalize sewage collection in order to achieve economies of scale, and achieve federal and state water pollution effluent standards. In order to induce the various local governmental units within its boundaries to participate in an intergovernmental cooperation agreement, the county placed a sphere of influence provision in said agreement. That provision essentially gave to the various municipalities the right to veto any development within their sphere of influence before it would be allowed to receive the regional sewer services available in the county.

The plaintiffs in this case owned some 585 developable acres in an unincorporated area of the county. For several years, starting in 1972, the plaintiff sought to obtain sewer services from various public and private entities, but his efforts were unsuccessful. He then sought an annexation agreement with another municipality which would have authorized a rather large-scale and intensive development on the property. That municipality did not have a public sewage treatment facility. Simultaneously, the county was developing its regional sewage treatment plan and soon began negotiations with its resident municipalities regarding their mutual needs. These negotiations concluded with the signing of the intergovernmental agreement which included the sphere of influence provision.

Plaintiff alleged that there was a conspiracy to deprive the plaintiff of his right to develop his property and to displace competition through such acts as the gerrymandering of the Grayslake's sphere of influence territory, Grayslake's veto of the plaintiff's proposal to connect to one of the available regional sewer interceptor lines, and the defendants' objections to plaintiff's plan to build his own "package plant" to service the proposed development. The jury agreed and awarded damages which were trebled by the judge in accordance with the Sherman Antitrust Act, 15 USC § 15. On appeal, the defendants are arguing that intergovernmental cooperation agreements to provide services, or otherwise engage in police power functions such as franchising, are not covered by the conspiracy provisions of the Sherman Antitrust Act. Since it has been reported that at least 60 per cent of the local governmental units in the United States have entered into some form of intergovernmental cooperation agreement, the ramifications of this case are obvious.

If this decision should remain intact, the franchising and contracting powers of local governmental units would be more circumscribed than before. In addition, intergovernmental agreements covering franchising and contracting would immediately become suspect under the antitrust laws. The potential for substantial recoveries against local governmental bodies would undoubtedly have a chilling effect on their activities, especially as it might relate to franchising or contracting with individuals providing innovative products or services.

The appellate courts have been liberally interpreting the two-prong *Boulder-Lafayette* test for the application of the *Parker* exemption for substate unit action. In *Town of Hallie v City of Eau Claire,* 700 F2d 381 (7th Cir 1983), the court in another sewage case held that the city does not have to point to a particular and specific state statute in order for there to be a "clearly articulated and affirmatively expressed" state policy. The court held that if the anticompetitive effects are a reasonable and foreseeable consequence of the state statute, the *Parker* exemption will apply to the substate unit. *Id* 381. *See also Central Iowa Refuse Systems, Inc v Des Moines Metropolitan Solid Waste Agency,* 715 F2d 419 (8th Cir 1983); *Gold Cross Ambulance & Transfer v City of Kansas City,* 750 F2d 1005 (8th Cir 1983).

Intensive lobbying efforts of local governmental and other interest groups led to the enactment, in October 1984, of the Local Government Antitrust Act of 1984 (LOGATA), Pub L No 98-544, 98 Stat 2750 (to be codified at 15 USC §34). LOGATA essentially prohibits prevailing plaintiffs in antitrust suits from recovering damages, interest, costs, or attorney's fees from "any local government, or official or employee thereof acting in an official capacity" or "in any claim against a person based on any official action directed by a local government, or official or employee thereof acting in an official capacity." Pub L No 98-544, §3(b), 98 Stat 2750. Congress was clearly influenced by the $28.5 million judgment resulting from the *Unity Ventures* case. HR Rep 965, 98th Cong, 2d Sess 10, *reprinted in* 1984 US Code Cong & Ad News 4611. Congress sought to balance the strong federal interest in preventing anticompetitive activity and the strong state and local interest in protecting their fiscal integrity by adopting an approach which did not affect the substantive antitrust principles, but merely limited the available remedies. HR Rep No 965, 98th Cong, 2d Sess 18-19, *reprinted in* 1984 US Code Cong & Ad News 4619. The principal remaining remedy for potential antitrust plaintiffs is the injunctive relief authorized by §16 of the Clayton Act, 15 USC §26. See **§3.04.**

In addition to the limits on remedies against local governmental entities, LOGATA contains two other important provisions. While primarily aimed at protecting local governments and their officers, Congress also sought to deal with the problem of *private party* immunity from damages where the private parties are acting pursuant to any official action directed by a local government. Pub L No 98-544, 98 Stat 2750, §4(a), 15 USC §34. The Conference Committee Report reflects a congressional intent to shield persons from damage actions, if the conduct would have shielded had it been executed by a state government under the *Parker v Brown* doctrine (*Parker v Brown,* 317 US 341 (1943)). HR Rep No 1158, 98th Cong, 2d Sess, *reprinted in* 1984 US Code Cong & Ad News 4627. The judicial approach to private party immunity under the *Parker* doctrine has recently been clarified and will be discussed below. *See, e.g., Southern Motor Carriers Rate Conference, Inc v United States,* 105 S Ct 1721 (1985); *California Retail Liquor Dealers Assn v Midcal Aluminum, Inc,* 445 US 97 (1980); *New Motor Vehicle Board v Orrin W Fox Co,* 439 US 96 (1978).

The issue of retroactive or prospective application of LOGATA was a troublesome one for Congress. The House bill was intentionally silent on the issue. HR Rep No 965, 98th Cong, 2d Sess 22 (1984), *reprinted in* 1984 US Code Cong & Ad News 4623. The final version, as it emerged from the Conference Committee, was designed to allow the district court to determine on a case-by-case basis its application to pending cases. The local government would have the burden to show it would be inequitable not to apply LOGATA's remedial limitations. Where the case was in the prelitigation stage, LOGATA would more readily apply.

Another statutorily mandated factor a district court must consider is the availability and effectiveness of injunctive relief. H.Conf. Rep 1158, 98th Cong, 2d Sess 3-4 (1984) *reprinted in* 1984 US Code Cong & Ad News 4626-27.

There have already been two cases applying LOGATA. In *TCI Cablevision, Inc v City of Jefferson, Missouri,* 604 F Supp 845 (WD Mo 1984), the court held that LOGATA was applicable where the case had been filed only 17 days prior to its passage. Applying both of the factors contained in LOGATA—availability of alternative relief and stage of litigation—the court had no problem on the facts of the case finding that it would be inequitable not to apply LOGATA. *Id* 866.

In *Jefferson Disposal Co v Parish of Jefferson, Louisiana,* 603 F Supp 1125 (ED La 1985), the court, after an extensive look at the legislative history of LOGATA, also concluded that it would be inequitable not to apply LOGATA to the pending lawsuit. This case was in the pretrial discovery stage at the time of enactment of LOGATA. Although considerable expense had been incurred by the plaintiffs, the court felt that the legislative history of LOGATA reflected a strong congressional concern for immunizing local governments from monetary liability for antitrust violations, except in rare cases. The court also minimized a rather strong argument by the plaintiffs that injunctive relief was unavailable and, even if available, would not be effective. Notwithstanding the express placement of the burden of proof on the local government, the court seemingly shifted the burden to the antitrust plaintiff because of the strong congressional intent not to invade local governmental treasuries. *Id* 1133-35. The court also upheld the retroactivity of LOGATA against a constitutional argument. *See, e.g., Bradley v School Board,* 416 US 696 (1974).

In addition to the legislative response to *Lafayette* and *Boulder,* the United States Supreme Court, in *Town of Hallie v City of Eau Claire,* 105 S Ct 1713 (1983) and *Southern Motor Carriers Rate Conference, Inc v United States,* 105 S Ct 1721 (1985), clarified the application of the antitrust laws to local governmental activities and activities of private individuals taken pursuant to state or local governmental action. The result of these two decisions, along with LOGATA, is to give local governments greater leeway in acting in concert with the private sector without fear of antitrust liability.

In *Hallie,* several townships had sued the City of Eau Claire for refusing to extend sewage treatment services to a township unless it also purchased sewage collection and transportation services or agreed to be annexed to the city. The Seventh Circuit dismissed the antitrust claim, 700 F2d 376 (7th Cir 1983), finding sufficient state authority to meet the first prong of the *Lafayette-Boulder* test. The Seventh Circuit thus avoided the second prong of the *Lafayette-Boulder* test that *Parker* immunity is only available to local governments if there was active state supervision. 700 F2d at 384.

The Supreme Court thus confirmed its adherence to the first prong of the *Lafayette-Boulder* test. Local governmental actions, to be immune under *Parker,* must be taken pursuant to a "clearly articulated and affirmatively expressed . . . state policy." *Hallie,* 105 S Ct at 1717. The court, in analyzing the state statutes dealing with the provision of services, found that anticompetitive conduct was a foreseeable result of giving local governments to power to refuse to serve unannexed areas. This specific regulatory power differentiates *Hallie* from *Boulder* where the state grant of authority was the broad and nondescript grant of general home rule powers. Here Wisconsin had adopted a specific policy towards sewage treatment and collection services, while in *Boulder,* Colorado had not expressed any specific policy regarding CATV regulation. 105 S Ct at 1718-19.

The court also rejected the claim, based on *Cantor v Detroit Edison,* 428 US 579 (1976) and *Goldfarb v Virginia State Bar,* 421 US 773 (1975) that the city must show that the state must compel, not merely authorize, the city to engage in anticompetitive activities. The court rejected the application of *Cantor* and *Goldfarb* because both concerned actions of private parties. Where a municipality is the actor, following a clearly articulated state policy, authorization is all that the city must prove. 105 S Ct at 1719-20.

The court had more difficulty with the "active state supervision" second prong of *Lafayette-Boulder.* No case involving a local governmental defendant had reached the active state supervision test, and the issue had been specifically left open in *Boulder.* In *Hallie,* the court held that *no* active state supervision need be shown by the local government where the municipality is the alleged actor. 105 S Ct at 1720. The requirement was deemed to be redundant of the clearly articulated state policy requirement where the actions of a local government are involved. As the court concluded: "Once it is clear that state authorization exists, there is no need to require the State to supervise actively the municipality's execution of what is a properly delegated function." 105 S Ct at 1721.

After *Hallie,* most state enabling legislation dealing with a specific power will probably immunize the local government from antitrust liability. Only the broadest delegation of power to local governments, such as home rule statutes, will fail the new single-prong test of existence of a "clearly articulated state policy" which could envision the exercise of that policy in an anticompetitive manner.

In *Southern Motor Carriers Rate Conference, Inc v United States,* ___ US ___, 105 S Ct 1721 (1985), the Supreme Court dealt with the problem of what test to apply when a private party acts pursuant to a state or local governmental policy which involves alleged anticompetitive behaviour. The antitrust defendants in this case met together, within the various states covered by the litigation, and submitted single requests for intrastate trucking rates with the state regulatory body. The state statutes did not compel this kind of uniform ratemaking, but they did not prohibit it. The private associations which developed the uniform and singular rate request are made up of the actual motor carriers engaged in intrastate trucking. None of the members of the associations is bound by the decision to file a single rate. Each is free, by virtue of the state statutes and the rules of the association, to file their own rates if they desire.

The Fifth Circuit found that the collective ratemaking violated the Sherman Anti-Trust Act and refused to apply the two-prong test of *California Retail Liquor Dealers Association v Midcal Aluminum, Inc,* 445 US 97 Ct (1980), because in that case a public official was the named defendant. The *Midcal* test is essentially the same as the *Lafayette-Boulder* two-prong test. Immunity can only be found if the anticompetitive activity is clearly articulated and affirmatively expressed as state policy and if the policy is actively supervised by the state itself. *Id* 105. In *Midcal,* the court concluded that a state statute allowing for resale price maintenance in the wine industry, while taken pursuant to a clearly articulated state policy, was not actively supervised by the state, since it was the private parties who determined the prices that would be maintained at the retail level. *Id* 105-06. A state cannot give private individuals immunity from antitrust liability merely by authorizing them to violate the Sherman Act.

The Fifth Circuit nonetheless went on in dictum to hold that even if *Midcal* was applicable, the activities of the conferences were not immune because there was no active state supervision. It concluded that since the private associations determine the rate they were going to submit, even though the state regulatory body had overall supervisory and approval power, there was insufficient state supervision of the decision-making process that went into the actual rate-seeking petition which the state regulatory commission passed on.

The *Southern Motor Carriers* decision clearly extended the *Parker* cloak of immunity to private parties who were engaging in activities that would violate the antitrust laws. It also adopted the *Midcal* test to determine the applicability of *Parker.*

In defining the clearly articulated policy requirement of *Midcal,* the Court, as in *Hallie,* had to deal with the argument that only action compelled by state statutes would confer *Parker* immunity. As in *Hallie,* the Supreme Court rejected the compulsion argument and concluded that the various state statutes in question provided a clearly articulated state policy to provide for rate regulation over intrastate trucking commerce. If the states chose to articulate a policy which is permissive of anticompetitive behavior, the first prong of the *Midcal* test has been met. *Hallie,* 105 S Ct at 1728. While a policy which compels anticompetitive behavior is the best evidence of a clearly articulated policy, it is not the only evidence. To have a more restrictive definition of the first prong of the test is unnecessarily to hamstring the regulatory alternatives available to states to deal with public problems. 105 S Ct at 1729. The clearly articulated state policy must be a policy approved by a state legislature, as in *New Motor Vehicle Board v Orrin W Foxx Co,* 439 US 96 (1978), or by a state supreme court, as in *Bates v State Bar,* 433 US 350 (1977). In interpreting each of the state statutes creating the state regulatory and pricing mechanism for intrastate truck hauling rates, the Supreme Court concluded that the conduct of the private associations was taken pursuant to a clearly articulated state policy.

The court emphasized the requirement of active state supervision of the private party in order to prevent states from frustrating the national policy in favor of competition by casting a "gauzy cloak of state involvement" over the private anticompetitive behavior. *Hallie,* 105 S Ct at 1727 *citing Midcal,* 445 US at 106. There was no need for further analysis of the active state supervision prong of the test since the plaintiff had conceded that there was active supervision. It was clear that the states, through their regulatory bodies which set the rates, were engaged in activity supervising the private parties. This is in contrast to the *Midcal* situation, where no state agency was involved with the setting of the resale prices for the wine. Those agreements were between the wholesalers and the retailers with no actual state supervision. Therefore, the court concluded that, since the two prongs of the *Midcal* test had been met, there was antitrust immunity for the private association.

As with *Hallie,* the liberalization of the two-prong test for finding immunity should give greater flexibility to state and local governments in dealing with innovative solutions to modern problems. As long as there is some clearly articulated policy which authorizes or envisions private anticompetitive behavior, which is in some manner supervised by the state, the *Parker* immunity doctrine will apply to private parties.

The combination of *Hallie, Southern Motor Carriers,* and LOGATA will have a marked influence on state and local regulation of innovative products and systems. It may encourage state and local governments to regulate the private sector through the use of anticompetitive methods, since the fear of huge monetary judgments had been eliminated. That may lead, however, to closer scrutiny over state and local regulatory mechanisms, because courts will no longer have to choose between tolerating antitrust violations or imposing large damage awards against governmental treasuries.

§8.29B —Development Agreements (New)

Whenever state or local governmental entities enter into a contract with a private party, several problems may arise. It is clear that governmental entities and private individuals may contract for the provision of goods and/or services, and the contract will be enforced as if it were between two private parties. In certain circumstances, however, governmental contracts may deal with restrictions on the governmental exercise of its legislative authority. It is with these type of contracts that substantial, and as yet unresolved, constitutional problems arise.

It may be to the advantage of both a local government and a marketer of an innovative product or system to enter into a bilateral contract which may restrict future governmental actions. Recently, a California statute, Cal Govt Code §§65864-65869.5 (West 1982 & Supp 1985), authorized local governments to enter into contracts with land developers whereby the developer would make concessions regarding the development in exchange for a governmental promise to freeze the extant land use regulatory standards that would apply to the development. In the absence of such an agreement, the local government is free to change the ordinances and regulations which impact the development as long as the developer has not achieved a "vested rights" status, which in California and other states requires the developer to have received all regulatory permits needed to develop. *Avco Community Developers, Inc v South Coast Regional Commission,* 17 Cal 3d 785, 553 P2d 546, 132 Cal Rptr 386, *appeal dismissed & cert denied,* 429 US 1083 (1977). This "development agreement," as it is called, essentially prohibits the local government's legislative body from enacting ordinances which would have an impact on the proposed development.

In fact, there is precedent in the annexation agreement cases for such a restriction on the exercise of local governmental legislative power. In annexation cases, courts have upheld contracts between land owners and municipalities whereby the municipality agrees to zone land in a certain way if the landowner will agree to, or at least not object to, the land being annexed by the municipality. *See, e.g., Geralnes BV v City of Greenwood Village,* 583 F Supp 830 (D Colo 1984); *Morrison Homes Corp v City of Pleasanton,* 58 Cal App 3d 724, 130 Cal Rptr 196 (1976). Where land use regulations might otherwise restrict the use of an innovative product or system, it might be possible to negotiate a contract using the same principles as those contained in development or annexation agreements, in order to allow the sale and use of the innovative product or system.

It is not clear, however, that agreements which tend to bind future legislative bodies from exercising their police powers are constitutional. The principal tension arises through the conflict in the "contracts" clause of the United States Constitution, US Const, art 1, §10, cl 1, and the judge-made "reserved powers" doctrine. The contracts clause essentially prohibits states from enacting laws which impair the obligation of existing public and private contracts. The "reserved powers" doctrine essentially prevents any legislative body from either binding future legislative bodies in exercising their legislative powers, or bargaining away to private individuals the legislative power. *See, e.g.,*

Stone v Mississippi, 101 US 814 (1879). *See also* Kramer, *Development Agreements: To What Extent Are They Enforceable?,* 10 Real Est LJ 29 (1981).

The doctrines are conceptually irreconcilable. The contracts clause seemingly prevents state and local governments from enacting laws which impair the obligation of earlier governmental contracts. *See, e.g., United States Trust Co v New Jersey,* 431 US 1 (1977). The reserved powers doctrine basically prohibits the government from entering into that contract which restricts future legislative prerogatives. The cases which have dealt with the dichotomy have not come to grips with the essence of the problem. In *United States Trust Co,* the court reaffirmed the coexistence of both the contracts clause prohibition and the reserved powers doctrine. It rejected the earlier test that had been applied, namely what type of power was being contracted away, to determine if the contracts clause or reserved powers doctrine was applicable. For example, earlier cases held that governments can bargain away the spending and taxing powers, *Georgia Railway & Banking Co v Redwine,* 342 US 299 (1952), while it cannot bargain away the police power to protect the public health and general welfare, *Stone v Mississippi,* 101 US 814 (1879). The court articulated as a replacement a test which requires the court to inquire as to whether or not an essential attribute of sovereignty has been bargained away. However, in applying the new test, the *United States Trust* decision returned to the categorization approach of earlier decisions. 431 US at 23-24.

Notwithstanding the uncertainty about the enforcement of these bilateral agreements between governmental entities and private individuals, they may provide another tool or mechanism to encourage the dissemination of innovative products or technologies. The need to avoid the regulatory morass of state and local governments is important for innovators. Bilateral agreements based on the experience of land developers with annexation and development agreements may provide a way to clear these regulatory hurdles.

9 State and Local Promotion of Innovation

Frank J. Mauro

The Science and Technology Path to Economic Growth

§9.06 —Present Status

In note 54, page 336, fourth line, delete "1963" and insert:

1973

Add to note 60, page 337:

One specific manifestation of this linkage between the State of North Carolina and the federal government is the North Carolina Science and Technology Research Center (NCSTRC). The NCSTRC is a statutorily created state agency (NC Gen Stat §§143B-442 to 143B-445) which has operated since 1964 under a continuing contract with the National Aeronautical and Space Administration, and functions as one of NASA's regional industrial applications centers (ICAs). It is, however, the only IAC that is also a state agency and therefore it plays a dual role. As a NASA IAC, it is available to provide technical assistance to businesses in states throughout the southeastern United States. As a state agency, administered by the North Carolina Department of Commerce and receiving state funds, it provides a more intensive level of services to businesses located in North Carolina. See note 62 *infra* for additional information on the NCSTRC.

In note 62, page 337, last line, delete "909-549-0671" and substitute:

919-549-0671 or 800-334-8561

Page 339, in first full paragraph, second line, delete "1974 and 1975" and substitute:

1964 and 1975

Add to note 66, page 339:

The program of support for local technology development agencies (officially referred to as regional technology development organizations) has been continued beyond the initial three-year test period. Start-up funding has now been provided to technology development organizations in eight areas of the state. These include the Capital Region High Technology Development Council, P.O. Box 1907, Albany, New York 12201, 518-455-2200; Central New York High Technology Council, c/o Knowledge Systems and Research, Inc., 500 South Salina Street, Syracuse, New York 13202, 315-470-1350; High Technology of Rochester, Inc., c/o The University of Rochester, 276 Administration Building, Rochester, New York 14627, 716-275-3322; Mid-Hudson High Technology Council, 61 Livingston Street, Poughkeepsie, New York 12601, 914-452-4860; Mohawk Valley Applied Technology Commission, 209 Elizabeth Street, Utica, New York 13501, 315-724-9805; Northern Technology Council, c/o Clarkson College, Potsdam, New York 13676, 315-268-7727; Regional Industrial Technical Education Council (RITEC) of Long Island, Inc., 80 Hauppauge Road, Commack, New York 11725, 516-751-3329; and the Western New York Technology Development Center, 2211 Main Street, Buffalo, New York 14214, 716-831-3472.

Examples of the activities undertaken by the first three regional technology development organizations funded under this program included the coordination of the technically oriented education activities in an area with economic development programs relating to high technology industries operating in that region, the undertaking of a comprehensive job inventory and follow-up biannual surveys to improve projections of regional employment needs by discipline and degree, and the preparation and distribution of an inventory of a region's assets relevant to high technology development.

In order to ensure that state funds go as far as possible, the appropriations made for this program have included a priority for grants to organizations in areas of the state not previously funded and a provision for a greater local match after the first year of funding. New York State Science and Technology Foundation, Technology Development Organizations (Mar 1985). Since 1981, responsibility for the administration of a number of new programs has also been assigned to the foundation. These include the Corporation for Innovation Development (see **§9.14** notes 114 & 115), the Centers for Advanced Technology program (see supplement at **§9.16,** note 134), the Productivity Improvement Program (see supplement at end of **§9.15**), and the Industrial Innovation Extension Service (see supplement at end of **§9.11**).

Page 339, add at end of section:

The first new state science foundation in 13 years was established by Indiana in 1982. *See* Ind Code Ann §§4-3-11-1 to -5 (Burns). While the powers and duties of the new Indiana Corporation for Science and Technology are quite similar to a number of the earlier foundations, it is unique in one regard. While it was created by statute, and is governed by a 24-member board of directors appointed by the Governor, it was chartered as a private not-for-profit corporation.

The enabling statute authorizes the Governor acting "on behalf of the state" to "request" the establishment of "a private not-for-profit corporation named the Indiana corporation for science and technology." Ind Code §4-3-11-1. Both the proposed governance of the corporation and its powers and duties indicate that the underlying theme of the statute is enhanced public-private cooperation. While all 24 members of the board of directors are to be appointed by the Governor, eight are to be representatives of the public sector, eight are to be representatives of the private business and labor sector, and eight are to be educators.Ind Code §4-3-11-2(2). In addition to being charged with identifying problem areas and targets of opportunity, the corporation's powers and duties include "assist(ing) in establishing cooperative associations of universities in Indiana and of private enterprises for the purpose of coordinating research and development programs that will, consistent with the primary educationsl function of the universities, aid in the creation of new jobs in Indiana"; "assist(ing) in financing the establishment in Indiana of technology-intensive businesses"; and "determin(ing) how public universities can increase income derived from the sale or licensure of products or processes having commercial value that are developed as a result of university-sponsered research programs." Ind Code §4-3-11-3.

In 1983, Kansas enacted legislation establishing a multi-purpose Advanced Technology Commission to "promote, develop and coordinate education, research and economic development programs in the field of advanced technology." This commission was given several assessment and planning responsibilities but was also charged with both establishing priorities for the distribution of the funds available to it and distributing those funds among the state's institutions of higher education on the basis of its established priorities. 1983 Kan Sess Laws ch 450. In 1983, the state legislature appropriated $610,000 to this new commission for research project grants at the state's four public universities, with the proviso that any such grants be matched on a one and one-half to one basis by the institutions involved. 1983 Kan Sess Laws ch 24. As of March 1, 1984, 11 research projects had been funded with approximately $370,000 in state resources and

$570,000 in matching funds from 16 businesses. The Advanced Technology Commission is also responsible for overseeing the state's Centers of Excellence program. In 1983-84, three grants of $130,000 each were made, under similar one and one-half to one matching requirements, to assist in the establishment of Centers for Productivity Enhancement (at Wichita State University), Bioanalytical Research (at Kansas University), and Artificial Intelligence and Automated Control Systems (at Kansas State University). The commission recommended second-year state support of $215,000 for each of these three centers. Kansas Advanced Technology Commission, 1984 Report to the Legislature (1984).

§9.07 Tax Incentives

Add to note 67, page 340:

A 1983 survey listed 14 states (California, Indiana, Maine, Maryland, Massachusetts, Minnesota, New Jersey, New York, Rhode Island, Vermont, Virginia, Washington, West Virginia, and Wisconsin) as having tax incentives designed to encourage research and development. These incentives include deductions and credits against corporate income taxes, exemptions from sales and use taxes for materials consumed in research and development activities, and exemptions and abatements from property taxes for plant and equipment engaged primarily in research and development activities. National Association of State Development Agencies, National Council for Urban Economic Development, and The Urban Institute, Directory of Incentives for Business Investment and Development in the United States (1983).

Page 341, add at end of section:

In 1981, New York and Minnesota both enacted new research and development tax credits to supplement or complement the accelerated depreciation options already offered to taxpayers making investments of this type in their respective states. New York authorized an investment tax credit, against its corporate franchise tax and its personal income tax, of 10 per cent of the cost (or other basis used for federal income tax purposes) of investments in buildings and equipment used for research and development purposes. 1981 NY Laws ch 103, §§14,16, codified as NY Tax Law §§210(18), 606(h)(McKinney). This credit was allowed as an alternative to the one-year writeoff deduction which continues to be available to taxpayers. The importance placed on this type of economic activity was underscored by the fact that even though the state's regular investment tax credit was increased by that same legislation to 6 per cent, the credit for research expenditures is more than 50 per cent greater.

A new credit authorized by Minnesota in 1981 differed from the New

York credit in that it was allowed an addition to the previously authorized accelerated depreciation It was also different in that the amount of the credit was equal to a portion (10 per cent) of the amount, if any, by which the research and experimental expenditures paid or incurred by the taxpayer in Minnesota during the taxable year exceeded the average of the taxpayer's research and experimental expenditures in Minnesota during the three preceding taxable years. 1981 Minn Laws (3d special sess) ch 2, codified as Minn Stat Ann §290.068(West).

Minnesota amended its tax credit provision the following year to base its calculation of increased research expenditures on the calculation of "qualified research and experimental expenditures" for federal income tax purposes, with a few slight modifications. *See* IRC §44F, as added by §221 of the Economic Recovery Tax Act of 1981. Also see **§§5.11, 5.12,** and **5.13**; Collins, An Early Assessment of Three R&D Tax Incentives Provided by the Economic Recovery Tax Act of 1981 (National Science Foundation, April 1983); Subcommittee on Oversight, House Ways and Means Committee and Joint Committee on Taxation, 98th Cong, 2d Sess, Description of Present Law and Issues Relating to the Research and Experimentation Credit (Joint Comm Print 1984); Bucy, *Meeting the Competitive Challenge: The Case for R&D Tax Credits,* 1 Issues in Sci & Tech 4, 69 (Summer 1985); Eisner, *The R&D Tax Credit: A Flawed Tool,* 1 Issues in Sci & Tech 4, 79 (Summer 1985) for discussions of federal tax treatment of research and development expenditures. The 1982 Minnesota amendments also changed the amount of the credit from 10 per cent of the increase in qualified research expenses to 12.5 per cent of the first $2 million of such increase and 6.25 per cent of any increase beyond $2 million. In addition to qualified research expenses as defined in the Internal Revenue Code, Minnesota added an interesting variation to this credit. It included as a qualified research expense any contributions which the taxpayer makes to a nonprofit corporation established under Minnesota law for the purpose of promoting the establishment and expansion of business in the state, provided that the contributions are used by the nonprofit corporation to provide financing to small, technologically innovative enterprises in the early stages of their development. 1982 Minn Laws ch 523, amending Minn Stat Ann §290.068 (West), as originally enacted by 1981 Minn Laws (3d special sess) ch 2.

Indiana and Wisconsin have also adopted incremental research and development credits similar to the federal and Minnesota credits. The Indiana credit, which is authorized through 1985, is based on the increases in the amount by which a firm increases its research and development expenditures in Indiana minus the amount of the federal credit received for those expenditures. The credit is equal to 2 per cent of the difference between these two amounts for expenditures made in 1982 and 1983 and 5 per cent of the difference for expenditures made in 1984 and 1985. Unused credits may be carried forward. Ind Code Ann §§6-3.1-4-1 through 6-3.1-4-6 (Burns 1984). In 1984, Wisconsin enacted a credit of 5 per cent of the incremental amount of research and development expenditures in Wisconsin and a credit of 5 per cent of the cost of the construction and equipping of new and expanded research facilities in Wisconsin. Wis Stat §§71.04(2)(b)(5), 71.05(1)(a)(16), 71.09(12r) & 71.09(12rf) (1984), as enacted by 1983 Wis Laws 27.

At least one state, Washington, has adopted a tax preference designed to promote innovation in a particular industry. Specifically, Washington exempts from its sales and use tax the fuel used in aircraft by aircraft manufacturers during research, development, and testing activities. Wash Rev Code Ann §82.08.0255(1) (1981).

In addition to these tax incentives, which are designed to promote industrial innovation by encouraging research and development, a number of states have provided tax incentives as a means of helping to meet the capital requirements of innovative firms. Some states have formed state financing entities and have made tax credits available to taxpayers who help to capitalize these financing intermediaries (see **§9.14** regarding the Indiana Corporation for Innovation Development, the Maine Capital Corporation, and the Iowa Venture Capital Fund). Other states have provided tax incentives to taxpayers who invest directly in certain types of small high-technology businesses (see **§9.14,** notes 116 and 117 regarding incentives of this type in New York and California), or who invest in private investment companies that are formed to provide capital for such samll businesses (see **§9.14** following note 117 regarding programs of this latter kind in Montana and Louisiana).

§9.08 Advisory Councils

Page 342, add at end of section:

In 1981, Governor Edmund G. Brown Jr. of California established a California Commission on Industrial Innovation by Executive Order No B-91-81. This commission conducted a series of studies and recommended policies for maintaining California's lead in industrial innovation. Among the results of this commission's studies was the Microelectronics Innovation and Computer Research Opportunities (MICRO) program, discussed in **§9.09.** In September 1982, Governor Brown approved legislation establishing a Commission on Industrial Innovation on a more permanent, statutory basis. 1982 Cal Stat ch 1527. This statute charged the commission with "developing and recommending effective and aggressive strategies for expanding the state's economy and employment rate" and gave it a number of specific assignments, including the following:

1. To advise in the establishment of programs in scientific and technical education and job training which will prepare Californians for employment in industrially innovative firms and for life in an increasingly technological society. This advice shall particularly address policies to aid in the transition of workers into mature innovative industries
2. To assess and evaluate programs which improve productivity while maintaining regard for worker involvement and satisfaction
3. To identify means by which industry and universities can cooperate on research and development projects and in the establishment or expansion of cooperative research and development institutes
4. To identify means of financing innovative industry including, but not limited to, public and private pension funds and industrial development bonds
5. To identify and solicit financial support for the activities of the commission, such as foundation grants and individual donations
6. To identify management techniques necessary to successful industrial innovation
7. To examine and evaluate the effectiveness of any state programs assisting or hindering industrial innovation
8. To identify shortcomings in the transportation, sewer, water, energy, and waste disposal systems which inhibit existing or future industrially innovative firms

§9.09 Promotion of Innovation in Particular Industries and Technologies

Add to note 83, page 343:

See also Bureau of National Affairs, *New York Energy Research Agency Spends, Saves Millions for State,* 13 Energy Users Rep (BNA) 482 (May 23, 1985); Regens, *A Role for State Government?.* 27 Envt 18 (May 1985). Regens also cites a 1984 study (Randolph, *Energy Conservation Programmes: A Review of State Initiatives in the USA.* 12 Energy Pol 425 (1984)) which found that 28 states were performing at least some energy research, development, and demonstration function using state and other funds.

Add to note 84, page 343:

Other coal-producing states (i.e., Illinois, West Virginia, and Ohio) have also undertaken research, development, and demonstration programs to spur the use of coal produced in those states. These efforts are frequently directed at technologies designed to eliminate or ameliorate some of the negative consequences of coal usage such as sulfur emissions. Regens, *A Role for State Government?*, 27 Envt 18 (May 1985).

In note 86, page 344, sixth line, delete "1982-83" and insert:

1981-82.

Add to note 86, page 344:

The MICRO program is now administered by an executive committee of five faculty members from five of the University of California campuses under the overall guidance of a nine-member policy board, composed of three representatives each from industry, the state government, and the University of California. See Microelectronics Innovation and Computer Research Opportunities (MICRO) Programs, 1983-84 Progress Report (Jan 31, 1984). In the program's second year of operations (1982-83), 51 projects were supported with $1,656,000 from the state and $3,215,000 in matching grants from 33 participating businesses. For further information, contact the Officer of Associate Vice President for Academic Affairs, University of California, Berkeley, California 94720, 415-642-9786.

Page 344, add at end of section:

Other recent efforts have focused on society's need to find solutions to particularly complex and troubling problems. One example of this emphasis is provided by the State of New Jersey, which has joined with the National Science Foundation and 10 industrial firms to establish the National Center for Research in Hazardous/Toxic Substances. The center, which is operated by a consortium of five universities, is focusing on such areas as incineration, biological and chemical treatment, and physical treatment. For further information, contact the National Center for Research in Hazardous/Toxic Substances, 323 High Street, Newark, New Jersey 07102, 201-596-3465.

Renewed attention is also being paid to the current viability of state efforts to promote innovation in particular industries and technologies in which there has been a long-standing state presence. For example, a recent study of the New York State food industry by a consortium of state agencies called attention to the role played by the agricultural land grant colleges in making the United States the greatest food-producing country in the world and to what it characterized as a dwindling research commitment which is negatively affecting the competitive position of this country's food industries. This report concluded that New York State should pursue the following approaches to the support of research and technological development in food science:

1. The institution of a matching fund program to encourage both public and private monies for both new product development and market research.
2. The provision of research and development support to smaller food processing companies that cannot otherwise afford the team of technically trained personnel and facilities needed to maintain a competitive R&D program
3. The maintenance of continued support for both basic and targeted research on the physical and chemical properties of New York food commodities through the Center for Advanced Technology located at Cornell University
4. Encouragement of the adoption of new technology through technical and financial support, technology demonstration, and extension efforts
5. Encouragement of a strong program of aquaculture research and development in the State University System with emphasis on disease control, genetics, nutrition, reproduction, and engineering of culture systems

New York State Department of Agriculture and Markets, New York State Department of Commerce, New York State College of Agriculture and Life Sciences and New York State Sea Grant, The New York State Food Industry: Economic Development Opportunities for the Future (October 1984).

§9.10 Research and Technology Parks

Add to note 87, page 345:

See notes 60 and 62 for information on the North Carolina Science and Technology Research Center. See **§9.09** at note 85 for information on North Carolina State Energy Institute. See **§9.16** at notes 128 and 129 for information on the Microelectronics Center of North Carolina. The North Carolina Biotechnology Center was established in 1981 by Governor James B. Hunt, Jr., to stimulate the development of North Carolina's biotechnology resources. The center is located in the Research Triangle Park and works with North Carolina's universities, both public and private, the Research Triangle Institute, and interested industries "to develop collaborative research and education programs addressing the biomedical, agricultural and engineering applications of biotechnology." North Carolina Biotechnology, informational brochure published by the North Carolina Biotechnology Center and the Industrial Development Division of the North Carolina Department of Commerce (undated). For further information, contact the North Carolina Biotechnology Center, P.O. Box 12235, Research Triangle Park, NC 27709, 919-549-0671.

Add to note 88, page 345:

Under the Florida statute, codified as Fla Stat §§23.145-23.148, four parks have been initiated. These include the Tampa Bay Research Park at the University of Southern Florida, the Central Florida Research Park at the University of Central Florida, the Florida Research and Technology Campus at the University of Florida, and the Innovation Park at Florida State University and Florida A&M University. US Office of Technology Assessment, Technology, Innovation and Regional Economic Development: Census of State Government Initiative for High-Technology Development 25 (1983).

Page 346, at end of §9.10, add:

In recent years, a variation on the theme of state-supported and university-based or university-related research and development parks has emerged. This variation—a concept referred to as incubator space or incubator facilities—has attracted enthusiastic support and offers significant promise as a positive sum approach (see **§9.17**) to the promotion of innovation. As opposed to the goals of many R&D or high tech parks, the incubator concept is oriented to the nurturing of new ventures that are in their earliest stages of development. The ventures involved are frequently, but not always, outgrowths of faculty or student research projects. Even for those ventures which do not have close university ties, the common characteristic is that these are firms which are at a stage in their development where they will benefit from the supportive, and to a significant degree protected, environment that the incubator facilities are intended to offer.

Perhaps the first high tech incubator building in this country was developed on the campus of Rensselaer Polytechnic Institute in Troy, New York, with planning assistance and financial support from the US Economic Development Administration and the New York State Urban Development Corporation. The rehabilitation and reconfiguration of an old dormitory into RPI's permanent incubator building was financed with industrial revenue bonds issued by the City of Troy's Industrial Development Agency.

The Advanced Technology Development Center, which was established by the State of Georgia and the Georgia Institute of Technology in 1980 to facilitate developing and expanding high technology industries, is currently constructing an incubator facility on the Georgia Tech campus. This facility, which is scheduled to open in September 1984, has been financed by a $3.4 million appropriation from the State of Georgia and a $1.7 million loan from a commercial bank in Atlanta, with the loan to be paid back out of the rents paid by the incubator firms. The Center had 11 small companies housed in an interim facility as of March 1983, with additional tenants in various stages of a rigorous screening process. Remarks by Jerry Birchfield,

Director of the Advanced Technology Development Center, to the Small Business High Technology Institute's conference, *The Billion Dollar Challenge—The States, The Professions, and The High Tech Companies,* Arlington, Virginia (Mar 18-19, 1983). For further information, contact the Advanced Technology Development Center, Georgia Institute of Technology, Atlanta Ga 30332, 404-894-3575. For additional information on this incubator program and on one established by the state of Illinois in Chicago's West Side Medical Center, see *America Rushes To High Tech for Growth,* Bus Week, Mar 28, 1983, at 87.

Under Florida's research and development park financing program, described above, plans are currently underway for the development of an incubator facility at the University of Florida. US Office of Technology Assessment, Technology, Innovation and Regional Development 25 (1983).

The recently enacted North Carolina New Technology Jobs Act requires the North Carolina Technological Development Authority (which is created by that statute) to establish one or more incubator facilities in North Carolina and authorizes it to make one-time grants of up to $200,000 to nonprofit corporations affiliated with higher education institutions and/or technical institutes for that purpose. Such grants must be matched by cash or real estate of an equal or greater amount from local government or private sources. NC Gen Stat §143B-471.4.

§9.11 Dissemination of Innovation

Page 346, at end of §9.11 add:

In recent years, a number of states have initiated dissemination or applications programs in the areas of industrial innovation, advanced technology, and related fields, which appear to provide the rudiments of a modern-day industrial innovation extension service similar in form and concept to the agricultural extension programs begun in the early part of the current century (see **§9.02**) but more general in scope and purpose.

In 1983, the University of Texas at Arlington initiated a program to help small technology-based companies in North Texas to expand and prosper, and "to encourage the formation and growth of other technology-based businesses in our community." With its major funding coming from the United States Small Business Administration, the university established the Technology Enterprises Development Center within its College of Engineering. According to the sponsors of this program, it is "the first technology-based Small Business Development Center in the United States. It is unique because it provides owners and operators of small businesses throughout North Texas with current technical information and engineering expertise, in addition to business-related assistance. These are services which many could not otherwise afford, yet such information is often critical to their success in the rapidly evolving high technology marketplace." Technology Enterprises Development Center, informational brochure published by the University of Texas at Arlington, College of Engineering. For further information contact the Technology Enterprises Development Center, UTA Station Box 19019, The University of Texas at Arlington, Arlington, TX 76019, 817-273-2571. This center's approach is similar to that of the agriculture extension programs in that it provides individualized, confidential counseling for small businesses that require technical information or engineering assistance, as well as more open workshops and seminars to inform small businesses of recent technology developments, use and availability of new products and techniques, and new market opportunities. The center is also a regional referral office for the Institute for Ventures in New Technology (INVENT) of the Texas Engineering Experiment Station, which provides services for entrepreneurs and inventors needing assistance with feasibility studies and business plans. See **§9.15,** note 119 for additional information on INVENT.

In both 1984 and 1985, the New York Legislature appropriated $150,000 to the New York State Science and Technology Foundation "for services and expenses related to the provision of assistance to small businesses in the application and utilization of productivity enhancing technologies." 1984 NY Laws ch 50 and 1985 NY Laws ch 53.

The foundation has utilized these funds for a series of productivity demonstration projects involving the application of new technologies in a variety of fields, including the fashion industry and packaging manufacturing. For further information on this program, see New York State Science and Technology Foundation. The Productivity Development Demonstration Project: Interim Status Report (February 27, 1985); New York State Science and Technology Foundation. The Productivity Development Demonstration Project: Second Interim Status Report (May 17, 1985). In addition, in 1985, the legislature appropriated $500,000 to establish an industrial innovation extension service. 1985 NY Laws ch 53. Such a service could perform a role analogous to that played by the various state agricultural extension services in bringing "scientific agriculture" to farmers earlier in this century. A variation of this approach is discussed in the [Massachusetts] Governor's Commission on the Future of Mature Industries, Final Report (June 1984).

The importance of the dissemination of innovation and the application of productivity-enhancing technologies in the entire spectrum of American industry is receiving increased recognition. A major study of technology, innovation, and economic development by the Office of Technology Assessment made this point very clearly. It concluded that while high technology industries (by whatever definition one uses) will probably grow somewhat faster than the economy generally over the next decade, they will, because of their small base, directly account for only a small portion of total employment growth in the United States. The major impact of these industries, however, is seen as coming in terms of "the diffusion and widespread application of their products by other industries, 'smokestack' and services alike." The report concludes by pointing out that "an emphasis on those industries per se, rather than on the process of technological innovation and diffusion, carries a risk of distorting public policy and ignoring real opportunities for promoting industrial competitiveness and sustainable economic development." US Office of Technology Assessment, Technology, Innovation and Regional Economic Development 3 (July 1984). For a similar perspective, see Botkin, Dimancescv, & Stata, The Innovators: Rediscovering America's Creative Energy (1984).

The High Tech Future: The 1970s, 1980s and Beyond

§9.12 Innovation as the Key to Economic Success

Add to note 93, page 347:

At least some officials in one state, California, are expressing concern over the possible negative effects of the current push for much greater defense spending. According to the California Office of Economic Policy, Planning and Research, as reported in Kindel & Teitelman, *Go East, Young Man,* Forbes, June 20, 1983, at 136, "Military spending is likely to attract skilled workers—craftsmen, engineers and scientists—into military production and thus away from civilian firms competing in world markets (particularly with nations with less military spending)." This study estimated that California, which now receives more than 17 per cent of all defense spending, "would actually lose 300,000 jobs through increased defense spending, because of the starvation of other, faster-growing sectors of the economy." A more traditional view of this subject is expressed in Gurwitz, Auda & Greer, Federal R&D and Defense Outlays in New York State (Federal Reserve Bank of New York, Feb 1984).

In note 95, page 347, delete "§§9.12, 9.13" and insert:

§§9.13, 9.14.

Page 348, add at end of section:

State governments are increasingly recognizing the essential role that human resources play in the innovation process and the part that skilled personnel, from scientists and engineers to qualified technicians and production workers, play in the success of high technology businesses and in the introduction of new technologies in traditional industries. *See,* for example, National Science Foundation, *Manufacturing Employment Becomes Increasingly Technological,* Sci Resource Studies Highlights (March 10, 1983).

The response has been an emphasis by states on both job training programs to prepare workers to meet the immediate needs of technologically oriented employers, on more long-range programs to improve the quality of elementary and secondary schools and thereby meet the challenges presented by the current technological revolution, and on programs to foster entrepreneurship by inspiring potential entrepreneurs to pursue such interests and by providing such individuals with the practical skills necessary to do so. See *Low-Tech Education Threatens The High-Tech Future,* Bus Week, Mar 28, 1983, at 95; Education Commission of the States, Task Force on Education for Economic Growth, Action For Excellence: A Comprehensive Plan To Improve Our Nation's Schools (1983); Education Commission of the States, Task Force on Education for Economic Growth, Action in the States: Progress Toward Educational Renewal (July 1984); State University of New York, Sampler of Training and Retraining Programs for Business and Industry (1983); California Work-Site Education And Training Act, Cal Unemp Ins Code §§9900-9910 (West); Pilcher, *Job Training in the States,* State Legislatures 9 (June 1983); California Office of Appropriate Technology, High Technology and High School: Preparing Students for California's Changing Economy (1982); Moser, *Meeting the Challenge of Technology Education,* 17 World 2 (Spring 1983), at 3; (Washington) High Technology Education and Training Act, 1983 Wash Laws (1st Ex Sess) ch 72; Bay State Skills Corporation Act, Mass Gen Laws Ann ch 40-I; Bluegrass State Skills Corporation, Ky Rev Stat §§157.710 to 157.750; Robinett, *What Schools Can Teach Entrepreneurs,* 7 Venture 2, 50 (Feb 1985); O'Connell & Hoerr, *There Really Are Jobs After Retraining,* Bus Week, Jan 28, 1985, at 76.

Regarding the provision of engineering education, particularly graduate-level engineering education in particular areas of states where such programs are not currently available but where their availability is seen as essential to realizing the development potential of such areas, *see* 1982 NY Laws ch 320 (development of an engineering school at the State University of New York at Binghamton) and 1984 NY Laws ch 50 (development of an engineering program at the State University of New York at New Paltz). Both branches of the university are located in areas of New York State (the Southern Tier and the Mid-Hudson Valley, respectively) where large and growing concentrations of high technology firms are located. *See also* 1984 Wash Laws ch 285, §517 which appropriated $80,000 to Washington State University "for needs assessment and planning by the higher education institutions in the Spokane area for programs in engineering and technology to meet community and industrial needs," and provides that the needs assessment and planning should be carried out by a consortium of institutions which "shall include, but not be limited to, Washington State University, Gonzaga University, Eastern Washington University, Whitworth College and the Spokane Community College District"; and 1984 Va Laws ch 755 §§1-72 to 1-76, which appropriated a little more than $5 million, in varying portions, to five different state institutions of higher education for the development and operation, on a consortium basis, of "an educational television project to provide graduate engineering education in the Richmond, Northern Virginia and Tidewater areas of the Commonwealth," with the provision that no allotment or expenditure of the funds appropriated for this purpose can be made until a joint plan is submitted by the participating institutions (University of Virginia, Virginia Commonwealth University, Virginia Polytechnic Institute and State University, Old Dominion University, and George Mason University) and approved by the Council on Higher Education.

§9.13 Helping to Meet the Capital Requirements of Innovative Firms

Add to note 103, page 349:

See Bearse, *A Comparative Analysis of State Programs to Promote New Technologically Based Enterprise,* 5 New Eng J Bus & Econ 49 (1979) for a comparison of the Connecticut Product Development Corporation and the Massachusetts Technology Development Corporation, discussed in **§9.14.** *See* Smart, *Venture Capital: Gambling on the Future,* 7 Inc 146 (Apr 1985) for a recent report on the activities of the Connecticut Product Development Corporation (CPDC). According to this report, the CPDC has committed $12.5 million to 59 new products in the past 10 years.

§9.14 —Recent Developments

Add to note 108, page 351:

The Main Capital Corporation is now independently established at One Monument Square, Portland, Me 04101, 207-772-1001.

Add to note 109, page 351:

In 1982, the corporation's purpose was broadened to include other segments of the state's economy, in addition to such renewable resources as timber, agriculture, fisheries, and renewable energy sources. *See* 1982 Alaska Sess Laws ch 142. At the same time, the corporation's name was changed to the Alaska Resources Corporation. For further information, contact the corporation at P.O. Box 828, Anchorage, Alaska 99510, 907-561-2210.

Add to note 112, page 351:

The Indiana Corporation for Innovation Development is now located at One North Capitol Avenue, Suite 520, Indianapolis, Ind 46204, 317-635-7325.

Page 351, at end of carryover paragraph add:

In 1980, Louisiana enacted two complementary statutes, the Louisiana Small Business Investment Law of 1980 (La Rev Stat Ann §§33:9061 to 33:9072) and the Louisiana Small Business Equity Corporation (LASBEC) Act of 1980 (La Rev Stat Ann §§33:9081 to 33:9092), designed to provide a mechanism by which the State of Louisiana could stimulate the flow of equity capital, long-term loans, and other financial assistance to small businesses. These statutes established a Louisiana Small Business Investment Fund, the monies in which may be invested by the LASBEC for a series of interrelated purposes including the purchase of debt instruments issued by small business financing companies, SBFCS, licensed pursuant to the provisions of these statutes, the provision of guarantee of funds to qualified financial institutions in order to encourage such institutions to provide financial assistance to licensed SBFCs, and the guarantee or consignment of funds to banks or other lending institutions to offset the costs of feasibility studies of new product or service development activities by *small business growth concerns,* which the statute defines as "a small business concern which contributes most directly to greater employment, productivity, innovation, and the production of wealth in the State." La Rev Stat Ann §§33:908(35), 33:9063(35).

While the LASBEC is generally charged with "stimulat[ing] the flow of private equity and venture capital and long term loans for the development, expansion, and retention of small business in the state," it is also given a more specific charge "to encourage the development of new products and services within the state through the infusion of financial aid for invention and innovation." La Rev Stat Ann §33:9085. *New product or service* is defined in these statutes as "any product, device, technique, process or service which is or may be exploitable commercially, the production or provision of which does not exist in the state or does not constitute a traditional activity, type of enterprise, or method of production or service within the state. Such terms shall not refer to pure research but shall be construed to apply to such products, devices, techniques, processes, or services, which have been advanced beyond the theoretical stage and are readily capable of being, or have been, reduced to practice." La Rev Stat Ann §§33:9068(25), 33:-9063(25).

A 1983 Utah statute, the Utah Technology and Innovation Act, established a Utah Technology Finance Corporation (UTFC) and authorizes that corporation to make direct loans and award research contracts to "emerging and developing technological and innovative small business." It may also make direct capital investments in such businesses provided that they are matched on a one-to-one basis by funds from private sources. Utah Code Ann §63-60-1 to 63-60-5. The UTFC is not only authorized to hold royalties, stock, or other ownership interests and proprietary rights in companies, patents, copyrights, etc., but its authorization to award research contracts includes the caveat that they must be made in return for royalty payments with the right to convert to equity. The statute does include the proviso that the property rights it holds in any company must be a noncontrolling minority interest. Another interesting aspect of the Utah statute is a requirement that the UTFC can only use its powers to "encourage the development and growth of business and technologies that are not detrimental to the quality of the land, water or general environment of Utah."

The State of Iowa, as part of an omnibus economic development bill enacted into law in 1983, created the Iowa Product Development Corporation (Iowa Code Ann §§28.81 to 28.95) for the purpose of "stimulat[ing] the development of new products within Iowa by the infusion of financial aid for invention and innovation in situations in which financial aid would not otherwise be reasonably available from commercial sources." Iowa Code Ann §28.87. The corporation is authorized "to enter into venture agreements with persons doing business in Iowa," and, pursuant to such agreements, to advance financial aid to such persons "for the development of specific products, procedures and techniques" in Iowa. Iowa Code Ann §28.87(2). This statute defines *financial aid* as "the infusion of risk capital to persons for use in the development and exploitation of specific inventions and products" Iowa Code Ann §28.82(2), and *venture agreement* as "a contractual agreement between a person and the corporation from which the corporation obtains rights, from or in an invention, product, or the proceeds from the product or invention in exchange for granting financial aid to the person." Iowa Code Ann §28.82(5). This statute appropriated $750,000 to the IPDC for the 1983-84 fiscal year and also gave the corporation the authority to issue bonds and notes as another means of providing sufficient funds for the achievement of its corporate purposes. Iowa Code Ann §§28.87(5), 28.90, 28.91, and 28.94.

This same omnibus economic development legislation provided for the incorporation by the Iowa Development Commission of a new entity to be called the Iowa Venture Capital Fund, with the initial board of directors of this new corporation to be appointed by the Governor. Iowa Code Ann §§28.40-.45. The fund is to be capitalized through a public offering of stock (Iowa Code Ann §28.42) with taxpayers investing in these securities being allowed a tax credit equal to five per cent of the amount of their investment. Iowa Code Ann §§422.10, 422.33(5). A limit of $5 million is placed on the amount of stock for which a tax credit will be allowed. Iowa Code Ann §28.45. The statute also sets forth an "investment policy" for the fund to ensure that it invests primarily in small businesses "which are principally engaged in the development or exploitation of inventions, technological improvements, new processes, or products not previously generally available in this state," or in small businesses which provide support to such technologically oriented small businesses. "The principal financial objective of the fund" is identified as the generation of "long-term capital appreciation by participating in the growth in equity value of Iowa-based companies in which the fund invests." Iowa Code Ann §28.43.

Another recent development of this type is the North Carolina Technological Development Authority (NC Gen Stat §143B-471), which is authorized to provide equity financing for the research activities of new and small business in North Carolina. NC Gen Stat §§143B-471.3A(6), 143B-471.5. The financing which the authority can provide takes the form of research contracts with the proviso that the "awards per research project shall not exceed fifty thousand dollars ($50,000) per fiscal year." NC Gen Stat §143B-471.5(d). It will be interesting to see if the research grant-funding mechanism and equity financing arrangements can be effectively combined, as this statute and the Utah statute envision.

Add to note 113, page 352:

Through June 30, 1982, the MTDC had invested $2.95 million, leveraging another $17.8 million by other investors. In the year ending on that date, it approved loans of $650,000 and capital investments of $550,000 in eligible companies. The MTDC also packages financing for small technology-based companies completely form other sources, without investing any of its own funds. National Association of State Development Agencies, National Council for Urban Economic Development, and The Urban Institute, Directory of Incentives for Business Investment and Development in the United States (1983). *See also* Jarett, Fostering Capitalism: The Massachusetts Venture Capital Fund (Council of State Governments 1984).

Add to note 114, page 352:

In 1984, a state appropriation of $800,000 was made available to provide matching funds for any additional grants that might be available from the federal government. 1984 NY Laws ch 50.

Page 353, add at end of section:

Another approach that has been tried in a number of states involves the provision of state tax incentives to taxpayers who invest in private investment or capital companies organized for the purpose of investing in small high-tech businesses in the state providing the incentive. In 1981, Montana enacted a Small Business Investment Incentive Act. Mont Rev Code Ann §§15-33-101 to 15-33-106. As a means of encouraging private sector investment in small business investment corporations (SBICs) organized for the purposes of diversifying and strengthening employment opportunities in Montana, this legislation exempts any capital gains realized from such investments from state taxation. A second Montana initiative, enacted two years later, provides individual investors with a 25 per cent tax credit (up to $25,000) for investments in companies, licensed as Montana capital companies, which are established to make equity investments in Montana businesses. The total amount of tax credits available for all investments in all Montana capital companies may not exceed $1 million per year for each of the next four years. Montana Capital Company Act, 1983 Mont Laws ch 554.

A similar program was also enacted in Louisiana in 1983. Called the Louisiana Capital Companies Tax Credit Program, the statute establishing the program provides for the certification of companies that meet certain criteria set forth in the statute and makes a tax credit available for cash investments by individuals or corporations in certified Louisiana capital companies. The amount of the credit is either 20 per cent or 35 per cent, depending on the amount of the capital company's initial capitalization. A 20 per cent credit is available if the company's initial capitalization at the time it sought certification or within one year thereafter is between $3 million and $20 million. If the amount of capitalization is more than $20 million, a 35 per cent credit is available. La Rev Stat Ann §§51:1921 to 51:1931.

§9.15 Technical Assistance to Inventors and Entrepreneurs

Add to note 118, page 354:

The OPTI program lost its funding at the end of September 1982, with one recent study attributing its failure to the fact that it had a broad mandate, i.e., "it attempted to encourage everything from 'basement inventors' to sophisticated licensing arrangements," but little flexibility of funding. US Office of Technology Assessment, Technology, Innovation and Regional Economic Development: Census of State Government Initiatives for High-Technology Industrial Development 10 (1983). This study did note, however, that "a proposal to create a similar mechanism, with those errors corrected, has been introduced in the New Jersey Legislature."

Add to note 119, page 354:

Similar programs to assist inventors have also been established in Iowa and Wisconsin. The Iowa Program for Inventors, started in January 1983, is part of the Center for Industrial Research and Service of the Iowa State University Extension Service. It provides inventors with confidential, technical evaluations of their ideas or inventions and advice on how best to develop and market their inventions. At later stages in the product development process, the center is able to provide assistance in applying for loans and in such practical matters as plant layout. In addition to these aspects of the program's operations, which are similar to the other programs discussed in this section, the center also includes, in a newsletter it distributes to all Iowa manufacturers, a list of inventions available for lease or purchase.

The Wisconsin Innovation Service Center, while based at a state university (the University of Wisconsin at Whitewater), has a national scope of operations. Only 17 per cent of its clients are from Wisconsin, and two other states (California and Florida) serve as formal intermediaries. Like the other inventor programs, it evaluates inventions for commercial feasibility, but does so on the basis of a sophisticated rating scheme which it has developed and will refer those receiving high scores to businesses that have told the center that they are interested in ideas or products in particular areas. Inventions are evaluated on 33 criteria, which are weighted according to importance, and the inventors are given written and quantitative reports on the results. Grossman, *Iowa Assists Inventors To Create Jobs,* 26 State Govt News 16-19 (May 1983).

Similar programs have recently been established in a number of other states including the Institute for Ventures in New Technologies, which was set up as a division of the Texas Engineering Experiment Station which, in turn, is a part of the Texas A&M University system. The Texas Engineering Experiment Station is a state agency with responsibility for conducting engineering research and development for Texas. *Shaping New Ideas,* informational brochure published by the Institute for Ventures in New Technology (undated). For additional information contact INVENT, Engineering Research Center, Suite 310, College Station, TX 77843, 409-845-0538.

Page 354, add at end of section:

In 1981, after an extensive public relations and lobbying campaign premised on the results of studies which showed that small businesses were the source of a surprisingly large share of all technological innovations developed in this country, Congress enacted the Small Business Innovation Development Act of 1982, Pub L No 97-219, 96

Stat 217, amending 15 USC §638, requiring agencies with sizeable research budgets to set aside a fixed portion of those funds for award to small businesses. The program of federal research support established by this act is known as the Small Business Innovation Research (or SBIR) program. During the first year of this program, fiscal year 1983, over $20 million was awarded to small businesses. By fiscal year 1987, it is estimated that $450 million will be reserved for small businesses under the terms of this law. Many state governments have responded to this program by organizing workshops, seminars, and other activities designed to inform inventors and small high tech businesses about the program.

For example, the New York State Department of Commerce and the State's Science and Technology Foundation conducted four training workshops in different parts of the state with 460 small R&D firms participating. The Alabama Department of Economic and Community Affairs is using its state-wide network of Small Business Development field offices to assist small businesses in competing for federal SBIR research grants and, according to a report published by the National Governors' Association, at least 12 other states have initiated special outreach programs to encourage, and in some cases to assist, small businesses to participate in the SBIR program. National Governors' Association, Technology & Growth: State Initiatives in Technological Innovation 85-87 (1983).

While it might be argued that programs of this type are nothing more than efforts to increase a state's share of the federal R&D funds available to small businesses under the program, this effort will also result in an increase in the quality of applications and, therefore, in the quality of the research designs of the inventors, engineers, and scientists involved. With or without federal funding under this program, the result is likely to be an increase in the total number of research and development undertakings which have improved chances of making a positive contribution to the American economy.

Several states are taking the further step of making state funds available to complement the federal SBIR program. Pennsylvania has established a $300,000 small business research seed grant program that can be used in conjunction with the SBIR program and Utah has funded a Utah Small Business Innovation and Research Assistance Program under which additional financial awards, totalling $100,000 in the 1983-84 fiscal year, will be given to small innovative firms receiving federal SBIR support, as well as to those firms which are deemed worthy of federal support but are not funded due to the availability of SBIR funds. *Id* 87. In its 1984-85 state operations budget, New York State appropriated $500,000 "for Small Business Innovation Research Act matching grants." 1984 NY Laws ch 50. Later in 1984, legislation was enacted providing a framework for an ongoing grant program under which this $500,000 and any monies appropriated in future years would be made available to New York businesses that receive federal SBIR awards. 1984 NY Laws ch 448. This legislation amends the state's public authorities law, formally establishing a small business innovation research promotion program to be administered by the New York State Science and Technology Foundation.

The federal SBIR program involves a building-block approach designed to provide two review points at which the federal agency involved must conduct a review of the small business's work to date and then determine whether or not to provide funding for a continuation of that work. This procedure involves three phases. In Phase I, the federal government provides up to $50,000 for projects designed to evaluate the technical and economic feasibility of an idea. Phase I award recipients who successfully complete their initial contract work are eligible to apply for Phase II funding. It is during Phase II that intensive work on the development of the original idea is actually carried out. Only those Phase I projects that have the most potential are given Phase II awards, but the financial support during this phase is much more substantial: a project can be funded for one or two years at $500,000. Phase III is much less defined, with specific funding decisions and levels dependent upon the scope and potential of the specific idea or innovation involved. In some cases, the federal agency involved will provide a contract for the production of a particular product for use by the agency in its operations. In other cases, Phase III may entail private sector investment to exploit possible commercial application of the innovations derived from the Phase I and II research and development.

The New York legislation referred to above is designed to complement this federal funding procedure and is specifically aimed at assisting small firms in meeting the cash-flow problems that can occur once they have finished their Phase I work but have not yet received any funding under Phase II. State support under the New York legislation is specifically targeted to assist those small businesses that have (1) received federal SBIR Phase I awards, (2) successfully completed their Phase I work for the federal government, and (3) properly submitted a Phase II research proposal to the federal agency involved. NY Pub Auth Law §3102-b.

In addition to providing matching or complementing grants, the New York legislation also charges the Science and Technology Foundation with facilitating and promoting the establishment of joint small business-university technology councils at the state's Centers for Advanced Technology (see **§9.16**) and at other universities within the state. Such councils are seen as vehicles for encouraging both technology transfer and cooperative research and development undertakings by small firms and university experts. NY Pub Auth Law §3102-b(7).

§9.16 University-Based and University-Related Research and Education Centers

In note 124, page 355, delete note and insert:

As of November 1983, sponsorship for the Microelectronic and Information Sciences Center and its programs exceeded $10 million, with support coming from industrial grants and equipment gifts, University of Minnesota allocations, federal grants and contracts, and state funding. According to an informational brochure published by the center, it "provides an organizational structure which helps create and sustain the ongoing research efforts of the MEIS Center in the following ways:

— attracting visiting researchers from other institutions to participate in the programs of the MEIS Center

— providing state-of-the-art training in the microelectronic and information sciences for scientists and engineers, producing new degree recipients, and providing education for professionals already on the job

— establishing and maintaining state-of-the-art laboratories in the microelectronic and information sciences; and

— maintaining an effective vehicle for collaboration and communication among industry, university, and government to facilitate the transfer of technology and science.

MEIS: Microelectronic & Information Sciences Center, informational brochure published by the Center (Nov 1983). For further information contact the Microelectronic and Information Sciences Center, 227 Lind Hall, 207 Church Street S.E., Minneapolis, MN 55455.

Add to note 134, page 358:

See also 1982 NY Laws ch 561, "AN ACT in relation to providing for the construction of the center for industrial innovation on the campus of Rensselaer Polytechnic Institute in the city of Troy and to amend the urban development corporation act, in relation to the amount of bonds and notes such corporation is authorized to issue." In 1982, New York state also established a program under which the New York State Science and Technology Foundation (see **§§9.05 & 9.06** was authorized to designate and support Centers for Advanced Technology in fields "identified by the foundation as having significant potential for economic growth in New York, or in which the application of new technologies could significantly enhance the productivity and stability of New York businesses." NY Pub Auth Law §3102-a. Under this legislation, the foundation has designated centers for advanced technology in optics at the University of Rochester, in telecommunications at the Polytechnic Institute of New York, in biotechnology in agriculture at Cornell University, in medical biotechnology at the State University of New York at Stony Brook, in computers and information systems at Columbia University, in computer applications and software engineering at Syracuse University, and in health-care instruments and devices at the State University of New York at Buffalo. For the state's 1984-85 fiscal year, $7 million was appropriated for matching grants to these seven designated centers pursuant to the provisions of NY Pub Auth Law §3102-a(3), which requires that the amounts provided by the foundation must be matched by equal amounts from private and other (nonstate) governmental sources. 1984 NY Laws ch 50. Similar programs for the designation and support of centers of excellence have been established in Kansas (see **§9.06**), Pennsylvania (where that state's Ben Franklin Partnership "Challenge" grants have provided matching funds to establish four advanced technology centers at Lehigh University, the University of Pennsylvania, the University of Pittsburgh, and Pennsylvania State University), and New Mexico (*see* 1983 NM Laws ch 297 and 299).

Page 358, add at end of section:

In 1982, Massachusetts authorized the Massachusetts Technology Park Corporation, Mass Gen Laws Ann ch 40J, §§1-10 (West), to establish a Massachusetts Microelectronics Center containing design, fabrication, and testing equipment which will facilitate "the basic training of new engineers and technicians and the retraining of existing personnel to meet the constantly changing requirements for effective and productive work in the field of semi-conductor and microelectronic technologies." While the immediate focus of the legislation was on encouraging business and industry in the fields of semiconductor and microelectronics technologies to locate, develop, and expand in Massachusetts, it also concluded that "it may reasonably be anticipated that the pattern of events which has given rise to the need for a partnership of government and industry and education to create a training facility for semi-conductor and microelectronic technologies will recur in other technologies," and that "it is appropriate to utilize the skills and experience which will be developed in responding to the present need to meet such future needs." Consequently, the statute also authorized the Massachusetts Technology Park Corporation, after having satisfied itself that the microelectronic facility is operational, to establish further similar centers to support other technologies in which the corporation finds a need equal to that which was found to exist in the microelectronics field.

Cooperation among academic and industrial researchers and cooperative university-industry research centers are now widely accepted as useful approaches to the promotion of industrial innovation and virtually every state government is now involved, in one way or another, in trying to foster such cooperative arrangements. As the number of programs of this type grows, the need to learn from experience and to build upon past failures and successes becomes increasingly important. In one recent study evaluation of such efforts, the National Science Foundation reviewed 118 industry/university cooperative research projects that it has supported and published the results of a survey of 226 scientists involved in these projects. This study attempted to identify the project characteristics which contribute to successful technical and organizational outcomes and found that "in general the variables that seemed to contribute to all aspects of project success were those related to interpersonal interaction." Overall, the study found that "the theoretical orientation of the university reacted synergistically with the more practical concerns of industry scientists and yielded both scholarly outcomes and perception of commercial success." Johnson & Tornatzky, Cooperative Science: A National Study of University Industry Researchers: Assessment of the Industry/University Cooperative Research Projects Program (National Science Foundation, Nov 1984). *See also* Johnson, Tornatzky & Schlaaff, Cooperative Science: A National Study of University Industry Researchers (National Science Foundation, Nov 1984) for a series of case studies of "cooperative science as practical" in nine different fields (e.g., computer language systems, fluid dynamics, and nondestructive testing).

A practical approach to averting or effectively dealing with the problems inherent in university-industry relationships is set forth in Varrin & Kukich, *Guidelines for Industry-Sponsored Research at Universities,* 2272 Sci 385 (Jan 25, 1985), which recommends guidelines in such areas as patent ownership, publication rights, copyrights, confidentiality agreements, and international agreements. *See also* Holden, *New R & D Centers Will Test University Ties,* 227 Sci 150 (Jan 11, 1985). 1985).

Innovation and Products Liability
Kenneth Ross

10

Critical Issues in Historical Context

§10.03 Strict Liability

Page 365, in first line of section, delete "1962" and insert:

1963

In note 8, page 365, in first line delete "(1962)" and insert:

(1963)

§10.04 Section 402A

Add to note 14, page 366:

The main thrust of the Model Uniform Product Liability Act has been included in a Product Liability Act proposed in the United States Senate as S 100, 99th Cong, 1st Sess (1985). The balancing approach has been adopted in §5 of S 100; however, some courts are continuing to adopt the consumer expectations test in lieu of the risk/benefit analysis. *See* Lester v Magic Chef, Inc, 230 Kan 643, 641 P2d 353 (1982); Ward v Ford Motor Co, 657 P2d 95, 1 Prod Liab Rep (CCH) ¶9512 (Nev Sup Ct 1983).

Bases of Liability against a Manufacturer

§10.06 Manufacturing Defects

Add to note 17, page 367:

See also S 100, 99th Cong, 1st Sess, §5(a) (1985).

§10.07 Design Defects

Plaintiffs have asked courts to collaterally estop defendants from defending themselves on certain issues if such defendants have previously had such issues of fact adversely adjudicated. This procedure is called "offensive collateral estoppel" and is allowed by *Parklane Hosiery Co v Shore,* 439 US 322 (1979). Since 1979, a number of courts have ruled on the appropriateness of offensive collateral estoppel. While collateral estoppel has been allowed in asbestos lawsuits, *Borel v Fibreboard Paper Products Corp,* 493 F2d 1076 (5th Cir 1973), *cert denied,* 419 US 869 (1974), one court recently held that the *Borel* decision should not prevent asbestos manufacturers from litigating whether asbestos products are unreasonably dangerous or whether asbestos is in all products a producing cause of cancer. *Hardy v Johns-Manville Sales Corp,* 681 F2d 334 (5th Cir 1982). However, offensive collateral estoppel has been allowed in *Kaufman v Eli Lilly & Co,* 116 Misc 2d 351, 455 NYS2d 29 (1982) (DES) and in *Adducci v Skyworker,* 1 Prod Liab Rep (CCH) ¶9462 (DNY 1982) (man-lift).

Add to note 23, page 369:

See also S 100, 99th Cong, 1st Sess, §5(b) (1985), which includes many of the same criteria for determining a defective design as does the Model Uniform Product Liability Act.

Add to note 24, page 369:

No state other than Alaska has adopted the *Barker* concept of shifting the burden.

§10.08 Marketing Defects

In one of the most controversial decisions in 1982, the New Jersey Supreme Court held in *Beshada v Johns-Manville Products Corp,* 90 NJ 191, 447 A2d 539 (1982), that asbestos manufacturers who were sued in a strict liability action for failure to warn could not use a state-of-the-art defense. The rationale was that strict liability focuses only on the product and not on the culpability of the manufacturer. Therefore, the manufacturer could be held liable in such cases for failing to warn of dangers which were undiscoverable at the time of manufacture. If this opinion were adopted in many other courts, it could have a profound effect on industrial innovation.

An exception to the *Beshada* holding was created for drugs by the New Jersey Supreme Court in *Feldman v Lederle Laboratories,* 479 A2d 314 (1984). To date, only one court has adopted *Beshada: see Hayes v Ariens Co,* 391 Mass 404, 462 NE2d 273 (1984).

Add to note 28, page 370:

Courts are continuing to hold that a manufacturer has no duty to warn experienced product users: Strong v EI DuPont de Nemours Co, 667 F2d 682 (8th Cir 1981); Fierro v International Harvester Co, 127 Cal App 3d 862, 179 Cal Rptr 923 (1982).

Add to note 29, page 370:

S 100, 99th Cong, 1st Sess, §6(e)(1) (1985), has adopted the theory that there is no need for a manufacturer to warn of patent or obvious dangers. Also, while some courts are continuing to uphold the patent danger rule, Mach v Terex, 112 Mich App 158, 315 NW2d 561 (1982); Pressley v Sears-Roebuck Co, 738 F2d 1222 (11th Cir 1984) (interpreting Georgia law), two courts have recently rejected it, Holm v Sponco Mfg, Inc, 324 NW2d 207 (Minn 1982); Franchetti v Intercole Automation Inc, 529 F Supp 533 (D Del 1982).

Add to note 32, page 370:

This concept has been adopted in S 100, 99th Cong, 1st Sess, §6(d)(1) (1985).

Add to note 39, page 371:

S 100, 99th Cong, 1st Sess, §6(c) (1985) adopts the principle of a post-manufacturing duty to warn.

§10.10 Misrepresentation

Add to note 46, page 372:

This concept has been adopted in S 100, 99th Cong, 1st Sess, §7 (1985).

§10.11 Effect of Products Liability on Innovation

For a number of years, courts have awarded punitive damages against manufacturers who exhibited sufficient culpability to warrant their imposition. The increased risk of an award of punitive damages has had a great impact on industrial innovation. While there have been attempts to restrain the growth of such awards (*e.g.*, S 100, 99th Cong, 1st Sess, §12 (1985)), courts are continuing to find instances where punitive damages are appropriate. *See Airco, Inc v Simmons First National Bank,* 276 Ark 486, 638 SW2d 660 (1982) (artificial respirator); *Ford Motor Co v Nowak,* 638 SW2d 582 (Tex Civ App 1982) (transmission); *Hasson v Ford Motor Co,* 32 Cal 3d 388, 650 P2d 1171, 185 Cal Rptr 654 (1982) (brakes); *Moran v Johns-Manville Sales Corp,* 691 F2d 811 (6th Cir 1982) (asbestos); *Slough v JI Case Co,* 8 Kan App 2d 104, 650 P2d 729 (1982) (trencher); *Neal v Carey Canadian Mines, Ltd,* 548 F Supp 357 (ED Pa 1982) (asbestos); *Vossler v Richard Manufacturing Co,* 143 Cal App 3d 952, 192 Cal Rptr 219 (1983) (prosthetic device); *Cloroben Chemical Corp v Comegys,* 464 A2d 887 (Del Sup Ct 1983) (drain cleaner); *Toyota Motor Co, v Moll,* 438 So 2d 192 (Fla Dist Ct App 1983) (automobile); *Wooderson v Ortho Pharmaceutical Corp,* 235 Kan 387, 681 P2d 1038 (1984) (Ortho-Novum); *Palmer v AH Robins Co,* 684 P2d 187 (Colo 1984) (Dalkon Shield).

Add to note 49, page 373:

In 1981, Congress passed a law which would allow manufacturers to join together to form insurance cooperatives for product liability purposes. Product Liability Risk Retention Act, Pub L No 97-45, 95 Stat 949 (1981).

Add to note 50, page 373:

A number of courts have refused to adopt the market share liability doctrine as adopted by the California Supreme Court in *Sindell.* Mizell v Eli Lilly & Co, 526 F Supp 589 (DSC 1981) (DES); Starling v Seaboard Coast Line RR, 533 F Supp 183 (D Ga 1982) (asbestos); Payton v Abbott Labs, 386 Mass 540, 437 NE2d 171 (1982) (DES); Morton v Abbott Labs, 538 F Supp 593 (D Fla 1982) (DES). However, the New York Court of Appeals allowed a plaintiff to proceed without identifying the specific manufacturer under a concert of action theory in Bichler v Eli Lilly & Co, 55 NY2d 571, 436 NE2d 182, 450 NYS2d 776 (1982) (DES). Most recently, courts have more willingly adopted the market share or a related theory. *See* Morris v Parke, Davis & Co, 573 F Supp 1324 (DC Cal 1983) (DPT vaccine); Copeland v Celotex Corp, 447 So 2d 908 (Fla Dist Ct App 1984) (asbestos); Collins v Eli Lilly & Co, 116 Wis 2d 166, 342 NW2d 37 (1984) (DES).

Pre-Accident Prevention in the Production Process

§10.12 Introduction

Add to note 53, page 374:

See also Eads & Reuter, Designing Safer Products: Corporate Response to Product Liability Law and Regulation (1984).

§10.15 Considerations for a Safe Design

Add to note 56, page 376:

Owens v Allis-Chalmers Corp, 414 Mich 413, 326 NW2d 372 (1982).

Add to note 58, page 377:

See also Caterpillar Tractor Co v Ford, 406 So 2d 854 (Ala 1981).

§10.20 Advertising, Marketing and Distribution

Add to note 74, page 382:

Accord, Daberko v Heil Co, 681 F2d 445 (5th Cir 1982).

Prevention through Management Techniques

§10.25 Document Policy

Add to note 88, page 388:

See also American Bar Association, The Attorney-Client Privilege and the Work-Product Doctrine (1983).

Add to note 90, page 388:

The statute of repose has been declared unconstitutional in Alabama (Lankford v Sullivan, Long & Hagerty, 416 So 2d 996 (Ala 1982)), Rhode Island (Kennedy v Cumberland Eng Co, 471 A2d 195 (RI 1984)), and New Hampshire (Heath v Sears, Roebuck & Co, 123 NH 512, 464 A2d 288 (1983)), and declared constitutional in Tennessee (Petty v Vulcan Iron Works, Inc, 1 Prod Liab Rep (CCH) ¶9282 (Tenn Ct App 1982)); Mathis v Eli Lilly & Co, 719 F2d 134 (6th Cir 1983)); South Dakota (Van Den Hul v Baltic Farmers Elevator Co, 716 F2d 504

(8th Cir 1983)); Indiana (Pitts v Unarco Indus, 712 F2d 276 (7th Cir), *cert denied,* 104 S Ct 509 (1983); Scalf v Berkel, Inc, 448 NE2d 1201 (Ind Ct App 1983)), and North Carolina (Davis v Mobilift Equip Co, 1 Prod Liab Rep (CCH) ¶10,339 (NC Ct App 1984)).

§10.26 Attorney Involvement

Courts continue to hold contractual disclaimers and other limitations invalid. *Oldham's Farm Sausage Co v Salco, Inc,* 633 SW2d 177 (Mo Ct App 1982); *A & M Produce Co v FMC Corp,* 135 Cal App 3d 473, 186 Cal Rptr 114 (1982); *DCR, Inc v Peak Alarm Co,* 663 P2d 433, 1 Prod Liab Rep (CCH) ¶9566 (Utah 1983). It is important for an attorney to assist the client in drafting legally sufficient contractual disclaimers and limitations and to educate the client on ways to remedy any product problems in order to avoid a holding that the remedy failed of its essential purpose, thereby voiding all contractual protection.

Accident Prevention Programs

§10.28 Insurance Coverage and the Role of the Insurance Company

Add to note 93, page 390:

The United States Supreme Court has continued to refuse to decide this issue.

§10.30 Recalls, Retrofits and Repairs

Add to note 97, page 391:

See S 100, 99th Cong, 1st Sess, §13 (1985). The only exception contained in §13 to the inadmissibility of such evidence is if it is offered to impeach a witness for the manufacturer or product seller who has expressly denied the feasibility of such a measure.

§10.31 Spreading the Risk

In most states, the manufacturer or other product seller may be unable to obtain contribution or indemnity from a potential party if that party is the employer of the plaintiff. This is due to the interrelationship between products liability and the worker's compensation system. Most states, however, provide for a statutory or judicially created exception where there has been shown to be a deliberate intent to injure on the part of the employer. *See generally* Birnbaum & Wrubel, *Worker's Compensation and the Employer's Immunity Shield: Recent Exceptions to Exclusivity,* 5 J Prod Liab 119 (1982).

The Inside Counsel's Role in Innovation: A Conceptual Framework

Edward S. Binkowski

11

Overview: The Law and Innovation

§11.01 Introduction

Add to note 14, page 482:

This is often due to a technological bypass or leapfrogging of the economic rationale supporting the precedent as well. Rowe, *The Decline of Antitrust and the Delusions of Models: The Faustian Pact of Law and Economics,* 72 Geo LJ 1511 (1984).

Add to note 18, page 483:

Such changes place additional pressures on the legal profession to integrate technological competence with legal strategy. See Nyhart & Jones, *What You Don't Know About Technology Can Hurt You,* 69 ABAJ 1667 (Nov 1983); Freed, *The Demands of High-Tech Lawyering,* NY Times, Oct 2, 1983, at F3: Gemignani, *The Case for Computer Law,* 88 Case & Comment 3 (May/June 1983).

Add to note 19, page 483:

The rapid introduction of computers into all phases of industrial processes has been an especially fertile ground for creating such loopholes. See Pool, The Technologies of Freedom (1983) (First Amendment and communications law); Burnham, *Loophole in Law Raises Concern About Privacy in Computer Age,* NY Times, Dec 19, 1983, at A1 (privacy and theft of services); Lehman-Wilzig, *Frankenstein Unbound,* 13 Futures 442 (1981) (products liability and criminal law).

Page 483, add at end of section:

One of the latest phrases in vogue in current policy analysis and presidential politics is "industrial policy."

> Industrial policy remains an elusive concept. It encompasses ideas, some of them contradictory, ranging from the establishment of an industrial development bank to support winning industries to an industrial conversion agency to rehabilitate losing industries and their workers. Many proposals prescribe some type of three-way bargaining process in which business and labor make concessions in return for government help.

Arenson, *Debate Grows Over Adoption of National Industrial Policy,* NY Times, June 19, 1983, at A1. Most proponents see a coordinated federal response on these items, coupled with appropriate taxation and antitrust incentives, as necessary to meet international competition and ameliorate domestic decline of regional industries. Kilborn, *A Plan for Restoring U.S. Industrial Role Advocated in Study,* NY Times, Jan 12, 1984, at A1. Unfortunately, specifics of such a plan vary confusingly from advocate to advocate. See Arenson, *supra;* Kantrow, Eads, Levitt, McCraw, Reich & Thurow, *The Political Realities of Industrial Policy,* 83 Harv Bus Rev 76 (Sept/Oct 1983); Blumenthal, *Drafting a Democratic Industrial Plan,* NY Times, Aug 28, 1983, at VI28. The two approaches that have received the most public attention are Reich, the Next American Frontier (1983) and Rohatyn, The Twenty-Year Century (1984).

Many proponents of a national industrial policy see the Japanese Ministry of International Trade and Industry (MITI) as an appropriate model. An excellent discussion of MITI's structure and history can be found in Sawada, *Government Industrial Policy for a Health World Economy,* 24 Technological Forecasting of Soc Change 95 (Oct 1983). Critics of the transportability of the MITI paradigm point out its roots in the values of Japanese society, its unpublicized failures, and the extent to which successful marketing has been a factor in its achievements. Taylor & Tanakadate, *Why "Japan Inc" Might Not Work in the U.S.,* US News & World Rep, Jan 30, 1984, at 44. *See also* Lohr, *Japan's Trade Ministry Draws Praise and Ire,* NY Times, May 17, 1983, at A1; Lohr, *How Japan Guides Industries,* NY Times, May 18, 1983, at D1.

Serious political and operational objections have been raised to these plans. One former chairman of the Economic Council of Advisers believes that the industrial policy approach will not correct the most serious need of American industry: the gradual restoration of balanced economic prosperity and the reduction of huge structural budget deficits. Schultze, *Industrial Policy: A Solution in Search of a Problem,* 25 Cal Mgmt Rev 5 (Summer 1983). Another party holding the same post in a different administration finds the very notion of industrial policy vacuous, and its principal harm its impact on deficits. Stein, *Don't Fall for Industrial Policy,* Fortune, Nov 14, 1984, at 64. Less partisan writers argue that the reliance of the American political process on partisan politics and special interest groups would prevent any national industrial policy from being implemented, however sound it might be. Badaracco & Yoffie, *Industrial Policy: It Can't Happen Here,* 83 Harv Bus Rev 96 (Nov/Dec 1983). It can also be argued that apparent deindustrialization trends are the localized effects of an overstrong dollar in the world recession. Lawrence, Can America Compete? (1983). Alternatively, the greater volatility of recent business cycles due to inappropriate matching of fiscal and monetary policies of the last few years has been cited as the underlying cause of industrial malaise. Jasinowski, *Too Little Focus on the Business Cycle,* NY Times, June 12, 1983, at V2.

The most telling argument against the full-fledged industrial policy program of supporting "winners" and "culling" losers is the government's demonstrably poor track record in doing so (not that it would be necessarily worse than any other institutionalized attempt). See Kraus, *Can Creeping Socialism Cure Creaking Capitalism: One Man's MITI is Another Man's Poison,* 266 Harper's 17 (Feb 1983). Non-industry-specific encouragement of innovation, however, is the most legitimate portion of the industrial policy platform. For a narrower and more practical aspect of industrial policy focused on promotion of innovation generally, see **§§12.03-12.04** in text and this supplement.

This focus has been intensified by international factors as well. The President's Commission on Industrial Competitiveness reported that the ability of American business to compete with other countries has declined over the last 20 years, with significant slippage in high-tech industries such as pharmaceuticals, electrical equipment and components, plastics, industrial chemicals, and professional and scientific instruments. Kilborn, *Competitiveness Panel Finds 20-Year U.S. Lag,* NY Times, Feb 14, 1985, at D1. The effects of this decline show up dramatically in trade deficit estimates. In 1984, the United States was a net importer of electronics for the first time, with a deficit of nearly $7 billion. *America's High Tech Crisis,* Bus Wk, Mar 11, 1985, at 56. The electronics trade deficit with Japan alone was $15 billion, greater than the deficit associated with automobiles. Kilborn, *supra.*

§11.04 Extracorporate and Extralegal Background

Page 487, at end of first paragraph add:

Given scarce resources, accelerated technological innovation may be necessary to provide any growth at all. See Rothwell & Zegveld, Industrial Innovation and Public Policy, at 3-22 (1982); Gibson, *Revitalizing Old Industries,* 4 Venture 86 (Nov 1982).

Alternative Business Contexts of Innovation Having Legal Implications

§11.05 Economies of Scale

Page 488, at end of third paragraph add:

The very size of the firm, however, may make it easier to acquire access to new technology by other means. Link, *Alternative Sources of Technology: An Analysis of Induced Innovations,* 4 Managerial & Decisions Econ 40 (1983).

§11.06 Economies of Scope

Add to note 75, page 489:

The progress in computer-aided design and computer-aided manufacture (CAD/CAM) techniques has considerably broadened the possibilities for economies of scope in industrial processes, through enhanced flexibility in product design and mix, rapid response to environmental changes, greater control, accuracy, and repeatability of processes, reduced waste, greater predictability, faster throughput, and distributed processing capability. Goldhar & Jelinek, *Plan for Economies of Scope,* 83 Harv Bus Rev 141 (Nov/Dec 1983). *See also* Gunn, *The Mechanization of Design and Manufacturing,* 247 Sci Am 114 (Sept 1982).

§11.08 Radical Innovation

In note 97, page 492, delete "__US__" and insert:

454 US 810

Page 492, at end of carryover paragraph add:

In general, even radical innovation does not confer economic monopoly but instead raises additional barriers to entry. Lavey, *Patents, Copyrights, and Trademarks as Sources of Market Power in Antitrust Cases,* 27 Antitrust Bull 433 (1982).

Add to note 101, page 492:

cert denied, 454 US 810 (1982).

Page 492, at end of first full paragraph add:

The costs of and alternatives to this practice are a matter of some controversy. See Adelman, *Property Rights Theory and Patent-Antitrust: The Role of Compulsory Licensing,* 52 NYU L Rev 977 (1977).

§11.09 Expense-Capital Tradeoffs

Page 493, at end of fourth paragraph add:

These tradeoffs may also substantially increase employee quality of life, and thereby indirectly increase other efficiencies as well. Skinner & Chakraborty, The Impact of New Technology: People and Organizations in Manufacturing and Allied Industries (1982); Skinner & Chakraborty, The Impact of New Technology: People and Organizations in Service Industries (1982); and **ch 2,** especially **§2.07.**

§11.10 Innovation as Information

Page 494, at end of second paragraph add:

A former chief of the Antitrust Division of the Department of Justice argues that lack of adequate intellectual property and a failure to accept aggressive but still procompetitive licensing arrangements are the greatest failures of current law in promoting technological innovation. Baxter, *Antitrust Law and Technological Innovation,* 1 Issues in Sci & Tech 80 (Winter 1985)

Add to note 124, page 495:

See also Diamond v Diehr & Lutton, 450 US 175 (1981); Flewellen, *An Anomaly in the Patent System: The Uncertain Status of Computer Software,* 8 Rut J Computers & Tech 273 (1981). An excellent review of patent, copyright, and trade secret options is presented in Davidson, *Protecting Computer Software: A Comprehensive Analysis,* 23 Jurimetrics 337 (Summer 1983). *See also* Apple Computer, Inc v Franklin Computer, 714 F2d 1240 (3d Cir 1983), in which a computer program in object code embedded in ROM (read-only memory) was held copyrightable; Grogan & Kump, *The Broader Meanings of Apple v Franklin,* 1 Computer Law 15 (1984); Sanger, *The Gavel Comes Down on Computer Copycats,* NY Times, Oct 23, 1983, at F3; and **§4.02A,** this supplement.

Add to note 126, page 495:

See also **§4.20A,** this supplement.

Corporate Planning and Innovation

§11.11 Financing Options and Pay-Out Parameters

Add to note 140, page 496:

A review for a particular industry of the full range of governmental support mechanisms which may be brought to bear when foreign trade pressures are also an issue may be found in Joint Economic Committee, 97th Cong, 2d Sess, *International Competition in Advanced Industrial Sectors: Trade and Development in the Semiconductor Industry* (Feb 18, 1982); and Wilson, Ashton, & Egan, Innovation, Competition, and Government Policy in the Semiconductor Industry (1982).

Page 496, at end of second paragraph add:

Since the motivation for local governmental support is usually area development rather than technological advancement, the effects may be more immediate and successful. See **ch 12;** *see also* Subcommittee on Monetary and Fiscal Policy, Report of Joint Economic Committee, 97th Cong, 2d Sess (June 1, 1982); Brennan, *Advanced Technology Centers: Strategies for Corporate Growth,* 248 Sci Am E1 (May 1983); Pear, *States Fostering High Technology,* NY Times, Aug 17, 1983, at A1; see also **§9.10,** main text and this supplement.

Add to note 141, page 496:

See also Bronson, *U.S. Firms Plus Foreign Partners Equals Prosperity,* US News & World Rep, Mar 5, 1984, at 51, for a discussion of the usual trade-off in foreign joint ventures of technological expertise with market presence.

Add to note 142, page 497:

Long-standing proposals for favorable antitrust treatment of cooperative research and development efforts are receiving further extension by the Department of Justice and the Congress. The National Cooperative Research Act of 1984, Pub L No 98-462, 98 Stat 1815 (1984), 15 USC §§4301-4305, legislates a "rule of reason" standard for R&D joint ventures. For a detailed discussion of the Act, see **§3.10,** this supplement; *see also* Millstein & Millstein, *A Sound Approach on Industrial Policy,* NY Times, Aug 25, 1983, at A23; Hershey, *Stiff Antitrust Laws Called Rein on U.S.,* NY Times, Aug 1, 1983, at D1; HR Rep 656, 98th Cong, 2d Sess (1984); HR Rep 571, 98th Cong, 2d Sess (1983).

Page 497, text, following first carryover sentence insert:

International trade pressures are accelerating this trend. See Abelson, *International Competition in High Technology,* 219 Sci 1025 (1983); Joint Economic Committee, 97th Cong, 2d Sess, *International Competition in Advanced Industrial Sectors: Trade and Development in the Semiconductor Industry* (Feb 18, 1982).

§11.12 Mergers and Acquisitions

Page 498, at end of first full paragraph add:

In addition, familiar estimates of impact upon market share are rarely meaningful for a merger or acquisition predicated upon genuine innovation; such innovation constitutes 100 per cent of a market whose very existence is uncertain. Baxter, *The Definition and Measurement of Market Power in Industries Characterized by Rapidly Developing and Changing Technologies,* 53 Antitrust LJ 717 (1985).

§11.13 Advertising and Marketing

Add to note 174, page 500:

The "predatory innovation" argument may also be developed with respect to innovations achieved solely for the monopoly profit they induce and not for any genuine consumer benefit, especially in terms of product substitutability. Ordover & Willig, *An Economic Definition of Predation and Product Innovation,* 91 Yale LJ 8 (1981).

Page 500, at end of first full paragraph add:

The expenses of innovation will ensure that the typical innovative product should be priced above marginal cost, and the dilemma can be solved through protection of intellectual property approaches. Lavey, *Patents, Copyrights, and Trademarks as Sources of Market Power in Antitrust Cases,* 27 Antitrust Bull 433 (1982).

Finally, in industries with rapid product or production technology advances, the pace of innovation makes it difficult to rely on traditional product differentiation strategies (either on price-performance comparisons or non-performance-related consumer preferences) for any length of time. A focused, market-niche approach may be much more successful. See Lele & Maher, *Getting It Together: Building a Strategic Analysis Model,* 68 Bus Marketing 59 (Nov 1983) for a study with applications to the microcomputer market. Technological advances within a market niche will also raise strong barriers to entry for competitors. MacMillan, *Pre-emptive Strategies,* 4 J Bus Strategy 16 (Fall 1983).

Add to note 178, page 500:

Anticompetitive innovation in the computer industry as represented by unutilized patents is discussed in Sobel, *The Antitrust Interface with Patents and Innovation: Acquisition of Patents, Improvement Patents and Grant-Backs, Non-Use, Fraud on the Patent Office, Development of New Products and Joint Research,* 53 Antitrust LJ 681 (1985).

§11.14 Testing and Experimentation Liability

Add to note 185, page 501:

See also Feder, *Gene-Splicing: Legal Pitfalls,* NY Times, Dec 1, 1981, at D1. More recent feeling indicates that such fears were premature and overdrawn, Judson, *Thumbprints in Our Clay,* New Republic, Sept 19, 26, 1983, at 12.

§11.15 Internal Decision Structures

Add to note 203, page 502:

The same approach has been recommended as "internal entrepreneurship" and creation of a "skunkworks" organization for freewheeling innovation in Peters & Austin, In Search of Excellence (1985). The essential factors are given in Quinn, *Managing Innovation: Controlled Chaos,* 85 Harv Bus Rev 73 (May/June 1985) as avoidance of: top management isolation; intolerance of fanatics; short time horizons; restrictive accounting practices; excessive rationalism; excessive bureaucracy; and inappropriate incentives.

Page 502, add at end of section:

A recent popular study of successful American companies found the innovative success of even the largest contingent on the lack of corporate meddling in small internal research groups. Peters & Waterman, In Search of Excellence (1982). This view, however, is strenuously challenged by many authors who claim it is wasteful of resources and should be replaced by active and cost-justified "technology management." See Thackray, *The Technology Trap,* 17 Mgmt Today 86 (June 1983). Such a point of view clearly underestimates the necessity and the rewards of basic research. See **§11.28** in the main text. A more constructive compromise between the two points of view is to insist on clear strategic direction from upper management without any explicit involvement in the research process itself. See Petroni, *The Strategic Management of R&D, Part II: Organizing for Integration,* 16 Long Range Planning 51 (Apr 1983).

Measuring Innovation

§11.18 Choice of Surrogates for Innovation

Page 506, following first full paragraph add:

The granting of a patent is in its own right only a coarse and indirect measure of innovation. See Lavey, *Patents, Copyrights, and Trademarks as Sources of Market Power in Antitrust Cases,* 27 Antitrust Bull 433 (1982). Patterns in the granting of patents across an industry, however, provide a sounder guide. See Feder, *Technology: Measuring the "Cutting Edge,"* NY Times, Dec 3, 1981, at D2.

§11.20 Critical Mass Arguments

Add to note 261, page 508:

See also Bedell, *Give Money to the Real Innovators,* NY Times, Nov 29, 1981, at F2; Lentz, *Financing High-Technology: Congress Invests in a Better Idea,* 9 Rut J Computers & Tech 209 (1982). For a contrary interpretation of the same data, see Rothwell & Zegveld, Industrial Innovation and Public Policy, at 184-188 (1982).

§11.21 Rate of Innovation

Page 509, at end of second full paragraph add:

Governmental regulation also has alternatively retarding and accelerating effects on the rate of innovation, depending on whether the strictness of standards promulgated chokes or stimulates technological alternatives ("process" regulations generally accomplish the former and "end result" regulations the latter). For a literature review, see Rothwell & Zegveld, Industrial Innovation and Public Policy, at 116-146 (1982).

Innovation and Industry Structure

§11.23 Innovation and Competition

Page 512, at end of carryover paragraph add:

The surrounding ambiguity and debate continue. For review articles, see Adelman, *The Supreme Court, Market Structure, and Innovation: Chakrabarty, Rohm, and Hass,* 27 Antitrust Bull 457 (1982); Ginsburg, *Antitrust, Uncertainty, and Technological Innovation,* 24 Antitrust Bull 635 (1979). *See also* Kamien & Schwartz, Market Structure and Innovation (1982), for a technical summary.

A study attempting to reconcile the varied empirical results ends up itself so confused that it concludes with these metaphorically (and legally) strained platitudes:

> Innovation is inherently a chancy process that depends upon a wide variety of factors . . . What we need, therefore, is an equal amount of variety in firm size and industry structure. It is only through variety in one system that we can sustain variety in the other. Thus, technical progress is best facilitated by a judicious mix of both small and large firms. Smooth sailing in innovation requires islands of oligopoly in a sea of competition.

Sahal, *Technology, Productivity, and Industry Structure,* 24 Technological Forecasting & Soc Change 12 (Sept 1983).

Add to note 297, page 512:

See also Baumol, *Competition vs. Research,* NY Times, Aug 31, 1983, at D2.

§11.25 Innovation and Growth

Add to note 319, page 514:

See also Hayes & Garvin, *Managing as if Tomorrow Mattered,* 60 Harv Bus Rev 70 (May/June 1982).

Page 514, at end of third full paragraph add:

Growth may rationalize investment for innovation as well. One study shows that R&D decisions in growth businesses are responsive to competition, evolving strategies, and intra-firm resource flows, while in mature businesses, a pattern of stability and R&D equilibrium is evident, even in the face of changes by competitors. Hambrick, MacMillan & Barbosa, *Business Unit Strategy and Changes in the Product R&D Budget,* 29 Mmgt Sci 757 (July 1983).

Innovation and Decision Making

§11.27 Technological Forecasting

Add to note 338, page 516:

For the notable exceptions in which applications have been successfully anticipated, see Drucker, Innovation and Entrepreneurship (1985). Drucker sees the sources of corporate opportunities for innovation occurring in taking advantage of: unexpected occurrences (e.g., IBM's early shift to business rather than scientific applications); incongruities (e.g., land-based container technology revolutionizing the shipping industry); process needs (e.g., switchboard automation); and market changes (e.g., Ford Motors' introduction of the Mustang).

Add to note 344, page 517:

See also O'Keefe, Technology Assessment for State and Local Government (1982).

§11.28 Innovation and Pure Research

Page 519, at end of first full paragraph add:

Productivity following from applied research would decrease as well. See National Science Foundation, The 5-Year Outlook on Science and Technology, at 10-12 (1982).

§11.29 The Corporation and the University

Add to note 376, page 519:

See also Rothwell & Zegveld, Industrial Innovation and Public Policy, at 147-157 (1982).

Page 520, at end of first full paragraph add:

In addition, the corporate education system is a substantial alternative and sometimes competitor to the traditional educational process. Corporate-run education annually spends about the same ($60 billion) on the same number of students (8 million) as do the nation's four-year colleges and universities. The RAND Corporation even offers a Ph.D. program not only to its own employees but also to the public at large. Fiske, *Booming Corporate Education Efforts Rival College Programs Study Says,* NY Times, Jan 28, 1985, at A10.

Page 521, add at end of section:

Growth in the semiconductor industry has produced an analogous approach in directly embedding corporate research centers in graduate education. See Hechinger, *About Education: Looking Ahead to the Silicon Future,* NY Times, Dec 1, 1981, at C11. None of these forms of involvement has escaped criticism as compromising academic research. *Schools Study Ethics of Business-Aided Research,* NY Times, Apr 8, 1983, at A20.

Some initiatives are occurring from the academic side in terms of plans to fund, develop, and ultimately commercialize university research. Gibson, *Research for Sale,* 6 Venture 78 (Mar 1984). This approach, rather than the traditional sale of patents, will allow the university to exercise greater control over and receive greater rewards from the application. Schools with such plans include Baylor, with respect to virology and cardiovascular techniques; University of West Virginia, plastics and materials science; University of Delaware, marine engineering; Case Western Reserve, programmable machine tools; University of Utah, biomedical technology; and University of Pittsburgh, robotics. Other institutions, while countenancing such efforts on the part of individual faculty members, are reluctant to become directly involved due to fears of commercialization. *See also* Bouton, *Academic Research and Big Business: A Delicate Balance,* NY Times, Sept 11, 1983, at VI62; Reinhold, *Stanford and Industry Forge New Research Link,* NY Times, Feb 10, 1984, at A22.

Some initiatives that can be taken to preserve academic integrity in such associations are discussed in Varrin & Kukich, *Guidelines for Industry-Sponsored Research at Universities,* 227 Sci 385 (Jan 25, 1985), including: retention of publication rights; retention of patent ownership; copyright protection for university-developed software; minimization of confidentiality requirements; staffing of research units with faculty, students, and professional researchers; prohibition of faculty consulting with sponsors in sponsored research areas; prohibition of sponsorship by faculty entrepreneurs; avoidance of international agreements; sharing of personnel and equipment with the sponsor; and preparation of a model research agreement for potential industrial sponsors.

Particularly aggressive strategies are common in the computer industry, where manufacturers offer schools equipment and support in return for software development and hoped-for capture of students' interests. See Sanger, *Computer Makers Pursuing Campus Sales and Research,* NY Times, Jan 20, 1984, at A1; Fowler, *Information Systems Get Focus,* NY Times, Apr 11, 1984, at D24; *Brown and IBM Plan Computer Program,* NY Times, Nov 13, 1983, at 61. On this topic, see also **§9.16,** main text and this supplement.

§11.30 Lead Time and the Limits to Growth

Add to note 396, page 521:

National Research Council, Outlook for Science and Technology: The Next Five Years, at 1-19, 614-616 (1982).

Add to note 398, page 521:

Exacerbating these allocation problems, military spending, long a major sponsor of industrial innovation, is again on the rise. See Boffey, *R&D Gets a Boost from Washington,* NY Times, Feb 13, 1983, at E18. Resource constraints and issue visibility have also led to unprecedented corporate lobbying for R&D support. See Pollack, *The Birth of Silicon Statesmanship,* NY Times, Feb 27, 1983, at F1.

Major Policy Options

Richard A. Givens

12

§12.02 Need for Innovation from the Public Viewpoint

The need for innovation from the public standpoint was very effectively stated by the editors of *Business Week* in their cover story of July 4, 1983, at 62:

> When it comes to R&D, even conservative economists have little trouble justifying government help. It is generally agreed that the innovations resulting from R&D provide social returns—in improved quality of life, jobs, and expansion of knowledge—that far exceed the private investor's return on his outlay. R&D tends to be underfunded by the private sector, but it pays the public to make up the difference and thus capture that social spill-over. . . ."

See Weinschel, *Is There a Need for US Civilian Technology Planning?, National Technology Foundation Act of 1980, Hearings before the Subcommittee on Science, Research and Technology, House Committee on Science & Technology,* 96th Cong, 2d Sess No 179 at 499, 521-22 (1980) (statement on behalf of the American Assn of Engineering Societies).

On the disincentive of difficulty in properly assessing risks of research and development and like investments, see Hodder & Riggs, *Pitfalls in Evaluating Risky Projects,* 85 Harv Bus Rev 129 (Jan/Feb 1985).

On the increasing "lumpiness" of major research, see Broad, *Era of Big Science Diminishes Role of the Lonely Genius: Complex New Projects Require the Efforts of Teams of Researchers,* NY Times, Jan 1, 1985, at D1. *See also Cover Story,* Bus Wk Mar 11, 1985, at 58 (claiming that "the U.S. is frittering away its lead in high tech by spending too little on civilian research and development. Expenditures there have fallen behind those of our trading partners as a percentage of gross national product. Meanwhile, military R&D—half of the country's total outlay of $100 billion—no longer pushes civilian technology as it did in the 1950s and 1960s").

See also Terleckyj, Effects of R&D on the Productivity Growth of Industries (1974); Dickson, *Technology and Cycles of Boom and Bust,* 219 Sci 933 (Feb 25, 1983); Arrow, *The Allocation of Resources to Invention,* in R. Nelson, The Rate and Direction of Inventive Activity: Economic and Social Factors (1962); Freeman, Unemployment and Technological Innovation (1982); Pfeiffer, *The Basic Need for Basic Research,* NY Times Mag, Nov 24, 1957, at 23.

Definition of National Goals

Whether the appropriate means of promotion is for the legal system to get out of the way and allow private incentives to function, or to provide affirmative promotion to overcome disincentives due to high risk or postponed payout, defining the objectives to be sought may be important. These objectives may be: (1) General, such as to promote whatever will increase productivity in the economy, including unforeseeable individual innovations scattered throughout the sphere of activity, and (2) specific, such as to produce breakthroughs to overcome specific bottlenecks like scarcity or interruptibility of specific raw materials or adverse environmental effects of specific technologies.

Objectives are developed both interstitially by continuous private sector and public sector activity, and by declarations with high visibility at the national level, such as the announcement that the United States would put its citizens on the moon by the end of the decade of the 1960s—or the grand objectives of the Arsenal of Democracy (the details of which remained secret, of course, at the time!).

Conscious, specific articulation of key major objectives requiring additional encouragement might help to increase the contribution of the legal system and also help it not to interfere where this is what is needed, provided that no exclusion of other goals was implied and that no counterproductive compulsion was involved.

For these reasons, noncoercive and nonexclusive articulation of both general and specific national scientific and technological goals of both a short- and long-range nature has been proposed by the Committee on Federal Legislation of the New York County Lawyers Association as a means of helping the legal system to avoid obstruction of critical generalized and specific innovations, and to use its existing procedures for affirmative promotion without creating any additional authority in the public sector (Report No F-1, 1985).

This involves no creation of additional agencies nor additional spending or lending authority. Rather, the Bar Committee concept is to make better use of existing public sector authority. Industry, academic institutions, the scientific community, governmental agencies, legislative committees, state and local authorities, labor, consumer representatives, and the general public might all participate in public debate concerning what major goals are most vital to create additional options for the nation in the future. Once tentatively announced, such goals could be subject to revision by legislative action if justified.

While such definition can be achieved without new legislation, a legislative mandate would represent a national decision that the goals articulated be given importance. Similarly, existing agencies could use their existing powers to promote defined goals (general and specific) without further legislative direction. Such direction would help to secure the attention of diverse decisionmakers throughout the legal system, who have vast discretionary powers that may be used in the national interest without any additional authority being granted.

By way of historical example, the "Destroyer-Bases" deal of 1940, granting Great Britain over-age United States destroyers in return for United States custody (and guarding) of Caribbean bases, was accomplished without further legislation. Similarly, in the 1970s and 1980s, the discretionary persuasive power of banking supervisory agencies was influential in preventing a collapse of the silver market at a time when this might have affected financial stability, and in protecting uninsured stockholders of the Continental Illinois Bank. Such discretionary power is obviously employed in the national interest in wartime, as exemplified by the role of the RFC in World War II, described in Jones & Angly, Fifty Billion Dollars (1951), but need not be limited to periods of declared hostilities.

National goals, whether defined with resounding *éclat* through a process of public deliberation, by specific legislation, or more informally by consensus, can also be furthered by voluntary cooperation of all agencies which can expedite work needed to promote these goals. Legislation could formalize this process by directing all public sector entities (including state and local bodies under the Supremacy Clause) to use their best efforts to the extent appropriate to help to further defined national goals.

Through legislation or even without it, a "lead agency" for critical scientific and technological goals generally or relating to a specific effort could help guide an enterprise seeking to promote a critical goal through the maze of sometimes prohibitively complex legal barriers. Such "one-stop" service is sought to be provided to new activities in New York State by its Office of Business Permits and Regulatory Assistance. On a broader scale, the lead agency concept has been proposed in Richardson, *Cutting Regulatory Tape,* NY Times, Dec 13, 1984, at A31. The success of this concept obviously depends on securing the help of all relevant agencies through voluntary cooperation or by legislative direction.

§12.03 Primary Means of Promotion of Innovation

Furnishing the Basic Infrastructure and Allowing the Marketplace to Function

Page 527, text, after last paragraph add:

Efforts to remove excessive legal barriers to innovation can take the form either of direct modification of laws or regulations governing private activity, or of procedural change, such as the "lead agency" concept recommended in Richardson, *Cutting Regulatory Tape,* NY Times, Dec 13, 1984, at A31. If procedural reforms add to complexity, however, they can impede deregulation as well as slowing addition of new regulation. *See Motor Vehicle Manufacturers Association v State Farm Mutual Insurance Co,* 103 S Ct 2856 (1983) (airbag issue); *NAACP v FCC,* 682 F2d 993 (DC Cir 1982); *NRDC v United States Environmental Protection Agency,* 683 F2d 752 (3d Cir 1982); *Consumer Energy Council v FERC,* 673 F2d 425 (DC Cir 1982); *Office of Communication of United Church of Christ v FCC,* 560 F2d 529 (2d Cir 1977); *Building & Construction Trades Department v Donovan,* 543 F Supp 1282 (DDC 1982); 5 USC §551(5); Thomas, *Overregulating the Regulators,* NY Times, Jun 3, 1981 (Op-ed page); DeLong, *Repealing Rules,* 8 ABA Admin L News 2, at 1 (Winter 1983).

On developments in one industry, see Kessler, *Food Safety: Revising the Statute,* 223 Sci 1034 (Mar 9, 1984).

On state deregulation, see New York State Regulatory Reform: Report of Action Unit No 5 of the New York State Bar Association (1982) (Charles E Heming, Chair); *compare* Noam, *Divergent Goals for the Deregulators,* NY Times, Aug 6, 1981, at A23.

For the first time, research priorities themselves are under direct attack. See Sun, *Weighing the Social Costs of Innovation,* 223 Sci 1368 (Mar 30, 1984).

Affirmative Promotion

As discussed in Chapter 4, the patent system is a major means of affirmative promotion, but does not permit the innovator to recover more than what customers will pay for an item even with a monopoly. Thus, as explained in **§12.02,** external benefits to the economy justify further steps in specific instances, such as in the case of basic research and what the Business Roundtable has called "key technologies too remote, costly or long range to attract adequate private sector interest and investment," Strategy for a Vital U.S. Economy (May 1984) at 31.

Specific affirmative promotion may be effectuated by general-purpose agencies acting under their pre-existing powers, or by additional public sector activity specifically authorized for the purpose. Promotion requiring funding may be extended through contracts, grants, loans, or loan guarantees with requirements for private investment, and at times loans with requirements for "equity kickers" to compensate the lender for added risks.

Affirmative promotion of innovation was proposed as early as 1790 by President George Washington by means of creation of a national university. 1 State of the Union Messages of the Presidents 3 (1966); see Thomas Jefferson, *id* at 88; John Quincy Adams, *id* at 249.

The National Science Foundation was created in 1950 to promote this objective. *See* 42 USC §1861 *et seq;* Bronk, *The National Science Foundation: Origins, Hopes and Aspirations,* 188 Sci 409 (May 2, 1975); Fowler, *A Foundation for Research,* 188 Sci 414 (May 2, 1975). It has, however, no mechanism for generating a revolving fund, and in many instances absence of long-term commitments, due to annual current expense budget rather than investment funding, causes inefficiency. See note 26, text p 530; *compare* Congressional Budget Office, Federal Support for R&D and Innovation 93-94 (April 1984); Research & Development Relating to Sulfates in the Atmosphere: Report Prepared for the Subcommittee on Environment and the Atmosphere, House Committee on Science & Technology, 94th Cong, 1st Sess (1975), Serial F; *National Technology Foundation Act of 1980, Hearings before the Subcommittee on Science, Research and Technology, House Committee on Science and Technology,* 96th Cong, 2d Sess No 179, 521-22 (1980); Kennedy, *Government Policies and the Cost of Doing Research,* 227 Sci 480 (Feb 1, 1985); *U.S. Industrial Strategy, Hearing before the Subcommittee on Economic Stabilization,* House Committee on Banking, Finance & Urban Affairs, 97th Cong, 2d Sess 86 (1982). On the need for long-term commitments, *compare* Herbert Hoover, 2d Annual Message (Dec 2, 1930): ". . . any kind of construction requires, after its authorization, a considerable time before labor can be employed in which to make engineering, architectural, and legal preparations." 3 State of the Union Messages of the Presidents 2776 (Chelsea House 1966).

A requirement that a portion of government grants and contracts for research be set aside for small business was added by the Small Business Innovation Development Act of 1982, Pub L No 97-219, 96 Stat 217, amending 15 USC §638.

Contracts or grants function best for (1) pure basic research, (2) development of items that the contracting or granting agency expects to buy, and (3) items which are important to society but unlikely to be profitable without subsidization of the bulk of development costs. Examples of the latter are "orphan" products that are important to a very small number of potential customers or will only justify a small price, but carry an important public interest. *See, e.g.,* Orphan Drug Act, Pub L No 97-414, 96 Stat 2094 (1983); HR Rep No 840, 97th Cong, 2d Sess (1982), 1982 US Code Cong & Ad News 3577. "Generic technologies" for such types of purposes would be developed by federally established Centers for Industrial Technology authorized by the Stevenson-Wydler Technology Innovation Act of 1980, Pub L No 96-480, 94 Stat 2311, 15 USC §§3701-14.

These methods are likely to work *least* well in the far more frequent situation where technologies *can* profitably be marketed commercially. There, for best results to be obtained, industry needs to be involved in the development at the earliest stage possible and have a stake in the venture. It will then have the momentum to apply new technologies quickly and effectively. *Compare* Nelson & Langlois, *Industrial Innovation Policy: Lessons from American History,* 219 Sci 814 (Feb 18, 1983); Mencher, *Lessons for American Policy Making from the British Labor Government's 1964-70 Experiment in Applying Technology to Economic Objectives,* London Graduate School of Business, Report to National Science Foundation (Research Applied to National Needs (RANN) 1975) *with* Mendelssohn, The Search for Absolute Zero (1966); Katz, *Harvey and Medical Research,* 160 JAMA 1137 (1956); and Miller & Fox, *Lab Extraordinaire,* 123 Sci News 9 (Jan 1, 1983).

For precedents for use of specialized financial institutions, in addition to statutes cited in notes 15-17, text pp 528-529, *see* Federal Financing Bank Act, 12 USC §2281 *et seq;* Federal Farm Loan Act, Act of July 16, 1916, ch 245, 39 Stat 360 (1916); Federal Home Loan Bank Act, Act of July 22, 1932, ch 522, 47 Stat 725; Home Owners Loan Act, Act of June 16, 1933, ch 64 48 Stat 128; 12 USC §352a, repealed by Pub L No 85-699, §601, 72 Stat 697 (1958). Authority for loans and loan guarantees for defense-related industry has long existed under the Defense Production Act of 1950, as amended, 50 USC App §§2091-2097. A bipartisan proposal to adapt these precedents to technology development was made in S 3111, 94th Cong, 1st Sess (1976), Cong Rec S2981 (daily ed Mar 9, 1976).

Numerous proposals have recognized the need for affirmative promotion and for formulation of policy to that end, although not incorporating any long-term funding mechanism. *See, e.g.,* the proposed National Technology Foundation Act of 1983, HR 481, 98th Cong, 1st Sess (1983); Economically Strategic Industrial Research and Development Act, HR 1243, 98th Cong, 1st Sess (1983), 129 Cong Rec 2 (daily ed Feb 3, 1983).

The impact of an investment vehicle such as a national technology development bank or corporation, discussed in this section in the bound volume and **§12.04,** would depend on the scale of its operations. To prevent development of a bureaucracy, the staff of such an institution could be limited to, *e.g.,* 100 employees. Even more important would be the number of priority national goals for which loans or loan guarantees or direct investments could be given, and the number of such awards per area. If hundreds of categories were designated and hundreds of awards approved per area, control would be far more difficult, increasing the likelihood of abuses.

Herbert Hoover, in originally proposing the Reconstruction Finance

Corporation, asked that it be given authority to "finance modernization and construction of industrial plants and utilities so as to increase employment and plant efficiency" and to make loans "which could have been used for reproductive public works." 3 Memoirs of Herbert Hoover 98, 107 (1952). *Compare* Subcommittee on Economic Stabilization, House Committee on Banking, Finance & Urban Affairs, Committee Print 98-8, 98th Cong, 1st Sess (1983); HR 4360, 98th Cong, 1st Sess (1983); Muller, Revitalizing America (1980); Center for National Policy, Restoring America's Competitiveness: Proposals for an Industrial Policy (1984); the views of Felix Rohatyn in NY Times Magazine, Dec 5, 1982, at 12. But such a financing mechanism for broad objectives beyond research and development only would involve the following dangers:

1. Competition with other ongoing activities would cause unfairness and demands for subsidies for those hurt. See Schultze, *Industrial Policy: A Dissent,* Brookings Rev 3 (Fall 1983); *compare* Givens, Advocacy, ch 15, at 307-08 (Shepard's/McGraw-Hill 1980)
2. Pressure for political abuse of the power to grant lucrative credit assistance would most likely be difficult to resist. See Passell, *Banker of Last Resort: But Could an RFC Be Insulated Against Patronage?,* NY Times, June 4, 1982, at A30; Cloke, *Senate to Sift RFC Loans With View to Writing Curbs,* NY Times, Jan 22, 1950, at 1, col 6; *Editorial,* NY Times, Jan 23, 1984, at A20
3. Because a large number of ongoing activities not necessarily in core areas of national priorities would be involved, many "losers" would be likely to be picked. *Compare* Pasztor, *Synfuels Plant Sees $700 Million in Losses by 1995,* Wall St J, Apr 11, 1983, at 3; Lueck, *Synthetic Fuels' Appeal Fades,* NY Times, May 6, 1983, at D1

Now that the legislative veto has been eclipsed by *INS v Chadha,* 51 USLW 4409 (US June 23, 1983), control of categories considered national priorities and of issuance of obligations beyond a defined level could be accomplished by finely honed use of the power of the purse. Without involving the ongoing activities of the institution in annual appropriations, which would be counterproductive, a board of control could be established which would approve certain selected activities and which could function solely out of separate appropriations, which *would* be limited to a year or less in duration. The power to cut off further activities involving issuance of more obligations or new categories, or certain large loans or loan guarantees perhaps, would cause the institution to heed congressional disapprovals of such activities, even if technically only advisory and not binding as such.

Of course, congressional involvement in setting priority categories is important even if a specialized institution were involved in the literal sense—one with a special purpose, here to promote high-risk research in areas of vital national interest. A *general* purpose institution would create a risk of unbridled abuse of power of the kind which caused Andrew Jackson to veto the continuance of the Second Bank of the United States. A general purpose government bank would also be beset by so many demands for politically motivated loans that research or other true national priorities might wind up at the end of a very long line of applicants with their hats in hand and influential letters in their pockets. Here, as in many other instances, a limited if ambitious effort may be essential to success.

A Harris poll indicates that industry is strongly opposed to a general-purpose federal financing vehicle that would invest in ongoing economic activity. See *Executives Split on Saving Smokestack Industries,* Bus Week, Apr 18, 1983, at 18. A national technology development corporation dealing with vital high-risk research only, would not have these characteristics.

Another option is to establish a body representative of a broad spectrum of interests to guide policy. *See, e.g.,* Weil, *U.S. Industrial Policy: A Process in Need of a Federal Industrial Coordination Board,* 14 Law & Policy in Intl Bus 981 (1983); Eizenstadt, *Industrial Policy: Not If, But How,* Fortune, Jan 23, 1984, at 183.

Cooperation in emergencies can be vital, but coalitions *against* change can result from institutionalized consultation processes. See Olson, The Rise and Decline of Nations (1982), *reviewed in* 219 Sci 163 (Jan 14, 1983); *compare* Jacobs, the Economy of Cities (1969).

Collaboration of competitors in handling industrial problems was tried under the NRA, held invalid as an improper delegation of legislative power in *ALA Schechter Poultry Corp v United States,* 295 US 495 (1935), and recognized as a source of what were held illegal restraints of trade in numerous subsequent instances. *See, e.g., United States v Socony-Vacuum Oil Co,* 310 US 150 (1940); *Sugar Institute, Inc v United States,* 297 US 553 (1936); Galloway, Industrial Planning Under Codes (1935).

Use of a council containing competitors in industry as members to pass on investments in what could be competing businesses would also have obvious ill effects. *Compare* Bank Holding Company Act Amendments, Hearings before the House Committee on Banking & Currency, 91st Cong, 1st Sess 13, 15, 374-74, 383, 615-16 (1969). "If too many approvals are required before an idea can be born, the probabilities are overwhelming against it." Tribus, *Applying Science to Industry,* 70 US News & World Rep 35 (Jan 8, 1971).

According to the Business Roundtable, *Strategy for a Vital U.S. Economy* (May 1984) at 31:

> Technology is the key to new products and processes and, thus, to productivity and business growth. Technology is, therefore, a capability composed of science and engineering knowledge,

skills, innovative ideas, and experience deployed in a competitive environment for the production of goods and services. Technology is not science, is not information, is not research alone. It is a combination of all of these factors.

Technology is, of course, the by-product of basic and applied research and development, as well as industrial, engineering and entrepreneurial management. On the other hand, basic scientific and engineering research is typically aimed at exploring the art of the possible, at testing potentially valuable but high-risk ideas whose payoff lies out in the future. Strategic judgments on research directions depend on deep understanding of both science and the process by which the science may later be put to work. Industrial, government and academic scientists share the necessary knowledge and experience and must work together to make these judgments.

Commercial technology development also draws directly on scientific and engineering research and the education of the nation's pool of scientific and engineering talent. The ability of private companies to innovate is directly dependent on the vigor, availability and appropriateness of these knowledge and human resources and the education infrastructure that produces them.

Both government and industry have critical but distinct roles to play in the technology development and commercialization process. The appropriate role of government in this process is:

* * *

c. to limit its direct investment to basic research and engineering, the research infrastructure and those areas required for national defense, health and key technologies too remote, costly or long range to attract adequate private sector interest and investment . . .

Add to note 14, page 528:

See also Can OMB Cure Accountability Strife?, 218 Sci 1097 (Dec 10, 1982); David, *The Troglodyte Frieze,* 220 Sci 911 (May 27, 1983).

Add to note 23, page 529:

See also Nelson & Langlois, *Industrial Innovation Policy: Lessons from American History,* 219 Sci 814 (Feb 18, 1983); Worcester & Boggs, *Capital for Geothermal Energy Projects,* 14 Nat Resources Law 713 (1982). On private efforts, see Starr, *The Electric Power Research Institute,* 219 Sci 1190 (Mar 11, 1983).

Add to note 26, page 530:

See also House Committee on Banking, Finance & Urban Affairs, *U.S. Industrial Strategy, Hearing before the Subcommittee on Economic Stabilization,* 97th Cong, 2d Sess 86 (1982). On the negative impact of risk in deterring choices profitable on the average, see Roth, *Toward A Theory of Bargaining: An Experimental Study in Economics,* 220 Sci 687 (May 13, 1983).

Add to note 27, page 530:

See S 12, 98th Cong, 1st Sess (1983); S 20, 98th Cong, 1st Sess (1983) (proposing two-year budget cycle).

§12.04 Specific Policy Options

Consolidation of Approval Requirements

Add to note 30, page 531:

See People v Kerr-McGee Chem Corp, 677 F2d 571 (7th Cir 1982).

Encouragement of Research by Contractors

Page 532, at end of first full paragraph add:

Small businesses and nonprofit institutions were given the opportunity to obtain patent rights on inventions developed under federal grants and contracts by Pub L No 96-517, 94 Stat 3019 (1980), 35 USC §§200-211. Implementing regulations were issued in 1982, 47 Fed Reg 7556 (1982), specifying when licensing by patentees may be required and how to preserve the rights, but were not included in the Code of Federal Regulations. See Neuner, *Rules for Research Funded by US,* NYLJ (Dec 2, 1982), at 1.

On the importance of independent research and development by contractors, see Aerospace Industries Association of America, A Position Paper on Independent Research and Development (1974); *see also* Aerospace Industries Association of America, Federal Procurement Principles (1971); Federal Contracts Report (BNA) (looseleaf); Gallagher, the Law of Federal Negotiated Contract Formation (1981).

On a negative trend, see Kennedy, *Government Policies and the Cost of Doing Research,* 227 Sci 480 (Feb 1, 1985) (cost of lobbying and risks of loss of investments if grants or contracts are not renewed due to budgetary changes).

§12.05 Research Priorities

On the distinction between basic and applied research and the importance of both, see Baruch, *The Cultures of Science and Technology,* 224 Sci 9 (Apr 6, 1984).

On the possibility of creating materials to order from physical principles, see Cohen, Heine, & Phillips, *The Quantum Mechanics of Materials,* 246 Sci Am 82 (1982); *compare Now the Squeeze on Metals,* Bus Week, July 2, 1979, at 46-51; on water sources, see Hamilton, *What Will We Do When the Well Runs Dry,* 84 Harv Bus Rev 27 (Nov/Dec 1984); Johnson, *If We Could Take the Salt Out of Water,* NY Times Magazine, Oct 30, 1960, at 17; Marshall, *The Procrastinator's Power Source,* 224 Sci 268 (Apr 20, 1984) (fuel cell).

On identification of materials critically needed, *see* Economically Strategic Industrial Research and Development Act, HR 1243, 98th Cong, 1st Sess (1983), 129 Cong Rec 2 (daily ed Feb 3, 1983); *compare* Congressional Research Service, Library of Congress, Seventh Biennial Conference on National Materials Policy, *Report to House Committee on Science & Technology,* 98th Cong, 1st Sess (1983), Serial C.

On the need for affirmative encouragement of alternative resource research to prevent crises, see Douglas & Torman, *Understanding Nonrenewable Resource Supply Behavior,* 219 Sci 927 (Feb 25, 1983); *compare* Roxburgh, *Policy Responses to Resource Depletion,* 21 Contemp Stud in Econ & Fin Anal (1980).

For assertions that some long-term efforts are underfunded or subject to harmful uncertainties and changes in the level of support, see Kintner, *Casting Fusion Adrift,* 85 Tech Rev 64 (May/June 1982); Crease & Mann, *Gambling With the Future of Physics,* NY Times Magazine, Dec 5, 1982, at 66.

Add to note 45, page 534:

See also HR 478, 98th Cong, 1st Sess (1983).

Add to note 46, page 534:

See Buck & Goodwin, Alternative Energy: The Federal Role (Shepard's/McGraw-Hill 1982).

Add to note 48, page 534:

See also Seidels & Keyes, Can We Delay A Greenhouse Warning? (EPA 1983); Clark, Carbon Dioxide Review 1982, *reviewed,* 219 Sci 283 (Jan 21, 1983). *Compare* Heller, *Hydrogen-Evolving Solar Cells,* 233 Sci 1141 (Mar 16, 1984) *with* Shabecoff, *Broad Damage Found in Trees in Eastern US,* NY Times, Feb 26, 1984, at A1, col 1, A30, col 1.

§12.06 Long-Term Implications

While technological advance will create jobs by making an economy more competitive and creating new industries, some believe that the advent of robotics and other technological changes will eliminate more jobs than are added. Silk, *Structural Joblessness,* NY Times, Apr 6, 1983, at D2, goes so far as to suggest that this will end by threatening the viability of the work ethic; *compare* the earlier views discussed in NY Times, Feb 4, 1966, at 14, col 1. On the adverse consequences of any such development, see Givens, Advocacy, ch 22(B)(1)(b), at 126-31 (Shepard's/McGraw Hill, Supp 1983).

These consequences, however, would not follow if it were concluded that the amount of worthwhile work to be done were virtually unlimited, requiring not merely routinizable tasks entirely programmable in detail in advance but the "human use of human beings," in the phrase of Norbert Weiner in his book by that name in 1967. This would be so if such goals as space industrialization, cost-effective desalting of sea water, substitutes for scarce materials, research into the basic constituents of matter, and even efforts to prepare to avoid such future catastrophes as asteroid collision, to take just one example, were undertaken. See generally Sullivan, *Star May Cause Extinctions on Earth,* NY Times, Feb 21, 1984, at C1, C9; Thomas, *How Should Humans Pay Their Way,* NY Times, Aug 24, 1981, at A15; **§12.05.**

These scientific goals will require both highly trained experts and large-scale less skilled supporting efforts; see Sullivan, Build Brother Build (1969). Much hard hat and "smokestack industry" work may also be required for such goals as rebuilding blighted areas (*compare* HR 9431, 94th Cong, 1st Sess (1975)). Extensive expansion of human services in such fields as health care and education may also be possible and, indeed, necessary.

Development of alternatives to present techniques exhibiting serious side effects can also require a tremendous effort. An example of what occurs when such advance planning is ignored is the "Tris" fiasco, which led to enactment of a law authorizing suits by industry for compensation because the government required implementation of a technology that was later found to be hazardous and was subsequently banned. Pub L No 97-395, 96 Stat 2001. For debate, see Thurmond, *The Government was to Blame, in Part,* NY Times, Jan 9, 1983, at 2F; Nader, *A Bad Precedent,* NY Times, Jan 9, 1983, at 2F. For advance warning, see Richard L. Abbott, Vice President, Montgomery Ward & Co, Luncheon Address before the American Apparel Manufacturers Association, Feb 21, 1974.

At a more basic level, the goal of overcoming the balance of terror, through achieving a decisive margin of advantage for open societies and making totalitarian systems decisively obsolete in all fields, is vital to our survival, and will require more, not less, work. In light of these challenges, fears that people will have nothing worthwhile to do seem far-fetched. See Goeller & Zucker, *Infinite Resources: The Ultimate Strategy,*

223 Sci 456 (Feb 3, 1984); Freeman, Unemployment and Technological Innovation ch 10 (1982); Givens, Advocacy 126-31, 135-47, 198-200 (Shepard's/McGraw-Hill, Supp 1983). There may, however, be a net loss of blue collar jobs if current trends are not offset. See Hunt & Hunt, The Human Resources Implications of Robotics (1983); Kilborn, *Few on Layoffs May Be Rehired . . . Big Corporations Rely on Automation to Aid Output in a Recovery,* NY Times, May 16, 1983, at A1, A12.

This could be counteracted by, *e.g.*, hardware requirements for space development, as exemplified by Sullivan, Build Brother Build (1969), or systematic improvements in the infrastructure or rebuilding of blighted areas, as discussed in Federal Trade Commission, New York Regional Office, Report on Hearings on Obstacles to Expansion of the Building & Construction Industry (1973); *compare* HR 9341, 49th Cong, 1st Sess (1975).

The most important key to expanding the economic pie is quite evident if we consider why advanced industrial societies are economically wealthier than 100 or 1,000 years ago: the availability of technologies not then available. The most obvious critical question for economic development, although not the only one, then becomes how to promote such innovation while avoiding harmful side effects.

The means of doing so are equally evident in their broadest outlines, even if the details of how to achieve them may not be: (1) affirmative promotion of desirable new technologies, and (2) getting the public sector out of the way where it inhibits such technologies. Both kinds of activities are obviously desirable, if they can be achieved without overbalancing, counterproductive side effects.

In the case of the first activity, the more basic the research on a spectrum (ranging from discovering the underlying constituents of the universe to designing a better coffee cup), the more suboptimal investment is likely to become, because:

1. Investors, no matter how large, are averse to risk which is significant to their total portfolio *even if on the average the investment is cost/effective,* for the same reason individuals and institutions both buy insurance. Society as a whole can, of course, spread and thus carry larger risks, even though its decision-making processes are flawed
2. The "lumpiness" of the investment required—the minimum amount of resources needed to accomplish anything at all—tends to increase with the complexity of technology. The number of key innovations which can be developed in a garage tends to decline, and the costs and delays of accumulating capital become a major barrier to major investments even if profitable on the average
3. Much of the benefit of major innovations accrue to the entire economy and not merely the inventor or investor, as pointed out

by Business Week in its *Cover Story* of July 4, 1983 at 62, previously quoted

The Business Roundtable in *Strategy for a Vital U.S. Economy* stated in May 1984 at 31 that public investment would be justified for "basic research and engineering, the research infra structure and those areas required for national defense, health and *key technologies too remote, costly or long range to attract adequate private sector interest and investment* . . ." (emphasis added).

These are some of the reasons why society has always supported basic research, but today a vast area between pure basic and pure applied research has opened up, typified by such potential technologies as:

1. Cost/effective coal gasification
2. Substitutes or sources for scarce, interruptible critical materials (e.g., getting cobalt from coal)
3. Nonpolluting transportation and industrial power sources, such as hydrogen or methane engine fuel, and other substitutes for technologies producing environmental problems
4. Controlled nuclear fusion
5. Cost/effective desalting of sea water
6. The "supercomputer"
7. Space industrialization
8. Environmentally effective and power-producing waste disposal technologies

Certain existing policies not only do not solve these dilemmas but may add to them very significantly:

1. There is no public sector capital budget, so that investments in research are treated as current expenses, rarely return funds to a revolving account, and are subject to counterproductive uncertainty created by the annual budgeting process
2. In the private sector, anxiety about the quarterly bottom line causes short-run to become more important than long-term profitability in many instances
3. Private sector R&D must be expensed for accounting purposes even if a valuable investment, depressing the bottom line and increasing the vulnerability of management
4. "Technology assessment" focuses almost exclusively on stopping harmful technologies rather than developing desirable alternatives, and indeed often fails to stop objectionable activities because alternatives have not been developed

An example of the interaction of short-run perspectives is the Tris fiasco, in which the government required childrens' sleepwear to be flame-retardant without having developed a safe and effective means to that end; thereafter, Tris was developed and placed on many garments, only to cause the recall of those garments due to cancer risks, resulting in huge losses to the industry and such litigation as *Riegel Textile Corp v Celanese Corp,* 649 F2d 894 (2d Cir 1981). The imbroglio led to an ultimate bailout by Congress through Pub L No 97-496, 96 Stat 2001 (1982), discussed in Thurmond, *The Government Was to Blame, In Part,* NY Times, Jan 9, 1985, at F2 and Nader, *A Bad Precedent,* NY Times, Jan 9, 1983, at F2.

Alternatives

Proposals for massive governmental intervention through credit or subsidies for certain actual ongoing economic activities which compete with others (as contemplated by the Synfuels Corporation's assistance in building actual synthetic fuels plants to become operational in selling energy from that source) not only do not deal with these problems, but would worsen them:

1. Injury to competitors who might be more advanced in some cases would be caused
2. When (as always occurs) the subsidized technology itself becomes obsolete, bailouts would become necessary
3. Demands by other interest groups for assistance would become entwined with the projected intervention and become difficult or impossible to phase out later

An entirely different approach was developed during the "gas-line" crisis of the mid-1970s: a National Technology Development Bank which would provide loans or loan guarantees where private capital undertook a portion of the risk of research seeking to fulfill a *previously defined* major national scientific goal, but where additional help was needed. The private venturers could be required to pay a premium as well as repaying the loan if the effort succeeded, and would be able to get a private patent which was enforceable only if they exploited the patent. This would have had the advantage of a commitment by industry to investment in the project, making it more likely to be implemented if development was successful. *See* Nelson & Langlois, *Industrial Innovation Policy: Lessons from American History,* 219 Sci 814 (Feb 18, 1983).

By handling lumpy but vital research between the poles of pure and applied research in a businesslike manner, more of the funds available for current expenditures would have been released for critical basic research, which is, of course, the most critical of all, as illustrated by Mendelssohn, The Search for Absolute Zero (1966) (unforeseen and unforeseeable results of discovery of ability to liquify atmospheric gases); Katz, *Harvey and Medical Research,* 160 J AMA 1137 (1956).

The proposal for a National Technology Development Bank drew on a long tradition of use of specialized financial institutions to promote national objectives which could pay for themselves but required commitments extending beyond one year (see bound volume **§12.03,** nn15-17). Legislation along these lines was introduced by Senators Charles Mathias (R-Md), Jacob Javits (R-NY), and Hubert Humphrey (D-Minn) as S 3111, 94th Cong, 2d Sess (1976). Significant support was obtained for the concept from the legal and scientific communities (bound volume, **§12.03** n25), but interest evaporated with the gas-line crisis.

Under present circumstances, creating any new governmental agencies or adding any new spending or lending authority is out of the question. However, an even more powerful approach than that represented by the proposed bank may be available:

1. *Definition of priority national scientific and technological goals that require additional encouragement.* This could be sought by consultation embracing the scientific community, industry, academic institutions, relevant federal, state, and local governmental agencies, labor, and the public. Once defined, the goals would be nonexclusive and would, of course, change with conditions
2. *All public sector bodies at all levels could be directed to use their existing authority to the extent appropriate to promote the defined goals.* No expansion of public sector authority would be involved. No agency would be required to disregard its other obligations or act in a manner it considered inappropriate, but each would be requested to use its vast discretionary authority to the extent feasible to promote the defined goals. The reservoir of such authority is suggested by the "Destroyer-Bases" deal of 1940 in which over-age United States destroyers were traded for a United States obligation to guard leased British Caribbean bases against the Axis, or the persuasion used to bail out silver holdings during a silver crash threatening the stability of financial markets
3. *A lead agency could help private enterprises seeking to promote the defined goals in threading their way through the maze of legal restrictions which tend to stifle innovative efforts.* This approach is exemplified by the New York State Office of Business Permits and was suggested by Elliot Richardson in *Cutting Regulatory Tape,* NY Times, Dec 13, 1985, at A31

Ultimately, of course, a National Technology Development Bank or equivalent could also become part of the armory available for promotion of specifically and narrowly defined critical technologies of the sort identified by the Business Council in its May 1984 paper.

Protection of the lead obtained by the United States or cooperating nations can be sought by requiring that the research involved be carried on within the United States or a cooperating nation, which will insure that the expertise developed keeps us several jumps ahead of even those who may steal information, as exemplified by the Leiden low-temperature laboratory described in Mendelssohn, The Search for Absolute Zero (1966).

Options for State Action

Apart from these potential national initiatives, much can be done at the state level. The state can provide limited financial participation in new technological efforts to be located in the state which do not displace others, but its ability to do so is limited by the absence of congressional banking and currency powers, as well as the drawbacks of subsidies to ongoing economic activity discussed earlier.

Further state efforts to lessen the interference of the public sector regarding technological innovation could proceed along several lines:

1. *Duplication of regulation at federal, state, and local levels covering the same subject matter can be reduced except where the state feels strongly that federal requirements are seriously inadequate.*

Federal violations broadly defined by subject matter so as not to incorporate by reference specific statutes that may be amended, in violation of such provisions as New York State Const art III, §16, may be made state violations subject to state enforcement. This fulfills the intent of the Supremacy Clause that state enforcement of federal law be a basic feature of the federal system. This may be promoted by such concepts as that exemplified by New York State Administrative Procedure Act §206, enacted at the instance of New York State Bar Association Action Unit No 5, chaired by Charles E Heming, in 1982, which calls on state agencies to seek this result. It may also be promoted by continuous and conscious effort to reach this result where feasible, on the part of the legislative, executive, and judicial branches at all levels of state and local government.

Agencies such as the New York State Office of Business Permits and Regulatory Assistance should also be given authority to seek to guide economic enterprises through municipal, county, and other local requirements as well as those imposed at the state level. This is appropriate because all local entities are creatures of the state. Any politically or bureaucratically motivated opposition should be surmountable if it is recognized that bureaucratic turf can be a barrier to needed economic activity.

2. *Efforts to reduce the cost of government by pinpointing bureaucratic waste by new methods are possible.*

One such method is to conduct anonymous interviews with middle managers and working level employees in each agency, who know what is happening in that agency but cannot normally speak out if identified, because of the unpopularity of "whistle blowers," regardless of any formal protections purported to be given them.

Another method would be preparation of a flow chart showing each step at which a piece of paper is ever generated in accomplishing the 10 most common tasks performed by an agency. "Approval by the Commissioner," for example, would be three boxes, not one, if an assistant must prepare a memorandum covering the matter to be initialed and passed on by another assistant before action can be taken. Merely identifying such bureaucratic layering, a major source of delay, error and expense, would generate pressure for its elimination. The difference would be that such pressure could be pinpointed rather than wasted in generalized dissatisfaction.

3. *A comprehensive effort to simplify the legal system, since it may exert a counterproductive drag on economic activity at the state level, is important to attracting economic development to the state.*

This should include substantive as well as procedural and bureaucratic complexity. The Simplification Task Force of the New York State Bar Association is seeking to formulate recommendations of this kind and would welcome comments from both the public and private sectors.

An open society has tremendous advantages in long-run competition with rivals of all kinds: an open economy in which the marketplace can correct mistakes, and a First Amendment which permits us to correct some of our governmental mistakes. The one element we now lack is the ability to plan ahead to develop affirmatively scientific and technological options.

Developing the means to do so, combined with releasing energy still blocked by unnecessary bureaucratic barriers, may enable us not only to compete with trade rivals more successfully, but over a period of time to make totalitarian systems *decisively* obsolete. In that event, newly emerging leaders may at some future time decide that it is more in their interest to change their systems (as Portugal did in ousting Salazar) than to await a potential computer malfunction or human error. That would be a truly historic consequence, for which we, now in the last fifth of the twentieth century, can help to lay the foundation.

In the meantime, many more immediate advantages can also be obtained. By accelerating our scientific and technological efforts we may be able to provide more jobs, avoid environmental catastrophes (such as the projected cutting of forests to speed water runoff), and afford a society of compassion. To move toward these goals requires the cooperation of disparate interest groups, which can best be held together by a combination of wide consultation and arrangements to share the benefits of widening the economic pie—in other words, to create a "non-zero sum" game.

In particular, most new technologies will create jobs by engendering entirely new activities or by strengthening the competitive position of the nation, rather than destroying jobs; arrangements to cushion the effect of job-displacing advances can be built into the negotiations for encouragement of any such technologies, as discussed in Chapter 2. Similarly, any environmental hazards will more likely be present during full-blown implementation of a technology than at the research stage, and emphasis can be placed on research to develop techniques which will not only avoid hazards, but furnish substitutes for methods which now cause environmental problems such as acid rain.

Building coalitions for expanding the economy is thus a first order of business, requiring dedicated effort by both the public and private sectors.

Add to note 50, page 535:

See Weinberg, *The Forces of Nature,* 29 Bull Am Acad of Arts & Sci 4 (Jan 1976), at 1.

In note 55, page 536, fourth line, delete "15 USC §1021 et seq" and insert:

15 USC §2051 *et seq*

Add to note 55, page 536:

See also Mosteller, *Innovation and Evaluation,* 211 Sci 881 (Feb 27, 1981); S Rep No 625, 97th Cong, 2d Sess (1982); Congressional Research Service, Library of Congress, *A Review of Risk Assessment Methodologies, Report for the Subcommittee on Science, Research & Technology,* House Committee on Technology, 98th Cong, 1st Sess (1983), Serial B; National Research Council, Risk Assessment in the Federal Government (1983); *NRC Panel Urges Against Risk Assessment Superagency,* Wash Drug Letter 5 (Mar 21, 1983).

§12.07 Historical and Contemporary Context of Innovation

Page 540, add at end of section:

The bulk of the discussion here has consisted of suggestions of options for threading one's way through the labyrinth of legal barriers an innovator engaged in technological change must confront. The key in many cases is *early anticipation* of problems, before too many choices are foreclosed by events. This seems equally applicable to antitrust, labor, tax, environmental, products liability, and regulatory issues, as well as to use of existing opportunities for affirmative promotion.

Policies more favorable to innovation than have previously prevailed are already developing in a number of areas, based on public recognition that productivity growth must be sought and cannot be taken for granted. Numerous suggestions for further changes in policy to promote innovation have been made by many sources. Of those which deal with innovation and cut across industry lines, the following seem to have great potential impact:

1. Definition of major long- and short-term national scientific and technological goals justifying encouragement through a process of public discussion, preferably based on specific legislative authorization for the process, but if necessary without it, *provided* that the goals are clearly nonexclusive and not to form a basis for regulatory coercion of any kind. See **§12.02** of this supplement
2. Direction to all existing public sector agencies at all levels to use their best efforts to promote the defined goals by employing their existing powers, to the extent appropriate and consistent with their other responsibilities
3. Use of a single or differing "lead agencies" to assist private sector entities seeking to promote the defined goals to navigate the maze of legal obstacles to innovation, by providing one-stop service to the extent possible, as sought to be done in New York by its Office of Business Permits and Regulatory Assistance. See Richardson, *Cutting Regulatory Tape,* NY Times, Dec 13, 1984, at A31
4. Permitting capitalization of R&D expenditures and their amortization over an appropriate period for securities reporting purposes, with adequate disclosure where an enterprise wishes to do so, without affecting tax treatment of such expenditures (see **§§5.12** and **6.06,** this supplement; **§§5.11-5.12,** bound volume; *see also* Yorke, *What Killed Long-Term U.S. Business Planning,* NY Times, Jan 10, 1982, at A30). This is important because "money managers, bankers, accountants, stockholders, and business leaders should be challenged to deemphasize simple short-term financial measures, utilizing instead measures more consistent with long-range competitiveness and profitability." President's Commission on Industrial Competitiveness, Summary of Recommendations 52 (1985)
5. Focusing environmental review of innovation on actual impact rather than the detailed character of impact statement descriptions, to be supervised by a single agency in each instance, be it an environmental agency or one concerned with the specific project or industry, if fully responsible to see that environmental

review is adequate (see **§§7.57** and **12.04**, bound volume); thus, multiple agencies would not have to be involved

6. Use of the interstate commerce clause of the Federal Constitution to challenge state and local obstacles where the burden on commerce exceeds any benefit (see **§§3.15** and **8.29**, bound volume)
7. Clear, predictable inadmissibility of product improvements in products liability cases against the innovator (see **§10.30**, this supplement; *Cook v McDonough Power Equipment, Inc*, 720 F2d 829 (5th Cir 1983); Givens, *Product Improvement—Evidence of Liability?*, NYLJ (July 18, 1983))
8. Encouragement of research by contractors through favorable treatment of reasonable independent research as an overhead item, and through moderating paperwork burdens (**§12.04**, bound volume and this supplement)
9. Consideration of creation of a specialized financial institution to assist high-risk private sector research for vital national goals in a small number of areas defined in consultation with the scientific community industry and government. This would *not* subsidize *ongoing economic activity better regulated by the marketplace alone* (see **§§12.03-12.04**, bound volume and this supplement). Exchange of information and technological cooperation among competitors requested by such an institution could be made exempt from antitrust attack (see **§3.11**, bound volume and this supplement)
10. Adoption, consistent with budgetary exigencies and as a high priority among tax incentives allowed to survive, of permanent tax credits for research and development, including work done by new enterprises and to benefit business other than merely that of the taxpayer involved (see **§5.12**, this supplement; President's Commission on Industrial Competitiveness, Summary of Recommendations 52 (1985))
11. (a) Interpretation of the National Cooperative Research ("NCR") Act (discussed in **§3.10**, this supplement) to encourage and not inhibit joint research efforts where "lumpy" research costs are involved and restrictions on independent efforts outside the venture are absent, even if the venture is of wide scope, and
    (b) Avoidance of imposition of detailed reporting and procedural requirements upon joint research ventures which choose to file optional reports with enforcement authorities to obtain protection from treble damage suits under the NCR Act (see **§3.10**, this supplement)
12. Development of sound solutions to the problem of international strategy and economic technology piracy which are at the same

time not counterproductive by unduly limiting desirable exchanges of information

(a) Through effective but carefully honed and sensitive export controls as discussed in Storch, *Issues in Government Contracting and Secrecy Controls,* in ABA Nat Inst on Legal Problems of Innovation ch 11 (1983); Zonderman, *Policing High-Tech Exports,* NY Times Mag, Nov 27, 1983, at 100

(b) Through development of international consortiums where effective, as discussed in Norris, *Halting the Flow of High-Tech Bargains,* NY Times, Feb 26, 1984, at F3; Anders, *Europeans Offer Reasons for Research Lag,* Wall St J, Feb 1, 1984, at 28; Dickson, *New Push for European Science Cooperation,* 220 Sci 1134 (June 10, 1983)

(c) Through limiting the location where research supported by special assistance is carried on, which provides the advantage of lead time with facilities and expertise to keep ahead, as discussed in Mendelssohn, The Search for Absolute Zero (1966), and is deemed the most important way to retain the advantages of research, according to a preliminary report of a statistical study by Prof. Richard C. Levin of Yale, Interindustry Differences in R&D Appropriability and Technological Opportunity (1984), table 2

(d) Negotiations aimed at developing an international regime for protection of intellectual property in a more uniform manner, at least among nations with relatively open market economies. See Report of President's Commission on Industrial Competitiveness at 41 (1985)

13. Above all, recognition of the overarching importance of basic research and industrial innovation for the future of the nation and the open society (see **§§1.01** and **12.06,** this supplement, and **§12.07,** bound volume).

The ultimate results of a major increase in innovation may be far-reaching, along lines suggested as far back as 1953 by Harry S. Truman in his eighth annual message of January 7, 1953:

> The Communist rulers have a sense of time about these things wholly unlike our own. We tend to divide our future into short spans, like the 2-year life of this Congress, or the 4 years of the next Presidential term. They seem to think and plan in terms of generations. And there is, therefore, no easy, short-run way to make them see that their plans cannot prevail.
>
> This means there is ahead of us a long hard test of strength and stamina between the free world and the Communist domain; our politics and our economy, our science and technology against the best they can do; our liberty against their slavery. . . .

What, then, of the future? The answer, I believe, is this: As we continue to confound Soviet expectations, as our world grows stronger, more united, more attractive to men on both sides of the iron curtain, then inevitably there will come a time of change within the Communist world. We do not know how that change will come about, whether by deliberate decision in the Kremlin, by coup d'etat, by revolution, by defection of satellites, or perhaps by some unforeseen combination of factors such as these.

But if the Communist rulers understand they cannot win by war, and if we frustrate their attempts to win by subversion, it is not too much to expect their world to change its character, moderate its aims, become more realistic and less implacable, and recede from the cold war they began. . . .

The masters of the Kremlin live in fear their power and position would collapse were their own people to acquire knowledge, information, comprehension about our free society. Their world has many elements of strength, but this one fatal flaw: the weakness represented by their iron curtain and their police state. Surely a social order at once so insecure and so fearful, must ultimately lose its competition with our free society.

Provided just one thing—and this I urge you to consider carefully—provided that the free world retains the confidence and the determination to outmatch the best our adversary can accomplish and to demonstrate for uncertain millions on both sides of the iron curtain the superiority of the free way of life. . . .

Our resources are equal to the task. We have the industry, the skills, the basic economic strength. Above all we have the vigor of free men in a free society. We have our liberties. And, while we keep them, while we retain our democratic faith, the ultimate advantage in this hard competition lies with us. . . .

3 State of the Union Messages of the Presidents 3008-09 (Chelsea House 1966). On these advantages of an open society, see Sakharaov, Progress, Coexistence and Intellectual Freedom (1968); *compare* Orwell, Nineteen Eighty-Four (1949).

Cases

A

D

E

F

G

H

I

N

SEC v Ronson Corp, No 83-3030 (DNJ Aug 5, 1983 unreported) **§6.62**

SEC v Thayer, No 84 Civ 0066, [Current] Fed Sec 1 Rep (CCH) ¶99,718 (SDNY filed Jan 5, 1984) **§6.53**

Securities Indus Assn v Board of Governors of Fed Reserve, 104 S Ct 2979 (1984) **§6.47**

Shackelton v J Kaufman Iron Works, Inc, 689 F2d 334, 217 USPQ 98 (2d Cir 1982) **§§4.16, 4.17**

Shatterproof Glass Corp v Libbey-Owens Food Co, 728 F2d 613, 225 USPQ 633 (CAFC 1985) **§4.06**

Shop & Save Food Markets, Inc v Pneumo Corp, 693 F2d 27 (2d Cir), *cert denied,* 74 L Ed 2d 604 (1982) **§3.06**

Sierra Club v Gorsuch, 551 F Supp 785, 17 ERC 1748 (ND Cal 1982) **§7.23**

Sierra Club v Gorsuch, 17 Envt Rep Cas (BNA) 1748 (ND Cal 1982) **§7.25**

Sierra Club v Sigler, 695 F2d 957, 18 ERC 1649 (5th Cir 1983) **§§7.20, 7.22**

Sierra Club v United States Corps of Engrs, 701 F2d 1011, 18 ERC 1748 (2d Cir 1983) **§§7.15, 7.22**

Silver v New York Stock Exch, 373 US 341 (1963) **§3.11**

Slough v JI Case Co, 8 Kan App 2d 104, 650 P2d 729 (1982) **§10.11**

Smith, *In re,* 714 F2d 1127, 218 USPQ 976 (CAFC 1983) **§4.08**

Snow v Commissioner, 414 US 1128 (1974) **§5.12**

Sol Diamond, 56 TC 530 (1971) **§5.23**

South Corp v United States, 690 F2d 1368, 215 USPQ 657 (Fed Cir 1982) **§4.16**

Southern Motor Carriers Rate Conference, Inc v United States, 105 S Ct 1721 (1985) **§§3.11, 8.29A**

Southern Or Citizens Against Toxic Spraus, Inc v Clark, 720 F2d 1475, 20 ERC 1001 (9th Cir 1983) **§7.20**

Southern Pacific Co v Arizona, 325 US 761 (1945) **§3.11**

Starling v Seaboard Coast Line RR, 533 F Supp 183 (D Ga 1982) **§10.11**

Sterling Recreation Org Co v Segal, 537 F Supp 1024 (D Colo 1982) **§6.19**

Sticklke v Heublein, Inc, 716 F2d 1550, 219 USPQ 377 (Fed Cir 1983) **§4.25**

Stone v Mississippi, 101 US 814 (1879) **§8.29B**

Strong v EI DuPont de Nemours Co, 667 F2d 682 (8th Cir 1981) **§10.08**

Sugar Ins Inc v United States, 297 US 553 (1936) **§12.03**

Sunshine Books Ltd v Temple Univ, 697 F2d 90 (3d Cir 1982) **§3.07**

Sutter v Green, 667 F2d 197 (7th Cir 1982) **§6.19**

T

Tandy Corp v Personal Microspace Computers Co, Inc, 524 F Supp 171, 214 USPQ 178 (ND Cal 1981) **§4.02A**

TCI Cablevision, Inc v City of Jefferson, Mo, 604 F Supp 845 (WD Mo 1984) **§8.29**

Statutes

United States Code

See also Statutes at Large table for named act references

5 USC §551(5) **§12.03**
Commodities Exchange Act, 7 USC §1 *et seq* **§§6.50, 6.56**
7 USC §136a(c)(1)(D)(i) **§7.44**
7 USC §136a(c)(1)(D)(ii) **§7.44**
12 USC §133i **§6.36**
12 USC §352a **§12.03**
12 USC §404 **§6.38A**
Federal Financing Bank Act, 12 USC §2281 *et seq* **§12.03**
15 USC §6a **§3.12**
15 USC §15 **§§3.10, 3.12**
15 USC §18 **§3.03**
15 USC §26 **§3.11**
15 USC §34 **§§3.11, 8.29**
15 USC §77k **§6.42**
15 USC §77k(a) **§6.42**
15 USC §77k(c) **§§6.35, 6.47**
15 USC §77q(a) **§6.53**
15 USC §77s(a) **§6.34**
15 USC §78 **§6.50A**
15 USC §78(a) **§§5.50, 6.69**
15 USC §78c(a)(4) **§6.47**
15 USC §78c(a)(5) **§6.47**
15 USC §78c(a)(6) **§6.47**
15 USC §78(d) **§6.39**
15 USC §78(e) **§6.39**
15 USC §78(f) **§6.39**
15 USC §78m(d) **§§6.39, 6.50A, 6.69**
15 USC §78m(e) **§6.39**
15 USC §78m(f) **§6.50A**
15 USC §78m(g) **§§6.50A, 6.69**
15 USC §78p(b) **§§6.39, 6.69**
15 USC §78w(a) **§6.34**
Investment Company Act of 1940, as amended 54 Stat 789 (codified at 15 USC §§80a-1 to 80a-64) **§6.33**
15 USC §638 **§§9.15, 12.03, 12.04**
Small Business Act, 15 USC §638(d) **§3.11**
15 USC §775(c)(4) **§6.16**
15 USC §2051 *et seq* **§12.06**
Soft Drink Interbrand Competition Act, 15 USC §3501 *et seq* **§3.09**
Stevenson-Wydler Technology Innovation Act of 1980, Pub L No 96-480, 94 Stat 2311, 15 USC §3701 *et seq* **§§1.02, 12.03**
15 USC §§3701–3714 **§12.03**
15 USC §4013 **§3.12**
15 USC §4015 **§3.12**
15 USC §4016 **§3.12**
15 USC §4016(b)(5) **§3.12**

Statutes at Large

Internal Revenue Code

State Statutes

Administrative Materials

Code of Federal Regulations

Federal Register

Internal Revenue Service Treasury Regulations

Revenue Procedures

Revenue Rulings

Securities and Exchange Commission Rules

Securities and Exchange Commission Releases

Reports

Tender Offers (July 8, 1983) **§6.39**

Report of Special Study of Securities Markets of the Securities and Exchange Commission, HR Doc No 95, 88 Cong, 1st Sess (1963) **§6.69**

SEC Review of Enforcement Remedies Proposed Legislation on Administrative Proceedings, FOIA, Privilege, 16 Sec Reg L Rep (BNA) No 9, at 456 (Mar 2, 1984) (full text of SEC Statement to Congress) **§6.66**

Shark Repellants: The Role and Impact of Antitakeover Charter Amendments (Office of Chief Economist, SEC, Sept 1984) **§6.39**

Tender Offer Reform, 98th Cong, 2d Sess (Sept 7, 1984) **§6.39**

Final Report, 1984 SEC Government—Business Forum on Small Business Capital Formation (Jan 1, 1985) **§6.24**

Treasury Department Report to the President, Tax Reform for Fairness, Simplicity and Economic Growth (Nov 27, 1984) **§6.39**

SEC No Action and Interpretive Letters

City Stores Company (SEC no action letter available June 18, 1982) **§6.63**

New Jersey Economic Development Authority (avail Mar 1, 1984) **§6.14**

Price Communications Corporation (SEC no action letter available Feb 25, 1983) **§6.63**

STV Engineers, Inc (SEC no action letter avail Apr 28, 1983) **§§6.39, 6.49, 6.63**

Trust Indenture Act of 1939, Fed Sec L Rep (CCH) ¶77,345 **§6.40**

State Regulations

Blue Sky Reg, Reg 1, Art D, Rule 1.7.B **§6.24**

Or Admin R, *as amended,* Blue Sky Regulations, Rule 815-36-015(3)(19__)

Investment Act of 1940 Releases

Investment Advisers Act of 1940 Release No 961 (Mar 14, 1985) **§6.46**

Investment Advisers Act of 1940 Release No 956 (Feb 25, 1985) **§6.46**

Proposed Federal Reserve Board Rules

Board of Governors of the Federal Reserve System, No R-0538 (Feb 6, 1985), Fed Sec L Rep (CCH) ¶83,735 **§6.43**

NASD Releases

NASD Notice to Members 85-20 (Mar 28, 1985) **§6.39**

Legislative Materials (Documents and Legislative History)

Bills

S 44, 98th Cong, 1st Sess (1983), §14 **§10.30**
S 100, 99th Cong, 1st Sess (1985) **§§10.04, 10.06, 10.07, 10.08, 10.10, 10.30**
S 325, 99th Cong 1st Sess (1985) **§6.39**
S 409, 99th Cong, 1st Sess (1985) **§6.39**
S 411, 99th Cong, 1st Sess (1985) **§6.39**
S 414, Business Accounting and Foreign Trade Simplification Act, 98th Cong, 1st Sess, 129 Cong Rec S981–983 (daily ed Feb 3, 1983) **§6.64**
S 430, 99th Cong, 1st Sess (1985) **§6.64**
S 485, Lawyers Duty of Disclosure Act, 98th Cong, 1st Sess (1983) **§6.68**
S 568, 98th Cong, 1st Sess (1983) **§3.11**
S 737, 98th Cong, 1st Sess (1983) **§3.11**
S 1383, 98th Cong, 1st Sess (1983) **§3.11**
S 1561, 98th Cong, 1st Sess (1983) **§3.11**
S 2717, 97th Cong, 2d Sess (1982) **§3.11**
S 3111, 94th Cong, 1st Sess, Cong Rec S2981 (daily ed Mar 9, 1976) **§12.03**
S 3116, 97th Cong, 2d Sess (1982) **§3.11**

Reports

HR Conf Rep No 215, 97th Cong, 1st Sess, at 223–224 (Aug 1, 1981) **§5.12**
HR Rep 840, 97th Cong, 2d Sess (1982), 1982 US Code Cong & Ad News 3577 **§12.03**
HR Rep 571, 98th Cong, 2d Sess (1983) **§11.11**
HR Rep 656, 98th Cong, 2d Sess (1984) **§11.11**
HR Rep No 693, 98th Cong, 2d Sess (1984) **§1.01**
HR Rep No 656 98th Cong, 2d Sess (1984) **§3.10**
HR Rep No 965, 98th Cong, 2d Sess 22, *reprinted in* 1984 US Code Cong & Ad News 4623 **§8.29A**
H Conf Rep No 1158, 98th Cong, 2d Sess (1984) **§§3.11, 8.29A**
Joint Economic Committee, 97th Cong, 2d Sess, *International Competition in Advanced Industrial Sectors: Trade and Development in the Semiconductor Industry* (Feb 18, 1982) **§11.11**
Subcommittee on Monetary and Fiscal Policy, Report of Joint Economic Committee, 97th Cong, 2d Sess (June 1, 1982) **§11.11**
S Rep No 144, 97th Cong, 1st Sess, at 77–78 (July 6, 1981) **§5.12**
S Rep No 625, 97th Cong, 2d Sess (1982) **§12.06**

Hearings

Bank Holding Company Act Amendments, Hearings before the House Committee on Banking and Currency, 91st Cong, 1st Sess (1969) **§12.03**
House Committee on Banking, Finance & Urban Affairs, *U.S.*

Accounting Materials

Authorities

A

America Rushes to High Tech for Growth: Many Call It A Panacea, But Hopes for New Jobs Are Overly Optimistic, Bus Week, Mar 28, 1983, at 84, 87 **§§1.01, 9.10**

America's High Tech Crisis, Bus Week, Mar 11, 1985, at 56 **§11.01**

Anders, *Europeans Offer Reasons for Research Lag,* Wall St J, Feb 1, 1984, at 28 **§12.07**

Andewelt, Intellectual Property Section, Antitrust Division, *Basic Principles to Apply at the Patent-Antitrust Interface,* address to the Houston Patent Law Association (Dec 3, 1981) **§4.19**

Andewelt, *Competition Policy and the Patent Misuse Doctrine,* address to Bar Association for District of Columbia Patent, Trademark and Copyright Section (Nov 3, 1982) **§4.19**

Andrews, *Technology: Productivity or Innovation?* NY Times, Jan 31, 1979, at D4 **§1.01**

Areeda, *Antitrust Violations Without Damage Recoveries,* 89 Harv L Rev 1127 (1976) **§3.11**

Areeda, *Predatory Pricing,* 49 ABA Antitrust LJ 897 (1980) **§3.07**

Arenson, *Debate Grows Over Adoption of National Industrial Policy,* NY Times, June 19, 1983, at AL **§11.01**

A Technology Lag That May Stifle Growth, Bus Week, Oct 1, 1982, at 126 **§1.01**

Atleson, Values and Assumptions in American Labor Law (1984) **§12.07**

Auchter, *OSHA: A Year Later,* 33 Lab LJ 195 (1982) **§2.06**

Axinn, Fogg & Stoll, *Acquisitions Under the Hart-Scott-Rodino Antitrust Improvements Act* (Law Journal Seminars-Press 1980) **§6.39**

B

Bacon & Butler, Planned Innovation (2d ed 1981) **§§1.03, 11.03**

Badaracco & Yoffie, *Industrial Policy: It Can't Happen Here,* 83 Bus Rev 96 (Nov/Dec 1983) **§11.01**

Baderian & Goodman, *Plant Relocations and Union Contracts,* National LJ, Feb 27, 1984, at 24 **§2.02**

Baldridge, *Government Policies to Promote High Growth Industries Based on New Technologies and to Increase U.S. Competitiveness,* testimony before the Subcommittee on Savings, Pensions & Investment Policy, Senate Finance Committee, 98th Cong, 1st Sess 10 (1983) **§1.01**

Baldwin & McGovern, Preparation of a Product Liability Case (1981) **§1.04**

Bartholet, *Application of Title VII to Jobs in High Places,* 95 Harv L Rev 945 (1982) **§2.06**

Baumol, *Competition vs Research,* NY Times, Aug 31, 1983, at D2 **§11.23**

Baxter, *Antitrust Law and Technological Innovation,* 1 Issues in Sci & Tech 80 (Winter 1985) **§11.01**

Baxter, Interview, 52 ABA Antitrust LJ 1 (1983), at 32 **§3.10**

Baxter, Presentation before the National Association of Manufacturers, Washington, DC, 5 Trade Reg Rep (CCH) ¶50,447 (May 10, 1983) **§3.10**

Bearse, *A Comparative Analysis of State Programs to Promote New Technologically Based Enterprise,* 5 New Eng J Bus & Econ 49 (1979) **§9.13**

Bedell, *Give Money to the Real Innovators,* NY Times, Nov 29, 1981, at F2 **§11.20**

Bergman, *Business Development Companies,* 17 Rev Sec Reg 877 (June 20, 1984) **§6.45**

Biddle, *Crippling of 2d Space Shuttle Tied to Design Flaw,* NY Times Mar, 2, 1983, at 1 **§2.07**

Birchfield, Director of the Advanced Technology Development Center, *The Billion Dollar Challenge—The States, The Professions, and The High Tech Companies,* address to the Small Business High Technology Institute Conference, Arlington, Virginia (Mar 18-19, 1983) **§9.10**

Birnbaum & Wrubel, *Worker's Compensation and the Employer's Immunity Shield: Recent Exceptions to Exclusivity,* 5 J Prod Liab 119 (1982) **§10.31**

Blair, Economic Concentration (1972) **§3.13**

Boffey, *R&D Gets a Boost from Washington,* NY Times, Feb 13, 1983, at E18 **§11.30**

Boffey, *U.S. Priority on Technology is Urged,* NY Times, Apr 15, 1983, at A3 **§1.01**

Bolt, *Job Security: Its Time Has Come,* Harv Bus Rev 199 (Mar - Apr 1984) **§2.02**

Botkin, Dimancescu & Stata, The Innovators: Rediscovering America's Creative Energy (1984) **§9.11**

Bouton, *Academic Research and Big Business: A Delicate Balance,* NY Times, Sept 11, 1983, at VI 62 **§11.29**

Breakthrough: The Newsletter of Innovation by Boardroom Reports **§1.03**

Brennan, *Advanced Technology Centers: Strategies for Corporate Growth,* 248 Sci Am E1 (May 1983) **§11.11**

Broad, *Era of Big Science Diminishes Role of the Lonely Genius: Complex New Projects Require the Efforts of Teams of Researchers,* NY Times, Jan 1, 1985, at D1 **§12.02**

Brodley, *Joint Ventures and Antitrust Policy,* 95 Harv L Rev 1521 (1982) **§3.10**

Brodsky, *Timing of Corporate Disclosure,* NYLJ, Nov 7, 1984, at 1, col 1 **§6.52**

Bronk, *The National Science Foundation: Origins, Hopes and Aspirations,* 188 Sci 414 (May 2, 1975) **§12.03**

Bronson, *US Firms Plus Foreign Partners Equals Prosperity,* US News & World Rep, Mar 5, 1984, at 51 **§11.11**

Brown and IBM Plan Computer Program, NY Times, Nov 13, 1983, at 61 **§11.29**

Buck & Goodwin, Alternative Energy: The Federal Role (Shepard's/McGraw-Hill 1982) **§12.05**

Bucy *Meeting the Competitive Challenge: The Case for R&D Tax Credits,* 1 Issues in Sci & Techn 4 (Summer 1985) **§9.07**

Bureau of National Affairs, *New York Energy Research Agency Spends, Saves Millions for State,* 13 Energy Users Rep (BNA) 482 (May 23, 1985) **§9.09**

Burnham, *Loophole in Law Raises Concern About Privacy,* NY Times, Dec 19, 1983, at A1 **§11.01**

Business-Higher Education Forum, America's Competitive Challenge: The Need for a National Response (Apr 1983) **§1.01**

Business Roundtable, *Strategy for a Vital U.S. Economy* (May 1984) **§12.03**

3 Bus L Memo (Jan/Feb 1983) **§6.50**

Bus Week, Jan 10, 1983, at 45 **§1.01**

Bus Week, May 3, 1983, at 84 **§3.10**

Business Week—Harriss Poll, *Resounding Support for Price Competition,* Bus Week, Nov 14, 1984 at 24 **§3.03**

C

California Office of Appropriate Technology, High Technology and High School: Preparing Students for California's Changing Economy (1982) **§9.12**

Can OMB Cure Accountability Strife?, 218 Sci 1097 (Dec 10, 1982) **§12.03**

Caplan, *Social Support, Person-Environment Fit, and Coping,* in Mental Health and the Economy 89 (Ferman & Gorman eds 1979) **§2.01**

Carey, A Strategy for Developing Technology-Based Industry in New York (Governor's Office 1982) **§1.03**

Center for National Policy, Restoring America's Competitiveness: Proposals for an Industrial Policy (1984) **§12.03**

Civiletti, *The Fallout From Community Communications Co. v. City of Boulder: Prospects for a Legislative Solution,* 32 Cath UL Rev 379 (1983) **§8.29**

Clark, Carbon Dioxide Review 1982, *reviewed,* 219 Sci 283 (Jan 21, 1983) **§12.05**

Cloke, *Senate to Sift RFC with View to Writing Curbs,* NY Times Jan 22, 1950, at 1, col 6 **§12.06**

Cohen, Heine, & Phillips, *The Quantum Mechanics of Materials,* 246 Sci Am 82 (1982) **§12.05**

Coleman, *The Aging Mind Proves Capable of Lifelong Growth,* NY Times, Feb 2, 1984, at C1, C5 **§2.06**

Coleman & Seldin, *A Business Man's Guide to Capital-Raising Under the Securities Laws* (1982) **§16.13**

Collins, An Early Assessment of The R&D Tax Act of 1981 (National Science Foundation, Apr 1983) **§9.07**

Congressional Research Service, Library of Congress, *A Review of Risk Assessment Methodologies, Report for the Subcommittee on Science, Research & Technology,* House Committee on Technology, 98th Cong, 1st Sess (1983), Serial B **§12.06**

Congressional Research Service, Library of Congress, Seventh Biennial Conference on National Materials Policy, *Report to House Committee on Science & Technology,* 98th Cong, 1st Sess (1983), Serial C **§12.05**

Corporate Finance Activities, 17 Sec Reg & L Rep (BNA) No 10, at 403, (Mar 2, 1985) **§§6.24, 6.26**

Cover Story, Bus Week, July 4, 1983, at 62 **§§1.01, 12.02**

Cover Story, Bus Week, Mar 11, 1985, at 58 **§12.02**

Crawford, *The Problems of Causation of Private Legal Remedies for Damage from Acid Rain,* 17 Nat Resources Law 413 (1984) **§7.23**

Crease & Mann, *Gambling With the Future of Physics,* NY Times Magazine, Dec 5, 1982, at 66 **§12.05**

D

David, *Science Futures: The Industrial Connection,* 203 Sci 837 (Mar 2, 1979) **§1.02**

David, *The Troglodyte Frieze,* 220 Sci 911 (May 27, 1983) **§12.03**

Davidson, *Protecting Computer Software: A Comprehensive Analysis,* 23 Jurimetrics J 337 (Summer 1983) **§11.10**

Davis, *New Projects: Beware of False Economics,* 85 Harv Bus Rev 95 (Mar/Apr 1985) **§1.05**

Daykin, *Featherbedding,* 7 Lab LJ 699 (1956) **§2.03**

DeLong, *Repealing Rules,* 8 ABA Ad L News 2 (Winter 1983), at 1 **§12.03**

Dickson, *Britain Rises to Japan's Computer Challenge: In Departure from its Free-market Principles, the Government will Spend $300 Million to Subsidize Research on Information Science,* 220 Sci 799 (May 20, 1983) **§1.01**

Dickson, *New Push for European Science Corporation,* 220 Sci 1134 (June 10, 1983) **§12.07**

Dickson, *Technology and Cycles of Boom and Bust,* 219 Sci 933 (Feb 25, 1983) **§§1.01, 12.02**

Douglas & Torman, *Understanding Nonrenewable Resource Supply Bahavior,* 219 Sci 927 (Feb 25, 1983) **§12.05**

Drucker, Innovation and Entrepreneurship (1985) **§11.27**

Dula, *Regulation of Private Commercial Space Activities,* 23 Jurimetrics J 156 (1983) **§§1.03, 11.03**

E

Eads & Reuter, *Designing Safer Products: Corporate Response to Product Liability Law and Regulation* (1984) **§10.12**

Easterbrook, *The Court and the Economic System,* 98 Harv L Rev 4 (1984) **§3.15**

The Economist (London), Feb 12, 1983, at 67 **§2.07**

Editorial, NY Times Jan 23, 1984, at A20 **§12.03**

Education Commission of the States, Task Force on Education for Economic Growth, Action for Excellence: A Comprehensive Plan To Improve Our Nation's Schools (1983) **§9.12**

Education Commission of the States, Task Force on Education for Economic Growth, Action in the States: Progress Toward Educational Renewal (July 1984) **§9.12**

Egan, *To Err is Human Factors,* 85 Tech Rev 22 (Feb/Mar 1982) **§2.07**

Eisenstadt, *Access to Solar Energy: The Problem and its Current Status,* 22 Nat Resources J 21 (1982) **§8.19**

Eisenstadt, *Industrial Policy: Not If But How,* Fortune, Jan 23, 1984, at 183 **§12.03**

Eisner, *The R&D Tax Credit: A Flawed Tool,* 1 Issues in Sci & Tech 4, at 79 (Summer 1985) **§9.07**

EPA, The Acidic Deposition Phenomenon and Its Effects: Critical Assessment Review Papers (Public draft volumes May, 1983) **§7.23**

Equal Employment Advisory Council, Employee Selection (Monograph No 2, Potter ed 1983) **§2.06**

Evans, *Offerings of Securities Soley to Foreign Investors,* 40 Bus Law No 1 at 69 (Nov 1984) **§6.33**

Executives Split on Saving Smokestack Industries, Bus Week, Apr 18, 1983, at 18 **§12.03**

F

Farnsworth, *Implied Warranties of Quality in Non-Sales Cases,* 57 Colum L Rev 653 (1957) **§3.11**

FASB should Complete Work on Pension Project By End of 1985, Board Chairman Donald Kirk Says, 1 Acct Prac & Reg No 1 at 4 (Feb 11, 1985) **§6.64**

Fasman, *Legal Obstacles to Alternative Workforce Designs,* 8 Employee Rel LJ 256 (1982) **§2.06**

Fass & Wittner, Blue Sky Practice for Public and Private Limited Offerings (1985) **§6.16**

Feder, *Gene-Splicing: Legal Pitfalls,* NY Times, Dec 1, 1981, at D1 **§11.14**

Feder, *Technology: Measuring the "Cutting Edge,"* NY Times, Dec 3, 1981, at D2 **§11.18**

Federal Contracts Report (BNA) **§12.04**

Federal Trade Commission, New York Regional Office, Report on Hearings on Obstacles to Expansion of the Building & Construction Industry (1973) **§12.06**

Feders, Director, SEC Division of Enforcement, *Illegal Conduct-Materiality: An Enforcement Overview,* address to Federal Regulation of Securities Committee, Section on Corporation, Banking & Business Law, American Bar Association, Washington, DC, Nov 19, 1982, *reprinted in* 14 Sec Reg & L Rep (BNA) No 46, at 2057 **§§6.57, 6.64**

Ferrara & Phillips, *Opposition To 'Poison Pill' Warrants Is Mounting,* Legal Times of Wash, Oct 15, 1984, at 13, col 1 **§6.39**

Fishbein, Eishingrad & Garcia, *Only Congress Can Implement Tender Offer Proposals,* Legal Times of Wash, Feb 20, 1984, at §5, col 1 **§6.39**

Fiske, *Booming Corporate Education Efforts Rival College Program, Study Says,* NY Times, Jan 28, 1985, at A10 **§11.29**

Fiske, *Leaders of Companies and Educators Stress Need to Compete,* NY Times, May 16, 1983, at A1, A13 **§1.01**

Five Ways to Go Bust, The Economist (London), Jan 8, 1983, at 11 **§2.07**

Fleischer, Tender Offers: Defenses, Responses, And Planning (Law & Business, Inc, Harcourt Brace Jovanovich 1981) **§6.39**

Flewellen, *An Anomaly in the Patent System: The Uncertain Status of Computer Software,* 8 Rut J Computing & Tech 273 (1981) **§11.10**

"*Foreign 'Blocking Laws' Hinder US Enforcement, Fedders Says,*" 16 Sec Reg L Rep (BNA) No 13, at 578 Mar 30, 1984 **§6.66**

Fowler, *A Foundation for Research,* 188 Sci 414 (May 2, 1975) **§12.03**

Fowler, "*Information Systems Get Focus,*" NY Times, Apr 11, 1984, at D24 **§11.29**

Fox, *The New Merger Guidelines: A Blueprint for Microeconomic Analysis,* 27 Antitrust Bull 519 (1982) **§3.03**

Fox, *Updating the Antitrust Guide on International Operations—A Greener Light for Export and Investment Abroad,* 15 Vand J of Transnational L 709 (1982) **§3.12**

Freed, *The Demands of High-Tech Lawyering,* NY Times, Oct 2, 1983, at F3 **§11.01**

Freeman *et al,* Unemployment and Technological Innovation (1982) **§§1.02, 12.06**

G

Gallagher, The Law of Federal Negotiated Contract Formation (1981) **§12.04**

Gallogly, *Acid Precipitation: Can the Clean Air Act Handle It?* 9 B C Envtl Aff L Rev 687 (1981) **§7.23**

Galloway, Industrial Planning Under Codes (1935) **§12.03**

Gardner, Excellence XIV (1961) **§2.07**

Gemigani, *The Case For Computer Law,* 88 Case & Com 3 (May/June 1983) **§11.01**

Gene Therapy: Research in Public, 227 Sci 493 (Feb 1, 1985) **§7.14**

Gibbon, *Does the Law of Monopolization Inhibit Innovative Behavior?,* 49 ABA Antitrust LJ 925 (1980) **§3.04**

Gibney, Miracle by Design: The Real Reasons Behind Japan's Economic Success (1982) **§§1.01, 2.03, 2.07**

Gibson, *Research for Sale,* 6 Venture 78 (Mar 1984) **§11.29**

Gibson, *Revitalizing Old Industries,* 4 Venture 86 (Nov 1982) **§11.04**

Ginsburg, *Antitrust, Uncertainty, and Technological Innovation,* 24 Antitrust Bull 635 (1979) **§11.23**

Givens, Advocacy (Shepard's/McGraw-Hill 1980 & Supp 1983) **§§1.03, 1.05, 2.07, 12.03, 12.06**

Givens, Antitrust: An Economic Approach (1983) **§§3.03, 3.13**

Givens, *Dealing With National Emergency Labor Disputes,* 34 Temple LQ 17 (1960) **§2.03**

Givens, *Product Improvement - Evidence of Liability?,* NYLJ 1 (July 18, 1983) **§12.07**

Goeller & Zucker, *Infinite Resources: The Ultimate Strategy,* 223 Sci 456 (Feb 3, 1984) **§12.05**

Goldhar & Jelinek, *Plan for Economics of Scope,* 83 Harv Bus Rev 141 (Nov/Dec 1983) **§11.06**

Goldscheider, Technology Management Handbook (1984) **§1.03**

Gonson & Butler, *In Wake of 'Dirks', Courts Debate Definition of Insider,* Legal Times of Wash, Apr 2, 1984, at 16, col 1 **§6.53**

Grogan & Kump, *The Broader Meanings of Apple v Franklin,* 1 Computer Law 15 (1984) **§11.10**

Grossman, *Iowa Assists Inventors To Create Jobs,* 26 State Govt News 16-19 (May 1983) **§9.15**

Gunn, The *Mechanization of Design and Manufacturing,* 247 Sci Am 114 (Sept 1982) **§11.06**

Gurwitz, Auda & Green, Federal R&D and Defense Outlays in New York State (Federal Reserve Bank of New York, Feb 1984) **§9.12**

Guzzardi, *How to Foil Protectionism,* 107 Fortune 6, Mar 21, 1983, at 76 **§1.01**

H

Halloran, Venture Capital and Public Offering Negotiation (1982) **§1.03**

Hambrick, MacMillan & Barbosa, "Business Unit Strategy and Changes in the Product R&D Budget," 29 Mgmt Sci 757 (July 1983) **§11.25**

Hanan, How to Grow by Accelerating Profits on Existing Products, (McGraw-Hill Conference on Corporate Innovation Dec 6, 1979) **§3.10**

Hayes & Garvin, *Managing as if Tomorrow Mattered,* 60 Harv Bus Rev 70 (May/June 1982) **§11.25**

Heard & Pound, *Shark Repellants May Fail to Benefit Shareholders,* Legal Times, Mar 25, 1985, at 12, col 1 **§6.39**

Hechinger, *About Education: Looking Ahead to the Silicon Future,* NY Times, Dec 1, 1981, at C11 **§11.29**

Hedging Against New Accounting Pronouncements, 1984-2 Bus Rev 2 (Price Waterhouse 1984) **§6.63A**

Heilbroner, *Beyond Mass Production,* NY Times Book Rev, Jan 6, 1985, at 7 **§1.03**

Heller, *Hydrogen Evolving Cells,* 233 Sci 1141 (Mar 16, 1984) **§12.05**

Hershey, *Antitrust Chief Offers Guidelines for Research,* NY Times, May 11, 1983, at D1 **§3.10**

Hershey, *"Stiff Antitrust Laws Called Rein On U.S.,"* NY Times, Aug 1, 1983, at D1 **§11.11**

Hersey, Zest for Work (1955) **§2.07**

High Technology Jobs Going Overseas as U.S. Costs Rise, NY Times, Mar 19, 1983, at 1 **§1.01**

Hodder & Riggs, *Pitfalls in Evaluating Risky Projects,* 85 Harv Bus Rev 129 (Jan/Feb 1985) **§12.02**

Hoerr, *There Really are Jobs After Retraining,* Bus Week, Jan 28, 1985, at 76 **§9.12**

Holden, *New R&D Centers Will Test University Ties,* 227 Sci 150 (Jan 11, 1985) **§9.16**

Holmes, *Collected Legal Papers* 30-33, 270-71 (1921) **§2.07**

Hoover, 2d Annual Message 12/2/39, 3 State of the Union Messages of the Presidents 2776 (Chelsea House 1966) **§12.03**

3 Hoover, the Memoirs of Herbert Hoover 98 (1952) **§12.03**

Hudson, SEC Tightens Annual Meeting Proposal Rules, Wall St J, Aug 17, 1983 at 4 Cal 1 **§6.65**

The Human Factor in Innovation and Productivity, Report of Subcommittee, House Committee on Science & Technology, 97th Cong, 2d Sess (1982), Serial FF **§2.07**

Hunt & Hunt, The Human Resources Implications of Robotics (1983) **§12.06**

Hutlen, *On the Importance of Productivity,* NY Times, Jan 27, 1979, at 27 **§1.02**

I

Industry Committee, N.Y. Bar seek clarification of Insider Trading Bill, 15 Sec Reg & L Rep (BNA) No 10 (Mar 11, 1983), at 461 **§6.69**

Ingersoll, *SEC Asks Companies to volunteer in Test of Electronic Filings,* Wall St J, Mar 23, 1984, at 6, col 3 **§6.35**

Ingersoll, *SEC is Seeking to Punish Firms on False Data,* Wall St J, Feb 16, 1984, at 4 **§§6.35, 6.66**

The Institute for Ventures in New Technology, Texas A&M University, Shaping New Ideas **§9.15**

International Environment Reporter (BNA) **§1.04**

Intrapreneurial Now, The Economist (London), Apr 17, 1982, at 67 **§§2.07, 4.21**

Irving, *Closing and Sales of Businesses: A Settled Area?,* 33 Lab LJ 218 (1982) **§2.02**

J

Jackson, *Southland Probe is Ended by U.S. After Five Years and Some Convictions,* Wall St J, Mar 11, 1985, at 12, col 1 **§6.57**

Jacobs, The Economy of Cities (1969) **§12.03**

Janowitz, *Disarticulation,* NY Times, Apr 26, 1981, at E23 **§2.07**

Japan Engineering News, Patents and-Licensing (ISSN 0388-7081) **§1.03**

Jarett, Fostering Capitalism: The Massachusetts Venture Capital Fund (Council of State Governments 1984) **§9.14**

Javits, *For Labor, Employers,* NY Times, Jan 21, 1983, at A27 **§2.01**

Javits, Statement on Floor, *The Need for a National Productivity Improvement Policy,* Cong Rec S1100 (daily ed 2/6/79 **§1.02**

Jefferson, 1 State of the Union Messages of the Presidents 88 (1966) **§12.03**

Jente, *Assessing Recent Efforts to Challenge Aggressive Competition as an Attempt to Monopolize,* 49 ABA Antitrust LJ 961 (1980) **§3.04**

Johnson, *If We Could Take the Salt Out of Water,* NY Times, Oct 30, 1960 at 17 **§12.05**

Johnson, MITI and the Japanese Miracle (Stan Univ Press 1982) **§1.01**

Johnson, Organization for Economic Cooperation, The Industrial Policy of Japan (Paris 1972) **§1.01**

Johnson & Tornatzky, Cooperative Science: A National Study of University Industry Researchers: Assessment of the Industry/University Cooperative Research Projects Program (National Science Foundation, Nov 1984) **§9.16**

John, Tonatzky & Schlaaff, Cooperative Science: A National Study of University Industry Researchers (National Science Foundation, Nov 1984) **§9.16**

Jones & Angly, Fifty Billion Dollars (1951) **§12.02**

Judson, Thumbprints in Our Clay, New Republic, Sept 19/26, 1983, at 12 **§11.14**

Junior Stock, 8 Corporate Counsel 6 (Nov-Dec 1983) **§6.63A**

K

Kamien & Schwartz, Market Structure and Innovation (1982) **§11.23**

Kansas Advanced Technology Commission, 1984 Report to the Legislature (1984) **§9.06**

Kanter, The Change Masters: Innovation for Productivity in the American Corporation (1983) **§2.07**

Kanwit, Federal Trade Commission §11.05 (Shepard's/McGraw-Hill 1979) **§2.07**

Kartow, Eads, Levitt, McGraw, Reich & Thurow, *The Political Realities of Industrial Policy,* 83 Harv Bus Rev 76 (Sept/Oct 1983) **§11.01**

Katz, *Harvey and Medical Research,* 160 JAMA 1137 (1956) **§12.03**

Kelman, *Whatever Happened to U.S. Innovation?,* NY Times, Feb 14, 1979, at 14F **§1.02**

Kessler, *Food Safety: Revisiting the Statute,* 223 Sci 1034 (Mar 9, 1984) **§12.03**

Kilborn, *Competitiveness Panel Reports on US Lag,* NY Times, Feb 14, 1985, at D1 **§1.01**

Kilborn, *Few on Layoffs May Be Rehired . . . Big Corporations Rely on Automation to Aid Output in a Recovery,* NY Times, May 16, 1983, at A1, A12 **§12.06**

Kilburn, *A Plan for Restoring U.S. Industrial Role Advocated in Study,* NY Times, Jan 12, 1983, at A1 **§11.01**

Kindel & Teitelman, *Go East, Young Man,* Forbes, June 20, 1983, at 136 **§9.12**

Kintner, *Casting Fusion Adrift,* 85 Tech Rev 64 (May/June 1982) **§12.05**

Klein & Reese, Taking High Tech Companies Public (Packard Press Dec 1983) **§6.13**

Kobak, *Application of Antitrust Laws to Joint Research, Development, Part I,* NYLJ, Dec 10, 1984, at 1 **§3.11**

Konecci et al, Initiatives for Technology Venturing (IC2 Inst, Univ of Tex at Austin 1984) **§1.01**

Kossoris, *Work Rules in West Coast Longshoring,* 84 Monthly Lab Rev 1 (1961) **§2.03**

Kramer, *Development Agreements: To What Extent Are They Enforceable?* 10 Real Esate LJ 29 (1982) **§8.29B**

Kramer, *Transboundary Air Pollution and the Clear Air Act: An Historical Perspective,* 32 Kan L Rev 181 (1983) **§7.23**

Krasu, *Can Creeping Socialism Cure Creaking Capitalism: One Man's MITI is Another Man's Poison,* 266 Harper's 17 (Feb 1983) **§11.01**

Krotseng, *Plant Relocations and Consolidations: Employers' Obligations Regarding Employee Transfer and Union Recognition,* 33 Lab LJ 609 (1982) **§2.02**

Krueger & Yarema, *New Institutions for New Energy Technology: The Case of Ocean Thermal Conversion,* 54 S Cal L Rev 767 (1981) **§§1.03, 11.03**

L

Lab Extraordinaire, 123 Sci News 9, Jan 1, 1983 **§4.21**

Lavey, *Patents, Copyrights, and Trademarks as Sources of Market Power in Antitrust Cases,* 27 Antitrust Bull 433 (1982) **§§11.08, 11.13, 11.18**

Lawler & Mohrman, *Quality Circles After the Fad,* 85 Harv Bus Rev 31 (Jan/Feb 1985) **§2.01**

Lawrence, Can America Compete? (1983) **§11.01**

Lehman-Wilzig, *Frankenstein Unbound,* 13 Futures 442 (1981) **§11.01**

Lele & Maher, *Getting It Together: Building a Strategic Analysis Model,* 68 Bus Marketing 59 (Nov 1983) **§11.13**

Lentz, *Financing High-Technology: Congress Invests in a Better Idea,* 9 Rut J Computing & Tech 209 (1982) **§11.20**

Levenstein, Why People Work (1964) **§2.07**

Levi, Preliminary Report, Interindustry Differences in R&D Appropriability and Technological Opportunity (1984) **§12.07**

Levine & Ferrigno, *Corporate Image Suffers if Lesson of 'Dirks' is Ignored,* Legal Times of Wash, Apr 2, 1984, at 15, col 1 **§6.63**

Levinson, *General Partnership Interests and the Securities Act of 1933: Recent Judicial Developments,* 10 Ohio NUL Rev 463 (1983) **§6.06**

Lewis, *For Want of a Nail,* NY Times, Aug 31, 1981, at A17 **§2.07**

Lewis, *Technology, Enterprise and American Economic Growth,* 215 Sci 1204 (Mar 5, 1982) **§2.07**

Link, *Alternative Sources of Technology: An Analysis of Induced Innovations,* 4 Managerial & Decision Econ 40 (1983) **§11.05**

Lipman, The Personal Duties of Directors, Officers and Principal Stockholders Under the Securities Act of 1934 and Rule 144 of the Securities Act of 1933 (Packard Press Jan 1981) **§6.52**

Lipsky, *Current Antitrust Division Views on Patent Licensing Practices,* address to American Bar Association Antitrust Section (Nov 5, 1981) **§4.19**

Lipton, *Takeover Abuses Mortgage The Future,* Wall St J, Apr 5, 1985, at 16, col 4 **§6.39**

Lipton & Steinberger, Takeovers & Freezouts (Law Journal Seminare-Press 1978) **§6.39**

Lohr, *How Japan Guides Industries,* NY Times, May 18, 1983, at D1 **§11.01**

Lohr, *Japan's New Nonconformists: Technology Spurs Change,* NY Times, Mar 8, 1982, at D1 **§2.07**

Lohr, *Japan's Trade Ministry Draws Praise and Ire,* NY Times, May 17, 1983, at A1 **§11.01**

Long, Blue Sky Law Handbook (1985) **§6.16**

Low-Tech Education Threatens The High-Tech Future, Bus Week, Mar 28, 1983, at 95 **§9.12**

Lucas, *Acid Rain: The Canadian Position,* 32 Kan L Rev 165 (1983) **§7.23**

Lueck, *Synthetic Fuels' Appeal Fades,* NY Times, May 6, 1983, at D1 **§12.03**

Lundine, *The Jamestown Experience: A Case Study in Labor-Management Cooperation,* in Linstedt & Colglazier, Managing, ch 12 (1982) **§2.01**

Lutz, *Managing a Boundless Resource: United States Approaches to Transboundary Air Quality Control,* 11 Envtl L 321 (1980) **§7.23**

M

MacMillan, *Pre-emptive Strategies,* 4 J Bus Strategy 16 (Fall 1983) **§11.13**

Magaziner & Reich, Minding America's Business (1982) **§1.01**

Mansfield, *Basic Research and Productivity Increase in Manufacturing,* 70 Am Econ Rev 863 (Dec 1980) **§1.02**

Manual for the Handling of Applications for Patents, Design and Trade Marks Throughout the World (Octrooibureau Los En Steiger, Amsterdam, Holland) **§3.08**

Marshall, *Law of the Sea and the New U.S. Oceans Policy,* 55 NY St BJ (Apr 1983) **§1.03**

Maugh, *Acid Rain's Effects on People Assessed,* 226 Sci 1408 (Dec 21, 1984) **§7.23**

McCarroll, *Rehiring Retirees,* NY Times, Nov 22, 1981, at 8F **§2.02**

McDaniel, *A Simplified Indenture,* 13 Rev Sec Reg 16 (Sep 24, 1980), at 871 **§6.40**

McDowell, *Job Loyalty: Not the Virtue It Seems: Veteran Employees provide continuity, but all too often they are insecure, dull and dependent,* NY Times, Mar 3, 1985, §3, at Fl **§2.06**

McGraw-Hill, Biotechnology Newswatch (periodical) **§1.03**

McGregor, The Human Side of Enterprise (1960) **§2.07**

Mencher, *Lessons for American Policy Making from the British Labor Government's 1964-70 Experiment in Applying Technology to Economic Objectives,* London Graduate School of Business, Report to National Science Foundation (Research Applied to National Needs (RANN) 1975) **§12.03**

Mendelssohn, The Search for Absolute Zero (1966) **§§12.03, 12.07**

Microelectronic and Information Sciences Center, University of Minnesota, Microelectronic & Information Sciences (Nov 1983) **§9.16**

The Microelectronics Innovation and Computer Research Opportunities (MICRO) Program, 1983-84 Progress Report (Jan 31, 1984) **§9.09**

Midyear Meeting of American Bar Association, 51 USLW No 32 (Feb 22, 1983), at 2488 **§6.68**

Miller & Fox, *Lab Extraordinaire,* 123 Sci News 9 (Jan 1, 1983) **§12.03**

Millstein & Millstein, *A Sound Approach on Industrial Polocy,* NY Times, Aug 25, 1983, at A23 **§11.11**

Moore, *Junior Stock Plans Could Force Deadly Interpretations by IRS,* Legal Times of Wash, July 23, 1984, at 1, col 2 **§6.63A**

Moser, *Meeting the Challenge of Technology Education,* 17 World 2 (Spring 1983), at 3 **§9.12**

Mosteller, *Innovation and Evaluation,* 211 Sci 881 (Feb 27, 1981) **§12.06**

Mullaney, *The Spreading Malaise,* NY Times, July 7, 1978, at F15 **§1.62**

Muller, Revitalizing America (1980) **§12.03**

N

Nader, *A Bad Precedent,* NY Times, Jan 9, 1983, at 2F **§12.06**

NASAA Financial Planners Hearing Focuses on State Uniformity, 17 SEC L Rep (BNA) No 11, at 452 (Mar 15, 1985) **§6.46**

National Academy of Sciences, International Competition in Advanced Technology: Decisions for America (1983) **§1.01**

National Association of State Development Agencies, National Council for Urban Economic Development, & The Urban Institute, Directory of Incentives for Business Investment and Development in the United States (1983) **§§9.07, 9.14**

National Governors' Association, Technology & Growth: State Initiatives in Technological Innovation (1983) **§9.15**

National Research Council, Outlook for Science and Technology: The Next Five Years (1982) **§11.30**

National Research Council, Risk, Assessment in the Federal Government (1983) **§12.06**

National Science Foundation, *Manufacturing Employment Becomes Increasingly Technological,* Sci Resource Stud Highlights (Mar 10, 1983) **§9.12**

National Science Foundation, The 5-Year Outlook on Science and Technology (1982) **§11.28**

Nelson & Langlois, *Industrial Innovation Policy: Lessons from American History,* 219 Sci 814 (Feb 18, 1983) **§12.03**

Nelson, The Rate and Direction of Inventive Activity: Economic and Social Factors (1962) **§12.02**

Neuner, *Rules for Research Funded By US,* NYLJ (Dec 2, 1982), at 1 **§12.04**

New Data Explain Lag in Productivity, NY Times, Apr 7, 1983, at D6 New York State Regulatory Reform: Report of Action Unit No 5 of the New York State Bar Association (1982) (Charles E Heming, Chair) **§12.03**

New York State Department of Agriculture and Markets, New York State Department of Commerce, New York State College of Agriculture and Life Sciences and New York, State Sea Grant, The New York State Food Industry: Economic Development Opportunities for the Future (Oct 1984) **§9.09**

New York State Science and Technology Foundation, The Productivity Development Demonstration Project: Interim Status Report (Feb 27, 1985) **§9.11**

New York State Science and Technology Foundation, The Productivity Development Demonstration Project: Second Interim Status Report (May 17, 1985) **§9.11**

New York State Science and Technology Foundations, Technology Development Organizations (Mar 1985) **§9.06**

Noam, *Divergent Goals for the Deregulators,* NY Times, Aug 6, 1981 at A23 **§12.03**

Norris, Contract Compliance Under the Reagan Administration (1982) **§2.06**

Norris, *Halting the Flow of High-Tech Bargains,* NY Times, Feb 26, 1984, at F3 **§12.07**

North Carolina Biotechnology Center and North Carolina Department of Commerce Industrial Development Division, North Carolina Biotechnology **§9.10**

Note, 52 Colum L Rev 1020 (1952) **§2.03**

Note, 82 Yale LJ 258 (1972) **§3.11**

Note, *Community Communications Co v City of Boulder: The Emasculation of Municipal Immunity from Sherman Act Liability,* 32 Cath UL Rev 413 (1983) **§8.29A**

Note, *Injunctive Relief for State Violations of Federal Funding Conditions,* 82 Colum L Rev 1236 (1982) **§2.02**

Note, *Mandatory Disclosure of Corporate Projections and the Goals of Securities Regulation,* 81 Colum L Rev 1525 (Nov 1981) **§§6.59, 6.62**

Note, *Regulation of Toxic Substances Under the Occupational Safety and Health Act,* 33 Syracuse L Rev 887 (1982) **§2.06**

Note, *Technology Based Emission and Effluent Standards and the Achievement of Ambient Environmental Objectives,* 91 Yale LJ 792 (1982) **§1.04**

Note, *Title VII on Campus: Judicial Review of University Decisions,* 82 Colum L Rev 1206 (1982) **§2.06**

Nothstein, The Law of Occupational Safety and Health (1982) **§2.06**

NRC Panel Urges Against Risk Assessment Superagency, Wash Drug Letter 5 (Mar 21, 1983) **§12.06**

Nyhart & Jones, *What You Don't Know About Technology Can Hurt You,* 69 ABAJ 1667 (Nov 1983) **§11.01**

NY Times, Feb 4, 1966, at 14, col 1 **§12.06**

O

O'Keefe, Technology Assessment for State and Local Government (1982) **§11.27**

Olson, The Logic of Collective Action: Public Goods and the Theory of Groups (1971) **§1.05**

Olson, The Rise and Decline of Nations (1982), *reviewed in* 219 Sci 163 (Jan 14, 1983) **§1.05**

Olson, *SEC Enforcers Focus on Accounting Fraud in 1984,* Legal Times of Wash, Jan 23, 1984, at 12 **§6.62**

Ordover & Willig, *An Economic Definition of Predation and Product Innovation,* 91 Yale LJ 8 (1981) **§11.13**

O'Reilly, Gaynor, Carroll, & Cronin, Jr, Federal Regulation of the Chemical Industry (Shepard's/McGraw-Hill 1980) **§1.03**

Orwell, *Nineteen Eighty-four* (1949) **§12.07**

Ostrov, *Interboundary Stationary Source Pollution—Clean Air Act §126 and Beyond,* 8 Colum J Envtl L 37 (1982) **§7.23**

P

Passell, *Banker of Last Resort: But Could an RFC Be Insulated Against Patronage?,* NY Times, June 4, 1982, at A30 **§12.03**

Pasztor, *Synfuels Plant Sees $700 Million in Losses by 1995,* Wall St J, Apr 11, 1983, at 3 **§12.03**

Patent Law Association of Chicago, Tax Guide for Patents, Trademarks and Copyrights (1983) **§1.03**

Pear, *States Fostering High Technology,* NY Times, Aug 17, 1983, at A1 **§11.11**

Perkins, Managing Creation: The Challenge of Building A New Organization (Wiley Series on Organizational Assessment and Change 1983) **§2.06**

Personal Liability of Managers and Supervisors for Corporate EEO Policies and Decisions (Levinson ed 1982) **§2.06**

Peters & Austin, In Search of Excellence (1985) **§11.15**

Peters & Waterman, In Search of Excellence: Lessons from America's Best-Run Companies (1982) **§§2.07, 4.21**

Petillon & Hill, R&D Partnerships: Structuring the Transaction (1984) **§1.03**

Petron, *The Strategic Management of R&D, Part II: Organizing for Innovation,* 16 Long Range Planning 51 (Apr 1983) **§11.15**

Pilcher, *Job Training in the States,* State Legislatures 9 (June 1983) **§9.12**

Pione & Sabel, The Second Industrial Divide (1984) **§1.03**

Pollack, *The Birth of Silicon Statesmanship,* NY Times, Feb 27, 1983, §3, at 1 **§§1.01, 11.30**

Pool, Technologies of Freedom (1983) **§11.01**

Pope John XXIII, Mater et Magistra (1961) **§2.07**

Potts, *Join to Offer New High-Tech Tax Shelter Plan,* Wash Bus, Wash Post, June 27, 1983, at col 1 **§6.06**

Practising Law Institute, Environmental Law and Practice (1982) **§1.04**

Practising Law Institute, New Merger Guidelines (1982) **§3.03**

Practising Law Institute, Patent-Antitrust (1982) **§3.08**

Practising Law Institute, Product Design Litigation (1982) **§1.04**

Practising Law Institute, Product Liability of Manufacturers (1981) **§1.04**

Practising Law Institute, Protecting Trade Secrets (1982) **§2.06**

Practising Law Institute, Selected Problems in the Initial Public Offering for High Technology Companies (1984) **§1.03**

Practising Law Institute, State Regulation of Capital Formation and Securities Transactions (1983) **§1.03**

Practising Law Institute, Technology Licensing (1982) **§1.03**

Prentice, *The Sale of Business Doctrine: New Relief from Securities Regulation or a New Haven for Welshers?,* 44 Ohio St LJ 473 (1983) **§6.19**

President's Commission on Industrial Competitiveness, Summary of Recommendations 52 (1985) **§12.07**

R

Randolph, *Energy Conservation Programmes: A Review of State Initiatives in the USA,* 12 Energy Poly 425 (1984) **§9.09**

Regens, *A Role for State Government?* 27 Envt 18 (May 1985) **§9.09**

Reich, *Beyond Free Trade,* 61 Foreign Affairs 4 (Spring 1983), at 773 **§§1.01, 3.07**

Reich, The Next American Frontier (1983) **§§1.03, 11.01**

Reich, *The Next American Frontier: What's Really Wrong With the American Economy,* 251 Atlantic 43 (Mar 1983) **§1.03**

Reinhold, *Stanford and Industry Forge New Research Link,* NY Times, Feb 10, 1984, at A22 **§11.29**

Report, *Model Simplified Indenture,* 38 Bus Law 741 (Feb 1983) **§6.40**

Report of President's Commission on Industrial Competitiveness ch 4, at 41 (1985) **§12.07**

Research and Development Limited Partnerships, 38 Bus Law 1917 (Aug 1983) **§6.06**

Richardson, *Cutting Regulatory Tape,* NY Times, Dec 13, 1984, at A31 **§§1.02, 12.03, 12.07**

Rise of New Products: But How New Are They? NY Times, Dec 13, 1984 at A31 **§3.10**

Robinett, *What Schools Can Teach Entrepreneurs,* 7 Venture 2, (Feb 1985) **§9.12**

Robinson, The Accumulation of Capital (1966) **§1.01**

Rohatyn, *Junk Bonds and Other Securities Swill,* Wall St J, Apr 18, 1985, at 30, col 3 **§6.35**

Rohatyn, NY Times Magazine, Dec 5, 1982, at 12 **§12.03**

Rohatyn, The Twenty-Year Century (1984) **§11.01**

Roosevelt, 5th Annual Message Jan 3, 1938, 3 State of the Union Messages of the Presidents 2840-41 (Chelsea House 1966) **§2.02**

Ross, *Hill Panel Criticizes SEC for New Data System,* Wash Post, Mar 15, 1985, at D3, col 4 **§6.35**

Roth, *Toward A Theory of Bargaining: An Experimental Study in Economics,* 220 Sci 687 (May 13, 1983) **§12.03**

Rothwell & Zegveld, Industrial Innovation and Public Policy (1982) **§§11.04, 11.20, 11.21, 11.29**

Row, Disclosure Standards in SEC Proceedings, 16 Rev Sec Reg No 3 (Feb 2, 1983) at 1 **§6.66**

Rowe, *Disclosure of Negative News,* 17 Rev Sec Reg 804 (Nov 21, 1984) **§6.52**

Rowe, *How to Deal With Negative News—A Legal Analysis,* 5 Legal Notes & Viewpoints (Feb 1985) **§6.52**

Rowe, Settlement of an SEC Enforcement Action, Prentice-Hall Sec Reg Sec (1980) ¶1121 **§6.66**

Rowe, *The Decline of Antitrust and the Delusions of Models: The Faustian Pact of Law and Economics,* 72 Geo LJ 1511 (1984) **§11.01**

Rowe & Eppler, Preparation of Management's Discussion and Analysis of Financial Condition and Results of Operations Under Uncertain Economic Conditions, Prentice-Hall Sec Reg Sec (1983) **§11.06**

Roxburgh, Policy Responses to Resource Depletion, 21 Contemp Stud in Econ & Fin Anal (1980) **§12.05**

S

Sahal, *Technology, Productivity and Industry Structure,* 24 Technological Forecasting & Soc Change 1 (Sept 1983) **§11.20**

Sampson Says SEC Will Not Reconsider Decision To Eliminate Two Releases on Non-Audit Services, 1 Acct Prac & Reg No 5 at 2 (Mar 11, 1985) **§6.64**

Sanger, *Computer Makers Pursuing Campus Sales and Research,* NY Times Jan 20, 1984, at A1 **§11.29**

Sanger, *SEC's Computer Revolution,* NY Times, Apr 4, 1984, at D1, col 1 **§6.36**

Sanger, *The Gavel Comes Down on Computer Copycats,* NY Times, Oct 23, 1983, at F3 **§11.10**

Sansweet, *Hazardous Course: Carter Hawley's Trials Show Risk in Adopting Long-Range Strategy,* Wall St J, June 19, 1984, at 1, col 1 **§6.39**

Sawada, *Government Industrial Policy for a Healthy World Economy,* 24 Technological Forecasting & Soc Change 95 (1983) **§11.01**

Schaden & Heldman, Product Design Liability (1982) **§1.04**

Schneider, *Definition of a "Security"—1983/1984 Update,* 17 Rev Sec Reg 85 (Sept 12, 1984) **§6.17**

Schneider, Manko & Kant, Going Public: Practice Procedure and Consequences (Packard Press reprint June 1983) **§§6.13, 6.38**

Schools Study Ethics of Business-Aided Research, NY Times, Apr 8, 1983, at A20 **§11.29**

Schultze, *Industrial Polocy: A Dissent,* Brookings Rev (Fall 1983) **§12.03**

Schultze, *Industrial Policy: A Solution in Search of a Problem,* 25 Cal Mgmt Rev 5 (Summer 1983) **§11.01**

Science Indicators - 1980, Report of the National Science Board, National Science Foundation, Washington, DC 1981 **§1.01**

SEC, AICPA Defend Regulations; Legislators Hint of Congressional Action, 1 Acct Prac & Reg No 5, at 1 (Mar 11, 1985) **§6.62**

SEC Enforcement Chief Warns Companies, CPAs On Management Discussion, Analysis, 16 Sec Reg & L Rep (BNA) No 2, at 95 (Jan 13, 1984) **§6.54**

SEC Government-Business Forum on Small Business Capital Formation, Final Report (Nov 1982) **§§6.16, 6.22, 6.37, 6.44, 6.49, 6.54, 6.63, 6.69**

SEC, Hospital Corp Settle on Reporting of Insider Trading, Wall St J, Mar 27, 1984, at 53, col 3 **§6.50A**

SEC Staff Say Greenmail Payments Depress Price of Target's Stock, 16 Sec Reg & L Rep (BNA) No 39, at 1585 (Oct 5, 1984) **§6.39**

SEC Takes Dim View of Accounting 'Shenanigans', SEC Today vol 45, at 1 (Mar 6, 1984) **§6.62**

Sec To Go After Firms That Stretch Accounting Principles, Treadway Warns, 16 SEC Reg & L Rep No 9 (Mar 2, 1984), at 446 **§6.62**

SEC to Seek Expansion of Authority to Bring Administrative Proceedings, 16 Sec Reg & L Rep (BNA) No 7, at 267 (Feb 17, 1984) **§6.66**

SEC Urges Delaware Court To Void Poison-Pill-Takeover Defense, 17 SEC Reg & L Rep (BNA) No 12, at 481 (Mar 22, 1985) **§6.39**

SEC votes to Increase Enforcement Authority, SEC TODAY, No 84-33 (Feb 16, 1984) **§6.66**

Seidels & Keyes, Can We Delay A Greenhouse Warming? (EPA 1983) **§12.05**

Shabecoff, *Broad Damage Found in Trees in Eastern US,* NY Times, Feb 26, 1984, at A1, col 1, A30, col 1 **§12.05**

Shanklin & Ryans, *Organizing for High-Tech Marketing,* 84 Harv Bus Rev 1964 (Nov/Dec 1984) **§1.03**

Shenefield, *The Parker v. Brown State Action Doctrine and the New Federalism of Antitrust,* 51 ABA Antitrust LJ 337 (1983) **§3.11**

Silbergeld, *New Affirmative Action Regulations for Government Contractors,* 33 Lab LJ 230 (1982) **§2.06**

Silk, *Structural Joblessness,* NY Times, Apr 6, 1983, at D2 **§12.06**

Skinner & Chakraborty, The Impact of New Technology: People and Organizations in Manufacturing and Allied Industries (1982) **§11.09**

Slater, *Local Governments and State Action Immunity after City of Lafayette and City of Boulder,* 51 ABA Antitrust LJ 349 (1982) **§3.11**

Smart, *Venture Capital Gambling on the Future,* 7 Inc 146 (Apr 1985) **§9.13**

Smith, *Arbitrating Safety Grievances,* 33 Lab LJ 238 (1982) **§2.06**

Sobel, *The Antitrust Interface with Patents and Innovation: Acquisition of Patents, Improvement Patents and Grant-Back, Non-Use, Fraud on the Patent Office, Development of New Products and Joint Research,* 53 ABA Antitrust L J 681 (1985) **§11.13**

Special Report: A New Era for Management, Bus Week, Apr 25, 1983, at 50 **§1.03**

Spencer & Olson, *SEC's Quarterly Reporting Release Packs a Surprise,* Legal Times of Wash, Apr 2, 1984, at 18, col 1 **§6.63**

Stanger & Gunther, *"Big GAAP-Little GAAP": Should There be Different Financial Reporting for Small Businesses?,* 56 NYU L Rev 1207 (1981) **§6.63**

Starr, *The Electric Power Research Institute,* 219 Sci 1190 (Mar 11, 1983) **§§3.10, 12.03**

Starr & Rudman, *The Parameters of Technological Growth,* 182 Sci 358 (Oct 26, 1973) **§1.02**

State of the Union Message of the President, NY Times, Jan 26, 1983, at A14, col 5 **§1.01**

State University of New York, Sampler of Training and Retraining Programs for Business and Industry (1983) **§9.12**

Stein, *Don't Fall for Industrial Policy,* Fortune, Nov 14, 1983, at 64 **§11.01**

Stevens, *Two Southland Officials Cleared In Bribery Case,* Wall St J, Feb 28, 1985 **§6.57**

Stevenson & Gumpert, *The Heart of Entrepreneurship,* 85 Harv Bus Rev 85 (Mar/Apr 1985) **§2.07**

Storch, Issues in Government Contracting and Secrecy Controls, ABA National Institute, Legal Problems of Innovation (1983) **§12.07**

Stone, *The Common Law in the United States,* 50 Harv L Rev 4 (1936) **§3.11**

Subcommitee on Annual Review, *Federal Securities Regualtion (Case Law Section),* 40 Bus Law No 1 Nov 1984, at 159 **§6.17**

Subcommittee on Economic Stabilization, House Committee on Banking Finance to Urban Affairs Committee Print, 98th Cong, 1st Sess (1983) **§12.03**

Subcommittee on Oversight, House Ways and Means Committee and Joint Committee on Taxation, 98th Cong, 2d Sess, Description of Present Law and Issues Relating to the Research and Experimentation Credit (Joint Comm Print 1984) **§9.07**

Subcommittee on Partnerships, Trust and Unincorporated Association, Integration of Partnership Offerings: *A Proposal for Identifying a Discrete Offering,* 37 Bus Law No 4, at 1591 (July 1982) **§6.24**

Sullivan, *Advisor Warns US Lags in Scientific Growth Rate,* NY Times, Oct 6, 1973, at 10 **§1.02**

Sullivan, Build Brother Build (1969) **§12.06**

Sullivan, *Star May Cause Extinctions on Earth,* NY Times, Feb 21, 1984, at C1, C9 **§12.06**

Sweltz, *Generic Testing in the Workplace: An Analysis of the Legal Implications,* 19 Forum of ABA Tort to Ins Prac Sec No 2, at 323 (Winter 1984) **§2.06**

Symposium, *Current Antitrust Law and United States International Trade Practices,* 15 Vand J of Transnational L 4 (1982) **§3.12**

Symposium, *Employee Selection Procedures,* 8 Employee Rel LJ 572 (1983) **§2.06**

Symposium, *Law-Science Cooperation Under the National Environmental Policy Act,* 15 Nat Resources Law 569 (1983) **§7.12**

Symposium, *Permissible Cooperation Among Competitors,* 44 ABA Antitrust LJ 329 (1975) **§3.10**

T

Taylor & Tanakadate, *Why Japan Inc Might Not Work in U.S.*, US News & World Rep, Jan 30, 1984, at 44 **§11.01**

Technology Management Handbook (1982) **§1.03**

Technology and Trade: Some Indicators of the State of U.S. Industrial Innovation, Subcommittee on Ways and Means, US House of Representatives, US Government Printing Office, Washington, DC, 1980, at 2 **§1.01**

Terleckyj, Effects of R&D on the Productivity Growth of Industries (1974) **§1.02**

Term Loan Handbook (McCann ed 1983) **§1.03**

Thackray, America's New Technology Trap, Mgmt Today 86 (June 1983) **§11.15**

The (Massachusetts) Governor's Commission on the Future of Mature Industries, Final Report (June 1984) **§9.11**

Thomas, *How Should Humans Pay Their Way*, NY Times, Aug 24, 1981, at A15 **§12.06**

Thomas, *Overregulating the Regulators*, NY Times, June 3, 1981, (Op-ed page) **§12.03**

Thurmond, *The Government was to Blame, in Part*, NY Times, Jan 9, 1983, at 2F **§12.06**

Tomczak, Corporate and Commercial Finance Agreements (Shepard's/McGraw-Hill 1984) **§1.03**

Trade Secrets Handbook (1982) **§2.06**

Tribus, Applying Science to Industry, 70 US News & World Rep 35 (Jan 8, 1971) **§12.03**

Truman, Annual Message of Jan 7, 1953, 3 State of the Union Messages of the Presidents 309-09 (Chelsea House 1966) **§12.07**

Tsongas, *Op-ed*, NY Times, Mar 1, 1983, at A23 **§1.01**

Two Decades of Research on Innovation Selected Studies of Current Relevance, Prepared For the Use of the Special Study on Economic Change of the Joint Economic Committee of the Congress of the United States, US Government Printing Office, Washington, DC, 1980, at 138, 148 **§1.01**

U

University of Texas at Arlington, College of Engineering, Technology Enterprises Development Center **§9.11**

US Department of Justice, Vertical Restraints Guidelines, Jan 23, 1985, 4 BNA Antitrust & Trade Reg Rep No 1199 (Spec Supp Jan 24, 1985) **§3.03**

US Office of Technology Assessment, Technology, Innovation and Regional Economic Development (1983) **§§9.10, 9.11, 9.14**

V

Vanderstar, *Liability of Municipalities Under the Antitrust Laws: Litigation Strategies,* 32 Cath UL Rev 395 (1983) **§8.29**

Van der Water, *Industrial Productivity and the Law: A Study of Work Restrictions,* 43 Va L Rev 155 (1957) **§2.03**

Varrin & Kukich, *Guidelines for Industry-Sponsored Research at Universities,* 227 Sci 385 (Jan 25, 1985) **§§9.16, 11.29**

Vartanian, Schechter, West & Levine, *Leveraged ESOP Can Aid Thrift Takeover Defense,* Legal Times of Wash, Mar 25, 1983, at 18, col 1 **§6.38**

Vise, *Experts Disagree on Takeovers,* Wash Post, Mar 13, 1985, at F3, col 4 **§6.39**

W

Walton, *From Control to Commitment in the Workplace,* 85 Harv Bus Rev 77 (Mar/Apr 1985) **§2.07**

Warren, *Developments in State Takeover Regulation: MITE and its Aftermath,* 40 Bus Law No 2, at 671 (Feb 1985) **§6.39**

Washington, 1 State of the Union Messages of the Presidents 3 (1966) **§12.03**

Webster, *SEC Aims to Expedite Review,* Legal Times of Wash, Mar, 1984, at 1, col 2 **§6.35**

Weil, *US Industrial Coordination Board,* 14 Law & Policy in Int Bus 981 (1983) **§12.03**

Weinberg, *The Forces of Nature,* 29 Bull Am Acad of Arts & Sci 4 (Jan 1976), at 1 **§12.06**

Wheeler, *Product Liability: Civil or Criminal?,* 17 F of ABA Sec on Tort & Ins Prac 250 (1981) **§1.04**

Where Japan Has Research Edge, Bus Week, Mar 14, 1983, at 116 **§1.01**

Why Cap Cities is Sitting Pretty, Bus Week, Mar 21, 1983, at 121 **§2.07**

Why Small Companies Will Survive, Chem Week, Dec 1, 1982, at 60 **§4.21**

Williams, *Glory Days End for Pharmaceuticals,* NY Times Feb 23, 1985, §3, at F1, F11 **§1.01**

Williamson, *Commentary: The Reagan Administration's Position on Antitrust Liability of Municipalities,* 32 Cath UL Rev 371 (1983) **§8.29**

Wilson, Ashton, & Egan, Innovation, Competition, and Government Policy in the Semiconductor Industry (1982) **§11.11**

Worcester & Boggs, *Capital for Geothermal Energy Projects,* 14 Nat Resources Law 713 (1982) **§§1.03, 12.03**

Work in America: Report of a Special Task Force to the Secretary of Health Education and Welfare (1972) **§2.07**

Work Revolution in U.S. Industry: More Flexible Rules on the Job Are Boosting Productivity, Bus Week, May 16, 1983, at 100 **§2.06**

Y

York, *What Killed Long-Term US Business Planning,* NY Times, Jan 10, 1982, at A30 **§12.07**

Z

Zammit, *Computers, Software and the Law,* 68 ABAJ 970 (1982) **§1.03**

Zonderman, *Policing High-Tech Exports,* NY Times Magazine, Nov 27, 1983, at 100 **§12.07**

Index